Lecture Notes on Data Engineering and Communications Technologies

Volume 153

Series Editor

Fatos Xhafa, Technical University of Catalonia, Barcelona, Spain

The aim of the book series is to present cutting edge engineering approaches to data technologies and communications. It will publish latest advances on the engineering task of building and deploying distributed, scalable and reliable data infrastructures and communication systems.

The series will have a prominent applied focus on data technologies and communications with aim to promote the bridging from fundamental research on data science and networking to data engineering and communications that lead to industry products, business knowledge and standardisation.

Indexed by SCOPUS, INSPEC, EI Compendex.

All books published in the series are submitted for consideration in Web of Science.

Ning Xiong · Maozhen Li · Kenli Li · Zheng Xiao ·
Longlong Liao · Lipo Wang
Editors

Advances in Natural Computation, Fuzzy Systems and Knowledge Discovery

Proceedings of the ICNC-FSKD 2022

Volume 2

 Springer

Editors
Ning Xiong
Division of Intelligent Future Technologies
Mälardalen University
Västerås, Västmanlands Län, Sweden

Kenli Li
School of Information Science
and Technology
Hunan University
Changsha, Hunan, China

Longlong Liao
College of Computer and Data Science
Fuzhou University
Fuzhou, Fujian, China

Maozhen Li
Department of Electronic and Computer
Engineering
Brunel University London
Uxbridge, Middlesex, UK

Zheng Xiao
School of Information Science
and Technology
Hunan University
Changsha, Hunan, China

Lipo Wang
School of Electrical and Electronic
Engineering
Nanyang Technological University
Singapore, Singapore

ISSN 2367-4512   ISSN 2367-4520 (electronic)
Lecture Notes on Data Engineering and Communications Technologies
ISBN 978-3-031-20737-2   ISBN 978-3-031-20738-9 (eBook)
https://doi.org/10.1007/978-3-031-20738-9

This Springer imprint is published by the registered company Springer Nature Switzerland AG
The registered company address is: Gewerbestrasse 11, 6330 Cham, Switzerland

Organizing Committee

General Chairs

Lipo Wang   Nanyang Technological University, Singapore
Kenli Li   Hunan University, China

Organizing General Chairs

Wenzhong Guo   Fuzhou University, China
Yuanlong Yu   Fuzhou University, China

Program Chairs

Maozhen Li   Brunel University London, UK
Ning Xiong   Mälardalen University, Sweden

Organizing Chair

Longlong Liao   Fuzhou University, China

Publication Chairs

Xing Chen   Fuzhou University, China
Mingjian Fu   Fuzhou University, China
Guobao Xiao   Minjiang University, China

Publicity Chairs

Wenxi Liu   Fuzhou University, China
Chunyan Xu   Nanjing University of Science and Technology, China
Shanshan Fan   Beijing Language and Culture University, China

Finance Chair

Zheng Xiao   Hunan University, China

Sponsorship Chair

Xinqi Liu   University of Hong Kong, Hong Kong, China

Program Committee

Shigeo Abe	Kobe University, Japan
Henry N. Adorna	University of the Philippines, The Philippines
Davide Anguita	University of Genoa, Italy
Sabri Arik	Istanbul University, Turkey
Krassimir Atanassov	Bulgarian Academy of Sciences, Bulgaria
Sansanee Auephanwiriyakul	Chiang Mai University, Thailand
Philip Azariadis	University of the Aegean, Greece
Vladan Babovic	Singapore National University, Singapore
Thomas Bäck	Leiden Institute of Advanced Computer Science, Netherland
Emili Balaguer-Ballester	Bournemouth University, UK
Valentina Balas	Aurel Vlaicu University of Arad, Romania
Yaxin Bi	University of Ulster, UK
Federico Bizzarri	Politecnico di Milano, Italy

Tossapon Boongoen	Mae Fah Luang University, Thailand
Pierre Borne	Ecole Centrale de Lille, France
Hamid Bouchachia	Bournemouth University, UK
Ivo Bukovsky	Czech Technical University in Prague, Czech
Sujin Bureerat	KhonKaen University, Thailand
Godwin Caruana	Harvest Technology, Malta
Michele Ceccarelli	University of Sannio, Italy
Kit Yan Chan	Curtin University, Australia
Chen-Tung Chen	National United University, Taiwan
David Daqing Chen	London South Bank University, UK
Jianxia Chen	Washington University in St. Louis, USA
Syuan-Yi Chen	National Taiwan Normal University, Taiwan
Chi Tsun (Ben) Cheng	RMIT University, Australia
Jao Hong Cheng	National Yunlin University of Science and Technology, Taiwan
France Cheong	RMIT University, Australia
Jen-Shiun Chiang	Tamkang University, Taiwan
Panagiotis Chountas	University of Westminster, UK
Huey-Der Chu	Takming University of Science and Technology, Taiwan
Hung-Yuan Chung	National Central University, Taiwan
Alessandro Colombo	Politecnico di Milano, Italy
José Alfredo F. Costa	Universidade Federal do Rio Grande do Norte, Brazil
Keeley Crockett	Manchester Metropolitan University, UK
Zoltán Ernö Csajbók	University of Debrecen, Hungary
Darryl N. Davis	University of Hull, UK
Andre C. P. L. F. de Carvalho	University of Sao Paulo, Brazil
Marc de Kamps	University of Leed, UK
Mingcong Deng	Tokyo University of Agriculture and Technology, Japan
Minghua Deng	Peking University, China
Milena Djukanovic	University of Montenegro, Montenegro
Mustafa Dogan	Baskent University, Turkey
Prabu Dorairaj	Broadcom Inc, India
Giorgos Dounias	University of the Aegean, Greece
António Dourado	University of Coimbra, Portugal
Abdelali El Aroudi	Universitat Rovira i Virgili, Spain
Mohammed El Abd	The American University of Kuwait, Kuwait
Zuhal Erden	ATILIM University, Turkey
Geoffrey Falzon	STMicroelectronics (Malta) Ltd, Malta
Xiannian Fan	City University of New York, USA
Saeed Panahian Fard	Universiti Sains Malaysia, Malaysia
Elisabetta Fersini	University of Milan Bicocca, Italy
Zbigniew Galias	AGH University of Science and Technology, Poland

Ming Li	Nanjing University, China
Zhanhuai Li	Northwestern Polytechnic University, China
Steve Ling	University of Technology Sydney, Australia
Bin-Da (Brian) Liu	National Cheng Kung University, Taiwan
Lu Liu	University of Derby, UK
Xiangrong Liu	Xiamen University, China
Yong Liu	University of Aizu, Japan
Yubao Liu	Sun Yat-Sen University, China
José Manuel Molina López	Universidad Carlos III de Madrid, Spain
Jianquan Lu	Southeast University, China
Jinhu Lu	Chinese Academy of Sciences, China
Edwin Lughofer	Johannes Kepler University Linz, Austria
Jacek Mańdziuk	Warsaw University of Technology, Poland
Trevor Martin	University of Bristol, UK
Francesco Masulli	University of Genova, Italy
Masakazu Matsugu	Canon Research Center, Japan
Dinesh P. Mehta	Colorado School of Mines, USA
Hongying Meng	Brunel University, UK
Radko Mesiar	Slovak University of Technology Bratislava, Slovakia
Rym MHallah	Kuwait University, Kuwait
Hongwei Mo	Harbin Engineering University, China
Dusmanta Kumar Mohanta	MVGR College of Engineering, India
Robert Newcomb	University of Maryland, USA
Yoshifumi Nishio	Tokushima University, Japan
Yusuke Nojima	Osaka Prefecture University, Japan
Dimitri Ognibene	CNR-ISTC, Italy
Maciej Ogorzalek	Jagiellonian University, Poland
Kok-Leong Ong	LaTrobe University, Australia
Milos Oravec	Slovak University of Technology, Slovakia
Vasile Palade	Coventry University, UK
Linqiang Pan	Huazhong University of Science and Technology, China
Shaoning Pang	Auckland University of Technology, New Zealand
George Panoutsos	University of Sheffield, UK
Dong-Chul Park	Myong Ji University, Korea
Jessie Ju H. Park	Yeungnam Univ, South Korea
Petra Perner	Institute of Computer Vision and applied Computer Sciences, Germany
Valentina Plekhanova	University of Sunderland, UK
Petrica Pop	North University of Baia Mare, Romania
Man Qi	University of Canterbury, UK
Guangzhi Qu	Oakland University, USA
Rajesh Reghunadhan	Bharathiar University, India
Pedro Manuel Pinto Ribeiro	University of Porto, Portugal

Chan-Yun Yang	National Taipei University, Taiwan
Yingjie Yang	De Montfort University, UK
Zhijun Yang	Middlesex University London, UK
Yiyu Yao	University of Regina, Canada
Chung-Hsing Yeh	Monash University, Australia
Jian Yin	Sun Yat-Sen University, China
Wen Yu	CINVESTAV-IPN (National Polytechnic Institute), Mexico
Yuqing Zhai	Southeast University, China
Jie Zhang	Newcastle University, UK
Jinglan Zhang	Queensland University of Technology, Australia
Liming Zhang	Macau University, China
Liqing Zhang	Shanghai Jiao Tong University, China
Min-Ling Zhang	Southeast University, China
Zhongwei Zhang	University of Southern Queensland, Australia
Liang Zhao	University of Sao Paulo, Brazil
Wei Zheng	Xiamen University, China
Huiyu Zhou	Queen's University Belfast, UK
Ligang Zhou	Macau University of Science and Technology, Macau
Shangming Zhou	Swansea University, UK
Wenxing Zhu	Fuzhou University, China
William Zhu	Minnan Normal University, China
Jeffrey Zou	University of Western Sydney, Australia

Preface

The 2022 18th International Conference on Natural Computation, Fuzzy Systems and Knowledge Discovery (ICNC-FSKD 2022) was held from July 30 to August 1, 2022, online.

ICNC-FSKD is a premier international forum for scientists and researchers to present the state of the art of machine learning, data mining, and intelligent methods inspired from nature, particularly biological, linguistic, and physical systems, with applications to computers, systems, control, communications, and more. This is an exciting interdisciplinary area in which a wide range of theory and methodologies are being investigated and developed to tackle complex and challenging problems. We are delighted to receive many submissions from around the globe. After a rigorous review process, the accepted papers are included in this proceedings.

We have been looking forward to holding the conference in beautiful Fuzhou. However, the recent COVID-19 clusters in China have prompted the organizing committee to move this year's conference to online only. It is a pity that we cannot meet you all physically this year. But the pandemic will eventually be overcome, and we look forward to seeing you again next year in the sunny summer!

We would like to sincerely thank all organizing committee members, program committee members, and reviewers for their hard work and valuable contribution. Without your help, this conference would not have been possible. Special thanks go to the main organizer of this year's conference, College of Computer and Data Science of Fuzhou University, as well as the co-organizer, Science and Technology on Communication Information Security Control Laboratory. We thank Springer for publishing the proceedings. In particular, we thank Series Editors, Prof. Fatos Xhafa and Dr. Thomas Ditzinger, for their kind support. We are very grateful to the keynote speakers for their authoritative speeches. We thank all authors and conference participants for using this platform to communicate their excellent work.

August 2022

Ning Xiong
Maozhen Li
Kenli Li
Zheng Xiao
Longlong Liao
Lipo Wang

Contents

Deep Learning (34)

Multiple Layers Global Average Pooling Fusion 3
Silei Cao, Shun Long, Weiheng Zhu, Fangting Liao, Zeduo Yuan,
and Xinyi Guan

**Han Dynasty Clothing Image Classification Model Based
on KNN-Attention and CNN** 11
Guan Ziwei, Lv Zhao, and Teng Jinbao

**Self-Attention SSD Network Detection Method of X-ray Security
Images** 18
Hong Zhang, Baoyang Liu, and Yue Gao

**Cross Architecture Function Similarity Detection with Binary
Lifting and Neural Metric Learning** 27
Zhenzhou Tian, Chen Li, and Sihao Qiu

**Pain Expression Recognition Based on Dual-Channel
Convolutional Neural Network** 35
Xuebin Xu, Meng Lei, Dehua Liu, and Muyu Wang

**A Noval Air Quality Index Prediction Scheme Based on Long
Short-Term Memory Technology** 43
Lijiao Ding, Jinze Sun, Tingda Shen, and Changqiang Jing

**Code Summarization Through Learning Linearized AST Paths
with Transformer** 53
Zhenzhou Tian, Cuiping Zhang, and Binhui Tian

**Function Level Cross-Modal Code Similarity Detection
with Jointly Trained Deep Encoders** 61
Zhenzhou Tian and Lumeng Wang

Ease Solidity Smart Contract Compilation through Version
Pragma Identification 69
Zhenzhou Tian and Ruikang He

Towards Robust Similarity Detection of Smart Contracts
with Masked Language Modelling 76
Zhenzhou Tian and Xianqun Ke

DeSG: Towards Generating Valid Solidity Smart Contracts
with Deep Learning 85
Zhenzhou Tian and Fanfan Wang

Combining AST Segmentation and Deep Semantic Extraction
for Function Level Vulnerability Detection 93
Zhenzhou Tian, Binhui Tian, and Jiajun Lv

A Novel Variational-Mode-Decomposition-Based Long
Short-Term Memory for Foreign Exchange Prediction 101
Shyer Bin Tan and Lipo Wang

Multi-modal Scene Recognition Based on Global Self-attention
Mechanism 109
Xiang Li, Ning Sun, Jixin Liu, Lei Chai, and Haian Sun

Sheep Herd Recognition and Classification Based on Deep
Learning 122
Yeerjiang Halimu, Zhou Chao, Jun Sun, and Xiubin Zhang

A Deep Learning-Based Innovative Points Extraction Method 130
Tao Yu, Rui Wang, Hongfei Zhan, Yingjun Lin, and Junhe Yu

Deep Embedded Clustering with Random Projection Penalty 139
Kang Song, Wei Han, Chamara Kasun Liyanaarachchi Lekamalage,
and Lihui Chen

Multi-title Attention Mechanism to Generate High-Quality
Images on AttnGAN 147
Pingan Qiao and Xiwang Gao

Hyperspectral Remote Sensing Images Terrain Classification
Based on LDA and 2D-CNN 157
Jing Liu, Yang Li, Meiyi Wu, and Yi Liu

Design and Implementation of Vehicle Density Detection
Method Based on Deep Learning 165
Jiale Yi, Xia Zhang, Zhili Mao, Huimin Du, and Yu Ma

Rotated Ship Detection with Improved YOLOv5X 173
Xuanhong Wang, Shuai Gao, Jingchen Zhou, and Yun Xiao

Contents

Application of End-To-End EfficientNetV2 in Diabetic Retinopathy Grading 182
Xuebin Xu, Dehua Liu, Muyu Wang, and Meng Lei

Sternal Fracture Recognition Based on EfficientNetV2 Fusion Spatial and Channel Features 191
Xuebin Xu, Muyu Wang, Dehua Liu, Meng Lei, and Xiaorui Cheng

Three-Segment Waybill Code Detection and Recognition Algorithm Based on Rotating Frame and YOLOv5 201
Jiandong Shen and Wei Song

A Method for Classification of Skin Cancer Based on VisionTransformer 210
Xuebin Xu, Haichao Fan, Muyu Wang, Xiaorui Cheng, and Chen Chen

MOOC Courses Recommendation Algorithm Based on Attention Mechanism Enhanced Graph Convolution Network 218
Xiaoyin Wang and Xiaojun Guo

Network Traffic Anomaly Detection Based on Generative Adversarial Network and Transformer 228
Zhurong Wang, Jing Zhou, and Xinhong Hei

A Deep Learning-Based Early Patent Quality Recognition Model 236
Rongzhang Li, Hongfei Zhan, Yingjun Lin, Junhe Yu, and Rui Wang

Defect Detection of Exposure Lead Frame Based on Improved YOLOX 244
Wanyu Deng, Dunhai Wu, Jiahao Jie, and Wei Wang

Deep Neural Network for Infrared and Visible Image Fusion Based on Multi-scale Decomposition and Interactive Residual Coordinate Attention 254
Sha Zong, Zhihua Xie, Qiang Li, and Guodong Liu

An Adaptive Model-Free Control Method for Metro Train Based on Deep Reinforcement Learning 263
Wenzhu Lai, Dewang Chen, Yunhu Huang, and Benzun Huang

Neighborhood Graph Convolutional Networks for Recommender Systems 274
Tingting Liu, Chenghao Wei, Baoyan Song, Ruonan Sun, Hongxin Yang, Ming Wan, Dong Li, and Xiaoguang Li

Emotion Recognition from Multi-channel EEG via an Attention-Based CNN Model 285
Xuebin Xu, Xiaorui Cheng, Chen Chen, Haichao Fan, and Muyu Wang

**Multi-feature Short-Term Power Load Prediction Method
Based on Bidirectional LSTM Network** 293
Xiaodong Wang, Jing Liu, Xiaoguang Huang, Linyu Zhang,
and Yingbao Cui

Natural Computation Application (29)

**Prediction of Yak Weight Based on BP Neural Network
Optimized by Genetic Algorithm** 307
Jie He, Yu-an Zhang, Dan Li, Zhanqi Chen, Weifang Song,
and Rende Song

**Hybrid Sweep Algorithm and Modified Ant System
with Threshold for Travelling Salesman Problem** 317
Petcharat Rungwachira and Arit Thammano

**Determining All Pareto-Optimal Paths for Multi-category
Multi-objective Path Optimization Problems** 327
Yiming Ma, Xiaobing Hu, and Hang Zhou

**Improved Whale Optimization Algorithm Based on Halton
Sequence** 336
Wenyu Zhang, Bingchen Zhang, Yongbin Yuan, Changyou Zhang,
and Xining Jia

**A Security-Oriented Assignment Optimization Model of Main
Equipment and Facilities in Prefabricated Building** 344
Chunguang Chang, Zhuo Zuo, and Hongbo Hou

**Mineral Identification in Sandstone SEM Images Based
on Multi-scale Deep Kernel Learning** 353
Mei Wang, Simeng Fan, Fei Han, Zhigang Liu, and Kejia Zhang

**Optimal Selection of Left and Right Hand Multi-channel Pulse
Features Based on Neighbourhood Component Analysis** 361
Lin Fan, Yan Li, Jinsong Wang, Rong Zhang, and Ruiling Yao

**A Dissolving P System for Multi-objective Gene Combination
Selection from Micro-array Data** 369
Fan Liu, Shouheng Tuo, and Chao Li

**Design and Implementation of Scalable Power Load Forecasting
System Based on Neural Networks** 377
Shu Huang, Ze-san Liu, Hong-min Meng, Zhe-nan Xu, Ai-jun Wen,
Shan Li, Di Liu, and Ge Ding

**Heave Compensation Sliding Mode Predictive Control Based
on an Elman Neural Network for the Deep-Sea Crane** 386
Zhimei Chen, Yingbin Lu, Zhenyan Wang, Xuejuan Shao,
and Jinggang Zhang

Multi-feature Fusion Flame Detection Algorithm Based on BP Neural Network 395
Jin Wu, Ling Yang, Yaqiong Gao, and Zhaoqi Zhang

Single Infrared Image Non-uniformity Correction Based on Genetic Algorithm 402
Gaojin Wen, Changhai Liu, Hongmin Wang, Pu Huang, Can Zhong, Zhiming Shang, and Yun Xu

Load Forecasting of Electric Vehicle Charging Station Based on Power Big Data and Improved BP Neural Network 410
Hao Sun, Shan Wang, and Chunlei Liu

A Probabilistic Fuzzy Language Multi-attribute Decision Making Method Based on Heronian Operator and Regret Theory 419
Wenyu Zhang, Xue Gao, Yongbin Yuan, Weina Luo, and Jiahao Zeng

TODIM Multi-attribute Decision-Making Method Based on Spherical Fuzzy Sets 428
Wenyu Zhang, Weina Luo, Xue Gao, Changyou Zhang, and Keya Wang

Modelling of Fuzzy Discrete Event Systems Based on a Generalized Linguistic Variable and Their Generalized Possibilistic Kriple Structure Representation 437
Shengli Zhang and Jing Chen

Design of AUV Controllers Based on Generalized S-Plane Function and AFSA Optimization 447
Chunmeng Jiang, Jiaying Niu, Shupeng Li, Fu Zhu, and Lanqing Xu

Online Tuning of PID Controllers Based on Membrane Neural Computing 455
Nemanja Antonic, Abdul Hanan Khalid, Mohamed Elyes Hamila, and Ning Xiong

Fuzzy Logic Based Energy Management Strategy for Series Hybrid Bulldozer 465
Cong feng Tian, Jia jun Yang, Ru wei Zhang, Jin dong Xu, and Yong Zhao

A Hybrid Improved Multi-objective Particle Swarm Optimization Feature Selection Algorithm for High-Dimensional Small Sample Data 475
Xiaoying Pan, Jun Sun, and Yufeng Xue

Exploiting Inhomogeneities of Subthreshold Transistors as Populations of Spiking Neurons 483
Etienne Mueller, Daniel Auge, and Alois Knoll

Backpropagation Neural Network with Adaptive Learning Rate
for Classification 493
Rujira Jullapak and Arit Thammano

Bidirectional Controlled Quantum Teleportation of Two-Qubit
State via Eight-Qubit Entangled State 500
Jinwei Wang

Incremental Bayesian Classifier for Streaming Data
with Concept Drift 509
Peng Wu, Ning Xiong, Gang Li, and Jinrui lv

DDoS Detection Method Based on Improved Generalized
Entropy 519
Jiaqi Li, Xu Yang, Hui Chen, Haoqiang Lin, Xinqing Chen,
and Yanhua Liu

Subway Train Time-Energy Alternative Set Generation
Considering Passenger Flow Data Based on Adaptive NSGA-II
Algorithm 527
Benzun Huang, Dewang Chen, Yunhu Huang, and Wenzhu Lai

Research on FPGA Accelerator Optimization Based on Graph
Neural Network 536
Jin Wu, Xiangyang Shi, Wenting Pang, and Yu Wang

A Weighted Naive Bayes for Image Classification Based
on Adaptive Genetic Algorithm 543
Zhurong Wang, Qi Yan, Zhanmin Wang, and Xinhong Hei

Quantum Voting Protocol Based on Blind Signature 551
Qiang Yuwei, Chen Sihao, Li Na, and Bai Qian

Computer Vision (17)

Face Detection and Tracking Algorithm Based on Fatigue Driving 561
Zhe Li and Jing Ren

Algorithm Application Based on YOLOX Model in Health
Monitoring of Elderly Living Alone 568
Zhe Li and Jing Dang

Abdominal Multi-organ Localization with Adaptive Random
Forest in CT Images 575
Ruihao Wang, Jiaxin Tan, Laquan Li, and Shenhai Zheng

An Indoor Floor Location Method Based on Minimum Received
Signal Strength (RSS) Dynamic Compensation and Multi Label
Classification 584
Mingzhi Han and Yongyi Mao

An Improved Multi-dimensional Weighted K-nearest Neighbor Indoor Location Algorithm 592
Yongyi Mao and Rong Liu

Image Caption Description Generation Method Based on Reflective Attention Mechanism 600
Qiao Pingan, Li Yuan, and Shen Ruixue

Research on Image Description Generation Method Based on Visual Sentinel 610
Pingan Qiao, Ruixue Shen, and Yuan Li

Improved Weakly Supervised Image Semantic Segmentation Method Based on SEC 621
Xingya Yan, Zeyao Zheng, and Ying Gao

A Document Image Quality Assessment Algorithm Based on Information Entropy in Text Region 629
Zongrui Zhang, Jian Qiu, and Hao He

RetinaHand: Towards Accurate Single-Stage Hand Pose Estimation 639
Zilong Xiao, Luojun Lin, Yuanxi Yang, and Yuanlong Yu

Land Use/Cover Change Estimation with Satellite Remote Sensing Images and Its Changing Pattern of Chebei Creek in the Rapid Urbanization Process 648
Junxiang Liu, Hongbin Wang, Zhong Xu, Weinan Fan, Wenxiong Mo, and Lin Yu

Image Fusion Method Based on Improved Framelet Transform 656
Weiwei Kong, Yang Lei, and Chi Li

Video Stream Forwarding Algorithm Based on Multi-channel Ring Buffer Technology 663
Yingbao Cui, Xiaoguang Huang, Jing Liu, Linyu Zhang, and Xiaodong Wang

An Intelligent Annotation Platform for Transmission Line Inspection Images 673
Jing Liu, Xiaodong Wang, Yingbao Cui, Xiaoguang Huang, Linyu Zhang, and Yang Zhang

Image Interpolation Algorithm Based on Texture Complexity and Gradient Optimization 682
Yinbo Wang and Huimin Du

A Defect Detection Method for Semiconductor Lead Frame Based on Gate Limite Convolution 693
Wanyu Deng, Jiahao Jie, Dunhai Wu, and Wei Wang

**YOLOx-M: Road Small Object Detection Algorithm Based
on Improved YOLOx** 707
Jiaze Sun and Di Luo

Knowledge Discovery (35)

Cross-view Geo-localization Based on Cross-domain Matching 719
Xiaokang Wu, Qianguang Ma, Qi Li, Yuanlong Yu, and Wenxi Liu

**Adaptive-Impulsive Synchronization of Uncertain Complex
Networks with Heterogeneous Nodes** 729
Qunjiao Zhang, M. A. Aziz-Alaoui, Cyrille Bertelle, and Li Wan

**Robustness Analysis and Optimization of Double-Layer Freight
Relationship Network in "The Silk Road Economic Belt"** 739
Fengjie Xie, Yuwei Cao, and Jianhong Yan

**Optimization of Safety Investment in Prefabricated Building
Construction Based on SSA-SVR** 754
Chunguang Chang and Xiaoxue Ling

**An Efficient Heuristic Rapidly-Exploring Random Tree
for Unmanned Aerial Vehicle** 764
Chunping Yin, Meijin Lin, Qun Liu, and Hongmei Zhu

A Review of Research on Automatic Scoring of English Reading 773
Xinguang Li, Xiaoning Li, Xiaolan Long, Shuai Chen, and Ruisi Li

Methods for Solving the Change Data Capture Problem 781
Liang Hao, Tao Jiang, Yatuan Lin, and Yitong Lu

**A Multi-attribute Decision-making Method for Probabilistic
Language VIKOR Based on Correlation Coefficient and Entropy** 789
Wenyu Zhang, Weina Luo, Xue Gao, Chuanqiang Zhang,
and Siyuan Zhao

**The Transfer of Perceptual Learning Between First-
and Second-Order Fine Orientation Discriminations** 798
Mingliang Gong, Tingyu Liu, and Lynn A. Olzak

**Some New Characterizations of Ideals in Non-involutive
Residuated Lattices** 810
Chunhui Liu

**An Analysis of College Students' Behavior Based on Positive
and Negative Association Rules** 819
Feng Hao, Long Zhao, Haoran Zhao, Tiantian Xu, and Xiangjun Dong

**Identification and Analysis on Factors in Establishing the Green
Supply Chain Contractual Relationship: Literature Review
Based on NVivo** 833
Qian Zhang, Hongmei Shan, Jiapan Wang, and Mengmeng Miao

**A Data-Based Approach for Computer Domain Knowledge
Representation** 841
Lin Zhou, Qiyu Zhong, and Shaohong Zhang

**No Reference Image Quality Assessment Based
on Self-supervised Learning** 849
Zhen Wei, Wanyu Deng, Qirui Li, and Huijiao Xu

**Research and Implementation of Remote Sensing Image Object
Detection Algorithm Based on Improved ExtremeNet** 859
Xia Zhang, Lanxin Wang, Yulin Ren, and Zhili Mao

**A Review Focusing on Knowledge Graph Embedding Methods
Exploiting External Information** 869
Yuxuan Chen and Jingbin Wang

**A Survey of Spectrum Sensing Algorithms Based on Machine
Learning** 888
Youyao Liu and Juan Li

**Terrain Classification of Hyperspectral Remote Sensing Images
Based on SC-KSDA** 896
Jing Liu, Yinqiao Li, Yue Ye, and Yi Liu

**A Survey of Counterfactual Explanations: Definition,
Evaluation, Algorithms, and Applications** 905
Xuezhong Zhang, Libin Dai, Qingming Peng, Ruizhi Tang,
and Xinwei Li

**Retrospective Characterization of the COVID-19 Epidemic
in Four Selected European Countries Via Change Point Analysis** 913
Carmela Cappelli

**Analyzing the Time-Delay Correlation Between Operational
and State Parameters of Blast Furnace: A DTW and Clustering
Based Method** 921
Wenji Wang, Denghui Hao, and Yin Zhang

Multi-scale Algorithm and SNP Based Splice Site Prediction 930
Jing Zhao, Bin Wei, and Yaqiong Niu

**ACP-ST: An Anticancer Peptide Prediction Model Based
on Learning Embedding Features and Swin-Transformer** 939
YanLing Zhu, Shouheng Tuo, Zengyu Feng, and TianRui Chen

Research on Genome Multiple Sequence Alignment Algorithm
Based on Third Generation Sequencing 947
Zhiyu Gu, Junchi Ma, Xiangqing Meng, and Hong He

Psychological Portrait of College Students from the Perspective
of Big Data 956
Wen Man, Yu Zhu, Tingting Zhu, and Jie Zhu

A Multi-stage Event Detection Method 968
Xiaoshuo Feng, Zeyu Lv, Wandong Xue, Zhengping Sun,
and Dongqi Wang

Influencing Factor Analysis for Information Technology
Training Institutions 974
Zhenzhen Li, Peng Yang, Zhenhua Yuan, Yan Chen,
Shaohong Zhang, and Deying Liu

Predicting Impact of Published News Headlines Using Text
Mining and Classification Techniques 983
Parikshit Banerjee, Usha Ananthakumar, and Shubham Singh

Public Opinion Analysis for the Covid-19 Pandemic Based
on Sina Weibo Data 993
Feng Wang and Yunpeng Gong

A Survey on Temporal Knowledge Graphs-Extrapolation
and Interpolation Tasks 1002
Sulin Chen and Jingbin Wang

Analysis of Abnormal User Behavior on the Internet Based
on Time Series Feature Fusion 1015
Long Zhao, Yanyan Wang, Wenbin Fan, DeXuan Wang, Yuan Zhou,
Yekun Fang, and Enhong Chen

A Product Design Scheme Based on Knowledge Graph
of Material Properties 1024
Pengcheng Ding, Hongfei Zhan, Yingjun Lin, Junhe Yu, and Rui Wang

Blockchain Application in Emergency Materials Distribution
Under Disaster: An Architectural Design and Investigation 1034
Hongmei Shan, Yu Liu, Jing Shi, and Yun Zhang

Design and Implementation of Virtual Power Plant System
Based on Equipment-Level Power and Load Forecasting 1045
Xu Zhenan, Liu Zesan, Meng Hongmin, Huang Shu, Wen Aijun,
Li Shan, Jin Siyu, and Cui Wei

Intrusion Detection Method Based on Minkowski Distance
Negative Selection 1056
Yang Lei and Wenxuan Bie

Machine Learning and Data Science (19)

Multi-scale Prototypical Network for Few-shot Anomaly
Detection 1067
Jingkai Wu, Weijie Jiang, Zhiyong Huang, Qifeng Lin,
Qinghai Zheng, Yi Liang, and Yuanlong Yu

Distributed Nash Equilibrium Seeking with Preserved Network
Connectivity 1077
Qingyue Wu

Heterogeneous Network Representation Learning Guided
by Community Information 1087
Hanlin Sun, Shuiquan Yuan, Wei Jie, Zhongmin Wang,
and Sugang Ma

Prediction and Allocation of Water Resources in Lanzhou 1095
Lei Tang, Simeng Lin, Zhenheng Wang, and Kanglin Liu

A Deterministic Effective Method for Emergency Material
Distribution Under Dynamic Disasters 1105
Yingfei Zhang, Xiangzhi Meng, Ruixin Wang, and Xiaobing Hu

Unsupervised Concept Drift Detectors: A Survey 1117
Pei Shen, Yongjie Ming, Hongpeng Li, Jingyu Gao,
and Wanpeng Zhang

Music Video Search System Based on Comment Data and Lyrics 1125
Daichi Kawahara, Kazuyuki Matsumoto, Minoru Yoshida,
and Kenji Kita

Data Analysis of Nobel Prizes in Science Using Temporal Soft Sets 1136
Feng Feng, Jing Luo, Jianke Zhang, and Qian Wang

Few-Shot Learning for Aspect-Based Sentiment Analysis 1146
Heng Ruan, Xiaoge Li, Xianliang Li, Huikai Jiang, and Yingchao Li

Multimodal Incremental Learning via Knowledge Distillation
in License Classification Tasks 1158
Jian Qiu, Zongrui Zhang, and Hao He

FERTNet: Automatic Sleep Stage Scoring Method Based
on Frame Level and Epoch Level 1167
Xuebin Xu, Chen Chen, Kan Meng, Xiaorui Cheng, and Haichao Fan

Container Load Prediction Based on Extended Berkeley Packet
Filter 1176
Zhe Qiao, LiJun Chen, and YuXuan Bai

A Review on Pre-processing Methods for Fairness in Machine
Learning 1185
Zhe Zhang, Shenhang Wang, and Gong Meng

Analysis of Test Scores of Insurance Salesman Based
on Improved K-means Algorithm 1192
Wei Bai and Jianhua Liu

The Difference Between the Impact of Intellectual Capital
on Corporate Market Value and Book Value in the Taiwan
Electronics Industry GMM Method 1202
Li-Wei Lin, Xuan-Ze Zhao, Shu-Zhen Chen, and Kuo-Liang Lu

A Synchronous Secondary Index Framework Based
on Elasticsearch for HBase 1210
Xiaohui Lin, Wenzhong Guo, and Kun Guo

MOOC Resources Recommendation Based on Heterogeneous
Information Network 1219
Shuyan Wang, Wei Wu, and Yanyan Zhang

Model Inversion-Based Incremental Learning 1228
Dianbin Wu, Weijie Jiang, Zhiyong Huang, Qinghai Zheng,
Xiaodong Chen, WangQiu lin, and Yuanlong Yu

Knowledge Representation by Generic Models for Few-Shot
Class-Incremental Learning 1237
Xiaodong Chen, Weijie Jiang, Zhiyong Huang, Jiangwen Su,
and Yuanlong Yu

Information Systems (28)

A Medical Privacy Protection Model Based on Threshold
Zero-Knowledge Protocol and Vector Space 1251
Yue Yang, Rong Jiang, Chenguang Wang, Lin Zhang, Meng Wang,
Xuetao Pu, and Liang Yang

Hierarchical Medical Services Data Sharing Scheme Based
on Searchable Attribute Encryption 1259
Chenguang Wang, Rong Jiang, Lin Zhang, Meng Wang, Yue Yang,
Xuetao Pu, and Liang Yang

Spatial and Temporal Distribution Analysis of Regional
Industrial Patents of Injection Molding Machine 1267
Chengfan Ye, Hongfei Zhan, Yingjun Lin, Junhe Yu, and Rui Wang

Design of Frequency Reconfigurable Antenna Based on Liquid
Metal 1274
Jingjing Ren and Xiaofeng Yang

About the Parsing of NMEA–0183 Format Data Streams in GPS 1282
Siyao Dang, Haisheng Huang, and Xin Li

Dynamic Computing Offloading Strategy for Multi-dimensional Resources Based on MEC 1290
Jihong Zhao, Zihao Huang, Xinggang Luo, Gaojie Peng, and Zhaoyang Zhu

A Novel Ultra Wideband Circularly Polarized Antenna for Millimeter Wave Communication 1304
Zhong Yu, Zhenghui Xin, Yanli Cui, Li Shi, and Dixuan Liu

Calculation of Phase Difference and Amplitude Ratio Based on ADRV9009 1313
Jin Wu, Heng Wen, Xiangyang Shi, and Ling Yang

Cross-Polarized Directional Antenna with Y-Shaped Coupling Feed Structure for Ultra-wideband 1320
Zhong Yu, Li Shi, and ZhengH ui Xin

Scattering Properties of High Order Bessel Vortex Beam by Perfect Electrical Conductor Objects 1329
Zhong Yu, Yanli Cui, and Li Shi

A Zero Trust and Attribute-Based Encryption Scheme for Dynamic Access Control in Power IoT Environments 1338
Wenhua Huang, Xuemin Xie, Ziying Wang, and Jingyu Feng

A Parallel Implementation of 3D Graphics Pipeline 1346
Wenjiong Fu, Tao Li, and Yuxiang Zhang

From Source Code to Model Service: A Framework's Perspective 1355
Jing Peng, Shiliang Zheng, Yutao Li, and Zhc Shuai

An Efficient Software Test Method for the Autonomous Mobile Robot Control Program 1363
Chuang Cao, Xiaoxiao Zhu, and Xiaochen Lai

Forward Kinematics and Singularity Analysis of an Adjusted-DOF Mechanism 1371
Junting Fei, Qingxuan Jia, Gang Chen, Tong Li, and Yifan Wang

A New Visual and Inertial and Satellite Integrated Navigation Method Based on Point Cloud Registration 1385
Ping'an Qiao, Ruichen Wu, Jinglan Yang, Jiakun Shi, and Dongfang Yang

Path-Following Formation Control of Multi-mobile Robots Under Single Path 1398
Zhangyi Zhu, Wuxi Shi, and Baoquan Li

FPGA Prototype Verification of FlexRay Communication
Controller Chip 1407
Zejun Liu and Xiaofeng Yang

Design of Extended Interface of the UAV Flight Control
Computer 1414
Qintao Wang and Xiaofeng Yang

A Dynamic Selective Replication Mechanism for the Distributed
Storage Structure 1423
Siyi Han, Youyao Liu, and Huinan Cai

An Effective Algorithm for Pricing Option
with Mixed-Exponential Jump and Double Stochastic
Volatility 1433
Sumei Zhang, Panni Liu, and Zihao Liao

Data Mining and Knowledge Management for O2O Services,
Brand Image, Perceived Value and Satisfaction Towards
Repurchase Intention in Food Delivery 1441
Alfred Tjandra and Zhiwen Cai

Analysis of Data Loss and Disclosure in the Process of Big Data
Governance in Colleges and Universities in the Smart Era 1453
Wen Man, Yu Zhu, and Jingyi Zhang

Digital Twin System Design for Textile Industry 1463
Qingjin Wu, Danlin Cai, and Daxin Zhu

Design of FPGA Circuit for SHA-3 Encryption Algorithm 1471
Yuxiang Zhang, Wenjiong Fu, and Lidong Xing

L-Band Broadband Dual-Polarized Dipole Phased Array 1479
Jin Wu, Ruiqing Guo, Yu Wang, and Heng Wen

Design and FPGA Implementation of Numerically Controlled
Oscillators with ROM Look-Up Table Structure 1486
Han Zhang, Haisheng Huang, and Xin Li

Parallelization Designs of SpMV Using Compressed Storage
for Sparse Matrices on GPU 1494
Jianxin Wei

Author Index 1503

Knowledge Discovery (35)

Cross-view Geo-localization Based on Cross-domain Matching

Xiaokang Wu[✉], Qianguang Ma, Qi Li, Yuanlong Yu[✉], and Wenxi Liu

College of Computer and Data Science, Fuzhou University, Fuzhou, China
200320081@fzu.edu.cn, 200327074@fzu.edu.cn, 200320074@fzu.edu.cn,
yu.yuanlong@fzu.edu.cn, wenxiliu@fzu.edu.cn

Abstract. As a recently emerging problem, cross-view geo-localization aims at finding image pairs captured from different views (e.g., drone and satellite views) or domains yet same location, which can be widely employed in various applications. However, unlike traditional scene classification problem, it faces several challenges, including large intra-class distance and small inter-class distance caused by domain gap, as well as redundant contextual information and visual distractors across views. To address the concerns, we propose a novel cross-domain matching framework to handle this task, which measures the similarity for query and candidate images from two different domains. Comparing to prior classification based framework, our matching based framework is better suited for the task by forcing the model to learn discriminative features for scenes. Moreover, to aid cross-domain matching, we propose a matching-oriented feature modulation scheme, in which we not only apply a large-view attention module to enhance spatial features but also employ channel shuffling to loose the correlation of key feature semantics and distractors in the respective domains. Last, we conduct experiments to show that our model achieves the state-of-the-art performance and surpasses the competing method by a large margin on the public benchmarks.

Keywords: Drone · Geo-localization · Cross-domain image matching

1 Introduction

Cross-view geo-localization aims to address the recently emerging problem of finding image pairs captured from different view angles yet the same location in a geo-tagged image gallery. In practical application, cross-view geo-localization can exist in many field, e.g., autonomous driving, event detection, robot navigation and so on. [1–4].

Cross-view geo-localization faces several major challenges. The major challenge is that, the cross-view localization task suffers from large intra-class distance caused by domain gap and small inter-class distance especially in satellite views. As shown in Fig. 1, different buildings in the satellite views appear to be similar, while there exists a significant gap between the same building from the drone view and satellite view. Thus, it leads to a major challenge to distinguish them across views.

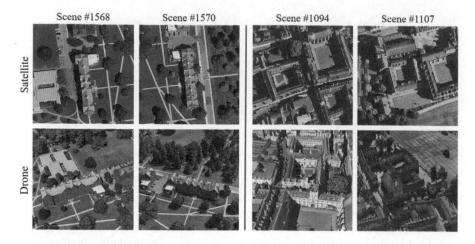

Fig. 1. The cross-view geo-localization task suffers from a large intra-class distance and a small inter-class distance.

Second, in most images, geographic targets (e.g., landmarks or buildings) only occupy small areas of images, while the rest of images is background or distractive visual information. Another challenge is the misaligned scale and mismatched image content between the query image and candidate images, which are captured from two significantly different views.

To remedy the aforementioned shortcomings, we propose a novel cross-domain matching framework to handle this task, which measures the similarity for query and candidate images from two different domains (i.e., drone view or satellite view). Comparing to the existing classification based framework, performing a cross-domain image matching encourages the model to extract discriminative features for recognizing scenes while preventing the distraction of irrelevant background contexts. To facilitate cross-domain matching in our framework, we propose a matching-oriented feature modulation scheme to emphasize on critical features and reduce the negative impact of visual distractors. To relieve this concern, we propose a simple yet effective practice, i.e., shuffling the feature channels, so as to loose the channel correlation and reduce the co-dependency of semantics, and then we employ another large-view spatial attention module to enhance and yield the domain sharing features.

In the experiment, our proposed model was fully tested and proved that our model can achieve the most advanced performance of this task.

In summary, the main contributions of this work are below:

- In order to resolve the cross-view geo-localization task, we propose a novel cross-domain matching framework, instead of a scene classifier, which effectively encourages to learn discriminative features.
- We present a matching-oriented feature modulation scheme that obtains the domain-sharing visual features via feature shuffling and large-view spatial attention for cross-domain scene matching.
- We conduct experiments to show the state-of-the-art performance of the proposed model on the public benchmarks.

2 Related Work

2.1 Cross-View Geo-localization

In cross-view geo-localization, a query image of a particular scene is given, such as an image of a landmark building and the purpose of the task is to find images containing this scene in a series of image sets that may be captured from different platforms [5,6]. Cross-view geo-localization is very promising and challenging because it can be implemented in many applications, such as person re-identification [7], remote sensing [8], medical image search [9], among many others.

Cross-perspective geo-location has attracted more and more research in recent years, Many works [2,10–12] have been dedicated to addressing this problem. Benefited from recent advance of deep learning, it provides a more accurate alternative for cross-view image geo-localization and has recently dominated this area [12–17]. However, these methods typically ignore context information and focus only on global information. Unlike existing works, the proposed approach includes not only the relevant context, but also the landmark architecture. We propose a large-view spatial attention module that can focus on extracting features of landmark buildings and their contexts occupying large image regions, and a channel shuffling module that increases model robustness, misalignment between query and candidate images.

3 Method

In this section, we first introduce the motivation and principles of our proposed cross-domain matching framework. Next, we describe our proposed matching-oriented feature modulation scheme.

3.1 Our Cross-Domain Matching Framework

Network Architecture As illustrated in Fig. 2, given a query image I_Q and a candidate image I_C from two separate domains, our proposed model F extracts their visual features through a shared network as X_Q and X_C, which follows the practice of Siamese network architecture. Next, the extracted features are passed through our proposed matching-oriented feature modulation scheme (please refer to Sec. 3.2 for details), which intends to modulate the features via spatial feature enhancement and disentangling channel correlation to make it suited for matching task. Then, inspired by [18], the modulated features are re-organized through a square-ring partition (SRP) module, which divides the high-level semantic features into four "square-rings" to yield four vectors of 1×2048 dimension on each network branch. Following the practice of [18], SRP is dedicated to encode the structural information of the images from satellite view or drone view for scene classification. Formally, the generated vectors or feature embeddings of the query and candidate images along the four square-rings can be denoted as $\mathbf{x}_Q^i$ and $\mathbf{x}_C^i$ ($i = \{1, \ldots, 4\}$), respectively. In final, for each network branch, a fully connected layer f is employed to classify the vectors into K scene classes, which is supervised by a cross-entropy loss. More importantly, to accomplish the matching of the query and candidate images, we introduce MSE losses to measure the distance between the feature embeddings of the query and the candidate images. Hence, the overall objective function is below:

$$\mathcal{L} = \sum_i \mathcal{L}_{ce}(f(\mathbf{x}_C^i), \mathbf{y}) + \sum_i \mathcal{L}_{ce}(f(\mathbf{x}_Q^i), \mathbf{y}) + \lambda \sum_i \mathcal{L}_{mse}(\mathbf{x}_C^i, \mathbf{x}_Q^i), \quad (1)$$

where $f(\cdot)$ represents the fully-connected layer as classifier, and $\mathbf{y}$ refers to the ground-truth label of scene class. $\mathcal{L}_{ce}$ denotes the cross entropy loss, while $\mathcal{L}_{mse}$ is the MSE loss that encourages the model to reduce the distance between the feature embeddings of I_C and I_Q, i.e., $\mathcal{L}_{mse}(\mathbf{x}_C^i, \mathbf{x}_Q^i) = \|\mathbf{x}_C^i - \mathbf{x}_Q^i\|_2^2$. λ refers to the balance weight.

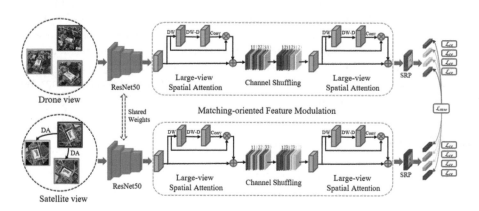

Fig. 2. Overview of the proposed matching based framework. The framework based on a Siamese network architecture is composed of two branches.

3.2 Matching-Oriented Feature Modulation

We propose the matching-oriented feature modulation scheme that refines the CNN-based deep features to aid the task of image matching. Its structure is illustrated in Fig. 2. In specific, it first employs a large-view spatial attention (LSA) module f_{LSA} to strengthen the features with the reference to rich contextual information, so it is able to emphasize on important features. However, the strong bond among semantics in the respective domains does not always benefit visual matching across domains. To hurdle this problem, inspired by [19], we propose to incorporate channel shuffling f_{CS} to loose the bond among channels, and then apply another LSA module f'_{LSA} to strengthen the domain-agnostic features. The entire process can be simplified as $\hat{X}_z = f'_{LSA}(f_{CS}(f_{LSA}(X_z)))$ $(X_z \in \{X_Q, X_C\})$. In the following, we will describe the main structure of our proposed scheme.

Large-View Spatial Attention. We propose to use a large-kernel convolutional attention module that pays more attention to the contextual information of local regions. The large-kernel convolutional attention module can better discriminate the key building area and context via the perception of a large image area. Yet, the large kernel may significantly increase the computation overhead. Thus, we can decompose the standard convolution into two cascaded DepthConv that performs computation along each channel, so as to reduce the computation burden.

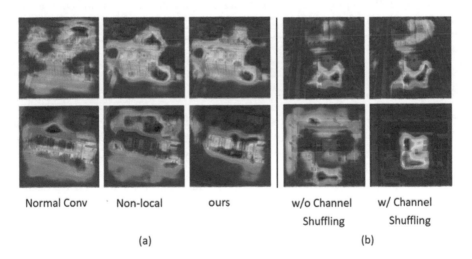

Normal Conv   Non-local   ours   w/o Channel   w/ Channel
                                      Shuffling   Shuffling

(a)   (b)

Fig. 3. Fig a show that visualizations for the spatial attention module using the normal convolution with 3×3 kernel (left), non-local attention (middle), and our large-view attention module (right). Fig b show that visual comparison produced by the model with or without channel shuffling.

As shown in Fig. 2, the residual branch of LSA can be divided into three components: (1) a spatial local convolution, (2) a spatial large-view convolution, and (3) a channel convolution. The underlying principle is that, the $K \times K$ convolution can be approximated as the combination of several layers, i.e., a $\lceil \frac{K}{d} \rceil \times \lceil \frac{K}{d} \rceil$ depth-wise dilation convolution with dilation d, a $(2d-1) \times (2d-1)$ depth-wise convolution, and a 1×1 convolution. Hence, the LSA module is able to capture longer-range dependency with extra slight computational cost and parameters, and produce a large-view attention map that can be further adopted to enhance features.

Purpose of Two LSA Modules. As shown in Fig. 2, we deploy the LSA module twice. For the first LSA, it receives the features of different domains, and thus LSA is able to enhance the features via contextual information to roughly adapt for matching. For the second LSA that is placed after the channel shuffling module, it receives the features whose channel-wise correlation is distengled, and its goal is to further consolidate the features.

Channel Shuffling Based Feature Enhancement. As shown in Fig. 2, the features enhanced through f_{LSA} are adapted for matching. However, due to the features are extracted by a shared network backbone, there remains strong correlation among semantics in the respective domains, which may not be suited for matching. To relieve the concern, we propose a simple yet effective approach by incorporating a feature channel shuffling module. It re-orders the feature channels in a certain pattern, which can effectively loose the bond of the correlated channels. Comparing to the purpose of the shuffling operation in [19] that is originally deployed to compensate the limitation of group convolution, the incorporation of channel shuffling in our model has a different purpose.

4   Experiments

4.1   Implementation Details

Model training. We employ ResNet-50 [20] with pre-trained weights on ImageNet [21] as backbone of siamese network to extract visual features. In practice [3,18], During training, our model employ Stochastic Gradient Descent (SGD) as the optimizer with a momentum of 0.9, a weight decay of 0.0005, and a mini-batch of 32. We trained 120 epochs, and the learning rate decayed 0.1 for every 80 epochs.

4.2   Comparison with the State-of-the-Arts

As shown in Table 1, we compare the proposed method with other competing methods on *University-1652*. In [3], they propose a simple two-branch network and try different loss functions including instance loss [3], contrastive loss [13], triplet loss [22], and soft margin triplet loss [12]. Besides, we also compare against LPN [18] for our task. LPN proposes to divide high-level features into several parts in square ring partitions, and then utilizes a classifier module to predict the geo-label of each part. Note that, in Table 1, LPN* indicates that an additional training set collected from Google Images is not deployed in the training phase on top of the LPN. Comparing to these methods, our approach achieves the state-of-the-art performance for both tasks, improving both R@1 and AP by at least about 3% advantage.

Table 1. Comparison with the state-of-the-art results.

Method	Drone → Satellite		Satellite → Drone	
	R@1	AP	R@1	AP
Instance loss [3]	58.49	63.31	71.18	58.74
Contrastive loss [13]	52.39	57.44	63.91	52.24
Triplet loss [22]	55.18	59.97	63.62	53.85
Soft margin triplet loss [12]	53.21	58.03	65.62	54.47
LPN* [18]	74.18	77.39	85.16	73.68
LPN [18]	75.93	79.14	86.45	74.79
Ours	79.19	82.07	88.30	78.56

4.3 Ablation Studies

We designed several ablation studies to verify the validity of the components in our model

Modules of our framework. As shown in Table 2, we analyze the effectiveness of our matching-based framework and the matching-oriented feature modulation, dubbed MFM.

As the baseline, we employ the vanilla Siamese classification network. For comparison, we first upgrade it into a cross-domain matching based framework by adding matching loss and we adopt data resampling strategy to set up image pairs for matching. As observed, it gains more than 2% improvements in terms of R@1 and AP in both tasks and surpasses the other comparing methods. After incorporating MFM, our model is improved by 0.9–1.7%, which fully demonstrates the effectiveness of our MFM for our network.

Table 2. Effectiveness of our proposed modules.

Matching	MFM	Drone → Satellite		Satellite → Drone	
		R@1	AP	R@1	AP
		75.48	78.69	84.96	73.94
✔		78.29	81.27	86.59	77.22
✔	✔	79.19	82.07	88.30	78.56

Matching-oriented feature modulation. Our proposed MFM module consists of three sub-modules: the first large-view spatial attention f_{LSA}, channel shuffling f_{CS}, and the second large-view spatial attention f'_{LSA}. We show the ablation study on the effectiveness of these sub-modules in the Table 3.

In experiments, our baseline model is the vanilla matching framework without LSA and channel shuffling. As observed in the Table 3, incorporating LSA or CS leads to performance gain to a certain extent, but combining them can reach the optimal performance thanks to the collaboration of spatial attention and domain-specific feature adaptation.

In the following, we first study the effectiveness of the first LSA. We can observe that, without f_{LSA}, the performance of the model obviously drops in both tasks, comparing to the baseline model. This is because, the first LSA plays the most important role to adapt features for matching. Besides, without f_{CS}, the feature modulation is essentially based on two cascaded LSA structures, which also demonstrates slightly worse performance than baseline. Comparing to the above structures, our model without the second LSA surpasses the baseline, which implies that channel shuffling plays an important role for the cross-domain matching task. Note that, for the second task "Satellite → Drone", this model can achieve obvious improvement over baseline with nearly 1% boost in R@1 and more than 1.4% gain in mAP. Last, our proposed structure achieves the optimal performance for both tasks in most metrics, which reflects the importance of f'_{LSA}.

Table 3. Ablation study on our feature modulation module.

f_{LSA}	f_{CS}	f'_{LSA}	Drone → Satellite		Satellite → Drone	
			R@1	mAP	R@1	mAP
			78.29	81.27	86.59	77.22
✓			78.72	81.62	85.88	77.65
	✓		78.47	81.45	87.16	77.97
	✓	✓	77.31	80.37	86.59	77.14
✓		✓	78.16	81.16	85.45	76.67
✓	✓		78.75	81.93	87.45	*78.68*
✓	✓	✓	*79.19*	*82.07*	*88.30*	78.56

5   Conclusion

In this work, we propose a novel cross-domain matching framework to handle the cross-view geo-localization task, which measure the similarity of query and candidate images from two different domains. Comparing to prior classification based framework, our matching based framework is better suited for the task by encouraging the model to learn discriminative features for scenes. Moreover, we propose a matching-oriented feature modulation scheme to refine features, in which we not only apply a large-view attention module to enhance spatial features but also employ channel shuffling to disentangle the correlation of key feature semantics and distractors. Last, we conduct experiments to show that our model achieves the state-of-the-art performance and surpasses the competing method by a large margin on the public benchmarks.

Acknowledgements. This work was supported by National Natural Science Foundation of China (NSFC) under grant 61873067, and University-Industry Cooperation Project of Fujian Provincial Department of Science and Technology under grant 2020H6101.

References

1. Liu, L., Li, H.: Lending orientation to neural networks for cross-view geo-localization. In: CVPR
2. Shi, Y., Liu, L., Yu, X., Li, H.: Spatial-aware feature aggregation for image based cross-view geo-localization. NIPS **32** (2019)
3. Zheng, Z., Wei, Y., Yang, Y.: University-1652: a multi-view multi-source benchmark for drone-based geo-localization. In: ACM MM
4. Workman, S., Jacobs, N.: On the location dependence of convolutional neural network features. In: Proceedings of the IEEE Conference on Computer Vision and Pattern Recognition Workshops
5. Babenko, A., Lempitsky, V.: Aggregating local deep features for image retrieval. In: IEEE
6. Zheng, L., Yang, Y., Tian, Q.: Sift meets cnn: a decade survey of instance retrieval. IEEE **40**(5)
7. Zheng, L., Shen, L., Tian, L., Wang, S., Wang, J., Tian, Q.: Scalable person re-identification: a benchmark. In: Proceedings of the IEEE International Conference on Computer Vision
8. Chaudhuri, U., Banerjee, B., Bhattacharya, A.: Siamese graph convolutional network for content based remote sensing image retrieval. Comput. Vision Image Understand. **184**
9. Nair, L.R., Subramaniam, K., Prasannavenkatesan, G.: A review on multiple approaches to medical image retrieval system. In: Intelligent Computing in Engineering
10. Shi, Y., Yu, X., Liu, L., Zhang, T., Li, H.: Optimal feature transport for cross-view image geo-localization. In: Proceedings of the AAAI Conference on Artificial Intelligence, vol. 34
11. Shi, Y., Yu, X., Campbell, D., Li, H.: Where am I looking at? joint location and orientation estimation by cross-view matching. In: Proceedings of the IEEE/CVF Conference on Computer Vision and Pattern Recognition
12. Hu, S., Feng, M., Nguyen, R.M., Lee, G.H.: Cvm-net: cross-view matching network for image-based ground-to-aerial geo-localization. In: Proceedings of the IEEE Conference on Computer Vision and Pattern Recognition
13. Lin, T.Y., Cui, Y., Belongie, S., Hays, J.: Learning deep representations for ground-to-aerial geolocalization. In: Proceedings of the IEEE Conference on Computer Vision and Pattern Recognition

14. Tian, Y., Chen, C., Shah, M.: Cross-view image matching for geo-localization in urban environments. In: Proceedings of the IEEE Conference on Computer Vision and Pattern Recognition

15. Vo, N.N., Hays, J.: Localizing and orienting street views using overhead imagery. In: European Conference on Computer Vision

16. Workman, S., Souvenir, R., Jacobs, N.: Wide-area image geolocalization with aerial reference imagery. In: Proceedings of the IEEE International Conference on Computer Vision

17. Zhai, M., Bessinger, Z., Workman, S., Jacobs, N.: Predicting ground-level scene layout from aerial imagery. In: Proceedings of the IEEE Conference on Computer Vision and Pattern Recognition

18. Wang, T., Zheng, Z., Yan, C., Zhang, J., Sun, Y., Zheng, B., Yang, Y.: Each part matters: Local patterns facilitate cross-view geo-localization. IEEE (2021)

19. Zhang, X., Zhou, X., Lin, M., Sun, J.: Shufflenet: an extremely efficient convolutional neural network for mobile devices. In: Proceedings of the IEEE Conference on Computer Vision and Pattern Recognition

20. He, K., Zhang, X., Ren, S., Sun, J.: Deep residual learning for image recognition. In: Proceedings of the IEEE Conference on Computer Vision and Pattern Recognition, pp. 770–778 (2016)

21. Deng, J., Dong, W., Socher, R., Li, L.J., Li, K., Fei-Fei, L.: Imagenet: a large-scale hierarchical image database. In: 2009 IEEE Conference on Computer Vision and Pattern Recognition, pp. 248–255. IEEE (2009)

22. Chechik, G., Sharma, V., Shalit, U., Bengio, S.: Large scale online learning of image similarity through ranking. J. Mach. Learn. Res. 11(3) (2010)

Adaptive-Impulsive Synchronization of Uncertain Complex Networks with Heterogeneous Nodes

Qunjiao Zhang[1]([⊠]), M. A. Aziz-Alaoui[2], Cyrille Bertelle[2], and Li Wan[1]

[1] School of Mathematical and Physical Sciences, RCNS, Wuhan Textile University,
Wuhan 430073, China
`qjzhang@wtu.edu.cn`, `2006040@wtu.edu.cn`
[2] UNIHAVRE, LMAH-LITIS, FR-CNRS-3335, ISCN, Normandie University, 76600
Le Havre, France
`aziz.alaoui@univ-lehavre.fr`, `cyrille.bertelle@univ-lehavre.fr`

Abstract. Based on adaptive-impulsive control method, synchronization of uncertain complex networks with heterogeneous nodes is considered in this paper. A novel control protocol is derived for the proposed uncertain complex networks model, which guarantee the achievement of synchronization and identification of uncertain node's parameters. Further, a simple corollary is given for adaptive-impulsive synchronization of networks with homogeneous nodes. Finally, take a scale-free complex network as an example, some numerical experiments illustrate the validity and feasibility of our proposed results.

Keywords: Adaptive-impulsive synchronization · Complex network · Heterogeneous nodes

1 Introduction

Complex dynamical networks widely exist in many man-made and natural systems, such as neural networks, biological networks, metabolic networks, citation networks, computer networks and social networks, which are greatly prosperous in ecology, physics, mathematics and social sciences [1,2].

Synchronization is a common phenomenon exhibited in practical complex networks, many kinds of synchronization were studied and many important results have been reported in the existing literatures [3–6]. In the same period, various of control methods, including pinning control, adaptive feedback control, impulsive control, have been explored to realize the synchronization of complex dynamical networks [7–11].

Different from continuous time control methods, impulsive control protocol is popular and efficient because it applies control on nodes only at discrete moments. Therefore, it is still important and widely used to synchronize chaotic systems and complex networks via impulsive control up to now. Wang et al.

© The Author(s), under exclusive license to Springer Nature Switzerland AG 2023
N. Xiong et al. (Eds.): ICNC-FSKD 2022, LNDECT 153, pp. 729–738, 2023.
https://doi.org/10.1007/978-3-031-20738-9_82

discussed the synchronization of coupled networks under mixed impulsive controllers based on a delayed inequality approach [12]. Combining with pinning and impulsive control, Ding et al. investigated the exponential synchronization of nonlinearly coupled complex networks with time-varying delays and stochastic disturbances [13]. Based on adaptive and impulsive control, Chen et al. studied adaptive-impulsive synchronization for uncertain chaotic systems [14]. Further, Wan and Sun proposed a new strategy to realize adaptive impulsive synchronization of chaotic systems [15].

In practical, complex networks are much more universal than systems, which can describe realistic situations more accurately. Consequently, Li et al. reported some results of uncertain networks with nonidentical topological structures, in which parameter identification and adaptive-impulsive synchronization were discussed [16]. In the next year, based on generalized Barbalat's Lemma, they further studied pinning and adaptive-impulsive synchronization of fractional-order complex networks [17]. Recently, quantized adaptive control and quantized impulsive control were combined to synchronize the considered complex networks with given goal dynamics [18]. An adaptive impulsive control scheme was presented for the chaotic-oscillation synchronous of generators networks [19]. Based on impulsive-adaptive control protocols, some criteria for cluster synchronization of nonlinearly delayed coupled Lur'e networks were derived [20].

In real-world complex networks, different individuals often display different dynamics, thus a complex network consisting of heterogeneous nodes can describe the real situation more accurately. Additionally, there exist a various of uncertain factors caused by internal or external noises. Motivated by the above reasons, adaptive-impulsive synchronization of general uncertain complex dynamical networks with heterogeneous nodes is further investigated in this paper.

2 Synchronization of Complex Networks with Heterogeneous Nodes via Adaptive-Impulsive Control

Let the isolate node be an n-dimensional dynamical system, which is given by

$$\dot{x}_i(t) = F_i(t, x_i(t), \Theta_i), \tag{1}$$

where $x_i(t) = (x_{i1}(t), x_{i2}(t), \ldots, x_{in}(t))^{\mathrm{T}} \in \mathbf{R}^n$ is the state vector, $\Theta_i \in \mathbf{R}^p$ is the parameter vector. If one separate the state variables and parameters, it can be expressed as

$$\dot{x}_i(t) = f_i(t, x_i(t)) + g_i(t, x_i(t)) \cdot \Theta_i, \tag{2}$$

where $f_i(t, x_i(t)) : \mathbf{R}^+ \times \mathbf{R}^n \to \mathbf{R}^n$ is a continuous nonlinear function vector, $g_i(t, x_i(t)) : \mathbf{R}^+ \times \mathbf{R}^n \to \mathbf{R}^{n \times p}$ is a continuous function matrix, and $\Theta_i \in \mathbf{R}^p$ is an unknown parameter vector.

In this paper, the vector-valued function $F_i(t, x_i(t), \Theta_i)$ is assumed to satisfy the uniform Lipschitz condition about Θ_i, that is

Assumption 1 (A1). *For any* $x_i(t) = (x_{i1}(t), x_{i2}(t), \ldots, x_{in}(t))^{\mathrm{T}}$ *and* $y_i(t) = (y_{i1}(t), y_{i2}(t), \ldots, y_{in}(t))^{\mathrm{T}}$, *there exists a positive constant* $L_i > 0$, *such that*

$$\|F_i(t, y_i(t), \Theta_i) - F_i(t, x_i(t), \Theta_i)\| \le L_i \|y_i(t) - x_i(t)\|.$$

Actually, **(A1)** is easily satisfied when the solution of Eq. (1) is bounded.

Consider a general complex network, which consists of N heterogeneous dynamical nodes. Each isolate node of this network is an n-dimensional system satisfying **(A1)**. Then the drive network is described by the following equations:

$$\dot{x}_i(t) = F_i(t, x_i(t), \Theta_i) + \sum_{j=1}^{N} c_{ij} A x_j(t), \quad i = 1, 2, \ldots, N, \tag{3}$$

where $x_i(t) = (x_{i1}(t), x_{i2}(t), \ldots, x_{in}(t))^{\mathrm{T}} \in \mathbf{R}^n$ is a state vector of ith node, $C = (c_{ij})_{N \times N}$ is a weighted configuration matrix, and $c_{ij} \ne 0$ when there is a directed coupling from node i to j ($j \ne i$); otherwise, $c_{ij} = 0$. The matrix $A = (a_{ij}) \in \mathbf{R}^{n \times n}$ is an inner connecting matrix for different components of each node. Another slaved network with impulsive controllers is designed as

$$\begin{cases} \dot{y}_i(t) = F_i(t, y_i(t), \hat{\Theta}_i) + \sum\limits_{j=1}^{N} c_{ij} A y_j(t), \ t \ne t_k, \\ \Delta y_i(t^+) = d_{ik}(y_i(t) - x_i(t)), \qquad t = t_k, \\ y_i(t_0) = y_{i0}, \end{cases} \quad k = 1, 2, \ldots, \tag{4}$$

where $y_i(t) = (y_{i1}(t), y_{i2}(t), \ldots, y_{in}(t))^{\mathrm{T}} \in \mathbf{R}^n$ is a response state vector of ith node. $\hat{\Theta}_i$ is an estimation of unknown parameters Θ_i, and the impulsive feedback gain received by the ith node at impulsive moment t_k is noted as d_{ik}. $\Delta y_i(t_k^+) = y_i(t_k^+) - y_i(t_k^-)$, $y_i(t_k^+) = \lim\limits_{t \to t_k^+} y_i(t)$, and at each t_k, the solution of (4) is left continuous, that is, $y_i(t_k^-) = y_i(t_k)$. The impulse time series satisfy $t_1 < t_2 < \cdots < t_k < t_{k+1} < \cdots$, $\lim\limits_{k \to \infty} t_k = \infty$, and $\tau_k = t_k - t_{k-1} < \infty$.

Let the error vectors be

$$e_i(t) = y_i(t) - x_i(t), \quad \tilde{\Theta}_i = \hat{\Theta}_i - \Theta_i, \quad i = 1, 2, \ldots, N. \tag{5}$$

According to the Eqs. (3)–(5), the error systems can be described by

$$\begin{aligned} \dot{e}_i(t) &= F_i(t, y_i(t), \hat{\Theta}_i) - F_i(t, x_i(t), \Theta_i) + \sum_{j=1}^{N} c_{ij} A y_j(t) - \sum_{j=1}^{N} c_{ij} A x_j(t) \\ &= \left[g_i(t, y_i(t)) \cdot \hat{\Theta}_i - g_i(t, y_i(t)) \cdot \Theta_i \right] + \left[F_i(t, y_i(t), \Theta_i) - F_i(t, x_i(t), \Theta_i) \right] \\ &\quad + \sum_{j=1}^{N} c_{ij} A e_j(t), \end{aligned} \tag{6}$$

$$e_i(t_k^+) = y_i(t_k^+) - x_i(t_k^+) = (1 + d_{ik}) e_i(t_k). \tag{7}$$

Theorem 1. *Suppose that **(A1)** holds and λ is the largest eigenvalue of $(C \otimes A) + (C \otimes A)^{\mathrm{T}}$, if there exists a constant $\gamma > 0$ such that*

$$2\alpha_k \tau_k + \ln \bar{d}_k + \gamma < 0, \qquad k = 1, 2, \ldots, \tag{8}$$

where $\alpha_k = \frac{1}{2} + \max\limits_i \{\hat{L}_i(t_k)\}$, $\bar{d}_k = \max\limits_i \{(1 + d_{ik})^2\}$. With the following updating laws

$$\begin{aligned}
\dot{\hat{\Theta}}_i &= -g_i^{\mathrm{T}}(t, y_i(t)) e_i(t), \\
\dot{\hat{L}}_i &= e_i^{\mathrm{T}}(t) e_i(t),
\end{aligned} \tag{9}$$

then the network (3) and slaved network (4) is adaptive-impulsive synchronous. Moreover, $\hat{\Theta}_i \to \Theta_i$ for any $1 \le i \le N$.

Proof. Construct the Lyapunov function, as has been done in [14]:

$$V(t) = \frac{1}{2} \sum_{i=1}^{N} e_i^{\mathrm{T}}(t) e_i(t) + \frac{1}{2} \sum_{i=1}^{N} (1 + d_{ik})^2 \left[(\hat{\Theta}_i - \Theta_i)^{\mathrm{T}} (\hat{\Theta}_i - \Theta_i) + (\hat{L}_i - L_i)^2 \right]. \tag{10}$$

One can notice that, the feedbacks $d_{ik} \ne 0$ only at the impulse moments t_k^+, but they remain zero in any impulse intervals. So for any $t \in (t_{k-1}, t_k)$, we have

$$\dot{V}(t) = \sum_{i=1}^{N} e_i^{\mathrm{T}}(t) \dot{e}_i(t) + \sum_{i=1}^{N} \dot{\hat{\Theta}}_i^{\mathrm{T}} (\hat{\Theta}_i - \Theta_i) + \sum_{i=1}^{N} \dot{\hat{L}}_i (\hat{L}_i - L_i),$$

along with Eq. (6) and **(A1)**, one further has

$$\dot{V}(t) \le \sum_{i=1}^{N} e_i^{\mathrm{T}}(t) \left[g_i(t, y_i(t)) \cdot (\hat{\Theta}_i - \Theta_i) \right] + \sum_{i=1}^{N} L_i e_i^{\mathrm{T}}(t) e_i(t) + \sum_{i=1}^{N} \sum_{j=1}^{N} e_i^{\mathrm{T}}(t) c_{ij} A e_j(t)$$

$$+ \sum_{i=1}^{N} \dot{\hat{\Theta}}_i^{\mathrm{T}} (\hat{\Theta}_i - \Theta_i) + \sum_{i=1}^{N} \dot{\hat{L}}_i (\hat{L}_i - L_i).$$

From Eq. (9), we know that $\hat{L}_i(t)$ is increasing monotonically, so $\hat{L}_i(t) \le \hat{L}_i(t_k)$ for $t \in (t_{k-1}, t_k]$. Substitute Eq. (9) into the above inequality and denote $\bar{L}_k = \max\limits_i \{\hat{L}_i(t_k)\}$, which can be written as

$$\dot{V}(t) \le E^{\mathrm{T}}(t)(C \otimes A) E(t) + \sum_{i=1}^{N} \hat{L}_i(t_k) e_i^{\mathrm{T}}(t) e_i(t)$$

$$= E^{\mathrm{T}}(t) \frac{(C \otimes A) + (C \otimes A)^{\mathrm{T}}}{2} E(t) + \bar{L}_k \sum_{i=1}^{N} e_i^{\mathrm{T}}(t) e_i(t)$$

$$\le \left(\frac{\lambda}{2} + \bar{L}_k \right) \sum_{i=1}^{N} e_i^{\mathrm{T}}(t) e_i(t) \le 2\alpha_k V(t), \tag{11}$$

where $E(t) = (e_1^{\mathrm{T}}(t), e_2^{\mathrm{T}}(t), \ldots, e_N^{\mathrm{T}}(t))^{\mathrm{T}}$ and $\alpha_k = \frac{1}{2} + \bar{L}_k$. Integral the above inequality, there is

$$V(t) \le V(t_{k-1}^+) e^{2\alpha_k(t - t_{k-1})}, \qquad t \in (t_{k-1}, t_k]. \tag{12}$$

Specially, for $t = t_k^+$, from Eqs. (7) and (10), then

$$V(t_k^+) = \frac{1}{2} \sum_{i=1}^{N} (1 + d_{ik})^2 e_i^{\mathrm{T}}(t_k) e_i(t_k)$$

$$+ \frac{1}{2} \sum_{i=1}^{N} (1 + d_{ik})^2 \left[(\hat{\Theta}_i - \Theta_i)^{\mathrm{T}}(\hat{\Theta}_i - \Theta_i) + (\hat{L}_i - L_i)^2 \right]$$

$$\leq \max_i (1 + d_{ik})^2 V(t_k^-) = \bar{d}_k V(t_k^-), \tag{13}$$

where $\bar{d}_k = \max_i \{(1 + d_{ik})^2\}$. Let $k = 1$ in the inequality (12), for any $t \in (t_0, t_1]$, we have

$$V(t) \leq V(t_0^+) e^{2\alpha_1(t - t_0)}.$$

Specially,

$$V(t_1) \leq V(t_0^+) e^{2\alpha_1(t_1 - t_0)}.$$

Since $V(t)$ is left continuous at t_k, from (13), we have

$$V(t_1^+) \leq \bar{d}_1 V(t_1) \leq V(t_0^+) \bar{d}_1 e^{2\alpha_1(t_1 - t_0)} = V(t_0^+) \bar{d}_1 e^{2\alpha_1 \tau_1}.$$

Repeating the similar process, there is

$$V(t_k^+) \leq V(t_0^+) \prod_{q=1}^{k} \bar{d}_q e^{2\alpha_q \tau_q}. \tag{14}$$

In terms of the inequality (8) in **Theorem** 1, we further know that

$$\bar{d}_q e^{2\alpha_q \tau_q} < e^{-\gamma}, \quad q = 1, 2, \ldots$$

Thus, the inequality (14) can be rewritten as

$$V(t_k^+) < V(t_0^+) e^{-\gamma k},$$

therefore, $V(t_k^+) \to 0$ as $k \to \infty$.

For $t \in (t_{k-1}, t_k]$, from Eq. (12), we have $V(t) \leq V(t_{k-1}^+) e^{2\alpha_k(t - t_{k-1})} \leq V(t_{k-1}^+) e^{2\alpha_k \tau_k}$, thus, one can derive that $V(t) \to 0$ as $k \to \infty$. Furthermore, along with Eq. (10), it follows that the state errors $e_i(t) \to 0$, the estimated parameters $\hat{\Theta}_i \to \Theta_i$ and $\hat{L}_i \to L_i (i = 1, 2, \ldots, N)$ when $t \to \infty$. That's to say, the adaptive-impulsive synchronization between the networks (3) and (4) occurs. The proof is done. $\qquad \square$

Corollary 1. *If the complex network is composed with N homogenous nodes, which is given by*

$$\dot{x}_i(t) = f(t, x_i(t)) + g(t, x_i(t)) \cdot \Theta + \sum_{j=1}^{N} c_{ij} A x_j(t), \tag{15}$$

then the unknown parameters Θ can be identified by the estimated $\hat{\Theta}$, with the impulsively controlled response network

$$
\begin{cases}
\dot{y}_i(t) = f(t, y_i(t)) + g(t, y_i(t)) \cdot \hat{\Theta} + \sum_{j=1}^{N} c_{ij} A y_j(t), \ t \neq t_k, \\
\Delta y_i(t^+) = d e_i(t), \quad t = t_k, \\
y_i(t_0) = y_{i0}.
\end{cases}
\qquad k = 1, 2, \ldots.
$$

(16)

If there exist a positive constant γ satisfying

$$
2\alpha_k \tau_k + 2\ln(1 + d) + \gamma < 0, \qquad k = 1, 2, \ldots,
$$

(17)

where d is the constant impulsive feedback gain which is independent on i and k, $\alpha_k = \frac{\lambda}{2} + \hat{L}(t_k)$, and λ is the largest eigenvalue of $(C \otimes A) + (C \otimes A)^{\mathrm{T}}$. Under the following updating laws

$$
\begin{aligned}
\dot{\hat{\Theta}} &= -\sum_{i=1}^{N} g_i^{\mathrm{T}}(t, y_i(t)) e_i(t), \\
\dot{\hat{L}} &= \sum_{i=1}^{N} e_i^{\mathrm{T}}(t) e_i(t),
\end{aligned}
$$

(18)

then the network (15) and the network (16) with delayed impulsive controllers is adaptive-impulsive synchronous. Furthermore, $\hat{\Theta} \to \Theta$.

3   Numerical Simulations

Example 1. Let's consider an uncertain complex network without coupling delay, which is composed with identical nodes and the inner-coupling matrix A is an identity matrix, i.e. $A = I_{n \times n}$. The node's dynamical function is given by the chaotic Lorenz system, which is described as

$$
F(t, x_i(t), \Theta) = \begin{pmatrix} a(x_{i2} - x_{i1}) \\ c x_{i1} - x_{i1} x_{i3} - x_{i2} \\ x_{i1} x_{i2} - b x_{i3} \end{pmatrix}, \quad i = 1, 2, 3 \ldots, 200,
$$

(19)

where the parameters $a = 10, b = 8/3, c = 28$ and $\Theta = (a, b, c)^{\mathrm{T}}$. For any state vectors $x(t) \in \mathbf{R}^3$ and $y(t) \in \mathbf{R}^3$ of Lorenz system, the **(A1)** is easily satisfied because the Lorenz attractor is bounded in certain region. Here, all the parameters a, b, c are assumed to be unknown in prior.

A B-A scale-free network with $N = 200, m_0 = 2, m = 2$ is taken as the network model, one can know that $\lambda = 0$ from this generated network. Choose the impulsive feedback gain constant $d = -0.99$, impulsive interval $\tau_k = 0.02$ and exponent positive constant $\gamma = 2$ respectively. For a suitable choice for $\bar{L}(t_k) = 160$, the inequality (17) in **Corollary** 1 can be easily verified with the all given parameters. That is, $2\alpha_k \tau_k + 2\ln(1+d) + \gamma = -0.8103 < 0$. More details about the relationship among $\hat{L}, d, \tau_k$ and γ, can be found in [14].

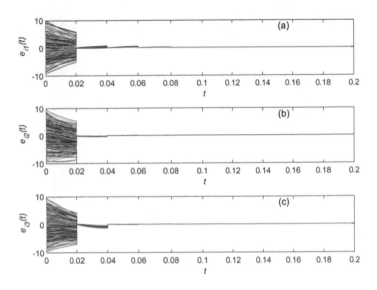

Fig. 1. The synchronization errors $e_{i1}(t), e_{i2}(t), e_{i3}(t)$ $(i = 1, 2, \ldots, 200)$.

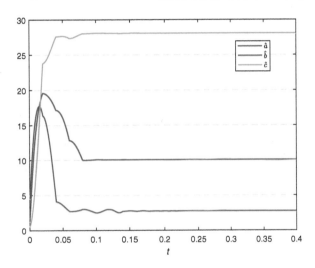

Fig. 2. The estimation of $\hat{a}$, $\hat{b}$, $\hat{c}$.

Figure 1 displays the adaptive-impulsive synchronization errors of $e_{i1}(t)$, $e_{i2}(t)$, $e_{i3}(t)$ under the updating laws (18). Obviously, all synchronization errors are converging to zero rapidly. Figure 2 shows the tracking process of unknown system parameters a, b, c during the synchronization occurs. Figure 3

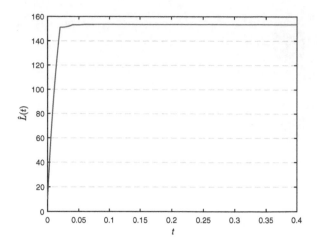

Fig. 3. The estimation of $\hat{L}(t)$.

allows us to check that $\hat{L}(t_k) \leq \bar{L}(t_k) = 160$. Actually, Eq. (19) is rewritten as

$$F(t, x_i(t), \Theta) = \begin{pmatrix} -a & a & 0 \\ c & -1 & 0 \\ 0 & 0 & -b \end{pmatrix} \begin{pmatrix} x_{i1} \\ x_{i2} \\ x_{i3} \end{pmatrix} + \begin{pmatrix} 0 \\ -x_{i1}x_{i3} \\ x_{i1}x_{i2} \end{pmatrix} \doteq Bx_i(t) + W(x_i(t)).$$

(20)

From Fig. 4, there exists a constant $M = 50$ satisfying $|x_{ij}(t)| \leq M$ and $|x_{ij}(t)| \leq M (j = 1, 2, 3)$. Similar to [21], then

$$\|F(t, y_i(t), \Theta) - F(t, x_i(t), \Theta)\| \leq (\|B\| + 2M)\|y_i(t) - x_i(t)\|. \qquad (21)$$

After calculating, $\|B\| + 2M = 30.0731 + 100 = 130.0731 < 140 < \bar{L}(t_k) = 160$. That's to say, an upper bound estimation of Lipchitz constant L is effective.

4   Conclusion

This paper presented an impulsive and adaptive control protocol for the synchronization of a class of unknown general complex networks with heterogeneous nodes. By applying the impulsive Lyapunov stability theory, some sufficient conditions have been derived to guarantee the synchronization for the proposed impulsive networks after strict mathematical induction. Furthermore, a simple corollary is given for adaptive-impulsive synchronization of uncertain networks with homogeneous nodes. Finally, a numerical example has been provided to illustrate the correctness of theoretical analysis and control schemes.

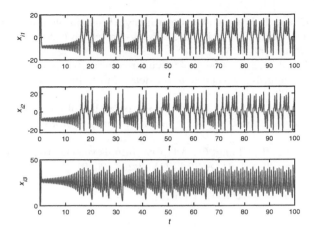

Fig. 4. The evolution of state variables $x_{i1}(t), x_{i2}(t), x_{i3}(t)$.

Acknowledgements. This work was supported by National Natural Science Foundation of China under Grant No. 61901308.

References

1. Strogatz, S.H.: Exploring complex networks. Nature **410**, 268–276 (2001)
2. Albert, R., Barabási, A.: Statistical mechanics of complex networks. Rev. Mod. Phys. **74**, 47–97 (2002)
3. Li, C., Sun, W., Kurths, J.: Synchronization between two coupled complex networks. Phys. Rev. E **76**, 046204 (2007)
4. Tang, H., Chen, L., Lu, J., Tse, C.K.: Adaptive synchronization between two complex networks with nonidentical topological structures. Physica A **387**(22), 5623–5630 (2008)
5. Chen, J., Jiao, L., Wu, J., Wang, X.: Adaptive synchronization between two different complex networks with time-varying delay coupling. Chin. Phys. Lett. **26**(6), 060505 (2009)
6. Wu, Z., Chen, G., Fu, X.: Outer synchronization of drive-response dynamical networks via adaptive impulsive pinning control. J. Franklin Inst. **352**(10), 4297–4308 (2015)
7. Yu, W., Chen, G.: On pinning synchronization of complex dynamical networks. Automatica **45**(2), 429–435 (2009)
8. Zhang, Q., Wu, X., Liu, J.: Pinning synchronization of discrete-Time complex networks with different time-varying delays. J. Syst. Sci. Complex **32**, 1560–1571 (2019)
9. Tu, Z., Ding, N., Li, L., Feng, Y., Zou, L., Zhang, W.: Adaptive synchronization of memristive neural networks with time-varying delays and reaction-diffusion term. Appl. Math. Comput. **311**(15), 118–128 (2017)
10. Hayakawa, T., Haddad, W., Volyanskyy, K.: Neural network hybrid adaptive control for nonlinear uncertain impulsive dynamical systems. Nonlinear Anal. Hybrid Syst. **2**(3), 862–874 (2008)

11. Zhang, Q., Chen, G., Wan, L.: Exponential synchronization of discrete-time impulsive dynamical networks with time-varying delays and stochastic disturbances. Neurocomputing **309**, 62–69 (2018)
12. Wang, Y., Lu, J., Li, X., Liang, J.: Synchronization of coupled neural networks under mixed impulsive effects: a novel delay inequality approach. Neural Netw. **127**, 38–46 (2020)
13. Ding, D., Tang, Z., Wang, Y., Ji, Z.: Synchronization of nonlinearly coupled complex networks: distributed impulsive method. Chaos Solitons Fractals **133**, 109620 (2020)
14. Chen, Y., Hwang, R., Chang, C.: Adaptive impulsive synchronization of uncertain chaotic systems. Phys. Lett. A **374**, 2254–2258 (2010)
15. Wan, X., Sun, J.: Adaptive impulsive synchronization of chaotic systems. Math. Comput. Simulat. **81**, 1609–1617 (2011)
16. Li, H., Jiang, Y., Wang, Z., Zhang, L., Teng, Z.: Parameter identification and adaptive-impulsive synchronization of uncertain complex networks with nonidentical topological structures. Optik **126**(24), 5771–5776 (2015)
17. Li, H., Hu, C., Jiang, Y., Wang, Z., Teng, Z.: Pinning adaptive and impulsive synchronization of fractional-order complex dynamical networks. Chaos Solitons Fractals **92**, 142–149 (2016)
18. Zhang, W., Li, C., Yang, S., Yang, X.: Exponential synchronization of complex networks with delays and perturbations via impulsive and adaptive control. IET Control Theory Appl. **13**(3), 395–402 (2019)
19. Zhu, D., Wang, R., Liu, C.: Synchronization of chaotic-oscillation permanent magnet synchronous generators networks via adaptive impulsive control. IEEE Trans. Circuits Syst. II **67**(10), 2194–2198 (2020)
20. Xuan, D., Tang, Z., Feng, J., Park, J.: Cluster synchronization of nonlinearly coupled Lur'e networks: delayed impulsive adaptive control protocols. Chaos Solitons Fractals **152**, 111337 (2021)
21. Zhou, J., Lu, J.: Topology identification of weighted complex dynamical networks. Physica A **386**, 481–491 (2007)

Robustness Analysis and Optimization of Double-Layer Freight Relationship Network in "The Silk Road Economic Belt"

Fengjie Xie, Yuwei Cao$^{(\boxtimes)}$, and Jianhong Yan

School of Modern Post, Xi'an University of Posts and Telecommunications, Xi'an 710061, China
xiefengjie@xupt.edu.cn, 786370994@stu.xupt.edu.cn,
yanjianhong@stu.xupt.edu.cn

Abstract. The "Silk Road Economic Belt" effectively promotes trade connectivity among countries and regions along the Eurasian continent. The construction of cross-border freight infrastructure is the realistic foundation of trade interconnection. As a part of infrastructure construction, the rational layout and efficient operation of freight relationship networks are very important for the development of trade. Therefore, we study the robustness analysis and optimization method of the double-layer freight relationship network of "The Silk Road Economic Belt". Considering the geospatial characteristics of the freighted relationship and the speed difference between railway and air transportation, a double-layer freight relationship network is constructed, and transportation distance is taken as weight. On this basis, deliberately attack different proportions of railway and airlines in the double-layer freight relationship network from large to small according to the weight of the edge. From the practical significance, taking transportation efficiency and transportation cost as important indicators to measure the robustness of the freight relationship network, the robustness of the double-layer freight relationship network is analyzed. Finally, a multi-objective model is constructed to optimize the robustness of the double-layer freight relationship network. To provide a reference for the cross-border transportation planning of "The Silk Road Economic Belt".

Keywords: Complex network · The silk road economic belt · Multimodal transport · Network robustness · Multi-objective model

1 Introduction

The Silk Road Economic Belt has effectively promoted trade connectivity among countries and regions along the Eurasian continent. Cross-border freight infrastructure provides a realistic basis for the trading connectivity of the "Silk Road

Foundation items: Project supported by the National Social Science Foundation, China (20BGL282)

N. Xiong et al. (Eds.): ICNC-FSKD 2022, LNDECT 153, pp. 739–753, 2023.
https://doi.org/10.1007/978-3-031-20738-9_83

Economic Belt", and is a prerequisite for the trade development of the "Silk Road Economic Belt". As part of cross-border freight transport infrastructure construction, the reasonable layout and efficient operation of the freight network is crucial to the trade development of "the Silk Road Economic Belt". At present, domestic and foreign studies on the freight network of "the Silk Road Economic Belt" focus on the freight network of a single mode of transportation, which can be mainly divided into railway freight network and air freight network according to different modes of transportation. The research of railway freight relationship networks focuses on the analysis of network structure characteristics [1], network quality [2] and service network design [3,4], Air cargo network focuses on the analysis of network structure characteristics [5], network layout and optimization [6].

A freight relation network is a freight network system composed of node cities and transportation routes between them. In the process of cargo transportation, when facing natural disasters such as extreme weather, will directly lead to the failure of freight lines, which will lead to the failure of cargo transportation on time. When these sudden disturbances lead to the obstruction of the operation of some freight routes, it is easy to cause the joint effect of related freight routes, and the whole freight network will face the danger of paralysis. From the perspective of complex networks, these problems belong to the network robustness problem. That is, the tolerance of the network to faults and attacks [7], it is also often called "network elasticity" [8]. At present, literature [9,10] have studied the robustness of air freight networks and railway express freight networks along the "Belt and Road", it is found that both air freight networks and rail freight networks show strong robustness under random attack and weak robustness under deliberate attack. In addition, literature [10] finds that the selection of node importance evaluation indexes based on line grade has a greater impact on the robustness of railway express freight networks than those based on degree and betweenness centrality.

The freight relations of the "Silk Road Economic Belt" are mainly composed of cross-border air and rail transport. The existing studies on the robustness analysis of freight networks only focus on the single transport mode of air or railway, which cannot fully reveal the freight function paralysis caused by the failure of multiple freight lines caused by natural disasters such as extreme weather. In addition, the existing freight relationship network robustness analysis [9,10] is based on the analysis method of complex relationship network robustness, without considering the geographic distance information between node cities contained in the freighted relationship. In the railway freight and air freight network of "the Silk Road Economic Belt", the node cities and freight relations are in a European space with geographical constraints, and the realization of the freight transport function has an inseparable relationship with the geographical distance between the node cities. The analysis based on the robustness of the complex relationship network only focuses on the information of the transport relationship between node cities, ignoring the geographical distance information between them, and fails to reflect the structural and functional characteristics of the freight network. Finally, a large number of studies on the robustness of transport networks from the perspective of complex network theory [11–13] focus on the analysis of robustness, and few studies further discuss how to improve and optimize the robustness of networks.

Given this, this paper will focus on the robustness analysis and optimization of the double-layer freight relationship network of the "Silk Road Economic Belt". Based on the actual data of cross-border railway and cross-border air freight routes between cities along the "Silk Road Economic Belt", consider the geospatial characteristics embodied in freight relations and the speed differences between rail and air transport, build a two-layer freight relationship network with transportation distance as weight. On this basis, deliberately attack different proportions of railway and airlines in the double-layer freight relationship network from large to small according to the weight of the edge. Taking transportation efficiency and transportation cost as important indicators to measure the robustness of the freight relationship network, and then analyze the robustness of the double-layer freight relationship network. Finally, by constructing a multi-objective model, the robustness of the double-layer freight relationship network of the "Silk Road Economic Belt" is optimized and adjusted.

2   The "Silk Road Economic Belt" Freight Relation Network Topology Description

2.1   Definition of Data Sources and Network Relationships

The data of railway freight lines along the "Silk Road Economic Belt" comes from China Railway Corporation, China railway supply chain logistics website, and major train platform companies. A total of 108 railway freight cities along the "Silk Road Economic Belt" were collected, except deleting several independently connected freight routes, including 53 domestic cities and 55 foreign cities. The data on air cargo routes along the "Silk Road Economic Belt" comes from the website of the International Air Transport Association (IATA), check out the 2020 global cargo airline rankings on the website, and check the flight schedule and flight data from the official websites of the top 15 airlines, and collect a total of 113 air cargo cities, excluding independent cities that have no route connection with any city along the "Silk Road Economic Belt", 78 aviation cities with mutual navigation are finally obtained.

Take 108 railway freight cities and 78 aviation cities as nodes, take the freight lines existing between node cities in the railway layer and aviation layer as the edge, take the transportation distance between node cities as the weight of the edge, build the freight relationship network of the "Silk Road Economic Belt". Because the original railway freight line data includes entry-exit rail changing cities, However, there is no freight relationship between the cities that change rails, so the data of cities that change rails in the original railway data are excluded.

2.2 Construction of Freight Relationship Matrix

Firstly, according to the definition of network relation, the cross-border railway freight relation matrix $w_r = [w_{ij}]$ and cross-border air freight relation matrix $w_a = [w_{mn}]$ are constructed respectively. The weight of edges in the cross-border railway and cross-border aviation matrix is the transportation distance between cities and represents the transportation distance between railway city i and city j. w_{mn} represents the transportation distance between aviation city m and city n. $w_{ij} > 0\,(w_{mn} > 0)$ indicates that there is a freight relationship between city $i(m)$ and city $j(n)$; $w_{ij} = 0\,(w_{mn} = 0)$ indicates that there is no freight relationship between city $i(m)$ and city $j(n)$. The freight transportation distance between cities is calculated by the formula of large circle distance [14],

$$
\begin{aligned}
GC_{ij} = &R \cdot \arccos\left(\cos\left(lat_i \cdot \frac{\pi}{180}\right) \cdot \cos\left(lat_j \cdot \frac{\pi}{180}\right)\right. \\
&\left. \cdot \cos\left(Ing_i \cdot \frac{\pi}{180} - Ing_j \cdot \frac{\pi}{180}\right)\right) + \sin\left(lat_i \cdot \frac{\pi}{180} \cdot \sin\left(lat_j \cdot \frac{\pi}{180}\right)\right)
\end{aligned}
\tag{1}
$$

In Formula (1), GC_{ij} is the transportation distance (km) between node city i and node city j; R is the radius of the earth, which is a constant value, value is 6370.99681 km; Ing_i represents the longitude value of node city i and represents the latitude value of node city i.

To analyze the robustness of the double-layer freight relationship network, this paper uses the weighted super adjacency matrix [12] to construct the double-layer freight relationship matrix w, as follows:

$$
w = \begin{bmatrix}
\cdots & \cdots & \cdots & \cdots & \cdots & \cdots \\
\cdots & w_r & \cdots & k_{ra} & \cdots \\
\cdots & \cdots & \cdots & \cdots & \cdots & \cdots \\
\cdots & k_{ra} & \cdots & w_a & \cdots \\
\cdots & \cdots & \cdots & \cdots & \cdots & \cdots
\end{bmatrix}
\tag{2}
$$

Matrix w is the freight relation matrix with the weight of the transportation distance between railway node cities corresponding to row and column $1 \sim 108$. w_a is the freight relation matrix with the weight of the transportation distance between air node cities corresponding to row $109 \sim 186$ and column $109 \sim 186$. The freight matrix of the two layers establishes the connection between the layers through the common city nodes. In the weighted super adjacency matrix set the value of the corresponding position to represent the transit distance from the railway freight station to the air freight station within the city. If there is only one freight mode of railway or aviation in a city, $k_{ra} = 0$ in the corresponding position of the city.

2.3 Freight Transport Network Topology

Nodes and edges in the two-layer freight network represent different types of objects and multiple relationships respectively. Figure 1. shows a simple example of a multi-layer freight relationship network. Moscow, Chongqing, and Hamburg have both railway freight and air freight, and the freight relations at the railway and air levels are different, The railway freight and air freight are connected by the road transport represented by the dotted line, forming a double-layer freight network.

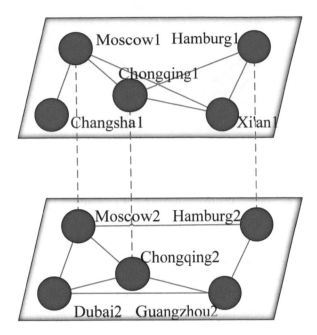

Fig. 1. Schematic diagram of the network structure of the railway-air double-deck freight relationship.

To analyze the distribution characteristics of the double-layer freight network in each region, the actual distribution of the network is depicted on the map, and the network is visualized in HTML language. The results are shown in Fig. 2. In the figure, N_r and N_a respectively represent the total number of nodes in the railway and air freight network layer; E_r, E_a, respectively represent the amount of connections between nodes in the railway and air freight relationship network layer; E_{ra} is the number of connections between rail freight relationship network layer and air freight relationship network layer.

It can be seen from Fig. 2. that the railway freight relationship network and air freight relationship network show obvious differences in regional distribution and the number of nodes and sides, as follows:

The node cities of the railway layer are mainly located in Central Asia, Eastern Europe, and China, while the node cities of the aviation layer are mainly located in West Asia, South Asia, Eastern Europe, and China except in Central Asia. Through statistical analysis of the number of nodes and edges of the railway layer and the aviation layer, it is found that the ratio of the quantity of edges

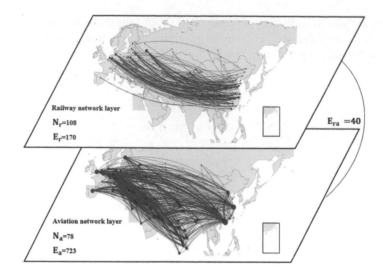

Fig. 2. Rail-air double-decker freight relationship network.

to the amount of nodes in the railway layer and the aviation layer is 1.574 : 1 and 9.269 : 1, respectively. This indicates that each node city in the railway layer has freight relations with only 1 to 2 cities on average, while each node city in the aviation layer has freight relations with 9 cities on average, indicating that the connection between cities in the aviation layer is closer than that in the railway layer. In addition, there are a total of 40 identical node cities in the railway layer and the aviation layer, of which there are 108 node cities in the railway layer and 78 node cities in the aviation layer. This shows that more than 50% of the node cities in the aviation layer have two cross-border transportation modes: Railway and aviation, but only about 37% of the node cities in the railway layer have two cross-border transportation modes: Railway and aviation, and about 63% of the cities can only carry out freight exchanges with the areas along the "Silk Road Economic Belt" through cross-border railway transportation.

3   Robustness Analysis of the Freight Relationship Network of the "Silk Road Economic Belt"

3.1   Robustness Metrics and Attack Strategies

Robustness Metrics In the field of complex networks, the relative size of the maximum connected subgraph of the network, the average path length, the network efficiency, and the network aggregation coefficient [15–17] are usually used to measure the robustness. This paper builds a freight relationship network based on cross-border freight transportation by rail and air, and pays more attention to examining freight efficiency and transportation costs [18–20]. Therefore, this paper uses the transportation efficiency of the maximal connected subgraph of the network and the transportation cost of the maximal connected subgraph to measure the robustness of the double-layer freight relationship network. The specific meanings of the two are as follows:

The robustness index of transport efficiency is defined as $CI = 1/AT$, where AT is the average shortest transport time of goods [21] (transport distance/transport speed), including railway transport time, air transport time and air-railway transfer time. The larger the CI, the higher the transport efficiency of the network. The transportation efficiency index CI adopted in this paper not only considers the actual transportation time of the goods, but also considers the transit time of the goods, which is more in line with the actual situation than the traditional evaluation method. Among them, the transit time includes the handling time, the changing and packing time, the road transportation time between the sites within the same city, etc. Through relevant investigation and analysis, the values of rail transport speed v_r, ir transport speed v_a and road transport speed v_t are respectively: $v_r = 80$ km/h, $v_a = 800$ km/h, $v_t = 100$ km/h [22].

Transportation cost C includes fixed cost and variable cost. Fixed cost refers to the customs declaration fee, labor fee, voucher preparation fee, weighing fee and other fixed expenses after the goods arrive at different types of stations; The variable cost is the actual transportation cost of the goods during the transportation process, which is determined by the transportation distance between cities and the unit transportation cost. The specific calculation formula is as follows:

$$C = \sum (w_r \cdot c_r + w_a \cdot c_a + w_t \cdot c_t + F_r + F_a) \tag{3}$$

In Formula (3), w_r and w_a are the transportation distance between railway and aviation node cities; w_t is the road transportation distance between the railway station and the aviation station in the city (it can be calculated by the great circle distance formula); c_r is the railway transportation cost; c_a is the air transportation cost; c_t is the road transportation cost; F_r is the fixed cost of the railway freight station; F_a is the fixed cost of the air freight station. Based on the regulations of the former Ministry of Railways on freight rates, statistics on the official websites of each airline, and the "International Container Vehicle Transportation Fee Collection Rules", the values of railway, air and road transportation costs and fixed costs are as follows: $c_r = 0.65\,CNY/(t \cdot km)$, $c_a = 2\,CNY/(t \cdot km)$, $c_t = 1\,CNY/(t \cdot km)$, $F_r = 3000CNY$, $F_a = 3500CNY$.

Robust Attack Strategy When attacking the double-layer freight relationship network, considering the differences in transportation speed and transportation cost between the railway and air transportation, the edges in the railway layer and the air layer are heterogeneous. To reflect the heterogeneity between layers, this paper selects different proportions of railways and airlines to conduct deliberate attacks and then analyzes the robustness of the double-layer freight relationship network. In the pre-experiment, it is found that the edge failure ratio of the double-layer freight relation network is about 11%, and the network is completely paralyzed. Therefore, let the proportion of air-railway attack edges be Q, select a certain proportion to attack 1% of the edges in the double-layer freight relationship network each time, and continuously adjust the ratio of railway transportation lines and air transportation lines in these 1% edges until Network crashes.

3.2   Robust Results Analysis

According to the robust attack strategy of this paper, the double-layer freight relationship network of the "Silk Road Economic Belt" is deliberately attacked. In each attack, extract the maximum connected subgraph of the remaining network, calculate the network transportation efficiency CI and transportation cost C indicators, and observe the changing trend of transportation efficiency CI and transportation cost C with different Q values under different attack proportions, as shown in Fig. 3.

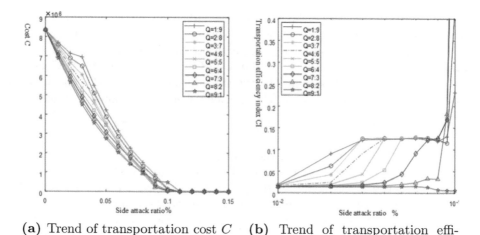

(a) Trend of transportation cost C

(b) Trend of transportation efficiency index CI

Fig. 3. Variation trend of transport cost C, and transport efficiency indexes CI when the edge is deliberately attacked.

It can be seen from the changing trend of transportation cost in Fig. 3a that under the deliberate attack, with the increase of the proportion of failed edges, the overall transportation cost of the network shows a downward trend. The value of Q is different for each attack, and the cost of the double-layer freight relationship network is also different. The analysis found that under the deliberate attack, the smaller the proportion of air transportation lines Q, the less the failure of air transportation lines, and the higher the cost of the network. This is because air freight is more than three times as expensive per unit of distance as rail freight, and cities on the air tier are more interconnected than on the rail tier. Therefore, when the proportion of air routes failing is reduced, the cost of the double-layer freight network will rise.

As can be seen from Fig. 3b variation trend of transport efficiency, the larger proportion of railway transport lines Q is under deliberate attack, namely, the more failures of railway transport lines, the faster growth of the transport efficiency index CI and the higher network transport efficiency. This is because more line failures on the railway layer lead to an increase in the number of isolated nodes on the railway layer, making the aviation layer dominate. The calculation shows that the average shortest transportation time of the railway freight relationship network is 249.3532 h, while the average shortest transportation time of the air freight relationship network is 7.8705 h. The average shortest transportation time of the air layer is much smaller than that of the railway layer, so the transportation efficiency of the whole network can be improved. It can be seen that the larger the proportion of air freight lines is, the more favorable the transportation efficiency of the double-layer freight relationship network is so that the double-layer freight relationship network can maintain higher transportation efficiency.

In addition, in the case of attacks with different Q values, tends to infinity when the failure edge ratio of the two-layer freight network is about, CI tends to infinity when the failure edge ratio of the two-layer freight network is about 11%, and the network breaks down and loses the transportation function. When the proportion of failed edges is less than, the transportation efficiency index CI has an obvious plateau period. Due to the different values Q, the proportion of failed edges in the transportation efficiency index CI reaching the platform during deliberate attacks is also different. Specifically, in each attack, the value of Q is different, and the degree of change of the transport efficiency index is different. When $Q - 1 : 9$, the transportation efficiency index CI reaches the plateau first, and the transportation efficiency index CI of the Q value changes the most when the failure edge ratio is 1%; When $Q = 2 : 8$, $Q = 3 : 7$ and $Q = 4 : 6$, the network transportation efficiency index CI changes the most when the proportion of failure edges is 1%~2%, and the order is: $Q = 5 : 5$, $Q = 3 : 7$, $Q = 2 : 8$ according to the degree of change from large to small; When $Q = 5 : 5$, transportation efficiency index CI changes the most when the failure edge ratio is 2%~3%; When $Q = 6 : 4$, $Q = 7 : 3$ and $Q = 8 : 2$, the transportation efficiency index CI changes the most when the failure edge ratio is 3%~4%, 4%~5%, and respectively; When $Q = 9 : 1$, the network transportation efficiency index CI does not increase. When the proportion of side attacks increases, the network transportation efficiency shows a downward trend.

By sorting and comparing the attacked edges when the network transportation efficiency index CI changes the most under different Q values, it is found that the same edges appear many times, indicating that these edges have a great impact on the transportation efficiency of the whole double-layer freight relationship network. The primary reason is that the edges of the railroad layer appear more frequently. The actual transportation routes of these sides are: Chengdu-Moscow, Tianjin-Moscow, Shijiazhuang-Moscow, Chifeng-Moscow, Harbin- Bikliang, Urumqi-Naples, Urumqi-Duisburg, Changsha-Tehran, Yinchuan-Aktau, Qingdao-Atenkoli, Qingdao-Ashgabat, Yiwu-Chelyabinsk, Yiwu-Mazar-e-Sharif,

Xi'an-Baku, Xi'an-Tehran, Hohhot-Bam, etc. By analyzing the above actual transportation lines, it is found that these lines are mainly from cities with large node degrees, such as Moscow, Duisburg, Yiwu, Xi'an, etc. In addition, the transportation distance of these lines is long and they are long-distance transportation.

4   Optimization of the Network Structure of Freight Relations in the "Silk Road Economic Belt"

4.1   Analysis Strategy Considering Network Transportation Efficiency and Transportation Cost

Through the analysis of the robustness of the double-layer freight relationship network of the "Silk Road Economic Belt" in Chap. 2, it is found that when the network is attacked, the overall scale and transportation cost of the network under different Q values show a downward trend, but the overall transportation efficiency of the network shows an upward trend.

When analyzing the double-layer freight relationship network, if only the network transportation efficiency is considered, the network transportation cost will be higher, and if only the network transportation cost is considered, the network transportation efficiency will be reduced. For example, the smaller the proportion of air transportation lines in Q value, the transportation efficiency can be improved, but at the same time, the transportation cost will also increase. On the contrary, the smaller the proportion of railway transportation lines in Q value, the transportation cost can be reduced, but at the same time, the transportation efficiency will also be reduced. Therefore, by constructing a multi-objective model, this paper discusses how to adjust the distribution ratio of the railway layer and the aviation layer in the double-deck freight relationship network, so that the transportation cost and transportation efficiency of the double-deck freight relationship network can be Pareto optimal, so that the double-deck freight relationship network can achieve Pareto optimality. The layout of the relationship network is more reasonable.

4.2   Multi-objective Model Construction

Taking the adjustment idea of considering both network transportation efficiency and network transportation cost in Sect. 3.1, this paper uses the efficacy coefficient function [23] to construct multiple objective functions into one expression, and obtains the following objective function:

$$f = \left(\frac{CI - g_2}{g_1 - g_2} \right)^\omega * \left(\frac{C - h_2}{h_1 - h_2} \right)^\delta \tag{4}$$

In Formula (4), ω and δ are the weight coefficients of network transportation efficiency and network transportation cost, respectively, and both are 0.5, indicating that network transportation efficiency and network transportation cost are equally important in this function; CI is the value of the transportation efficiency index, g_1 and g_2 are the max and min values of the transportation efficiency index, separately; C is the value of the network transportation cost, and h_1 and h_2 are the max and min values of the network transportation cost, respectively.

The specific restrictions of the f function are shown in Formula (5):

$$\begin{cases} AN - 1 \le AE \le 1484 \\ RN - 1 \le RE \le 340 \\ 0 \le AN \le 78 \\ 0 \le RN \le 108 \end{cases} \tag{5}$$

(Note: The maximum number of connected edges between nodes in the actual air cargo network is currently 1484; The maximum number of connected edges between nodes in the current rail freight network is 340)

In Formula (5), AN is the amount of air freight nodes; RN is the quantity of railway freight nodes; AE is the number of air freight lines; RE is the number of railway freight lines.

The model aims at maximizing the value of the function f. By continuously adjusting the ratio of railway freight lines and air freight lines, it analyzes under which distribution ratio the function f can take the maximum value, so as to adjust the double-layer freight relationship network, the ratio of the rail layer to the air layer edge. In addition, the network transportation efficiency index CI and the transportation cost index C are both used as the robustness indexes of the entire network when analyzing different proportions of railway and air-lines, so these two indexes also have constraints in the f function. Finally, this paper constructs a multi-objective model of a double-layer freight relationship network with the constraints of network transportation efficiency and network transportation cost.

4.3   Multi-objective Result Analysis

By analyzing the changing trend of the multi-objective function f when adjusting the ratio of railway freight lines and air freight lines under different Q values, as shown in Fig. 4.

Figure 4 shows that the value of the function f will be different when the adjustment ratio of the railway freight line and the air freight line is different. Moreover, when the value of Q is different, the function f will have a maximum value under different adjustment ratios. If the proportion of air transportation routes in the Q value is smaller, the optimization result of the network will be better, and the function f will appear the at maximum value first. In other words, when $Q = 1 : 9$, the optimal result appears when the edge adjustment ratio is 3%, the multi-objective result f is greater than 0.6, and the transportation efficiency and transportation cost of the network reach the Pareto optimal result. That is, the combination ratio of railway freight lines and air freight lines in the two-layer freight network is relatively optimal.

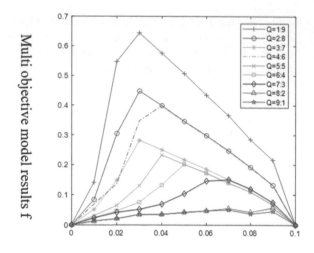

Side (aviation, railway) adjustment %

Fig. 4. change trend of multi-objective model result when adjusting the network edge.

At this time, the network matrix when the function f value of the multi-objective model is the largest is extracted, and the UCINET software is imported to describe a simple network topology structure, and the overall connection of the double-layer freight relationship network is observed. The results are shown in Fig. 5.

Figure 5 shows that the overall connection of the double-layer freight relationship network is good, but the connection between 21 freight node cities and the main network is completely disconnected, so it is impossible to continue to play the function of cross-border freight forwarding. The numbers of these 21 node cities are shown in Fig. 5. According to the numbering index, they all belong to the node cities in the railway layer, specifically Wuhan, Changsha, Suzhou, Xi'an, Hefei, Yiwu, Shihezi, Harbin, Kunming, Lanzhou, Yinchuan, Jinan, Linyi, Shenyang, Nanchang, Lianyungang, Jinhua, Ganzhou, Qinzhou, Xuzhou and Baiyin and other cities. These cities belong to China's domestic cross-border railway nodes, but it is found in Fig. 5 that when these 21 node cities fail, the overseas cities connected to these cities do not lose connection with the whole network and still play a freight role in the network. This shows that the failure of these 21 cities will not affect the overall connectivity of the double-layer freight relationship network.

It can be seen from the above analysis that when adjusting the edges of the double-layer freight relationship network, the transportation efficiency and transportation cost of the network can be Pareto optimal at the same time when the air-rail attack edge ratio is 1 : 9, and the function f value of the multi-objective model is the largest. Therefore, the aforementioned 21 node cities can be closed. The closure of these cities will not only not affect the overall

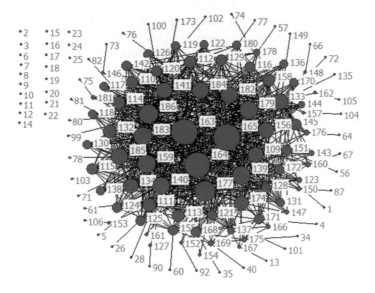

Fig. 5. Overall connection of double-layer network when the multi-objective result f is maximum.

connectivity of the double-layer freight relationship network, but will also reduce the transportation cost of the network. At the same time, most of the edges that have a greater impact on network transportation efficiency from the robust attack in Sect. 3.2 are sent from these node cities. Therefore, the optimization scheme can effectively solve the problems found in the previous paper. After these domestic cities are closed, the goods of these cities can be integrated and transferred to China. And select other domestic railway freight node cities or air freight node cities that are closer in distance, and ship overseas through these cross-border freight cities, which not only ensures the connectivity of the double-layer freight relationship network, but also improves the transportation efficiency of the network, and saves money at the same time. The transportation cost of the double-layer freight relationship network is calculated.

5   Conclusions and Suggestions

By analyzing the geographical distribution of nodes in the double-layer freight relationship network, it is found that the node cities of the railway layer are mainly located in Central Asia, Eastern Europe, and China, while the node cities of the aviation layer are mainly located in West Asia, South Asia, Eastern Europe, and China except central Asia; More than half of the node cities in the aviation layer have both railway transport and air transport, while most of the cities in the railway layer can only transport goods with the regions along the "Silk Road Economic Belt" through a single railway transport. According to the current distribution of railway and aviation node cities, it is suggested that various transportation modes should be appropriately added in future development, and the layout should be carried out according to the geographical characteristics of different regions. For example, the layout of railway transportation should be developed in South Asia, to expand its coverage and promote overland trade. At the same time, node cities that only involve railway transportation should develop a variety of transportation modes as far as possible to improve the transportation efficiency of the double-layer freight relationship network of the "Silk Road Economic Belt"; Air transport should comprehensively refer to the geographical advantages of Central Asia and play a bridge role in central Asia to promote the development of air transport.

By observing the overall change trend of robustness, it is found that under deliberate attack, the fewer side failures in the aviation layer, the higher the transportation efficiency of the double-layer freight relationship network. However, due to the dense air transportation lines and high unit transportation cost, the transportation cost of a double-layer freight relationship network is high.

According to the results of a multi-objective analysis, after removing 21 railway freight node cities in China, such as Wuhan, Changsha, Suzhou, Xi'an, Hefei, Nyi'u, Shihezi, Harbin, Kunming, Lanzhou, Yinchuan, Jinan, Linyi, Shenyang, Nanchang, Lianyungang, Jinhua, Ganzhou, Qinzhou, Xuzhou and Baiyin, the overseas railway freight cities connected with these cities can still play the role of freight transportation in the railway freight relationship network. In addition, the transportation efficiency of the double-layer freight relationship network can be improved, and the transportation cost will not be too high. According to this, we can adjust the pattern of China's external transportation mode. First, package and integrate the goods of these 21 node cities, and then select a reasonable transportation mode to transfer to other domestic cities, which will be transported to overseas destination cities, which can effectively reduce unnecessary transportation and reduce costs.

References

1. Zhao, Y., Zhu, L., Ma, B., Xu, Y., Jiang, B.: Characteristics of inter-provincial network connection based on railway freight flow in China, 1998–2016. Sci. Geogr. Sin. **40**(10), 1671–1678 (2020)
2. Feng, Z.: Research on railway freight safety risk assessment based on BP neural network. China Saf. Sci. J. **28**(S1), 178–185 (2018)
3. Zhang, Y.-z.: Review of service network design for express freight transportation. J. Transp. Syst. Eng. Inf. Technol. **21**(03), 1–12 (2021)
4. Lixia, H., Bin, S., Wenxian, W.: Optimization of hub-and-spoke service network design for railway express freight train. Appl. Res. Comput. **32**(07), 1974–1978 (2015)
5. Li, H.q., Yuan, J.l., Zhao, W.c., Zhang, L.: Statistical characteristics of air cargo-transport network of China. J. Beijing Jiaotong Univ. (Soc. Sci. Edn.) **16**(02), 112–119 (2017)

6. Yan-ling, C.H.U., Zhong-zhen, Y.A.N.G., Ru-mei, F.E.N.G.: The optimal distribution of air cargo transportation network based on accessibility index. J. Transp. Syst. Eng. Inf. Technol. **17**(05), 214–220 (2017)

7. Albert, R., Jeong, H., Barabasi, A.L.: Error and attack tolerance of complex networks. Nature **406**(6794), 378–382 (2000)

8. Newman, M., Newman, M.E.J.: The structure and function of complex networks. SIAM Rev. **45**(2), 167–256 (2003)

9. Zhiqiang, Z.H.U.O., Hongguang, Y.A.O.: Study on the structure and the robustness in the aviation network along "the belt and road initiative". Logistics Sci.-Tech. **41**(05), 78–84 (2018)

10. Zhang, Z., Zhang, Y., Wang, X.: Robustness analysis of railway express freight network. China Saf. Sci. J. **30**(03), 150–156 (2020)

11. Feng, X.U., Jingfu, Z.H.U., Jianjun, M.I.A.O.: The robustness of high-speed railway and civil aviation compound network based on the complex network theory. Complex Syst. Complex. Sci. **12**(01), 40–45 (2015)

12. Xiao-xuan, S.U.N., Ye, W.U., Xin, F.E.N.G., Jing-hua, X.I.A.O.: Structure characteristics and robustness analysis of multi-layer network of high speed railway and ordinary railway. J. Univ. Electron. Sci. Technol. China **48**(02), 315–320 (2019)

13. Ma, X., Cai, Y.m.: Study on the robustness of Chinese railway and airline multilayer networks based on complex network theory. Shandong Sci. **30**(05), 70–78 (2017)

14. Fu, Z.: Calculation of the great-circle distance and heading. J. Civil Aviat. Univ. China **8**(1), 1–10 (1990)

15. Zeng, X., Tang, X., Jiang, K.: Measure of China airline networks invulnerability based on complex networks. Syst. Simul. Technol. **8**(02), 111–116 (2012)

16. Ya-ru, D.A.N.G., Su-zhen, S.O.N.G.: Invulnerability analysis of chinese air passenger flow network based on centrality. Complex Syst. Complex. Sci. **10**(01), 75 82 (2013)

17. Lordan, O., Sallan, J.M., Simo, P., et al.: Robustness of the air transport network. Transp. Res. Part E Logist. Transp. Rev. **68**(68), 155–163 (2014)

18. Liu, H.-W., Wu, J., Liang, W.: Efficiency of highway cargo transportation and its effect factors based on perspective of Wanjiang demonstration zone. J. Transp. Syst. Eng. Inf. Technol. **18**(02), 27–32 (2018)

19. Xu, J.: Research on the relationship between China's railway transport industry price reform and freight transport efficiency. Price Theory Pract. **10**, 163–166+179 (2020)

20. Huang, J.: Six strategies to reduce the operating cost of air cargo. CAAC NEWS, 2015-08-05(005)

21. Peng, T., Zhang, Y-p., Cheng, S.-w.: Robustness analysis of hierarchical airline network based on penalty factor. J. Transp. Syst. Eng. Inf. Technol. **16**(03), 187–193 (2016)

22. Fengjie, X.I.E., Wentian, C.U.I.: Analyzing and optimizing the robustness of weighted express networks. Syst. Eng. Theory Pract. **36**(09), 2391–2399 (2016)

23. Peng, T.: Structural Performance Optimization of Hierarchical Airline Network. Harbin Institute of Technology (2017)

Optimization of Safety Investment in Prefabricated Building Construction Based on SSA-SVR

Chunguang Chang$^{(\boxtimes)}$ and Xiaoxue Ling

School of Management, Shenyang Jianzhu University, Shenyang 110168, China
{ccg,LingXX}@sjzu.edu.cn

Abstract. For reducing the losses caused by safety accidents in the construction of prefabricated building (PB), the sparrow search algorithm (SSA) and support vector regression (SVR) were used to determine the quantitative relationship function between safety control investment and economic loss of accidents in the construction process. Accordingly, a safety investment optimization model was constructed, which was based on the minimization of economic losses from accidents as the objective and the safety input resources as the constraint. A case study proved that the SSA-SVR-based construction safety investment optimization method can reduce the loss of safety accidents in an effective way during the construction of PB and reduce the capital invested by construction companies.

Keywords: Prefabricated building · Safety investment · Sparrow search algorithm · Support vector regression

1 Introduction

Relevant data shows that China's new prefabricated building (PB) area has accounted for 20.5% of the country's new building area in 2020. PB has become an inevitable trend in the transformation and development of the construction industry. However, its construction characteristics are obviously different from those of traditional buildings, such as complex construction processes and frequent heavy tower crane operations, which bring new challenges for construction site safety management.

The state of construction safety is closely related to safety inputs. Many scholars have conducted a lot of research on safety inputs. Smith et al. [1] conducted a study to prove the correlation between safety investment and accident rate. Jiang et al. [2] constructed a production safety investment program evaluation model by entropy weight-TOPSIS method. Li [3] and Yang [4] developed an optimization model of safety investment. However, most research on construction safety management of PB is focused on safety risk evaluation and risk factor identification, etc., and relatively few research results are aimed at construction safety investment. In this context, the author draws on more mature research methods to optimize safety input in PB construction, establish a support vector regression (SVR) model based on the sparrow search algorithm (SSA), and verify the feasibility.

© The Author(s), under exclusive license to Springer Nature Switzerland AG 2023
N. Xiong et al. (Eds.): ICNC-FSKD 2022, LNDECT 153, pp. 754–763, 2023.
https://doi.org/10.1007/978-3-031-20738-9_84

2 Establishment of Safety Investment Model by SSA-SVR

2.1 Designing of SVR

SVR maps of training sample set $D = \{(X_1, Y_1), (X_2, Y_2) (X_m, Y_m)\}$ $(X \in R^n, Y \in R)$ to the high-dimensional feature space by function $\varphi(x)$, which determines linear regression model. Where, w is weight vector, b is a constant term:

$$f(X_i) = [w^T \cdot \varphi(X_i)] + b \tag{1}$$

Assuming that there is a maximum allowable error ε between $f(X_i)$ and actual value Y_i, the loss is calculated only when the difference value between $f(X_i)$ and Y_i is greater than ε. Then the original SVR model is obtained, where, σ is the penalty factor, l_ε is the loss function of ε.

$$\min_{w,b} \frac{1}{2}\|w\|^2 + \sigma \sum_{i=1}^{m} l_\varepsilon [f(X_i) - Y_i \tag{2}$$

$$l_\varepsilon = \begin{cases} 0, & |u| \leq \varepsilon \\ |u| - \varepsilon, & |u| > \varepsilon \end{cases} \tag{3}$$

$$|u| = |f(X_i) - Y_i| \tag{4}$$

In practical application, it is difficult to determine the suitable ε. Therefore, the relaxation variables ξ and ξ^* are introduced to relax the function as follows.

$$\min_{w,b,\xi,\xi^*} \frac{1}{2}\|w\|^2 + \sigma \sum_{i=1}^{m} (\xi, \xi^*) \tag{5}$$

$$\text{s.t.} \ [w^T \varphi(X_i) + b] - Y_i \leq \varepsilon + \xi_i \tag{6}$$

$$Y_i - [w^T \varphi(X_i) + b] \leq \varepsilon + \xi_i^* \tag{7}$$

$$\xi_i \geq 0, \xi_i* \geq 0, i = 1, 2, \ldots, m \tag{8}$$

The Lagrange multiplier α is introduced to obtain the pair-wise problem.

$$\min \frac{1}{2} \sum_{i=1}^{m} \sum_{j=1}^{m} (\alpha_i^* - \alpha_i)(\alpha_j^* - \alpha_j)[\varphi(X_i) \cdot \varphi(X_j)]$$

$$+ \varepsilon \sum_{i=1}^{m} (\alpha_i^* + \alpha_i) - \sum_{i=1}^{m} Y_i(\alpha_i^* - \alpha_i) \tag{9}$$

$$\text{s.t.} \ \sum_{i=1}^{m} (a_i - a_i^*) = 0, \ 0 \leq \alpha_i, \alpha_i^* \leq \sigma, \ i = 1, \ldots, m, \tag{10}$$

Then radial basis function is introduced as kernel functions to realize the spatial mapping [5], and w is represented by α to obtain decision function of SVR model, where, X_i and X are respectively the inputs to the training and prediction sample sets, and g is the kernel function parameter.

$$f(X) = \sum_{i=1}^{m} (\alpha_i - \alpha_i^*) K(X_i, X) + b \tag{11}$$

$$K(X_i, X) = \exp\left(-g \|X_i - X\|^2\right) \tag{12}$$

2.2 Establishing of Optimization Model for Safety Investment

The objective function of optimization model for safety investment in PB construction is obtained as follows. It minimizes the economic loss by safety accidents.

$$\min f(X) = \sum_{i=1}^{m} (\alpha_i - \alpha_i^*) \exp\left(-g \|X_i - X\|^2\right) + b \tag{13}$$

The main risk factors in the construction segment of PB can be divided into the technical level of operators, the safety awareness of operators, the safety status of construction machinery, and the environmental conditions in the construction site. Assuming that the total safety investment is divided into d subsections for controlling construction risks, the model constraints are analyzed as follows.

Firstly, because of the requirements of corporate profits and industry norms, it is necessary to set upper and lower limits on total security investment. Secondly, each indicator needs a minimum input ratio as a guarantee. Finally, because of the interaction between some of the safety input indicators, therefore, a lower input limit needs to be set to constrain the input of these safety indicators.

Supposing $r_{\min}$ and $r_{\max}$ is respectively the minimum and maximum share of safety performance in project construction cost C; v denotes the input indicator number, r_v denotes the minimum share of performance for indicator v, r_s denotes the minimum share of performance for certain safety performance indicators to be considered, Ω is the quantity consisting of the sequence numbers of certain safety performance indicators to be considered, and $X = [x_1, x_2, \ldots, x_d]$. In summary, the optimization model of safety investment in PB construction can be established as follows.

$$\min f(x) = \sum_{i=1}^{m} (\alpha_i - \alpha_i^*) \exp(-g \|X_i - X\|^2) + b \tag{14}$$

$$\text{s.t. } r_{\min} \leq \sum_{v=1}^{d} x_v / C \leq r_{\max} \tag{15}$$

$$r_v \leq x_v / \sum_{v=1}^{d} x_v \leq 1 \quad v = 1, 2, \ldots, d \tag{16}$$

$$r_s \leq \sum_{v \in \Omega} x_v / \sum_{v=1}^{d} x_v \leq 1 \tag{17}$$

2.3 Designing of SSA

SSA has a simple structure, performs better in avoiding local optima compared to other swarm intelligence algorithms [6]. Inspired by the foraging behavior of sparrows, SSA divides individuals into three roles (finder, follower and watcher) in the process of finding the optimal solution more efficiently. Assuming that there are n sparrows in the population and the dimension of the problem to be optimized is d, , the population of sparrows is shown in (18). Denoting the fitness value by f, , the fitness values of all sparrows can be expressed as (19).

$$
G = \begin{bmatrix}
x_{1,1} & x_{1,2} & \cdots & x_{1,d} \\
x_{2,1} & x_{2,2} & \cdots & x_{2,d} \\
& & \cdots & \\
x_{n,1} & x_{n,2} & v & x_{n,d}
\end{bmatrix}
\tag{18}
$$

$$
F = \begin{bmatrix}
f(x_{1,1} \ x_{1,2} \cdots x_{1,d}) \\
f(x_{2,1} \ x_{2,2} \cdots x_{2,d}) \\
\cdots \\
f(x_{n,1} \ x_{n,2} \cdots x_{n,d})
\end{bmatrix}
\tag{19}
$$

In SSA, the discoverer is responsible for finding the area with the most abundant food, and the higher f is the easier for the discoverer to obtain food. When the safety value R of the foraging area is lower than the danger value ST, the discoverer will lead the flock of finches to fly to other areas for foraging. Accordingly, the iterative process of the finder's location is expressed as (20), where t is the current number of iterations, $X_{i,j}$ is the location information of individual i, a is a random number on (0, 1], $item_{max}$ is the upper limit of iteration amount, Q is a normal distribution random value, and L is a matrix with all elements of 1.

$$
X_{i,j}^{t+1} = \begin{cases}
X_{i,j}^t \cdot \exp(-\frac{i}{a \cdot item_{max}}), & R_2 < ST \\
X_{i,j}^t + Q \cdot L, & R_2 \geq ST
\end{cases}
\tag{20}
$$

The follower obtains food by monitoring the discoverer, and the two identities can be transformed into each other, and when the follower's fitness value is too low to satisfy the demand, it will fly to other places to continue foraging. Accordingly, the location update of the follower is denoted as (21), where X_P denotes the optimal location of the current discoverer, X_{worst} denotes the current worst foraging location in the whole domain, and A denotes the matrix whose elements are assigned to 1 or -1 in random. If $i > n/2$, the follower i has a low fitness value and needs to change the foraging area. Otherwise, the follower i will forage near the optimal discoverer.

$$
X_{i,j}^{t+1} = \begin{cases}
Q \cdot \exp\left(\frac{X_{worst}^t - X_{i,j}^t}{i^2}\right), & i > n/2 \\
X_P^{t+1} + \left| X_{i,j}^t - X_P^{t+1} \right| \cdot A^+ \cdot L, & i \leq n/2
\end{cases}
\tag{21}
$$

The sparrow that chirps to indicate when it encounters danger is called a vigilant, and its position is updated as (22), where, X_{best}^t denotes the current optimal position in total domain, B is the control step, k is a random number on $[-1, 1]$, e is a constant to avoid denominator 0, f_b and f_w is respectively the optimal and worst fitness value in the current total domain. When $f_i \neq f_b$, individual i is at the edge of the population and will move to other regions. Otherwise, the individual is at the center of the population and will randomly approach other individuals to shift the danger.

$$
X_{i,j}^{t+1} = \begin{cases} X_{\text{best}}^t + B \cdot \left| X_{i,j}^t - X_{\text{best}}^t \right|, f_i \neq f_b \\ X_{\text{best}}^t + k \cdot \left(\frac{X_{i,j}^t - X_{\text{best}}^t}{|f_i - f_w| + e} \right), \quad f_i = f_b \end{cases}
\tag{22}
$$

2.4 SSA Optimization Parameter Flow

Detailed steps are as follows:

Step1: After the initialization of the algorithm parameters, the training set samples are fitted by cross-validation to find the better individual fitness.

Step2: Find the global optimum by continuously iterating over the positions of discoverers, followers and vigilantes.

Step3: After satisfying the maximum number of iterations, the optimal parameters are output, and then the test sample set data are input into the resulting model to check the fitting results and obtain the optimal SVR model.

3   Model Validation and Case Analysis

3.1   Training and Testing of SSA-SVR Model

20 different PB projects within the last 5 years are selected, combined with the actual situation of PB construction, the safety investment subsections of the construction site are reclassified, and five safety investment subsections are finally determined as the input indexes of this model. They are investment in communication and education, civilized construction, labor protection, security management and security facilities.

After recalculating the safety inputs of 15 PB construction projects according to the above indicators, the original data samples are obtained (as Table 1), which are used as the basis for fitting training and testing of the SSA-SVR model.

Set the maximal iteration times of SSA as 200; the population size is set to be 30, of which 70% are discoverers, 20% are vigilantes and 10% are followers; the safety queue is 0.6; the initial ranges of σ and g are [0.001, 100] and [0.001, 1000], respectively. Then the SSA-SVR model was tested and validated using the 5-fold cross-validation method. After running the code of the SSA-SVR model, the values of the optimal parameters are obtained as $\sigma = 77.7549$, $g = 0.11171$. The population fitness function curve is shown in Fig. 1. The optimal parameters are brought back to the model for data sample training and testing, and the fitting curves of expected and predicted values are obtained, as shown in Fig. 2. The fitted curves of expected and predicted values for the test sample set are shown in Fig. 3.

Table 1. Original data of construction safety investment and accident loss (unit: 10000 yuan).

No.	Investment in communication and education	Investment in civilized construction	Investment in labor protection	Investment in security management	Investment in security facilities	Economic loss of the accident
1	62.97	14.59	39.73	16.69	48.99	87.48
2	82.56	18.34	58.46	24.01	60.56	107.96
3	50.94	15.24	40.77	16.66	38.92	83.56
4	51.39	14.12	23.65	13.75	40.26	75.56
5	63.04	12.02	37.25	16.98	42.94	83.54
6	60.36	17.2	47.21	18.18	44.84	89.25
7	92.64	23.45	61.8	23.74	63.22	115.74
8	63.13	11.17	40.26	17.18	46.65	86.57
9	94.58	24.49	67.32	29.2	78.29	122.13
10	92.84	22.55	60.85	26.72	71.57	117.58
11	109.53	81.88	70.66	32.34	80.84	135.06
12	87.83	22.19	59.61	30.62	64.57	117.95
13	108.52	30.72	75.34	31.63	88.03	133.88
14	121.93	31.99	89.42	40.76	105.03	144.43
15	76.78	20.2	56.1	19.4	63.95	104.36

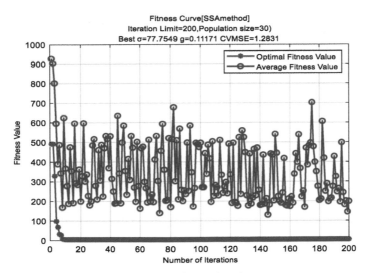

Fig. 1. Population fitness function curve.

The results show that the SSA-SVR model has a mean square error MSE of 1.2831 and a correlation coefficient of about 0.96 for the prediction results of all samples.

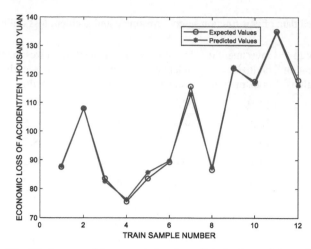

Fig. 2. The fitted curve of the expected and predicted values of the training set.

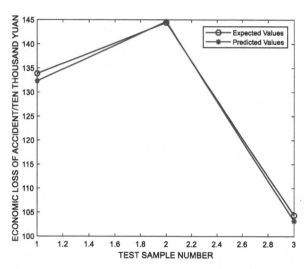

Fig. 3. The fitted curve of the expected and predicted values of the test set.

After training and testing, 11 support vectors are obtained in original data sample, as shown in Eq. (23). The optimal parameters and support vector-related data obtained from the model are brought into Eq. (14) to obtain the objective function of the optimization problem as shown in Eq. (24), where $(\alpha_i - \alpha_i^*) = [-0.3774, 77.7549, -77.7549, -$

77.7549, $-77.7549, 77.7549, -77.7549, -17.5341, 77.7549, 17.9115, 77.7549$].

$$
\begin{bmatrix} X_1 \\ X_2 \\ X_3 \\ X_4 \\ X_5 \\ X_6 \\ X_7 \\ X_8 \\ X_9 \\ X_{10} \\ X_{11} \end{bmatrix} = \begin{bmatrix} 0.1695 & 0.0484 & 0.2445 & 0.1088 & 0.1523 \\ 0.0000 & 0.0576 & 0.2603 & 0.1077 & 0.0000 \\ 0.0063 & 0.0417 & 0.0000 & 0.0000 & 0.0203 \\ 0.1704 & 0.0120 & 0.2068 & 0.1196 & 0.0608 \\ 0.1327 & 0.0853 & 0.3582 & 0.1640 & 0.0895 \\ 0.5874 & 0.1737 & 0.5801 & 0.3699 & 0.3676 \\ 0.1717 & 0.0000 & 0.2525 & 0.1270 & 0.1169 \\ 0.6147 & 0.1884 & 0.6640 & 0.5720 & 0.5955 \\ 0.5902 & 0.1609 & 0.5656 & 0.4802 & 0.4939 \\ 0.8253 & 1.0000 & 0.7148 & 0.6883 & 0.6341 \\ 0.5197 & 0.1558 & 0.5468 & 0.6246 & 0.3880 \end{bmatrix} \tag{23}
$$

$$
\min f(x) = \sum_{i=1}^{11} (\alpha_i - \alpha_i^*) \exp(-0.11171 \|X_i - X\|^2) + 102.0463 \tag{24}
$$

3.2 Case Analysis

PB construction project H is selected to test the feasibility and validity of the above model. Its construction and installation project cost is 427.86 million RMB. Here, $r_{max} = 0.03$, i.e. 3% of the construction and installation project cost is used as construction safety investment to ensure the safe conduct of the project. In addition, the lower limit of the total safety investment is set at 2.5%, i.e. $r_{min} = 0.025$, according to the regulations of safety production cost extraction. Considering the safety management requirements of the construction site and the restrictions of the project itself on the safety decision, the lower limit of the input is set for each safety input sub-item and the combination of safety input indicators that have a greater impact on the safety state, and the safety input optimization model is constructed as follows:

$$
\min f(x) = \sum_{i=1}^{11} (\alpha_i - \alpha_i^*) \exp(-0.11171 \|X_i - X\|^2) + 102.0463 \tag{25}
$$

$$
\text{s.t. } 0.025 \le \sum_{v=1}^{5} x_v / 42786 \le 0.03 \tag{26}
$$

$$
0.28 \le x_1 / \sum_{v=1}^{5} x_v \le 1 \tag{27}
$$

$$
0.09 \le x_2 / \sum_{v=1}^{5} x_v \le 1 \tag{28}
$$

$$0.2 \leq x_3 / \sum_{v=1}^{5} x_v \leq 1 \qquad (29)$$

$$0.2 \leq x_3 / \sum_{v=1}^{5} x_v \leq 1 \qquad (30)$$

$$0.09 \leq x_4 / \sum_{v=1}^{5} x_v \leq 1 \qquad (31)$$

$$0.25 \leq x_5 / \sum_{t=1}^{5} x_t \leq 1 \qquad (32)$$

$$0.6 \leq (x_1 + x_5) / \sum_{v=1}^{5} x_v \leq 1 \qquad (33)$$

The minimum value of economic loss of H project accident is 4.792 million RMB, the optimal combination of input is X = [371.38248, 119.8008, 251.58168, 107.82072, 347.42232], and the total security investment is about 11.980 million RMB. Table 2 shows the comparison between the results obtained from the safety investment optimization model and the original safety investment of the project.

Table 2. Comparison of safety investment and accident economic loss data (unit: 10,000 yuan).

Indicator	Value of the optimized index	Value of the original index	Differences
Investment in communication and education	371.38	346.57	24.81
Investment in civilized construction	119.80	116.87	2.93
Investment in labor protection	251.58	320.90	−69.32
Investment in security management	107.82	128.36	−20.54
Investment in security facilities	347.42	320.90	26.52
Total security investment	1198.01	1283.58	−85.57
Economic loss of the accident	479.20	551.94	−72.74

By comparison, it can be seen that after the reallocation of safety input optimization model, the economic loss of safety accidents in project H is reduced by RMB 727,400, and the total safety investment is reduced by about RMB 855,700. There are different degrees of waste and shortage of investment in each sub-project of the original project, which indicates that managers lack scientific basis in making safety investment decisions.

4 Conclusion

Through the SSA-SVR model fitting training, the quantitative relationship function between safety investment and accident economic loss of PB construction is obtained. The safety investment optimization model constructed according to this can effectively save safety investment, reduce accident economic losses, and provide theoretical support for construction safety investment decision. However, due to the limitation of sample size and financial data, the model parameter search structure is subject to random influence, and the model prediction accuracy can be improved by increasing the sample size.

Acknowledgments. This work is supported by National Natural Science Foundation of China (51678375), Liaoning Provincial Colleges and Universities' Innovative Talents Support Plan (LR2020005), Liaoning Provincial Natural Science Foundation's Guiding Plan (2019-ZD-0683).

References

1. Smith, N.M., Ali, S., Bofinger, C., et al.: Human health and safety in artisanal and small-scale mining: an integrated approach to risk mitigation. J. Clean. Prod. **129**(5), 43–52 (2016)
2. Jiang, F., Zhou, S., Zengtong, W., et al.: Analysis of coal mine safety investment decision based on entropy weight-TOPSIS method. J. China Saf. Sci. **31**(7), 24–29 (2021)
3. Li, Z., Cao, Q., Yang, T.: Research on safety investment model of coal mine enterprises based on support vector machine and continuous ant colony algorithm. Min. Saf. Environ. Protect. **46**(1), 109–113 (2019)
4. Yang, Y., Zhang, Q., Yang, J., et al.: Optimization of shipping safety investment based on GA-SVR-PSO. Saf. Environ. Eng. **27**(1), 146–151 (2020)
5. Wang, J., Lanyi, H., Qi, C.: Prediction of bulk commodities based on internet concerns. Syst. Eng. Theory Pract. **37**(5), 1163–1171 (2017)
6. Hongzhi, H., Qin, C., Guan, F., et al.: Tool wear recognition based on sparrow search algorithm optimized support vector machine. Sci. Technol. Eng. **21**(25), 10755–10761 (2021)

An Efficient Heuristic Rapidly-Exploring Random Tree for Unmanned Aerial Vehicle

Chunping Yin[1], Meijin Lin[2(✉)], Qun Liu[1], and Hongmei Zhu[2]

[1] School of Aerospace Engineering, Xiamen University, Xiamen 361102, China
{yin_chunping,lqyclq}@xmu.edu.cn
[2] Department of Marine Physics and Engineering, Xiamen University, Xiamen 361102, China
{meijinlin,flyzhu323}@xmu.edu.cn

Abstract. Autonomous unmanned aerial vehicle (UAV) can be utilized to replace humans to do hard work or work in dangerous environment. Path planning is one of the crucial technologies for intelligent flight of UAV. The Rapidly-exploring Random Tree (RRT) has a wide range of applications in path planning with the advantage of the sampling-based path planning which avoids complex construction of the configuration space. However, these methods perform a uniform random sampling and thus do not lead to the most efficient solution. In this paper, in order to improve the efficiency of the traditional RRT, a heuristic strategy with goal-bias was exploited and incorporated in the traditional RRT. A simulation platform was built to assess the performance of the improved algorithm and analyze the influences of the parameters, and a flight test platform was established based on the DJI Matrice 100 to carry out the flight experiments of UAV path planning. The simulations proved that the heuristic RRT is much higher efficient than the standard version. The experiments validated the feasibility to apply the new RRT in the navigation for UAVs in reality.

Keywords: Heuristic strategy · Rapidly-explore random tree · Sensor-based random tree · Unmanned aerial vehicle · Path planning

1 Introduction

Unmanned aerial vehicle (UAV) can relieve human from hard work and dangerous environments. In recent years, the research and development of applications for autonomous UAVs and robotic systems have increased tremendously [1]. What is more, during pandemic periods, i.e. the current outbreak of COVID 19, the UAV delivery service can reduce the risk of infection of plagues. Path or motion planning is one of the most crucial technical development for UAV, especially in automatic navigation. An effective and feasible autonomous path planning control strategy is a key indicator of the degree of autonomy and intelligence of an unmanned mobile system [1–4].

Path planning means finding a collision-free and optimal path from the start point to the target point judged by certain evaluation criteria, such as the shortest path, the shortest planning time etc. [5]. Sampling-based motion planning methods, especially rapidly-exploring random tree (RRT) proposed by LaValle [6, 7], have gained the popularity for

© The Author(s), under exclusive license to Springer Nature Switzerland AG 2023
N. Xiong et al. (Eds.): ICNC-FSKD 2022, LNDECT 153, pp. 764–772, 2023.
https://doi.org/10.1007/978-3-031-20738-9_85

the capability of efficient search. Sampling-based path planners obtain the information of obstacles in the environment through collision detection of sampling points. They avoid the complex construction of the configuration space and can better solve the path planning problem with many degrees of freedom [8].

RRT has been proven to be efficient for path planning in environments populated with obstacles. It leads the search in a blank region by random sampling, so it is not necessary to model the space, which has better optimization efficiency than grid-based planners [9]. The searchability of RRT is powerful. However, the randomness and blindness caused by random sampling have obvious defects in execution time, and path length as well. Therefore, the efficiency of the traditional RRT can be further improved.

In this work, in order to ameliorate the efficiency of the basic RRT, a heuristic strategy with goal-bias was exploited. The simulations were executed to verify the efficiency enhancements of the improved algorithm compared to the original RRT, and the flight experiments were carried out to validate the performance of the new RRT on UAVs in reality.

2   Methods

2.1   Basic RRT

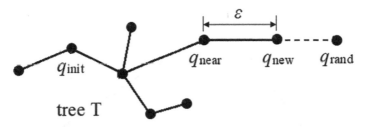

Fig. 1. Extension of RRT.

The extension of an standard RRT [6] is shown in Fig. 1. The construction of the tree T is initiated by the setting a point as the tree root from an obstacle-free region. In this case, it is the point q_{init}, i.e. the initial position of a UAV. Samples are then randomly generated over the obstacle-free region. At each iteration, a random sampling point q_{rand} is generated. Next, T is searched to find the node q_{near}, which is closest to q_{rand}. The algorithm proceeds by selecting a point q_{new} that belongs on the virtual line that connects q_{near} and q_{rand}. Q_{new} is located at a random distance ε from q_{near}, and ε is bounded to [0, D_{MAX}], where D_{MAX} denotes the maximum expansion distance possible in the sampling area. If the connectivity between q_{new} and q_{near} is possible without obstacles, an edge or a new node is created and q_{near} becomes the parent of q_{new}, otherwise, q_{rand} is rejected and the next algorithmic iteration is performed. By repeating the operation above within the maximum number N of the times of iterations, RRT can quickly search the obstacle-free space until getting a collision-free path from the initial point to the target point. The algorithm ends when the path is successful founded when q_{new} reaches the goal position q_{end}.

2.2   Improvement with Heuristic Probability Strategy

RRT typically performs a uniform sampling of the state space and q_{rand} is selected totally randomly with no contribution of the goal position. The uniform sampling treats all points in the obstacle-free region equally, which guarantees the algorithm's completeness but pays a price of longer execution time and path length [10]. In order to avoid searching the invalid areas as much as possible, the improved RRT has higher probability to grow toward the goal [3]. A heuristics probability strategy illustrated in Fig. 2 is incorporated into the RRT to modify the sampling distribution in this work.

In order to relate the sampling with the goal position, in this work, a goal-bias heuristic factor p $(0 \leq p \leq 1)$ is introduced to determine the node q_{heur} which replaces q_{rand} in the basic RRT algorithm according to Eq. (1):

$$\overrightarrow{q_{near}q_{heur}} = p\overrightarrow{q_{near}q_{end}} + (1-p)\overrightarrow{q_{near}q_{rand}} \qquad (1)$$

When $p = 0$, the algorithm serves as the basic RRT. When $p = 1$, the tree will grow the branches only in the direction toward the goal position for every sample, which will lead to high probability of failure of the path planning when there are obstacles on the straight line connecting the start point and the target point.

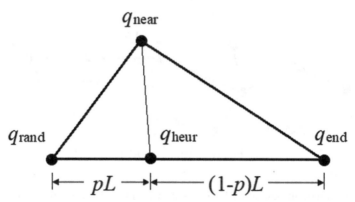

Fig. 2. The heuristic strategy incorporated into RRT. L represents the length between q_{rand} and q_{end}, and p $(0 \leq p \leq 1)$ is the goal-bias heuristic factor.

3   Simulations and Experiments

3.1   Simulations

TO evaluate the improved algorithm in the complex environments of high obstacle density, e.g. long narrow corridors or mazes, a simulation platform was built with C++ utilizing the Microsoft Foundation Class (MFC) library in Visual studio (VS) 2013. In addition, the effects of some pivotal parameters on the algorithm performance were analyzed, which provides guidance for experiments to set up the parameters.

For the global path planning, there are four input parameters: (1) ε is the advance distance between q_{new} and q_{near}, (2) N is the threshold value for the number of iterations to search q_{new}, (3) p is the goal-bias heuristic factor, and (4) d is the maximum radius of a UAV size. And the main output parameters are (1) planning time length t, (2) number of nodes of a tree m, and (3) the path length L. In the sampling window of the rectangular block, the start location is set at the most top-left (0, 0), and the target position the most bottom-right (483, 410).

Figure 3 shows the simulation results and compares the improved RRT with the standard RRT. The input parameters were set up as following: $\varepsilon = 20$ pixels, N $= 1000$ and $d = 10$ pixels for both the traditional and the improved RRT, while p is random between 0 and 1 for the improved RRT, and $p = 0$ for the basic RRT. The size of the sampling window is 483 * 410 (pixel * pixel).

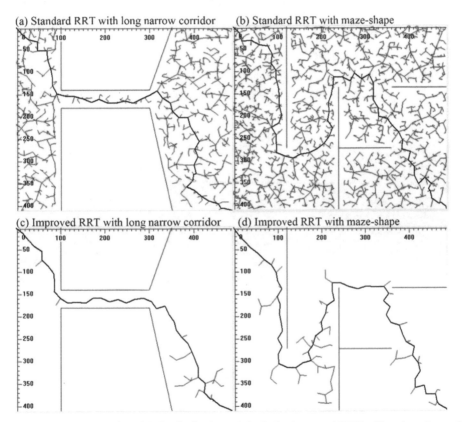

Fig. 3. Simulation results of (a-b) the basic and (c-d) the improved RRTs. The obstacle environment for (a) and (c) is long narrow corridor, and (b) and (d) maze-shape. Red line represents obstacle, black line is the planned trajectory. The branches of pink lines and the trajectory constitute the whole tree. The start position of UAV is the most top-left, and the goal position is the most bottom-right.

Figure 3 shows that, compared to the standard RRT, the improved RRT introduced here can greatly enhance the path planning efficiency. It is obvious that, compared to the improved RRT in the environments of corridor and maze-shape showed in Fig. 3(c) and (d), respectively, the sampling branches or nodes in the basic RRT showed in Fig. 3(a) and (b) uniformly distribute much more densely over all the obstacle-free regions. Table 1 lists the data averaged over 50 times of repeated trials. It demonstrates quantitatively that the improved RRT consumes much less computing resource and time significantly (by an order of magnitude), and generates relatively shorter paths as well than the basic RRT.

Table 1. Statistics of the path planning simulation in two typical complex scenarios

Scenario	Algorithm	Planning time t (s)	Number of nodes m	Path length L (Pixel)
Long narrow corridor	Basic RRT	0.164	6714	830
	Improved RRT	0.014	67	770
Maze-shape	Basic RRT	0.3185	1193	1194
	Improved RRT	0.0299	168	1164

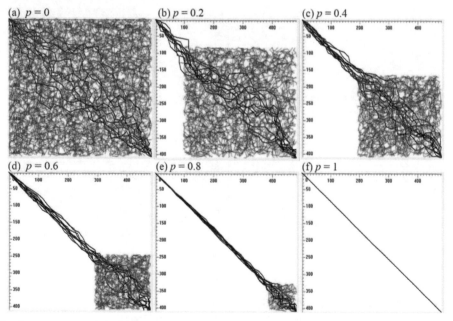

Fig. 4. The effect of goal-bias heuristic factor p on the path planning. The black lines are the paths, and the pink lines are the branches not included in the paths

The effect of the goal-bias heuristic factor p is showed in Fig. 4. To execute the path planning without failure, no obstacle is placed in the sampling block. There are 10 trials for each p value. In Fig. 4(a), when $p = 0$, the improved algorithm is same as basic RRT, and the sampling is totally random without any bias, so the branches or nodes of the trees distribute uniformly all over the simulation block. In Fig. 4(f), when $p = 1$, all the ten trials get the same path which is a straight line directly from the start to the goal, and there is no randomness at all. As a result, there is no branch other than the trunk of the path, which means that the path cannot be planned successfully once there are obstacles on the strait line connecting the points of the start and the goal. In Fig. 4(b-e), when $0 < p < 1$, the branches (in pink) not included in the paths only appear in the last part of path planning, and in the first part of the planning, there is only the trunks included in the paths. The ratio of length of the trunk without branch to the length of the whole path is close to the p value. It can be concluded that the efficiency of path planning increases with the p value, but at the cost of higher probability of failure.

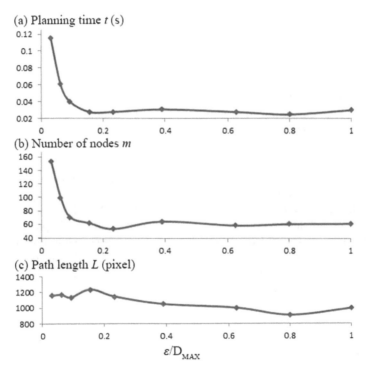

Fig. 5. The effects of the ratio of advance distance ε to D_{MAX} on (a) planning time, (b) number of nodes, and (c) Path length.

The effects of the advance distance ε between q_{new} and q_{near} is shown in Fig. 5, where ε is normalized to D_{MAX}, which in this case is the length of the diagonal of the sampling window, as the ratio ε/D_{MAX}. Advance distance ε means the maximum distance a branch can grow for a sampling step. The data are the averages of 10 times of trials with the scenario of maze-shape. The effects on path planning time t in Fig. 5(a) and the number of nodes m in a tree in Fig. 5(b) are similar. The reason is that the planning time is proportional to the number for sampling to grow a tree. Both t and m

decrease quickly as ε/D_{MAX} increases from 0 to circa 0.2, and then are quite stable after ε/D_{MAX} exceeds circa 0.3. The effect on the path length L is not very distinct, and it seems that L decreases slightly as ε/D_{MAX} increases from 0 to circa 0.8. Since there is only one optimal path, the weak effect of the advance distance on the path length shows that the improved RRT can usually find a quasi-optimal path close to the optimal one with all possible advance distances.

3.2 Flight Experiments

To further illustrate the adaptivity and feasibility of the improved RRT in real world applications, the flight experiments were carried out with the application of the following devices: (1) a drone of DJI Matrice 100; (2) a visual sensing system of DJI Guidance, which provides the coordinates of the drone; (3) an airborne computer of Intel NUC5I7RYH; (4) a ground station served by a laptop of ASUS Y481C; (5) a wireless router of TP-LINK WDR8500, which creates a local area network (LAN) to link the ground station and the airborne computer. UAV system were constructed with the first three device. A ground station and a LAN are utilized to monitor the actual flight of the UAV.

A 7 m * 6 m polystyrene (KT) board was used to provide the coordinates for the drone. The start point is at the most top-left with the coordinates (0, 0) of the KT board, and the goal point the most bottom-right with the coordinates (7 m, 6 m). A plate of 1.5 m long, 0.1 m thick, and 1.6 m high was located at the coordinates from (3 m, 2 m) to (3 m, 3.5 m) on the KT board as an obstacle. The flight altitude was fixed at 1.2 m. Before each flight, the drone was placed at the start position, so the coordinate systems of the drone and KT board can coincide. The input parameters for the experiments were set up as follow: p is random between 0 and 1, $\varepsilon = 0.4$ m, N = 1000, and $d = 1.2$ m.

Figure 6 shows the experiment on UAV in reality with the improved RRT. The whole UAV flight was recorded by a camera (as the image shown in Fig. 6(a)) and the ground station (shown in Fig. 6(b)). In Fig. 6(b), the flight track was recorded in pink on the monitor of the ground station to compare to the path (in blue) planed with the algorithm. It difficult to differentiate the two trajectories from each other, because they almost perfectly match and overlap together with each other, except a little deviation at the beginning of the flight, which is due to the inaccuracy of the coordinates given by DJI Guidance. Therefore, the flight experiments show that the heuristic RRT can be implanted in a UAV in reality for the navigation to avoid obstacles.

4   Conclusions

The RRT has been widely employed because it avoids the complex construction of the configuration space and offers low computational complexity. However, the RRT typically perform a uniform sampling of the state space without any preference, which lower the efficiency of the path search. In this work, an efficient RRT algorithm combined with a goal-bias heuristic strategy was designed to address the weakness of the basic RRT. The sampling can be bias in the direction toward the goal at a probability determined by the goal-bias heuristic factor. The simulation demonstrated the great enhancement

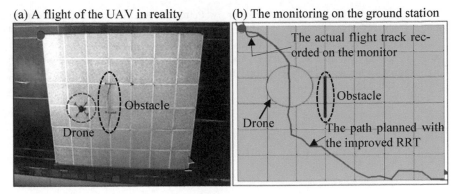

Fig. 6. The UAV flight carried out according to the improved RRT. (a) An image from the video to record the UAV flight in reality, and (b) the synchronous monitoring of the actual flight from the ground station. On the monitor screen, the location of the UAV is denoted as a circle by the monitor, the blue line is the path planned with the improved RRT before the flight, and the pink line is the actual flight track recorded on the monitor.

of the highly efficient heuristic RRT with much shorter planning time and much fewer nodes by an order of magnitude, and shorter path length as well. The effects of some parameters were analyzed with the simulations. UAV experiments were carried out to validate the feasibility of applying this improved RRT in UAV navigation in reality.

Acknowledgement. This work was supported by the NNSF of China under Grant 81401405.

References

1. Matlekovic, L., Juric, F., Schneider-Kamp, P.: Microservices for autonomous UAV inspection with UAV simulation as a service. Simul. Model. Pract. Theory **119**, 102548 (2022)
2. Shi, K.J., Wu, P., Liu, M.S.: Research on path planning method of forging handling robot based on combined strategy. In: 2021 IEEE International Conference on Power Electronics, Computer Applications (ICPECA), pp. 292–295. IEEE (2021)
3. Yuan, C.R., Zhang, W.Q., Liu, G.F., et al.: A heuristic rapidly-exploring random trees method for manipulator motion planning. IEEE Access **8**, 900–910 (2020)
4. Zhao, Y.Q., Liu, K., Lu, G.H., et al.: Path planning of UAV delivery based on improved APF-RRT* algorithm. J. Phys: Conf. Ser. **1624**(4), 042004 (2020)
5. Alexopoulos, C., Griffin, P.M.: Path planning for a mobile robot. IEEE Trans. Syst. Man Cybern. **22**(2), 318–322 (1992)
6. LaValle, S.M.: Rapidly-exploring random trees: a new tool for path planning. Iowa State University (1998)
7. LaValle, S.M., Kuffner, J.J.: Rapidly-exploring random trees: progress and prospects. In: 4th International Workshop on the Algorithmic Foundations of Robotics (WAFR), pp. 293–308. A K Peters, Ltd., Dartmouth College, Hanover, New Hampshire, USA (2001)
8. Lindemann, S.R., LaValle, S.M.: Current issues in sampling-based motion planning. pp. 36–54. Springer, Berlin, Heidelberg (2005)

9. Tsardoulias, E.G., Iliakopoulou, A., Kargakos, A., Petrou, L.: A review of global path planning methods for occupancy grid maps regardless of obstacle density. J. Intell. Rob. Syst. **84**(1–4), 829–858 (2016). https://doi.org/10.1007/s10846-016-0362-z
10. Xue, Y., Zhang, X., Jia, S., et al.: Hybrid bidirectional rapidly-exploring random trees algorithm with heuristic target graviton. In: 2017 Chinese Automation Congress (CAC), pp. 4357–4361 (2017)

A Review of Research on Automatic Scoring of English Reading

Xinguang Li[1], Xiaoning Li[1(✉)], Xiaolan Long[1], Shuai Chen[1], and Ruisi Li[2]

[1] Guangdong University of Foreign Studies, Guangzhou 510006, China
{lxg,20191003155,20181010003,20191010005}@gdufs.edu.cn
[2] South-Central Minzu University, Wuhan 430072, China
202021061031@mail.scuec.edu.cn

Abstract. With the growing demand for English as a second language across the world, there has been growing interest in automatic assessment of spoken language proficiency. This paper introduces the current research status and the technology development of the automatic assessment of English reading in China. This paper presents the scoring scales for the manual assessment of English reading and demonstrates the feature selection methods of the automatic assessment system. Specifically, this paper elaborates the methods of assessing segmental features and prosodic features. By analyzing the current research status and practical application of automatic speech quality assessment, this paper concludes that the theory and technology of automatic assessment of English reading quality are improving constantly, but there are still some problems that need to be further solved, such as the reliability of the assessment system, the selection methods of scoring features and the feedback function of the system.

Keywords: English reading · Feature selection method · Automatic assessment · Feedback and guidance

1 Introduction

Among the four basic language learning abilities of "listening, speaking, reading and writing", "speaking" is in second place. Therefore, the evaluation for speaking proficiency is an important part of language proficiency test. The earliest test for speaking proficiency is one-to-one communication between examinees and examiners, which is difficult to carry out in a large scale. At the end of the twentieth century, Heaton [1] believed that it was almost impossible to define a credible and objective measurement of the spoken language skills due to the subjectivity of human grading. Bachman and Palmer [2] proposed that how to score during the test was the core issue, as score was the main indicator of a learner's competence. Through interviews and surveys, Chen [3] found that in the real practice of scoring, human raters tended to pay more attention to the word diversity, grammar, text organization and the correlation with the topics in the candidates' responses, which validated John's view [4] that the pronunciation was relatively unimportant in determining speaking proficiency.

N. Xiong et al. (Eds.): ICNC-FSKD 2022, LNDECT 153, pp. 773–780, 2023.
https://doi.org/10.1007/978-3-031-20738-9_86

With the development of linguistics and natural language processing technology, scholars in education, language and computer fields began to pay attention to computer-assisted language learning (CALL). The Programmable Logic/Learning for Automated Teaching Operation (PLATO), which came out in 1959, is recognized as the first large-scale CALL system in the world [5]. Its appearance significantly promoted the use of computers in foreign language learning. The second-generation CALL system focused on educational field, represented by MIT's Athena Language Learning Project (ALLP) [6]. The third-generation CALL system has been focusing on human-computer inter-action and multimedia in language acquisition since the 1990s. Researchers attempt to quantify and digitize different spoken language features when designing assessment systems. Setting a reasonable machine scoring rule is critical for improving the validity and reliability of machine scoring.

2 Selection and Research Status of Feature Parameters for Automatic Scoring of English Reading

In the practice of oral language teaching, reading aloud is a traditional training method. Many scholars believe that reading quality can directly reflect a learner's oral ability. Heaton [1] proposed that reading with preparation was an effective way to assess a speaker's pronunciation quality. Gao [7] evaluated the pronunciation quality in terms of pronunciation, vocabulary, fluency, comprehension and other features, and demonstrated the importance of reading skills in the comprehensive language abilities.

The automatic scoring system of spoken English is an artificial intelligence sys-tem that can assess the spoken English quality. Generally, it includes the following processes: speech preprocessing, speech feature extraction and comparison with the standard speech. The final score is obtained by fusing the pronunciation accuracy score and prosodic feature score. The framework of an automatic scoring system for English reading quality is shown in Fig. 1.

Fig. 1. Framework diagram of automatic scoring system for English reading quality

2.1 Manual Scoring Method for English Reading

According to Wang [8], reading aloud in the college entrance examination mainly tests students' reading competence, which includes the mastery of word pronunciation, word stress, sentence stress, pauses, rhythm, fluency and intonation. The traditional human scoring method is mainly based on the human raters' overall impression and the scor-ing scales only include some abstract and vague indicators such as fluency, coherence, correct rhythm, etc. As a result, the score is determined only by the rater's knowledge and prior experience, which diminishes the test results' reliability. In the reading score scheme introduced by Gao [9], pronunciation accuracy was evaluated in terms of syntax, vocabulary and phoneme pronunciation; fluency was evaluated based on the speech speed and the occurrence of disfluent words. Ying and Zeng [10] conducted an in-depth study of several phonetic features that affect the reading quality and proposed 12 scoring indicators, including splutter, adding phoneme, phoneme error, word error or word block error, stress error and intonation error in terms of phonology; repetition, self-correction, pauses and filler words in terms of fluency; words addition and words omission in terms of word accuracy. They recommended that the intonation score be increased and that the speech tempo and content scores be appropriately reduced.

In general, the scoring for reading quality can be categorized into two parts: segment scoring and suprasegment scoring. In machine scoring, segment scoring refers to the scoring of pronunciation accuracy, also known as pronunciation confidence, which includes vocabulary, pronunciation, adding, and swallowing phonemes; suprasegment scoring refers to prosody scoring, which includes fluency, stress, intonation, and rhythm. The following will sort out and summarize recent domestic research on automatic scoring of English reading, with a focus on segment and prosody evaluation methods, and introduce the development of an automatic English reading scoring system.

2.2 Feature Selection for Automatic Scoring of English Reading

Speech assessment technology in the CALL system refers to the computerization of distinguishing features of spoken language. The selection of scoring features, which can be categorized into segmental and prosodic features, is the most significant part of speech evaluation technology.

Segmental Features. The features of the phonological units of the speech are segmental features, also known as pronunciation accuracy features. The common approach to evaluate segmental features is to compute the similarity of the reference speech and test speech based on an acoustic model. The likelihood algorithm, log posterior probability algorithm and segmental classification algorithm are some of the methods to calculate pronunciation accuracy, with the log posterior probability algorithm is the most widely used due to its insensitivity to rapid changes of individual features and sound intensity.

Yan et al. [11] from the University of Science and Technology of China (USTC) proposed to obtain phoneme-related transformation parameters by learning the manual scoring and adopted a phoneme-based log posterior probability transformation method to evaluate the pronunciation quality, so as to reduce the interference of phonemes with similar pronunciation on the scoring results. Liu et al. [12] used linguistic knowledge of Mandarin pronunciation to constrain the value of the log posterior probability algorithm and improved the performance of the scoring model of phonetic segmental pronunciation. Liang et al. [13] from Tsinghua University proposed the PASS (phone-based automatic score for l2 speech quality) algorithm. The PASS algorithm used the phoneme-level speech recognizer to calculate the log-likelihood ratio and normalized speed segment duration, and defined a new scoring method based on the Mahalanobis distance.

From the above research, it can be found that pronunciation accuracy was mainly rated based on the text-dependent methods. Text-dependent approaches have the advantage of improving recognition rate and lowering system computing burden, but they also have significant drawbacks. As a result, scholars began to investigate text-independent approaches. People from different places will have distinct accents, so different ways to evaluate varied accented speech should be used. Therefore, linguistic knowledge should be employed to improve the scoring performance of the system. In the field of speech evaluation, the application of linguistic knowledge to automatic assessment systems will become a major research focus.

Prosodic Features. The prosodic features mainly include the fundamental frequency (F0), energy and duration features, which are responsible for intonation, fluency, and rhythm. In the teaching practice, the learner's mastery of stress, rhythm and intonation is crucial to the success of phonics learning.

In the current automatic scoring system for reading, fluency is an important evaluation index. However, in most research, fluency is evaluated using simple methods, such as calculating the average duration of each phoneme to get the fluency score. For reading aloud task, a more accurate evaluation method of fluency should be explored. Huang et al. [14] developed a speech fluency grading system to evaluate the speaker's reading skills as well as more advanced reading skills related to coherent expression and rhythmic proficiency. It first calculated the score of pronunciation hesitation, pause, continuous reading, loss of blasting and overall speaking rate, and then used a hierarchical scoring fusion strategy in sentence and passage level. This is relatively comprehensive research on fluency scoring methods that can be found so far. Li et al. [15] conducted research on sentence stress and rhythm evaluation, using a two-threshold method based on energy and duration for stress evaluation and improved dPVI (distinct Pairwise Variability Index) method for rhythm evaluation. Many studies have shown that the pitch in the suprasegment could be used to represent intonation, so scholars started to focus on extracting the intonation-related features from the F0 contour. Jia and Tao [16] built an intonation model based on the F0 contour. Li et al. [17] used cepstrum analysis to obtain the pitch period and adopted DTW (Dynamic Time Warping) algorithm to compute the pitch distance between reference sentence and test sentence. Xu [18] innovatively brought in the scoring index of speech emotion in the automatic assessment system, using the fuzzy classifier and evaluating the speech emotion quality based on the similarity between test speech and reference speech.

Prosody evaluation plays a vital role in speech evaluation. Prosody can directly reflect the speaker's attitude, intention, mood, and expectations. As computer technology and cross-border researchers' language skills have improved, prosody evaluation has caught the attention of more and more researchers. The main goal of language learning is to communicate. In real-life conversation, the speaker changes their intonation to express different emotions, so the evaluation of speech emotion will also be a research hotspot in the future.

3   The Assisted-Learning Function of Automatic Scoring System of English Reading

The automatic scoring system for English reading can help students' learning as well as lowers the instructor's workload. The system should be able to provide students with instant feedback on their weakness. For further guidance, personalized resources should be recommended based on students' English level. Figure 2 shows the system with feedback function that allows students to keep track of their progress and receive helpful guidance.

The FLUENCY system developed by Carnegie Mellon University's Institute of Language Technology could detect syllable and rhythm errors in learners' speech, but rhythm feedback was only about speech duration [19]. The PLASER system, proposed by the Hong Kong University of Science and Technology, was able to provide accurate feedback for speakers with dialectal accents and visualize their errors during practice [20]. Liang and Liang [21] from South China Normal University presented an intelligent tutor system which could not only provide feedback on students' performance in terms of accuracy, fluency and completeness, but also could recommend personalized resources to learners. However, the study did not evaluate the validity and reliability of the feedback function. Wang [22] from Nanjing University of Posts and Telecommunications developed an intelligent English pronunciation training system that provided students with formant charts of their pronunciation and standard pronunciation and assisted them in correcting their mouth shape, but its feedback functions should be validated as well.

Feedback-guided learning is an effective strategy to improve students' spoken English skills, and it will be a major trend of computer-assisted language learning.

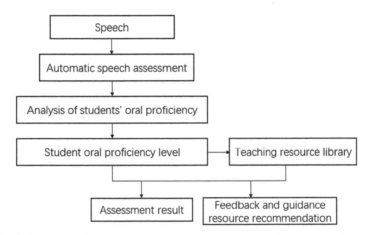

Fig. 2. A framework for an automatic computer-assisted learning system with feedback

4   Summary and Prospect

This paper summarizes and analyzes the related research on automatic scoring of English reading in China. It can be found that most of the prior work in this area focus on the evaluation method of pronunciation accuracy, which includes feature selection and algorithm improvement using linguistic knowledge. From the number of research on pronunciation accuracy evaluation, we can know its vital role in the automatic scoring of English reading. Table 1 summarizes the selection of features in previous studies. After the scoring results of pronunciation accuracy were recognized by language experts, scholars then began to explore prosody features to improve the system performance. Among the multiple features of prosody, the speech rate is widely used due to the ease of implementation. The research direction of scholars has transitioned from single-feature pronunciation quality evaluation to multi-feature fusion pronunciation quality evaluation.

By analyzing the related research, we can know that there are numerous systems developed for automatic assessment for English reading quality, but most of them lacked analysis of validity and reliability. A number of the systems only showed the human-machine correlation coefficient, but no inter-human correlation coefficient was illustrated. Some research only introduced the system design and did not evaluate the system performance. Since these studies used unpublished and self-built corpora, it is difficult to do horizontal comparisons between these systems. Therefore, future research on automatic scoring of English reading quality can be improved from the following aspects:

1. Improve system reliability. In the existing research, many evaluation algorithms have been proposed. Future research should focus on verifying the effectiveness of the algorithms and selecting the optimal scoring features.
2. Set reasonable weight to each scoring feature and consider studying a multi-feature fusion algorithm involving emotion feature.
3. Study the feedback and guidance function to better help learners improve their English reading proficiency.

Table 1. Comparison of research on scoring feature selection

References	Pronunciation accuracy	Fluency	Stress	Rhythm	Intonation	Emotion
[12]	Y					
[13]	Y	Y				
[15]	Y	Y	Y	Y	Y	
[16]	Y	Y		Y	Y	
[17]	Y	Y	Y	Y	Y	
[18]	Y	Y		Y	Y	Y
[22]	Y					
[23]	Y	Y	Y		Y	
[24]	Y	Y				
[25]	Y	Y	Y		Y	
[26]	Y	Y	Y		Y	
[27]	Y	Y	Y	Y	Y	

Acknowledgment. This research was financially supported by a National Natural Science Foundation (No.61877013), Special fund for scientific and technological innovation strategy of Guangdong Province (pdjh2021a0170, pdjh2021b0176 and pdjh2022b0174) and a Guangdong University of Foreign Studies Graduate Research Innovation Project (No.21GWCXXM-42).

References

1. Heaton, J.B.: Writing English Language Tests (1988)
2. Bachman, L.F., Palmer, A.S.: Language Testing in Practice: Designing and Developing Useful Language Tests, vol. 1. Oxford University Press (1996)
3. Chen, H., Li, J.N.: Reflections on the current situation of phonological assessment: a survey among raters of standardized oral tests in China. Foreign Lang. Their Teach. **5**, 81–87 (2017)
4. Levis, J.M.: Pronunciation and the Assessment of Spoken Language. Spoken English, TESOL and Applied Linguistics, pp. 245–270. Palgrave Macmillan, London (2006)
5. Franco, H., et al.: Automatic pronunciation scoring for language instruction. In: 1997 IEEE International Conference on Acoustics, Speech, and Signal Processing, vol. 2. IEEE (1997)
6. Neumeyer, L., et al.: Automatic text-independent pronunciation scoring of foreign language student speech. In: Proceeding of Fourth International Conference on Spoken Language Processing. ICSLP'96, vol. 3. IEEE (1996)
7. Gao Xia, Zhu Zhengcai, and Yang Huizhong. "The role of reading aloud in foreign language teaching and testing." Foreign Language Science2(2006):8
8. Wang, G.: The scoring requirements of reading aloud questions in the college entrance examination. Mod. Foreign Lang. **3**, 5(1991)
9. Gao, X.: Evaluation and design of reading score scheme. Foreign Lang. Teach. 4, 6 (2007)
10. Ying, X., Zeng, Y.: A study of pronunciation features and score predictors for the read-aloud task. J. Xi'an Int. Stud. Univ. **2**, 14 (2017)
11. Yan, K., et al.: Pronunciation evaluation based on a phoneme-dependent posterior probability transformation. J. Tsinghua Univ. (Sci. Technol.) (2011)
12. Liu, Q., et al.: The linguistic knowledge based improvement in automatic Putonghua pronunciation quality assessment algorithm. J. Chin. Inf. Process. **21**(4), 92–96 (2007)
13. Liang, W., et al.: Phoneme-based pronunciation quality evaluation algorithm. J. Tsinghua Univ.: Nat. Sci. **45**(1), 5–8 (2005)
14. Huang, S., et al.: Automatic assessment of speech fluency in computer aided speech grading system. Tsinghua Univ. (Sci. Tech.) **49**, 1349–1355 (2009)
15. Li, X., Wang, G., Yang, S.: Research on objective evaluation system of English sentences based on stressed syllables and prosody. Jisuanji Gongcheng yu Yingyong (Comput. Eng. Appl.) **49**(8) (2013)
16. Jia, H., Tao J.: A preliminary study on prosodic variability analysis and automatic prosodic evaluation of intonation. In: Proceedings of the 8th Chinese Phonetics Conference and International Conference on Frontiers of Phonetics in Celebration of Wu Zongji's 100th Birthday (2008)
17. Li, X.-G., et al.: A study of assessment model of oral English Imitation reading in college entrance examination. In: 2018 11th International Congress on Image and Signal Processing, BioMedical Engineering and Informatics (CISP-BMEI). IEEE (2018)
18. Xu J.: Research and Application System Design of English Pronunciation Quality Evaluation Based on Phonetic Emotion. Guangdong University of Foreign Studies (2016)
19. Eskenazi, M., Hansma, S.: The Fluency pronunciation trainer. In: Proceedings of the STiLL Workshop (1998)

20. Chen, G., Parsa, V.: Output-based speech quality evaluation by measuring perceptual spectral density distribution. Electron. Lett. **40**(12), 783–785 (2004)

21. Liang, Y., Liang, Y.: Speech assessment-based intelligent tutoring system for Spoken English learning. Mod. Educ. Technol. (2012)

22. Wang, S.: Design and Implementation of Intelligent English Pronunciation Training System Based on Android Platform. Nanjing University of Posts and Telecommunications (2013)

23. Li C.: An Objective Evaluation Method for Pronunciation Quality in Interactive Language Learning System (2007)

24. Yan, K.: Research on Automatic Evaluation of English Recitation and Retelling Test. A dissertation for master's degree at USTC (2008)

25. Teng, H., Liu, X., Wang, L.: The research of English pronunciation evaluation system based on the speech recognition technology. J. Yancheng Ins. Tech.(Nat. Sci. Ed.) 3, 17–22 (2016)

26. Li, G., Weiqian, L., Yuguo, D.: Feasibility study and practice of machine scoring of repetition question in large-scaled English oral test. Comput.-Assist. Foreign Lang. Educ. **2**, 10–15 (2009)

27. Chen, H., Kui, W., Li, J.: The automatic scoring system of Chinese learners' spoken English—A new system to assess oral English proficiency. Technology Enhanced Foreign Language Education (Chinese) **185**, 72–77 (2019)

Methods for Solving the Change Data Capture Problem

Liang Hao[1], Tao Jiang[1], Yatuan Lin[1], and Yitong Lu[2](✉)

[1] HBIS Digital Technology Co., Ltd., Hebei, Shijiazhuang 050022, China
{haoliang,hbjiangtao,linyatuan}@hbisco.com
[2] School of Computer Science and Technology, Xidian University, Xi'an, Shaanxi, China
21031211582@stu.xidian.edu.cn

Abstract. Change data capture (CDC) is an important strategic part of the data integration infrastructure. Obtaining change data from the source data system is the key to the incremental maintenance of data warehouse and business intelligence data. It is also the focus and difficulty of ETL (Extraction Transform Load). CDC first identifies and captures changes made to data in the source database, then records these changes in the order in which they occur, and finally delivers them to downstream processes or systems in real-time through messaging middleware. CDC currently uses various methods, including timestamps, differential snapshots, triggers, and archive logs. Many database vendors provide their own CDC (Change Data Capture) products, but these CDC products are developed based on their own database systems and are expensive. This paper lists several standard CDC technologies, introduces their fundamental principles and typical applications, analyzes the shortcomings of current CDC technologies, and finally provides a future research direction.

Keywords: CDC (Change Data Capture) · ETL (Extract Transform and Load) · Data warehouse

1 Introduction

With the vigorous development of information technology, the amount of data carried by information technology is also increasing, which means that the data we capture and need to analyze will also increase rapidly. Visible application scenarios in life, from driverless cars to daily transaction systems, all rely on capturing changing data.

Change Data Capture (CDC) is the process of identifying when data in a data source system has changed and prompting the system to act on the changes based on the changes in downstream processes or reminders. An everyday use case is to react to data changes in different target source systems so that data in related systems stay in sync.

ETL (Extract, Transform and Load) is essential to data warehouse and business intelligence. It is a crucial step in realizing the data warehouse and is responsible for completing the data transformation process from data source to target data warehouse. The essential role of ETL work is to recognize the system operation and management

N. Xiong et al. (Eds.): ICNC-FSKD 2022, LNDECT 153, pp. 781–788, 2023.
https://doi.org/10.1007/978-3-031-20738-9_87

of global enterprise data and to integrate the data sources of various business systems to establish a shared data warehouse. Among them, the data extraction part is divided into two types, one is total extraction, and the other is incremental extraction. Incremental extraction is the extraction of data changes in the source system, a process known as change data capture (CDC).

2 Related Work

The change data capture algorithm is generally divided into two methods: query-based and log-based methods.

Query-based methods can be subdivided into the following: snapshot method, trigger method, timestamp method, shadow table method, and control table change method [1]. In 2018, Harry Chandra completed the analysis and comparison of the existing implementations of change data capture by conducting experiments to simulate the execution of queries according to the specified number and period. The research and analysis have drawn the following conclusions. Each implementation algorithm has its optimal algorithm for different database system structures. The data structures applicable to the CDC algorithm based on timestamps include the data structure of the flat file model, the hierarchical model, and the mesh model data structure. The database systems applicable to the trigger-based CDC algorithm include relational model data structures and binary relational data structures [2]. In 2013, Mayuri B. Bokade et al. researched and analyzed the overall framework of change data capture and real-time data warehouse. They proposed the importance and significance of a data warehouse [3].

2.1 The CDC Method Based on Snapshot

A snapshot is an image of a storage system in a database that is used for data backup and recovery. The realization idea is to extract all the data in the source database at one time and load it into the buffer of the data warehouse. Every time it is necessary to synchronize and then extract all the data from the source database as a new version and compare the previous version with the current version each time to obtain the data changes.

In 2015, Du Wei et al. adopted two techniques to improve the efficiency of the differential snapshot algorithm [4]. The first technology implements the query with a summary in the open-source database MySQL (An open-source relational database management system), which reduces the differential snapshot algorithm's I/O (Input/Output) cost. The second technology adopts the parallel programming structure based on Hadoop Map/Reduce, which significantly improves the computing efficiency of the differential snapshot to avoid errors.

In implementing the new MySQL query syntax, the critical issue is the generation of summaries. When it is necessary to compare the corresponding lines of two snapshot files to find differences, the usual method is to compare the attribute values of the related lines one by one. The overhead of this method is higher when the number of attributes is large. An improved way is to perform the summary calculation on the attribute value of a row, directly compare the summary value of the corresponding row, and finally obtain the differential result after the comparison. Previously used methods of this type required three times of database visits to generate a snapshot file with summary values or summaries. Wei Du et al. simplified this process and proposed a SQL query language with summaries. The language implements query statements in the open-source database MYSQL and generates summaries by the MD5 algorithm.

In the implementation of the introduction of the MapReduce parallel processing framework, the critical issue is the acquisition process of the difference. Existing research proposes two methods to reduce the difficulty of obtaining the differences between snapshot files, namely the hash partition algorithm and the window algorithm, which may cause errors due to the limitation of memory size. The differencing algorithm merges and sorts two snapshot files. Although the accuracy of the results is guaranteed, the processing efficiency of a single CPU is low, which causes the low processing efficiency of the CDC process. Therefore, Wei Du et al. combined the MapReduce parallel processing framework and the merge sort algorithm into the differential algorithm, which ensured the correctness of the CDC results and the efficiency of the processing.

SQL Server 2008 provides two ways to capture change data: Change Tracking and Change Data Capture. Paul S. Randal conducted an in-depth comparison of these two ways in 2008. The changes provided by SQL Server Data capture is a snapshot-based capture method [5].

2.2 The CDC Method Based on Trigger

Trigger is a commonly used CDC implementation method. It is a storage mechanism invoked by the database system under specific conditions or when a particular event occurs. Triggers are implemented using SQL statements such as "BEFORE UPDATE" or "AFTER INSERT" when database data is changed. The general idea of the trigger-based CDC method is to create three triggers of insert, modify and delete in the source database that needs to be extracted to identify the corresponding changes in the source data. When the source data table changes, the trigger will capture these changes, write the changed data to the intermediate table, call the thread to extract the data through the intermediate table, and propagate the changed information to the destination database for subsequent operations.

In 2007, Ziyu Lin et al. conducted a study on the implementation of actively acquired change data capture. They concluded that the extraction efficiency and real-time performance of data capture based on triggers are higher than other implementations because the entire extraction work is performed. It is done by the DBMS itself [6]. But the disadvantage is that there is a relatively significant burden on the operating system, and the DBMS that requires the data source and target must be the same.

Zhejiang Synergy Data System Co., Ltd. Implemented a trigger-based incremental data extraction method in 2010 [7]. This method extracts the corresponding complete change record by extracting the change data recorded in the intermediate table. This extraction process is done in the incremental extraction process in the ETL. The extraction process is completed, aiming to solve the deficiency that most ETL commercial products do not realize the automatic generation of incremental data.

In 2013, Carlos Roberto Valêncio et al. proposed a zero-latency ETL extraction technique that executes task structures through triggers and creates these triggers using a tool that automatically generates SQL code [8]. The capture of changing data by this zero-latency ETL tool occurs concurrently with the use of the system. It is therefore also referred to as online extraction, as shown in Fig. 1. The CDC process was added to the data extraction because it needs to be persistently active to store changed data. This ensures that other steps can only operate on this changed information, not all the information stored in control.

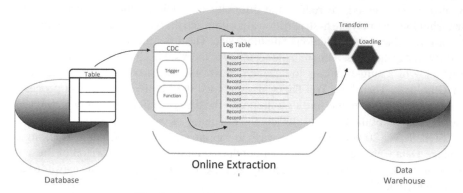

Fig. 1. A model of an online extraction.

In the task structure of this technique, when a user operates (such as an insert, edit, or delete) on a source database table that must be regenerated in the data warehouse, a trigger is activated to begin the change data extraction. When the trigger is activated, it executes a function that points to the table on which the operation is performed. This function stores the changed metadata information and ends the extraction.

2.3 The CDC Method Based on Timestamp

The principle of the CDC technology based on the timestamp of the source system is relatively simple. It captures data changes through the timestamp flag of the source system table. Changes in the database system can be added to the timestamp to find the specific change time of the data and then passed to the target system to make the same changes to ensure the consistency of the target system and the source system.

In 2009, Fuliang Xu and others researched various mainstream change data capture technologies. They concluded that the main advantages of timestamp-based CDC technology are that the technical difficulty and cost ratio is relatively low, and the intrusiveness and impact on the source system are moderate. Adding timestamps when extracting data consumes limited system resources and has little effect on the data increment and performance of the entire database [9].

The timestamp-based CDC method requires a timestamp field to be attached to each table in the source database to record the modification time of each table. This means a table scan is required unless an index is defined on the timestamp attribute in the table [10]. Under normal circumstances, the extraction technology based on timestamps has poor real-time performance and cannot obtain process data changes. Because only the final changes in the database before the extract operation are detectable, state changes in the source system cannot be captured. This makes timestamp-based methods challenging to apply to source systems that do not natively support timestamps or that frequently change table structures.

2.4 The Log-Based CDC Method

Another commonly used CDC method is to capture database change data by acquiring and parsing logs. The log-based method is the focus of research in change data capture. The analysis of the log is relatively complex, and for each database management system, the structure, content, and usage of the log are different. Therefore, the technical realization of this method is difficult and high cost.

In 2012, Xianxia Zou and others conducted a log-based change data capture study. They proposed that the current database architecture almost all includes database logs, and this logging method does not affect the performance of the source system and is highly low-invasive [11]. Therefore, log-based change data capture methods are also more common.

In 2022, JingGang Shi et al. proposed a framework for change data capture and data extraction based on log analysis. The design structure of the method is shown in Fig. 2. The structure includes log-based change data capture components, scheduling controllers, and data loading and transformation components.

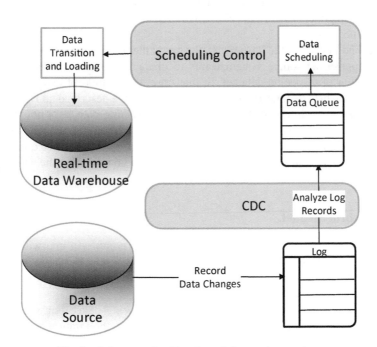

Fig. 2. A framework of log-based change data capture.

The method analyzes the online log and pushes the analyzed captured data to the data queue for data scheduling processing. The process of log analysis and change data capture includes initializing logs, building a data dictionary, loading log files, analyzing logs, and collecting data [12].

2.5   The CDC Method Based on Publish and Subscribe Queues

The publish and subscribe pattern was the first to use push data changes instead of pulling. Solutions like versioning database table rows and updating timestamps require the target system to pull changed data. Still, the source system actively pushes the changed data in the publish-subscribe model.

Typically, the publish-subscribe model requires an intermediary broker between the source database and the target system that pushes changes to a queue when data changes in the original system. The target system keeps listening to this queue and then decides to use them when changes reach a set threshold.

Such push data solutions offer many benefits, primarily scalability. For example, in a heavily loaded metadata system, where thousands of pieces of data are updated every second, a proactive approach to pulling changed data would require pulling many changes from the source system at once and consuming them all. This means an unavoidable time overhead, likely leading to high network latency before the target system requests new data. On the other hand, the push method allows the source system to send as many updates as possible to the intermediate queue. The target system can scale the number of applications using the queue's data, allowing the data to be processed faster and without errors when necessary [13].

The second benefit of the push data solution is the decoupling of source and target systems. Suppose the source system wants to change its underlying database or the storage location of a specific data set. In that case, the method of extracting changed data based on snapshots, timestamps, and triggers we have described above needs to be adjusted accordingly. In contrast, the technique of pushing data does not. As long as the source system continues to push the changed data to the intermediate queue in the same data format as before, the target queue can continue to listen for queue updates without making any changes. The target system is also unaware that the source system has changed.

2.6   The Hybrid Method

In order to obtain a change data capture solution that is more in line with the actual business scenario, many scholars have proposed two or more methods based on the above introduction, combined with their characteristics, and learn from each other to achieve better algorithms to solve more needs.

Beijing JinHe Software Co., Ltd. proposed an incremental data capture and extraction method based on timestamps and logs in 2012 [14]. This method improves the efficiency of data extraction and reduces the pressure of extraction on the existing business system and the technical complexity of extraction.

2.7 How to Choose CDC Algorithm

There is no right or wrong which CDC method to use. This article discusses the above methods' advantages, disadvantages, and application scenarios. The choice of method depends entirely on the requirements for capturing the changes and what the data in the target system will be used for.

If the target system's use case relies on always up-to-date data, we should consider implementing a push-based CDC solution. Even if the current use case is not real-time based, this approach may still need to be considered instead of using a pull-based CDC system.

If a push-based CDC solution is not possible, there are also many factors to consider when choosing a pull-based solution. First, if we can modify the source schema, adding update timestamps or row versions should be pretty straightforward by creating some database triggers. The overhead of managing an update timestamp system is much less than a row versioning system, so we should prefer to use update timestamps whenever possible.

If the source system cannot be modified, the only options are probably to take advantage of the source database's built-in change logging capabilities or change scanning. If the source system that provides the data in the file cannot accommodate change scanning, then the change scanning method needs to be used at the table level. This means fetching all the data in the table each time and determining what has changed by comparing it with the data stored in the target. This expensive method only works on source systems with relatively small datasets, so it should be used as a last resort.

Finally, when faced with complex or uncommon business requirement scenarios, we may consider combining multiple CDC technologies. But DIY CDC implementation isn't always easy, so it's safer to use an off-the-shelf CDC option in many cases.

3   Future Work

Many studies have adopted methods based on logs, triggers, and replication. However, most of the research done so far in the field has only tested each method on one type of data source. This results in the method being best on some types of data sources tested but not necessarily working equally well on other data sources. We need to analyze the test results of the best change data capture method in each type of data source at the data structure level to determine the best change data capture method for future new database applications with similar data structures.

Nowadays, the data sources that many companies use for operations come from disparate sources. Therefore, proposing a general change data capture method to meet the synchronization requirements between heterogeneous databases is the next issue that needs to be considered and studied in depth. The change capture method and synchronization strategy in the synchronization of heterogeneous databases are closely related. In order to solve the problems in the above solutions and realize the synchronization between heterogeneous databases, follow-up research can be carried out on the change data capture method and synchronization strategy of heterogeneous data sources simultaneously.

Acknowledgments. This work is supported by the National Natural Science Foundation of China under Grant No. 62172316 and the Key R&D Program of Hebei under Grant No. 20310102D. This work is also supported by the Key R&D Program of Shaanxi under Grant No. 2019ZDLGY13-03-02, and the Natural Science Foundation of Shaanxi Province under Grant No. 2019JM-368.

References

1. Jing, Z.: Research and Implementation of Open Heterogeneous Database Replication Framework. Institute of Software, Chinese Academy of Sciences, Beijing (2004)
2. Chandra, H.: Analysis of change data capture method in heterogeneous data sources to support RTDW. In: 2018 4th International Conference on Computer and Information Sciences (ICCOINS) (2018)
3. Bokade, M.B., Prof. Dhande, S.S., Prof. Vyavahare, H.R.: Framework of change data capture and real time data warehouse. Int. J. Eng. Res. Technol. (IJERT) **2**(4), 1418–1425 (2013)
4. Du, W., Zou, X.: Differential snapshot algorithms based on hadoop MapReduce. In: International conference on fuzzy systems and knowledge discovery (FSKD), pp. 1203–1208 (2015). https://doi.org/10.1109/FSKD.2015.7382113
5. Randal, P.S.: Tracking Changes in Your Enterprise Database. Microsoft TechNet Magazine (2008)
6. Ziyu, L., Dongqing, Y., Guojie, S., Tengjiao, W.: A review of research on change data capture in real-time active data warehouse. Comput. Res. Dev. **44**, 447–451 (2007)
7. Zhejiang Collaborative Data System Co., Ltd.: A trigger-based data incremental extraction method. Chinese Invention Patent, CN101923566.2010-12-22
8. Valêncio, C.R., Marioto, M.H., Zafalon, G.F.D., Machado, J.M., Momente, J.C.: Real time delta extraction based on triggers to support data warehousing. In: 2013 International Conference on Parallel and Distributed Computing, Applications and Technologies, pp. 293–297 (2013)
9. Fuliang, X., Zude, Z.: Research on change data capture technology. J. Wuhan Univ. Technol. Inf. Manage. Eng. Edn. **31**(5), 740–743 (2009)
10. Ram, P., Do, L.: Extracting delta for incremental data warehouse maintenance. In: Proceedings of 16th International Conference on Data Engineering (Cat. No.00CB37073), pp. 220–229 (2000). https://doi.org/10.1109/ICDE.2000.839415
11. Xianxia, Z., Weijia, J., Jiuhui, P.: Research on change data capture based on database log. Small Microcomput. Syst. **33**(3), 531–536 (2012)
12. Shi, J., Bao, Y., Leng, F., Yu, Y.: Study on log-based change data capture and handling mechanism in real-time data warehouse. In: 2008 International Conference on Computer Science and Software Engineering, pp. 478–481 (2008)
13. Gustriansyah, R.: The change data capture and the web application messaging protocol on the real time dashboard. Int. J. Eng. Adv. Technol.
14. Beijing Jinhe Software Co., Ltd.: An incremental data capture and extraction method based on timestamp and log. Chinese Invention Patent, CN102915336.2013-02-06

A Multi-attribute Decision-making Method for Probabilistic Language VIKOR Based on Correlation Coefficient and Entropy

Wenyu Zhang[1,2], Weina Luo[1(✉)], Xue Gao[1], Chuanqiang Zhang[1],
and Siyuan Zhao[3]

[1] Xi'an University of Posts and Telecommunications, Xi'an 710016, China
zwy888459@xupt.edu.cn, 17392189719@stu.xupt.edu.cn, gx876645@stu.xupt.edu.cn,
Chuanqiang0706@xupt.edu.cn
[2] China Research Institute of Aerospace Systems Science and Engineering, Beijing
100048, China
[3] Xi'an University of Finance and Economics, Xi'an 710100, China
1831040360@xaufe.edu.cn

Abstract. This paper proposes extending the VIKOR technique to the Probabilistic Linguistic Term Set (PLTS) to address the issue of retaining even more assessment information as possible while reasonably calculating the weights of experts and characteristics in the multi-attribute decision-making issue. Initially, a new PLTS aggregation method is recommended; then the model PLTS is constructed through the correlation coefficient weights and the weights based on entropy measures, and the objective weights of the experts and the objective weights of the attributes are obtained; secondly, the VIKOR method is extended to PLTS, The alternatives are ranked and the best one is selected considering the subjective preferences of decision makers; finally, through comparison analysis and further green supplier instances, the method's viability and efficacy are confirmed.

Keywords: Probabilistic linguistic term set · Correlation coefficient · Entropy · VIKOR

1 Introduction

People need to make more and more decisions as society grows, and the complexity of decision-making is becoming larger and larger. We can't make decisions only by relying on the experience and intuition of decision-makers. Therefore, it is more urgent to use reasonable decision-making methods to solve problems in real life [1]. MADM has become more and more common. There are two key issues that need to be resolved: how to keep the original assessment information more thorough, and how to decide on the expert's attribute and weight

N. Xiong et al. (Eds.): ICNC-FSKD 2022, LNDECT 153, pp. 789–797, 2023.
https://doi.org/10.1007/978-3-031-20738-9_88

in a reasonable manner [2].Focused on uncertainty and ambiguity of valuation information. Zadeh [3] proposed the theory of vague sets, since then, there has been a lot of interest in and research on the theory of ambiguous sets. Owing to the continuous innovation of the theory, the set of probabilistic linguistic terms (PLTS) is proposed [4,5], This is a more flexible and convenient rendering method. Numerous scholars have also looked into how to fairly weigh experts and qualities. The ideal expert weight is created using an adaptive consensus approach, according to Pang et al. [6]. Building on the fundamentals of the conventional TOPSIS method, Wang et al. [7] develops an optimization model that may be used to calculate the attribute weight. However, most experts' or qualities' relative importance in the judgment process is set arbitrarily. To solve this problem, the correlation coefficient weight determination model between the PLTs and the weight determination model based on the entropy measure are constructed, which incorporate both the DM's personal preference and the evaluation data's objective information. In MADM, the selection of the sorting technique is a crucial decision. For various decision settings, a number of MADM techniques, including TOPISIS [7], TODIM [8], PROMETHEE [9], and VIKOR [10], have been developed. The VIKOR method, out of the three previously described approaches, produces a more logical decision result because it maximizes group utility while minimizing individual regret. This document presents an extended version of VIKOR in the PLT environment.

2   Preliminaries

Definition 1. [4] LTS $S = \{s_i \mid i = -\tau, \ldots, -1, 0, 1, \ldots, \tau\}$, a set of probabilistic linguistic concepts (PLTS) can be defined as follows [4]:

$$L(p) = \left\{ \left(s_i^l, p^l\right) \mid p^l \geq 0, l = 1, 2, \ldots, \#L, \sum_{l=1}^{\#L} p^k = 1 \right\} \tag{1}$$

Where $\left(s_i^l, p^l\right)$ denotes the l linguistic term s_i^l associated with the corresponding probability value p^l, and $\#L$ represents the number of linguistic terms with $p \neq 0$ in $L(p)$.

Definition 2. $L_1(p) = \left\{\left(s_i^{1l}, p^{1l}\right) \mid l = 1, 2, \ldots, L\right\}$ and $L_2(p) = \left\{\left(s_i^{2l}, p^{2l}\right) \mid l = 1, 2, \ldots, L\right\}$ be two different PLTSs. The hamming distance between $L_1(p)$ and $L_2(p)$:

$$d\left(L_1(p), L_2(p)\right) = \frac{1}{L} \sum_{l=1}^{L} \left(g\left(s_i^{1l}\right) p^{1l} - g\left(s_i^{2l}\right) p^{2l}\right) \tag{2}$$

Definition 3. Let $L(p) = \left\{ (s_i^l, p^l) \mid p^l \geq 0, l = 1, 2, \ldots, L, \sum_{l=1}^{L} p^l = 1 \right\}$ be a PLTS, then the mean and variance of the PLTS of L (p) is given as follows:

$$\overline{L(p)} = \frac{1}{L} \sum_{l=1}^{L} \left(g\left(s_i^l\right) p^l \right) \tag{3}$$

$$Var(L(p)) = \frac{1}{L} \sum_{l=1}^{L} \left(g\left(s_i^l\right) p^l - \overline{L(p)} \right)^2 \tag{4}$$

Definition 4. $L_1(p) = \left\{ (s_i^{1l}, p^{1l}) \mid l = 1, 2, \ldots, \#L_1 \right\}$ and $L_2(p) = \left\{ (s_i^{2l}, p^{2l}) \mid l = 1, 2, \ldots, \#L_2 \right\}$ be two different PLTSs. The PLTS that has been adjusted can be labeled as $\widehat{L_1(p)} = \left\{ (s_i^{1k}, p^{k*}) \mid \sum_{k=1}^{K} p^{k*} = 1, \; k = 1, 2, \ldots, K \right\}$ and $\widehat{L_2(p)} = \left\{ (s_i^{2k}, p^{k*}) \mid \sum_{k=1}^{K} p^{k*} = 1, k = 1, 2, \ldots, K \right\}$, The correlation coefficient between $L_1(p)$ and $L_2(p)$ is then defined as:

$$\rho\left(L_1(p), L_2(p)\right) = \frac{c\left(L_1(p), L_2(p)\right)}{\sqrt{Var\left(L_1(p)\right) \cdot Var\left(L_2(p)\right)}}$$

$$= \frac{\sum_{k=1}^{K} \left[g\left(s_i^{1k}\right) p^{k*} - \overline{L_1(p)} \right] \left[g\left(s_i^{2k}\right) p^{k*} - \overline{L_2(p)} \right]}{\sqrt{\sum_{l=1}^{L} \left(g\left(s_i^{1k}\right) p^{k*} - \overline{L_1(p)} \right)^2 \cdot \sum_{l=1}^{L} \left(g\left(s_i^{1k}\right) p^{k*} - \overline{L_2(p)} \right)^2}} \tag{5}$$

Therefore, the correlation coefficients between matrix A and matrix B is:

$$\rho(A, B) = \sum_{i=1}^{m} \sum_{j-1}^{n} \left(\frac{c\left(L_1(p), L_2(p)\right)}{\sqrt{Var\left(L_1(p)\right) \cdot Var\left(L_2(p)\right)}} \right) \tag{6}$$

Definition 5. $L_1(p), L_2(p), \ldots, L_n(p)$ are n PLTSs, and $W = (w_1, w_2, \ldots, w_n)^T$ is the weight that PLTSs relate to $w_j \geq 0$. Next, give the following definition of the Probabilistic Linguistic Weighted Average (PLWA) operator:

$$PLWA\left(L_1(p), L_2(p), \ldots, L_n(p)\right)$$

$$= \left\{ \left[g^{-1} \left(\frac{(1 + g\left(s_i^1\right))^{w_1} - (1 - g\left(s_i^1\right))^{w_1}}{(1 + g\left(s_i^1\right))^{w_1} + (1 - g\left(s_i^1\right))^{w_1}} \right), p^1 \right] \right\}$$

$$\oplus \left\{ \left[g^{-1} \left(\frac{(1 + g\left(s_i^2\right))^{w_2} - (1 - g\left(s_i^2\right))^{w_2}}{(1 + g\left(s_i^2\right))^{w_2} + (1 - g\left(s_i^2\right))^{w_2}} \right), p^2 \right] \right\} \oplus$$

$$\cdot \oplus \left\{ \left[g^{-1} \left(\frac{(1 + g\left(s_i^n\right))^{w_n} - (1 - g\left(s_i^n\right))^{w_n}}{(1 + g\left(s_i^n\right))^{w_n} + (1 - g\left(s_i^n\right))^{w_n}} \right), p^n \right] \right\} \tag{7}$$

Definition 6. $L(p) = \left\{ (s_i^l, p^l) \mid p^l \geq 0, l = 1, 2, \ldots, L, \sum_{l=1}^{L} p^k = 1 \right\}$ be a PLTS, then the defined entropy measure of PLTS is as follows:

$$E(L(p)) = -$$

$$\frac{1}{2L \ln 2} \sum_{l=1}^{L} \left[p^l \ln p^l + (1 - p^l) \ln (1 - p^l) + \frac{g(s_i^l) + g(s_i^{L-l+1})}{2} \ln \frac{g(s_i^l) + g(s_i^{L-l+1})}{2} \right. $$
$$\left. + \frac{2 - g(s_i^l) - g(s_i^{L-l+1})}{2} \ln \frac{2 - g(s_i^l) - g(s_i^{L-l+1})}{2} \right] \tag{8}$$

3   Extended VIKOR Method with PLTS Information

Let $E = \{e_1, e_2, \ldots, e_t\}$ represent the expert group set, $W = \{w_1, w_2, \ldots, w_t\}$ represent the corresponding weight, $0 \leq w_k \leq 1$ and $\sum_{k=1}^{t} w_k = 1; A = \{A_1, A_2, \ldots, A_m\}$ be the alternative set; $C = \{c_1, c_2, \ldots, c_n\}$ be the decision attribute set, $\lambda = \{\lambda_1, \lambda_2, \ldots, \lambda_n\}$ be each attribute's weight vector, $0 \leq \lambda_j \leq 1$ and $\sum_{j=1}^{n} \lambda_j = 1$.

Step 1 The evaluation information of experts is collected to form the decision matrix of probabilistic linguistic.

Step 2 Compute the combined weight information for experts.

Step 2.1 According to the DMs' experience and knowledge to determine the subjective weight w_k^s of experts.

Step 2.2 Calculate the correlation coefficients between e_k and e_l according to Eqs. (5) and (6).

Step 2.3 Calculate the overall correlation coefficients between expert e_k and other expert decision-making information CC_k as:

$$CC_k = \sum_{l=1}^{t} \rho(e_k, e_l) \tag{9}$$

Step 2.4 The greater the overall correlation coefficients CC_k of the expert e_k, the higher the importance of the expert. Therefore, the objective weight w_k^o of the expert e_k is calculated as:

$$w_k^o = CC_k / \sum_{k=1}^{t} CC_k \tag{10}$$

Finally, the combined weights of experts $w_k = (w_1, w_2, \ldots, w_k)$ could be defined:

$$w_k = \alpha w_k^s + (1 - \alpha) w_k^o \tag{11}$$

Step 3 Using the PLWA, According to the Eq. (7), each expert's PLTs decision matrix is combined into a single PLTs matrix.

Step 4 Compute the combined weight information for attributes.

Step 4.1 Using the AHP method, the subjective weight of attributes are determined.

Step 4.2 Compute the entropy of each attribute C_j under different alternatives.

$$E_j = \frac{1}{m} \sum_{i=1}^{m} E\left(L_k^{ij}(\mathrm{p})\right) \tag{12}$$

Step 4.3 The obtained entropy information is used to calculate the attribute's objective weight.

$$\lambda_j^o = \frac{1 - E_j}{\sum_{j=1}^{n}(1 - E_j)} = \frac{1 - E_j}{n - \sum_{j=1}^{n}(E_j)} \tag{13}$$

Finally, the comprehensive weight vector of the attribute is $\lambda = \{\lambda_1, \lambda_2, \ldots, \lambda_n\}$:

$$\lambda_j = \beta \lambda_j^s + (1 - \beta)\lambda_j^o \tag{14}$$

β is a coefficient that indicates the choice for subjective weights.

Step 5 Determine the positive P_i^+, and the negative N_i^- qualities for every criterion within all alternatives. Eqs. (3) and (4) are employed.

Step 6 Calculate the group utility worth GU_i and the individual regret worth IR_i form every alternative A_i.

$$GU_i = \sum_{j=1}^{n} \lambda_j \cdot \frac{d\left(P_j^+, L^{ij}(\mathrm{p})\right)}{d\left(P_j^+, N_j^-\right)} \tag{15}$$

$$IR_i = {}_1 \le j \le n \frac{\lambda_j d\left(P_j^+, L^{ij}(\mathrm{p})\right)}{d\left(P_j^+, N_j^-\right)} \tag{16}$$

Step 7 Calculate the comprehensive evaluation value Q_i of each alternative:

$$Q_i = \delta \frac{(GU_i - GU^-)}{(GU^+ - GU^-)} + (1 - \delta)\frac{(IR_i - IR^-)}{(IR^+ - IR^-)} \tag{17}$$

δ is a predetermined coefficient used to express DMs' subjective opinions when they make their final decision. In addition, $GU^+ = \max_i\{GU_i\}, GU^- = \min_i\{GU_i\}, IR^+ = \max_i\{IR_i\}$, and $IR^- = \min_i\{IR_i\}$.

Step 8 According to Q_i, GU_i, and IR_i. The value of generates three different ranking tables, ranging from small to large. The lower the value, the better the alternative.

4   A Case Study and Comparative Analysis

4.1   A Case Study

At this time, one of the four logistics service provider $A = \{A_1, A_2, A_3, A_4\}$ has been chosen as the partner who best complies with the requirements of green development. Relying on three experts $E = \{e_1, e_2, e_3\}$ using the same linguistic term $\{s_{-3}$: nothing; s_{-2} :verylow; s_{-1} : low; s_0 : medium; s_{-1} : highs s_2 :veryhigh;s s 3 :perfect $\}$ the four suppliers are evaluated on the basis of the following four evaluation attributes: product price (C_1), product quality (C_2), delivery time (C_3), and environmental performance (C_4).

Step 1 Gather scoring information for each alternative against each attribute reported by each expert. Table 1 displays the PLTS assessed value of e_1, as well as the PLTS assessed values of other experts.

Table 1. Evaluation matrix of e_1.

	C_1	C_2	C_3	C_4
A_1	$(s_{-1},0.4)\,(s_1,0.6)$	$(s_0,0.7)\,(s_1,0.3)$	$(s_2,0.6)\,(s_3,0.4)$	$(s_3,1)$
A_2	$(s_0,0.3)\,(s_1,0.7)$	$(s_2,1)$	$(s_{-2},1)$	$(s_{-3},0.4)\,(s_{-2},0.6)$
A_3	$(s_0,0.7)\,(s_1,0.3)$	$(s_{-2},0.6)\,(s_{-1},0.4)$	$(s_{-1},1)$	$(s_0,0.3)\,(s_1,0.7)$
A_4	$(s_{-1},0.6)\,(s_0,0.4)$	$(s_{-2},0.5)\,(s_{-1},0.5$	$(s_2,0.6)\,(s_3,0.4)$	$(s_0,0.7)\,(s_1,0.3)$

Step 2.1 According to the DMs' experience and knowledge to determine the subjective weight of experts is $w_k^s = (0.4, 0.3, 0.3)$.

Step 2.2 According to Definition 4, calculate the correlation coefficients between e_k and e_l. Table 2 shows them.

Table 2. The correlation coefficients between e_k and e_l.

	e_1	e_2	e_3
e_1	–	6.71	11.84
e_2	6.71	–	10.79
e_3	11.84	10.79	–

Step 2.3 Establish the objective weight $w_k^o = (0.34, 0.36, 0.3)$ of the expert e_k by the Eqs. (9) and (10).

Step 2.4 Determine the total expert weights using Eq. (11). To depict a neutral opinion, we set $\alpha = 0.5$ to represent a neutral attitude. The global expert weights vector is $w_k = (0.33, 0.36, 0.31)$.

Step 3 Reconstruct the group PL score matrix based on the full expert weights obtained using the PLWA operator proposed in Definition 5 are shown in Table 3.

Step 4.1 Using the AHP method to determine the subjective weight of attributes. $\lambda_j^s = (0.4, 0.1, 0.3, 0.2)$.

Step 4.2 Determine the exponent within each attribute c_j under different alternatives based on entropy measure according to Eq. (12).

Table 3. Aggregation PLTS decision matrix.

	c_1	c_2	c_3	c_4
A_1	$(s_{0.66}, 0.5), (s_2, 0.5)$	$(s_{-0.23}, 0.5)(s_{0.64}, 0.1)$ $(s_{1.05}, 0.4)$	$(s_{-0.29}, 0.5)(s_{0.29}, 0.1)$ $(s_3, 0.4)$	$(s_2, 1)$
A_2	$(s_{-0.34}, 0.5)(s_{0.78}, 0.2)$ $(s_{1.35}, 0.5)$	$(s_3, 1)$	$(s_{-0.9}, 0.4)(s_{0.1}, 0.6)$	$(s_{0.52}, 0.4)(s_{1.36}, 0.6)$
A_3	$(s_{-0.69}, 0.6)(s_{-0.47}, 0.1)$ $(s_{0.54}, 0.1)(s_{0.97}, 0.2)$	$(s_{-1.74}, 0.6)(s_{-0.93}, 0.4)$	$(s_{0.1}, 1)$	$(s_{1.34}, 0.3)(s_{2.75}, 0.7)$
A_4	$(s_{0.86}, 0.6)(s_3, 0.4)$	$(s_{-0.33}, 0.5)(s_{2.3}, 0.5)$	$(s_{1.47}, 0.6)(s_3, 0.4)$	$(s_{0.08}, 0.4)(s_{1.13}, 0.6)$

Step 4.3 Based on the entropy information, compute the objective weights of the attributes using Eq. (13). The vector of the objective weights of the attributes $\lambda_j^o = (0.36, 0.27, 0.22, 0.15)$.

Step 4.4 Determine the total weights according to Eq. (14). To represent a neutral setting, we set level $\beta = 0.5$. The total weights vector of attributes is $\lambda_k = (0.39, 0.185, 0.261, 0.175)$

Step 5 Determine the positive P_j^+ and negative N_j^- values of each attribute across all alternatives. Based on Eqs. (3–4) and Table 3, we have the following positive and negative values for each attribute among all alternatives:

$P_j^+ = \{(s_{0.66}, 0.5)(s_2, 0.5); (s_3, 1); (s_{1.47}, 0.6)(s_3, 0.4); (s_2, 1)\}$

$N_j^- = \{(s_{-0.69}, 0.6)(s_{-0.47}, 0.1)(s_{0.54}, 0.1)(s_{0.97}, 0.2); (s_{-1.74}, 0.6)(s_{-0.93}, 0.4);$
$(s_{-0.9}, 0.4)(s_{0.1}, 0.6); (s_{0.08}, 0.4)(s_{1.13}, 0.6)\}$

Step 6 Calculate and rank the group utility value GU_i, individual regret value IR_i, and qualitative assessment value $Q_i(\delta = 0.5)$ of each alternative. Table 4 displays the results of the above aspects based on Eqs. (15–17).

Table 4. Calculation process and results of GU_i, IR_i, and Q_i.

	GU_i	Ranking	IR_i	Ranking	Q_i	Ranking
A_1	0.32	1	0.2	1	0	1
A_2	0.67	3	0.34	4	0.88	4
A_3	0.78	4	0.25	2	0.68	3
A_4	0.52	2	0.26	3	0.43	2

Step 7 Generate three rankings of alternatives based on the value of Q_i, GU_i. From Table 4, clearly shoes that the alternative $Q_4 - Q_1 = 0.43 > 1/(4-1)$, A_4 is still the best when referring to GU_i and IR_i. Therefore, according to the evaluation conditions, A_4. Is the best alternative.

4.2 Comparison Analysis

To test the method's feasibility and effectiveness, the method in this paper is compared with PL-TOPSIS [2], PL-TODIM ($\theta = 0.5$), PL-PROMETHEE [8], the final ranking is obtained through the selected method. As shown in Table 5.

Table 5. Order by using diverse methods.

	Order	Optimal alternative	Bad alternative
PL-TOPSIS [10]	$A_1 > A_4 > A_2 > A_3$	A_1	A_3
PL-TODIM [10]	$A_1 > A_4 > A_3 > A_2$	A_1	A_2
PL-PROMETHEE [10]	$A_1 > A_3 > A_4 > A_2$	A_1	A_2
PL-VIKOR	$A_1 > A_4 > A_3 > A_2$	A_1	A_2

When the above four results are compared, it is clear that the ranking results generated by various decision-making methods are not significantly different, and the optimal solutions are all the same, which shows the efficiency and practicability of decision-making method.

5   Conclusions

This paper proposes a new MADM method based on PLTS as evaluation information. The following three aspects reflect the main work of this paper:

1. To preserve the language information of the original PLTS, this paper proposes a new PLTS aggregation method, which can avoid the loss of probability language and make the aggregated results better reflect the opinions of experts.
2. Using correlation coefficient and entropy measure, a new comprehensive weighting method is proposed, which can deduce the objective weight of experts and attributes, and the calculation is simple.
3. Express the opinions of experts through PLTS, and extend the VIKOR method to MADM, which can consider the subjective preference of DM. It only needs to maximize the group utility and minimize personal regret. The calculation is simple and easy to get the results quickly.

References

1. Gou, X.J., Deng, F.M., Xu, Z.S.: Research on large-scale group consensus decision-making method based on self-confidence double-layer language preference relationship and its application. Chinese Management Science (2021)
2. Mou, N.Y.: MADM method based on generalized hesitation triangular fuzzy power mean operator. Control Decis. Mak. **02**, 282–292 (2018)
3. Zadeh, L.A.: Fuzzy sets, information and control. Inf. Control **8**(3), 338–353 (1965)
4. Chen, Y., Wang, Y.M.: Multi attribute group decision making based on power operator of interval probability language term set. Fuzzy Syst. Math. **03**, 91–107 (2021)
5. Shen, L.L., Pang, X.D., Zhang, Q., Qian, G.: TODIM method based on probabilistic language term set and its application. Stat. Decis. Mak. **18**, 80–83 (2019)
6. Pang, J.F., Liang, J.Y., Song, P.: An adaptive consensus method for multi-attribute group decision making under uncertain linguistic environment. Appl. Soft Comput. **58**(58), 339–353 (2017)

7. Wang, Z.P., Fu, M., Wang, P.W.: Multi attribute group decision making model based on prospect theory and TOPSIS method in probabilistic fuzzy environment. Sci. Technol. Eng. **22**(04), 1329–1337 (2022)
8. Zhang, Y., Yang, W.: TODIM method based on hesitation in Pythagorean fuzzy environment. Fuzzy Syst. Math. **02**, 85–92 (2020)
9. Geng, X.L., Zhou, Q.C.: Multi criteria decision making method based on probability language BWM and PROMETHEE II. Oper. Res. Manage. **29**(06), 124–129 (2020)
10. Yu, Q., Cao, J., Hou, F.J., Tan, L.: Hesitation triangle fuzzy language multi-attribute decision-making method based on VIKOR and correlation coefficient. Practice Understand. Math. **08**, 22–33 (2020)
11. Wang, J., Wei, G., Wei, C., Wei, Y.: MABAC method for multiple attribute group decision making under q-rung orthopair fuzzy environment. Defence Technol. **01**, 208-216 (2020)

The Transfer of Perceptual Learning Between First- and Second-Order Fine Orientation Discriminations

Mingliang Gong[1,2(✉)], Tingyu Liu[1], and Lynn A. Olzak[2]

[1] School of Psychology, Jiangxi Normal University, Nanchang, China
{gongml,202040100260}@jxnu.edu.cn
[2] Department of Psychology, Miami University, Oxford, OH 45056, USA
olzakla@miamioh.edu

Abstract. FirsT- and second-order systems have been proposed to explain visual information processing. With regard to the communications between the two systems, mixed results have been shown. The transfer of perceptual learning between first- and second-order systems was examined in fine orientation discrimination tasks. Observers were either trained with luminance-modulated (LM) orientation and tested with contrast-modulated (CM) orientation (Experiment 1) or trained with CM orientation and tested with LM orientation (Experiment 2). The difficulty of the discrimination of the two types of orientations was equalized. Learning curves were tracked and compared between observers who had training and those who had no training. Results showed that the performance of observers trained with LM orientation improved rapidly in CM task and vice versa, while the performance of untrained observers tended to stay low. This two-way transfer suggests that there are bidirectional communications between first- and second-order systems wherein higher-level cortical areas might be involved and the recruitment of common population of neurons might be playing an important role.

Keywords: Perceptual learning · First-order · Second-order · Transfer · Orientation discrimination · Contrast-modulated

1 Introduction

Performance in perceptual tasks improves significantly after training or practice. This type of learning, widely known as perceptual learning, suggests that neurons involved in the tasks are plastic [1–5]. Many studies have demonstrated that the learning cannot transfer to untrained features such as motion direction [6], orientation [7], location [2, 7] and contrast [8]. This is believed to reflect that the neuronal plasticity occurs in early visual cortex wherein receive fields still retain fine selectivity to basic visual features [2, 5, 9]. Supporting this claim, perceptual learning has been found to increase the BOLD signal in V1 [10, 11].

However, the specificity of perceptual learning has been challenged by studies showing that the degree of learning specificity is modulated by task difficulty [12, 13], stimulus

© The Author(s), under exclusive license to Springer Nature Switzerland AG 2023
N. Xiong et al. (Eds.): ICNC-FSKD 2022, LNDECT 153, pp. 798–809, 2023.
https://doi.org/10.1007/978-3-031-20738-9_89

complexity [14], task precision [15], training procedure [16–19] and training duration [20, 21]. For instance, in an odd-element detection task that requires the detection of a possible oddly oriented element in an array of light bars, Ahissar and Hochstein [12] manipulated task difficulty during the training session by varying orientation differences between target and distractors, number of possible target locations, and SOAs. They found that when the training task was easy, the learning effect in one condition (e.g., one specific orientation difference) transferred to other conditions (e.g., another orientation difference). However, the transfers did not occur when the training task was difficult. This study suggests that task difficulty affects the amount of transfer.

Liu and Weinshall [13] trained participants to distinguish between the directions of moving dots, and it was demonstrated that if the task was challenging, learning transfer from a trained direction to a new one did not occur. When tracking their performance in the task of the new direction, however, they only needed half the number of sessions to reach the plateau, which means that their learning speed doubled although their initial performance was poor. The authors proposed that this was because observers had learned some knowledge in the initial training task that was used in the second task. This study, together with studies showing the transfer of perceptual learning (e.g. [12]), suggests that high-level cognitive processes can be involve in perceptual learning [22, 23].

Disparate results have been shown with regarding to the transfer of perceptual learning between these two types of stimuli. First-order stimuli are those that defined by differences in luminance from their backgrounds (i.e., luminance-modulated, LM), while second-order stimuli differ in contrast or texture (e.g., contrast-modulated, CM). The sensitivity of the visual system to second-order stimuli is significantly lower than first-order stimuli [24–26] and more importantly, they are processed in different ways. The mechanism of first-order system can be explained by a linear model whereas the second-order system includes both linear and nonlinear mechanisms (linear-nonlinear-linear model) [27–31].

Given the differences between the first- and second-order systems, it is intriguing to know whether learning to discriminate one stimulus transfers to the other. Many studies have shown that the two systems involve two distinct mechanisms [32–36]. Does this mean there is no communication between them? According to the explanation models (i.e., linear model vs. linear-nonlinear-linear model), the two systems share an early linear process that includes luminance filters [37, 38], which makes the communication a possibility. A wealth of studies have probed this question by the investigation of learning transfer between the two systems [37–40], yet it remains open questions whether there is a transfer and what is the nature of the transfer if it did exist.

Whereas a few studies have not found transfer between first- and second-order systems [41, 42], most studies have reported such transfer, regardless of the direction of the transfer. The most common finding is asymmetric transfer [37–40, 43]. In Petrov and Hayes [38] study, for instance, observers were first pretested in either a CM or a LM motion judgment task and then trained in the other task (i.e., LM or CM motion judgment task), and finally posttested for the task that had pretested. They showed a full transfer of perceptual learning from CM motion to LM motion, but no significant transfer was observed vice versa. Others have shown asymmetric transfer in the opposite direction e.g. [39], or even both-way transfer [44, 45]. For example, Cruickshank and Schofield

[44] showed two-way partial transfer of tilt after-effects between LM and CM cues using a tilt-after-effect paradigm. The transfer, though not due to perceptual learning, at least suggests the existence of a link between first- and second-order systems. In sum, results from earlier research on perceptual learning transfer between first- and second-order systems have been mixed.

Intriguingly, previous research regarding perceptual learning transfer between first- and second-order visual systems has mostly been focusing on motion; orientation, as a fundamental visual feature, has rarely been explored. In the present investigation, we explored the transfer of perceptual learning between first- and second-order orientation discriminations. Given that task difficulty modulates perceptual learning transfer [12], we minimized the influence of this potential confound by attempting to equalize the visibility of LM and CM stimuli. By comparing performance of trained and untrained observers in the transfer task, we aimed to examine whether there was any transfer or at least saving in the transfer task due to previous training, like the one shown in Liu and Weinshall study [13]. To this end, we tracked the performance of both groups for at least 8 sessions. If there is a transfer or saving, trained observers should need fewer sessions to reach plateau than untrained observers; if there is no transfer or saving, the same amount of sessions are needed to reach plateau.

2 General Methods

2.1 Observers

Five observers (ACT, CJL, KEC, NRM and TLB), four female and one male, participated in Experiment 1. NRM[1] and another four observers (ADL, KLS, XHL and YQH), 1 male and 4 female, participated in Experiment 2. All were undergraduate students of Miami University, naive to the purpose of the study. Before running the experiments, two observers in each experiment (TLB and CJL in Experiment 1, NRM and KLS in Experiment 2) received considerable training with LM stimuli (Experiment 1) and CM stimuli (Experiment 2). The other three observers in each experiment (ACT, KEC and NRM in Experiment 1, ADL, XHL and YQH in Experiment 2) were not trained and served as a baseline group. All observers had normal or corrected-to-normal vision. The study was approved by the IRB at Miami University and was carried out following the tenets of the Declaration of Helsinki.

2.2 Apparatus and Stimuli

The stimuli were display on a gamma corrected 17-inch CRT monitor. The mean display luminance was 19.8 cd/m^2, and the resolution was 2 pixels/minute of visual arc. In order to make up for the huge pixel size, the viewing distance was set to 2.74 m.

The LM stimuli were circular sinusoidal grating patches, with a diameter of 40 min of arc and a spatial frequency of 4 cycles/degree. They were presented at a contrast of

---

[1] NRM only finished 12 experimental sessions with CM stimulus and did not run any sessions with LM stimulus in Experiment 1. This experience only affected the time she needed in the training task (CM stimulus) but not the transfer task (LM stimulus) in Experiment 2.

0.1 and tilted slightly to the left or right of vertical. The contrast modulated second-order stimuli had a binary noise carrier, presented at a modulation depth of 0.4. Other parameters of the CM stimuli were identical to that of LM stimuli. See Fig. 1 for example stimuli.

Fig. 1. LM grating (left) and CM grating (right) used in the experiments.

3   Procedure

Both experiments started with a training session. In Experiment 1, observers were trained with LM gratings. The fixation point, which was 9 degrees left or right of the stimuli, was displayed on the screen throughout the whole trial. All along, observers had to keep their attention on the fixation. Training began with an easy orientation discrimination task, with the grating either tilted to the left or to the right. A simple orientation discrimination test with a grating that was either slanted to the left or the right served as the training's first task. The orientation magnitude of tilting was then gradually and systematically reduced as training progressed. On each trial, a grating, accompanied by a middle-pitch tone, with an exposure time of 500 ms. Observers were asked to determine whether the grating was leftward-tilted or rightward-titled on a 6-point scale. Feedback (a low or high tone) was given after the response. If no respond was made within 5000 ms, or an invalid key was pressed, the trial was repeated after being mixed in with the other trials.

It is well established that the visual system was less sensitive to second-order stimuli than to first-order stimuli (e.g. [26]). As result, CM stimuli are less visible and thus more difficult to discriminate than LM stimuli. Consistent with this conclusion, we found that, in the pilot study of Experiment 2, the task was too difficult for observers when the CM stimuli had a modulation depth of 0.1 and presented 9 degree peripherally during the training. Thus we made some adjustments to equalize the visibility of the two types of stimuli: First, the contrast of the CM stimuli was raised to 0.4; second, the target stimulus was presented in the center and observers used foveal viewing to do the task. Through the adjustments, the performance of the observers for first- and second-order orientations reached approximately the same plateaus, i.e., sensitivity index (d') lying between 1.5 and 2.0, when they reached their discrimination thresholds.

Transfer tasks followed the training sessions. In the transfer tasks, two trained observers together with three untrained ones in each experiment completed several sessions of experiment with CM (Experiment 1) or LM stimuli (Experiment 2). The procedure in each experiment was identical to the training session. To examine whether there was any transfer or saving in the transfer task due to previous training, observers' performance was tracked for at least 8 sessions of 240 trials each until their performance became stable.

4 Results

4.1 Experiment 1: LM to CM

To test whether perceptual learning with LM stimuli could transfer to CM stimuli, performance of observers who received training with LM stimuli (TLB and CJL) was tracked in the transfer task (i.e., CM task), and more importantly, their performance was compared to the performance of those who did not receive any training (ACT, NRM and KEC).

Figure 2(a) shows the performance of observers trained with LM gratings ("-trained") and those without any training ("-untrained") in the CM task as a function of sessions. It is clear that all observers performed poorly in the first and second sessions. After that, however, performance rapidly diverged, with trained observers rapidly achieved asymptote performance with d' larger than 2.0 for most sessions while performance remained low ($d' < 1.5$) for untrained observers. Pairwise comparisons showed that performance of the trained observers (TLB and CJL) was significantly better than the untrained observers (ACT, KEC and NRM), ps < 0.01, whereas there was no significant difference between the two trained observers or between any pair of the untrained observers (See Fig. 2(b)). This result suggested that observers had learned some knowledge in the LM task which partially transferred to the CM task. Together, the experiment showed that perceptual learning with LM stimuli did transfer to CM task.

4.2 Experiment 2: CM to LM

Similarly, to examine whether perceptual learning with CM stimuli could transfer to LM stimuli, observers including those who received training with CM gratings (NRM and KLS) and untrained ones (YQH, ADL and XHL) participated in the transfer task (i.e., LM task). Figure 2(c) shows the performance of observers trained with CM gratings ("-trained") and untrained ("-untrained") observers in the LM task as a function of sessions. The results revealed that while the performance of the trained observers stayed at high level (d' around 1.5) from the first session on, the performance of untrained observers was inconsistent: two out of three (YQH and ADL) stayed at low level (d' around 0.5), while the third observer (XHL) rapidly achieved asymptote ($d' = 1.24$) and performed almost as good as the trained observers ($d' = 1.52$), t (17) = 1.33, p > .05. Planned comparisons confirmed that the performance of the trained observers (NRM and KLS) was significantly better than the two untrained observers (YQH and ADL), ps < .001. When averaged over observers and sessions, performance of trained observers ($d' = 1.52$) was still significantly better than untrained observers ($d' = .72$), t (17) = 5.57, p < 0.001. See Fig. 2(d). Thus the result suggests that significant transfer of learning effect in CM task to LM task can occur.

To summarize, Experiments 1 and 2 together showed that training significantly improved observers' performance, even though individual difference was observed in Experiment 2. This finding indicates that there is two-way transfer of perceptual learning between first- and second-order orientations.

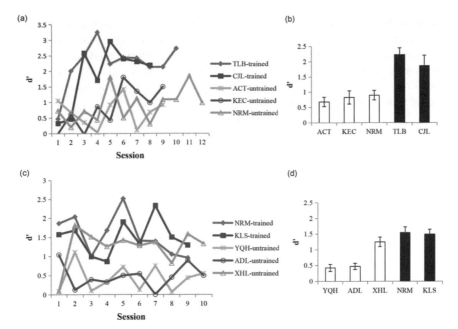

Fig. 2. Orientation discrimination sensitivity (d′) in Experiments 1 (a, b) and 2 (c, d). The left two panels show performance in the transfer task with CM stimulus (a) and with LM task (c), respectively. The right two panels show performance of each observer averaged across sessions in Experiment 1 (b) and Experiment 2 (d), respectively. In (a) and (c), "-trained" signifies observers received training with LM and CM, respectively; "-untrained" signifies observers with no perceptual learning experience. In (b) and (d), the first three columns (hollow) and the last two columns (solid) signify untrained and trained observers. Error bars indicate 1 ± SEM.

5   Discussion

5.1   Learning Transfer Between First- and Second-Order System

While many studies suggest that different pathways are involved in the process of first- and second-order stimuli (e.g.[32]), a growing body of evidence showing transfers in perception learning of motion such as [38, 40, 46], motion aftereffect [43] and tilt aftereffects [44] between the two systems indicates that they share some common mechanisms rather than being completely separated. Here we probed the perceptual learning transfer between first- and second-order of a basic visual feature, i.e., orientation. The results showed two-way transfer, which suggests a communication between the two systems.

A dual-pathway model has been put forth to explain the separation of first- and second-order systems [29, 37, 38, 47] (See Fig. 3). At the first layer, a bank of first-order linear filters, such as spatial-frequency filters at V1, process both first- and second-order inputs [47]. Then the two pathways are separated: whereas the first-order signal goes directly to the template analysis, the second-order signal goes through additional processes before being analyzed by the second-order template. Specifically, the second-order signal is processed by a non-linear rectification which generates components not presented in the first-order pathway. Then this rectified signal is further smoothed by a group of second-order texture filters at V2 or higher level cortical areas [26]. After the template analysis, signals in both pathways reach its final decision layer. Although the dual-pathway model is proposed to explain asymmetric transfer between second- and first-order systems, we propose that it can also be accommodated to explain the symmetric transfer observed in our study.

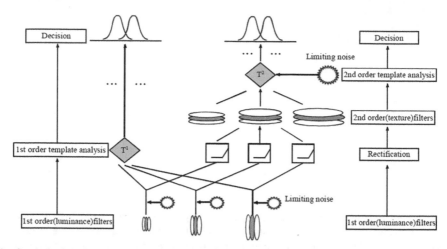

Fig. 3. A dual-pathway model. The left part shows the pathway for first-order signal processing. The right part shows the pathway for second-order signal processing, which includes additional non-linear processes (non-linear rectification and second-order texture filters) (Adapted from [37])

A widely accepted theory concerning the transfer of perceptual learning is re-weighting [47–49]. According to this theory, signal processing is largely limited by internal and external noises, as well as by the strength of the connections between layers. Perceptual learning can mitigate these limits (See also [1, 50]) and strengthen the connections between the early neural representations and the areas responsible for making decisions (See also [51]). More importantly, with appropriate training (e.g. double training), new connections may form that enable the use of learned rules in previously untrained contexts (e.g., new contrast) [52]. When observers are trained with CM stimuli, both linear and non-linear processes are strengthened or re-weighted, it follows that the perception of LM stimulus that only involves linear process is also strengthened [37]. This explains the transfer from CM to LM. When observers are trained with LM stimulus, the overlapping layer, linear filters, is strengthened, yet other layers in the second-order pathway (i.e., non-linear rectification and second-order filters), which is vital to processing second-order stimuli, are not strengthened. According to Chen et al. [37] and Petrov and Hayes [38], there will be no transfer from LM to CM. However, this transfer was observed in the current study.

The differences in task stimuli could be one explanation for the inconsistent results between the present study and earlier ones. In the present study, an orientation discrimination task was used. Orientation is primarily processed in V1 [53, 54], thus perceptual learning of both first- and second-order orientation discriminations (as in our study) can rely heavily on the improvement of the sensitivity of neurons in V1. Thus when observers were trained with LM orientation discrimination, improvement mostly occurs in V1 (or first-order linear filtering layer). When they are presented with CM gratings, a more reliable signal from the first-order linear filtering layer in V1 would feed onto the second filtering layer, which may facilitate the processing of CM orientation discrimination [39]. By contrast, most studies showing asymmetry transfer used a motion discrimination task [37, 38, 40] Motion is mainly processed in V3 and V5 [55, 56], thus the perception of first- and second-order motion relies on the improvement of the sensitivity of neurons not only in V1, but also in V3 and V5. However, training with LM only improves the sensitivity of neurons in V1, which restricts learning transfer from LM motion to CM motion. Taken together, non-linear processes (i.e., non-linear rectification and second-order texture filters) taking place in V2 and higher level cortex are more likely to set a bottleneck on motion discrimination than on orientation discrimination.

Psychophysical studies suggest that the improvement of performance caused by perceptual learning can be attributed to neural synaptic modifications that are produced by repeated representation of the same stimulus pattern during training [57, 58]. We assume that the occurrence of learning transfer between different tasks depends on the amount of overlapping representations involved in these tasks. In accordance with this view, perceptual learning has been shown to transfer across different combinations of task and stimulus configuration as long as the same spatial axis of the positional judgment (e.g., the vertical axis) is employed [59]. Another study showing perceptual learning transfer across several tasks appears to support this view [52]. In this study, participants were underwent training on three different perceptual tasks—curvature discrimination, orientation discrimination, and "global form" coherence—before the degree of transfer between them was assessed. Results showed that learning improvements in one task did transfer to other tasks. The authors suggested that this was because all three tasks required the identification of the orientation of the elements [51]. Thus the specific form of the tasks may differ; however, they may involve overlapping representation (e.g., the same spatial axis or the orientation of the elements) and therefore recruit common population of neurons. This might be a critical factor that leads to perceptual learning transfer [59]. According to the dual-pathway model, common early luminance filters are shared by first- and second-order pathways [38, 39], which can lead to "common recruitment": both tasks recruit common population of neurons that are responsible for the shared early luminance filters. This could be the brain mechanism underlying the current study's observed transfer of perceptual learning across first- and second-order orientation discriminations.

5.2 Locus of Perceptual Learning

Perceptual learning is extremely particular to training features and locations, according to numerous studies (e.g. [2]), which suggests that it takes place in the primary visual cortex. This postulation is supported by an electrophysiological study showing the connection between V1 activity and perceptual learning [60] and imaging studies showing the enhancement of BOLD signals in V1 due to training [61, 62]. More recent evidence, however, suggests that the degree of learning specificity is influenced by factors such as task difficulty [12] and training procedure [17]. In our study, we equalized the task difficulty in the first- and second-order tasks and showed bidirectional transfer between the LM and CM orientation discriminations although different locations were used in the two tasks. This finding is informative because it shows that perceptual learning can not only generalize across locations as showed by Zhang et al. [19], but also generalize across location and type of stimuli simultaneously. Thus the finding further challenges the specificity of perceptual learning.

The locus of perceptual learning is still inconclusive, yet the current study, together with others [16–19, 63], indicating that higher-level brain areas beyond the retinotopic regions are engaged in perceptual learning. In accordance with this assumption, neurophysiological studies have shown that top-down signal modulates perceptual learning [64]. This top-down signal is likely to be from higher-level cortical areas [18, 65, 66]. Evidence from imaging studies suggests that these higher level brain areas include the frontal eye field (FEF), intraparietal sulcus (IPS), and supplementary eye field [67]. Therefore, perceptual learning not only changes activity in primary visual cortex (e.g., [61]) and the weight of connections between the visual cortex and decision units [48], but also it is modulated by the activities in higher order cortical areas.

6 Conclusion

The present study examined perceptual learning transfer between first- and second-order systems in fine orientation discrimination tasks, with the difficulty of the discrimination of the two types of orientations equalized. Results showed that the performance of observers trained with LM orientation improved rapidly in CM task and vice versa, whereas the performance of untrained observers remained low. In other words, the first- and the second-order systems are symmetrically transferred between. This finding suggests that the two systems are not separate, but rather they can communicate, most likely in higher-level cortical areas.

Acknowledgement. The authors gratefully acknowledge supports from Jiangxi Provincial Educational Science Planning Project (Grant No. 21QN007) and the Science and Technology Research Project of Jiangxi Provincial Department of Education (Grant No. GJJ210312).

References

1. Hua, T., et al.: Perceptual learning improves contrast sensitivity of V1 neurons in cats. Curr. Biol. **20**(10), 887–894 (2010)
2. Karni, A., Sagi, D.: Where practice makes perfect in texture discrimination: evidence for primary visual cortex plasticity. Proc. Natl. Acad. Sci. U. S. A. **88**(11), 4966–4970 (1991)
3. Herpers, J., Arsenault, J. T., Vanduffel, W., Vogels, R.: Stimulation of the ventral tegmental area induces visual cortical plasticity at the neuronal level. Cell Rep. **37**(6) (2021)
4. Maniglia, M., Seitz, A.R.: Towards a whole brain model of perceptual learning. Curr. Opin. Behav. Sci. **20**, 47–55 (2018)
5. Watanabe, T., Nanez, J.E., Sr., Koyama, S., Mukai, I., Liederman, J., Sasaki, Y.: Greater plasticity in lower-level than higher-level visual motion processing in a passive perceptual learning task. Nat. Neurosci. **5**(10), 1003–1009 (2002)

6. Ball, K., Sekuler, R.: A specific and enduring improvement in visual motion discrimination. Science **218**(4573), 697–698 (1982)
7. Crist, R.E., Kapadia, M.K., Westheimer, G., Gilbert, C.D.: Perceptual learning of spatial localization: Specificity for orientation, position, and context. J. Neurophysiol. **78**(6), 2889–2894 (1997)
8. Yu, C., Klein, S.A., Levi, D.M.: Perceptual learning in contrast discrimination and the (minimal) role of context. J. Vis. **4**(3), 169–182 (2004)
9. Bejjanki, V.R., Beck, J.M., Lu, Z.L., Pouget, A.: Perceptual learning as improved probabilistic inference in early sensory areas. Nat. Neurosci. **14**(5), 642–648 (2011)
10. Furmanski, C.S., Schluppeck, D., Engel, S.A.: Learning strengthens the response of primary visual cortex to simple patterns. Curr. Biol. **14**(7), 573–578 (2004)
11. Yotsumoto, Y., Watanabe, T., Sasaki, Y.: Different dynamics of performance and brain activation in the time course of perceptual learning. Neuron **57**(6), 827–833 (2008)
12. Ahissar, M., Hochstein, S.: Task difficulty and the specificity of perceptual learning. Nature **387**(6631), 401–406 (1997)
13. Liu, Z., Weinshall, D.: Mechanisms of generalization in perceptual learning. Vis. Res. **40**(1), 97–109 (2000)
14. Bakhtiari, S., Awada, A., Pack, C.C.: Influence of stimulus complexity on the specificity of visual perceptual learning. J. Vis. **20**(6), 1–19 (2020)
15. Jeter, P.E., Dosher, B.A., Petrov, A., Lu, Z.L.: Task precision at transfer determines specificity of perceptual learning. J. Vis. **9**(3), 1–13 (2009)
16. Xie, X.Y., Yu, C.: Perceptual learning of Vernier discrimination transfers from high to zero noise after double training. Vis. Res. **156**, 39–45 (2019)
17. Xiao, L.Q., Zhang, J.Y., Wang, R., Klein, S.A., Levi, D.M., Yu, C.: Complete transfer of perceptual learning across retinal locations enabled by double training. Curr. Biol. **18**(24), 1922–1926 (2008)
18. Zhang, J.Y., Zhang, G.L., Xiao, L.Q., Klein, S.A., Levi, D.M., Yu, C.: Rule-based learning explains visual perceptual learning and its specificity and transfer. J. Neurosci. **30**(37), 12323–12328 (2010)
19. Zhang, T., Xiao, L.Q., Klein, S.A., Levi, D.M., Yu, C.: Decoupling location specificity from perceptual learning of orientation discrimination. Vis. Res. **50**(4), 368–374 (2010)
20. Hung, S.C., Seitz, A.R.: Prolonged training at threshold promotes robust retinotopic specificity in perceptual learning. J. Neurosci. **34**(25), 8423–8431 (2014)
21. Jeter, P.E., Dosher, B.A., Liu, S.H., Lu, Z.L.: Specificity of perceptual learning increases with increased training. Vis. Res. **50**(19), 1928–1940 (2010)
22. Mastropasqua, T., Galliussi, J., Pascucci, D., Turatto, M.: Location transfer of perceptual learning: passive stimulation and double training. Vis. Res. **108**, 93–102 (2015)
23. Wang, R., et al.: Perceptual learning at a conceptual level. J. Neurosci. **36**(7), 2238–2246 (2016)
24. Hess, R.F., Achtman, R.L., Wang, Y.Z.: Detection of constrast-defined shape. J. Opt. Soc. Am. A **18**(9), 2220–2227 (2001)
25. Lewis, T.L., Kingdon, A., Ellemberg, D., Maurer, D.: Orientation discrimination in 5-year-olds and adults tested with luminance-modulated and contrast-modulated gratings. J. Vis. **7**(4), 9 (2007). https://doi.org/10.1167/7.4.9
26. Lin, L.M., Wilson, H.R.: Fourier and non-Fourier pattern discrimination compared. Vis. Res. **36**(13), 1907–1918 (1996)
27. Ellemberg, D., Allen, H.A., Hess, R.F.: Investigating local network interactions underlying first- and second-order processing. Vis. Res. **44**(15), 1787–1797 (2004)
28. Schofield, A.J., Yates, T.A.: Interactions between orientation and contrast modulations suggest limited cross-cue linkage. Perception **34**(7), 769–792 (2005)

29. Wilson, H.R., Ferrera, V.P., Yo, C.: A psychophysically motivated model for two-dimensional motion perception. Vis. Neurosci. **9**(1), 79–97 (1992)
30. Zhou, Y.X., Baker, C.L.: A processing stream in mammalian visual cortex neurons for non-Fourier responses. Science **261**(5177), 98–101 (1993)
31. Graham, N.V.: Beyond multiple pattern analyzers modeled as linear filters (as classical V1 simple cells): useful additions of the last 25 years. Vis. Res. **51**(13), 1397–1430 (2011)
32. Ashida, H., Lingnau, A., Wall, M.B., Smith, A.T.: fMRI adaptation reveals separate mechanisms for first-order and secondorder motion. J. Neurophysiol. **97**, 1319–1325 (2007)
33. Larsson, J., Landy, M.S., Heeger, D.J.: Orientation-selective adaptation to first- and second-order patterns in human visual cortex. J. Neurophysiol. **95**(2), 862–881 (2006)
34. Ledgeway, T., Smith, A.T.: Evidence for separate motion-detecting mechanisms for first- and second-order motion in human vision. Vis. Res. **34**(20), 2727–2740 (1994)
35. Lu, Z.L., Sperling, G.: The functional architecture of human visual motion perception. Vis. Res. **35**(19), 2697–2722 (1995)
36. Pavan, A., Campana, G., Guerreschi, M., Manassi, M., Casco, C.: Separate motion-detecting mechanisms for first- and second-order patterns revealed by rapid forms of visual motion priming and motion aftereffect. J. Vis. **9**(11), 27, 21–16 (2009). https://doi.org/10.1167/9.11.27
37. Chen, R., Qui, Z.P., Zhang, Y., Zhou, Y.F.: Perceptual learning and transfer study of first- and second-order motion direction discrimination. Progr. Biochem. Biophys. **36**, 1442–1450 (2009)
38. Petrov, A.A., Hayes, T.R.: Asymmetric transfer of perceptual learning of luminance- and contrast-modulated motion. J. Vis. **10**(14), 11 (2010)
39. Chung, S.T., Li, R.W., Levi, D.M.: Learning to identify near-threshold luminance-defined and contrast-defined letters in observers with amblyopia. Vis. Res. **48**(27), 2739–2750 (2008)
40. Zanker, J.M.: Perceptual learning in primary and secondary motion vision. Vis. Res. **39**(7), 1293–1304 (1999)
41. Morgan, M.J., Mason, A.J., Baldassi, S.: Are there separate first-order and second-order mechanisms for orientation discrimination? Vis. Res. **40**(13), 1751–1763 (2000)
42. Vaina, L.M., Chubb, C.: Dissociation of first- and second-order motion systems by perceptual learning. Atten. Percept. Psychophys. **74**(5), 1009–1019 (2012)
43. Schofield, A.J., Ledgeway, T., Hutchinson, C.V.: Asymmetric transfer of the dynamic motion aftereffect between first-and second-order cues and among different second-order cues. J. Vis. **7**(8), 1–1 (2007)
44. Cruickshank, A.G., Schofield, A.J.: Transfer of tilt aftereffects between second order cues. Spat. Vis. **18**(4), 379–397 (2005)
45. Georgeson, M.A., Schofield, A.J.: Shading and texture: separate information channels with a common adaptation mechanism? Spat. Vis. **16**(1), 59–76 (2002)
46. Zhang, J.Y., Yang, Y.X.: Perceptual learning of motion direction discrimination transfers to an opposite direction with TPE training. Vis. Res. **99**, 93–98 (2014)
47. Dosher, B.A., Lu, Z.L.: Level and mechanisms of perceptual learning: learning first-order luminance and second-order texture objects. Vis. Res. **46**(12), 1996–2007 (2006)
48. Dosher, B.A., Lu, Z.L.: Perceptual learning reflects external noise filtering and internal noise reduction through channel reweighting. Proc. Natl. Acad. Sci. U. S. A. **95**(23), 13988–13993 (1998)
49. Tyler, B., Takeo, W.: Modeling visual perceptual learning of contrast discrimination with integrated reweighting. J. Vis. **21**(9), 2613 (2021)
50. Sagi, D.: Perceptual learning in vision research. Vis. Res. **51**(13), 1552–1566 (2011)
51. McGovern, D.P., Webb, B.S., Peirce, J.W.: Transfer of perceptual learning between different visual tasks. J. Vis. **12**(11) (2012). https://doi.org/10.1167/12.11.4

52. Green, C.S., Kattner, F., Siegel, M.H., Kersten, D., Schrater, P.R.: Differences in perceptual learning transfer as a function of training task. J. Vis. **15**(10), 5 (2015). https://doi.org/10.1167/15.10.5

53. Hubel, D.H., Wiesel, T.N.: Receptive fields, binocular interaction and functional architecture in the cat's visual cortex. J. Physiol. **160**, 106–154 (1962)

54. Hubel, D.H., Wiesel, T.N.: Receptive fields and functional architecture of monkey striate cortex. J. Physiol. **195**(1), 215–243 (1968)

55. Braddick, O.J., O'Brien, J.M., Wattam-Bell, J., Atkinson, J., Hartley, T., Turner, R.: Brain areas sensitive to coherent visual motion. Perception **30**(1), 61–72 (2001)

56. Dubner, R., Zeki, S.M.: Response properties and receptive fields of cells in an anatomically defined region of the superior temporal sulcus in the monkey. Brain Res. **35**(2), 528–532 (1971)

57. Bao, S., Chan, V.T., Merzenich, M.M.: Cortical remodelling induced by activity of ventral tegmental dopamine neurons. Nature **412**(6842), 79–83 (2001)

58. Dinse, H.R., Ragert, P., Pleger, B., Schwenkreis, P., Tegenthoff, M.: Pharmacological modulation of perceptual learning and associated cortical reorganization. Science **301**(5629), 91–94 (2003)

59. Webb, B.S., Roach, N.W., McGraw, P.V.: Perceptual learning in the absence of task or stimulus specificity. PLoS ONE **2**(12), e1323 (2007)

60. Pourtois, G., Rauss, K.S., Vuilleumier, P., Schwartz, S.: Effects of perceptual learning on primary visual cortex activity in humans. Vis. Res. **48**(1), 55–62 (2008)

61. Schwartz, S., Maquet, P., Frith, C.: Neural correlates of perceptual learning: a functional MRI study of visual texture discrimination. Proc. Natl. Acad. Sci. U. S. A. **99**(26), 17137–17142 (2002)

62. Yotsumoto, Y., Chang, L.H., Watanabe, T., Sasaki, Y.: Interference and feature specificity in visual perceptual learning. Vis. Res. **49**(21), 2611–2623 (2009)

63. Sasaki, Y., Nanez, J.E., Watanabe, T.: Advances in visual perceptual learning and plasticity. Nat. Rev. Neurosci. **11**(1), 53–60 (2010)

64. Li, W., Piech, V., Gilbert, C.D.: Perceptual learning and top-down influences in primary visual cortex. Nat. Neurosci. **7**(6), 651–657 (2004)

65. Gilbert, C.D., Sigman, M., Crist, R.E.: The neural basis of perceptual learning. Neuron **31**(5), 681–697 (2001)

66. Jing, R., Yang, C., Huang, X., Li, W.: Perceptual learning as a result of concerted changes in prefrontal and visual cortex. Curr. Biol. **31**(20), 4521–4533 (2021)

67. Mukai, I., Kim, D., Fukunaga, M., Japee, S., Marrett, S., Ungerleider, L.G.: Activations in visual and attention-related areas predict and correlate with the degree of perceptual learning. J. Neurosci. **27**(42), 11401–11411 (2007)

Some New Characterizations of Ideals in Non-involutive Residuated Lattices

Chunhui Liu[✉]

School of Educational Science, Chifeng University, Chifeng 024000, P. R. China
chunhuiliu1982@163.com

Abstract. In this paper, we further study the properties and structural characteristics of ideals in non-involutive residuated lattices. On the one hand, some new equivalent characterizations of ideals are obtained in non-involutive residuated lattices. On the other hand, the relationship between the concepts of ideal and filter is discussed.

Keywords: Fuzzy logic · Non-involutive residuated lattice · Ideal · Filter

1 Introduction

Following the developments of computer science and mathematics, non-classical mathematical logics has been deeply studied [1]. At present, It has been proved to be a powerful and practical technique and tool in the field of computer science to deal with fuzzy and uncertain problems. At the same time, Various logical algebras have been introduced as the corresponding algebraic systems of non-classical mathematical logic systems. For instance, residuated lattices, MV algebras [2], BL algebras [3], MTL algebras [4], Gödel algebras, lattice implication algebras [5] and NM-algebras, which is also called R_0-algebras [6]. Among these logical algebras, residuated lattices proposed by Ward and Dilworth [7] are very important and basic algebraic system because the other logical algebras are all subalgebras of residuated lattices. They have been investigated by many scholars (see [8–10]).

Filter and ideal are two applicable and effectively tools for researching of logical algebraic structures and logical systems. In terms of logic, different filters and ideals corresponding to different sets of provable formulas. It not hard to find that in logical algebras satisfying involutive (regular) properties, because ideals and filters are dual to each other, most people pay more attention to the problem of filters (see [11–13]). But, if the negation operation of logical algebra not have involution, the duality between ideals and filters will also disappear. Thus, it will be a valuable work to discuss the topic of ideals and its application of non-involutive logical algebras. Based on this, the notions of ideals were introduced of non-involutive residuated lattices in [14,15], and some conclusions with theoretical height and application value have been given.

© The Author(s), under exclusive license to Springer Nature Switzerland AG 2023
N. Xiong et al. (Eds.): ICNC-FSKD 2022, LNDECT 153, pp. 810–818, 2023.
https://doi.org/10.1007/978-3-031-20738-9_90

In order to further study the the structural characteristics of residuated lattices, in this paper, As the deepening and continuation of the work in [14,15], we study the new properties and structural characteristics of ideals in non-involutive residuated lattices. some significative and interesting conclusions are obtained.

2 Preliminaries

Firstly, we give some existing knowledge and conclusions about residuated lattices [7–10], which are used to this paper.

By a residuated lattice we mean an algebra $\mathcal{L} = (L, \vee, \wedge, \otimes, \rightarrow, 0, 1)$ of type $(2, 2, 2, 2, 0, 0)$ such that:

(R1) $(L, \vee, \wedge, 0, 1)$ is a bounded lattice with the least element 0 and the greatest element 1,

(R2) $(L, \otimes, 1)$ is a commutative monoid,

(R3) $(\otimes, \rightarrow)$ is an adjoint pair on L, i.e.,

$$(\forall x, y, z \in L)(x \otimes y \leqslant z \Longleftrightarrow x \leqslant y \rightarrow z). \tag{1}$$

In a residuated lattice $\mathcal{L}$, we denote $x^* = x \rightarrow 0$ for all $x \in L$, and $\mathcal{L}$ is said to be a non-involutive residuated lattice if $\exists x \in L$ such that $x^{**} \neq x$. In a non-involutive residuated lattice $\mathcal{L}$, the following conclusions hold:

$$(\forall x, y \in L)(x \leqslant y \Longleftrightarrow x \rightarrow y = 1) \tag{2}$$

$$(\forall x \in L)(x \rightarrow x = 1, x \rightarrow 1 = 1, 1 \rightarrow x = x) \tag{3}$$

$$(\forall x, y \in L)(x \leqslant y \rightarrow x \text{ and } ((x \rightarrow y) \rightarrow y) \rightarrow y = x \rightarrow y) \tag{4}$$

$$(\forall x, y, z \in L)(x \rightarrow y \leqslant (y \rightarrow z) \rightarrow (x \rightarrow z)) \tag{5}$$

$$(\forall x, y, z \in L)(y \rightarrow z \leqslant (x \rightarrow y) \rightarrow (x \rightarrow z) \leqslant x \rightarrow (y \rightarrow z)) \tag{6}$$

$$(\forall x, y \in L)(x \leqslant (x \rightarrow y) \rightarrow y \text{ and } y \leqslant (x \rightarrow y) \rightarrow y \tag{7}$$

$$(\forall x, y, z \in L)(x \leqslant y \text{ implies } z \rightarrow x \leqslant z \rightarrow y \text{ and } y \rightarrow z \leqslant x \rightarrow z)) \tag{8}$$

$$(\forall x, y, z \in L)(x \rightarrow (y \wedge z) = (x \rightarrow y) \wedge (x \rightarrow z)) \tag{9}$$

$$(\forall x, y, z \in L)((y \vee z) \rightarrow x = (y \rightarrow x) \wedge (z \rightarrow x)) \tag{10}$$

$$(\forall x, y, z \in L)(x \leqslant y \rightarrow z \Longleftrightarrow y \leqslant x \rightarrow z) \tag{11}$$

$$(\forall x, y, z \in L)(x \rightarrow (y \rightarrow z) = (x \otimes y) \rightarrow z = y \rightarrow (x \rightarrow z)) \tag{12}$$

$$(\forall x, y, z \in L)(x \rightarrow (y \vee z) \geqslant (x \rightarrow y) \vee (x \rightarrow z)) \tag{13}$$

$$(\forall x, y, z \in L)((y \wedge z) \rightarrow x \geqslant (y \rightarrow x) \vee (z \rightarrow x)) \tag{14}$$

$$(\forall x, y, z \in L)(x \otimes (x \rightarrow y) \leqslant y \text{ and } (x \rightarrow y) \otimes (y \rightarrow z) \leqslant x \rightarrow z) \tag{15}$$

$$(\forall x, y \in L)(x \otimes y \leqslant x \wedge y) \tag{16}$$

$$(\forall x, y, z \in L)(x \leqslant y \text{ implies } x \otimes z \leqslant y \otimes z) \tag{17}$$

$$(\forall x, y \in L)(x \leqslant y \text{ implies } y^* \leqslant x^* \text{ implies } x^{**} \leqslant y^{**}) \tag{18}$$

$$(\forall x \in L)(x \otimes x^* = 0, x \leqslant x^{**} \text{ and } x^{***} = x^*) \tag{19}$$

$$(\forall x, y \in L)(x^* \leqslant x \to y \leqslant y^* \to x^*) \tag{20}$$

$$(\forall x, y \in L)(x \to y^* = y \to x^* = (x \otimes y)^*) \tag{21}$$

$$(\forall x \in L)(x \vee x^* = 1 \iff x \wedge x^* = 0) \tag{22}$$

$$(\forall x, y \in L)((x \vee y)^* = x^* \wedge y^* \text{ and } (x \wedge y)^* \geqslant x^* \vee y^*) \tag{23}$$

$$(\forall x, y \in L)((x \to y^*)^{**} = x \to y^*) \tag{24}$$

$$(\forall x, y \in L)((x \to y^{**})^{**} = x \to y^{**} = x^{**} \to y^{**}) \tag{25}$$

$$(\forall x, y \in L)(x^{**} \to y^* = y^{**} \to x^*). \tag{26}$$

Let $\mathcal{L}$ be a residuated lattice. A nonempty subset F of L is said to be a filter of $\mathcal{L}$, if it satisfies:

$$(\forall x, y \in L)(x \leqslant y \text{ and } x \in F \text{ imply } y \in F), \tag{27}$$

$$(\forall x, y \in L)(x \in F \text{ and } y \in F \text{ imply } x \otimes y \in I). \tag{28}$$

The set of all filters of $\mathcal{L}$ is denoted by $\mathbf{Fil}(L)$. It is obvious that $\{1\} \in \mathbf{Fil}(L)$ and $L \in \mathbf{Fil}(L)$.

Lemma 1. [11,12] *Let $\mathcal{L}$ be a residuated lattice and $\emptyset \neq F \subseteq L$. Then $F \in \mathbf{Fil}(L)$ if and only if it satisfies $1 \in F$ and:*

$$(\forall x, y \in L)(x \in F \text{ and } x \to y \in F \text{ imply } y \in F). \tag{29}$$

3   On Ideals in Non-involutive Residuated Lattices

In this section, we give some new properties and equivalent representations of ideals in non-involutive residuated lattices.

In a non-involutive residuated lattice $\mathcal{L}$, the binary opearation $\oplus$ is defined by $x \oplus y = x^* \to y$ for all $x, y \in L$. It is emphasized that $\oplus$ is not commutative in general.

Definition 1. [15] *Let $\mathcal{L}$ be a non-involutive residuated lattice. A nonempty subset I of L is called an ideal of L, if it satisfies:*

$$(\forall x, y \in L)(x \leqslant y \text{ and } y \in I \text{ imply } x \in I), \tag{30}$$

$$(\forall x, y \in L)(x \in I \text{ and } y \in I \text{ imply } x \oplus y \in I). \tag{31}$$

We denote the set of all ideals of L by $\mathbf{Id}(L)$.

Note 1. Let $\mathcal{L}$ be a non-involutive residuated lattice. It is obvious that $\{0\} \in \mathbf{Id}(L)$ and $L \in \mathbf{Id}(L)$.

Lemma 2. [14,15] *Let $\mathcal{L}$ be a non-involutive residuated lattice and $\emptyset \neq I \subseteq L$. Then $I \in \mathbf{Id}(L)$ if and only if it satisfies $0 \in I$ and:*

$$(\forall x, y \in L)(x \in I \text{ and } (x^* \to y^*)^* \in I \text{ imply } y \in I). \tag{32}$$

Lemma 3. [14,15] *Let $\mathcal{L}$ be a non-involutive residuated lattice and $I \in \mathbf{Id}(L)$. Then the following conclusion holds:*

$$(\forall x \in L)(x \in I \Longleftrightarrow x^{**} \in I). \tag{33}$$

Theorem 1. *Let $\mathcal{L}$ be a non-involutive residuated lattice and $\emptyset \neq I \subseteq L$. Then $I \in \mathbf{Id}(L)$ iff it satisfies the following conditions:*

$$(\forall x, y \in L)(x \in I \text{ and } y \in I \text{ imply } \downarrow (x \oplus y) := \{z \in L | z \leqslant x \oplus y\} \subseteq I). \tag{34}$$

Proof. Assume that $I \in \mathbf{Id}(L)$, $x \in I$ and $y \in I$, then $x \oplus y \in I$ by (31). For any $z \in \downarrow x \oplus y$, we have $z \leqslant (x \oplus y)$, it follows from (30) that $z \in I$. Hence $\downarrow x \oplus y \subseteq I$.

Conversely, assume that I satisfies (34). On the one hand, since $0 \leqslant x \oplus y$ for any $x, y \in I$, we have that $0 \in \downarrow (x \oplus y)$. Hence $0 \in I$ by (34). On the other hand, let $x \in I$ and $(x^* \to y^*)^* \in I$, then $\downarrow (x \oplus (x^* \to y^*)^*) \subseteq I$ by (34). Since

$$
\begin{aligned}
y \to (x \oplus (x^* \to y^*)^*) &= y \to (x^* \to (x^* \to y^*)^*) \quad &\text{[by definition } \oplus] \\
&= y \to ((x^* \to y^*) \to x^{**}) \quad &\text{[by (21)]} \\
&= (x^* \to y^*) \to (y \to x^{**}) \quad &\text{[by (12)]} \\
&= (y \to x^{**}) \to (y \to x^{**}) \quad &\text{[by (21)]} \\
&= 1, \quad &\text{[by (2)]}
\end{aligned}
$$

we have $y \leqslant x \oplus (x^* \to y^*)^*$ by (2). Thus $y \in \downarrow (x \oplus (x^* \to y^*)^*) \subseteq I$, this shows that I also satisfies (32). Therefor $I \in \mathbf{Id}(L)$ by Lemma 2. $\qquad\square$

In order to give more new properties and equivalent characterization of ideals, in a non-involutive residuated lattice $\mathcal{L}$, we define the binary opearation $\oslash$ by $x \oslash y = x^* \to y^{**}$ for all $x, y \in L$.

Proposition 1. *Let $\mathcal{L}$ be a non-involutive residuated lattice. Then the following conclusions hold:*

$$(\forall x, y \in L)(x \oslash y = y \oslash x), \tag{35}$$
$$(\forall x, y, z \in L)((x \oslash y) \oslash z = x \oslash (y \oslash z)), \tag{36}$$
$$(\forall x, y, z \in L)(x \leqslant y \Longrightarrow x \oslash z \leqslant y \oslash z), \tag{37}$$
$$(\forall x, y \in L)(x \oslash y = (x \oslash y)^{**} = x^{**} \oslash y^{**} = x^{**} \oplus y^{**}), \tag{38}$$
$$(\forall x, y \in L)(x \oplus y \leqslant x \oslash y). \tag{39}$$

Proof. (i) Let $x, y \in L$, then by (21) we can obtain that

$$x \oslash y = x^* \to y^{**} = y^* \to x^{**} = y \oslash x.$$

(ii) Let $x, y, z \in L$, then by (21), (25) and (12) we have that

$$
\begin{aligned}
(x \oslash y) \oslash z &= (x^* \to y^{**})^* \to z^{**} \\
&= z^* \to (x^* \to y^{**})^{**} \\
&= z^* \to (x^* \to y^{**}) \\
&= x^* \to (z^* \to y^{**}) \\
&= x^* \to (y^* \to z^{**}) \\
&= x^* \to (y^* \to z^{**})^{**} = x \oslash (y^* \to z^{**}) = x \oslash (y \oslash z).
\end{aligned}
$$

(iii) Let $x, y \in L$ and $x \leqslant y$, then $y^* \leqslant x^*$ by (18). It follows from (8) that

$$
x \oslash z = x^* \to z^{**} \leqslant y^* \to z^{**} = y \oslash z.
$$

(iv) Let $x, y \in L$, on the one hand, by (25) we have

$$
x \oslash y = x^* \to y^{**} = (x^* \to y^{**})^{**} = (x \oslash y)^{**}.
$$

On the other hand, by (19) we have

$$
x \oslash y = x^* \to y^{**} = x^{***} \to (y^*)^{***} = (x^{**})^* \to (y^{**})^{**} = x^{**} \oslash y^{**},
$$

and

$$
x \oslash y = x^* \to y^{**} = x^{***} \to y^{**} = (x^{**})^* \to y^{**} = x^{**} \oplus y^{**}.
$$

(v) Let $x, y \in L$, since $y \leqslant y^{**}$ by (19), it follows from (8) that

$$
x \oplus y = x^* \to y \leqslant x^* \to y^{**} = x \oslash y.
$$

□

Theorem 2. *Let $\mathcal{L}$ be a non-involutive residuated lattice and $\emptyset \neq I \subseteq L$. Then $I \in \mathbf{Id}(L)$ iff it satisfies (30) and the following condition:*

$$
(\forall x, y \in L)(x \in I \text{ and } y \in I \text{ imply } x \oslash y \in I). \tag{40}
$$

Proof. Assume that $I \in \mathbf{Id}(L)$, then I satisfies (30). On the other hand, let $x \in I$ and $y \in I$, we have $x^{**} \in I$ and $y^{**} \in I$ by (33). Hence by (38) and (31) we can obtain that $x \oslash y = x^{**} \oplus y^{**} \in I$. Therefore I satisfies (40).

Conversely, assume that I satisfies (30) and (40). For proving $I \in \mathbf{Id}(L)$, it is sufficient to prove that I satisfies (31) by Definition 1. In fact, let $x \in I$ and $y \in I$, then $x \oslash y \in I$ by (40). Since $x \oplus y \leqslant x \oslash y$ by (39), we obtain that $x \oplus y \in I$ by (30). Therefore I satisfies (31). □

Theorem 3. *Let $\mathcal{L}$ be a non-involutive residuated lattice and $\emptyset \neq I \subseteq L$. Then $I \in \mathbf{Id}(L)$ iff it satisfies the following conditions:*

$$
(\forall x, y \in L)(x \in I \text{ and } y \in I \text{ imply } \downarrow (x \oslash y) := \{z \in L \mid z \leqslant x \oslash y\} \subseteq I). \tag{41}
$$

Proof. Assume that $I \in \mathbf{Id}(L)$, $x \in I$ and $y \in I$, then $x \oslash y \in I$ by (40). For any $z \in\downarrow x \oslash y$, we have $z \leqslant (x \oslash y)$, it follows from (30) that $z \in I$. Hence $\downarrow x \oslash y \subseteq I$.

Conversely, assume that I satisfies (41). On the one hand, since $0 \leqslant x \oslash y$ for any $x, y \in I$, we can have that $0 \in\downarrow (x \oslash y)$, Hence $0 \in I$ by (41). On the other hand, let $x \in I$ and $(x^* \to y^*)^* \in I$, then $\downarrow (x \oslash (x^* \to y^*)^*) \subseteq I$ by (41). Since

$$
\begin{aligned}
y \to (x \oslash (x^* \to y^*)^*) &= y \to (x^* \to (x^* \to y^*)^{***}) && \text{[by definition of } \oslash] \\
&= y \to (x^* \to (x^* \to y^*)^*) && \text{[by (19)]} \\
&= y \to ((y \to x^{**}) \to x^{**}) && \text{[by (21)]} \\
&= (y \to x^{**}) \to (y \to x^{**}) && \text{[by (12)]} \\
&= 1, && \text{[by (3)]}
\end{aligned}
$$

we have that $y \leqslant x \oslash (x^* \to y^*)^*$ by using (2). Thus $y \in\downarrow (x \oslash (x^* \to y^*)^*) \subseteq I$, this shows that I also satisfies (32). Therefor $I \in \mathbf{Id}(L)$ by Lemma 2.   □

4   Relationship Between Ideals and Filters

In this section, we investigate the relationship between ideals and filters in non-involutive residuated lattices.

Definition 2. [14,15] Let $\mathcal{L}$ be a residuated lattice and $\emptyset \neq I \subseteq L$. The least ideal containing A is said to be the ideal of $\mathcal{L}$ generalized by A, written $\langle A \rangle$.

Let $\mathcal{L}$ be a non-involutive residuated lattice and $A \subseteq L$. Define two subsets of L as follows:

$$A^* = \{x^* \in X \mid x \in A\} \text{ and } N(A) = \{x \in X \mid x^* \in A\}. \tag{42}$$

About the relationship between ideals and filters, we know that $F \in \mathbf{Fil}(L)$ if and only if $F^* \in \mathbf{Id}(L)$ in MV-algebras. But this result may not be true in non-involutive residuated lattices (see Examples 3.3 and 3.4 in [15]). Nextly, we will analyze the relationship between ideals and filters by using the subset $N(A)$ of L as following theorem.

Theorem 4. Let $\mathcal{L}$ be a non-involutive residuated lattice, $I \in \mathbf{Id}(L)$ and $F \in \mathbf{Fil}(L)$. Then the following conclusions are valid:

$$N(I) \in \mathbf{Fil}(L) \text{ and } I^* \subseteq N(I), \tag{43}$$

$$N(F) \in \mathbf{Id}(L) \text{ and } N(F) = \langle F^* \rangle, \tag{44}$$

$$I = N(N(I)), \tag{45}$$

$$F \subseteq N(N(F)), \tag{46}$$

$$N(I) = N(N(N(I))). \tag{47}$$

Proof. (i) Since $I \in \mathbf{Id}(L)$, we have $1^* = 0 \in I$, thus $1 \in N(I)$. Now let $x \in N(I)$ and $x \to y \in N(I)$, then $x^* \in I$ and $(x \to y)^* \in N(I)$, it follows from $I \in \mathbf{Id}(L)$ and (40) that

$$x^* \oslash (x \to y)^* \in I.$$

Since

$$
\begin{aligned}
y^* \to (x^* \oslash (x \to y)^*) &= y^* \to (x^{**} \to (x \to y)^{***}) && \text{[by definition of } \oslash] \\
&= y^* \to (x^{**} \to (x \to y)^*) && \text{[by (19)]} \\
&= y^* \to ((x \to y) \to x^{***}) && \text{[by (21)]} \\
&= (x \to y) \to (y^* \to x^*) && \text{[by (12) and (21)]} \\
&= 1, && \text{[by (20) and (2)]}
\end{aligned}
$$

we can obtain that $y^* \leqslant x^* \oslash (x \to y)^*$. Thus $y^* \in I$ by (30). Hence $y \in N(I)$ and $N(I)$ satisfies (29). Therefore $N(I) \in \mathbf{Fil}(L)$. Finally, it follows from (33) that $x \in I$ if and only if $x^{**} \in I$ for all $x \in L$, hence $x^* \in N(I)$ and $I^* \subseteq N(I)$.

(ii) Since $F \in \mathbf{Fil}(L)$, we have $0^* = 1 \in F$, thus $0 \in N(F)$. Now let $x \in N(F)$ and $y \in N(F)$, then $x^* \in F$ and $y^* \in F$. Since

$$
\begin{aligned}
x^* \to (y^* \to (x \oslash y)^*) &= x^* \to ((x \oslash y) \to y^{**}) && \text{[by (21)]} \\
&= (x \oslash y) \to (x^* \to y^{**}) && \text{[by (12)]} \\
&= (x \oslash y) \to (x \oslash y) && \text{[by definition of } \oslash] \\
&= 1, && \text{[by (3)]}
\end{aligned}
$$

by (29) we have $(x \oslash y)^* \in F$. Hence $x \oslash y \in N(F)$ and $N(F)$ satisfies (40). Let $x \leqslant y$ and $y \in N(F)$, then $y^* \leqslant x^*$ and $y^* \in F$ by (18), thus $x^* \in F$ by (27). Hence $x \in N(F)$ and $N(F)$ satisfies (30). It follows from Theorem 2 that $N(F) \in \mathbf{Id}(L)$.

Now we prove that $N(F) = \langle F^* \rangle$. In fact, let $x \in F$, since $x \leqslant x^{**}$ by (19), we have $x^{**} \in F$ by $F \in \mathbf{Fil}(L)$, thus $x^* \in N(F)$ and $F^* \subseteq N(F)$. Let $J \in \mathbf{Id}(L)$ and $F^* \subseteq J$, if $x \in N(F)$, then $x^* \in F$, thus $x^{**} \in F^* \subseteq J$, and thus $x \in J$ by (33). Hence $N(F) \subseteq J$ and the proof is completed by Definition 2.

(iii) It follows from the face that $I \in \mathbf{Id}(L)$ and (33).

(iv) It follows from that $x \leqslant x^{**}$ for all $x \in L$.

(v) Since $x \in N(I)$ iff $x^* \in I$ iff $x^{***} \in I$, we have that $N(I) = N(N(N(I)))$.
$\square$

Example 1. Let $L = \{0, a, b, c, 1\}$ and $0 < a < b < c < 1$, the binary operators $\to$ and $\otimes$ on L be defined as Tables 1 and 2. and $x^* = x \to 0$ for every $x \in L$, then $(L, \leqslant, \wedge, \vee, \otimes, \to, 0, 1)$ is a negative non-involutive residuated lattice. Let $I = \{0, a\}$, we easy to know that $I \in \mathbf{Id}(L)$. It is easy to verify that $N(I) = \{b, c, 1\} \in \mathbf{Fil}(L)$, but by

$$b \in I^* \text{ and } b \to c \in I^*, \text{ but } c \notin I^*,$$

we know that $I^* = \{b, 1\} \notin \mathbf{Fil}(L)$.

Table 1. Def. of "→".

→	0	a	b	c	1
0	1	1	1	1	1
a	b	1	1	1	1
b	a	a	1	1	1
c	0	a	b	1	1
1	0	a	b	c	1

Table 2. Def. of "⊗".

→	0	a	b	c	1
0	0	0	0	0	0
a	0	0	0	a	a
b	0	0	b	b	b
c	0	a	b	c	c
1	0	a	b	c	1

5   Concluding Remarks

As well known, the concepts of filters and ideals occupy an important position in reflecting the structural characteristics of logical algebras. In [14,15], the present author and Liu ea al. have proposed the notion of ideals in non-involutive residuated lattices and given some their important properties. In this study, as a continuation of [14,15], we have given some new properties and equivalent characterizations of ideals and investigated the relationship between ideals and filters. Because the various logical algebras such as MTL algebras, BL algebras, Gödel algebras, NM algebras, lattice implication algebras and MV algebras and so on, are all particular cases of non-involutive residuated lattices, the conclusions have obtained in this study regarded as common characteristic of those algebras. We hope that more interesting and valuable results be obtained by the stipulating of the work of this paper.

Acknowledgments. This work is supported by Research Program of science and technology at Universities of Inner Mongolia Autonomous Region of China(Grant No. NJZY21138, NJZY22146)

References

1. Wang, G.J.: Nonclassical Mathematical Logic and Approximate Reasoning. Science Press, Beijing (2003)
2. Chang, C.C.: Algebraic analysis of many value logic. Trans. Amer. Soc. **87**, 1–53 (1958)

3. Hájek, P.: Metamathematics of Fuzzy Logic. Kluwer Academic Publishers, Dordrecht (1998)
4. Esteva, F., Godo, L.: Monoidal t-norm based logic: towards a logic for left-continuous t-norms. Fuzzy Sets Syst. **124**, 271–288 (2001)
5. Xu, Y., Ruan, D., Qin, K.Y., et al.: Lattice-Valued Logic. Springer, Berlin (2004)
6. Wang, G.J., Zhou, H.J.: Introduction to Mathematical Logic and Resolution Principle, 2nd edn. Science Press, Beijing (2009)
7. Ward, M., Dilworth, R.P.: Residuated lattices. Trans. Am. Math. Soc. **45**, 335–354 (1939)
8. Pavelka, J.: On fuzzy logic II. Enriched residuated lattices and semantics of propositional calculi. Zeitschrift Für Mathematische Logik und Grundlagen der Mathematik, **25**, 119–134 (1979)
9. Pei, D.W.: The characterization of residuated lattices and regular residuated lattices. Acta Math Sinica **42**, 271–278 (2002)
10. Belohlavek, R.: Some properties of residuated lattices. Czechoslovak Math J. **53**, 161–171 (2003)
11. Kondo, M.: Filters on commutative residuated lattices. Adv. Intell. Soft. Comput. **68**, 343–347 (2010)
12. Zhu, Y.Q., Xu, Y.: On filter theory of residuated lattices. Inf. Sci. **180**, 3614–3632 (2010)
13. Dumitru, B., Dana, P.: Some types of filters on residuated lattices. Soft Comput. **18**, 825–837 (2014)
14. Liu C.H.: LI-ideals theory in negative non-involutive residuated lattices. Appl. Math. J. Chin. Univ. (Ser. A) **30**, 445–456 (2015)
15. Liu, Y., Qin, Y., Qin, X.Y., Xu, Y.: Ideals and fuzzy ideals on residuated lattices. Int. J. Mach. Learn. Cybern. **8**, 239–253 (2017)

An Analysis of College Students' Behavior Based on Positive and Negative Association Rules

Feng Hao[1], Long Zhao[1], Haoran Zhao[2], Tiantian Xu[1], and Xiangjun Dong[1(✉)]

[1] Department of Computer Science and Technology, Qilu University of Technology (Shandong Academy of Sciences), Jinan 250300, China
haof2@sda.cn, {zhaolong,ttxu,dxj}@qlu.edu.cn
[2] University of Ottawa (Engineering), Ottawa, Canada
hzhao108@uottawa.ca

Abstract. Positive association rule (PAR) mining has been often used in campus data analysis, while negative association rule (NAR) mining has not, which will result in a lot of valuable information missing. This paper collects real campus data that contains the student's academic performance, e-card consumption behavior, book lending records and mental health status and analyzes them with NAR and PAR techniques. We first preprocess the data to obtain a suitable format. Then we use a method called Positive and Negative Association Rules on Correlation (PNARC) to mine NARs and PARs from these data. Finally, we obtain a lot of valuable information, for example, there is a strong negative correlation between depression and academic performance. These results are very helpful for educators to improve college students' performance.

Keywords: Academic performance · Campus data · Negative association rules · PNARC model

1 Introduction

The construction of the digital campus system generates a large amount of campus data [1]. All these data, gathered from diverse and usually heterogeneous sources, contains a wealth of information about student life and learning. However, in fact these data may be no help to students in its raw form [2]. Therefore, how to extract instructive information from such complex data, especially extract information that can improve students' academic performance becomes more important.

Some studies have been done on analyzing the campus data to establish students' management models or predict students' performance by different methods such as statistical analysis [3], decision trees [4], clustering [5], and machine learning algorithms [6]. However, they didn't pay attention to the analysis of behavioral data, which is very necessary because it has certain guiding significance for improving student performance. In references [7–12], the algorithms of association rules mining which can easily discover the related rules hidden in the datasets were used to analyze the correlation between

© The Author(s), under exclusive license to Springer Nature Switzerland AG 2023
N. Xiong et al. (Eds.): ICNC-FSKD 2022, LNDECT 153, pp. 819–832, 2023.
https://doi.org/10.1007/978-3-031-20738-9_91

student performance and behavior. However, they only focused on mining positive association rules (PARs) and did not consider the negative association rules (NARs) [13]. In fact, sometimes, it is important to find NARs in some situations because it can provide more actionable information and play roles irreplaceable by PARs alone.

Therefore, in this paper, we focus on the application of positive and negative association rules (PNARs) in analysis of campus data, aiming to find valuable information to improve college students' performance. The data and basic information collected from one Chinese college during the period of September 2016 to January 2017. It contains the students' final grade, e-card consumption behavior, book lending records and mental health status. The PNARC model [14] which can not only mine PARs, but also mine NARs is introduced to discover the hidden relationships that exist among the e-card consumption behavior, book lending records, students' performance and mental health status. Relevant experiments and analysis show that we can easily get the relationships between academic performance and other influencing factors. The discovery of PNARs can provide guidance for educators to improve college students' performance.

The main contributions in this paper are summarized as follows:

1. The real campus data collected from one Chinese college is applied to analyze the factors affecting the students' performance.
2. Due to the heterogeneity and complexity of the data, we have done a lot of data preprocessing work including but not limited to reduction processing, discretization and integration processing.
3. The PNARC model which can not only mine PARs, but also mine NARs is introduced to discover the hidden relationships that exist among the e-card consumption behavior, book lending records, mental health status and students' performance.
4. The PNARs we discovered can provide guidance for educators to improve college students' performance.

The remaining parts of this paper are organized as follows. Section 2 introduces related work. Section 3 shows preprocessing of campus data. Section 4 presents the experimental results and analysis, followed by conclusions in Sect. 5.

2 Related Work

In this section, we first introduce the related work on applying the association rules mining in analysis of campus data. Then, we put our focus on introducing the PNARC model, which is the main method applied in mining PNARs from campus data in this paper.

2.1 Application of Association Rules in Analysis of Campus Data

Zhang et al. Found the association rules between different course scores through research to help students improve their college [7]. They proposed a k-means clustering algorithm based on sample distribution density, which optimizes the initial clustering center and preprocesses outliers. Experiments show that using this improved k-means algorithm to assist association rule mining can effectively reduce invalid and wrong rules.

In Ref. [8], the author proposed a mining algorithm based on fuzzy association rules to mine the hidden relationship between students' academic performance and enrollment. The mining rules provided guidance for the prediction indicators of freshmen's enrollment quality.

Banswal et al. designed a system called SPACS which combines fuzzy logic and the association rules mining method to analyze the students' academic performance by identifying all the critical factors. The experimental results show that this system has teaching and research significance [9].

In 2014, Wang et al. proposed an efficient association rule mining method by improving the basic Apriori algorithm. And applied it to the data relationship mining of the curriculum management system, and finally achieved good results.

Jiang et al. found that there are three strong association sets, scholarships, the low canteen consumption amount and the greater number of consumption times by using Apriori association rule algorithm [11].

All studies mentioned above, however, just consider PARs, and do not take NARs into account. In fact, sometimes, NARs can provide more useful information. L. Zhao et al. used the method named PNARC to mine both PARs and NARs from one real college dataset [12]. But they only analyzed the relationship among mental health factors and did not analyze the impacts of mental health on student performance.

2.2   The Positive and Negative Association Rules on Correlation Model

Let $I = \{i_1, i_2, ..., i_n\}$ be a set of n distinct items. D is a database of transactions. An association rule is an implication of the form $X \Rightarrow Y$, where $X \cap I$, $Y \cap I$, and $X \cap Y = \Phi$. The calculation methods of support and confidence of rule $X \Rightarrow Y$ are shown in formula 1 and 2 respectively.

$$s(X \Rightarrow Y) = s(X \cup Y) = (X \cup Y).count/|D| \qquad (1)$$

$$c(X \Rightarrow Y) = s(X \cup Y)/s(X) = (X \cup Y).count/X.count \qquad (2)$$

An association rule is valid if its support and confidence are greater than the minimum support (*ms*) and the minimum confidence (*mc*) respectively. It can also be expressed by the following inequality (3) and inequality (4):

$$s(X \Rightarrow Y) \geq ms \qquad (3)$$

$$c(X \Rightarrow Y) \geq mc \qquad (4)$$

Apriori algorithm is the basis of all association rule mining algorithms. It uses an iterative approach of layer-by-layer search to mine *FIS*. It generates $(k + 1)$-*FIS* with k-*FIS*. It mainly generates candidate sets by self-joining steps and pruning steps, and then picks out the itemset that meets the *ms* from the candidate set. Finally, mining association rules that meet the *mc* from *FIS*. The specific process can be found in the Ref. [15].

However, Apriori algorithm is only suitable for mining PARs, and it will generate contradictory rules when the NARs are considered.

PNARC model can effectively solve this problem, because it uses the correlation test method to remove the contradiction association rules. The correlation between the two itemsets can be expressed by the following equation.

$$corr_{X,Y} = s(X \cup Y)/(s(X)s(Y)) \tag{5}$$

$corr_{X,Y}$ has three possible situations: if $corr_{X,Y} > 1$, X and Y are positively correlated; if $corr_{X,Y} = 1$, X and Y are independent of each other; if $corr_{X,Y} < 1$, X and Y are negatively correlated.

The pseudo code of the PNARC algorithm is as shown in Table 1.

In PNARC model, if $corr = 1$, it means that A and B are contradictory and do not generate rules. Obviously, PNARC can not only effectively mine the PNARs, but also eliminate the contradiction rules and ensure the accuracy of the results.

3 Experiment Data Processing

The real data is collected from one Chinese college from September 2016 to January 2017. The data producers are second-year university students who major in the same profession. Two main reasons can be accounted for that. One is that the data for second-year students is highly stable and representative. The second reason is that there are different course difficulty and course arrangement for different majors. Comparing data from the same profession can improve the accuracy of experimental results. The data includes students' academic performance, e-card consumption records, book borrowing records and mental health status. Although these data are derived from the same campus databases, their data structures, data types, and dimensions are all different. Therefore, dealing with these four kinds of data separately is necessary.

3.1 Academic Performance Data

Student grades are composed of public and professional classes. In this paper, in addition to the total score, we also calculate the professional course score as an important research factor. The total score and professional course score are calculated by Eqs. (6) and (7), respectively. The total score is the weighted average score of all grade scores which are weighted by credits, and the professional course score is the weighted average score of all professional course scores which are weighted by credits.

$$\sum \text{total grade} = \sum \text{single grade score} \times \text{credits}/ \sum \text{total credits} \tag{6}$$

$$\sum \text{professional course grade} = \sum \text{single professional course grade score}$$
$$\times \text{credits}/ \sum \text{total professional credits} \tag{7}$$

As shown in Table 2, the student's grades are divided into two categories.

Table 1. The pseudo code of the PNARC algorithm

Algorithm:	PNARC
Input:	*ms:* minimum support *mc:* minimum confidence; *L:* Frequent ItemSets;
Output:	*NAR:* set of NARs; *PAR:* set of PARs;
(1)	$NAR = \Phi; PAR = \Phi;$
(2)	Generate *FIS* with the same steps as Apriori
(3)	**for** any *FIS X* in *L* **do begin**
(4)	**for** any itemset $A \cup B = X$ and $A \cap B = \Phi$ **do begin**
(5)	$corr = supp(A \cup B)/(\ supp(A)supp(B));$
(6)	$conf(A \Rightarrow B) = supp(A \cup B)/\ supp(A);$
(7)	$conf(\neg A \Rightarrow \neg B) = 1 - conf(\neg A \Rightarrow B);$
(8)	$conf(\neg A \Rightarrow B) = (\ supp(B) - supp(A \cup B))/(1 - supp(A))$
(9)	$conf(A \Rightarrow \neg B) = 1 - conf(A \Rightarrow B)$
(10)	**if** $corr = 1$ **then** continue
(11)	**if** $corr > 1$ **then**
(12)	**if** $conf(A \Rightarrow B) \geq mc$ **then**
(13)	$PAR = PAR \cup \{\ A \Rightarrow B\}$
(14)	**if** $conf(\neg A \Rightarrow \neg B) \geq mc$ **then**
(15)	$NAR = NAR \cup \{\neg A \Rightarrow \neg B\}$
(16)	**if** $corr < 1$ **then**
(17)	**if** $conf(\neg A \Rightarrow B) \geq mc$ **then**
(18)	$NAR = NAR \cup \{\neg A \Rightarrow B\};$
(19)	**if** $conf(A \Rightarrow \neg B) \geq mc$ **then**
(20)	$NAR = NAR \cup \{\ A \Rightarrow \neg B\};$
(21)	**end;**
(22)	**end;**
(23)	**Return** *PAR* and *NAR;*

Table 2. Grade after classification

Student ID	Total score	Professional course score
101	≥ 80	≥ 80
102	≥ 80	≥ 80
103	<80	<80

3.2 College Student Consumption Data

The student's consumption records at the school are kept in e-card database. In this paper, the students' consumption behaviors of breakfast, lunch and dinner will be extracted from e-card database. According to the specific situation of our school, we do the following analysis.

Breakfast regularity: if the student eats breakfast before 8:00 am and the number of breakfasts in this semester is greater than or equal to 75 times (remove statutory holidays and weekends), we think that the student has a regular breakfast, otherwise, we think that the breakfast is not regular.

Lunch regularity: if the student has lunch between 11:30 and 12:30, and the number of lunches in this semester is greater than or equal to 75, we believe that the lunch habit is regular, otherwise, if the number of lunches was less than 75, the lunch is not regular.

Dinner regularity: If the student has dinner between 5:00 pm and 6:30 pm, and the number of dinners in this semester is greater than or equal to 75 times, we believe that the dinner is regular, otherwise, if the number of dinners is less than 75, we think that the dinner is irregular.

Consumption level: We average the student's consumption amount during the semester. If the consumption is greater than the average (¥2344.50), it belongs to the high-consumption group. If it is less than the average, it belongs to the low-consumption group. The records after classification and discretization are shown in Table 3.

Table 3. Discretized data

Student ID	Breakfast habits	Lunch habits	Dinner habits	Consumption
101	Regularity	Regularity	Regularity	High
102	Irregularity	Irregularity	Regularity	Low
103	Irregularity	Regularity	Irregularity	High

3.3 Book Borrowing Data

Book borrowing data is also a criterion for judging whether a student is studying at school. To this end, we also collected and analyzed the book borrowing data for this semester. Like the campus card source data table, the book borrowing source data table also stores the student's book borrowing information in a record-by-record manner. We will translate it into all the borrowing records of each student in the current semester. The key steps include traversing the database to get the number of borrowings for each student during the semester. "Professional course books" means that books are closely related to the majors of the students. If they borrowed the book during the semester, they will be recorded as "yes", otherwise recorded as "no". The processed data format is shown in Table 4.

Table 4. Book borrowing data after discretization

Student ID	Books	Professional course books
101	Yes	No
102	Yes	Yes
103	Yes	No

3.4 Mental Health Data

The mental health data is collected from students' symptom self-rating scale (SCL_90) test which is one of the most famous mental health test scale.

This test contains a total of ten factors including somatization, obsessive, fear and so on. The converted result is showed in Table 5.

3.5 Data Integration

After the data processing of the above steps, we need to integrate the various attributes in a database.

The final data consists of 19 items including gender, breakfast regularity, lunch regularity, dinner regularity, consumption level, total score, professional class score, borrowing books, borrowing professional books, somatization, obsessive, interpersonal sensitivity, depression, anxiety, hostility, fear, paranoid, psychoticism and the mental health status.

4 Experiment Results and Analysis

The obtained results based on PNARC algorithm are given in Sect. 4.1. Some suggestions for the improvement of college students' performance are given in Sect. 4.2.

4.1 The Obtained Results

In order to obtain more rules with high confidence, ms and mc are set to be 0.2 and 0.5 respectively. The PARs whose antecedents contain the items related to the regularity of eating as well as consequents contain the items related to the academic performance are shown in Table 6.

Rule 1 implies among all the students who often eat breakfast, 92.20% of students have a total score greater than or equal to 80. Rule 2 means among all the students who often eat breakfast, students whose professional course score greater than or equal to 80 accounted for 90.55%. Rules 3 and 4 imply that among all the students who often have lunch, 52.25% of students with a total score greater than or equal to 80, and 58.30% of students with a professional course score greater than or equal to 80. Rule 5 means that if students often eat dinner, the probability that the total score is greater than or equal to 80 is 62.88%. Rule 6 means that if students have regular meals for breakfast, lunch, and

Table 5. Mental health data after discretization

Gender	somatization	Obsessive	Interpersonal sensitivity	Depression	Anxiety	Hostility	Fear	Paranoid	Psychoticism	Mental health status
Male	Yes	Yes	Yes	Yes	Yes	Yes	Yes	Yes	Yes	Yes
Male	No	Yes	Yes	Yes	Yes	Yes	Yes	Yes	Yes	Yes
Male	No	Yes	Yes	Yes	Yes	Yes	No	No	Yes	Yes
Male	No	Yes	Yes	Yes	Yes	Yes	Yes	Yes	Yes	Yes
Female	No	No	No	No	No	No	No	No	No	No
Female	Yes	No	No	Yes	No	No	No	No	Yes	Yes
Female	No	Yes	No	No	No	No	No	No	Yes	Yes

Table 6. PARs between regularity of eating and academic performance

Number	Association rule	Confidence (%)
1	Breakfast regularity ⇒ total score ≥ 80	92.20
2	Breakfast regularity ⇒ professional course score ≥ 80	90.55
3	Lunch regularity ⇒ total score ≥ 80	52.25
4	Lunch regularity ⇒ professional course score ≥ 80	58.30
5	Dinner regularity ⇒ total score ≥ 80	62.88
6	Breakfast regularity, lunch regularity, dinner regularity ⇒ total score ≥ 80	94.50
7	High ⇒ total score ≥ 80	60.88
8	Low ⇒ total score ≥ 80	55.45

dinner, they have a 94.50% chance of having a total score higher than or equal to 80. And the rules 7 and 8 mean that among high-spending students, 60.88% will have a total score higher than or equal to 80, and among low-spending students, 55.4% will have a total score higher than or equal to 80, which indicate that the level of consumption has little effect on student achievement.

Table 7 shows the partial NARs between students' academic performance and regularity of eating.

We find that there is a significant negative correlation between the irregularity of breakfast and the student's performance. Among all the students who have irregular meals at breakfast, 87.43% of the students have a total score of less than 80. In addition, if the students' breakfast, lunch and dinner are irregular, they have a 94.75% probability of having a total score less than 80.

Table 7. NARs between regularity of eating and academic performance

Number	Association rules	Confidence (%)
1	Breakfast regularity ⇒ (total score < 80, professional course score < 80)	88.47
2	¬Breakfast regularity ⇒ score < 80	87.43
3	¬(Breakfast regularity, lunch regularity, dinner regularity) ⇒ score < 80	94.75

Our research results show that good eating habits are positively related to students' academic performance. Breakfast in three meals a day is the most influential factor on students' academic performance.

The PNARs whose antecedents contain the items related to the book lending as well as consequents contain the items related to the academic performance are shown in Table 8.

Table 8. PNARs between book borrowing habit and academic performance

Number	Rule	Confidence (%)
1	Book $\Rightarrow$ total score $\geq$ 80	85.57
2	Book $\Rightarrow$ professional course score $\geq$ 80	82.28
3	Professional book $\Rightarrow$ professional course score $\geq$ 80	95.56
4	No book $\Rightarrow \neg$ (total score $\geq$ 80)	83.38

By analyzing the PARs, we can get that among the students who borrowed books frequently, 85.57% of the students had a total score higher than or equal to 80, and 82.28% of the students had a professional course score higher than or equal to 80. Among all students who have borrowed professional books, 95.56% of them had a professional course score higher than or equal to 80. By analyzing the NARs, we can find that 83.38% of the students who did not borrow books related to the major did not perform well in professional courses. Through the above analysis, we can find that there is a positive correlation between student academic performance and borrowing habits. Usually, students who borrow more books can reflect that they are more serious in their studies. There is a strong correlation between professional books and professional achievements. If students are interested in their majors, they usually borrow books from related majors. Cultivating students' interests in professional learning is crucial.

The rules whose antecedents contain the items related to the mental illness as well as consequents contain the items related to the academic performance are shown in Table 9.

As we can see that almost all mental illnesses have an impact on student performance, depression is the mental illness that has the greatest impact on student performance. Among the students suffering from depression, 85.45% of the students have a professional course score of less than 80, and 92.55% of the students have a total score of less than 80. 91.36% of students with anxiety disorders have a total score of less than 80. 88.47% of students with fear have a total score of less than 80. Among the students with sensitive interpersonal relationships, 91.35% of them do not achieve satisfactory results. In addition, we can see that if a student suffers from depression, anxiety and phobia at the same time, it has a 95.69% chance of having a total score below 80. Conversely, among all the students with mental health, 93.45% of them had a total score higher than or equal to 80. We checked the relevant information and learned that students with poor academic performance have a greater emotional change. They are more depressed and pessimistic than ordinary students. They are not interested in many things, have no spirit, are prone to fatigue, and even have an anxiety tendency. Students with poor academic performance feel inferior to others in the process of interpersonal communication. They often feel shy in the crowd. In view of the above analysis, it is particularly important for schools to strengthen their mental health education. Our research results can provide better decision support for schools.

We also explored the impact of various behavioral habits on academic performance. Table 10 shows some PARs, and Table 11 provides NARs.

Table 9. PNARs between mental illness and academic performance

Number	Rule	Confidence (%)
1	Depression ⇒ professional course score < 80	85.45
2	Depression ⇒ total score <80	92.55
3	Anxiety ⇒ total score <80	91.36
4	Fear ⇒ total score <80	88.47
5	Interpersonal sensitivity ⇒ total score <80	91.35
6	Healthy ⇒ total score ≥ 80	93.45
7	Depression, anxiety, fear ⇒ total score <80	95.69
8	Depression, anxiety ⇒ total score <80, professional course score <80	89.92
9	Unhealthy ⇒ ¬(total score ≥ 80)	93.67

Table 10. PARs between behavioral habit and academic performance

Number	Rule	Confidence
1	Breakfast regularity, book ⇒ total score ≥ 80, lunch regularity	90.50
2	Healthy, book ⇒ total score ≥ 80	92.70
3	Breakfast regularity, lunch regularity, dinner regularity, book ⇒ total score ≥ 80	94.60
4	Breakfast regularity, lunch regularity, dinner regularity, healthy ⇒ total score ≥ 80, professional course score ≥ 80	93.52
5	Breakfast regularity, lunch regularity, dinner regularity, healthy, book, professional book ⇒ total score ≥ 80, professional course score ≥ 80	95.60
6	Total score ≥ 80, breakfast regularity ⇒ book, healthy	90.25
7	Total score ≥ 80, book ⇒ breakfast regularity, healthy	91.56
8	Total score ≥ 80, healthy ⇒ breakfast regularity, book	91.79

By analyzing those rules, we can get that the habits of having breakfast, mental health status and book borrowing behavior all have an impact on students' academic performance. 94.60% of all students who have regular breakfast, regular lunch, regular dinner, and frequent borrowing have achieved good academic performance. 93.52% of students with regular breakfast, regular lunch, regular dinner and mental health can achieve good academic performance.

Among the students who have regular breakfast habits, regular lunch habits, regular dinner habits, the habits of borrowing professional class books, and no mental illness, 95.60% of them have a total score higher than or equal to 80. Rules "Good, Breakfast regularity ⇒ book, healthy", "Good, book ⇒ Breakfast regularity, healthy" and "Good, healthy ⇒ Breakfast regularity, book" show that there is a strong correlation between good grades, good breakfast habits, borrowing books and mental health, which means that the regularity of breakfast, borrowing behavior and mental health can promote students' performance. Conversely, the NARs in Table 11 indicate mental illness, irregularities in breakfast, and lack of extracurricular learning can have a negative impact on students' performance.

Table 11. NARs between behavioral habit and academic performance

Number	Rule	Confidence (%)
1	Breakfast regularity, book ⇒ ¬(total score <80)	90.85
2	Healthy, book ⇒ ¬(total score <80)	92.70
3	¬(Breakfast regularity, book, healthy) ⇒ (total score <80)	89.56
4	¬(Breakfast regularity, book, professional book) ⇒ professional course score <80	88.70
5	Depression, anxiety, fear ⇒ ¬(total score ≥ 80)	95.69
6	Depression, anxiety, interpersonal sensitivity ⇒ ¬(total score ≥ 80)	94.66
7	Depression, breakfast irregularity, no book ⇒ ¬(total score ≥ 80)	93.55

4.2 Some Suggestions Based on Rule Analysis

According to the analysis of the rules in Sect. 4.1, we propose some suggestions for improving the performance of college students in this section.

First of all, students should develop good eating habits, especially ensure good breakfast habits, because according to the obtained rules whose antecedents contain the items related to the eating habits as well as consequents contain the items related to the academic performance, breakfast has the greatest impact on academic performance (breakfast regularity ⇒ total score ≥ 80, confidence = 92.20%). Secondly, through the analysis of the book borrowing behavior, we find that the professional class books are very important for the professional class results (professional book ⇒ professional course score ≥ 80, confidence = 95.56%), so it is recommended that students read more books related to the major. Last but not least, mental health has a great impact on student achievement, so it is recommended that students maintain good mental health as much as possible, especially away from depression (depression ⇒ total score < 80, confidence = 92.55%).

5 Conclusions

There is a lot of valuable information hidden in the campus data. The purpose of this paper is to discover which factors would affect students' performance by analyzing campus data collected from one Chines college school. The processed data consists of 19 items including gender, breakfast habits and the mental health status, etc. The PNARs mining method called PNARC is introduced to mine PNARs from these items. Relevant experiments and analysis show that we can easily get the factors that are good for academic performance and factors that are detrimental to academic performance. The discovery of PNARs can provide guidance for educators to improve college students' performance.

Acknowledgment. This paper was supported by the National Natural Science Foundation of China (62076143, 61806105, 61906104) and the Natural Science Foundation of the Shandong Province (ZR2019BF018).

References

1. Shi, L.: A study on the strategy of integrating knowledge resources in college students education. Comput. Simul. **4**(6), 25–32 (2017)
2. Jiang, T., Cao, J., Dan, S., Yang, X.: Analysis and data mining of students' consumption behavior based on a campus card system. In: 2017 International conference on smart city and systems engineering, pp. 58–60. IEEE Computer Society, Changsha (2017)
3. Yang, W.: Network mining of students' friend relationships based on consumption data of campus card. IOP Conf. Ser.: Earth Environ. Sci. **632**(5), 1–5 (2021)
4. Xu, L., Mclean, S., Meadows, K.N., Heffernan, A., Campbell, N.: Using online decision trees to support students' self-efficacy in the laboratory. AJP Adv. Physiol. Educ. **44**(3), 430–435 (2020)
5. Yang, X., Wu, S., H, Xia., Li, Y., Li, X.: Campus economic analysis based on K-means clustering and hotspot mining. Summary Educ. Theory **3**(2), 1–9 (2020)
6. Bertolini, R., Finch, S., Nehm, R.: Enhancing data pipelines for forecasting student performance: integrating feature selection with cross-validation. Int. J. Educ. Technol. Higher Educ. **18**(1), 1–23 (2021)
7. Zhang, T., Yin, C., Pan, L.: Improved clustering and association rules mining for university student course scores. In: 2017 12th International Conference on Intelligent Systems and Knowledge Engineering, pp. 1–6. IEEE, Nanjing (2017)
8. Oladipupo, O.O., Oyelade, O.J., Aborisade, D.O.: Application of fuzzy association rule mining for analysing students academic performance. Int. J. Comput. Sci. Issues **9**(6), 216–223 (2012)
9. Banswal, R., Madaan, V.: SPACS: students' performance analysis and counseling system using fuzzy logic and association rule mining. Int. J. Comput. Appl. **134**(6), 732–742 (2016)
10. Wang, H., Liu, P., Li, H.: Application of improved association rule algorithm in the courses management. In: IEEE 5th International Conference on Software Engineering and Service Science, pp. 804–807. IEEE, Beijing (2014)
11. Jiang, N., Xu, W.: Student consumption and study behavior analysis based on the data of the campus card system. Microcomput. Appl. **31**(2), 35–38 (2015)
12. Zhao, L., Hao, F., Xu, T., Dong, X.: Positive and negative association rules mining for mental health analysis of college students. Eurasia J. Math. Sci. Technol. Educ. **13**(8), 5577–5587 (2017)
13. Wu, X., Zhang, C., Zhang, S.: Efficient mining both positive and negative association rules. ACM Transa. Inf. Syst. **22**(3), 381–405 (2004)

14. Dong, X., Wang, S., Song, H., Lu, Y.: Study on negative association rules. J. Beijing Inst. Technol. **24**, 978–981 (2004)
15. Agrawal, R., Swami, A.: Mining association rules between sets of items in large databases. In: ACM SIGMOD International Conference on Management of Data, pp. 207–216. ACM, Washington (1993)

Identification and Analysis on Factors in Establishing the Green Supply Chain Contractual Relationship: Literature Review Based on NVivo

Qian Zhang[1](✉), Hongmei Shan[1], Jiapan Wang[2], and Mengmeng Miao[1]

[1] Xi'an University of Posts and Telecommunications, Xi'an, Shannxi 710061, China
zhangqian@stu.xupt.edu.cn, shmxiyou@xupt.edu.cn, 2011210005@stu.xupt.edu.cn
[2] University of British Columbia, V6T 1Z4 Vancouver, BC, Canada
wjiapan@student.ubc.ca

Abstract. Under the dual constraints of limited resources and environmental protection pressure, contract has become an important means to coordinate the relationship between upstream and downstream firms in the green supply chain (GSC). Which factors should be considered in establishing the GSC contract relationship is always a hard nut to crack faced by both theoretical and practical circles. This paper conducts coding analysis of relevant literatures on GSC contractual relationship in recent ten years based on Grounded Theory, and puts forward a five-factor model for establishing GSC contractual relationship. The study findings are: The obstacle factors and members' behavioral preferences is core category factors, the government incentives policy is sub-core category factor, another, the requirements of upstream and down-stream partners, and market pressure is support category factors. The managers should not only pay attention to the core and sub-core essential factors, but also comprehensively consider other subsidiary factors when establishing the contract relationship with the green supply chain partners in the future. The results will provide useful reference for the decision-making of green supply chain practice activities under the complex and changeable environment.

Keywords: Green supply chain · Coordination · Contract · Grounded theory · Literature review

1 Introduction

With the increasingly serious problems of environmental pollution and resource shortage, green supply chain (GSC) management has become an important means for enterprises to reduce operating costs and improve resource utilization.

Foundation items: supported by the National Social Science Foundation, China (No. 21BJY216)

Handfield and Srivastava found that establishing inter-firm GSC coordination is the best way of operating with the least negative environmental impact and the highest resource utilization, which is conducive to the greening of the whole supply chain [1,2]. As early as 1985, Paste put forward the concept of supply chain contract and believed that supply chain contract is an important means to promote cooperation between enterprises [3]. Jiang and Li [4] established a supply chain game model considering the factor of product greenness, and optimized the profits between manufacturers and retailers through revenue sharing contract. Zhang et al. [5] established the contract with cost-sharing and revenue-sharing to coordinate green supply chain on the basis of fairness preference and government subsidy.

It can be seen from the above that scholars have achieved fruitful research results on the establishment of GSC contract relationship. However, the existing results do not systematically study which factors are should be considered for the establishment of GSC contract relationship, it cannot clearly provide detailed decision-making reference for theoretical research and enterprise practice decision-making. This study attempts to use grounded theory to solve the following two main questions: (1) what factors do scholars consider when using contract to establish GSC coordination relations? (2) Which factors are the core and sub-core factors in numerous factors?

2   Data Sources and Research Methodology

2.1   Literature Collection

Fig. 1. Word cloud of the sample dataset.

This paper mainly collects relevant literatures on contract for supply chain coordination in green Supply Chain (GSC) from 2011 to 2020 through the CNKI database in China (all literatures are journals) and Web of Science data platform (Article or Review). Finally, 92 valid samples were determined in this paper.

This paper uses NVivo 12.0 software to conduct visualization analysis on confirmed valid dataset samples. Figure 1 indicates that the Green and Supply Chain are highly concerned words of the literatures; the main participants of GSC are Manufacturer, Retailer and Supplier. The GSC contract coordination relationship focuses on price, profit, revenue, sharing and cost. To precisely identify the importance of factors, this paper further introduces the coding number of "reference points" to reflect the importance of the factors considered by GSC.

2.2 Research Methodology and Tools

Grounded theory was proposed by Anselm Strauss and Barney Glaser, from Columbia University. The purpose is to use a systematic procedure to generalize a theory based on a phenomenon. This method is currently considered one of the most scientific qualitative research tools [6]. The coding process was completed with NVivo 12.0 to shorten the research cycle and improve efficiency.

3 The Coding Analysis of Considerations

3.1 Open Coding

The research is based on the principles of Grounded theory to analyze 92 sample datasets carefully, sort the considerations in texts comprehensively, and code and conceptualize the involving factors. A total of 156 reference nodes were sorted out. Finally, 33 conceptual categories (tertiary nodes), such as risk appetite, equity concern, and green reputation, were identified through modification, deletion, and merging processes. These nodes are located at the bottom of the subordination hierarchy and are the direct factors in designing the GSC contract.

3.2 Axial Coding

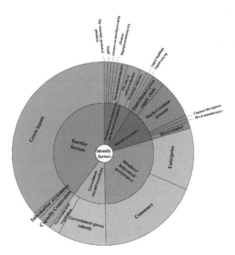

Fig. 2. The node coding hierarchy.

Axial coding is the process of implementing cluster analysis and comparison to the open coding results, and refining the main category [7]. This paper groups 33 tertiary nodes obtained from open coding into 21 secondary nodes. Subsequently, the 21 secondary nodes are further grouped into 5 primary nodes. Figure 2 shows that barrier factors (40.0%) and member behavioral preferences (29.1%) are the mainstream factors considered by researchers so far.

3.3 Selective Coding

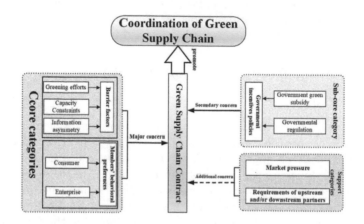

Fig. 3. The five-factor model established by the GSC contract.

Selective coding must further summarize and refine the five factors derived from the primary axial coding with the research purpose [8]. Firstly, in terms of coding reference points, the barrier factors and members' behavioral preferences are at 69.1% of all the nodes, which are overarching to the whole situation and belong to the "core categories". Secondly, although the market pressure and government incentives are external factors, government subsidies directly affect the games among supply chain participants, which affects the establishment of contractual relationships and belongs to the "sub-core category". Finally, the upstream and downstream partner requirements and market pressure nodes account for 20.6%, which are essential supplements of the "core" and "sub-core" categories but will be treated as "support" because they do not directly impact the contract design. Based on the above analysis, the five-factor model for establishing green supply chain contractual relationship is constructed as shown in Fig. 3.

4   The Analysis Results of Considerations

4.1   Core Categories

Barrier Factors. Barrier factors are the general term of all factors that lead to the incoherence (overall benefit reduction) of GSC due to supply chain members' inputs, capacity constraints, information asymmetry, etc. Fig. 2 evidently show that green inputs are the most concerned by current researchers, followed by capacity constraints and information asymmetry factors.

Green Inputs. The greenness of products is the first concern in green input. Swami and Shah [9] believe that tariff contracts composed of two parts have greater potential in terms of profitability and environmental benefits when considering the greenness of products. Rong and Xu [10] analyzed the effect of retailers bearing manufacturers' green development costs on GSC stability.

Capacity Constraints. Capacity constraints are a barrier factor in GSC partnerships which can lead to inefficiency of the overall supply chain. Chang et al. [11] analyzed the interaction between manufacturers' capacity constraints and retailers' green marketing efforts. They combined cost-sharing and benefit-sharing to coordinate the supply chain for the inefficiencies under decentralized GSC.

Information Asymmetry. The information asymmetry is also part of the barrier factors that can lead partners to make wrong decisions and cause profit loss. Li and Zhu [12] propose a new contract scheme that combines wholesale price contracts with cost-sharing contracts to improve information asymmetry between food supply chains.

Members' Behavioral Preferences. Members' behavioral preference is that supply chain participants tend to make decisions under the domination of the subconscious mind. It can be seen from Fig. 2 that existing studies have focused more on consumers' green preferences and risk preferences of corporate.

Consumers' green preferences. From the consumers' perspective, green preference is the main consideration, which refers to the degree of consumers' recognition of the new consumption model that is conducive to the harmonious development of human beings and nature [13]. Gao and Wu [14] found that establishing a cost-sharing contract based on consumers' green preferences increased product greenness and supply chain members' profits.

Enterprises' Risk Preference. From the enterprises' perspective, the risk appetite has a crucial influence on decision making. Decision-makers can only make the best decisions to obtain the maximum investment return if they precisely understand their risk preference. Ql et al. [15] found the optimal strategy for GSC by considering the transportation strategy and default risk to establish an appropriate agreement.

4.2   Sub-core Categories

Government Incentive Policy. Government incentive policy is a range of policy tools such as green subsidies or rewards or legal means, which enforce constraints on GSC participants to protect the environment and promote resource-efficient development. According to Fig. 2, the governments' green subsidies are the most concern in government incentive policies. Hu and Dai [16] studied the government's subsidy behavioral of loan interest rate to manufacturers and the coordination effect of different contracts in the management of the green supply chain and conducted a comparative analysis. Two pricing contracts proposed by Zhang et al. and Wang [17] can coordinate GSC under government subsidies and output uncertainty.

4.3 Supporting Categories

Requirements of Upstream and Downstream Partners. The requirement of partner is the sum of all factors that the firms from upstream and downstream of supply chain propose additional demand in GSC coordination strategies. Zhang [18] considered the rate of carbon reduction while designing GSC contracts. He explored how cost-benefit sharing contracts affect the members' carbon emission strategies and coordination. Guo [19] focused on whether upstream and downstream supply chain members can simultaneously stabilize economic benefit goals to achieve GSC coordination. Besides, researchers have also focused on factors such as deferred payment [20], environmental labelling strategies [21], and transportation strategies [18] as seen in Fig. 2.

Market Pressure. Market pressure is a general term of all factors that lead to uncoordinated supply chains due to market uncertainty. Under the consideration of market pressure, Hua and Ding [22] found that a combination contract not only reduces the selling price of green agricultural products but also increases greenness and the market demands of green agricultural. In addition, the existing studies also concern on channel disruption [23], competing manufacturers [24], and competing retailers [25] as seen in Fig. 2.

5   Conclusion and Outlook

5.1   Conclusion

Based on the grounded theory research method with NVivo 12.0 software, this research conducted a systematic coding analysis of the literatures related to the GSC contractual coordination relationship in the past ten years. The five-factor model for establishing a GSC contractual relationship is proposed, including 5 primary nodes, 22 secondary nodes, and 32 tertiary nodes. The research results show that: the core categories factors include obstructive factors and members' behavioral preference in the five-factor model, sub-core category factor is government incentive policy, and support categories factors have requirements of upstream and downstream partners and market pressure. The existing literature mainly considers the barrier factors and the members' behavioral preferences. However, the researches concerned on the requirements of upstream and downstream partners, market pressure and government policies are still insufficient. This study result improves the theoretical knowledge system of the GSC contractual relationship and provides a decision-making reference for the cooperation between supply chain enterprises.

5.2 Outlook

Although fruitful research on establishing the GSC coordination relationship has been achieved, some issues still exist, such as incomplete considerations and imperfect contract design. Further research on the GSC contract needs to be completed.

(1) Explicit the multifaceted consideration factors. The research finds that the current theoretical community has exhaustive studies on the core genetic factors for establishing GSC contractual relationships. However, the sub-core factors and support factors give less attention. Due to supply chain enterprises' complex and volatile operating environment, partners' requirements from upstream and downstream and external market pressure will be critical challenges for researchers and should be the focus in future.
(2) Focus on the dynamic nature of the various consideration. With complication of the environment, the same factor in the supply chain environment can constantly change in different period when establishing a contractual relationship.
(3) Explore the unknown influencing factors actively. Due to the increasing complexity of the supply chain network, there is a growing number of impactors that affect the establishment of the GSC contractual relationship.

References

1. Handfield, R.B., Walton, S.V., Seegers, L.K., et al.: 'Green' value chain practices in the furniture industry. J. Oper. Manage. **15**(4), 293–315 (1997)
2. Srivastava, S.K.: Green supply-chain management: a state of the art literature review. Int. J. Manage. Rev. **9**(1), 53–80 (2007)
3. Pasternack, B.A.: Optimal pricing and returns policies for perishable commodities. Mark. Sci. **4**, 166–176 (1985)
4. Jiang, S., Li, S.: Green supply chain Game model and revenue sharing contract with product green degree. Chin. J. Manage. Sci. **23**(06), 169–176 (2015)
5. Zhang, H., Huang, J., Cui, Y.: Game models and contract of green Supply chain considering fairness preference and government subsidies. J. Ind. Technol. Econ. **37**(01), 111–121 (2018)
6. Strauss, A.: Qual. Anal. Soc. Sci. Cambridge University Press, New York (1987)
7. Wu, D., Liu, H., et al.: Research on formation mechanism of tourism public opinion based on Grounded Theory: the perspective of psychological contract violation. Manage. Rev. **33**(04), 170–179 (2021)
8. Wang, N.: Identification and analysis of the development resistance and restriction factors of Maker space in university libraries: the Grounded Theorical analysis based on NVivo. New Century Library **10**, 34–41 (2020)
9. Swami, S., Shah, J.: Channel coordination in green supply chain management: the case of package size and shelf-space allocation. Technol. Oper. Manage. **2**(1), 50–59 (2011)
10. Rong, L., Xu, M.: Impact of revenue-sharing contracts on green supply chain in manufacturing industry. Int. J. Sustain. Eng. **13**(4), 316–326 (2020)

11. Chang, S., Hu, B., He, X.: Supply chain coordination in the context of green marketing efforts and capacity expansion. Sustainability **11**(20), 5734–5734 (2019)
12. Li, X., Zhu, Q.: Contract design for enhancing green food material production effort with asymmetric supply cost information. Sustainability **12**(5), 2119–2119 (2020)
13. Zhu, X., Sun, X.: Analysis on the source of Chinese consumers' green consumption motivation: functional need or symbolic need. Enterp. Econ. **12**, 68–75 (2015)
14. Gao, G., Wu, Q.: Green supply chain contract coordination from the perspective of cost sharing. J. Railway Sci. Eng. **15**(10), 2700–2706 (2018)
15. Ql, S., Zhou, Y., et al.: Optimal strategy for a green supply chain considering shipping policy and default risk. Comput. Ind. Eng. **131**, 172–186 (2019)
16. Hu, G., Dai, P.: Study on green supply chain coordination for manufacturer quality improvement under loan discount. J. Xihua Univ. (Philos. Soc. Sci.) **39**(04), 72–82 (2020)
17. Zhang, S., Wang, S.: Contracts analysis of green supply chain: with a consideration of government subsidies and yield uncertainty. Ecol. Econ. **36**(11), 66–74 (2020)
18. Guo, X., Cheng, L., Liu, J.: Green supply chain contracts with eco-labels issued by the sales platform: profitability and environmental implications. Int. J. Prod. Res. **58**(5), 1485–1504 (2020)
19. Zhao, L., Li, L., et al.: Research on pricing and coordination strategy of a sustainable green supply chain with a capital-constrained retailer. Complexity **2018**, 1–12 (2018)
20. Cao, Y., Hu, H., Li, Q.: Research on environmental labeling strategy selection of green supply chain under cost sharing contract. Chin. J. Manage. Sci. 1–13 (2021)
21. Zhang, H., Xu, H., Pu, X.: A cross-channel return policy in a green dual-channel supply chain considering spillover effect. Sustainability **12**(6), 2171–2171 (2020)
22. Hua, J., Ding, X.: Three-level supply chain coordination of green agricultural products under stochastic demand. Ind. Eng. J. **23**(03), 51–58 (2020)
23. Aslani, A., Heydari, J.: Transshipment contract for coordination of a green dual-channel supply chain under channel disruption. J. Clean. Prod. **223**, 596–609 (2019)
24. Ma, P., Zhang, C., et al.: Pricing decisions for substitutable products with green manufacturing in a competitive supply chain. J. Clean. Prod. **183**, 618–640 (2018)
25. Wang, W., Liu, X., et al.: Coordination of a green supply chain with one manufacturer and two competing retailers under different power structures. Discrete Dyn. Nat. Soc. **2019**, 1–18 (2019)

A Data-Based Approach for Computer Domain Knowledge Representation

Lin Zhou, Qiyu Zhong, and Shaohong Zhang$^{(\boxtimes)}$

School of Computer Science and Cyber Engineering, Guangzhou University,
Guangzhou, Guangdong, China
2112106079@e.gzhu.edu.cn, zimzsh@qq.com

Abstract. Representation learning is a method to compute the corresponding vectorized representations of entities or relationships. It is one of the most basic and essential natural language processing tasks. Current computer domain knowledge modeling techniques have two flaws: (1) the neglect of fine-grained knowledge hierarchies, and (2) the lack of a unified reference standard for modeling domain information. The fine-grained knowledge hierarchy includes knowledge domains, units, and topics. We use the Computer Science Guidelines as a standard to annotate an unstructured and unlabeled corpus in the computer domain with knowledge annotation and topic mapping. We organise the corpus into a computer domain knowledge system with a three-level hierarchy. We propose a knowledge representation method that incorporates contextual semantic information and topic information. The method can be applied to discover connections between knowledge of entities of different granularity. We compare it with several existing textual representation methods. Experimental results on extracting knowledge representations in computer domains show that combining contextual semantic information and topic information methods are more effective than single ones.

Keywords: Domain knowledge modeling · Knowledge representation and quantification · Multi-granularity semantic similarity

1 Introduction

With the rapid development of information technology and the advent of the industrial Internet era, computer technology is increasingly being used in various fields. Furthermore, computer domain knowledge is abundant and widely disseminated. It can be found in various technical forums, scientific and technical papers, computer-related education material, and job search information. In contrast, standard organization and administration of computer domain knowledge are critical to the industry's success. In 2013, the Association for Computing Machinery and the Institute of Electrical and Electronics Engineers Computer Society (ACM/IEEE-CS) published the Computer Science Curricula 2013 (CS2013) [1], a guide for undergraduate computer science courses for computer

N. Xiong et al. (Eds.): ICNC-FSKD 2022, LNDECT 153, pp. 841–848, 2023.
https://doi.org/10.1007/978-3-031-20738-9_93

science undergraduates. It specifies what a computing course must include and the syllabus for each course.

The language model (LM), a technique used to assess the plausibility of textual data, i.e., to quantify whether a phrase makes sense. It is a concept strongly connected to a text representation. Language models, primarily via learning semantics from large-scale corpus texts, may anticipate the next language unit based on context. Text representation converts text into computable numerical vectors, i.e., a vectorized representation of the extracted knowledge. The datasets we utilize contain entities at various degrees of granularity. This includes fine-grained knowledge with small semantic units, such as terminology, concepts, algorithms, technologies, expertise, and high-granularity knowledge, such as knowledge domains, knowledge units, courses, syllabi, and jobs. Furthermore, large-grained things are collections of knowledge. We quantified these entities to produce a representation of various granularity entities in the same vector space to compute and characterize links between multi-granularity domain knowledge.

In this paper, we propose knowledge representation methods that combine contextual semantic information and topic information. It can be used for extracting, quantitative representing, and topic mapping computer domain knowledge to discover connections between the knowledge of different entities. It can also provide some thoughtful approaches for subject teaching and scientific research. The remainder of the paper is structured as follows. Section 2 summarizes some relevant work, Sect. 3 outlines our technique, Sect. 4 conducts the experiments, and Sect. 5 summarizes our study.

2   Related Work

From the late 1980s to 2010, statistical linguistic modeling (SLM), modeled statistically, became the dominant approach [2]. In 2003, Bengio et al. [3] proposed the Neural Network Language Model (NNLM), which uses low-dimensional, dense, real-valued vectors to represent the components of language. In 2010, Mikolov et al. [4] proposed the first recurrent neural network (RNN)-based language model, the RNNLM. In 2012, Sundermeyer et al. [5] proposed the LSTM-RNNLM, a language model based on long short-term memory recurrent neural networks, which further improved the performance of language modeling. In addition, the convolutional neural network (CNN)-based language model (CNNLM) has also achieved success [6]. In recent years, neural network-based language models have become the dominant approach to representing text.

The vector space model is a simple and effective model for text representation, first proposed by Salton et al. [7] at Harvard University, and is a discrete model. In 2003, Blei et al. [8] proposed the Latent Dirichlet Allocation (LDA) model, one of the most representative topic models. The use of graphs for text representation was first proposed by Schenker et al. [9]. The most representative application model of graph-based text representation is the TextRank algorithm [10], which is an improvement of the PageRank algorithm [11]. In recent years, the research and application of deep learning based on neural networks have become more and more widespread. There are various neural network models for modeling text, such as using convolutional neural networks to extract some words as input features, similar to the idea of n-grams, and using recurrent neural network models with temporal feature memory to input word vector features in a sequential manner. Currently, representative works on text representation based on deep neural networks include ELMo [12], Transformer [13], BERT [14] and XLNet [15].

3   Methodology

This section focuses on our knowledge representation method that incorporates subject information and contextual meaning for the characteristics of computer domain knowledge.

Our knowledge documents contain not only large-grained knowledge, i.e., knowledge domains, knowledge units, courses, course outlines, and recruitment information, but also small-grained knowledge such as terminology, concepts, algorithms, and technologies. At the same time, data in computer verticals, especially knowledge domains and units, are often strongly topic-related, so we want to characterize knowledge by considering both contextual semantic information and topic information of the text.

The LDA topic model can identify the document's topic well among the existing text representation methods. In contrast, the BERT model can obtain contextual semantic information well. We combine these two models to characterize knowledge to obtain both subject matter information and contextual semantic information. The method flow block diagram is shown in Fig. 1.

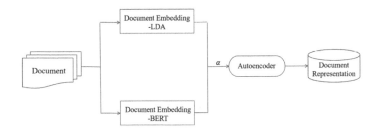

Fig. 1. Flowchart of the knowledge representation method

Here we used the distilbert-base-nli-stsb-mean-tokens model from the pre-training model SentenceTransformers [13], which performs best in the text-similarity task. To balance the relative importance of topic information and contextual semantics, we use a weight hyperparameter to stitch together the document vector obtained with the LDA topic model and the document vector obtained with the BERT model. Then a self-encoder is used to learn the representation of the vectors in the low-dimensional space, embedding the text into the vector space to obtain a knowledge representation of the topic with contextual semantics. The specific algorithm is shown in Algorithm 1. The self-encoder neural network we used in this process contains only one hidden layer. The aim is to map a high-dimensional document vector to a vector on a low-dimensional space, resulting in a document-level embedded vector containing contextual semantic information and topic information.

Algorithm 1 Converged knowledge representation algorithm

Input:
    Documents $D = D_1 D_2 \ldots D_n$, $D_i \rightarrow sentence\ i$.
Output:
    Document representation E.
1: **for** $i = 1$ to n **do**
2:     Vectorising the input document D_i with the LDA algorithm, v_{i1}
3:     Vectorising the input document D_i with the BERT algorithm, v_{i2}
4:     $v_i = \alpha \times v_{i1} + (1 - \alpha) \times v_{i2}$
5:     $e_i = Autoencoder\ (v_i)$
6: **end for**
7: **return** $E \leftarrow e_1 e_2 \ldots e_n$

---

4 Experiments

In this section, we do experiments to compare the clustering effectiveness of our proposed method with the TF-IDF model, the LDA model, and the BERT model. We evaluate these models on three internal metrics and four external metrics. Internal metrics include the Topic Coherence, the Silhouette Coefficient, and the Calinski Harabasz. External metrics include Adjusted Rand Index (ARI), the Adjusted Mutual Information (AMI), the V measure, and the Fowlkes Mallows Index (FMI) [16,17]. The dataset we used was selected from CS2013. It contains 871 large-grained entity knowledge at the document level and 3393 small-grained knowledge at the phrase level. Firstly, we preprocessed the data into the KA-KU-Topics structure.

4.1 Experimental Results

Large Granularity Knowledge Representation The results of our knowledge representation clustering of 871 document-level and large-granularity knowledge unit KU entities are shown in Fig. 2. The values assessed in terms of internal and external metrics are shown in Table 1.

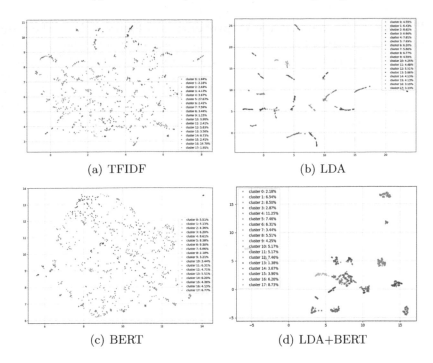

Fig. 2. Clustering results based on large-grained entity knowledge representations

We can see that the clusters of the same color in Fig. 2(a) and (c) are relatively close but without clear boundaries. Because TF-IDF model is a bag-of-words-based model, it loses contextual information. Whereas BERT model captures the contextual information but ignores the topic features. The clustering of the graphs in Fig. 2(b) and (d) is better because it incorporates both contextual semantic and topic information, both of which produce separated clusters.

Table 1. Clustering results based on large-grained entity knowledge representations

Method	Internal indicators			External indicators			
	Topic coherence	Silhouette coefficient	Calinski Harabasz	ARI	AMI	V measure	FMI
TFIDF	**0.5590**	0.0349	5.5636	0.1018	0.2803	0.3401	0.1822
LDA	0.4183	**0.6957**	**486.0739**	0.0660	0.1788	0.2281	0.1285
BERT	0.4349	0.0643	25.4074	**0.1583**	**0.3307**	**0.3702**	**0.2169**
LDA_BERT	0.4328	0.3142	102.8948	0.1054	0.2294	0.2729	0.1642

Small Granularity Knowledge Representation Figure 3 shows the results of our tests on clustering 3393 phrase and sentence level, small granularity knowledge representations, and Table 2 shows the values evaluated for internal and external metrics. Compared to large-grained knowledge, small-grained knowledge is more fragmented and incoherent. Therefore, the LDA model is less effective in representing small-grained knowledge than large-grained knowledge. However, our representation model, which incorporates contextual semantic information and topic information, still achieves good clustering separation, as shown in Fig. 3(d).

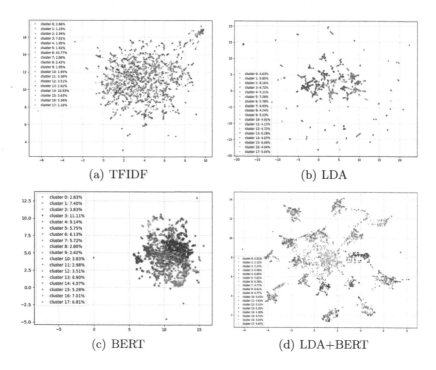

Fig. 3. Clustering results based on small-grained knowledge representations

We note that the LDA model approach performs better numerically under the assessment of internal metrics. This is because the internal metrics for clustering are mainly assessed jointly by combining the closeness of clusters and the dispersion between clusters, reflecting the model's identification of different categories of topics. In contrast, the values of the BERT model approach performed better under the evaluation of external metrics. This is because external metrics are evaluated based on the actual category labels of the text, i.e., reflecting the accuracy of the delineation. On the other hand, the BERT model is modeled precisely by the semantics of the text context. Its ability to accurately identify the meaning of the text itself allows for the best accuracy of the clustering classification.

Table 2. Clustering results based on small-grained knowledge representations

Method	Internal indicators			External indicators			
	Topic coherence	Silhouette coefficient	Calinski Harabasz	ARI	AMI	V measure	FMI
TFIDF	**0.4126**	0.0186	11.6680	0.0012	0.1326	0.1735	0.1356
LDA	0.3594	**0.4629**	**744.4613**	0.0385	0.0749	0.0875	0.1031
BERT	0.4063	0.0387	70.1212	**0.0870**	**0.1901**	**0.2015**	**0.1529**
LDA_BERT	0.3917	0.2204	233.4825	0.0478	0.0860	0.0979	0.1110

In summary, the knowledge representation method based on word embedding model and topic model can obtain contextual semantic features and topic features of knowledge and better characterize knowledge at different granularities.

5   Conclusion and Future Work

We propose a knowledge representation method for the computer vertical domain that integrates contextual semantic information and subject information. It can be utilized to uncover links between the knowledge of different entities. Compared to existing knowledge representation methods, our method can effectively characterize and quantify computer domain knowledge. Our present study focuses on semantic knowledge classification in knowledge domains, units, and subject concepts. In the future, we will refine the current knowledge of computer vocabulary into types of research challenges, solutions, resources, tools, methods, and datasets.

Acknowledgment. The work described in this paper was partially supported by grants from the funding of Guangzhou education scientific research project [No. 1201730714], and the Guangdong Basic and Applied Basic Research Foundation [No. 2022A151501-1697].

References

1. ACM/IEEE-CS Joint Task Force on Computing Curricula: Computer science curricula 2013. Technical Report. ACM Press and IEEE Computer Society Press (2013). https://doi.org/10.1145/2534860
2. Bendersky, M., Croft, W.B.: Modeling higher-order term dependencies in information retrieval using query hypergraphs. In: Proceedings of the 35th International ACM SIGIR Conference on Research and Development in Information Retrieval, pp. 941–950 (2012)
3. Bengio, Y., Ducharme, R., Vincent, P.: A neural probabilistic language model. In: Advances in Neural Information Processing Systems, vol. 13 (2000)
4. Mikolov, T., Karafiát, M., Burget, L., Cernockỳ, J., Khudanpur, S.: Recurrent neural network based language model. In: Interspeech, vol. 2, pp. 1045–1048. Makuhari (2010)

5. Sundermeyer, M., Schlüter, R., Ney, H.: LSTM neural networks for language modeling. In: Thirteenth Annual Conference of the International Speech Communication Association (2012)
6. Wang, P., Xu, J., Xu, B., Liu, C., Zhang, H., Wang, F., Hao, H.: Semantic clustering and convolutional neural network for short text categorization. In: Proceedings of the 53rd Annual Meeting of the Association for Computational Linguistics and the 7th International Joint Conference on Natural Language Processing, vol. 2: Short Papers, pp. 352–357 (2015)
7. Salton, G., Wong, A., Yang, C.S.: A vector space model for automatic indexing. Commun. ACM **18**(11), 613–620 (1975)
8. Blei, D.M., Ng, A.Y., Jordan, M.I.: Latent dirichlet allocation. J. Mach. Learn. Res. **3**(Jan), 993–1022 (2003)
9. Schenker, A., Last, M., Bunke, H., Kandel, A.: Graph representations for web document clustering. In: Iberian Conference on Pattern Recognition and Image Analysis, pp. 935–942. Springer (2003)
10. Mihalcea, R., Tarau, P.: Textrank: Bringing order into text. In: Proceedings of the 2004 Conference on Empirical Methods in Natural Language Processing, pp. 404–411 (2004)
11. Page, L., Brin, S., Motwani, R., Winograd, T.: The pagerank citation ranking: Bringing order to the web. Technical Report. Stanford InfoLab (1999)
12. Peters, M.E., Neumann, M., Iyyer, M., Gardner, M., Clark, C., Lee, K., Zettlemoyer, L.: Deep contextualized word representations. In: Proceedings of the 2018 Conference of the North American Chapter of the Association for Computational Linguistics: Human Language Technologies, vol. 1 (Long Papers), pp. 2227–2237. Association for Computational Linguistics, New Orleans, Louisiana (2018). https://doi.org/10.18653/v1/N18-1202. https://aclanthology.org/N18-1202
13. Vaswani, A., Shazeer, N., Parmar, N., Uszkoreit, J., Jones, L., Gomez, A.N., Kaiser, L., Polosukhin, I.: Attention is all you need. In: Advances in Neural Information Processing Systems, vol. 30 (2017)
14. Kenton, J.D.M.W.C., Toutanova, L.K.: BERT: pre-training of deep bidirectional transformers for language understanding. In: Proceedings of NAACL-HLT, pp. 4171–4186 (2019)
15. Yang, Z., Dai, Z., Yang, Y., Carbonell, J., Salakhutdinov, R.R., Le, Q.V.: XLNet: generalized autoregressive pretraining for language understanding. In: Advances in Neural Information Processing Systems, vol. 32 (2019)
16. Liu, Y., Li, Z., Xiong, H., Gao, X., Wu, J.: Understanding of internal clustering validation measures. In: 2010 IEEE International Conference on Data Mining, pp. 911–916. IEEE (2010)
17. Zhang, S., Yang, Z., Xing, X., Gao, Y., Xie, D., Wong, H.S.: Generalized pair-counting similarity measures for clustering and cluster ensembles. IEEE Access **5**, 16904–16918 (2017). https://doi.org/10.1109/ACCESS.2017.2741221

No Reference Image Quality Assessment Based on Self-supervised Learning

Zhen Wei[✉], Wanyu Deng, Qirui Li, and Huijiao Xu

Xi'an University of Posts and Telecommunications, Xi'an 710121, China
{weizhenkk,dengwanyu,li123,210}@stu.xupt.edu.cn

Abstract. No-Reference Image Quality Assessment (NR-IQA) is an underlying research topic that has been of interest to the research field of computer vision for a long time. Image quality assessment has a major role in image manipulation systems for algorithm analysis and verification, as well as system capability evaluation. In recent, so much researches already raised solve this subject with deep neural networks, but this problem is still a great challenge because the absence of labeled train samples. We propose a self-supervised learning-based research methods to solve the subject of lack of labeled training samples. We first use the classification prediction of the type and degree of image distortion as an auxiliary task of self-supervised learning. To enable better learning of representations, realistic distorted images and synthetic distorted images are included in the dataset selection. We use a contrastive learning approach to train a network framework to solve the proposed auxiliary task and add a mixed domain attention module to improve local distortion prediction. After training, we use a linear regressor to map the learned features to a no-reference environment to evaluate the quality for the score. We validate our findings through extensive experiments, which is indicated that the method present in this paper can also achieve high quality representations with subjective consistency without a significant amount of artificial labeled datasets.

Keywords: No reference image quality assessment · Self-supervised learning · Attention model

1 Introduction

With the rapid development of image information technology, the assessment of image quality has become a widespread and fundamental issue. Because image information has advantages that cannot be matched by other information, the rational treatment of image information has been an essential instrument in various fields. However, in the process of acquiring, processing, compressing, transmitting and recording images, the current image imaging technology, image processing methods, image signal transmission media and recording equipment are not perfect. This has caused a great obstacle for people to know the objective world in reality and to study and analyze to solve these problems. Therefore, it is very important to study the method of reasonable evaluation of image quality for application [1].

© The Author(s), under exclusive license to Springer Nature Switzerland AG 2023
N. Xiong et al. (Eds.): ICNC-FSKD 2022, LNDECT 153, pp. 849–858, 2023.
https://doi.org/10.1007/978-3-031-20738-9_94

The no-reference image quality assessment method does not use any original reference image information, but only uses the features of the image itself for evaluation. Over the past decades, scholars have made great efforts to propose a variety of evaluation datasets and objective models. The KangCNN [2] proposed by Kang et al. is one of the earliest representative works using CNN to solve the NR-IQA problem. The author implemented the feature extract and score regression into a unified framework. The Deep-IQA [3] by Bosse et al. is as well an IQA model based on an end-to-end framework. The model uses an extended VGG16 as the backbone network, which has deeper layers than other IQA models and achieves better performance in multiple datasets. Lin et al. innovatively added Generative Adversarial Networks (GAN) to the NR-IQA domain [4], proposing a new perspective to solve the discomfort problem of NR-IQA. Their network structure mainly has three components, which include a qualitatively-aware generation network, a discriminative network, and a phantom-guided quality conditioning network.

Most of the above approaches use supervised learning for image quality assessment, and most of the models rely on a large number of manually labeled datasets that contain a raw high quality image, and distorted images obtained by common distortions. From the point of view of no-reference model design, we would like to obtain a model that does not require additional manual annotation and can handle both synthetic and real distortion, so that it can be applied to any image and many more scenes.

In this paper, we present a self-supervised learning based image quality evaluation method to analyze the above mentioned methods and the problems in the practical process. This model builds a network based on ResNet50 and Multilayer Perceptron (MLP), and incorporates a attention model to improve local feature distortion prediction. Distortion type and distortion degree prediction are used as auxiliary tasks.

2   Related Work

2.1   Self-supervised Learning

Self-supervised learning focuses on using the auxiliary tasks to derive their own supervised information from the massive un-supervised data and use this constructed supervised of information to train the network so that it can learn the representation that is valuable for the downstream task. Recent state-of-the-art approaches are reliant on the task of instanced based recognition, where instead image and its enhancements are considered as a separate class Liu et al. present a NR-IQA model using image-ranking as an auxiliary task [5]. In this paper using the Identify the type of distortion and distortion levels as a self-supervised auxiliary task, we use representational learning for image quality prediction.

2.2   No Reference Image Quality Assessment

Depending on framework used, the approaches can be categorized into two separate groups. The first type uses a "feature extraction + regression/fitting" framework, where the features of the image are first extracted and then a mapping model between the features and the quality scores is built using machine learning methods; the second type uses an end-to-end framework, where the mapping model between the image to be observed and the quality scores is built directly by deep learning.

NR-IQA based on the feature extraction + regression/fitting framework. There are two parts to this evaluation method: feature extraction and fitting/regression. Where features are used to express the content of the image and directly determine the merit of the algorithm. Depending on the features used, they can be classified as artificial features and depth features. The fitting/regression methods include support vector regression, random forest, and deep neural network. The most representative ones, CORINA [6] and BRISQUE [6], have a similar feature extraction approach [12], both of which use artificial features-MSCN coefficients-and then use SVR fitting methods to obtain quality assessment scores. Although conventional models show good performance when assessed on synthetic distorted images, their performance is usually limited when real distorted images and their combinations are incorporated.

Based on the above studies, this paper accomplishes the following: (1) employs the training of a deep neural network with respect to distortion type and distortion level as an auxiliary task; (2) adds a mixed domain attention module in the middle of the encoder and mapping layers to improve local distortion prediction; (3) uses a multi-scale transformation approach to improve the robustness of learning; (4) mapping the extracted representations by means of linear regression functions to obtain quality scores; (5) evaluating the algorithm using the most commonly used PLCC and SROCC.

3 The Proposed Method

This paper presents an image quality assessment model. The method is used contrastive learning and attention model and requires only unlabeled images for training. After this section, the article will introduce each module of the framework. The overall framework of the model in this paper consists of an encoder module, an attention module, and a projection layer module (Fig. 1).

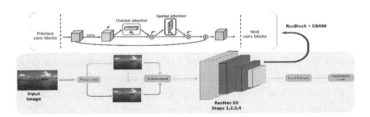

Fig. 1. The model framework figure

3.1 Auxiliary Tasks

In self-supervised learning, it is very important and challenging to propose an efficient auxiliary task. For this research, the effect we want to achieve is to be able to identify varying amounts of distortion and different types of distortion. Therefore, we convert the problem of image quality assessment to classifying the types of image distortion, which includes different types of distortion and different degrees of distortion in the classification. The goal of our model is to obtain features that can distinguish images as distortion types. So we use the cross entropy loss function commonly used in classification problems to achieve this goal during training.

We design a model: a Resnet50 network with the fully connected layer removed is used as the encoder and a multilayer perceptual network is used as the projection layer. We use $d(.)$ and $f(.)$ to Express the depth encoding network and the projection network.

$$\mathbb{Z} = d(x), \quad v = f(y) = f(d(x)) \quad \mathbb{Z} \in = R^D, v \in R^K \tag{1}$$

The loss function chosen is the a normalized temperature-scaled cross entropy (NT-Xent) [7], defined as follows:

$$\mathcal{L} = \frac{1}{|P(i)|} \sum_{j \in P(i)} -\log \frac{\exp(\phi(z_i, z_j)/\tau)}{\sum_{k=1}^{N} \mathbf{1}_{k \neq i} \exp(\phi(z_i, z_k)/\tau)} \tag{2}$$

N is the number of images that appear in the batch, t is the temperature parameter, and $P(i)$ is a set containing the index of images that belong to the same class as x_i. $|P(i)|$ is its base. In our design, we use the synthetic distortion prior knowledge as class labels and in function (2), the similarity of each pair of images is measured.

3.2 Multilayer Perceptron

Since the parameters of the fully connected layer alone can account for about 80% of the network parameters due to the redundancy of the fully connected layer, the encoder in our framework uses the resnet50 network with the fully connected layer removed. Due to the existence of features with very high relevance to the target in the actual data of image quality evaluation, we choose a multilayer perceptron to achieve the classification work. The multilayer perceptron is able to solve nonlinear problems that cannot be solved by a single-layer perceptron and can be very effective in the task of reducing the dimensionality of the representations generated by the encoder. We use the representations learned by the encoder as the input to the multilayer perceptron, where the input layer in the multilayer perceptron does not do anything but provide data. A linear transformation is computed once for each node in the hidden and output layers, the hidden layer neurons use the function sigmoid as the activation function, the output layer uses the softmax function, and the amount of nodes in the output layer is relative to the amount of classification.

3.3 Attention Model

The mechanism of attention works by enabling the network to acquire attention—the ability to ignore unrelated information and concentrate on key information. Attention mechanisms are classified according to the domain of attention: spatial domain, channel domain and hybrid domain. In this paper, we use the Convolutional Block Attention Module (CBAM) [8], which enables the network to learn image features while paying more attention to the specific details of local distortion in the image, highlighting the features of important targets in the image and thus predicting the degree of distortion more effectively. This module partitions the attention course in two separate fractions, a channel attention module and a spatial attention module. The channel attention module aims to focus on which of the input feature maps are meaningful, i.e., to judge the importance of the input feature channels and then assign appropriate weights, enabling the network to extract more useful and detailed feature maps. The spatial attention module, as a complement to the channel attention module, it is designed to attend to the most informative portions of the feature map. And for each passing module, the generated attentional feature maps need to be multiplied with the input feature maps of that module for adaptive feature refinement.

3.4 Data Enhancement

To obtain a more diverse sample, we cropped the input image to a constant random size. We assume that the cut image inherits the distortion type of the original image and keeps the distortion type unchanged. For each cut image we will have two sizes, full size and half size. For the half-size image, the whole image is zero-filled to maintain the same resolution. The usual graphical transformations are color dithering, Gaussian blur, and random resizing, but all these enhancement techniques modify the original image quality and will modify the distortion information, which is obviously not compatible with the purpose of image quality evaluation. We have performed a horizontal flip transformation of the image in order to expand the data set without modifying the image quality for the purpose of complementary quality information in different domains.

3.5 Evaluation Metrics

To acquire the quality fraction of the image, a learned representation is applied to the quality prediction problem to evaluate the image quality. After training, we remove the projector and use the output of the encoder network as the image representation, and we draw on the linear evaluation protocol used by x to evaluate the accuracy of self-supervised classification [9], using ridge regression training to obtain the quality score, with the expression:

$$\mathbb{Z} = LK, \quad L = \underset{L}{\operatorname{argmin}} \sum_{i=1}^{N} (HG_i - \mathbb{Z}_i)^2 + \lambda \sum_{j=1}^{M} L_j^2 \tag{3}$$

HG express the true score, Z is the estimation score, L express a vector having the same dimension as K, λ is the regularization parameter, M is the number of dimensions of K, and n is the number of images appearing.

4 Experiment

4.1 Data Set

In this paper, we used two public datasets, TID2013 [10] and LIVE [11], and a hybrid dataset with 2000 images in coco. TID2013 included 25 reference images and 3000 distorted images, there are 24 types of distortion. 29 reference images are available in the LIVE standard database, to which 5 different types and levels of distortion are added, generating 779 distorted images.

The LIVE database provides the Difference Mean Opinion Score (DMOS) corresponding to each image as its subjective quality score in [0, 100]. A larger DMOS of an image indicates its poorer quality, as shown in Fig. 2

(a) DMOS:75.70   (b) DMOS:72.38   (c) DMOS:70.14   (d) DMOS:21.23   (e) DMOS:13.40

Fig. 2. Images and DMOS of the same distortion type with different degrees in the LIVE database

The TID2013 dataset also provides Mean Opinion Scores (MOS) in the range of [0, 9]. Higher MOS means better image quality, as shown in the Fig. 3

(a) MOS:5.91   (b) MOS:5.68   (c) MOS:4.69   (d) MOS:3.68   (e) MOS:1.46

Fig. 3. Images and MOS of the same distortion type with different degrees in the TID2013 database

4.2 Experimental Platform and Parameter Settings

The experiments in this paper are based on the PyTorch deep learning framework, using a NVIDIA GeForce 2060 GPU with Windows 10, CUDA version 10.1, Pytorch version 1.8, and Python version 3.7. We use RESNET50 as the encoder network and a two-layer multilayer perceptron as the projection network, with each hidden layer of the projection network containing 512 neurons. The temperature parameter chosen in function (2) is fixed at $t = 0.1$. The initial learning rate of the network is 0.6. Based on this, the epoch of each training is 250 rounds. Multiple training operations are performed and tested to obtain the algorithm performance index value each time, and finally the mean value is taken as its final performance evaluation.

4.3 Experimental Evaluation Index

In the experiments, we chose two common general performance metrics to evaluate our algorithm, which are SSROCC and PLCC. Both metrics are used to measure the consistency of the subjective evaluation of the algorithm and the Human visual system (HVS), and SROCC evaluates the monotonicity of the algorithm output and the subjective evaluation; PLCC evaluates the linear correlation of the algorithm output and the subjective evaluation [12].

Spearman rank-order correlation coefficient (SROCC) is as follows:

$$\text{SROCC} = 1 - \frac{6\sum_{i=1}^{n} d_i^2}{n(n^2 - 1)} \tag{4}$$

d_i is the difference between the subjective quality score and the predicted score of the i image.

Pearson linear correlation coefficient (PLCC) is as follows:

$$\text{PLCC} = \frac{\sum_{i=1}^{n}\left(s_i - \overline{s}^-\right)(q_i - \overline{q})}{\sqrt{\sum_{i=1}^{n}\left(s_i - s^2\right)}\sqrt{\sum_{i=1}^{n}(q_i - \overline{q})^2}} \tag{5}$$

s_i is the subjective quality score of the i image, and x_i is its prediction score.

4.4 Results

During the experiments, some representative full-reference and no-reference algorithms were selected for comparison in order to verify the accuracy and effectiveness of the algorithms in this paper. The full-reference methods include SSIM, PSNR and other methods; the no-reference evaluation methods include CORNA, BRISQUE, DII-VINE, VI-IQA, MS-C and other methods. We used the LIVE dataset to test the effectiveness of the algorithm against synthetic distortion and also used LIVE In the Wild Image Quality Challenge Dataset [13] established by the LIVE lab at the University of Texas at Austin to verify the effectiveness of the algorithm against natural distortion.

Table 1. PLCC comparison on LIVE database

Method	WN	JPEG	JP2k	GBl	FF	aLL
PSNR	0.942	0.885	0.870	0.763	0.847	0.866
SSIM	0.964	0.946	0.939	0.907	0.941	0.913
CNN-IQA	0.980	0.960	0.934	0.934	0.892	0.953
CORNIA	0.976	0.955	0.943	0.969	0.906	0.942
DIICVNE	0.984	0.910	0.913	0.923	0.863	0.916
VI-IQA	0.984	0.964	0.941	0.937	0.89	0.960
MS-C	0.987	0.966	0.940	0.936	0.905	0.958
FRIQUEE	0.947	0.919	0.983	0.937	0.844	0.957
Proposed	0.991	0.960	0.954	0.932	0.912	0.962

Table 2. SROCC comparison on the LIVE database

Method	WN	JPEG	JP2k	GBl	FF	aLL
PSNR	0.942	0.885	0.870	0.763	0.847	0.866
SSIM	0.964	0.946	0.939	0.907	0.941	0.913
CNN-IQA	0.980	0.960	0.934	0.934	0.892	0.953
CORNIA	0.976	0.955	0.943	0.969	0.906	0.942
DIICVNE	0.984	0.910	0.913	0.923	0.863	0.916
VI-IQA	0.984	0.964	0.941	0.937	0.89	0.960
MS-C	0.987	0.966	0.940	0.936	0.905	0.958
FRIQUEE	0.947	0.919	0.983	0.937	0.844	0.957
Proposed	0.991	0.960	0.954	0.932	0.912	0.962

From Tables 1 and 2, it can be seen that the method proposed in this paper basically obtains good subjective and objective agreement for different distortion types, and the performance indexes for SROCC and PLCC are basically improved by more than 0.5%. In the case of comprehensive evaluation of the dataset, the self-supervised learning-based method proposed in this paper shows significant improvement compared with other full-reference and no-reference methods, with SROCC and PLCC improving by at least 0.9% or more.

To verify that the algorithm has good subjective consistency against natural distortion as well, we also perform experimental validation on the LIVE In the Wild Image Quality Challenge Database dataset. This dataset includes 1162 images. The distortion type of these images was not generated by computer simulation, but was verified using real images captured by various mobile cameras such as smartphones and tablets.

Table 3. SROCC and PLCC comparison on the LIVE in the wild image quality challenge dataset

Method	SROCC	PLCC
FRIQUEE	0.6820	0.7066
BRISQUE	0.6020	0.6100
DIQaM-NR	0.6060	0.6010
WaDIQaM-NR	0.6710	0.6800
DB-CNN	0.8510	0.8690
DeepBIQ	0.8894	0.9082
Proposed	0.8930	0.9190

From Table 3, both for the real distortion type, the SROOC and PLCC indexes of the proposed method in this paper have been improved compared with other algorithms. Based on the experimental results of the above two data sets, the generality and effectiveness of the algorithm in this paper can be fully illustrated.

5   Conclusion

In this paper, we proposed a method for no-reference image quality assessment based on self-supervised learning. We train our network model using only unlabeled distorted images using a contrast learning approach and improve local feature extraction by adding a attention mechanism. By comparing the improved algorithm model with several classical image quality evaluation methods, the SROCC and PLCC metrics are synthetically validated on the LIVE dataset with 0.960 and 0.957, and experiments on the LIVE In the Wild Image Quality Challenge Database dataset also demonstrate that the algorithm is also effective for realistic distorted images. The algorithm improves the deficiency due to the lack of marker data and can accurately evaluate various types and degrees of distorted images, and the prediction results are in good agreement with human visual perception.

References

1. Lin, K.Y., Wang, G.: Self-supervised deep multiple choice learning network for blind image quality assessment. In: BMVC, p. 70 (2018)
2. Kang, L., Ye, P., Li, Y., et al.: Convolutional neural networks for no-reference image quality assessment. In: Proceedings of the IEEE Conference on Computer Vision and Pattern Recognition, pp. 1733–1740 (2014)
3. Bosse, S., Maniry, D., Müller, K.R., et al.: Deep neural networks for no-reference and full-reference image quality assessment. IEEE Trans. Image Process. **27**(1), 206–219 (2017)
4. Lin, K.Y., Wang, G.: Hallucinated-IQA: no-reference image quality assessment via adversarial learning. In: Proceedings of the IEEE Conference on Computer Vision and Pattern Recognition, pp. 732–741 (2018)
5. Liu, X., Van De Weijer, J., Bagdanov, A.D.: Rankiqa: learning from rankings for no-reference image quality assessment. In: Proceedings of the IEEE International Conference on Computer Vision, pp. 1040–1049 (2017)
6. Ye, P., Kumar, J., Kang, L., et al.: Unsupervised feature learning framework for no-reference image quality assessment. In: 2012 IEEE Conference on Computer Vision and Pattern Recognition, pp. 1098–1105. IEEE (2012)
7. Khosla, P., Teterwak, P., Wang, C., et al.: Supervised contrastive learning. Adv. Neural. Inf. Process. Syst. **33**, 18661–18673 (2020)
8. Woo, S., Park, J., Lee, J.-Y., Kweon, I.S.: CBAM: Convolutional block attention module. In: Ferrari, V., Hebert, M., Sminchisescu, C., Weiss, Y. (eds.) ECCV 2018. LNCS, vol. 11211, pp. 3–19. Springer, Cham (2018). https://doi.org/10.1007/978-3-030-01234-2_1
9. Murray, N., Marchesotti, L., Perronnin, F. AVA: a large-scale database for aesthetic visual analysis. In: 2012 IEEE Conference on Computer Vision and Pattern Recognition, pp. 2408–2415. IEEE (2012)
10. Ponomarenko, N., Ieremeiev, O., Lukin, V., et al.: Color image database TID2013: peculiarities and preliminary results. In: European Workshop on Visual Information Processing (EUVIP), pp. 106–111. IEEE (2013)

11. Sheikh, H.R., Sabir, M.F., Bovik, A.C.: A statistical evaluation of recent full reference image quality assessment algorithms. IEEE Trans. Image Process. **15**(11), 3440–3451 (2006)
12. Zhang, L., Zhang, L., Mou, X., et al.: A comprehensive evaluation of full reference image quality assessment algorithms. In: 2012 19th IEEE International Conference on Image Processing, pp. 1477–1480. IEEE
13. Ghadiyaram, D., Bovik, A.C.: Massive online crowdsourced study of subjective and objective picture quality. IEEE Trans. Image Process. **25**(1), 372–387 (2015)

Research and Implementation of Remote Sensing Image Object Detection Algorithm Based on Improved ExtremeNet

Xia Zhang[✉], Lanxin Wang, Yulin Ren, and Zhili Mao

Xi'an University of Posts and Telecommunications, Xi'an 710121, China
zhangxia@xupt.edu.cn, 972962238@qq.com, 2749863169@qq.com,
mzl000@163.com

Abstract. According to the characteristics of rich edge information of remote sensing images, DOTA dataset is annotated with edge keypoints on the objects to adapt to the object detection algorithm ExtremeNet. Aiming at the problem of large differences in object scales in remote sensing images, the depthwise separable convolutions and CBAM attention mechanism are introduced to improve ExtremeNet. Experimental results show that the average detection accuracy of the improved ExtremeNet is better.

Keywords: Object detection · Remote sensing image · ExtremeNet

1 Introduction

Remote sensing images are widely used in military and civil fields because of their large monitoring range, wide dynamic range, and a huge amount of information.

Object detection in remote sensing images refers to searching, locating, and identifying interested objects in remote sensing images. It is a key technology in the field of remote sensing. However, it often faces several challenges, including the obvious changes in the visual appearance of objects caused by viewpoint variation, background clutter, illumination, shadow and so on [1].

There are great differences between remote sensing image object detection and natural image object detection. There is large background noise in remote sensing images, and the object scale changes greatly, which is easy to cause the missed detection of small objects. The remote sensing image is taken from the top view, so the detailed features of the objects are not obvious, but the edge information of the object is richer.

In recent years, with the continuous development of deep learning, many excellent deep learning networks that perform well in natural image object detection have been applied to remote sensing image object detection. Combined with the characteristics and difficulties of remote sensing image object detection, researchers put forward some effective methods to improve the performance of remote sensing image object detection. In order to solve the large shape difference between different classes of objects in remote sensing images, researchers have successively proposed introducing variability convolution, shape attention mechanism, and object detection methods based on key points

[2–4]. In order to solve the large scale difference in remote sensing images, the method of multi-scale feature fusion is introduced [5]. In order to adapt to the random direction of objects in remote sensing images, some deep learning networks based on rotating bounding boxes are proposed [6]. Various attention mechanisms are used to enhance object information and reduce the interference of complex background [7].

ExtremeNet is an anchor free object detection network proposed by UT Austin at the 2019 CVPR International Conference [8]. ExtremeNet extracts four extreme points (top-most, bottom-most, left-most, right-most) and one center point of an object using a keypoint estimation network and performs very well on COCO dataset. Because there is little coverage and occlusion between the objects in remote sensing images, it is easy to extract the edge points of objects under the top view. Therefore, ExtremeNet can deal with the remote sensing image object detection well, but there are still the following limitations: first, the image annotation information of the existing remote sensing image dataset does not contain the edge information of objects; second, it is difficult to detect small objects; third, ExtremeNet has a large number of parameters and long detection time.

Aiming at remote sensing image object detection, this paper carries out research work based on ExtremeNet. Because the annotation method of DOTA dataset is not suitable for the training of ExtremeNet based on keypoints, we have annotated the edge keypoints of four categories of objects in DOTA dataset. Because of the different scales of objects in remote sensing images and the missed detection of small objects, CBAM attention mechanism module is added to the feature extraction network of ExtremeNet to strengthen the feature learning ability of the middle channel and reduce the feature loss of small objects. Meanwhile, the residual block in ExtremeNet is optimized by depthwise separable convolution, which can effectively reduce the number of network parameters.

2   Related Work

2.1   ExtremeNet

The framework of ExtremeNet is shown in Fig. 1 [8]. The network uses hourglass-104 as the feature extraction network, and then predicts the heatmaps and offset maps of four edge extreme points (top-most, bottom-most, left-most, right-most), and a center heatmap of the object. There is no offset prediction for the center map. The network produces five $C \times H \times W$ heatmaps and four $2 \times H \times W$ offset maps, where C is the number of classes, W is the weight and H is the height of the output channel. First, a 3×3 window is used to collect all the peaks of four extreme points for each category. The peaks are greater than the peak selection threshold. Then, the brute enumeration is carried out. For each combination of extreme points, the geometric center of the composed bounding box is computed. If the geometric center has a high response in the predicted center map, the extreme points are committed as a valid detection.

2.2   DOTA Dataset

DOTA [9] is one of the most common remote sensing image object detection datasets. It contains 2806 remote sensing images. The pixel size of each image is in the range of

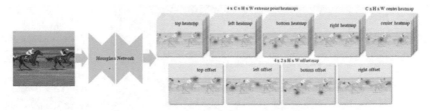

Fig. 1. Illustration of ExtremeNet [8]

800×800 to 4000×4000. It contains 15 object classes, with a total of 188282 object instances. It has the characteristics of different object scales, random directions, and large shape differences.

There are two types of bounding boxes in DOTA dataset. One is the horizontal bounding box which uses the horizontal rectangular box to frame the object. The other is the oriented bounding box, which can better enclose the objects and differentiate crowded objects from each other.

3 Proposed Method

3.1 Edge Keypoints Annotation on Objects

ExtremeNet uses the extreme points of the object to locate the object, which is very beneficial for remote sensing image object detection with rich edge information. There are two ways of annotation methods in DOTA dataset. The horizontal bounding box annotation method discards the rich edge information of objects in remote sensing images. The oriented bounding box annotation method only considers the edge information of some objects, such as vehicles, tennis courts, etc., because the bounding box is just the same as the edge of the object. The edge information of other objects such as planes, storage tanks, harbors, and so on is also ignored. Therefore, in order to make ExtremeNet use the edge information of objects for training, it is necessary to relabel DOTA dataset.

The original sizes of images in DOTA dataset are different, and some images have a huge amount of data. If these images are sent to the network directly, the network will have a long training time because of the huge amount of data. Most of the networks using DOTA dataset will cut the original image by a specific step before training. In this paper, the original image is cropped to 512×512. We select the cropped images which contain planes, storage tanks, harbors, and tennis courts for annotation. The reason for choosing the four categories is that the edge characteristics of these objects are different and have their own characteristics. A total of 7862 objects are annotated in this paper. Figure 2 shows the annotation results.

3.2 Improved ExtremeNet for Remote Sensing Image Object Detection

Based on ExtremeNet, CBAM attention mechanism module, lightweight residual module with depthwise separable convolution and center offset are introduced into the network. The overall structure of the improved network is shown in Fig. 3.

Fig. 2. The annotation results of four categories of objects

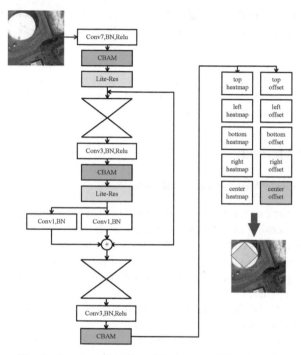

Fig. 3. Structure diagram of the improved ExtremeNet

The input image passes through the feature extraction network first. In the feature extraction network, the image passes through 7×7 convolution, CBAM attention mechanism module, and a lightweight residual module named as Lite-Res. Then it passes through two hourglass modules, and there is a 3×3 convolution and a CBAM module after each hourglass module. The idea of spanning connection is adopted between the two hourglass modules to fuse the shallow features with the middle features. After the feature extraction network, there is the detection module, which determines the position of the object through the four extreme point heatmaps, the center heatmap, and the offsets of each position.

Feature Extraction Network with CBAM Attention Mechanism The features of some small objects in remote sensing images will disappear with the increase of convolution operation, which will result in missed detection. Therefore, this paper introduces Convolutional Block Attention Module (CBAM) to optimize the feature extraction ability of the original ExtremeNet.

CBAM combines the channel attention module and the spatial attention module, as shown in Fig. 4 [10]. It can be widely applied to boost the representation power of CNNs.

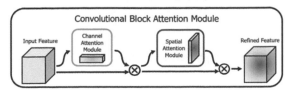

Fig. 4. The overview of CBAM [10]

The main work of the channel attention module includes: first step, two global pooling operations on the input feature map are carried out; second step, both features are forwarded to a shared network to produce a channel attention map; third step, the output feature vectors are merged by using element-wise summation, and the final channel attention feature is generated after sigmoid function; Finally, the original input feature map is element-wise multiplied by the output feature map of the channel attention module, and then the result is sent to the spatial attention module.

The main work of the spatial attention module includes: first step, the average-pooling and the max-pooling operations are applied to the channel-refined feature and an efficient feature descriptor is generated by concatenating the two feature maps; second step, a convolution operation and sigmoid function are used to generate spatial attention map; Finally, the spatial attention map is element-wise multiplied by the input of the spatial attention module to produce the final output of CBAM.

In this paper, CBAM attention mechanism is introduced into three positions in ExtremeNet to improve the feature extraction ability of the network for small objects.

Lightweight Residual Network with Depthwise Separable Convolution In order to reduce the parameters and speed up the reasoning speed of the network, a lightweight residual module named Lite-res is proposed in this paper, as shown in Fig. 5.

The lightweight residual network maintains the original jump connection structure and replaces the two 3 × 3 convolutions with two depthwise separable convolutions, which significantly reduces the number of network parameters. The parameters of the original residual network are about 1.4 times that of Lite-res.

Center Offset The original ExtremeNet has no center offset. It uses four extreme points to enumerate and calculate the highest score to determine the position of the center point. In some cases, it will cause missed detection. In order to reduce the occurrence of missed detection, the center offset is added to change the confirming way of the center point, as shown in Fig. 3.

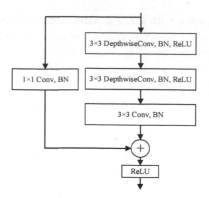

Fig. 5. Lightweight residual module Lite-res

Loss Function Because the center offset is introduced into the improved ExtremeNet, the loss function needs to be adjusted. ExtremeNet follows the basic structure and loss function of CornerNet [11]. In order to balance the positive and negative locations, a modified focal loss is used for training, as shown in Formula (1).

$$L_{\text{det}} = -\frac{1}{N} \sum_{i=1}^{H} \sum_{j=1}^{W} \begin{cases} (1 - \hat{Y}_{ij})^\alpha \log(\hat{Y}_{ij}) & \text{if } Y_{ij} = 1 \\ (1 - Y_{ij})^\beta (\hat{Y}_{ij})^\alpha \log(1 - \hat{Y}_{ij}) & \text{o.w} \end{cases} \tag{1}$$

$\hat{Y}_{ij}$ represents the score at location (i, j) in the predicted heatmaps, Y_{ij} represents the ground-truth heatmap augmented with the unnormalized Gaussians, and N represents the number of objects in the image. α and β are hyper-parameters.

The offset map is trained with the smooth L1 Loss [12] at ground-truth extreme point locations and center location. The offset loss function in the improved ExtremeNet is shown in Formula (2).

$$L_{off} = \frac{1}{5}(L_{off-top} + L_{off-left} + L_{off-bottom} + L_{off-right} + L_{off-center}) \tag{2}$$

The final loss function is shown in Formula (3):

$$L_{loss} = L_{\text{det}} + \omega_{off} L_{off} \tag{3}$$

ω_{off} is the weight coefficient of offset loss.

4   Experiment

4.1   Experimental Environment and Parameter Setting

In this paper, the platform used for training and testing the network is Intel(R) Core (TM) i5-6500 K @ 3.20 GHz, and the graphics card is NVIDIA Tesla P40. The whole experiment is carried out under the Ubuntu 16.04 operating system, by using Python 3.6 combined with PyTorch1.0.1 and CUDA10.0.

Our implementation is based on ExtremeNet. The loss function is focal loss function, and Adam algorithm is selected for network optimization. The number of iterations of the algorithm is fixed at 100,000, and the initial learning rate is 0.000125. Every 30,000 iterations, the learning rate is adjusted to one-tenth of the previous one.

We use average precisions (APs) at different IoUs and Aps for different object sizes as the main evaluation metrics.

4.2 Verification of DOTA Dataset with Edge Keypoints Annotation

In order to verify the validity of the annotated DOTA dataset, we train and test ExtremeNet using the dataset without edge keypoints annotation and the dataset with edge keypoints annotation respectively. AP_{50} is used as the evaluation metric. The experimental results are shown in Table 1.

Table 1. AP50 of four categories of objects (ExtremeNet)

Dataset	Plane	Storage tank	Harbor	Tennis court	mAP
AP_{50} on original dataset	56.4	58.2	50.4	54.4	54.8
AP_{50} on annotated dataset	80.6	65.1	52.4	74.3	68.1

Table 1 shows that the AP50 of ExtremeNet on annotated dataset is higher than the AP50 of ExtremeNet on original dataset. In particular, the AP50 of plane has been increased by 24.2%. However, for the harbor whose edge feature is not very obvious, the average precision is slightly improved. Figure 6 shows the examples of experimental results. It can be seen that the mask formed by extreme points can match the object as much as possible.

4.3 Object Detection Performance of Improved ExtremeNet

The experimental environment and parameter setting are the same as those in Sect. 4.1. In order to verify the impact of CBAM attention mechanism module on the network, a series of comparative experiments have been completed in this paper. Table 2 shows the experimental results. AT, AM, AB, and A mean adding CBAM modules at the top, middle, bottom, and all positions of the hourglass network respectively. The results show that the average precision of ExtremeNet-A is the best. Adding CBAM attention mechanism module in three positions can enhance the expression ability of object features.

In order to verify the effectiveness of the improved ExtremeNet in remote sensing image object detection, the following comparative experiments are completed in this paper. The experimental results are shown in Table 3. ExtremeNet with different improved modules is compared with the baseline experiment of the original ExtremeNet. A means adding CBAM attention mechanism module, L means adding Lite-res module, and C means adding a center offset.

It can be seen that the improved ExtremeNet obtains better APs than the original ExtremeNet on the annotated dataset. The Aps of ExtremeNet-ALC are the best. AP_{50}

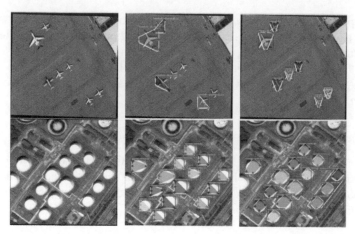

Fig. 6. Qualitative results comparison. First column: original remote sensing image. Second column: object detection results of ExtremeNet on original dataset. Third column: object detection results of ExtremeNet on annotated dataset.

Table 2. Improved ExtremeNet APs with CBAM in different locations

Method	AP	AP_{50}	AP_{75}	AP_{small}	AP_{mid}	AP_{large}
ExtremeNet-AT	59.0	68.8	65.9	29.7	53.0	67.9
ExtremeNet-AM	58.7	68.0	65.7	25.3	52.6	67.6
ExtremeNet-AB	58.9	67.9	65.0	26.6	52.5	67.8
ExtremeNet-A	59.4	69.3	66.3	30.6	53.3	68.0

Table 3. APs comparison of different networks

Method	AP	AP_{50}	AP_{75}	AP_{small}	AP_{mid}	AP_{large}
ExtremeNet	58.7	68.1	65.2	24.0	52.1	67.1
ExtremeNet-A	59.4	69.3	66.3	30.6	53.3	68.0
ExtremeNet-L	59.2	69.1	66.5	28.2	54.1	67.6
ExtremeNet-C	60.2	69.5	67.5	30.2	53.9	68.0
ExtremeNet-AL	60.5	70.4	67.0	34.5	54.2	68.5
ExtremeNet-ALC	61.7	71.6	67.8	35.4	56.2	69.8

has increased by 3.5%, and AP_{small} has increased by 11.4%, which proves that the method is effective and has an improvement in small object detection. Figure 7 shows the experimental results. Table 4 shows the parameters of different networks. It can be seen that when the three improvements are all introduced into ExtremeNet, the number of parameters is the least. The total number of parameters of ExtremeNet-ALC is 64.7% of the total number of parameters of the original ExtremeNet.

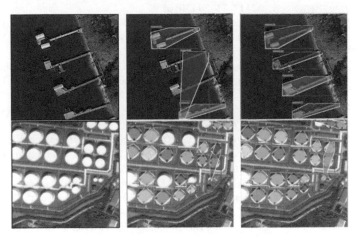

Fig. 7. Qualitative results comparison. First column: original remote sensing image. Second column: object detection results of ExtremeNet. Third column: object detection results of ExtremeNet-ALC.

Table 4. Parameters of different network

Method	Parameters
ExtremeNet	198,338,754
ExtremeNet-A	198,340,900
ExtremeNet-L	128,370,882
ExtremeNet-C	199,509,662
ExtremeNet-AL	127,200,646
ExtremeNet-ALC	128,371,474

5 Conclusion

In summary, we annotated the edge key points of four categories of objects in DOTA dataset and proposed an improved ExtremeNet which included CBAM attention mechanism module, lightweight residual module, and center offset. This method not only improves the APs, but also reduces the number of network parameters. In the future, we will annotate other categories of objects in DOTA dataset and carry out further research to improve the APs.

References

1. Cheng, G., Han, J.: A survey on object detection in optical remote sensing images. ISPRS J. Photogramm. Remote. Sens. **117**, 11–28 (2016)
2. Ren, Y., Zhu, C., Xiao, S.: Deformable faster R-CNN with aggregating multi-layer features for partially occluded object detection in optical remote sensing images. Remote Sens. **10**(9), 1470 (2018)
3. Li, C., Xu, C., Cui, Z., et al.: Feature-attentioned object detection in remote sensing imagery. In: 2019 IEEE International Conference on Image Processing (ICIP), pp. 3886–3890. IEEE, Taipei, China (2019)
4. Hwabc, D., Yue, Z., Zcabc, D., et al.: Oriented objects as pairs of middle lines. ISPRS J. Photogramm. Remote. Sens. **169**, 268–279 (2020)
5. Yao, Q., Hu, X., Lei, H.: Geospatial object detection in remote sensing images based on multi-scale convolutional neural networks. In: IGARSS 2019—2019 IEEE International Geoscience and Remote Sensing Symposium, pp. 1450–1453. IEEE, Yokohama, Japan (2019)
6. Cao, L., Zhang, X., Wang, Z., et al.: Multi angle rotation object detection for remote sensing image based on modified feature pyramid networks. Int. J. Remote Sens. **42**, 5253–5276 (2021)
7. Li, Q., Mou, L., Jiang, K., Liu, Q., et al.: Hierarchical region based convolution neural network for multiscale object detection in remote sensing images. In: IGARSS 2018—2018 IEEE International Geoscience and Remote Sensing Symposium, pp. 4355–4358. IEEE, Valencia, Spain (2018)
8. Zhou, X., Zhuo, J., Krahenbuhl, P.: Bottom-up object detection by grouping extreme and center points. In: 2019 IEEE/CVF Conference on Computer Vision and Pattern Recognition (CVPR), pp. 850–859. IEEE, Long Beach, CA, USA (2019)
9. Xia, G. S., Bai, X., Ding, J., et al.: DOTA: a large-scale dataset for object detection in aerial images. In: 2018 IEEE/CVF Conference on Computer Vision and Pattern Recognition, pp. 3974–3983. IEEE, Salt Lake City, USA (2018)
10. Woo, S., Park, J., Lee, J.-Y., Kweon, I.S.: CBAM: Convolutional block attention module. In: Ferrari, V., Hebert, M., Sminchisescu, C., Weiss, Y. (eds.) ECCV 2018. LNCS, vol. 11211, pp. 3–19. Springer, Cham (2018). https://doi.org/10.1007/978-3-030-01234-2_1
11. Zhou, X.Y., Wang, D.Q., Philipp, K.P., et al.: Objects as points. In: IEEE Conference on Computer Vision and Pattern Recognition, IEEE Computer Society, Washington (2019)
12. Girshick, R.: Fast R-CNN. In: IEEE Conference on Computer Vision and Pattern Recognition, pp. 1440–1448. IEEE Computer Society, Washington (2015)

A Review Focusing on Knowledge Graph Embedding Methods Exploiting External Information

Yuxuan Chen [ID] and Jingbin Wang[(✉)] [ID]

Fuzhou University, Fuzhou Fujian 350108, China
wjb@fzu.edu.cn

Abstract. Knowledge Graph (KG) is a structural way to represent knowledge. Many applications in the industry rely on KG, such as recommendation systems, relationship extraction, and question answering. However, most existing knowledge graphs are incomplete, so tackling KG completion becomes a crucial problem. Knowledge Graph Embedding (KGE) is an effective method for KG completion. Based on the literature published in recent years, we review existing KGE methods, including traditional approaches and approaches exploiting external information. Traditional methods only utilize triplet information and ignore the more informative external information. Therefore, our work mainly focuses on the methods that utilize external information, including textual description, relation paths, neighborhood information, entity types, and temporal information. Experimental results show that methods exploiting external information generally outperform traditional methods.

1 Introduction

Knowledge Graph (KG) is a structured way to store knowledge of the world. A KG could be regarded as a directed graph, where each vertex stands for an entity and each edge represents a relationship. A fact is represented by a triple (h, r, t), where h, r, t stands for head entity, relation, and tail entity, respectively. A series of triples shown above represent the facts existing in a KG. So far, KG has extensive industrial uses, such as Recommender Systems and Question Answering. The state-of-the-art KGs available currently include WikiData [1], FreeBase [2], Yago [3], DBPedia [4]. However, even the most advanced KG is still suffering from incompleteness. For example, in Freebase, the nationality of 75% of people is missing, and 99% of people have no ethnicity [5–7]. Consequently, knowledge graph completion has become an active research field. Learning the Knowledge Graph Embedding (KGE) is a productive way to complete the KG. Various ways that leverage machine learning or deep learning techniques have been proposed to learn the KGE.

Regarding the existing approaches of KGE, we can divide them into two categories. One is the models that solely use triples for learning, and the other is those that exploit triples and external information for learning. The external information here may be textual description, relation paths, entity types, temporal information, and neighborhood

© The Author(s), under exclusive license to Springer Nature Switzerland AG 2023
N. Xiong et al. (Eds.): ICNC-FSKD 2022, LNDECT 153, pp. 869–887, 2023.
https://doi.org/10.1007/978-3-031-20738-9_96

information. However, most existing methods belong to the first category. The predictions of those approaches may not be accurate enough because they only require the learned embedding to be compatible with every triplet in KG [8]. Approaches that exploit external information consider more helpful information. Thus, more accurate predictions can be provided.

The contribution of this paper lies in explaining common approaches of KGE and discussing the advantages of KGE methods exploiting external information.

Outline: Sect. 2 provides the definition of knowledge graph embedding and knowledge graph completion. Section 3 introduces different KGE methods in a taxonomic way. Section 4 discusses the advantages of approaches exploiting external information. Section 5 gives concluding remarks.

2   Definition of Knowledge Graph Embedding and Knowledge Graph Completion

This section provides a detailed explanation of Knowledge Graph Embedding (KGE) and Knowledge Graph Completion.

2.1   Knowledge Graph Embedding

KGE is a machine learning task to project relations and entities into low-dimensional space while preserving the semantic meanings [8].

In KGE, the following characteristics play an essential role:

1. Representation space: a low-dimensional space where relations and entities are mapped to the form of a vector [9].
2. Encoding model: a model which defines how relations and entities interreact in the representation space [9].
3. Scoring function: a measurement of the plausibility of a given triple to be a factual triple [9]. The scoring function is closely related to the encoding model, and to some extent, the encoding model is defined by the scoring function.

2.2   Knowledge Graph Completion

Most knowledge graphs are built semi-automatically or manually, and many implicit relationships and entities have yet to be discovered, so the incompleteness of KGs has become a common phenomenon [10].

Thus, knowledge graph completion is dedicated to alleviating the sparsity and incompleteness due to the missing entities and relations. We can divide the knowledge graph completion problem into two sub-problems:

1. Entity prediction: given the head entity and relation $(h, r, ?)$ or tail entity and relation $(?, r, t)$, predict the missing tail or head entity, respectively.
2. Relation prediction: given the head and tail entities, predict the missing relation $(h, ?, t)$.

Knowledge graph completion can be classified into open environment completion and closed environment completion [10]. Closed environment completion, also called static knowledge graph completion, predicts the missing entities and relations belonging to the original KG. In contrast, open environment completion serves relations and entities that may not exist in the original KG. The approaches we discuss only focus on static knowledge graph completion in this review.

3   Knowledge Graph Embedding Approaches

In this section, we will explain different KGE methods. In the first part, we review the traditional methods. These methods are subdivided into the translational distance model, tensor decomposition model, and neural network model. The second part describes the models exploiting external information and subdivides them by the types of external information (textual description, relation paths, neighborhood information, entity type, temporal information).

3.1   Traditional Approaches

We will outline the ideas of each model before going into detail. For translational distance models, the plausibility of a fact is calculated by the distance between the head entity translated by relation and the tail entity. Regarding tensor decomposition models, they treat KG as a three-way tensor and then decompose it into a series of vectors, which are the embeddings of relations and entities. By optimizing the scoring function through the training set, embeddings that capture the semantic similarities can be learned. Neural Network Models compute a semantic matching score by feeding relations and entities into deep neural networks [9].

Distance Model
Distance Model is the most classic model in knowledge graph embedding. Although the complexity is not high, these models are still powerful and perform well in knowledge graph completion tasks. Here, we will show three paradigms of the translational distance model.

The first one is TransE [11] proposed in 2013. The algorithm of TransE first projects relations and entities into the low-dimensional space and then applies stochastic gradient descent to the scoring function:

$$f_r(h, t) = -\|h + r - t\|. \tag{1}$$

The embeddings of the head and tail entities will be close to each other after finally adding relation embedding to the head entity.

However, even though TransE is powerful and performs well in link prediction, it still has drawbacks. Since the relation embeddings and entity embeddings are determined, N-N, N-1, 1-N relationships cannot be expressed.

In 2014, TransH [12] was proposed. Compared with TransE, TransH does the translation by first projecting the embeddings of entities into a relation-specific hyperplane, then adding the relation embedding to the projected head entity embedding. The scoring function is defined as:

$$f_r(h, t) = -\left\| \left(h - w_r^T h w_r\right) + d_r - \left(t - w_r^T t w_r\right) \right\|, \tag{2}$$

where w_r is the normal vector for projecting entities. One breakthrough of TransH is that it overcomes the flaws of TransE concerning the reflexive/one-to-many/many-to-one/many-to-many relations while inheriting its efficiency [12].

TransR [13] separates embeddings of relations and entities into two different representation spaces. For a triplet (h, r, t), TransR projects the embedding of h and t to the relation space by a transformation matrix $M_r \in R^{k \times d}$, where k is the dimension of entity embeddings, d is the dimension of relation embedding. Then embedding of the relation is added to the head entity for translation. The scoring function of TransR is:

$$f_r(h, t) = -\|h M_r + r - t M_r\|. \tag{3}$$

Regarding link prediction, TransR performs significantly better than TransE and TransH [13]. However, since a matrix needs to be learned regarding each relation, the space complexity is relatively large.

Tensor Decomposition Models

A three-way tensor can represent a knowledge graph. Each slice, an adjacency matrix, corresponds to one specific relation and indicates whether that relation exists between entities. The idea behind these models is to decompose the 3-way tensor into vectors or matrices. These vectors or matrices, which capture the latent representation of relations and entities, are learned by the defined scoring function and are supposed to predict the unseen facts in the graph.

RESCAL [14] is one of the most classic tensor decomposition models. RESCAL aims to factorize X_k to $AR_k A^T$, where A is an n × r matrix representing the entities and R_k is an asymmetric r × r matrix that models the interactions of the latent components in the k-th predicate [16]. Here, r is the dimensionality, and n is the number of entities. Each row of A corresponds to one entity embedding. RESCAL's scoring function is defined as:

$$f_r(h, t) = h^t M_r r, \tag{4}$$

which is in a bilinear form.

DistMult [15] is one simplified model of RESCAL. Compared with RESCAL, DistMult limits the representation matrix of a relation to a diagonal matrix. Consequently, the complexity of the model is dramatically decreased while improving the performance.

However, the models above cannot well model the asymmetric relations since their scoring functions have a property of commutativeness. In 2016, ComplEx [16] was proposed to address this issue. Similar to DistMult, the relation representation matrix is diagonalized in ComplEx, but the embedding space is extended from real numbers to complex numbers. By defining the scoring function:

$$f_r(s, r) = Re\left(e_s^T W \overline{e_o}\right), \tag{5}$$

ComplEx disables the commutativeness and successfully models the asymmetric relationships.

Neural Network Models
A neural network consists of multiple layers involving one input layer, one or many hidden layers, and an output layer. Inputs from one layer are sent to the next layer by first combining the weights and biases and then are activated by a non-linear function, where the weights and biases can be learned through training. In recent studies, neural networks have achieved excellent predictive performance [9].

ConvE [17] takes the idea of the convolutional neural network. It applies 2D convolution on embeddings. To compute a score for a triplet (h, r, t), it first reshapes the embeddings of h and r to two matrices respectively and concatenates them. The obtained "image" is sent to the convolutional layer, which will return a series of characteristics mapping matrices. Those matrices are reshaped to one vector and transposed to the dimension of the embeddings before processing. The final score is computed via a dot product between the resulting vector and the tail embedding. ConvE defines the scoring function as:

$$\psi_r(e_h, e_t) = f(vec(f([\overline{e_h}; \overline{r_r}] * \omega))W)e_t. \tag{6}$$

Here, $[\overline{e_h}; \overline{r_r}]$ means the concatenation of the reshaped head and relation embeddings, $*$ is the convolution operator, ω is the convolution filters, and W is the transformation matrix.

However, the extent of interactions captured by ConvE is not enough. A new approach, InteractE [18], was proposed to address this issue. On the basis of ConvE, InteractE proposed three novel ideas: feature permutation, different ways of reshaping, and circular convolution. Feature permutation aims to capture more interactions by generating different input orders. Permutations for embeddings e_s and e_r is denoted by:

$$P_t = \left[\left(e_s^1, e_r^1 \right); \ldots; \left(e_s^t, e_r^t \right) \right]. \tag{7}$$

There are three ways of reshaping mentioned in InteractE, which are stack, alternate, and chequer. Regarding stack, e_s and e_r are first transformed into two 2-dimensional matrices, and then one is stacked over another to get the result. Alternate also needs to transform e_s and e_r into 2-dimensional matrices, but each row of two matrices is stacked alternately. For chequer, it creates a matrix where the same embedding occupies no two adjacent cells. Circular convolution applied in InteractE enhances interactions compared with the convolution used in ConvE. Circular convolution on an input $I \in \mathbb{R}^{m \times n}$ with a filter $\omega \in \mathbb{R}^{k \times k}$ is defined as:

$$[I * \omega]_{p,q} = \sum_{i=-\lfloor k/2 \rfloor}^{\lfloor k/2 \rfloor} \sum_{j=-\lfloor k/2 \rfloor}^{\lfloor k/2 \rfloor} I_{[p-i]_m, [q-j]_n} \omega_{i,j}, \tag{8}$$

where $[x]_n$ denotes x modulo n and $\lfloor \cdot \rfloor$ denotes the floor function [18]. The scoring function of InteractE is defined as:

$$\psi_r(e_h, e_t) = g(vec(f(\phi(P_k) * \omega))W)e_t, \tag{9}$$

where ϕ is the reshaping function, $*$ denotes depth-wise circular convolution, g and f are sigmoid and ReLU, respectively.

3.2 Approaches Exploiting External Information

Besides triples, the knowledge base often contains a wealth of external information, including textual descriptions, relation paths, neighborhood information, entity types, and temporal information. Semantic information can be better captured if we can take full advantage of external information. In detail, we will introduce ways to leverage different external information for the rest of this section.

Textual Description
The majority of knowledge graphs involve concise and abundant descriptions of entities, which are ignored by the traditional methods. It is non-trivial to utilize textual descriptions since it can alleviate the structure sparseness of knowledge graphs and low performance on one-to-many/many-to-one/many-to-many relations [19]. One challenge of knowledge graph embedding utilizing textual description is to embed both unstructured textual information and structured knowledge in the same space [9]. Here, we will introduce two approaches, DKRL [20] and StAR [21].

The idea of DKRL is to enhance TransE to consider entity descriptions further. Regarding entity e, DKRL learns two representations: structure-based representation e_s, and description-based representation e_d. The structure-based representation, which captures structural information in KG, is the same as the representation learned from the traditional approaches. The description-based representation is derived from external textual descriptions of entities. For encoding textual descriptions, DKRL mentions two methods, which are continuous Bag-of-words Encoder and Convolutional Neural Network (CNN) Encoder [20]. The energy function defined by DKRL is:

$$E = E_S + E_D, \tag{10}$$

E_S is the structure-based energy function, which shares the same equation with TransE, and E_D is the description-based energy function. To make the learning process of E_D to be compatible with E_S, E_D is defined as follows:

$$E_D = E_{DD} + E_{DS} + E_{SD}, \tag{11}$$

$$E_{DD} = \|h_d + r - t_d\|, \tag{12}$$

$$E_{DS} = \|h_d + r - t_s\|, \tag{13}$$

$$E_{SD} = \|h_s + r - t_d\|. \tag{14}$$

StAR consists of a structure-aware triple encoder and a structure-augmented scoring module. Given the text of a triplet $\left(x^{(h)}, x^{(r)}, x^{(t)}\right)$, StAR first feeds $\left(x^{(h)}, x^{(r)}\right)$, and $x^{(t)}$ to the transformer encoder, obtaining the representation across head and relation $\boldsymbol{u}$, and the representation of tail v, respectively. These two representations are then passed to the structure-augmented scoring module. 2 parallel scoring strategies are applied in the structure-augmented scoring module to learn both contextualized and structured knowledge. The scoring strategies are named deterministic representation learning and spatial structure learning. For the former strategy, an interactive concatenation is applied to u and v. The result is written as:

$$c = [u : u \times v; u - v; v], \tag{15}$$

where c represents the semantic relationship between two parts of a triplet. c is then sent to a Multilayer Perceptron (MLP) classifier which could produce the score of the input triple, denoted s^c. The latter strategy scores a triple to augment structured knowledge in the encoder, where the scoring function for this strategy is defined as:

$$s^d = - \|u - v\|. \tag{16}$$

By leveraging these two scores in training, contextualized and structured knowledge can be well modeled.

Relation Paths

Relation paths contain rich semantic information. Considering relation paths in models can better utilize global information, which is beneficial for knowledge graph embedding. In this subsection we are going to introduce three models: PTransE [22], RTransE [23], TransE-Compose [24].

PTransE models relation paths based on TransE. Given a triple (h, r, t), the energy function is defined as:

$$G(h, r, t) = E(h, r, t) + E(h, P, t) \tag{17}$$

where

$$P = P(h, t) = \{p_1, p_2, \ldots, p_n\}, \tag{18}$$

and p_i is the path from h to t $(r_1, \ldots, r_l)$. Here, $E(h, r, t)$ is exactly the same with the energy function of TransE, while $E(h, P, t)$, which is the part considering relation paths, is defined as:

$$E(h, P, t) = \frac{1}{Z} \sum_{p \in P(h,t)} R(p|h, t)E(h, p, t), \tag{19}$$

where $R(p|h, t)$ indicates the reliability of the path p, and it is calculated through a network-based resource allocation mechanism [25], Z is the normalization factor:

$$Z = \sum_{p \in P(h,t)} R(p|h, t), \tag{20}$$

and $E(h, p, t)$ has the same form of the energy function of TransE. Regarding the representation of path $p = (r_1, \ldots, r_l)$, PTransE proposes three ways to model it:

$$\text{Addition:} \, p = r_1 + \cdots + r_l, \tag{21}$$

$$\text{Multiplication: } p = r_1 \circ \ldots \circ r_l, \tag{22}$$

$$\text{RNN: } c_i = f\left(W[c_{i-1}; r_i]\right), \tag{23}$$

where f is the non-linear function, $[\bullet]$ is the concatenation operation, W is the composition and c_i is the vector representation of aggregated relation.

RTransE improves TransE in the training stage to accurately reproduce composition, such that given a path (h, l, t), (t, l', t'), we should have $h + l + l' \approx t'$. RTransE claims that the loss function of TransE:

$$\sum_{(h,l,t)\in S,(h',l,t')\in S_{(h,l,t)}} \left[\gamma + d(h, l, t) - d(h', l, t')\right]_+, \tag{24}$$

where S is the set of facts, $S_{(h,l,t)}$ is the corrupted facts set, $[x]_+ = \max(x, 0)$ is the positive part of x, γ is the margin hyperparameter and $d(h, l, t)$ is defined as:

$$d(h, l, t) = \|h + l - t\|, \tag{25}$$

can train the tail embedding being the nearest neighbor of the translated head, but it cannot make sure the distance between the tail and translated head is small, which is significant for composition. Thus, it adds two regularization terms to account for the compositionality:

$$\lambda \sum_{(h,l,t)\in S} d(h, l, t)^2, \tag{26}$$

$$\alpha \sum_{(h,\{l_1,l_2\},t)\in S,} N_{l\rightarrow\{l_1,l_2\}} d(h, \{l_1, l_2\}, t)^2. \tag{27}$$

The first term is applied to the knowledge of the original KG, whereas the latter is to the relational path. Here, $N_{l\rightarrow\{l_1,l_2\}}$ indicates the number of paths that contain relations $\{l_1, l_2\}$. More weight could be given to the reliable paths by applying this criterion.

TransE-Compose proposes a novel training objective that improves the composable models' abilities to answer queries. During the training stage, it does not only consider triples that contain entities connected by relation but also those connected by relation paths. In TransE-Compose, let us take TransE as the composable model being improved, given an entity pair (h, t), (h, p, t) will be used as the external triple for training, where p is the relation path $(r_1, \ldots, r_l)$, which will translate h to t. The representation of p is $r_1 + \ldots + r_l$. Similarly, it can enhance RESCAL by feeding the external triple (h, p, t) in training, while in this case, the representation of p becomes $M_1 \circ \ldots \circ M_l$, where M_i represents i-th relation in the path.

Neighborhood Information

The local neighborhood surrounding an entity usually provides plenty of valuable and inherent information. However, traditional models treat the triples independently without leveraging them. Models exploiting neighborhood information will be introduced for the rest of this subsection.

Models leveraging neighborhood information are generally based on GNN (Graph Neural Network). In the task of knowledge graph completion, the GNN-based models act as an encoder that produces entity representation and cooperates with a decoder (score function) for link prediction.

In the literature [26], the proposed model: R-GCN, which extends GCN (Graph Convolutional Network) to model relational data, acts as an encoder to learn entity representations. The equation of one layer of the network is:

$$h_{(i)}^{l+1} = \sigma \left(\sum_{r \in R} \sum_{j \in N_i^r} \frac{1}{c_{i,r}} W_r^{(l)} h_j^{(l)} + W_0^{(l)} h_i^{(l)} \right), \tag{28}$$

where N_i^r denotes the set of nodes that are direct neighbors of node i under relation r, $c_{i,r}$ is the normalization constant, h is the entity representation and W is the weighted matrix. The entity representations are then sent to DistMult, which acts as the decoder to predict the labeled edge.

RGHAT [27] is an encoder that views an entity's neighborhood as a two-layer hierarchy. The first layer of the hierarchy contains different relations associated with the entity, while the second layer contains different sets of neighborhood entities. Each set of the neighborhood entities in the second layer corresponds to one specific relation that appeared in the first layer. RGHAT utilizes a two-level mechanism to aggregate neighborhood information. The first level is named relation-level. It corresponds to the first layer in the hierarchy, giving different weights for different relations. While the second level, the entity level, is dedicated to the second layer, emphasizing the weight of different neighboring entities under the same relation. The final output of this encoder is computed by the information aggregator, which takes both entity-level and relation-level attention into consideration.

Entity Types
Knowledge graph usually contains rich information on entity types. Accurately leveraging the information of entity types can make knowledge graph embedding better at capturing semantic information and thus produce better results. We are going to introduce three approaches, TKRL [28], TypeComplex [35], and AutoETER [29] that exploit entity type.

In TKRL, hierarchical type c implies one entity's different roles in different scenarios. Two different encoders: Recursive Hierarchy Encoder (RHE) and Weighted Hierarchy Encoder (WHE) were proposed to encode c into representation learning. The projection matrix derived from RHE is:

$$M_c = \prod_{i=1}^{m} M_{c^{(i)}}, \tag{29}$$

where m is the number of layers for type c in the hierarchical structure, and $M_{c^{(i)}}$ stands for the projection matrix of i-th sub-type $c^{(i)}$. WHE produces the result by summing up the projection matrices of sub-types with different weights:

$$M_c = \sum_{i=1}^{m} \beta_i M_{c^{(i)}}, \tag{30}$$

where β_i is the corresponding weight of $c^{(i)}$. Additionally, the relation-specific information constrains the final type projection matrix applied to entities, which is shown in Eq. (31):

$$M_{re} = \frac{\sum_{i=1}^{n} a_i M_{c_i}}{\sum_{i=1}^{n} a_i}, a_i = \begin{cases} 1, c_i \in C_{re} \\ 0, c_i \notin C_{re} \end{cases}, \quad (31)$$

where n is the number of types entity e possesses, c_i is the i-th type e belongs to, C_{re} represents the set of types of e in relation r given by the relation-specific type information, M_{c_i} is the projection matrix of c_i. This transformation matrix projects the head and tail entities into the corresponding type spaces. The energy function of TKRL is defined as:

$$E(h, r, t) = \|M_{rh}h + r - M_{rt}t\|. \quad (32)$$

TypeComplex modifies ComplEx so as to consider type information. TypeComplex combines two terms $C_v(s, r)$, $C_w(o, r)$ on ComplEx's scoring function $f_r(s, r, o)$ to further measure type compatibility. The scoring function is defined in Eq. (33):

$$f_r'(s, r, o) = f_r(s, r, o) C_v(s, r) C_w(o, r), \quad (33)$$

$$C_v(s, r) = \sigma(v_r \cdot u_s), \quad (34)$$

$$C_w(o, r) = \sigma(w_r \cdot u_o). \quad (35)$$

Here, v_r, w_r are vectors associated with relation to encode type for the subject entity, and object entity, respectively. Whereas u_s and u_o are vectors directly related to the subject and object entities to encode type.

AutoETER utilizes two encoders for modeling KG with entity types: an entity-specific triple encoder and a type-specific triple encoder. Entity-specific triple encoder defines an energy function:

$$E_1(h, r, t) = \left\| \left(h - h^T w_r h \right) \circ r - \left(t - t^T w_r t \right) \right\| \quad (36)$$

to learn entity-specific embeddings in complex space with a hyper-projection strategy. The type-specific triple encoder uses a relation-aware projection mechanism to learn the type-specific embeddings, aiming to make the type embedding of the head entity as close as possible to the translated tail type embedding, where the corresponding energy function is:

$$E_2(h, r, t) = \|M_r y_h + y_r - M_r y_t\|. \quad (37)$$

In addition, AutoETER claims that type embeddings of head entities (or tail entities) are supposed to be similar if they share the same relations. Thus, it defines an equation:

$$E_3((h_1, r, t_1), (h_2, r, t_2)) = \frac{1}{2}(\|M_r y_{h1} + y_r - M_r y_{t1} + M_r y_{h2} + y_r - M_r y_{t2}\|), \quad (38)$$

which is expected to have low value. The training objective of AutoETER is to minimize the value of the three Eqs. (36–38) mentioned above.

Temporal Information

Temporal information is of great significance to the knowledge graph since most facts in a knowledge graph are only valid within a specific interval. Thus, it will not be accurate enough to consider the static knowledge graph model only. This part discusses the model of learning knowledge graph embedding using temporal information.

HyTE [30] combines temporal information in entity-relationship space by associating a hyperplane with the corresponding timestamp. For each timestamp in KG, there would be a normal vector w_t representing it. Triples at that timestamp would be projected to the corresponding hyperplane. The equations of the projected representation are:

$$P_\tau(e_h) = e_h - \left(w_\tau^T e_h\right) w_\tau, \tag{39}$$

$$P_\tau(e_t) = e_t - \left(w_\tau^T e_t\right) w_\tau, \tag{40}$$

$$P_\tau(e_r) = e_r - \left(w_\tau^T e_r\right) w_\tau. \tag{41}$$

The scoring function is similar to TransE:

$$f_t(h, r, t) = \|P_\tau(e_h) + P_\tau(e_r) - P_\tau(e_t)\|. \tag{42}$$

TA-DistMult [31] exploits temporal information by utilizing a recurrent neural network, which could learn the time-aware representations of relation types. These representations learned are used in conjunction with the factorization method DistMult for link prediction. During training, predicate sequence, which contains relation type and possibly temporal information, is first extracted from the fact, then mapped to the embedding space via a linear layer, and finally sent to the recurrent neural network: LSTM (long short-term memory) to obtain its embedding $e_{p_{seq}}$. This predicate sequence embedding can now be used in conjunction with subject and object embeddings in DistMult. The final scoring function is:

$$f\left(s, p_{seq}, o\right) = (e_s \circ e_o) e_{p_{seq}}^T. \tag{43}$$

CyGNet [32] investigates the phenomenon of recurrent temporal facts and aims to infer future facts by exploiting facts in history in the temporal knowledge graph. CyGNet is based on a time-aware copy-generation mechanism, which possesses two modes of inference to make predictions, named copy mode and generation mode. Copy mode only considers historical entities, whereas the whole entity vocabulary does the generation mode. Take NBA as an example, in copy mode, the 2024 NBA champion is predicted by only considering the previous champion teams, whereas all NBA teams would be considered in generation mode. By combining the predictions of two modes, better predictions can be made.

4   Discussion of Pros for Models Exploiting External Information

This section will discuss the advantages of models exploiting external information. The advantages discussed here might be lower time complexity, stronger expressivity, novel ways to solve problems, and better performance than other models. Again, we will classify the models by different categories of external information and discuss the merits of models under each category.

4.1   Textual Description

Traditional models that solely use triplets are sensitive to the structure sparseness of KG. However, models exploiting textual descriptions are relatively robust to this problem since those enriching descriptions can expand KG's semantic structure.

In DKRL, the representation of textual descriptions is learned through an encoder. The existing continuous Bag-of-words encoder falls the shortage of ignoring word orders since it produces the embedding of textual description by simply summing up all of the embeddings of keywords. However, to alleviate the issue, DKRL proposes a CNN encoder, which takes the word orders into consideration and is also more robust towards the noise. In the experiment of DKRL, dataset FB15K [11] was adopted. The optimal learning rate, margin, embedding dimension, and convolutional size of CNN encoder were 0.001, 1.0, 100, and 2, respectively [20]. As shown in Table 1, a union model of CNN and TransE consistently outperforms TransE on entity prediction.

Table 1. Evaluation results of DKRL and TransE on entity prediction taken from [20]

	Hits@10 (Raw)	Hits@10 (Filter)
TransE	48.5	66.1
DKRL (CBOW)	38.5	51.8
DKRL (CNN)	44.3	57.6
DKRL (CNN) + TransE	**49.6**	**67.4**

Traditional models that learn structural information only, such as TransE, are hardly generalizable to elements that are not visited during training. The textual encoding approach KG-BERT alleviates the problems above nicely. However, it has high computational complexity in calculating contextualized representation, and structural information is not taken into account as well. StAR reduces the complexity of calculating contextualized representation by partitioning triple into a concatenation of head and relation versus tail. Meanwhile, structural information is also taken into account through its structure-augmented scoring module.

4.2   Relation Paths

Relation paths contain rich semantic information for inference. Models exploiting relation paths have a better ability to avoid confusing inference, leading to more accurate predictions.

While maintaining the same model parameter size ($N_e K + N_{rK}$), PTransE achieves considerably better performance than TransE [22]. Taking the relation paths into consideration makes it easy for PTransE to infer confusing facts in TransE. For example, Donald Trump and Barack Obama were presidents of the United States, so their embeddings are pretty similar in TransE. Therefore, predicting the spouse of Donald Trump becomes confusing. PTransE addresses this issue by exploiting the relation paths between Donald Trump and Melania Trump.

RTransE and TransE-Compose all contribute to the reduction of cascading error. Even though reducing cascading error aims to improve path query answering, it can naturally lead to more accurate prediction for a single edge, thus beneficial for knowledge graph completion. For instance, consider the Horn clause: sister(x, y) $\land$ father(y, z) $\Rightarrow$ aunt(x, z), which means that if x is the sister of y and y is the father of z, then x is the aunt of z. The Horn clause body indicates a path from x to z. If the path is well modeled, it would be spontaneous to infer the head of the horn clause [24].

In the experiment, both PTransE and RTransE set embeddings dimension to 100, the learning rate to 0.001, and limited the number of epochs over training to 500 [22, 23]. Table 2 presents a comparison of the performance between RTransE and PTransE regarding entity prediction on FB15k. PTransE outperforms RTransE under the metric of filtered Hits@10, whereas RTransE outmatches PTransE under the evaluation of filtered mean rank.

Table 2. Experimental results of RTransE and PTransE. Results of PTransE are taken from [22]. Results of RTransE are taken from [23]

	Mean rank	Hits@10
RTransE	**49.5**	76.2
PTransE (ADD, 2-step)	54	83.4
PTransE (MUL, 2-step)	67	77.7
PTransE (RNN, 2-step)	92	82.2
PTransE (ADD, 3-step)	58	**84.6**

4.3 Neighborhood Information

Neighborhood information greatly supports knowledge graph completion because they provide plenty of inherent information. However, the importance of the information afforded by different neighbors varies. Thus, it is of great significance for models that utilize neighborhood information to perceive different weights for different neighbors.

The significance of a neighborhood entity is related to the relation to which it is connected. For example, a neighborhood entity connected by the relation "lover" is supposed to provide more information than a neighborhood entity connected by the relation "enemy" since a person is intuitively similar to the lover but not to the enemy.

R-GCN improves GCN by assigning different weights to different relations and thus being able to perceive different relations during aggregation.

RGHAT raises the attention mechanism to the triple level. Compared with R-GCN, the hierarchical attention mechanism adopted considers both relation-level attention and entity-level attention, where triple-level attention is the combination of these two attentions. The hierarchical attention mechanism makes the weights of neighboring triples with the same relations can be computed collectively, making RGHAT perform more stably and more consistent with human intuition. In the experiment, a two-layer RGHAT was adopted to train the embeddings, and the dimension of which was set to 100. The learning rate was set as 0.005. In the test phase, the head and tail entities were replaced in turn with all entities in KG for every triple in the test set [27]. Comparatively experimental results on FB15k are listed in Table 3, indicating that RGHAT outperforms both DistMult and R-GCN.

Table 3. Evaluation results of R-GCN and RGHAT taken from [27]

	MRR	Hits @1	Hits @3	Hits @10
DistMult	0.789	–	–	0.893
R-GCN + DistMult	0.696	0.601	0.760	0.842
RGHAT	**0.812**	**0.760**	**0.843**	**0.898**

4.4 Entity Types

Entity types contain rich semantic information. Models using them generally have intense expressivity because they allow entities to have different representations in different contexts.

The energy function of TKRL [Eq. (32)] indicates that each entity can have different representations when that entity is associated with different relations. The transformation matrix M_{re}, which plays a crucial role here, results from the hierarchical type encoder. In addition, due to utilizing entity types, entities with the same type will tend to have similar representations, leading to more errors in entity prediction [20]. A negative sampling approach during training, called Soft Type Constraint (STC) was proposed to alleviate this issue. STC increases the probabilities of selecting entities that have the same types constrained by relation-specific type information and consequently enlarges the distances between entities with the same type. As shown in Table 4, regarding entity prediction on FB15k, TKRL trained with STC performs better than the counterpart without STC and TransR under the metric Hits@10, indicating the balance between diversity and similarity contributes to better performances.

Observing that entity prediction failure often accompanies an incompatible type required by the relation, TypeComplex improves ComplEx by combining two type-compatibility functions with the scoring function. Without explicit supervision from a type catalog, TypeComplex obtains up to 7% MRR gains over base models [35]. Table 5 shows the corresponding experimental results on FB15K.

Table 4. Evaluation results of TKRL on entity prediction taken from [28]

	Hits@10 (Raw)	Hits@10 (Filter)
TransR	47.2	67.2
TKRL (RHE)	49.2	69.4
TKRL (WHE)	49.2	69.6
TKRL (RHE + STC)	**50.4**	73.1
TKRL (WHE + STC)	50.3	**73.4**

Table 5. Evaluation results of TypeComplex on entity prediction taken from [35]

	MRR	Hits@1	Hits@10
ComplEx	70.50	61.00	86.09
TypeComplex	**75.44**	**66.32**	**88.51**

AutoETER proposes both entity-specific triple encoder and type-specific triple encoder to learn the KG embeddings with entity types. The entity-specific triple encoder in AutoETER embeds entities and relations in complex space, regarding relation as a rotation similar to RotatE [33], which enables the ability to model symmetry and inversion relation patterns, meanwhile projecting entities into a hyperplane associated with a specific relation to further model the complex N-N, N-1, 1-N relations. At the same time, the type-specific triple encoder can also model those relation patterns by utilizing a relation-aware projection mechanism. In the experiment of AutoETER, the learning rate was set as 0.0001, the batch size was 1024, the dimension of the entity and relation embeddings in entity-specific triples was set as 1000, and the dimension of the type and relation embeddings in type-specific triples was set to 200 [29]. Table 6 compares the performance regarding link prediction between AutoETER and TKRL on FB15K, indicating AutoETER is better at leveraging the explicit types.

Table 6. Evaluation results of AutoETER and TKRL taken from [29]

	MR	MRR	Hits@10
TKRL	68	–	0.694
TypeComplex	–	0.753	0.869
AutoETER	**33**	**0.799**	**0.896**

4.5 Temporal Information

Most facts in the world are only valid during their time scopes, which are ignored by the traditional methods. Clearly, considering these time scopes could help yield better knowledge graph embeddings. Apart from that, embedding temporal information makes it possible to make future inferences from history.

The sparsity and heterogeneity of temporal expressions usually make representation learning for temporal information a challenging task. However, TA-DistMult shows outstanding robustness towards these challenges. The LSTM network leveraged by TA-DistMult can utilize digits and modifiers such as "since", thus facilitating the transfer of information across similar timestamps. Apart from that, LSTM can use triples both with and without temporal information in the training stage, indicating its robustness to the sparsity of temporal information.

The author of CyGNet observed that many facts repeatedly occurred throughout history, such as the Olympics every four years, financial crises every 7–10 years [34], and so on. The above phenomenon indicates that historical facts are meaningful for inferring future facts. Nevertheless, the temporal KGE model for perceiving such evolution patterns was not well studied. Thus, CyGNet was proposed. Time-aware copy-generation mechanism proposed in CyGNet combines two modes of copy and generation to make the inference, in which the copy mode predicates future events by historical facts, taking the evolution pattern into account. In the experiment of CyGNet, embedding dimension, learning rate, and batch size was set to 200, 0.001, and 1024, respectively. The training epoch was limited to 30 [32]. Evaluation results on link prediction are listed in Table 7, where CyGNet consistently performed better than TA-DistMult on dataset YAGO.

Table 7. Evaluation results of TA-DistMult and CyGNet taken from [32]

	MRR	Hits@3	Hits@10
DistMult	59.47	60.91	65.26
TA-DistMult	61.72	65.32	67.19
CyGNet	**63.47**	**65.71**	**68.95**

5 Conclusion

This review summarizes the models for Knowledge Graph Completion and divides them into two categories: traditional models utilizing triples alone and models exploiting external information. Since external information is very informative, leveraging them in training is an excellent way to improve the performance of KGE. However, external information is not as highly regarded as traditional models. So, we focus on models of external information and discuss their corresponding advantages. In general, models exploiting external information outperform traditional models under various metrics. In the future, we will continue to study models exploiting other types of external information, such as visual information and uncertainty of relations, etc.

References

1. Vrandečić, D., Krötzsch, M.: Wikidata: a free collaborative knowledgebase. Commun. ACM **57**(10), 78–85 (2014)
2. Bollacker, K., Evans, C., Paritosh, P., Sturge, T., Taylor, J.: Freebase: a collaboratively created graph database for structuring human knowledge. In: Proceedings of the 2008 ACM SIGMOD International Conference on Management of Data, pp. 1247–1250. Association for Computing Machinery, New York, NY, USA (2008)
3. Suchanek, F.M., Kasneci, G., Weikum, G.: Yago: a core of semantic knowledge. In: Proceedings of the 16th International Conference on World Wide Web, pp. 697–706. Association for Computing Machinery, New York, NY, USA (2007)
4. Auer, S., Bizer, C., Kobilarov, G., Lehmann, J., Cyganiak, R., Ives, Z.: DBpedia: a nucleus for a web of open data. In: Aberer, K., et al. (eds.) ASWC/ISWC -2007. LNCS, vol. 4825, pp. 722–735. Springer, Heidelberg (2007). https://doi.org/10.1007/978-3-540-76298-0_52
5. Dong, X., Gabrilovich, E., Heitz, G., Horn, W., Lao, N., Murphy, K., Strohmann, T., Sun, S., Zhang, W.: Knowledge vault: a web-scale approach to probabilistic knowledge fusion. In: Proceedings of the 20th ACM SIGKDD International Conference on Knowledge Discovery and Data Mining, pp. 601–610. Association for Computing Machinery, New York, NY, USA (2014)
6. West, R., Gabrilovich, E., Murphy, K., Sun, S., Gupta, R., Lin, D.: Knowledge base completion via search-based question answering. In: Proceedings of the 23rd International Conference on World Wide Web, pp. 515–526. Association for Computing Machinery, New York, NY, USA (2014)
7. Rossi, A., Barbosa, D., Firmani, D., Matinata, A., Merialdo, P.: Knowledge graph embedding for link prediction: a comparative analysis. ACM Trans. Knowl. Discov. **15**(2), 1–49 (2021)
8. Wang, Q., Mao, Z., Wang, B., Guo, L.: Knowledge graph embedding: a survey of approaches and applications. IEEE Trans. Knowl. Data Eng. **29**(12), 2724–2743 (2017)
9. Ji, S., Pan, S., Cambria, E., Marttinen, P., Philip, S.Y.: A survey on knowledge graphs: representation, acquisition, and applications. IEEE Trans. Neural Netw. Learn. Syst. **33**(2), 494–514 (2021)
10. Chen, Z., Wang, Y., Zhao, B., Cheng, J., Zhao, X., Duan, Z.: Knowledge graph completion: a review. IEEE Access **8**, 192435–192456 (2020)
11. Bordes, A., Usunier, N., Garcia-Duran, A., Weston, J., Yakhnenko, O.: Translating embeddings for modeling multi-relational data. In: Proceedings of the 26th International Conference on Neural Information Processing Systems, pp. 2787–2795. Curran Associates Inc., Red Hook, NY, USA (2013)
12. Wang, Z., Zhang, J., Feng, J., Chen, Z.: Knowledge graph embedding by translating on hyperplanes. In: Proceedings of the Twenty-Eighth AAAI Conference on Artificial Intelligence (AAAI'14), pp. 1112–1119. AAAI Press, Palo Alto, California, USA (2014)
13. Lin, Y., Liu, Z., Sun, M., Liu, Y., Zhu, X.: Learning entity and relation embeddings for knowledge graph completion. In: Proceedings of the Twenty-Ninth AAAI Conference on Artificial Intelligence (AAAI'15), pp. 2181–2187. AAAI Press, Palo Alto, California, USA (2015)
14. Nickel, M., Tresp, V., Kriegel, H.-P.: A three-way model for collective learning on multi-relational data. In: Proceedings of the 28th International Conference on Machine Learning, pp. 809–816. Omnipress, Madison, WI, USA (2011)
15. Yang, B., Yih, W.T., He, X., Gao, J., Deng, L.: Embedding entities and relations for learning and inference in knowledge bases. arXiv:1412.6575 (2014)
16. Trouillon, T., Welbl, J., Riedel, S., Gaussier, É., Bouchard, G.: Complex embeddings for simple link prediction. In: Proceedings of the 33rd International Conference on International Conference on Machine Learning, pp. 2071–2080. JMLR.org (2016)

17. Dettmers, T., Minervini, P., Stenetorp, P., Riedel, S.: Convolutional 2D knowledge graph embeddings. In: Proceedings of the Thirty-Second AAAI Conference on Artificial Intelligence and Thirtieth Innovative Applications of Artificial Intelligence Conference and Eighth AAAI Symposium on Educational Advances in Artificial Intelligence (AAAI'18/IAAI'18/EAAI'18), Article 221, pp. 1811–1818. AAAI Press, Palo Alto, California, USA (2018)

18. Vashishth, S., Sanyal, S., Nitin, V., Agrawal, N., Talukdar, P.: InteractE: Improving convolution-based knowledge graph embeddings by increasing feature interactions. In: Proceedings of the AAAI Conference on Artificial Intelligence, vol. 34, no. 03, pp. 3009–3016. AAAI Press, Palo Alto, California USA (2020)

19. Wang, Z., Li, J., Liu, Z., Tang, J.: Text-enhanced representation learning for knowledge graph. In: Proceedings of the Twenty-Fifth International Joint Conference on Artificial Intelligence (IJCAI'16), pp. 1293–1299. AAAI Press, Palo Alto, California, USA (2016)

20. Xie, R., Liu, Z., Jia, J., Luan, H., Sun, M.: Representation learning of knowledge graphs with entity descriptions. In: Proceedings of the Thirtieth AAAI Conference on Artificial Intelligence (AAAI'16), pp. 2659–2665. AAAI Press, Palo Alto, California, USA (2016)

21. Wang, B., Shen, T., Long, G., Zhou, T., Wang, Y., Chang, Y.: Structure-augmented text representation learning for efficient knowledge graph completion. In: Proceedings of the Web Conference 2021 (WWW'21), pp. 1737–1748. Association for Computing Machinery, New York, NY, USA (2021)

22. Lin, Y., Liu, Z., Luan, H., Sun, M., Rao, S., Liu, S.: Modeling relation paths for representation learning of knowledge bases. In: Proceedings of the 2015 Conference on Empirical Methods in Natural Language Processing, pp. 705–714. Association for Computational Linguistics, Stroudsburg, PA, USA (2015)

23. Garcia-Duran, A., Bordes, A., Usunier, N.: Composing relationships with translations. In: Proceedings of the 2015 Conference on Empirical Methods in Natural Language Processing, pp. 286–290. Association for Computational Linguistics, PA, USA (2015)

24. Guu, K., Miller, J., Liang, P.: Traversing knowledge graphs in vector space. In: Proceedings of the 2015 Conference on Empirical Methods in Natural Language Processing, pp. 318–327. Association for Computational Linguistics, PA, USA (2015)

25. Zhou, T., Ren, J., Medo, M., Zhang, Y.C.: Bipartite network projection and personal recommendation. Phys. Rev. E **76**(4), 046115 (2007)

26. Schlichtkrull, M., Kipf, T.N., Bloem, P., Berg, R.V.D., Titov, I., Welling, M.: Modeling relational data with graph convolutional networks. In: European semantic web conference, pp. 593–607. Springer, Cham (2018)

27. Zhang, Z., Zhuang, F., Zhu, H., Shi, Z., Xiong, H., He, Q.: Relational graph neural network with hierarchical attention for knowledge graph completion. In: 34th AAAI Conference on Artificial Intelligence, pp. 9612–9619. AAAI Press, Palo Alto, California, USA (2020)

28. Xie, R., Liu, Z., Sun, M.: Representation learning of knowledge graphs with hierarchical types. In: Proceedings of the twenty-fifth international joint conference on artificial intelligence (IJCAI'16), pp. 2965–2971. AAAI Press, Palo Alto, California, USA (2016)

29. Niu, G., Li, B., Zhang, Y., Pu, S., Li, J.: AutoETER: Automated entity type representation for knowledge graph embedding. In: Findings of the Association for Computational Linguistics: EMNLP 2020, pp. 1172–1181. Association for Computational Linguistics, Stroudsburg, PA, USA (2020)

30. Dasgupta, S.S., Ray, S.N., Talukdar, P.: Hyte: Hyperplane-based temporally aware knowledge graph embedding. In: Proceedings of the 2018 conference on empirical methods in natural language processing, pp. 2001–2011. Association for Computational Linguistics, Stroudsburg, PA, USA (2018)

31. García-Durán, A., Dumančić, S., Niepert, M.: Learning sequence encoders for temporal knowledge graph completion. In: Proceedings of the 2018 Conference on Empirical Methods in Natural Language Processing, pp. 4816–4821. Association for Computational Linguistics, Stroudsburg, PA, USA (2018)
32. Zhu, C., Chen, M., Fan, C., Cheng, G., Zhan, Y.: Learning from history: modeling temporal knowledge graphs with sequential copy-generation networks. In: Proceedings of the AAAI Conference on Artificial Intelligence, pp. 4732–4740. AAAI Press, Palo Alto, California US (2021)
33. Sun, Z., Deng, Z. H., Nie, J.Y., Tang, J.: Rotate: knowledge graph embedding by relational rotation in complex space. arXiv:1902.10197 (2019)
34. Korotayev, A.V., Tsirel, S.V.: A spectral analysis of world GDP dynamics: Kondratieff waves, Kuznets swings, Juglar and Kitchin cycles in global economic development, and the 2008–2009 economic crisis. Struct. Dyn. **4**(1) (2010)
35. Jain, P., Kumar P., Mausam, Chakrabarti, S.: Type-sensitive knowledge base inference without explicit type supervision. In: Proceedings of the 56th Annual Meeting of the Association for Computational Linguistics, pp. 75–80. Association for Computational Linguistics, Melbourne, Australia (2018)

A Survey of Spectrum Sensing Algorithms Based on Machine Learning

Youyao Liu and Juan Li[(✉)]

School of Electronics and Engineering, Xi'an University of Posts and Telecommunications,
Xi'an, Shanxi 710121, China
lyyao2002@xupt.edu.cn, 68691_j@stu.xupt.edu.cn

Abstract. Due to the lack of spectrum resources required for information trans-mission, Spectrum perception is one of the key technologies in cognitive radio that can solve this problem. The traditional spectrum sensing algorithm has poor detection performance when the signal-to-noise ratio is low. For higher accuracy of spectrum perception. The combination of neural network and spectral sensing algorithm is proposed, which can effectively improve the perceptual accuracy. This paper reviews the spectrum sensing algorithms based on machine learning in recent years, including spectrum sensing based on convolutional neural network, spectrum sensing based on residual neural network and spectrum sensing based on long short-term memory neural network.

Keywords: Spectrum sensing · Cognitive radio · Machine learning · Detection performance

1 Introduction

With the acceleration of the information process, the demand for wireless communi-cation services is also increasing, and the existing spectrum resources and spectrum allocation methods cannot meet the needs of users, which is an important factor hinder-ing the progress of wireless communication technology. The fixed spectrum allocation method results in a low utilization of spectrum resources, with only 15–85% of spec-trum utilization in licensed frequency bands. To make full use of spectrum resources, spectrum sensing technology is proposed [1]. Common spectrum sensing techniques include energy detection [2], cyclic stationary feature detection [3], covariance matrix detection [4]. Collaborative spectrum sensing is an increased interaction of information with other second users to determine the usage status of the frequency band. However, in the actual communication environment, the traditional spectrum sensing cannot be quickly perceived and the detection performance is susceptible to the influence of noise, so a spectrum sensing algorithm based on machine learning is proposed.

Machine learning is the core technology of artificial intelligence, which collects environmental information and user states in cognitive radio (CR) networks for modeling and inference learning, and then determines the parameters of the classifier through training, and uses the trained classifier to classify the signal. In the literature [5], the

© The Author(s), under exclusive license to Springer Nature Switzerland AG 2023
N. Xiong et al. (Eds.): ICNC-FSKD 2022, LNDECT 153, pp. 888–895, 2023.
https://doi.org/10.1007/978-3-031-20738-9_97

characteristic parameters of the extracted signal are judged by random forest (RF), but this method is a single point of perception, which cannot overcome the effects of multipath and fading. In the literature [6], using energy detection the local sensing, and the sensing results of each collaboration points are formed into a training sequence, and the Back propagation neural network (BPNN) is used in the fusion center to adjust the fusion weight of each collaboration node, Perception results are more accurate. The literature [7] proposed a collaborative spectrum sensing method based on the support vector machine (SVM) and the k-nearest neighbor (KNN) supervised learning algorithm, which does not require the design of fusion rules, but uses energy vectors as features for training and recognition, and the recognition effect is poor at low SNR. In this paper, we summarize the spectrum sensing algorithms based on machine learning in recent years.

2 Spectrum Sensing Model

Assume that the CR consists of 1 primary user and M second users. $y(n)$ is the received signal of the second user, $x(n)$ is the transmitted signal of the primary user, $h(n)$ is the channel coefficient between the primary and second users, and the noise obeys the Gaussian distribution of the mean of 0 and the variance of σ^2, denoted as $u(n)$. Binary hypothesis testing [8] performed by an unauthorized user for spectrum sensing is expressed as:

$$\begin{cases} y(n) = u(n) & H_0 \\ y(n) = x(n) + u(n) & H_1 \end{cases} \tag{1}$$

where H_0 indicates that there is no primary user and the spectrum is free; H_1 indicates that there is a primary user and the spectrum is occupied. $s(n) = x(n)h(n)$ indicates the useful signal received by the second user.

The performance indicators of the spectrum sensing algorithm include: detection probability (P_d) and false alarm probability (P_f), P_d indicates the probability that the SU detects the existence of the PU when there is signal transmission; P_f indicates that the PU did not transmit the signal, but the SU misjudged the probability of the existence of the PU. The defined equation is:

$$P_d = P\{H_1|H_1\} \tag{2}$$

$$P_d = P\{H_1|H_0\} \tag{3}$$

3 Machine Learning Models

3.1 Convolutional Neural Networks

Convolutional neural networks (CNNs) are deep learning models or multilayers similar to artificial neural networks (ANNs) that are commonly used to analyze visual images. CNNs are mainly composed of input layers, convolution layers, pooling layers, fully-connected layers, and output layers [9]. The convolutional layer can extract some of the more complex features in the image through convolution [10]. The main function of the pooling layer is to reduce the dimensionality of the characteristics extracted by the convolutional layer, reduce the number of parameters and calculations, and avoid overfitting. The fully connected layer maps the extracted high dimensional feature map into one-dimensional feature vectors [11]. The advantages of CNN are [12]:

(1) Sharing convolutional kernels makes it easier to work with high-dimensional data.
(2) Independent selection of features, feature classification results are accurate.

The structure of the CNN model as shown in Fig. 1.

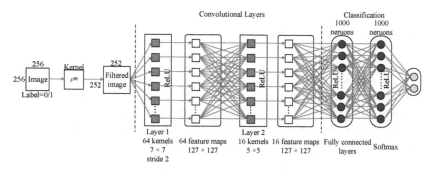

Fig. 1. Structure of the CNN model

3.2 Residual Neural Networks

Residual neural networks (ResNet) is an ANN whose structure is built on a structure known in the pyramid cells of the cerebral cortex. ResNet are implemented by skipping connections, or by jumping through some layers in shortcuts [13]. The typical ResNet model is implemented with a two-tier or three-tier skip.

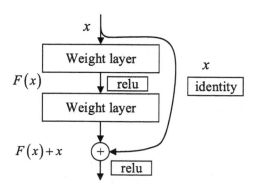

Fig. 2. ResNet model

As shown in Fig. 2, when the input x is convoluted and rectified linear unit (Relu), the result is set to $F(x)$, which is added to the original input x to have $H(x) = F(x) + x$. Compared with the output H of a traditional CNN, the residual network adds the convolutional output $F(x)$ to the input x, which is equivalent to calculating a small change in the input x, so that the output $H(x)$ is the superposition of x and the change. After the gradient propagation, the gradient transmitted to the previous layer has an x gradient, and it is precisely because of this shortcut that the gradient from the deep layer can be directly to the upper layer, so that the parameters of the shallow network layer are effectively trained. The advantages of ResNet are:

(1) It is easier to optimize and less time consuming.
(2) Because of these small changes, residual networks are more likely to propagate gradient information during back propagation.

3.3 Long Short-Term Memory Neural Networks

Long short-term memory (LSTM) neural networks are special recurrent neural networks (RNNs). LSTM has many of the same properties as gated loop cells, introducing memory cells, or cells for short [14]. In order to control the memory element, many gates are designed, such as input gates, output gates and forgetting gates, which are a mechanism used to reset the contents of the unit..; 0 is.

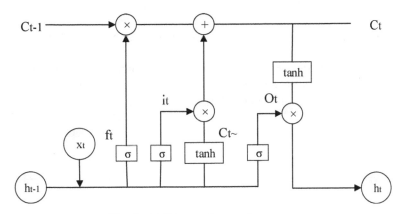

Fig. 3. LSTM neural network structure

In Fig. 3, x, h, and y denote the input sequence, hidden layer sequence, and output sequence, $t - 1$ and t denote the previous moment and the current moment, respectively, and f, i, o, and C denote the forgetting gate, input gate, output gate, and cell state, respectively.

4 Spectrum Sensing Based on Machine Learning

4.1 Spectrum Sensing Based on CNN

Zhang et al. proposed an OFDM spectrum perception method based on CNN [15]. This method first analyzes the cyclic autocorrelation and spectrum sensing model of OFDM signal, normalizes the grayscale of cyclic autocorrelation, and forms a cyclic autocorrelation grayscale map. Then, on the basis of LeNet-5 network, the CNN training data is used to learn hierarchically, abstract features are obtained, and finally the test data input is used to train the model for spectrum sensing.

In the literature [16], the threshold of the spectrum sensing algorithm based on the signal covariance matrix is difficult to obtain accurately, and the original signal information is not fully utilized. A collaborative spectrum perception algorithm using CNN and covariance is proposed. First, the normalized covariance matrix of the received I and Q quadrature signals is composed of a dual-channel input matrix, and then the CNN is used to directly extract the feature information of the covariance matrix, and the classifier is trained to obtain a classifier. Finally, the model is trained to complete the spectrum perception.

In the literature [17], in view of the problems of traditional CNN spectrum sensing methods, such as incomplete feature information extraction caused by simple network structure, deepening network and prone to gradient disappearance, etc. Using a collaborative spectral sensing method combined with deep convolutional neural network (DCNN), the spectral sensing problem is regarded as an image binary classification problem, and the covariance matrix of orthogonal QPSK is normalized, and uses it as an input to the deep convolutional neural network to train the DCCNN model through residual learning, at last, complete the spectrum sensing based on image classification.

A CNN-based spectrum sensing deep learning algorithm was proposed in [18]. This model-based spectrum sensing algorithm requires not only a primary user (PU) activity pattern model, but also a known signal-noise probability model. Our proposed deep learning method can better detect PU activity by learning the inherent PU activity pattern, that is, acquiring historical perception data and current perception data at the same time.

In deep learning, CNN has better performance in extracting input spatial features, while LSTM network is more suitable for extracting temporal features. In order to learn the PU activity pattern, the energy-related features of the perceptual data are first obtained by the CNN-LSTM detector [19] using the CNN to generate the covariance matrix from the perceptual data, and then the LSTM will receive the extracted multiple perceptual cycles The corresponding series of energy-related features. The CNN-LSTM detector can change the performance in the scene of noise uncertainty.

4.2 Spectrum Sensing Based on ResNet

In the literature [20], Proposing a spectrum sensing algorithm based on ResNet. The algorithm uses the signal real part sequence to construct a two-dimensional matrix, maps the matrix to a grayscale image, and then extracts the abstract features in the grayscale image by training ResNet, and uses the trained ResNet as the final classifier to achieve spectrum perception. When the SNR is −20 dB in the Gaussian channel, its detection accuracy of the ResNet is 89.60%, and when the SNR is −18 dB in the Rayleigh fading channel, the detection probability is 91.90%. In the low SNR, Experiments show that it has good detection performance.

In the literature [21], dense connections are added to the CNN spectrum sensing method, the feature map information is reused, and shortcut connections are added at both ends of the dense unit, and method for combining residual dense network (ResDenNet) and spectral sensing has been proposed. This method maps the problem of spectrum sensing to the problem of image binary classification. Take advantage of this approach, When the SNR is −19 dB, the result can reach 0.96, the false alarm probability is 0.1.

Dynamic spectrum allocation requires the minimum spectrum sensing cost to detect spectrum usage within a certain time, so the NN-ResNet prediction model is proposed in [22] to solve this challenge. To improve the utilization of deployed sensors and reduce the perception cost, the spatiotemporal spectrum usage in the region is predicted by a deep learning prediction model based on ResNets and CNN [23], and the spectrum usage data in the unperceived region is recovered by neighbor (NN) interpolation. Compared with CNN and ConvLSTM prediction models, this model has lower error rates in various sparse sensor scenarios.

4.3   Spectrum Sensing Based on LSTM

In the literature [24], LSTM technique combined with deep learning concepts was used to recalculate the prediction errors present in future WiFi band master and GSM user estimates. Simulation results show that this technique has the characteristics needed to improve channel usage prediction compared to other techniques. Although LSTM performs better in generating time series forecasts, it has high computational complexity due to its neural structure [25]. Therefore, it is feasible to implement in a cognitive radio network based on a centralized network topology.

Reference [26] proposes an LSTM-based Spectrum Sensing (LSTM-SS), which can learn the correlation between current and past timestamps from spectral data. In addition, the CR system can also improve the sensing performance through spectrum sensing based on PU activity statistics (PAS-SS), which mainly improves performance by calculating information such as PU duty cycle. This sensing scheme is validated on spectral data obtained from different radio technologies using an experimental testbed setup, compared to the proposed LSTM-SS scheme. It improves detection performance under low SNR conditions.

In the literature [27], a blind spectrum sensing method combined with deep learning has been proposed, which combines one-dimensional CNN, LSTM and fully connected neural network (FCNN) three kinds of neural networks to conduct end-to-end signal detection requires no prior knowledge of the detected signals and can be adapted to most current modulation types. The effects of different layers of LSTM on the detection performance were experimentally verified. Experimentally detect performance when the number of LSTM network layers is 2.

5 Concluding Remarks

This paper first summarizes the research status of spectrum sensing algorithms, and then briefly introduces the spectrum sensing model and three learning models widely used in the spectrum sensing research literature of machine learning: the structure of CNN, the basic structure of ResNet, and the hidden layer cell structure of LSTM. Finally, the spectrum sensing methods based on CNN, LSTM and ResNet are reviewed.

Acknowledgements. The work supported by the Natural Science Foundation of China (grants nos. 61874087, 61834005).

References

1. Haykin, S.: Cognitive radio: brain-empowered wireless communications. IEEE J. Sel. Areas Commun. **23**(2), 201–220 (2005)
2. Gaiera, B., Patel, D. K., Soni, B.: Performance evaluation of improved energy detection under signal and noise uncertainties in cognitive radio networks. In: IEEE International Conference on Signals and Systems, pp. 131–137 (2019)
3. Zhu, Y., Liu, J., Feng, Z.: Sensing performance of efficient cyclostationary detector with multiple antennas in multipath fading and lognormal shadowing environment. J. Commun. Netw. **16**(2), 162–171 (2014)
4. Charan, C., Pandey, R.: Double threshold based spectrum sensing technique using sample covariance matrix for cognitive radio networks. In: 2017 2nd International Conference on Communication Systems, Computing and IT Applications (CSCITA), pp. 150–153 (2017)
5. Wang, X., Wang, J.K., Liu, Z.G.: Random forest-based cognitive network spectrum sensing algorithms. J. Instrum. Instrum. **34**(11), 2471–2477 (2013)
6. Chen, Y., Zhang, H., Hu, H.: Cooperative spectrum sensing technology based on BP neural network. Comput. Sci. **42**(2), 43–45 (2014)
7. Thilina, K.M., Choi, K.W., Saquib, N.: Machine learning techniques for cooperative spectrum sensing in cognitive radio networks. IEEE J. Sel. Areas Commun. **31**(11), 2209–2221 (2013)
8. Hochreiter, S., Schmidhuber, J.: Long short-term memory. Neural Comput. **9**(8), 1735–1780 (1997)
9. Sundriyal, A., Baghel, A.: CNN based cognitive spectrum sensing with optimization. In: 2020 4th International Conference on Electronics, Communication and Aerospace Technology (ICECA), pp. 1537–1540 (2020)
10. Yadav, S.S., Jadhav, S.M.: Deep convolutional neural network based medical image classification for disease diagnosis. J. Big Data **6**(1), 1–18 (2019). https://doi.org/10.1186/s40537-019-0276-2
11. Badza, M.M., Marko, C.: Classification of brain tumors from MRI images using a convolutional neural network. Appl. Sci. **10**, 6, 1999 (2020)
12. Satya P.S., Lipo P., Wang, S.G., Balazs, G., Parasuraman, P.: Shallow 3D CNN for detecting acute brain hemorrhage from medical imaging sensors. IEEE Sens. J. **21**(13) 14290–14299 (2021)
13. Ren, X., Mosavat-Jahromi, H., Cai, L., Kidston, D.: Spatio-temporal spectrum load prediction using convolutional neural network and ResNet. IEEE Trans. Cogn. Commun. Networ. 1–1 (2021)
14. Greff, K., Srivastava, R.K., Koutnik, J.: LSTM: a search space odyssey. IEEE Trans. Neural Netw. Learn. Syst. **28**(10), 2222–2232 (2017)

15. Zhang, M.B., Wang, L.W., Feng, Y.Q.: OFDM spectrum sensing method based on convolutional neural network. Syst. Eng. Electron. Technol. **41**(01), 178–186 (2019)
16. Lu, H.C., Zhao, Z.J., Shang, J.N.: Collaborative spectrum sensing algorithm using convolutional neural networks and covariance. Signal Process. **35**(10), 1700–1707 (2019)
17. Gai, J.X., Xue, X.F., Wu, J.Y.: Collaborative spectrum sensing method based on deep convolutional neural network. J. Electron. Inf. **43**(10), 2911–2919 (2021)
18. Xie, J.D., Liu, C., Liang, Y.C.: Activity pattern aware spectrum sensing: a CNN-based deep learning approach. IEEE Commun. Lett. **23**(6), 1025–1028 (2019)
19. Xie, J., Fang, J., Liu, C.: Deep learning-based spectrum sensing in cognitive radio: a CNN-LSTM approach. IEEE Commun. Lett. **24**(10), 2196–2200 (2020)
20. Chen, Y., Zhang, X., Ai, W.B.: Spectrum sensing algorithm based on residual neural network. Mod. Electron. Technol. **45**(07), 1–5 (2022)
21. Gai, J.X., Xue, X.F., Nan, R.X.: Spectrum sensing method based on residual dense network. J. Commun. **42**(12), 182–191 (2021)
22. Ren, X., Mosavat-Jahromi, H., Cai L.: Spatio-temporal spectrum load prediction using convolutional neural network and ResNet. IEEE Trans. Cogn. Commun. Netw. 1–1 (2021)
23. Justin, K., Lipo, P.W., Jai, R., Tchoyoson, L.: Deep Learning applications in medical image analysis. IEEE Access **6**, 9375–9389 (2018)
24. Hernández, J., López, D., Vera, N.: Primary user characterization for cognitive radio wireless networks using long short-term memory. Int. J. Distrib. Sens. Netw. **14**(11), 1–20 (2018)
25. Satya, P.S., Lipo, P.W., Sukrit, G., Haveesh, G., Parasuraman P., Balazs, G.: 3D deep learning on medical images: a review, Sensors **20**(5097) (2020)
26. Balwani, N., Patel, D.K., Soni, B.: Long short-term memory based spectrum sensing scheme for cognitive radio. In: 2019 IEEE 30th Annual International Symposium on Personal, Indoor and Mobile Radio Communications, pp. 1–6 (2019)
27. Yang, K., Huang, Z.T., Wang, X.: A blind spectrum sensing method based on deep learning. Sensors **19**(10), 1–17 (2019)

Terrain Classification of Hyperspectral Remote Sensing Images Based on SC-KSDA

Jing Liu[1]([✉]) [iD], Yinqiao Li[1] [iD], Yue Ye[1] [iD], and Yi Liu[2] [iD]

[1] Xi'an University of Posts and Telecommunications, Xi'an 710121, China
jingliu@xupt.edu.cn, {yinqiaoli,851807613}@stu.xupt.edu.cn
[2] School of Electronic Engineering, Xidian University, Xi'an 710071, China
yiliu@xidian.edu.cn

Abstract. To solve the problem of "same object but different spectrum" of hyperspectral remote sensing images and further extract more effective nonlinear separability features, in this paper, spectral clustering and subclass discriminant analysis (SC-SDA) is generalized to the kernel space, spectral clustering and kernel subclass discriminant analysis (SC-KSDA) is proposed. The SC-SDA method can automatically distinguish the number of clusters for the same class of ground objects that can be divided, and divide them into subclasses, and calculate the final feature subspace by subclass discriminant analysis (SDA). In this paper, the results of subclass division of various classes are obtained based on the SC-SDA method, and then the criterion function on data separability and kernel parameters is designed to map the data divided by subclasses into high-dimensional kernel space. Finally, each subclass is treated as a separate class, and linear discriminant analysis (LDA) method is used to extract nonlinear separable features in kernel space. The experimental results of the three measured hyperspectral remote sensing images using the minimum distance classifier show that proposed SC-KSDA can effectively improve accuracy of feature classification.

Keywords: Hyperspectral remote sensing images · Feature extraction · Kernel subclass discriminant analysis

1 Introduction

Hyperspectral remote sensing images are characterized by multiple wavebands, wide spectral coverage and large amount of data. However, in hyperspectral remote sensing images, there is a common phenomenon of "same object but different spectrum", that is, the same ground objects have the same species but different spectral curves under the influence of external factors such as light, pests, diseases and radioactive substances, which makes various samples have multi subclass distribution, and linear methods cannot be used for effective classification and discrimination. Linear discriminant analysis (LDA) is a typical supervised extraction method. However, LDA is only suitable for samples of Gaussian distribution. Kernel linear discriminant analysis (KLDA) [1] is a nonlinear generalization of LDA, which is based on the principle of extracting the separability characteristics of samples in kernel space using the LDA method after projecting

© The Author(s), under exclusive license to Springer Nature Switzerland AG 2023
N. Xiong et al. (Eds.): ICNC-FSKD 2022, LNDECT 153, pp. 896–904, 2023.
https://doi.org/10.1007/978-3-031-20738-9_98

samples into high-dimensional kernel space. However, KLDA cannot solve the feature extraction problem of hyperspectral data distributed by multiple subclasses within a class. Subclass discriminant analysis (SDA) method proposed by Manli Zhu et al. [2] can solve the problem of "same object but different spectrum" in hyperspectral remote sensing images, but SDA does not consider the separability of various samples, which will cause the problem of overdivision [3–5]. Kernel subclass discriminant analysis (KSDA) [6] is the result of SDA nonlinearization, which can effectively extract nonlinear separable features from data distributed in multiple subclasses within a class. The literature [7] proposed spectral clustering and subclass discriminant analysis (SC-SDA), considered the separability of various samples, proposed a cluster similarity criterion for determining spectral clustering (SC) subspace dimension, and automatically selected the appropriate number of clusters according to the intraclass separability, but it is difficult to directly extract the nonlinear separable features.

In this paper, the SC-SDA method is generalized to the kernel space, and a nonlinear method is proposed, spectral clustering and kernel subclass discriminant analysis (SC-KSDA). Firstly, based on the principle of improving the separability of the overall data after subclass division, it is judged whether the various classes of training samples are separable. The cluster similarity criterion [7] and the segregation within each class are used to select the optimal SC subspace dimension and the number of subclasses. Then, the SC method is used to subclass classes that need to be classified in the original space. Finally, the subclass-divided data is mapped to the high-dimensional kernel space, and each subclass remains independent in the kernel space, and LDA is used in the kernel space to extract the nonlinear separable features of the hyperspectral remote sensing images. Experimental results show that compared with the LDA, KLDA, KSDA and SC-SDA, proposed SC-KSDA method can effectively improve accuracy of terrain classification.

2 Proposed Method

In the original space $\mathbf{R}^D$, given a set of data sets with C classes $\mathbf{X} = [\mathbf{X}_1, \mathbf{X}_2, \ldots, \mathbf{X}_C]$. Then, nonlinear mapping $\phi(\cdot)$ can map samples from the original space into the high-dimensional kernel space $\mathbf{F}$, namely $\phi : \mathbf{x} \in \mathbf{R}^D \rightarrow \phi(\mathbf{x}) \in \mathbf{F}$. Order H_i represents the number of subclasses of class i. When class i is not a molecular class, $H_i = 1$.

SC-KSDA is maximized

$$\mathbf{W} = \arg\max_{\mathbf{W}} \mathrm{tr} \left(\frac{\mathbf{W}^{\mathrm{T}} \mathbf{S}_{\mathrm{B}}^{\phi} \mathbf{W}}{\mathbf{W}^{\mathrm{T}} \mathbf{S}_{\mathrm{W}}^{\phi} \mathbf{W}} \right) \tag{1}$$

to solve the projection matrix $\mathbf{W}$ of the feature subspace. $\mathbf{S}_{\mathrm{B}}^{\phi}$ and $\mathbf{S}_{\mathrm{W}}^{\phi}$ represent the interclass scatter matrix and intraclass scatter matrix of the training samples in kernel space $\mathbf{F}$. The purpose of mapping data to the kernel space is to improve the divisibility of the data, and often in practical problems where a direct high-dimensional mapping is not possible, it can be solved by transforming the inner product in the kernel space into a kernel function in the sample space.

$$k\left(\mathbf{x}_i, \mathbf{x}_j\right) = < \phi(\mathbf{x}_i), \phi(\mathbf{x}_j) > \tag{2}$$

$< \phi(\mathbf{x}_i), \phi(\mathbf{x}_j) >$ represents the inner product of two samples in the kernel space. $k(\mathbf{x}_i, \mathbf{x}_j)$ is a kernel function, the kernel matrix $\mathbf{K}$ can be obtained by the kernel function:

$$\mathbf{K} = [\mathbf{K}_{lh}]_{\substack{l=1,\ldots,H \\ h=1,\ldots,H}} = \mathbf{\Phi}\mathbf{\Phi}^{\mathrm{T}} \tag{3}$$

$\mathbf{\Phi}^{\mathrm{T}}=[\phi(\mathbf{x}_1), \phi(\mathbf{x}_2), \ldots, \phi(\mathbf{x}_N)]$ represents a sample set in kernel space. $[\mathbf{K}_{lh}]$ is the inner product matrix with the class h sample, its size is $N_l \times N_h$. From the regenerative kernel theory, it can be seen that the solution vector $\mathbf{w}$ in kernel space $\mathbf{F}$ is a linear combination of training sample $\mathbf{\Phi}^{\mathrm{T}}$, i.e.

$$\mathbf{w}=\sum_{i=1}^{N} \alpha_i \phi(\mathbf{x}_i) = \mathbf{\Phi}^{\mathrm{T}}\boldsymbol{\alpha} \tag{4}$$

$\boldsymbol{\alpha}$ is a coefficient vector. According to the $\boldsymbol{\alpha}$ vectors, a matrix of coefficients can be obtained $\mathbf{A}= [\boldsymbol{\alpha}_1, \boldsymbol{\alpha}_2, \ldots, \boldsymbol{\alpha}_d]$. Then the projection matrix $\mathbf{W}$ of the SC-KSDA subspace can be expressed as

$$\mathbf{W}=\mathbf{\Phi}^{\mathrm{T}}\mathbf{A} \tag{5}$$

From the above, it is clear that the optimal projection matrix $\mathbf{W}$ can be calculated by solving the coefficient matrix $\mathbf{A}$. Assuming that the subclasses in the kernel space are independent of each other, we can calculate the mean vectors of each subclass in the kernel space $\boldsymbol{\mu}_l^{\phi}$ and the overall mean vector $\boldsymbol{\mu}^{\phi}$, respectively:

$$\begin{aligned}
\boldsymbol{\mu}_l^{\phi} &= \frac{1}{N_l} \sum_{t=1}^{N_l} \phi\left(\mathbf{x}_t^l\right) \\
&= \frac{1}{N_l}\left[\phi\left(\mathbf{x}_1^l\right) + \phi\left(\mathbf{x}_2^l\right) + \cdots + \phi\left(\mathbf{x}_{N_l}^l\right)\right] \\
&= \frac{1}{N_l}\left[\phi\left(\mathbf{x}_1^l\right), \phi\left(\mathbf{x}_2^l\right), \ldots, \phi\left(\mathbf{x}_{N_l}^l\right)\right]\mathbf{1}_{N_l \times 1} \\
&= \frac{1}{N_l}\mathbf{\Phi}_l^{\mathrm{T}}\mathbf{1}_{N_l \times 1} \tag{6}
\end{aligned}$$

$$\begin{aligned}
\boldsymbol{\mu}^{\phi} &= \frac{1}{N} \sum_{l=1}^{H}\sum_{t=1}^{N_l} \phi\left(\mathbf{x}_t^l\right) \\
&= \frac{1}{N}\left[\phi\left(\mathbf{x}_1^1\right), \ldots, \phi\left(\mathbf{x}_{N_1}^1\right), \ldots, \phi\left(\mathbf{x}_1^H\right), \ldots, \phi\left(\mathbf{x}_{N_H}^H\right)\right]\mathbf{1}_{N \times 1} \\
&= \frac{1}{N}\mathbf{\Phi}^{\mathrm{T}}\mathbf{1}_{N \times 1} \tag{7}
\end{aligned}$$

$\mathbf{\Phi}_l^{\mathrm{T}}(l = 1, 2, \cdots, C)$ is the l-th sample in kernel space, $\mathbf{1}_{N \times 1}$ is a row vector with all values of 1. In the kernel space, interclass scatter matrix and intraclass scatter matrix obtained from the mean vector are:

$$\mathbf{S}_{\mathrm{B}}^{\phi} = \sum_{l=1}^{H} P_l(\boldsymbol{\mu}_l^{\phi} - \boldsymbol{\mu}^{\phi})(\boldsymbol{\mu}_l^{\phi} - \boldsymbol{\mu}^{\phi})^{\mathrm{T}} = \mathbf{\Phi}^{\mathrm{T}}\mathbf{B}\mathbf{\Phi} \tag{8}$$

$$\mathbf{S}_W^\phi = \sum_{l=1}^{H} \frac{P_l}{N_l} \sum_{t=1}^{N_t} \left(\phi\left(\mathbf{x}_t^l\right) - \mathbf{\mu}_l^\phi\right)\left(\phi\left(\mathbf{x}_t^l\right) - \mathbf{\mu}_l^\phi\right)^T = \mathbf{\Phi}^T \mathbf{V} \mathbf{\Phi} \qquad (9)$$

where, $\mathbf{1}_{N \times N}$ is a matrix with a value of all 1 and a size of $N \times N$.

$$\mathbf{B} = \frac{1}{N}\left(\mathrm{diag}\left(\frac{1}{N_1}\mathbf{1}_{N_1 \times N_1}, \ldots, \frac{1}{N_H}\mathbf{1}_{N_H \times N_H}\right)\right) - \frac{1}{N}\mathbf{1}_{N \times N} \qquad (10)$$

$$\mathbf{V} = \mathrm{diag}\left(\frac{N_1}{N}, \ldots, \frac{N_l}{N}\right) - \mathrm{diag}\left(\frac{1}{N}\mathbf{1}_{N_1 \times N_1}, \ldots, \frac{1}{N}\mathbf{1}_{N_l \times N_l}\right) \qquad (11)$$

Substituting Eqs. (8) and (9) into Eq. (1), after collation, we can get a criterion function about the coefficient matrix $\mathbf{A}$:

$$\frac{\mathbf{W}^T \mathbf{S}_B^\phi \mathbf{W}}{\mathbf{W}^T \mathbf{S}_W^\phi \mathbf{W}} = \frac{\mathbf{A}^T \mathbf{K} \mathbf{B} \mathbf{K} \mathbf{A}}{\mathbf{A}^T \mathbf{K} \mathbf{V} \mathbf{K} \mathbf{A}} = \frac{\mathbf{A}^T \mathbf{M}_B \mathbf{A}}{\mathbf{A}^T \mathbf{M}_W \mathbf{A}} \qquad (12)$$

By solving the eigenvector corresponding to the first d maximum eigenvalues of $(\mathbf{M}_W)^{-1}\mathbf{M}_B$ can obtain the optimal coefficient matrix $\mathbf{A}$. Further, bringing $\mathbf{A}$ into the Eq. (5) can be solved to obtain a projection matrix $\mathbf{W}$ of the SC-KSDA subspace, then any sample $\phi(\mathbf{x})$ in the original space can be projected into the SC-KSDA subspace, i.e.

$$\begin{aligned}
\mathbf{y} &= \mathbf{W}^T \phi(\mathbf{x}) \\
&= \mathbf{A}^T[\phi(\mathbf{x}_1)^T \phi(\mathbf{x}), \phi(\mathbf{x}_2)^T \phi(\mathbf{x}), \ldots, \phi(\mathbf{x}_N)^T \phi(\mathbf{x})] \\
&= \mathbf{A}^T[k(\mathbf{x}_1, \mathbf{x}), k(\mathbf{x}_2, \mathbf{x}), \ldots, k(\mathbf{x}_N, \mathbf{x})]^T \qquad (13)
\end{aligned}$$

The radial base kernel function is a universal kernel function with a wide range of applications.

$$k\left(\mathbf{x}_i, \mathbf{x}_j\right) = \exp\left(\frac{-\|\mathbf{x}_i - \mathbf{x}_j\|^2}{\sigma^2}\right) \qquad (14)$$

Aiming at the selection of kernel parameter σ, a kernel parameter optimization model based on data separability is established, and the optimal kernel parameters can be learned by using genetic algorithms, so that in the kernel space determined by the optimal kernel parameters, the sample has the greatest separability, that is, the ratio of distribution between heterogeneous subclasses and distribution within subclasses is the largest. First, we can calculate the distance square of the means of the two subclasses in the kernel subspace:

$$\begin{aligned}
\|\hat{\mu}_i - \hat{\mu}_j\|^2 &= \left(\hat{\mu}_i - \hat{\mu}_j\right)^T \left(\hat{\mu}_i - \hat{\mu}_j\right) \\
&= \hat{\mu}_i^T \hat{\mu}_i - 2\hat{\mu}_i^T \hat{\mu}_j + \hat{\mu}_j^T \hat{\mu}_j \qquad (15)
\end{aligned}$$

The squared distance between the sample and the mean within each subclass:

$$\frac{1}{N_i} \sum_{n=1}^{N_i} \|\mathbf{y}_n^i - \hat{\mu}_i\|^2 = \frac{1}{N_i} \sum_{n=1}^{N_i} \left(\mathbf{y}_n^i - \hat{\mu}_i\right)^T \left(\mathbf{y}_n^i - \hat{\mu}_i\right)$$

$$=\frac{1}{N_i}\sum_{n=1}^{N_i}\left(\mathbf{y}_n^{i\mathrm{T}}\mathbf{y}_n^i-2\mathbf{y}_n^{i\mathrm{T}}\hat{\mu}_i+\hat{\mu}_i^{\mathrm{T}}\hat{\mu}_i\right) \qquad (16)$$

where, $\mathbf{y}_n^i$ represents the nth sample of class i in the kernel space, and $\mathbf{Y}^i$ is the sample set of class i, N_i represents the number of the class i sample and $\hat{\mu}_i$ represents the mean of the class i.

$$\begin{cases} \mathbf{y}_n^i = \mathbf{W}^{\mathrm{T}}\phi\left(\mathbf{x}_n^i\right) \\ \mathbf{Y}^i = [\mathbf{y}_1^i, \mathbf{y}_2^i, \ldots, \mathbf{y}_{N_i}^i] \\ \mathbf{y}_n^{i\mathrm{T}}\hat{\mu}_i = \frac{1}{N_i}\mathbf{y}_n^{i\mathrm{T}}\mathbf{Y}^i\mathbf{1}_{N_i\times 1} \\ \hat{\mu}_i^{\mathrm{T}}\hat{\mu}_j = \frac{1}{N_iN_j}\mathbf{1}_{1\times N_i}\left(\mathbf{Y}^i\right)^{\mathrm{T}}\mathbf{Y}^j\mathbf{1}_{N_j\times 1} \end{cases} \qquad (17)$$

Then, the sum of the discretes between the heterogeneous subclasses D_b, and the sum of the discreteness of the samples within each subclass D_w can be represented as

$$D_b=\sum_{i=1}^{C-1}\sum_{n=1}^{H_i}\sum_{j=i+1}^{C}\sum_{t=1}^{H_n}p_{in}p_{jt}\left\|\hat{\mu}_{in}-\hat{\mu}_{jt}\right\|^2 \qquad (18)$$

$$D_w=\sum_{i=1}^{H}\frac{1}{N_i}\sum_{n=1}^{N_i}\left\|\mathbf{y}_n^i-\hat{\mu}_i\right\|^2 \qquad (19)$$

where, p_{in} is prior probability of the nth subclass of class i in the kernel subspace. $\hat{\mu}_{in}$ represents the mean of class i. Finally, criterion function based on data separability and kernel parameters can be represented as

$$J(\sigma)=\frac{D_b}{D_w} \qquad (20)$$

The genetic algorithm is used to solve the problem, so that the data separability under the optimal kernel parameter conditions is maximized.

3 Experimental Results

In the experiment, 20% of all kinds of samples are randomly selected as the training set, and 80% of all kinds of samples are used as the test set. The algorithm performance is evaluated by various recognition rates and the average recognition rate and Kappa coefficient after 10 classifications. The three measured hyperspectral remote sensing images are Indian Pines images, KSC images, and Botswana images.

Tables 1, 2 and 3 shows the experimental results of the feature classification of three hyperspectral images using the LDA, KLDA, KSDA, and SC-SDA subspace methods.

The experimental results of Tables 1, 2 and 3 show that SC-KSDA method has improved the accuracy of various classes of feature recognition compared with other methods.

Tables 4, 5 and 6 shows the average of 10 times of feature classification results of hyperspectral remote sensing images in LDA, SC-SDA, KLDA, KSDA and SC-KSDA subspace using the minimum distance classifier. It can be seen that the average accuracy of SC-KSDA subspace method is significantly improved compared with other methods.

Table 1. Experimental results of Indian Pines data under the minimum distance classifier.

Class	LDA	SC-SDA	KLDA	KSDA	SC-KSDA
1	69.67	76.74	74.42	81.39	74.42
2	73.57	86.21	63.99	62.51	81.78
3	62.07	71.21	66.86	64.62	83.06
4	72.73	74.33	77.01	82.89	70.05
5	86.43	87.49	84.42	87.94	90.95
6	97.66	98.33	90.13	89.63	96.82
7	23.81	71.43	90.48	90.48	90.48
8	99.49	97.19	92.58	82.61	97.70
9	37.50	62.50	100	100	96.25
10	68.35	86.69	76.74	73.90	87.86
11	55.37	76.39	69.15	68.54	86.73
12	73.93	91.24	80.45	82.28	82.28
13	98.82	99.41	97.05	98.83	98.23
14	87.63	90.53	95.94	94.39	95.56
15	79.61	80.59	57.56	60.19	78.75
16	86.84	86.84	82.89	85.53	88.95
OA(%) ± Std(%)	73.34 ± 20.54	83.57 ± 10.43	81.23 ± 12.30	81.61 ± 12.10	87.49 ± 8.30
Kappa	0.70	0.72	0.74	0.73	0.85

Table 2. Experimental results of Botswana data under the minimum distance classifier.

Class	LDA	SC-SDA	KLDA	KSDA	SC-KSDA
1	100	100	100	100	100
2	98.78	98.78	100	100	100
3	97.51	97.10	96.52	95.52	97.01
4	98.84	99.42	95.93	94.77	96.51
5	90.70	91.16	87.91	88.37	89.30
6	76.74	82.33	79.07	73.02	79.28
7	98.55	99.52	97.58	97.58	98.07
8	93.21	98.15	98.77	98.77	98.77
9	87.65	89.26	92.43	90.84	94.02
10	92.93	92.93	97.98	97.47	93.43
11	91.80	92.21	88.52	89.34	92.21
12	95.17	93.79	93.10	94.48	93.79
13	87.85	86.45	92.06	90.65	93.46
14	93.42	93.42	94.74	96.05	96.05
OA(%) ± Std(%)	93.08 ± 5.97	93.89 ± 5.16	93.90 ± 5.54	93.35 ± 6.73	94.42 ± 5.15
Kappa	0.91	0.66	0.92	0.91	0.93

Table 3. Experimental results of KSC data under the minimum distance classifier.

Class	LDA	SC-SDA	KLDA	KSDA	SC-KSDA
1	90.97	94.09	95.24	93.92	95.07
2	86.85	87.11	86.08	84.03	87.11
3	87.80	89.76	89.76	89.27	93.66
4	65.35	81.68	77.72	75.25	81.78
5	62.79	75.19	74.26	62.79	82.87
6	85.79	75.19	86.34	83.06	85.95
7	91.67	95.23	85.71	90.48	83.33
8	77.39	80.58	89.27	86.67	92.43
9	98.08	97.84	98.56	97.60	97.59
10	92.26	94.74	92.88	92.88	94.43
11	88.96	97.10	98.21	97.91	98.81
12	84.33	82.09	96.52	89.30	94.53

(*continued*)

Table 3. (*continued*)

Class	LDA	SC-SDA	KLDA	KSDA	SC-KSDA
13	97.57	97.57	97.44	98.65	97.44
OA(%) ± Std(%)	85.37 ± 10.47	88.32 ± 8.17	89.84 ± 7.44	87.83 ± 9.65	91.15 ± 5.86
Kappa	0.86	0.86	0.91	0.90	0.91

Table 4. Experimental results of Indian Pines data under the minimum distance classifier.

	LDA	SC-SDA	KLDA	KSDA	SC-KSDA
OA(%) ± Std(%)	73.70 ± 0.88	81.93 ± 0.80	82.13 ± 0.44	79.43 ± 0.71	86.41 ± 0.69
Kappa	0.70	0.72	0.74	0.70	0.83

Table 5. Experimental results of Botswana data under the minimum distance classifier.

	LDA	SC-SDA	KLDA	KSDA	SC-KSDA
OA(%) ± Std(%)	93.02 ± 0.67	94.48 ± 0.46	94.61 ± 0.48	93.85 ± 0.77	94.92 ± 0.51
Kappa	0.92	0.92	0.93	0.67	0.90

Table 6. Experimental results of KSC data under the minimum distance classifier.

	LDA	SC-SDA	KLDA	KSDA	SC-KSDA
OA(%) ⊥ Std(%)	88.14 ± 0.47	89.95 ± 0.39	88.75 ± 0.29	88.41 ± 0.69	92.43 ± 0.85
Kappa	0.87	0.88	0.90	0.75	0.91

4   Conclusion

In this paper, to solve the problem of "same object but different spectrum" in hyperspectral remote sensing images, SC-SDA is nonlinearly generalized and a nonlinear feature extraction method of SC-KSDA is proposed. From the experimental results based on three hyperspectral remote sensing images, it can be concluded that compared to the LDA, KLDA, KSDA and SC-SDA, the proposed SC-KSDA method can effectively extract the nonlinear separable features of hyperspectral remote sensing images, improve accuracy of feature classification, and obtain stable feature classification results.

References

1. Baudat, G., Anouar, F.: Generalized discriminant analysis using a Kernel approach. Neural Comput. **12**(10), 2385–2404 (2000)
2. Zhu, M., Martinez, A.M.: Subclass discriminant analysis. IEEE Trans. Pattern Anal. Mach. Intell. **28**(8), 1274 (2006)
3. Gkalelis, N., Mezaris, V., Kompatsiaris, I.: Mixture subclass discriminant analysis. IEEE Signal Process. Lett. **18**(5), 319–322 (2011)
4. Zhu, M., Aleix, Martinez, A. M.: Optimal Subclass Discovery for Discriminant Analysis. IEEE Computer Society (2004)
5. Nikitidis, S., Tefas, A., Nikolaidis, N., Pitas, I.: Subclass discriminant nonnegative matrix factorization for facial image analysis. Pattern Recogn. **45**(12), 4080–4091 (2012)
6. Bo, C., Li, Y., Liu, H., Zheng, B.: Kernel subclass discriminant analysis. Neurocomputing **71**(1–3), 455–458 (2007)
7. Liu, J., Guo, X., Liu, Y.: Hyperspectral remote sensing image feature extraction based on spectral clustering and subclass discriminant analysis. Remote Sens. Lett. (2020)

A Survey of Counterfactual Explanations: Definition, Evaluation, Algorithms, and Applications

Xuezhong Zhang[1], Libin Dai[1], Qingming Peng[1], Ruizhi Tang[1], and Xinwei Li[2(✉)]

[1] South Sulige Operation Branch of Changqing Oilfield Company, PetroChina, Xi'an 710018, Shaanxi, China
zhangxz01_cq@petrochina.com.cn, dailb_cq@petrochina.com.cn, pengqm_cq@petrochina.com.cn, tangrzh_cq@petrochina.com.cn
[2] School of Computer Science and Technology, Xidian University, Xi'an, Shaanxi, China
lixinwei@stu.xidian.edu.cn

Abstract. Explainable machine learning aims to reveal the reasons why black-box models make decisions. Counterfactual explanation is an example-based post-hoc explanation method. The counterfactual explanations aims to find the minimum perturbation that changes the model output with respect to the original instance. This study's goal is to review the literature that has already been written about counterfactual explanations and topics that are relevant to it. We provide a formal definition of counterfactual explanations and counterfactual explainer, and a summary and formulaic description of the properties of the generated counterfactual instances. In addition, we investigate the application of counterfactual explanations in two areas: model robustness, and generating feature importance. The findings demonstrate that the qualities necessary for counterfactual instances cannot be simultaneously satisfied by present methodologies. Finally, we go over potential future research directions.

Keywords: Explainable AI · Counterfactual explanations · Causability · Interpretable machine learning

1 Introduction

The interpretability of machine learning is mainly concerned with the interpretability of the model, in which the data is taken into account as the input of the model. At present, the interpretability of models is divided into two research ideas, one is to use Intrinsically interpretable models, and the other is to use various interpretable technologies or algorithms to explain complex models [1,2]. Intrinsically explainable models include decision trees, linear regression models, etc. However, the accuracy of model is often negatively correlated

© The Author(s), under exclusive license to Springer Nature Switzerland AG 2023
N. Xiong et al. (Eds.): ICNC-FSKD 2022, LNDECT 153, pp. 905–912, 2023.
https://doi.org/10.1007/978-3-031-20738-9_99

with the interpretability of the model, so the trade-off between model accuracy and model interpretability is a major challenge. Post-hoc interpretation can include instance-based interpretation and attribute-based interpretation, where attribute-based interpretation can generate information such as feature importance [3] and instance-based interpretation can generate representative data points [4].

Counterfactual explanations [5–7] are an instance-based interpretation technique that, in the context of interpretability in machine learning, turns the predictions of a given instance of interest for a particular machine learning model into a preset outcome by applying perturbations to it. Using the loan application task as an example, counterfactual explanations aids in identifying the adjustments that should be done by a loan applicant whose application has been rejected for it to be approved.

The rest of the paper is organized as follows. The associated domains are introduced in Sect. 2, counterfactual explanations is defined in Sect. 3, and its characteristics and metric formulas are covered in Sect. 4. Section 5 summarizes six classical counterfactual explanations algorithms, and two counterfactual explanations applications are described in Sects. 6: model robustness, and generating feature importance.

2   Related Work

2.1   Contrast Explanation

According to a study in cognitive science, people don't always want to know every reason that could have caused an event, which is usually irrational and unimportant [8]. A comparison, or an explanation of why this event occurred as opposed to one that people would prefer to occur, provides a more insightful justification for the explanation. A contrastive explanation is comparable to a human explanation of a decision and is typically more in line with the person being explained's mental expectations.

A contrastive explanation [9] gives an explanation by answering why did P happen rather than Q, where Q is another different decision, called a hypothesis, that is of interest to the explainee. Unlike the contrast explanation, The counterfactual explanation, aims to provide a variant of the explained instance that causes the machine learning model to make decision Q rather than decision P.

2.2   Causability

Calculating the feature importance of the data is one technique used in machine learning tasks to explain why decision P happens for a particular instance, however it is challenging for the explainee to establish a causal relationship with the final decision based on the feature importance. Causality is a key element in the interpretation of black-box models and reflects [10]. The notion of counterfactual reasoning, which involves substituting false situations or possibilities, has been employed extensively in the domains of psychology, statistics, cognitive science, and other fields. A consensus among cognitive scientists is that counterfactual reasoning is a key element in human cognitive modeling of causal relationships in complex environments. For machine learning systems, counterfactual explanations is an important aspect to achieve causal interpretability, and therefore, the design of counterfactual explanations algorithms should take relevant causal constraints, such as feasibility and plausibility, into account. Holzinger [11] argues that human-computer interfaces with counterfactual explanations facilitate causality.

3 Definition

For a given prediction model, the input of interest is altered as little as possible to generate a predefined output, and the altered input is the counterfactual explanations of that input.

Definition 1. *Counterfactual Explanations*
    Let $\mathcal{X} \subseteq \mathbb{R}^n$ denote the feature space, $\mathcal{Y}$ denote the label space, where $\mathcal{Y} \in \{1, \ldots, K\}$ or $\mathcal{Y} \subseteq \mathbb{R}$. Suppose there exists a prediction model $h_\theta : \mathcal{X} \mapsto \mathcal{Y}$, and the model parameters are θ. Map any feature vector $\boldsymbol{x} = (x_1, \ldots, x_n) \in \mathcal{X}$ to the label $h_\theta(\boldsymbol{x}) = y \in \mathcal{Y}$.

- *For the classification task, the goal of the counterfactual explanations is to convert the instance of interest x to x^{cf} in order to change the original prediction class to the target class y_T, s.t. $h_\theta(x^{cf}) = y_T$.*
- *For the regression task, set the threshold $\delta \in \mathbb{R} \backslash \{0\}$, and consider the counterfactual x^{cf} valid when $\left| h_\theta(x^{cf}) - h_\theta(x) \right| \geq \delta$.*

A basic requirement of counterfactual explanations is that the generated counterfactual explanations is as close as possible to the instance x on a given distance metric. Where the distance metric is related to the data type.

Definition 2. *Counterfactual Explainer*
    The counterfactual explainer g is defined as a mapping with the input being the instances of interest $x \in \mathcal{X}$, the known data set $X \subseteq \mathcal{X}$ and its corresponding label $Y \subseteq \mathcal{Y}$, given a prediction model h_θ, the output is the set $C = \{x_1^{cf}, x_2^{cf}, \ldots, x_k^{cf}\}, x_i^{cf} \in \mathcal{X}$, The explainer g can be represented as $g : (x, X, Y, h_\theta) \mapsto C$.

4 Properties and Metrics

The counterfactual explanations aim to find instances with predefined labels and closest to the original instance, but this alone is not sufficient. In addition, counterfactual explanations need to satisfy some additional qualities to ensure the generation of meaningful counterfactual explanations.

    Consider the case of N data points $\{x_i\}$ and generate counterfactual instances (CF) $\{x_i^{cf}\}$ for each data point. Let $x_{i,p}$ and $x_{i,p}^{cf}$ denote the pth feature of the ith data.

4.1   Validity

Validity is the most fundamental and critical requirement for counterfactual explanations. The percentage of predictions made by a given prediction model for the generated counterfactual instances $h_\theta(x^{cf})$ that agree with the target class y_T is known as validity and is calculated as follows

$$\frac{\sum_{i=1}^{N} I\left[f\left(x_i^{cf}\right) == y_T\right]}{N}$$

where I denotes the indicator function, and 1 if the condition holds, 0 otherwise.

4.2   Sparsity

Sparsity is defined as the ratio of the number of features that change from the instance of interest x to the counterfactual instance x^{cf} to the total number of features, and the number of features that change should be as few as possible. The metric formula is expressed as

$$\frac{1}{n * N} \sum_{i=1}^{N} \sum_{p=1}^{n} I\left(x_{i,p}, x_{i,p}^{cf}\right), I = \begin{cases} 1, \text{if} x_{i,p} = x_{i,p}^{cf} \\ 0, \text{if} x_{i,p} \neq x_{i,p}^{cf} \end{cases}$$

The final sparsity measure is obtained by averaging over all N data points.

4.3   Similarity

Similarity is defined as the distance between counterfactual instances x^{cf} and x under a given distance function. The formula used here comes from a metric proposed by Watcher [12] in the paper.

$$\begin{cases} \frac{1}{n_{num}*N} \sum_{i=1}^{N} \sum_{p=1}^{n_{num}} \frac{|x_{i,p}-x_{i,p}^{cf}|}{MAD_p}, & \text{if } x_{i,p} \text{ is numerical} \\ \frac{1}{n_{cat}*N} \sum_{i=1}^{N} \sum_{p=1}^{n_{cat}} \left(1 - I\left(x_{i,p}, x_{i,p}^{cf}\right)\right), & \text{if } x_{i,p} \text{ is categorical} \end{cases}$$

The measure used is $L1$ distance with the inverse of the median absolute deviation as the weight, where the formula for median absolute deviation is $MAD_p = \text{median}_{i \in \{1,...,N\}}\left(|x_{i,p} - \text{median}_{l \in \{1,...,N\}}(x_{l,p})|\right)$. The final similarity score is a weighted sum of numerical and discrete scores whose weights are $\frac{n_{num}}{n}$ and $\frac{n_{cat}}{n}$ respectively.

4.4   Plausibility

Plausibility measures whether the generated counterfactual is realistic, similar to the known dataset, and fits the correlations between observed features.

The way to determine the reasonableness of generating counterfactual instances is to manually define the real space representing the constraints for the dataset to determine whether the counterfactual instances are reasonable [20].

4.5 Actionability

Actionability considers whether the changed feature involves an infeasible feature, such as gender, race, etc. [21].

Given a set A of actionable features, a counterfactual x^{cf} is actionable if the difference $\delta_{x,x^{cf}}$ between the counterfactual x^{cf} and x involves only actionable features.

4.6 Speed

Speed represents the time to generate the counterfactual x^{cf}. In practice, the generation time of counterfactuals should be limited. The metric is $t_{final} - t_{init}$.

5 Algorithms

In this section, we summarize several classical counterfactual algorithms.

5.1 The Counterfactual Explanation Proposed by Wachter

By designing the loss function, the loss takes as input the instance of interest, the counterfactual and the desired (counterfactual) outcome, and then an optimization algorithm can be used to find the counterfactual explanation that minimizes this loss. Wachter et al. [12] propose to reduce the loss function as follows

$$L\left(x, x', y', \lambda\right) = \lambda \cdot \left(\hat{f}\left(x'\right) - y'\right)^2 + d\left(x, x'\right)$$

The first term is the squared distance between the model prediction counterfactual x' and the desired outcome y'. The second term is the distance d between the instance x to be explained and the counterfactual x'. The loss measures the distance between the predicted outcome of the counterfactual and the predefined outcome as well as the distance between the counterfactual and the instance of interest. The distance function d is defined as the Manhattan distance weighted using the inverse median absolute deviation (MAD) of each feature, and the formula is detailed in Sect. 4.3.

5.2 FACE

FACE [13] is concerned with the feasibility and actionability of counterfactual explanations, so under the constraint of the "closest" principle, counterfactual data points are required to be in the high-density region, and the connection path with the explained data point is continuous, and passes through the high-density region as much as possible. First, the data point graph is constructed based on one of the three methods based on KDE, k-NN or ϵ-graph, and then a set of candidate targets (data points) is obtained according to whether the subjective prediction confidence threshold (tp), density threshold (td), and whether the custom weight function (w) and the custom condition function (c) are met. The goal of applying FACE is to obtain more feasible and actionable data points.

5.3 SHAP-based

Combining SHAP and counterfactual explanations ideas, SHAP-C is utilized to implement counterfactual explanation of high-dimensional behavioral data and textual data prediction systems [14]. In order to implement counterfactual explanation, SHAP first maps the instances to a binary representation, then creates perturbed instances and assigns weights to them, creates important ranking lists through training, and finally applies the effective search algorithm for counterfactuals for linear models (lin-SEDC6). An extremely high degree of data imbalance currently exists for the model-neutral method SHAP-C.

5.4 C-CHVAE

C-CHVAE [15] is a general-purpose framework which is capable of generating faithful counterfactuals. Additionally, C-CHVAE uses an autoencoder to represent heterogeneous data and approximate the conditionally log-likelihood of the actionable attributes given the immutable features. C-CHVAE is suitable for tabular data, can generate fathful counterfactuals, is compatible with multiple AE architectures, and does not have to access the distance function for the input space.

5.5 Preserving Causal Constraints in Counterfactual Explanations

Mahajan et al. [16] focus on the challenge of feasibility in counterfactual explanation and formally define feasibility based on an underlying structural causal model between input features: a counterfactual example is feasible if the changes satisfy constraints entailed by the causal model. Based on constraints derived from the causal model, a causal proximity regularizer is proposed that can be added to any counterfactual generation method to obtain feasible counterfactual explana-tions. In practical situations, feasibility constraints are not always available, thus Mahajan et al. propose another VAE-based approach that can learn feasibility constraints from user feedback.

6 Applications

6.1 Model Robustness

Given two black-box models, one model is more robust if the original and counterfactual points are further apart on average compared to the other one [17]. Robustness score based on counterfactual explanations (CERScore) -

$$CERScore\,(\mathrm{model}\,) = \mathbb{E}_X\left[d\left(\mathbf{x}, \mathbf{c}^*\right)\right]$$

6.2   Feature Importance

The intuition behind the method of calculating feature importance based on counterfactual explanations is that features with high importance are more likely to be changed when generating counterfactual instances than features with low importance, which means that features that are changed more frequently when generating counterfactual instances are more important [18]. Thus the attribution score of a feature x_i can be defined as the proportion of the generated counterfactual instances in which the feature x_i is changed.

7   Conclusion

Counterfactual explanation is a novel form of explanation which is model-agnostic. The key to counterfactual is to make as few adjustments as necessary to have the desired consequences. By presenting factual and counterfactual, counterfactual explanation enables data subjects to quickly, and easily interpret information. Researchers argue that counterfactual explanations is consistent with GDPR legal requirements and contributes to causal interpretability, but existing counterfactual explanations algorithms are still unable to achieve this, simply by adding causal structural constraints to the optimization objective.

Combining counterfactual explanations with human-computer interaction is a meaningful research direction that can further enhance interpretability while helping to achieve causality. In addition, it makes sense to combine counterfactual explanations with traditional explanatory techniques. If, along with actionable feedback, the counterfactual explanation also gives the reason for the initial decision, this can help the user to further understand the model logic

Acknowledgement. This work is supported by the National Natural Science Foundation of China under Grant No. 62172316 and the Key R&D Program of Hebei under Grant No. 20310102D. This work is also supported by the Key R&D Program of Shaanxi under Grant No. 2019ZDLGY13-03-02, and the Natural Science Foundation of Shaanxi Province un-der Grant No. 2019JM-368.

References

1. Adadi, A., Berrada, M.: Peeking inside the black-box: a survey on explainable artificial intelligence (xai). IEEE Access **6**, 52 138–52 160 (2018)
2. Carvalho, D.V., Pereira, E.M., Cardoso, J.S.: Machine learning interpretability: a survey on methods and metrics. Electronics **8**(8), 832 (2019)
3. Lundberg, S.M., Lee, S.-I.: A unified approach to interpreting model predictions. Adv. Neural Inf. Process. Syst. **30** (2017)
4. Kim, B., Khanna, R., Koyejo, O.O.: Examples are not enough, learn to criticize! criticism for interpretability. Adv. Neural Inf. Process. Syst. **29** (2016)
5. Guidotti, R.: Counterfactual explanations and how to find them: literature review and benchmarking. Data Min. Knowl. Discov., pp. 1–55 (2022)

6. Karimi, A.-H., Barthe, G., Schölkopf, B., Valera, I.: A survey of algorithmic recourse: definitions, formulations, solutions, and prospects. *arXiv preprint* arXiv: 2010.04050 (2020)

7. Stepin, I., Alonso, J.M., Catala, A., Pereira-Fariña, M.: A survey of contrastive and counterfactual explanation generation methods for explainable artificial intelligence. IEEE Access **9**, 11 974–12 001 (2021)

8. Miller, T.: Explanation in artificial intelligence: insights from the social sciences. Artif. Intell. **267**, 1–38 (2019)

9. Lipton, P.: Contrastive explanation. R. Inst. Philos. Suppl. **27**, 247–266 (1990)

10. Pearl, J., et al.: Models, Reasoning and Inference, vol. 19(2). Cambridge University Press, Cambridge (2000)

11. Völkel, S.T., Schneegass, C., Eiband, M., Buschek, D.: "What is" intelligent "in intelligent user interfaces? a meta-analysis of 25 years of iui". In: Proceedings of the 25th International Conference on Intelligent User Interfaces, pp. 477–487 (2020)

12. Wachter, S., Mittelstadt, B., Russell, C.: Counterfactual explanations without opening the black box: automated decisions and the gdpr. Harv. JL & Tech. **31**, 841 (2017)

13. Poyiadzi, R., Sokol, K., Santos-Rodriguez, R., De Bie, T., Flach, P.: Face: feasible and actionable counterfactual explanations. In: Proceedings of the AAAI/ACM Conference on AI, Ethics, and Society, pp. 344–350 (2020)

14. Ramon, Y., Martens, D., Provost, F., Evgeniou, T.: A comparison of instance-level counterfactual explanation algorithms for behavioral and textual data: Sedc, lime-c and shap-c. Adv. Data Anal. Classif. **14**(4), 801–819 (2020)

15. Pawelczyk, M., Broelemann, K., Kasneci, G.: Learning model-agnostic counterfactual explanations for tabular data. In: Proceedings of The Web Conference, pp. 3126–3132 (2020)

16. Mahajan, D., Tan, C., Sharma, A.: Preserving causal constraints in counterfactual explanations for machine learning classifiers. *arXiv preprint* arXiv: 1912.03277 (2019)

17. Sharma, S., Henderson, J., Ghosh, J.: Certifai: counterfactual explanations for robustness, transparency, interpretability, and fairness of artificial intelligence models. *arXiv preprint* arXiv: 1905.07857 (2019)

18. Kommiya Mothilal, R., Mahajan, D., Tan, C., Sharma, A.: Towards unifying feature attribution and counterfactual explanations: Different means to the same end. In: Proceedings of the 2021 AAAI/ACM Conference on AI, Ethics, and Society, pp. 652–663 (2021)

19. Yousefzadeh, R., O'Leary, D.P.: Interpreting neural networks using flip points. *arXiv preprint* arXiv: 1903.08789 (2019)

20. Artelt, A., Hammer, B.: Convex density constraints for computing plausible counterfactual explanations. In: International Conference on Artificial Neural Networks, pp. 353–365. Springer (2020)

21. Ustun, B., Spangher, A., Liu, Y.: Actionable recourse in linear classification. In: Proceedings of the Conference on Fairness, Accountability, and Transparency, pp. 10–19 (2019)

Retrospective Characterization of the COVID-19 Epidemic in Four Selected European Countries Via Change Point Analysis

Carmela Cappelli[✉]

University of Naples Federico II, Via Porta di Massa 1, 80133 Naples, Italy
carcappe@unina.it

Abstract. The focus of this contribution is to show how the course of the pandemic can be retrospectively investigated in terms of change points detection. At this aim, an automatic method based on recursive partitioning is employed, considering the time series of the 14-day notification rate of newly reported COVID-19 cases per 100,000 population collected by the European Centre for Disease Prevention and Control. The application shows that the pandemic, at the individual country level, can be broken into different periods that do not correspond to the common notion of wave as a natural pattern of peaks and valleys implying predictable rises and falls. This retrospective analysis can be useful either to evaluate the implemented measures or to define adequate policies for the future.

Keywords: Change point analysis · COVID-19 pandemic · Atheoretical Regression Trees

1 Introduction

Automatic detection of change points in a signal, whose number and location are unknown, is useful in several fields of research and applications such as finance, medicine, meteorology, business, industry, especially when a large amount of time series is routinely collected and analyzed.

A change point breaks a time series into segments that differs from each other for a given statistical characteristics. As a consequence the change point is located where the underlying characteristics changes. Several methods have been proposed in the literature, an overview of the most common change point detection approaches can be found in Aminikhanghahi and Cook [1].

In the present paper a procedure for locating multiple changes occurring at unknown time points, that exploits the recursive partitioning approach of tree based methods, is applied to investigate the course of the coronavirus disease 2019 pandemic in some selected European countries focusing on time series of

N. Xiong et al. (Eds.): ICNC-FSKD 2022, LNDECT 153, pp. 913–920, 2023.
https://doi.org/10.1007/978-3-031-20738-9_100

the 14-day notification rate of newly reported COVID-19 cases per 100,000 population collected by the European Centre for Disease Prevention and Control (ECDC). Change point analysis has been employed in an early stage of the pandemic (up to June 2020) aiming at identifying possible turning point in the time series of daily confirmed cases in 20 selected countries all over the world [2]. In the present case, after more than two years of pandemic, a retrospective analysis is conducted.

Indeed, policymakers, media and researchers describe the COVID-19 pandemic in terms of common waves and tend to compare countries, but, from one side there's no strict definition of an epidemic wave (see the discussion in [3]) and moreover, as it will be shown, the upward and downward periods are not globally the same not only, as expected, in terms of number and dates of changes, but mostly because they are not homogeneous across the selected countries in terms of their duration, level and sing of the rates over time.

This retrospective analysis conducted at the individual country level, used in combination with the analysis of deaths, excess mortality, rates of hospitalization and of Intensive Care Unit admissions, can be useful both to evaluate the implemented measures and to define adequate preventing policies for the future.

2 Change Point Detection Via Atheoretical Regression Trees

In this section it is briefly illustrated a method named Atheoretical Regression Trees (ART) proposed by Cappelli et al. [4] that exploits Least Square Regression Trees (LSRT) methodology to partition a target variable Y maintaining its intrinsic ordering. Thus, in case the target variable is a time series, the partition will consist of intervals that retain the temporal order, in fact ART has been effectively employed in time series analysis to locate multiple changes occurring at unknown dates also considering non standard data as in [5,6].

In general, if we are dealing with a continuous target variable Y and a set of p covariates, $X_1, ..., X_p$ observed in a sample of n statistical units, a regression tree models the relationship between the target variable and the predictors by turning the covariates into dummy variables, known as splitting variables, and then it fits pice-wise constant functions to the data. The binary tree is the result of recursively splitting the training set $(y_i, x_{i1}, ..., x_{ij}, ..., x_{ip})_{i=1}^n$ into two subgroups by searching at any internal node h for the split s that provides the highest reduction in the sum of squares

$$\Delta SS(s, h) = SS(h) - [SS(h_l) + SS(h_r)] \tag{1}$$

where $SS(h)$ is the sum of squares at node h and $SS(h_l)$ and $SS(h_r)$ are the corresponding quantities at the left and right descendants. As the child nodes are an exhaustive partition of the father node, $SS(h)$ represents the total sum of squares $TSS_y(h)$ while $[SS(h_l) + SS(h_r)]$ is the within-group sum of squares $WSS_{y|s}(h)$. Therefore, the right side of Eq. 1 is the between-groups sum of squares $BSS_{y|s}(h)$ that, for a partition into two group, is defined as

$$BSS_{y|s}(h) = \frac{n(h_l)n(h_r)}{n(t)^2}(\hat{\mu}(h_l) - \hat{\mu}(h_r))^2. \qquad (2)$$

In other words, in LSRT, the best split of each internal node generates the child nodes that are as far as possible in terms of squared distance between the corresponding means.

After the partition of the father node, the splitting process is applied to each child node and so on, until either a minimum size is reached or no reduction of the node sum of squares can be achieved.

In case of ART, the continuous target variable is a univariate time series Y_t that is tree-regressed on the time itself, represented by an artificial covariate that is an arbitrary sequence of strictly increasing numbers, hence the name Atheoretical Regression Tree. The use of the artificial covariate transforms the tree growing procedure into a recursive application of the Fisher method [7] for grouping K elements into G contiguous subsets for maximum homogeneity where $G = 2$ at all internal nodes.

Then, in the binary tree generated by ART, the split points are interpreted as change points in the time series while the terminal nodes are homogeneous time segments.

The method is computationally effective as it requires $O(n(h))$ steps to identify the best split (change) point of a node h where $n(h)$ is the number of elements in the node, it produces a hierarchical structure providing information on the order of importance of the detected changes and further it can be easily adapted to study instability and find changes into parameters of time series models fitted to the data [8] .

3   Data Description and Application

The identification on Dec 31, 2019, of the first COVID-19 infections in Wuhan, China, was soon followed by the spread of the novel coronavirus all over the world. In Europe, that become the epicenter, first patients were diagnosed with the disease between January and February. Since then, national authorities started to record data about positive cases, number of tests performed, deaths, hospital and Intensive Care Unit (ICU) admissions and on 27 January, 2020, the ECDC and the World Health Organization Regional Office for Europe implemented the COVID-19 surveillance. As a consequence, various databases have been created over the time. In case of the present analysis the weekly updates on the 14-day notification rate of newly reported COVID-19 cases per 100,000 population collected by the ECDC Epidemic Intelligence (downloadable at [10]) has been considered.

It's worth noticing that testing policies, the tests performed per 100,000 population, laboratory capacity and the effectiveness of surveillance systems vary markedly across the countries and moreover, daily data may not correspond to the actual number of new cases on the given day but, instead, to the cases *reported* on that day, for this reason it is advisable to consider a longer time span that is less influenced by the variation in daily reporting and thus by missing data, delayed notifications, erroneous recording, etc.and overall provides more reliable information as also discussed by [9].

Two "small" countries, namely Austria and Sweden and 2 "big" countries, Italy and United Kingdom, have been selected; the corresponding time series all end in week 23 of 2022[1] while their starting points vary according to when first cases were diagnosed in each country and the data collection started. Moreover, for the purpose of the analysis, the time series of the 14-day rates of new cases per 100,000 population of the above mentioned countries, have been transformed into the first differences displayed in Fig. 1. Since the first differences show the change in the rate of confirmed cases per 100,000 population over 14 days relative to the number of cases in the previous 14 days, they provide a more vivid image of when the pandemic is accelerating, decelerating or staying the same and thus, searching for change points in these series fulfills the aim of the present analysis.

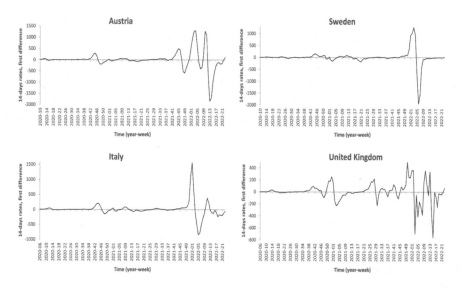

Fig. 1. 14-day notification rates of new cases per 100,000 population-first differences.

In order to identify the subperiods characterizing each country the ART approach has been applied, setting in the partitioning process, for the sake of homogeneity of the analysis across countries, the same minimum interval length, equal to 6 weekly notifications; another parameter that controls the size of a tree structure, namely the complexity parameter, has been also set to a common value very small ($cp = 0.001$) in such a way to let the procedure identify as much change points and subperiods as possible. Table 1 reports the results

---

[1] Weeks are numbered according to the European week numbering system based on the International Standard ISO 8601.

for each country, specifically the starting and ending week of each subperiod identified by the detected change points and the corresponding averages of the weekly confirmed new cases and of the change in the 14-day notification rate, respectively. In Fig. 2 the change points are depicted in the series of the weekly new cases.

Table 1. Subperiods identified by ART for each country: starting and ending date (format year-week) and corresponding average weekly cases and average 14-day rate change of new cases (below each subperiods in brackets).

	Austria Pop. 8,932,664	Sweden Pop. 10,379,295	Italy Pop. 59,236,213	United Kingdom Pop.68,059,863
1	2020–09;2020–40 (1583; +3.7)	2020–8;20–42 (2929;+2.6)	2020–5;2020–40 (9494; +1.5)	2020–05;2020–35 (9967;+0.9)
2	2020–41;2020–46 (26,652;+153.7)	2020–43;2020–51 (30,941;+85.5)	2020–41;2020–46 (157,851; +128.5)	2020–36;2020–48 (101,761;+24.5)
3	2020–47;2020–52 (23,648 ; −120.2)	2020–52;21–04 (31,784;-73.8)	2020–47;2020–52 (135,719;-81.0)	2020–49;2021–01 (260,000;+135.4)
4	2020–53 ;2021–40 (9908;-0.8)	2021–05;2021–15 (30,812;+33.6)	2020–53;2021–11 (110,773;+15.0)	2021–02;2021–07 (163,296;-152.2)
5	2021–41;2021–46 (51,066;-+281.2)	2021–16;2021–24 (19,267;-80.2)	2021–12;2021–22 (74,770; −41.0)	2021–08;2021–21 (26,752;-13.1)
6	2021–47;2021–52 (36,583;-258.1)	2021–25;2021–44 (4566;+1.8)	2021–23;21–21-42 (25,696;+0.4)	2021–22;2021–28 (149,971;+111.7)
7	2022–01;2022–11 (203,538;+588.2)	2021–45;2021–50 (13,627; +51.7)	2021–43;2021–48 (64,448;+41.7)	2021–29;2021–37 (223,014;-21.4)
8	2022–12;2022–23 (66,392; −532.3;)	2021–51;2022–04 (149,767;+768.0)	2021–49;2022–02 642,557;+631.2	2021–38;2021–48 (283,624;+26.8)
9	– –	2022–05;2022–10 (51752;-800.8)	2022–03;2022–08 (643,209;-504.1)	2021–49;2022–02 (819,987;+302.0)
10	– –	2022–11;2022–23 (3791;-18.5)	2022–09;2022–023 (326,543;-47.4)	2022–03;2022–08 (479,784;-327.6)
11	– –	– –	– –	2022–09;2022–23 (233,939;-41.2)

Notes and Comments. As a first observation, the selected countries share a common behavior mainly in the early period of the pandemic in fact no change points, for any of them, are detected by ART up to either the end of August 2020 (week 35 in Great Britain) or the beginning of autumn 2020 (week 42 in Sweden) confirming that, compared with what followed, the "first wave" of the pandemic, considered ended in June [11], is not actually distinguishable from the next two/three months.

Then, the partitioning process and the related change points enable to characterize, country by country, the course of the pandemic.

Austria presents the smallest number of time spans (8) and the longest interval (week 2020–53 starting on December 28, 2020, up to week 2021–41 ending on October 10) where the curve is almost flat (-0.8 average biweekly rate change) while it experienced the most intense acceleration ($+588.2$) during the interval between week 2022–01 (start January 3) and week 2022–11 (end March 20).

Sweden's response to the pandemic has been criticized because authorities kept the society open appealing to citizens' sense of responsibility. Its time series, divided into 10 intervals, shows, compared to the other countries, a delayed first change point occurring in week 2020–42 (start October 12) followed by 4 subperiods of alternate mild rises and incomplete falls never reaching a valley. Then, from week 2021 to 25 (start June 21) it has undergone a quite long interval of stability, lasted 23 weeks characterized by a low ($+1.8$) average change of the 14-day rate of new cases. Soon after, it's worth noticing, the presence of a very well defined leptocurtic wave started in week 2021–45 (start November 8) and ended in week 22–23 June that ART breaks into 4 distinct periods of similar length (6/8 weeks) corresponding to alternate slow/rapid accelerations/decelerations, respectively.

Italy, whose time series has been broken by ART into 10 subintervals, was the first European country to enact a lockdown nationwide and, in general, it has adopted the most restrictive measures. Nevertheless, after progressively making vaccination and COVID pass mandatory for a large part of the population, it was strongly hit by the more contagious new variant and between week 2021–49 and week 2022–02 the average biweekly rate change reached $+631.4$. After that, ART detects a similarly intense declining period (-504.6 average 14-day rate change) interrupted by a rise that overall decreased the deceleration.

Eventually, the United Kingdom time series is divided into 11 periods. It's particularly relevant that after a quite long interval of stable -under control-epidemic, from week 2021–08 (start March 8) up to week 2021–21 (end May 24) it shows a somewhat chaotic behavior that ART breaks into 6 segments characterized by varying accelerations and decelerations of the pandemic, due to fast-spreading variants compounded with the country's policy. Indeed, United Kingdom never pursued a zero-COVID goal in order to cushion the economic and social damage.

Overall, the location of the breaks detected in the series of the first differences of the biweekly rates of newly reported cases, depicted in the series of weekly new cases shows that the proposed approach is effective in partitioning the series into segments that, for each country, capture different phases of the pandemic.

As a final remark, when dealing with an unpredictable phenomenon, a retrospective analysis, like the one proposed, that retraces the past, may provide valuable information and contribute, in each country, to manage the present and designing appropriate guidelines for future outbreaks or epidemics. Indeed,

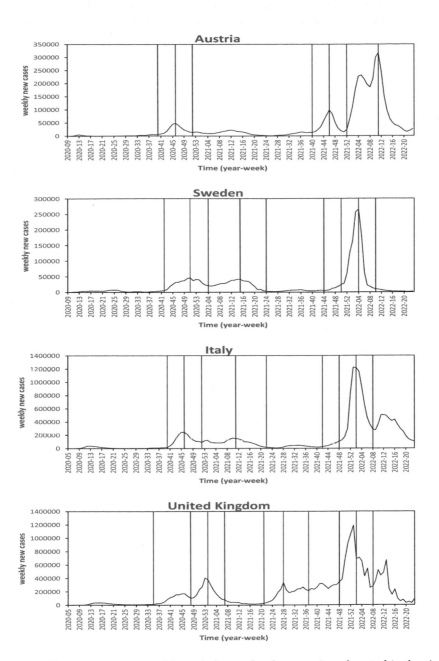

Fig. 2. Weekly new cases; vertical lines indicate the change points detected it the time series of the first differences of the 14-day notification rates.

a large scale study that includes all EU/EEA countries and encompasses the analysis of the series of deaths, hospitalizations and ICU admissions is the subject of ongoing research.

Acknowledgement. The author is profoundly grateful to her colleagues Prof. M. Corduas and Prof. D. De Caro for their support.

References

1. Aminikhanghahi, S., Cook, D.J.: A survey of methods for time series change point detection. Knowl. Inf. Syst. **51**, 339–367 (2017). https://doi.org/10.1007/s10115-016-0987-z
2. Coughlin, S.S., Yigiter, A., Hongyan, X., Berman, A.E., Chen, J.: Early detection of change patterns in COVID-19 incidence and the implementation of public health policies: a multi-national study. Public Health Pract. **2**, 100064 (2021). https://doi.org/10.1016/j.puhip.2020.100064
3. Zhang, S.X., Arroyo Marioli, F., Gao, R., Senhu, W.: A second wave? What do people mean by COVID waves? A working definition of epidemic waves. Risk Manag. Healthc. Policy **14**, 3775–3782 (2021). https://doi.org/10.2147/RMHP.S326051
4. Cappelli, C., Penny, R.N., Rea, W.S., Reale, M.: Detecting multiple mean breaks at unknown points in official time series. Math. Comput. Simul. **78**(2), 351–356 (2008). https://doi.org/10.1016/j.matcom.2008.01.041
5. Cappelli, C., D'Urso, P., Di Iorio, F.: Change point analysis of imprecise time series. Fuzzy Sets Syst. **225**, 23–38 (2013). https://doi.org/10.1016/j.fss.2013.03.001
6. Cappelli, C., D'Urso, P., Di Iorio, F.: Regime change analysis of interval-valued time series with an application to PM10. Chemom. Intell. Lab. Syst. **146**, 337–346 (2015). https://doi.org/10.1016/j.chemolab.2015.06.006
7. Fisher, W.D.: On grouping for maximum homogeneity. J. Am. Stat. Assoc. **53**, 789–798 (1958)
8. Cappelli C., Di Iorio, F.: Theoretical regression trees: a tool for multiple structural-change models analysis. In: Grigoletto, M., Lisi, F., Petrone, S. (eds) Complex Models and Computational Methods in Statistics, pp. 63–76. Springer, Milano (2013). https://doi.org/10.1007/978-88-470-2871-5_6
9. Our World in Data. https://www.ourworldindata.org
10. European Centre for Disease Prevention and Control Epidemic Intelligence. https://www.ecdc.europa.eu/en/publications-data/data-national-14-day-notification-rate-covid-19
11. Kontis, V., et al.: Magnitude, demographics and dynamics of the effect of the first wave of the COVID-19 pandemic on all-cause mortality in 21 industrialized countries. Nat. Med. **26**, 1919–1928 (2020). https://doi.org/10.1038/s41591-020-1112-0

Analyzing the Time-Delay Correlation Between Operational and State Parameters of Blast Furnace: A DTW and Clustering Based Method

Wenji Wang, Denghui Hao, and Yin Zhang[✉]

Software College, Northeastern University, Shenyang 110000, Liaoning, China
20195627@stu.neu.edu.cn, 20195355@stu.neu.edu.cn, zhangyin@mail.neu.edu.cn

Abstract. The complexity of the smelting process leads to many difficulties in accurately controlling the reaction of the blast furnace. It takes some time after the adjustment of the operational parameters to cause a change in the state parameters. The identification of time delay correlation between the operational parameters and state parameters of blast furnace is of great importance to optimization of blast furnace operation. In this paper, we propose a DTW and clustering based method to identify the time-delay characteristics between operational and state parameters. We propose to use the DTW algorithm as distance measure and use the k-means algorithm to clustering the pre-processed data to reduce the complexity of the calculation. We then use the Spearman correlation coefficient to measure time-delay correlation between operational and state parameters. We validated the proposed method using data collected from a real life blast furnace. Experimental results show that the proposed method can effective identify the time-delay characteristics.

Keywords: Blast furnace · Operational parameters · Time-delay characteristics

1   Introduction

In the study of blast furnace reactions, there are two types of parameters. (1) Operational parameters that are used to control and regulate the reaction process in the blast furnace, such as air velocity, oxygen enrichment flow, etc. (2) State parameters that refer to parameters such as top pressure and furnace temperature collected by the sensors in the blast furnace. These two types of parameters are characterized by a large amount of data and high dimensionality.

In production practice, there is a significant feature for the control of parameters: the operational parameters do not immediately lead to changes in the state

© The Author(s), under exclusive license to Springer Nature Switzerland AG 2023
N. Xiong et al. (Eds.): ICNC-FSKD 2022, LNDECT 153, pp. 921–929, 2023.
https://doi.org/10.1007/978-3-031-20738-9_101

parameters after manual adjustment, but after a period of time. This feature reflects the time delay characteristic of the operational parameters. In order to better control the reaction process, the time-delay characteristics of the operational parameters should be identified and the correlation between the factors should be determined.

When correlations are calculated between operational and state parameters, the large number of parameter attributes causes dimensionality inflation. Complex, high-dimensional data can increase the computational effort and slow down the computational speed. Meanwhile, there are many attributes in the operational and state parameters that are significantly correlated and their correlation has no sense of analysis. These attributes, if involved in the computation, also cause the problem of increasing the computational effort.

In this paper, the clustering method is used to analyze the pre-processed data, reduce the dimensionality of the data and the amount of computation and removes the data with obvious correlation. Correlation analysis algorithm of time series is used to analyze the time-delay correlation of the clustered data. By selecting representative clustering data and calculating the correlation coefficient, the magnitude of the correlation between two attributes can be measured. Through the above operation, a data-driven class analysis method reflecting the time-delay characteristics of operational parameters is established. Using the proposed method on a data set collected in a real blast furnace environment, better results were achieved to optimize the blast furnace operation.

2   Related Works

2.1   Blast Furnace Condition

There are two main approaches to the study of blast furnace conditions use computer science technology: data-driven modeling, and intelligent algorithms.

Data-driven model-based approach uses the experience and advice of experts in blast furnace-related fields, combined with artificial intelligence technology in the computer field to form a decision-making system to solve the problems. The first expert system for blast furnace condition prediction was developed by Kawasaki's "Go-Stop" system [1]. Junpeng Li et al. used a fuzzy model based on the data-driven Takagi-Sugenno algorithm to dynamically predict the silicon content of blast furnace molten iron [7]. It combines computational theory and empirical evaluation of blast furnaces to build a successful expert system.

Intelligent algorithm-based analysis method mainly uses data mining techniques and intelligent algorithms to analyze and model the blast furnace operation process, synthesizing the data from the blast furnace and discovering rules from it. Cheng Shi et al. used a distributed BP neural network model based on FCM (Fuzzy C-means) [2] to predict the furnace temperature of the blast furnace and used Particle Swarm algorithm for optimization. Yanjiao Li et al. used a hybrid predictive model based on a knowledge mining algorithm to evaluate the production status of the blast [6]. It provide reliable information for the field operators.

2.2 Clustering Based Blast Furnace Condition Analysis

Cluster analysis is an important class of data mining algorithms. Gu Dongyang proposed the use of similarity-based clustering algorithm and the use of time series correlation analysis for the relationship between operational parameters, state parameters and process indicators to improve the stability of the blast furnace production process and its automatic control level [3]. However, the clustering algorithms adopted in both studies use Euclidean distance as the basis for clustering. It does not fit well with the time-delayed nature of the operating parameters.

2.3 Time-Delay Correlation Analysis of Blast Furnace Conditions

Correlation coefficients between blast furnace operating parameters and state parameters can reflect the degree of correlation between the variables. Cui Guimei et al. used the autocorrelation function method [4] to determine the time lag of blast furnace process parameters on the silicon content of molten iron and other problems, and combined with the experience of relevant experts to determine the lag time of specific operation parameters [5]. These studies show that there is a time delay between the operational and state parameters of the blast furnace, and that studying this characteristic does make the operation of the blast furnace be optimized.

3 Proposed Method

In order to accurately identify the time delay characteristics of blast furnace operation parameters, a correlation analysis method is proposed. The method mainly includes: pre-processing of data, attribute clustering analysis based on k-means and correlation analysis considering time delay features. They are described in 3.1–3.3.

3.1 Data Pre-processing

Data pre-processing prior to data analysis ensures the accuracy and availability of the data. After manual collation, a database for blast furnace condition analysis and identification is constructed from the historical data related to the furnace condition obtained from different collection cycles of the blast furnace.

And then normalize the data to the most common value for different scales of data, mapping all data between 0 and 1. For missing data, if the value of a part of an attribute is missing, or the whole is missing, the whole attribute is eliminated.

Table 1. Operational parameters and state parameters of blast furnace

Operational parameters	State parameters
Atual blast volume	Multiple temperature
Oxygen-enriched flow	CO_2 content
Setting amount of coal injection	Oxygen-enriched rate
Blower humidity	Top pressure
Hot blast temperature	Pressure difference
Cold blast pressure cold blast flow	Breathability gas content
Theoretical combustion temperature	Gas index

3.2 Attribute Clustering Analysis Based on k-Means

After data acquisition by the sensors in the blast furnace, operational and status parameters related to the blast furnace status can be obtained. Some of these parameters are listed in Table 1.

It is necessary to calculate each of the hundreds of parameters two by two in order to perform a time delay correlation analysis of the operational parameters. This generates an extremely large amount of data, resulting in dimensional inflation.

More than 50,000 attribute pairs can significantly slow down the computation. Meanwhile, there exist obvious correlations between some attribute pairs in the operational and state parameters. If the above characteristics are ignored and the calculation is carried out directly, it will bring huge workload to the manual verification work of filtering out the significantly correlated and significantly uncorrelated results.

To solve the above problems, this paper uses an improved k-means clustering algorithm to cluster attributes, and selects representative attributes from each cluster to represent this cluster for correlation analysis. The purpose is to reduce the computational results, and increase the computational speed.

Traditional clustering algorithms usually use the Euclidean distance as the measure. Euclidean distance assumes independence between data. However, the blast furnace operational parameters data has a strong time dependence. So, to reflect the trend of data changes over time is important.

We propose to use DTW (Dynamic Time Warping) distance instead of Euclidean distance to calculate the similarity of two time series, and use k-means algorithm to perform clustering analysis. Spearman correlation coefficient is used to measure the similarity. The closer the value of this coefficient is to 1, the greater the correlation is.

DTW suppose there are two time series Q and C. The elements of the matrix (i, j) represent the distance $d(q_i, c_j)$ between q_i and c_j. Equations 1 and 2 can be used to calculate this distance.

$$DTW(Q, C) = \sqrt{\gamma(i, j)} \tag{1}$$

$$\gamma(i,j) = d(q_i, c_j) + min\gamma(i-1, j-1), \gamma(i-1, j), \gamma(i, j-1) \tag{2}$$

$\gamma(i,j)$ represents the cumulative distance from the point $(0,0)$ to match these two sequences Q and C. The whole algorithm can be represented as Algorithm 1.

K-means algorithm is used in the clustering process to group different objects according to their attributes in k groups. K is specified to represent the number of clusters to be generated.

Algorithm 1: Dynamic Time Warping algorithm

Input: Linear sequence Q,C,length of Q,C represent as n,m,
Output: distance of shortest path *shortest_distance*
1 **for** $i=0$ *to* $n-1$ **do**
2   **for** $j=0$ *to* $m-1$ **do**
3 | $Distance[i][j] \leftarrow Manhattan\_distances(Q[i], C[j])$
4   **end**
5 **end**
6 **for** $i=0$ *to* $n-1$ **do**
7   **for** $j=0$ *to* $m-1$ **do**
8 | $D_1 \leftarrow D_1 + min(D_0[i][j], D_0[i][j+1], D_0[i+1][j])$
9   **end**
10 **end**
11 $shortest\_distance \leftarrow D_1[-1][-1]$;
12 **return** $shortest\_distance$

The traditional k-means clustering algorithm uses the Euclidean distance measure to the centroid, in this paper, the DTW distance algorithm is chosen to be used instead based on the properties of the time series. Equation 3 describes the objective function of this process.

$$J = \sum_{i=1}^{n} \sum_{j=1}^{n} DTW(Q,C) \tag{3}$$

The smaller the value of the J-function of each point from the cluster centroid represents the closer it is to the centroid. It also means points should be grouped together. Where DTW(Q,C) represents the selected distance metric. The whole algorithm can be represented as Algorithm 2.

3.3 Correlation Analysis Based on Delay Characteristics

We calculate and compare the similarity distance between each type of data and other data, and finally determine the data with the smallest similarity distance in each type of data as the representative data for this type of data to participate in the correlation calculation.

$$\rho = \frac{\sum_i (x_{i-t} - \bar{x})(y_{i-t} - \bar{y})}{\sqrt{\sum_i (x_{i-t} - \bar{x})^2 \sum_i (y_{i-t} - \bar{y})^2}} \tag{4}$$

In the formula, t represents the offset, the value of ρ ranges from -1 to 1. The larger the absolute value of ρ is, the stronger the correlation between the two attributes are.

Algorithm 2: K-means clustering algorithm

Input: the number of clusters required k, training data X, number of training
        data sample m, maximum number of updates max_iter
Output: the centers of the cluster $centers$

1   $centers \leftarrow random(k), num\_iter \leftarrow 0$;
2   **while** $num\_iter < max\_iter$ **do**
3      $num\_iter \leftarrow num\_iter + 1$;
4   **for** $k = 0$ to $K - 1$ **do**
5          $current\_center \leftarrow center[k]$;
6          $dist \leftarrow calc\_disc(current\_center, current\_x)$;
7   **if** $dist < min\_dist$ **then**
8             $min\_dist \leftarrow dist$;
9             $center\_idx \leftarrow k$;
10   **end**
11         $center\_2\_x\_dict[center\_idx].append(i)$
12   **end**
13   **for** $k = 0$ to $K - 1$ **do**
14          $tmp\_x \leftarrow X[center\_2\_x\_dict[k]]$;
15          $centers[k] \leftarrow np.mean(tmp\_x)$;
16   **end**
17 **end**
18 **return** $centers$

---

4   Experiment and Analysis

This paper firstly obtains a part of long-running data from a steel mill as experimental data to be processed. Table 2 represents part of the data. Attribute names are substituted with "AA,AB,AC,AD,AE,AF".

In this paper, the DTW distance is used to replace the traditional Euler distance. Below are some of the clustering results. Figure 1 represent the first category, and Fig. 2 represent the second category

The data dimension before dimensionality reduction is as high as 234 attributes, and the high-dimensional data is converted into low-dimensional data, and the result after clustering is 36 categories.

Set the maximum number of offsets to 1000 and calculate the correlation coefficient between the representative data of various types and all other data

Table 2. Experimental data

AA	AB	AC	AD	AE	AF
−0.48493	0.465752	0.730796	−0.3925	0.57057	−0.9496
−0.47074	0.481117	0.737139	−0.37237	0.558971	−0.94992
−0.46454	0.481199	0.74771	−0.38518	0.554893	−0.94961
−0.46365	0.502488	0.72234	−0.36871	0.553639	−0.95032
−0.46631	0.508253	0.739253	−0.36322	0.543287	−0.95017

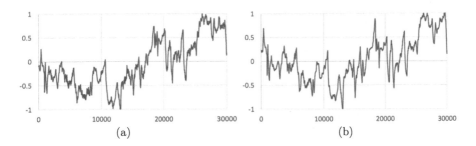

Fig. 1. Selected parameter in the first category

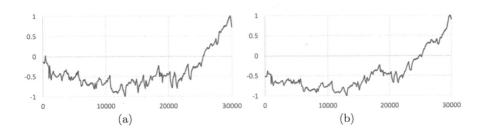

Fig. 2. Selected parameter in the second category

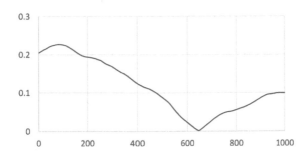

Fig. 3. Calculation results of partial correlation coefficients

with the change of offset, record the maximum value and make statistics. Figure 3 below shows some calculation results.

The abscissa represents the offset, and the ordinate represents the correlation coefficient of the two data. It can be seen from the figure that as the time-delay increases, the correlation of the two parameters increases, and gradually increases to the highest point. This means that the correlation of these two attributes is the largest at about time-delay 81. The correlation then rose again, possibly due to the cyclical nature of this attribute. When the correlation is reduced to the lowest point, the next cycle is entered, which makes the two data have a certain periodicity.

5   Conclusion

This paper studies the problem of analyzing time delay correlation between blast furnace operational and state parameters. To resolve the large amount of data when performing correlation delay analysis, this paper propose a DTW-based k-means clustering algorithm to cluster attributes with similar trends, and choose representative attributes from each cluster as properties to represent the class.

The time-delayed correlation between the representative data of each class and other data based on the delay are then calculated. Experimental results show that the proposed method can effective identify the time-delay characteristics. Our method can facilitate the better control of blast furnace reaction process and accurate selection of operating parameters. The results have certain significance for improving the economic benefit of blast furnace and reducing the energy consumption of blast furnace.

References

1. Ma, Z.W., Yang, F.Q.: Analysis and transplant of furnace condition prediction go-stop system for blast furnace. Metall. Ind. Autom. **1**(3–5), 19–63 (1992)
2. Chen, S., Cui. G.M.: Research on prediction of Blast Furnace Temperature Based on Fuzzy Distributed Model (2013) https://doi.org/10.3969/j.issn.1001-182X.2013.07.001
3. Gu, D.Y.: The operational parameters optimization based on data-driven for blast furnace ironmaking process. Master's thesis, Inner Mongolia University of Science & Technology (2014). https://doi.org/10.7666/d.D517251
4. Gao, C.H., Jian, L., Chen, J.M., Sun, Y.X.: Data-driven modeling and predictive algorithm for complex blast furnace ironmaking process. IEEE/CAA J. Automatica Sin. **35**(6), 725–730 (2009)
5. Cui, G.M., HOU, J.,Gao, C.L.: Optimization of operational parameters of bf iron-making based on multi-sensor. Transducer Microsyst. Technol. **34**(3), 21–23, 27 (2015). https://doi.org/10.13873/J.1000-9787(2015)03-0021-03
6. Li, Y., Li, H., Zhang, J., Zhang, S.: Data and knowledge driven approach for burden surface optimization in blast furnace. Comput. Electr. Eng. **92**, 107, 191 (2021). https://doi.org/10.1016/j.compeleceng.2021.107191

7. Li, J., Hua, C., Yang, Y., Guan, X.: Data-driven Bayesian-based Takagi-Sugeno fuzzy modeling for dynamic prediction of hot metal silicon content in blast furnace. IEEE Trans. Syst. Man Cybernet. Syst. **52**(2), 1087–1099 (2022). https://doi.org/10.1109/TSMC.2020.3013972

Multi-scale Algorithm and SNP Based Splice Site Prediction

Jing Zhao[1], Bin Wei[2(✉)], and Yaqiong Niu[2]

[1] Xijing University, Xi'an, China
zhaojing.83@163.com
[2] Engineering University of PAP, Xi'an, China
weibin82@126.com, yaqiongniu1994@163.com

Abstract. In the post-genomic era, gene function prediction get more and more attention. Splice site prediction is one of the most important part of those study. At present, a lot of algorithms have been proposed to this study, however, due to lack of understanding of the splicing mechanism, the performance of most of the methods has been influenced. This paper based on the relationship between the base and the codon designed a multi-scale algorithms for analysis of splice sites. In addition, the single nucleotide polymorphism has been used in this process, for exploration mutation on splicing mechanism by computation method. Finally, the experimental results showed that the proposed method can greatly improve the prediction accuracy. It is potentially interesting as an alternative tool in those studies.

Keywords: Bayesian · Genetic algorithm · SNP

1 Introduction

Gene function prediction and analysis is one of the hottest issues in current bioinformatics research [1], and eukaryotic gene splice site identification is not only an important and critical issue, but also the basis for our understanding of complex life processes [2].

In the past years, a large number of methods and software have been produced [3], such as artificial neural network (ANN) [4], variable Markov model [5], probability and statistical model [6], etc. However, due to the limitations of model methods and the lack of understanding of the splicing mechanism, most algorithms are not good in terms of prediction accuracy.

In summary, this paper designed a multi-scale algorithm to analyze splice sites based on the correspondence between bases and codons. In addition, single nucleotide polymorphism (SNP) is considered to be one of the most important discoveries in the post-genomic era [7], and it has a certain impact on splicing. In this paper the SNP data were integrated to explore the impact of variation on the splicing mechanism from a computational perspective for the first time. Finally, the effectiveness of the model designed in this paper is verified based on the comparison with the existing published methods by some experiments.

N. Xiong et al. (Eds.): ICNC-FSKD 2022, LNDECT 153, pp. 930–938, 2023.
https://doi.org/10.1007/978-3-031-20738-9_102

2 Multi-scale Locus Recognition Algorithm

Since this kind of research generally involves a relatively large amount of data and contains a lot of redundant information, it can be preprocessed by means of attribute reduction, which is also one of the most commonly used methods. At present, the most commonly used models are mainly divided into two categories: filtering and wrapper models [8].

However, almost methods treat introns and exons equally, that is, reduce and predict indiscriminately in the computational process. Obviously, such a treatment method does not fully consider the differences in functions and roles of introns and exons. Therefore, based on the fact that bases in exons correspond to codons, this paper designs a multi-scale prediction method for different characteristics of exons and introns (as shown in Fig. 1). That is, different scales are used for introns and exons in the processing of each model and algorithm. The intron part treats the base as an independent individual, while the exon part treats every three bases (with codon correspondence) as a whole.

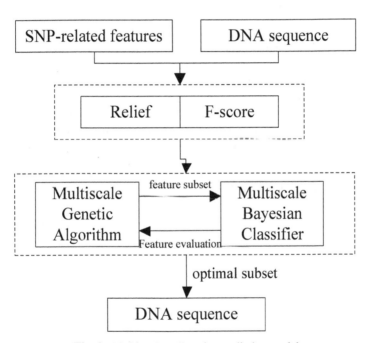

Fig. 1. Multiscale splice site prediction model

2.1 Filtering Module

F-score It is a simple and effective filtering algorithm, which defines the importance of features by their ability to distinguish between two categories. Given a set of $m * n$ dimensional data, where m and n are the number of samples and features, respectively. The number of samples in the positive and negative categories (true sites and false sites) are n_+ and n_- ($n_+ + n_- = m$), respectively, then the F-score of the i-th feature is defined as follows:

$$F(i) = \frac{(\bar{x}_i^{(+)} - \bar{x}_i)^2 + (\bar{x}_i^{(-)} - \bar{x}_i)^2}{\frac{1}{n_+-1} \sum_{k=1}^{n_+} (x_{k,i}^{(+)} - \bar{x}_i^{(+)})^2 + \frac{1}{n_--1} \sum_{k=1}^{n_-} (x_{k,i}^{(-)} - \bar{x}_i^{(-)})^2} \tag{1}$$

In the formula, $\bar{x}_i$, $\bar{x}_i^{(+)}$, $\bar{x}_i^{(-)}$ represents the average value of the i-th feature in the entire data set, positive class and negative class respectively; $x_{k,i}^{(+)}$ is the value of the i-th feature in the k-th positive class sample; $x_{k,i}^{(-)}$ is the value of the i-th feature in the k-th negative class sample. The numerator of Eq. 1 reflects the degree of distinction between the two types of samples, and the denominator reflects the distribution of the features within the two types of samples. The larger the value of F-score, the stronger the ability of the corresponding feature to distinguish two types of samples. Some redundant information can be filtered out by setting a threshold. In this paper, the median of all feature F values is selected as the threshold.

Relief The Relief algorithm is one of the most effective feature selection algorithms with high operating efficiency and is suitable for large-scale data processing. Relief evaluates the feature's ability to distinguish the nearest neighbor samples, and it believes that important features should make similar samples close and different between samples. The algorithm randomly selects a sample x in a given data set G, and according to the Euclidean distance, finds the most similar sample in the same sample and records it as $NH(x)$, and similarly finds the most similar sample in different types of samples and records it as $NM(x)$. And then for each dimension, the feature updates its weights according to Eq. 2.

$$W_i = W_i + \left| x^{(i)} - NM^{(i)}(x) \right| - \left| x^{(i)} - NH^{(i)}(x) \right|, i = 1, 2, ..., n \tag{2}$$

In the formula, $x^{(i)}$, $NM^{(i)}(x)$ and $NH^{(i)}(x)$ represent the value of the i-th feature of samples x, $NM(x)$ and $NH(x)$ respectively, and the initial weight of each feature is 0. It can be seen from the above formula that if $\left| x^{(i)} - NM^{(i)}(x) \right| > \left| x^{(i)} - NH^{(i)}(x) \right|$ indicates that the feature contributes to the classification, its weight increases; otherwise, the feature is not conducive to distinguishing two samples, and the weight should decrease. The algorithm repeats the above process L times, and finally outputs the average value of the feature weight. Larger the value is, stronger contribution will the feature commit towards classification. As in the previous section, the median is set as the threshold.

F-score and Relief algorithms have their own characteristics, and their filter (deletion) feature lists are necessarily not exactly the same, but there may be some overlap. Therefore, in this paper the features that are actually removed in the filtering part of the algorithm are the intersection of the above two lists (the purpose of this is to make the probability of potentially relevant features being removed in the filtering part as small as possible). In addition, the disjoint parts of the F-score and Relief filter feature lists are initialized to a small probability value at the corresponding MGA position (as shown in Fig. 2).

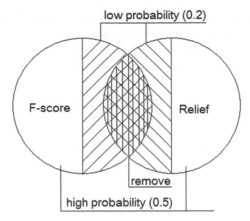

Fig. 2. Schematic diagram of the double filtering algorithm

2.2 Package Module

Multiscale Bayesian Classifier Naive Bayes classifiers have been widely used in biological information processing. It is known that in the process of coding, exons undergo transcription, translation and other operations according to the relationship between three bases and codon, and finally form the expression of proteins. Based on the that facts, this paper designs a multi-scale Bayesian classifier, that is, the algorithm divides the base sequence into two parts (exons and introns) during the training and prediction process. The bases are processed according to triplets (that is, the algorithm is improved according to the relationship of three bases corresponding to one codon), while the introns are still processed according to independent bases without any change.

Multiscale Genetic Algorithm Genetic Algorithm (GA) was proposed by Professor Holland [9]. It is one of the most typical and widely used optimization algorithm. In this paper, a multi-scale genetic algorithm is designed according to the corresponding relationship between bases and codons. In the process, the base sequence is divided into two parts in a similar way to the multi-scale Bayesian classifier. At this time, the triplet in the exon only needs one bit of the GA individual to correspond to it, that is to say the coding dimension in this part is only one-third of the original method, which greatly reduces the computational complexity and the demand for storage space.

3 Integration of SNP and Base Information

The data used in this paper were all obtained from Gene Bank Release 146, and Table 1 gives information on partial sequences. The commonly used methods for model training and prediction using DNA sequences are: (1) Use the complete DNA sequence and input sequentially using a sliding window; (2) According to the GT-AG rule, only short sequences containing GT or AG are used for training and testing data; (3) Use only sequences adjacent to the splice site, and then use a sliding window on these short sequences to input the network for training and testing. Comprehensively analyzing the advantages and disadvantages of the above methods, this paper designs a new method. Under the sliding window method, for the DNA sequence in the training sample, we use the short sequences of all splicing sites, and on this basis, randomly add a small number of complete sequences (less than 10% of the total number of sequences in the training set) are used to form training samples, and the test data can directly use complete sequences.

Table 1. List of sample data (Length: sequence length D: number of donor sites, A: number of acceptor sites)

GenBank code	Length	D	A	GenBank code	Length	D	A
HUMA1ATP	12,222	4	4	HUMHMG14A	8882	5	5
HUMA1GLY2	4944	5	5	HUMHSP90B	8210	11	11
HUMACTGA	3583	5	5	HUMHST	6616	2	2
HUMADAG	36741	11	11	HUMIL2	5737	3	3
HUMAK1	12,229	6	6	HUMIL1B	7824	6	6
HUMALBGC	19,002	14	14	HUMIL2B	5561	3	3
HUMALPHA	4556	10	10	HUMIL5	3230	3	3

SNP refers to the polymorphism of the DNA sequence caused by the variation of a single base in the genome with a probability of more than 1%, that is, any base in the DNA sequence is replaced by any one of the other three nucleotides, according to the two bases. Based on the sorting and analysis of more than 25 million human genome SNPs in the dbSNP database of the National Center for Biotechnology Information, we combined the SNPs with the DNA sequences involved in this paper, and using SNP information to annotate it, annotate the number of SNPs on each window, their location and type and other information, which are called SNP-related features, and combine this feature into a vector, which is used as the input of the algorithm together with the DNA sequence. In addition, the base feature in the original sequence will be replaced by the SNP feature to achieve the purpose of data integration.

4   Experimental Results and Analysis

4.1   Evaluation Indicators

Sensitivity and specificity are generally used in this field to evaluate the performance of the model.

$$Sn = \frac{TP}{TP + FN} \tag{3}$$

$$Sp = \frac{TN}{TN + FP} \tag{4}$$

where:
  TP—the number of correctly predicted splice sites;
  TN—the number of correctly predicted non-splicing sites;
  FN—the number of mispredicted splice sites;
  FP—number of mispredicted non-splicing sites.

If the sensitivity is high but the specificity is low, a high rate of false positives will be produced in practical applications; conversely, if the specificity is high and the sensitivity is low, a high rate of false negatives will be produced. Therefore, sensitivity and specificity need to be weighed. The comprehensive evaluation index Q^9 is given below.

$$q^9 = \begin{cases} \frac{TN-FP}{TN+FP}, & if : TP + FN = 0 \\ \frac{TP-FN}{TP+FN}, & if : TN + FP = 0 \\ 1 - \sqrt{2}\sqrt{\left(\frac{FN}{TP+FN}\right)^2 + \left(\frac{FP}{TN+FP}\right)^2}, \\ \quad if : TP + FN \neq 0, and, TN + FP \neq 0 \end{cases} \tag{5}$$

$$Q^9 = \frac{1+q^9}{2} \tag{6}$$

4.2 Results and Analysis

This paper discusses the donor data and the acceptor data separately, which is regarded as two classification problems.

Firstly, we tested the performance of the algorithm in the paper, and performed separate experiments on several elements used in the hybrid algorithm. Table 2 shows the experimental results (all experiments in the paper use five-fold cross-validation). From the table, we can see the importance of feature reduction in the study of such problems, that is, feature reduction can greatly improve the performance of the algorithm. In addition, the hybrid algorithm designed in this paper combines the advantages of the two models, so the result is the best.

Secondly, we tested the performance of the multi-scale algorithm designed according to the correspondence between bases and codons. Table 3 shows the performance of the multi-scale algorithm and the non-multi-scale algorithm (that is, all bases are regarded as independent features). From Table 3, it can be seen that the method proposed in this paper has been greatly improved the performance. In addition, compared with Table 2, it can be seen that the performance of the original method is even inferior to the multi-scale packaging algorithm. This also proves the effectiveness of this paper for algorithmic modification and the importance of designing algorithms according to actual biological phenomena.

Thirdly, we tested the impact of the newly involved SNP information on the performance. It can be seen from Table 4 that compared with the traditional data, the method of involved SNP-related features proposed in this paper greatly improves the identification performance of splicing sites (both donor and recipient sites have improved by nearly ten percentage points). Therefore, we can say that the SNP-related features involved in this paper to be successful in identifying splice sites.

Table 2. Hybrid algorithm performance verification

	Algorithm	Sn	Sp	Q9
Donor data	Multiscale Bayesian	0.8607	0.9971	0.9015
	F-score + relief + multi-scale Bayes	0.8861	0.9974	0.9194
	Multiscale genetic algorithm + multiscale Bayes	0.9240	0.9972	0.9462
	F-score + relief + multi-scale genetic algorithm + multi-scale Bayes	0.9620	0.9971	0.9730
Acceptor data	Multiscale Bayesian	0.7848	0.9940	0.8478
	F-score + relief + multi-scale Bayes	0.8481	0.9957	0.8925
	Multiscale genetic algorithm + multiscale Bayes	0.8481	0.9953	0.8925
	F-score + relief + multi-scale genetic algorithm + multi-scale Bayes	0.8987	0.9954	0.9283

Table 3. Multi-scale performance comparison

	Non-multiscale algorithm	Multiscale algorithm
Donor data	0.9397	0.9730
Acceptor data	0.8849	0.9283

Table 4. Influence of SNP information on algorithm performance

	Data	Sn	Sp	Q9
Donor data	No SNP information	0.9184	0.8814	0.8980
	There is SNP information	0.9620	0.9971	0.9730
Acceptor data	No SNP information	0.7468	0.9950	0.8210
	There is SNP information	0.8987	0.9954	0.9283

Finally, to further illustrate the performance of our method, Table 5 compares the prediction performance of our method and several existing commonly used splice site identification methods (GENIO [10] and FSPLICE [11]). It can be seen from the table that our method is obviously better than other methods on any kind of data, and this advantage is particularly obvious on the acceptor data. Overall, our method significantly outperforms these commonly used splice site prediction programs.

5 Conclusion

In this paper, SNP-related features are involved in the process of splice site prediction, and a multi-scale algorithm is designed according to the correspondence between bases

Table 5. Comparison of the effects of the method in this paper and several commonly used splice site prediction programs

	Algorithm	Sn	Sp	Q9
Donor data	GENIO	0.9241	0.9968	0.9463
	FSPLICE	0.9367	0.9976	0.9552
	Algorithm	0.9620	0.9971	0.9730
Acceptor data	GENIO	0.8577	0.8444	0.8509
	FSPLICE	0.8470	0.8541	0.8505
	Algorithm	0.8987	0.9954	0.9283

and codons. From a computational perspective, new splicing signals derived from SNPs are found. The effectiveness of our method and the fact that SNP data have a great influence on splice site prediction are proved by experiments results.

Acknowledgements. This study was supported by the Youth Fund for Humanities and Social Science Foundation of Ministry of Education of China (Grant No. 19XJC860006) and the National Natural Science Foundation of China (Grant No. 11974289/ A040506) and the Basic research plan of Natural Science in Shaanxi Province in 2021(Grant No. 2021JQ-878).

References

1. Sun, J.S.a.Y.: Predicting the hosts of prokaryotic viruses using GCN-based semi-supervised learning. BMC Biol. (2022)
2. Villemin, J.-P.: A cell-to-patient machine learning transfer approach uncovers novel basal-like breast cancer prognostic markers amongst alternative splice variants. BMC Biol., **19**(70) (2021)
3. Kuitche, E., Jammali, S., Quangraoua, A.O.: SimSpliceEvol: alternative splicing-aware simulation of biological sequence evolution. BMC Bioinf. **20**(Suppl 20)(640) (2019)
4. Moghimi, F., et al.: Two new methods for DNA splice site prediction based on neuro-fuzzy network and clustering. Neural Comput. Appl. **23**, S407–S414 (2013)
5. Quanwei, Z., et al.: Splice sites prediction of human genome using length-variable Markov model and feature selection. Expert Syst. Appl. **37**(4), 2771–2782 (2010)
6. LI Shaoyan, D.W.: Identification of splice sites based on probability statistical feature. Comput. Eng. Appl. **47**(31), 182–184 (2011)
7. Yao, Y.: CERENKOV2: improved detection of functional noncoding SNPs using data-space geometric features. BMC Bioinf. **20**(63) (2019)
8. Zhibo, Z., Qinke, P., Xinyu, G.: Personalized pagerank based feature selection for high-dimension data. In: 2019 11th International Conference on Knowledge and Systems Engineering (2019)
9. Zhang, P.: Selection of microbial biomarkers with genetic algorithm and principal component analysis. BMC Bioinf. **20**(Suppl 6), 413 (2019)

10. GENIO/splice: Splice Site and Exon Prediction in Human Genomic DNA.http://www.bio genio.com/sp-lice/splice.cgi
11. FSPLICE: FSPLICE 1.0, Prediction of potential splice sites in Homo_sapiens genomic DNA. http://sun1.softberry.com/berry.phtml?topic=fsplice&group=programs&%20subgroup=gfind

ACP-ST: An Anticancer Peptide Prediction Model Based on Learning Embedding Features and Swin-Transformer

YanLing Zhu[1,2], Shouheng Tuo[1,2(✉)], Zengyu Feng[1,2], and TianRui Chen[1,2]

[1] School of Computer Science and Technology, Xi'an University of Posts and Telecommunications, Xi'an 710121, Shaanxi, China
{zhuyanling,fengzengyu,tianruichen}@stu.xupt.edu.cn,
tuo_sh@xupt.edu.cn
[2] Shaanxi Key Laboratory of Network Data Analysis and Intelligent Processing, Xi'an 710121, Shaanxi, China

Abstract. Currently, cancer is becoming a common and serious disease that threatens people's health, and anticancer peptides (ACPs) has become a new technology for cancer treatment because of their special mechanism of action. A series of computer-based anticancer peptide detection methods have been proposed, but the problems of requiring manual feature representation, high computational effort and low recognition accuracy still exist. In this study, we propose an anticancer peptide prediction algorithm based Swin Transformer called ACP-ST, which avoids complex manual feature computation using the general word embedding, and Swin Transformer based multi-head self-attention mechanism is adopted for feature extraction by reshaping the feature encoding into picture form. The experimental results indicate that ACP-ST has higher accuracy and efficiency than existing excellent methods on several benchmark datasets. In addition, the prediction model constructed in this work has a linear computational complexity with the length of the peptide sequence, and the special data transformation can facilitate visual analysis and feature extraction.

Keywords: Anticancer peptide · Swin transformer · Learning embedding features

1 Introduction

Cancer is a disease caused by the uncontrolled proliferation of abnormal cells [1–3]. In recent years, anticancer peptides (ACPs) have attracted much attention from researchers as a new technology for cancer therapy due to their specific mechanism of action. However, only a small number of anticancer peptides can be identified in a large protein library by using traditional wet-lab experimentation, which is very time-consuming and expensive. Currently, computer-based identification methods have become an important approach for ACP identification [4, 5] and are much less costly and efficient. However, to identify ACPs based on computing technology, some technical challenges still need

© The Author(s), under exclusive license to Springer Nature Switzerland AG 2023
N. Xiong et al. (Eds.): ICNC-FSKD 2022, LNDECT 153, pp. 939–946, 2023.
https://doi.org/10.1007/978-3-031-20738-9_103

to be overcome: how to design a reasonable data representation so that it carries more possible features and how to design a model to fully extract the potential features of peptide sequences.

For those technical challenges, methods based on traditional machine learning (ML) and deep neural networks are used to identify ACPs. For example, Tyagi et al. [6] used amino acid composition and binary profile features to construct features and input them into the support vector machine (SVM) for feature extraction, in which the accuracy of the model using binary profile features was 91.44%. Chen et al. [7] combined that optimized feature with SVM to identify anticancer peptides, and the performance was proven by cross-validation. Wei et al. [8] proposed a model named ACPred-FL, in which the most informative 5-dimensional features selected by features were used as input of SVM for feature extraction, which was significantly improved compared with other models. Wei et al. [9] trained eight random forest prediction models by using adaptive features. Agrawal et al. [10] used various input features to develop models based on the Extra Tree classifier and obtained higher Matthews Correlation Coefficient values. Chen et al. [11] developed a data augmentation algorithm to increase the available training data when the amount of data is small, and it achieved excellent results in MLP.

All the above studies have contributed to ACP detection, but there are still problems that require manual feature representation, a large amount of calculation and low recognition accuracy, while the self-attention mechanism can well mine the potential association relationship between disjoint loci by learning contextual features. Based on inspiration, Wenjia et al. [12] proposed an ACP detection algorithm named ACPred-LAF, which adopted a learnable embedding feature representation and built a model using a multi-head attention mechanism, which increased the feature extraction ability without other feature engineering. Although it has achieved high accuracy, its feature representation is complex, its calculation complexity is high, and it is prone to information redundancy.

In view of the existing problems, we proposed a new anticancer peptide prediction model (named ACP-ST) based on a Swin Transformer [13] with the general embedding feature representation for ACP detection. ACP-ST has the following features:

(1) The general word embedding feature representation method is employed to characterize peptide sequences, which does not require additional work about feature representation.
(2) In the method, the feature matrix of the peptide sequence is treated as the picture pixel, which is a brand-new processing method for the ACP sequence to the best of our knowledge.
(3) The method has linear computational complexity with ACP length.

2 Related Work

The self-attention mechanism [14], a method of context modeling, was developed from the attention mechanism. It has been applied to reading comprehension, machine translation and other tasks due to its advantages in sequence modeling. The self-attention mechanism aims to represent the whole sequence by calculating the correlation of different positions in sequence data.

Swin Transformer was proposed to apply the Transformer [14] in visual tasks. It divides the pixels in a picture into equal-sized and non-overlapping patches and divides the original picture into different windows by hierarchically merging adjacent pixel patches. The hierarchical feature maps can be constructed by calculating the self-attention value in the different windows.

3 Proposed Method

3.1 Method

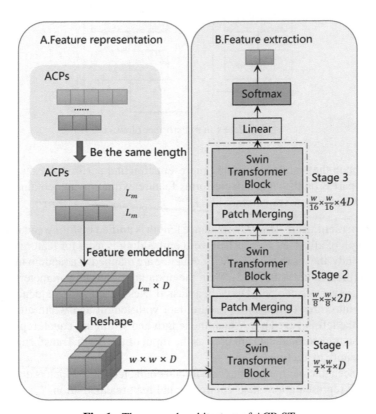

Fig. 1. The general architecture of ACP-ST

This study proposes a neural network method for the identification of anticancer peptides based on a Swin Transformer, named ACP-ST, as shown in Fig. 1, which goes through four procedures: (1) Unify ACP sequence length; (2) feature representation by word embedding and data reshaping; (3) feature extraction using a three-layer Swin Transformer structure; and (4) Use a linear layer and softmax function to obtain the prediction result. In Fig. 1, Section A on the left side aims to unify the length of the ACP

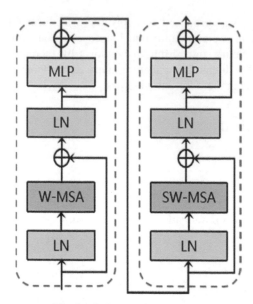

Fig. 2. Swin transformer block

sequence, obtain the feature embedding representation and perform the shape transformation, and part B on the right side performs feature extraction and obtains prediction results using three Swin Transformer blocks.

Feature Representation The ACP sequence length is unified by filling and intercepting until the length of all ACPs is L_m. Then word embedding is used for feature representation. In this study, the residues in the ACP sequence are mapped to a random initialization vector of D (the default value is 32). With the updating of model parameters, these initialized feature vectors can be adaptively adjusted. Next, each ACP sequence is further transformed into three-dimensional feature data with length and width attributes (see Reshape on the left in Fig. 1), in which the length and width are equal, represented by w. Finally, the transformed data are used as the input of the Swin Transformer structure for further feature extraction.

It can be seen that our method has end-to-end characteristics, which is different from most previous representations that require manual feature calculation.

Feature Extraction Based on Swin Transformer As shown in Section B of Fig. 1, the three layers of the Swin Transformer structure represent three stages, in which Stage 1 is composed of a Swin Transformer Block, and Stages 2 and 3 are composed of a Patch Merging and a Swin Transformer Block, respectively. Each feature vector is regarded as a "patch", and Patch Merging merges the fixed-size patches to form a non-overlapping window. Meanwhile, a linear layer is used to map the feature dimensions to other dimensions. In addition, the Swin Transformer Block improves the multi-head self-attention (MSA) used by the Transformer into window-based multi-head self-attention (W-MSA) and shifted-window-based multi-head self-attention (SW-MSA). Figure 2 depicts the Swin Transformer block, in which a Layer Norm (LN) is designed to increase the speed of convergence before each W-MSA layer, and there is a residual connection after each W-MAS layer. The last layer is a multilayer perceptron (MLP) with a GELU activation function. The structural details of W-MSA and SW-MAS are described in the literature [13].

The calculation process from stage 1 to stage 3 is as follows. In the 1st stage, adjacent patches are combined to form equal and non-overlapping windows, and the shape of the data is transformed from $w \times w \times D$ to $\frac{w}{4} \times \frac{w}{4} \times D$, where D represents the data dimension and $\frac{w}{4} \times \frac{w}{4}$ represents the number of windows. Then, the self-attention values in each window are calculated by the Swin Transformer block. In the 2nd stage, patch merging is first performed; that is, 2×2 merging is performed with windows. Then, the feature vectors are linearly mapped, and the data shape is transformed to a size of $\frac{w}{8} \times \frac{w}{8} \times 2D$. The self-attention value is calculated for the patches in $\frac{w}{8} \times \frac{w}{8}$ windows. After the same calculation of patch merging and Transformer block in the 3rd stage, the data with shape $\frac{w}{16} \times \frac{w}{16} \times 4D$ and the corresponding self-attention value can be obtained.

In the last linear layer, the outputs from stage 3 are mapped to a two-dimensional vector, and the final prediction probability is obtained by the softmax function.

The MSA used by the Transformer calculates the global self-attention score, while the improved W-MSA and SW-MSA in the Swin Transformer only calculate the self-attention score in the window. According to the literature [13], assuming that each window contains $M \times M$ patches, the computational complexity of the global MSA module and W-MSA module based on an ACP feature map with $h \times w$ patches can be expressed as:

$$\Omega(\text{MSA}) = 4hwC^2 + 2(hw)^2C \tag{1}$$

$$\Omega(W - \text{MSA}) = 4hwC^2 + 2M^2hwC \tag{2}$$

It can be seen that the global MSA has quadratic computational complexity with the size of the feature map, while W-MSA and SW-MSA (which have the same results as W-MSA) used in our method have linear computational complexity.

3.2 Performance Measurement

In this study, five common measurements were used to evaluate the performance of ACP-ST, including accuracy (ACC), specificity (SP), sensitivity (SN), Matthews correlation coefficient (MCC) and AUC (area under the Receiver Operating Characteristic curve). The measurements are as follows:

$$\begin{cases} \text{ACC} = \frac{TP+TN}{TP+TN+FP+FN} \\ \text{SE} = \frac{TP}{TP+TN} \\ \text{Sp} = \frac{TN}{TN+FP} \\ \text{MCC} = \frac{TP \times TN + FP \times FN}{\sqrt{(TP+FP)(TP+FN)(TN+FN)(TN+FP)}} \\ \text{AUC} = \frac{\sum_i^{n_{pos}} \text{rank}_i - \frac{n_{pos}(n_{pos}+1)}{2}}{n_{pos}n_{neg}} \end{cases} \tag{3}$$

Among them, TP, TN, FP and FN represent the numbers of true positive, true negative, false positive and false negative samples, respectively. And rank$_i$ denotes the ranking position of the i-th positive sample with respect to prediction probability, n_{pos} and n_{neg} are the number of positive samples and the number of negative samples, respectively and AUC are used to measure the overall prediction performance, while SE and SP measure the predictor's ability to predict positive samples and negative samples. In addition, training time is employed to measure the computational cost of the algorithm training process.

4   Results and Analysis

To evaluate the prediction performance of ACP-ST, this study tested LEE, Independent, ACPFL-500, ACPFL-164 [8], ACPred-Fuse 500, ACPred-Fuse 2710 [15], AntiCP 2.0 Main dataset and AntiCP 2.0 Alternate datasets [10], respectively.

The proposed ACP-ST is an improvement of ACPred-LAF. For a fair comparison, the AUC and training time were employed to compare ACP-ST and ACPred-LAF. As shown in Table 1, the proposed ACP-ST achieved higher accuracy in the four datasets compared to ACPred-LAF, with accuracy improvements of 3.49%, 4.01%, 0.87% and 4.28%, respectively. It also outperformed ACPred-LAF in other evaluation metrics, such as specificity and sensitivity. For the dataset AntiCP 2.0 Alternate, ACP-ST obtained 91.14, 91.15, 82.29, and 95.97% on ACC, SE, MCC and AUC, much higher than other comparative algorithms. In addition, the proposed ACP-ST algorithm takes the least time for training all datasets. The above results demonstrate that the window-based MSA adopted by the ACP-ST has comparable feature extraction capabilities with the global MAS adopted by ACPred-LAF, and the ACP-ST even has better performance and lower computational complexity.

We compared ACP-ST with eight models in the AntiCP 2.0 Alternate and AntiCP 2.0 Main datasets, namely, iACP-FSCM [16], ACPred [17], ACPred-FL [8], AntiCP, AntiCp_2.0 [10], iACP [7], PEPred-Suite [9] and ACPred-Fuse [15], are shown in Table 1, with some results from the literature [18]. ACP-ST obtained the best results in all metrics except specificity in the AntiCP 2.0 Alternate datasets. In the AntiCP 2.0 Main dataset, ACP-ST obtained a result only 0.14% lower than the highest accuracy of the AntiCP_2.0 and obtained the best results in MCC score, which were 0.51.

In conclusion, the above comparative analysis illustrates the practicality and versatility of ACP-ST as well as its prospects in the field of ACP identification.

5   Conclusion

In this study, we propose a new method for anticancer peptide identification called ACP-ST. Unlike most of the previous methods that required manual feature representation, the proposed method is more efficient and simpler by using learning embedding features, and the feature extraction model constructed in this way reduces the computational complexity of the original global MSA mechanism with a quadratic relationship with peptide sequence length to a linear relationship while still having comparable or even

Table 1. Comparison with existing models

Dataset	Model	ACC (%)	SE (%)	SP (%)	MCC	AUC	Training time(s)
LEE + independent	ACP-ST	**95.49**	**97.24**	93.70	**0.9102**	**0.9808**	**84.88**
	ACPred-LAF	92.00	84.00	**100.00**	0.8510	0.9717	328.19
ACPred-fuse 500 + ACPred-fuse 2710	ACP-ST	**97.34**	15.85	**99.89**	0.3519	0.7571	**136.87**
	ACPred-LAF	93.33	**91.33**	95.33	**0.8674**	**0.9661**	426.22
ACPLF-500 + ACPLF-164	ACP-ST	**85.63**	**88.75**	82.50	**0.7139**	**0.9219**	**70.48**
	ACPred-LAF	84.76	76.83	**92.68**	0.7040	0.8754	521.98
AntiCP 2.0 alternate	ACP-ST	**91.14**	**91.15**	91.15	**0.8229**	**0.9597**	**67.86**
	ACPred-LAF	86.86	86.08	87.63	0.7372	0.9078	199.77
	iACP-FSCM	88.90	87.60	90.20	0.779	–	–
	iACP	77.6	78.4	76.8	0.550	–	–
	PEPred-suite	57.5	40.2	74.7	0.160	–	–
	ACPred-fuse	78.9	64.4	**93.3**	0.60	–	–
	ACPred-FL	43.8	60.2	25.6	−0.15	–	–
	ACPred	85.3	87.1	83.5	0.71	–	–
	AntiCP	90.0	89.7	90.2	0.80	–	–
AntiCP 2.0 main	ACP-ST	75.30	82.04	68.64	**0.5111**	0.8285	**184.52**
	ACPred-LAF	72.09	79.65	64.53	0.447	**0.8293**	188.26
	AntiCP_2.0	**75.43**	77.46	73.41	0.51	–	–
	AntiCP	50.6	**100.0**	1.20	0.07	–	–
	ACPred	53.47	85.55	21.39	0.09	–	–
	iACP	55.1	77.9	32.2	0.11	–	–
	PEPred-suite	53.5	33.1	**73.8**	0.08	–	–
	ACPred-fuse	68.9	69.2	68.6	0.38	–	–
	ACPred-FL	44.8	67.1	22.5	−0.12	–	–

better feature extraction capability. The final comparative analysis demonstrated the good prospects of ACP-ST in the field of ACP detection.

However, the model in this study does not use the residue coding scheme, such as the physicochemical properties of amino acids, which may lead to the loss of residue specific information. In future research, we will try diverse feature coding schemes to find the best coding combination and adopt a more flexible method to feature extraction.

Acknowledgments. This work was supported by the Postgraduate Innovation Fund Project of Xi'an University of Posts and Telecommunications (No. CXJJYL2022051).

References

1. Jemal, A., et al.: Global cancer statistics. CA Cancer J. 69–90 (2011)
2. Torre, L.A., et al.: Global Cancer Statistics 2012, 87–108 (2015)
3. Ferlay, J., et al.: Estimates of worldwide burden of cancer in 2008: GLOBOCAN 2008. Cancer **127**, 2893–2917 (2010)
4. Feng, P., Wang, Z.: Recent advances in computational methods for identifying anticancer peptides. Curr. Drug Targets **20**, 481–487 (2019)
5. Wu, Q., et al.: Recent progress in machine learning-based prediction of peptide activity for drug discovery. Curr. Top. Med. Chem. **19**, 4–16 (2019)
6. Tyagi, A., Kapoor, P., Kumar, R., et al.: In silico models for designing and discovering novel anticancer peptides. Sci. Rep. **3**, 2984 (2013)
7. Chen, W., et al.: iACP: a sequence-based tool for identifying anticancer peptides. Oncotarget **7**(13), 16895 (2016)
8. Wei, L., et al.: ACPred-FL: a sequence-based predictor using effective feature representation to improve the prediction of anti-cancer peptides. Bioinformatics **34**, 4007–4016 (2018)
9. Wei, L., et al.: PEPred-Suite: Improved and robust prediction of therapeutic peptides using adaptive feature representation learning. Bioinformatics **35**, 4272–4280 (2019)
10. Agrawal, P., et al.: AntiCP 2.0: an updated model for predicting anticancer peptides. Briefings Bioinf. **22**(3) (2021)
11. Chen, X.G., Zhang, W., Yang, X., Li, C., Chen, H.: ACP-DA: Improving the prediction of anticancer peptides using data augmentation. Front. Genet. **12**, 698477 (2021)
12. He, W., et al.: Learning embedding features based on multisense-scaled attention architecture to improve the predictive performance of anticancer peptides. Bioinformatics **37**(24), 4684–4693 (2021)
13. Liu, Z., et al.: Swin transformer: hierarchical vision transformer using shifted windows. In: 2021 IEEE/CVF International Conference on Computer Vision (ICCV), pp. 9992–10002 (2021)
14. Vaswani, A., et al.: Attention is all you need. In: Proceedings of the 31st International Conference on Neural Information Processing Systems. Curran Associates Inc., Red Hook, pp. 6000–6010 (2017)
15. Rao, B., Zhou, C., Zhang, G., Su, R., Wei, L.: ACPred-fuse: fusing multi-view information improves the prediction of anticancer peptides.Briefings Bioinf. **21**(5), 1846–1855 (2020)
16. Charoenkwan, P., et al.: Improved prediction and characterization of anticancer activities of peptides using a novel flexible scoring card method. Sci. Rep. **11**, 3017 (2021)
17. Schaduangrat, N., et al.: ACPred: a computational tool for the prediction and analysis of anticancer peptides. Molecules **24**(10), 1973 (2019)
18. Lv, Z., et al.: Anticancer peptides prediction with deep representation learning features. Briefings Bioinf. **22**(5) (2021)

Research on Genome Multiple Sequence Alignment Algorithm Based on Third Generation Sequencing

Zhiyu Gu[1,3], Junchi Ma[2], Xiangqing Meng[1], and Hong He[1(✉)]

[1] School of Mechanical, Electrical and Information Engineering, Weihai 264209, China
`guzhiyu@mail.sdu.edu.cn`, `mengxiangqing@mail.sdu.edu.cn`, `hehong@sdu.edu.cn`
[2] School of Mathematics, Shandong University, Jinan 250100, China
`majunchi@mail.sdu.edu.cn`
[3] Institute of Software, Chinese Academy of Sciences, Beijing 100190, China

Abstract. Long reads produced by the third-generation gene sequencing technology need to be aligned to the reference genome using a more advanced aligner. To fully understand the current mainstream aligners, six mainstream aligners are selected to test their performance on datasets of a different magnitude to demonstrate and validate their ability to perform sequence alignment. According to the experimental results, the Minimap2 is the general aligner with the best all-around performance, which guarantees accuracy and consumes fewer resources. Graphmap2 and deSALT showed higher accuracy and sensitivity when resource usage was not restricted, especially deSALT identified all exons in exon calling. In addition, target detection is used to achieve sequence alignment and analyzes the reasons why target detection is useless for sequence alignment. Finally, deep learning and traditional algorithms are combined to solve the alignment problem, and the optimized algorithm performs better than Minimap2. ...

Keywords: Algorithm analysis · Sequence alignment · Deep learning · Object detection · LSTM

1 Introduction

Genetic research is an essential means for human beings to understand life. The genome obtained by sequencing contains the intrinsic correlation between all genes and life characteristics. The changes and breakthroughs in sequencing technology have brought a considerable impetus to the fields of genomics research, drug development, and breeding. The current third-generation sequencing technology is mainly the sequencing technology of companies such as pacbio [1] and ont [2]. Long sequencing reads can be as high as ten kbp and have an error rate of up to 15%, which brings more significant challenges to sequence alignment.

N. Xiong et al. (Eds.): ICNC-FSKD 2022, LNDECT 153, pp. 947–955, 2023.
https://doi.org/10.1007/978-3-031-20738-9_104

At present, most of the comparators of the third-generation sequencing technology use traditional algorithms, which are divided into two types of ideas: one principle is a hash table, such as Minimap [3], and the other is based on a suffix array or burrows Wheeler transform, such as BWA-mem [4].

The current deep learning achievements are remarkable, and some achievements have been made in applying sequence alignment, such as MiniScrub [5], which provides feasibility support for our research. Based on this, this paper tests the performance of mainstream aligners and uses the method of deep learning to meet this challenge. The contributions are as follows:

1. Selected six state-of-the-art aligners, conducted extensive comparative experiments, tested their performance on 9 data sets of different sizes and analyzed their advantages and disadvantages and the most suitable application scenarios.
2. The method of using target detection to solve the sequence alignment problem is explored, and the reasons for the poor performance of target detection on sequence alignment are analyzed.
3. This paper proposes a novel model that uses deep learning as a link in traditional approaches to sequence alignment to solve the alignment problem by combining the strengths of both strategies. Experimental comparisons with mainstream methods show that our method achieves competitive performance.

2   Related Work

At present, some alignment programs based on traditional algorithms specifically solve the alignment problem of long sequences, of which Heng Li has made outstanding contributions to the research of this problem.

In 2013, Heng Li proposed BWA-MEM recognition maximum match based on suffix array. In 2016, Minimap and minias proposed directly generating overlapping group sequences from the original sequence overlap through calibration-free assembly, with speed up to 50 times that of BWA-MEW. Heng Li [6] developed Minimap2 with additional functions in 2018, a sequence aligner specially designed for noisy long gene sequences and expanded the original algorithm by introducing a new heuristic method. It is 3–4 times faster than the mainstream short reading aligner in mapping short reading. It has the same accuracy, currently recognized as the general aligner with the best all-around performance.

In 2018, Sedlack [7] proposed NMGLR comparator as an aid to identify structural variations; The GraphMap proposed by Sovi ć et al. in 2017 improved the mapping sensitivity to a maximum of 80% and was able to map more than 95% of the bases, and then Graphmap2 [8] proposed in 2019 significantly improved the accuracy; In 2017, Kart using divide and conquer method proposed by Lin and Hsu [9] is faster in extended string reading and can still produce reliable mapping on data with high error rate. In 2019, Bo Liu [10] proposed deSALT, which has a good effect on exon recognition.

However, there are few types of research based on deep learning. For example, Nathan Lapierre proposed that MiniScrub process data using a convolutional neural network to remove low-quality segments in 2019 to improve accuracy.

3   Methodology

3.1   Traditional Comparator Performance Comparison Experiment

In this paper, six excellent aligners (minimap2, graphmap2, desalt, GMap, ngmlr, and kart), most used in gene sequencing, are selected from the multiple aligners. Testing on the same data set. The performance comparison of these aligners on different orders of magnitude and data sets is obtained on the supercomputer cluster of Shandong University.

3.2   Sequence Aligner Based on Object Detection

This paper attempts to solve the problem of sequence alignment through the target detection method based on Yolo v5 [11]. As shown in Fig. 1, the target sequence and reference sequence are converted into target sequence image and reference sequence image respectively. Then the target sequence image is detected on the reference sequence image. Finally, the recognition result is displayed in the image mark, and the target position, accuracy and other information are recorded. In converting the sequence into pictures, we will convert each letter into the picture of the corresponding letter, then splice it into the whole picture, and color each letter with different colors. Image coloring before target detection can increase the characteristics of image data and enhance the recognition effect.

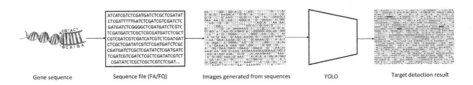

Fig. 1. YOLO identification flow chart

3.3   Traditional Algorithm Optimization Based on Deep Learning

In this paper, the sequence alignment problem is modeled as a classification problem of classifying subsequences of a target sequence and processed using deep neural networks. The general idea of this method is to train a corresponding neural network model for a specific reference genome to determine the region where the target sequence is located. The aligner is used for comparison; for the region where the model predicts that there is no target sequence, the traditional algorithm aligner with fast speed but slightly lower accuracy is used for comparison. Finally, the results are output uniformly.

First, the chromosomes in the reference genome are divided into several fragments and uniquely identified, and the divided fragments are used as the category of the classification problem. For the convenience of calculation, the target sequence is divided into multiple subsequences and input into the neural network for classification prediction. Adjacent subsequences will be classified as the same or adjacent segments. However, due to the sequence error of up to 15%, adjacent subsequences may be classified into different regions. After the classification prediction task is completed, the predicted results are aggregated into chains, and the target result is output, that is, the region where the target sequence is located.

In this paper, LSTM [12] is used for classification, and the segmented reference genome is used as the training data of the neural network, of which 80% of the data is used as the training set, and the remaining 20% of the data is used as the test set. The activation function of the output layer uses a sigmoid function for multi-label classification and a dropout layer to prevent overfitting. Graphmap2 was used for further alignment of regions where target sequences were likely to occur, and Minimap2 was used for detailed alignment of other regions.

4   Results

4.1   Experiment Design

Datasets This paper's data set consists of simulated and real data sets. The simulated data set is generated based on the reference genome of human, fruit fly and mouse (Table 1.) using nanosim and pbsim, while the real data set is obtained from na12878 samples through cDNA sequencing and directing sequencing, as shown in Table 2.

Table 1. Genome sequence sheet

Species	Single isoforms	Multiple isoforms	Short exons	All the generated transcripts
Fruit fly	3009	3065	2880	8954
Mouse	3072	4080	2783	9935
Human	2998	3603	3346	9947

Experimental Environment All traditional experiments in this paper are carried out on the supercomputing cluster of the school. Intel (R) Xeon (R) e5-2620 V3 CPU is used, which has 12 cores (24 threads), the size of memory is 32g, the running system is Linux RedHat, and each program runs with 20 threads.

Table 2. Genome data sheet

File type	#Runs	#Reads	Mean(b)	Read N50
Native RNA pass	30	10,302,647	1030.24	1334
Native RNA fail	30	2,686,736	430.96	840
cDNA pass	12	15,152,101	932.86	1072
cDNA fail	12	9,129,338	661.90	841

The evaluation criteria are:

Aligned reads%: the proportion of the number of bases correctly compared with their proper positions within the error range ($\sim$5%) in the logarithm of the total ratio.

Exact introns%: the proportion of exons correctly mapped by the algorithm within the error range ($\sim$5%).

Precision%: The proportion of the samples that the model thinks are correct that are positive.

Recall%:The proportion of the samples that the model thinks are correct that are positive.

$$Precision = \frac{TP}{TP + FP} \tag{1}$$

$$Recall = \frac{TP}{TP + FN} \tag{2}$$

4.2 Traditional Comparator Performance Comparison Experiment

The horizontal comparison diagram of the six programs is shown in Figs. 2 and 3.

According to the experimental results, minimap2 is the best comparator with the best all-around performance, less time and memory consumption, and accuracy reaching the average level. Graphmap2 has the best average accuracy, surpassing minimap2, but the resource consumption is very high. GMap takes less time to complete tasks, and its accuracy is slightly higher than the average, but its memory consumption is higher. The above is the performance of these comparators on conventional tasks. These comparators often perform better than other general comparators on the specific tasks they are good at. For example, the performance of ngmlr is like that of minimap2, and they are good at structure comparison tasks, while the performance of other tasks is slightly poor; Desalt is good at exon tasks and can identify all exons.

4.3 Sequence Aligner Based on Object Detection

The prediction accuracy result graph and confusion matrix graph are shown in Fig. 4.

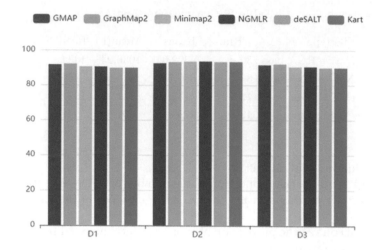

Fig. 2. Accuracy of mainstream aligners on different datasets

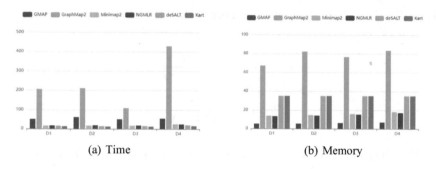

(a) Time   (b) Memory

Fig. 3. Time and memory consumption of mainstream aligners on different datasets

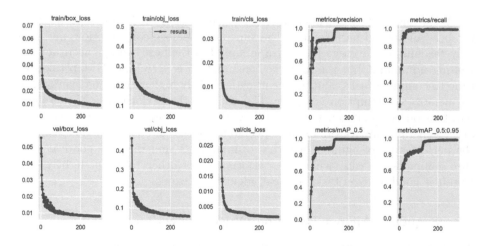

Fig. 4. Target detection accuracy result graph

In this chapter, target detection is applied to deal with sequence alignment problems, and the alignment tasks are completed by sequence mapping and letter coloring methods. However, some defects of this method are found in practical applications.

First, the purpose of the sequence alignment problem is to align the target gene sequence with the genome sequence. The target sequence cannot be determined each time. That is, it is difficult to predict the target sequence exhaustively, which means that if the method is practical, when applied in practice, the corresponding model needs to be retrained for different target sequences. This problem is challenging for current object detection, and the cost of each retraining is unacceptable.

Secondly, due to the data characteristics of gene sequences, it is currently a feasible idea to use deep learning methods to process them into graphs and perform target detection, which is more effective for short-string alignment of gene sequences. However, the data collected by the third-generation gene sequencing technology are long strings of K-level data volume. The cost of converting them into maps and performing target detection is too high, making it challenging to apply them practically.

Finally, the data collected by the third-generation gene sequencing technology has a 15% error. Deep learning inevitably has recognition and accuracy errors in the process of recognition, which is far from the traditional algorithm.

4.4 Traditional Algorithm Optimization Based on Deep Learning

This paper uses the same data set as the previous chapter, runs the test experiment on a GPU with a 1050TI machine, and selects LSTM Model as the model for training.

Tables 3 and 4 show this method's performance, time consumption, and memory consumption results compared with other mainstream methods. The results show that the time consumption of this method is almost the same as that of Minimap2 on the test data set, and the accuracy rate is a specific improvement.

This method is like other heuristic methods. The idea is to use a neural network to narrow the gene search range and perform sequence alignment. It is a method that combines deep learning with traditional methods. Before using traditional methods for alignment, the possible mapping of target sequences is determined. The regions were further accurately aligned using traditional methods. The region selection of the target sequence is performed with accurate and precise alignment equipment, and other regions are selected with fast and possibly accurate alignment equipment. Experiments show this method is better than other mainstream methods, including minimap2. Under the condition that the speed (excluding the time of model training on the genome) is almost the same, the accuracy is improved to a certain extent.

Innovation The sequence alignment is preprocessed and modeled as a classification problem, realizing the application of deep learning on sequence alignment.

Table 3. Comparison with other mainstream comparators

	Simulation dataset		Real dataset	
	Recall	Precision	Recall	Precision
Minimap2	0.9876	0.988	0.9695	0.9711
	0.9891	0.988	**0.9711**	0.9725
	0.9898	0.9898	**0.9698**	0.9697
deSALT	**0.9901**	0.988	**0.9721**	0.976
	0.9891	**0.9947**	**0.9711**	0.9713
	0.988	0.988	0.9654	0.9695
GraphMap2	0.9892	**0.9948**	0.9704	0.9809
	0.9875	0.9933	0.9643	**0.987**
	0.9876	0.991	0.9665	0.9827
LSTM	0.9866	0.9923	0.9577	**0.9815**
	0.9841	0.9891	0.9553	0.984
	0.9874	**0.9911**	0.9476	**0.9847**

Table 4. Compare the time and memory consumed by the test process

Time & memory					
Graphmap2	Wall time (M)	784.6	152.4	69	78.7
	CPU time (M)	743.4	624.7	624.8	969.3
	Memory (GB)	65.3	73	77	74.5
Minimap2	Wall time (M)	14.38	6.06	2.39	2.49
	CPU time (M)	14.9	19.7	14.05	24.43
	Memory (GB)	12.02	14.7	17.5	18
LSTM (Excluding training)	Wall time (M)	16.6	5.6	2.79	3
	CPU time(M)	22.23	24.57	23.14	26.73
	Memory (GB)	62.44	70.27	71.9	76.85
deSALT	Wall time (M)	18.53	5.18	3.16	3.07
	CPU time (M)	17.73	19.78	19.25	28.49
	Memory (GB)	30	34.85	34.18	38.81

Combining deep learning with traditional algorithms and combining the advantages of the two strategies, the resource consumption is almost unchanged, and the accuracy is improved.

Shortage At the same time, this method has some problems that can be used as a future improvement direction. For example, the reference genome needs to be trained in advance, and the 15% errors in the sequencing are not considered, and more advanced models such as Transformer can be used for training.

5   Conclusion

This paper introduces several mainstream aligners in the long reads sequence alignment field and tests their performance on real and synthetic datasets. The results show that Minimap2 is the general aligner with the best all-around performance, with high speed and low memory consumption; GraphMap2 has high accuracy and high resource consumption; other aligners such as deSALT perform well on specific tasks that they are good at. In addition to traditional methods, this paper argues that using deep learning to deal with sequence detection problems is a new problem-solving idea. This paper attempts to use the target detection method to solve the sequence alignment problem. The experimental effect is far from the traditional method, and the reasons are analyzed. Finally, this paper combines the deep learning method with the traditional method to solve the alignment problem, uses the neural network to judge the region where the target sequence exists and uses the traditional method for a detailed alignment. The combined algorithm performs better than Minimap2.

References

1. Rhoads, A., Au, K.F.: PacBio sequencing and its applications. Genomics Proteomics Bioinform. **13**, 278–289 (2015)
2. Jain, M., Olsen, H.E., Paten, B., Akeson, M.: The Oxford nanopore MinION: delivery of nanopore sequencing to the genomics community. Genome Biol. **17**, 239 (2016)
3. Li, H.: Minimap and miniasm: fast mapping and de novo assembly for noisy long sequences. Bioinformatics **32**, 2103–2110 (2016)
4. Li, H.: Aligning sequence reads, clone sequences and assembly contigs with BWA-MEM. arXiv:1303.3997 (2013)
5. LaPierre, N., Egan, R., Wang, W. Wang, Z.: MiniScrub: de novo long read scrubbing using approximate alignment and deep learning, 433573 (2018)
6. Li, H.: Minimap2: pairwise alignment for nucleotide sequences. Bioinformatics **34**, 3094–3100 (2018)
7. Sedlazeck, F.J., et al.: Accurate detection of complex structural variations using single-molecule sequencing. Nat. Methods **15**, 461–468 (2018)
8. Marić, J., Sović, I., Križanović, K., Nagarajan, N. Šikić, M.: Graphmap2—splice-aware RNA-seq mapper for long reads, 720458 (2019)
9. Lin, H.-N., Hsu, W.-L.: Kart: a divide-and-conquer algorithm for NGS read alignment. Bioinformatics **33**, 2281–2287 (2017)
10. Liu, B., et al.: deSALT: fast and accurate long transcriptomic read alignment with de Bruijn graph-based index. Genome Biol. **20**, 274 (2019)
11. Jocher, G., et al.: ultralytics/yolov5: v6.1—TensorRT, TensorFlow Edge TPU and OpenVINO Export and Inference. Zenodo (2022)
12. Donahue, J., et al.: Long-Term Recurrent Convolutional Networks for Visual Recognition and Description, pp. 2625–2634 (2015)

Psychological Portrait of College Students from the Perspective of Big Data

Wen Man, Yu Zhu[(✉)], Tingting Zhu, and Jie Zhu

Air Force Engineering University, Xi'an 710051, China
308754733@qq.com, 15162931@qq.com, 280774596@qq.com,
182179196@qq.com

Abstract. In view of the drawbacks of passive, lag and traditional mental health education for college students, this paper puts forward a psychological portrait strategy for college students based on big data. Through the use of large data network and campus application system to obtain all kinds of information of students, a complete "data acquisition, data analysis, data feedback" chain is formed, and the "random forest model based on the construction of data portrait evaluation model" is used to match the mental health measurement indicators, and finally the radar map is used to generate students' psychological portraits. Based on the mechanism of student portrait, the early warning of college students' mental health is realized, which provides an effective and feasible implementation path for the monitoring of college students' mental health, and has a certain practical reference value for guiding the mental health of college students.

Keywords: Big data · User portraits · Mental health

1 Introduction

With the rapid development of big data technology, it has been widely used in traditional scientific research, education, medical treatment, social governance and many other fields. Big data is changing people's life, work and thinking mode. At the university stage, the pressure of learning competition, career pressure, love distress, economic constraints, interpersonal difficulties, parents' high expectations and other pressures lead to the psychological overload of college students, and psychological problems become more and more prominent. However, at present, the mental health education of college students still stays in the traditional stage of propaganda education and ideological education, the management mode is mostly ex post education, the means of education are backward and passive, and there are many drawbacks, which make it difficult to adapt to the high intelligent control requirements of colleges and universities in the era of big data. How to use big data technology to achieve leapfrogging to solve the mental health education of college students has become a major problem to be solved in front of university administrators.

By using big data technology to analyze the data of existing application systems and extracting the psychological characteristics of students from the massive data of

N. Xiong et al. (Eds.): ICNC-FSKD 2022, LNDECT 153, pp. 956–967, 2023.
https://doi.org/10.1007/978-3-031-20738-9_105

students, this study makes it clearer to construct a clear psychological portrait from the family background, school performance, social networking, learning, community and interpersonal behavior of college students, and according to the portrait results. It will be able to improve the efficiency and pertinence of student management and realize the early prevention of student problems.

2 Related Research

2.1 User Portrait Research

In the field of academic research, user portrait technology has achieved some research results, mostly used in news information, e-commerce and other fields, such as news Application (App) or web pages through the analysis of users' browsing preferences, regular recommendation of similar information to users, Taobao, Tianmao and other e-commerce platforms, through the analysis of users' purchase records. Provide personalized service to improve customer satisfaction. Yang Xiong constructs multi-dimensional tags through student user portraits and applies them to personalized recommendation of educational resources [2]. In the article, Li Jieqiong proposed a precise advertising push method based on user portraits combined with collaborative recommendation [3]; Luo Biao and Cui Yanrong proposed to use a large number of student data in education App to build smart student portraits, and apply student portraits to education, so that teachers and parents can better understand the real-time dynamic and learning status of students [4]; Xu Man combed the shortcomings of current knowledge recommendation service in university libraries, and put forward optimization strategies for knowledge recommendation service in university libraries with the help of the construction process of user portraits [5]; Xu Chang, Mao Guifang and Zhou Yinjian elaborated the research and application of user portraits in the field of library and information, pointing out that user portraits are powerful tools for university libraries to understand user needs and realize personalized and intelligent services [6]; Shi Wenxing and Cao Shiyun proposed the application measures of user portrait technology in electric power enterprises based on the generation method and research object of user portrait [7]; Yin Jiajia predicted the label of users through clustering algorithm and built the user portrait system in the field of news [8]; Li Chunqiu applied user portraits to library services through in-depth characterization of user attributes and behaviors to improve the personalization and precision of library services [9]; Yu Yongli introduced user portraits and precision knowledge services into the study of academic new media knowledge services, and constructed a precision knowledge service model of academic new media based on user portraits [10].

2.2 Research on Mental Health of College Students

In recent years, many college students' mental health survey data show that mental health has become an important factor hindering the all-round development of college students. Liu (2011) investigated the mental health status of college students in 10 colleges and universities in Henan Province, and found that the proportion of abnormal psychology was more than 20%. In 2018, the Prospect of Mental Health Education for College Students: Research Evidence found that 30% of college students in China had psychological problems and 10% had psychological disorders [11]. The mental health status of college students is not optimistic and should be paid attention to.

What factors are related to the mental health of college students? How to monitor a student's mental health?

Through seemingly unrelated regression model analysis, Zeng et al. [12] found that their family background, campus life, work and rest, tobacco and alcohol intake, academic status, community and interpersonal interaction are all important factors affecting the mental health of college students.

Fang et al. [13] constructed three screening levels and 22 dimensional indicators of the scale for mental health screening of college students. The first-level screening was symptoms of severe psychological problems, including severe psychotic symptoms such as hallucinations, suicidal behavior and intention; The second-level screening was general psychological problem symptom screening, which was divided into seven indicators of internalized psychological problem symptoms, including anxiety, depression, paranoia, inferiority, sensitivity, social phobia and somatization, and eight indicators of externalized psychological problems, including dependence, hostile aggression, impulse, compulsion, Internet addiction, self-injury behavior, eating problems and sleep problems; The three-level screening is developmental distress screening, including five indicators of school adaptability difficulties, interpersonal distress, academic pressure, employment pressure and love distress. The first-level and second-level screening are the core of students' mental health screening, while the third-level screening mainly reflects students' adaptability problems and indicates the source of potential psychological problems.

To analyze 22 dimensional indicators and extend them to all aspects of university life, if we want to make early warning, we can only do analysis from the existing data, as far as possible to obtain the data of students' learning, life, social interaction, network and other aspects after they enter the school, combined with the comprehensive analysis of students' family background, through the psychological portrait of students to predict, for the school to do a good job of students ideological work to provide a basis.

3 Data Sources, Collection, Mining and Filtering

3.1 Data Source and Overview

All aspects of college students' life are basically concentrated in the school, and all kinds of information extraction are relatively concentrated, which can be roughly divided into three aspects: learning life, network life and daily consumption, as shown in Fig. 1.

This study holds that the essence of the student's mental health portrait is the student's psychological model, which extracts and depicts the student's mental health information panorama comprehensively and meticulously from different dimensions after enrollment.

Through the campus information system, the static and dynamic characteristics of students are collected [14]. The static attributes include name, gender, major, grade, dormitory, age, native place, family background, hobbies and other characteristics; Dynamic attributes include classroom attendance information, online and offline learning, library

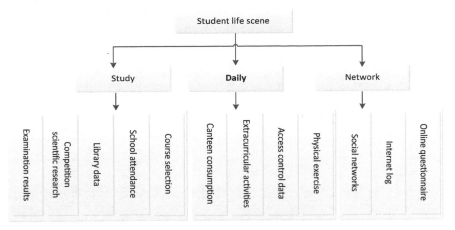

Fig. 1. Scene of college students' life in school

borrowing, dormitory return rate, access control system, campus card system, canteen dining, campus Internet access, participation in competitions, community situation, bathing frequency and other data.

The data sources in this study are mainly obtained from the following aspects:

(1) Personal information data of student status system
(2) Data of course selection and examination results in the educational administration system
(3) Canteen consumption data, library borrowing data and attendance data
(4) Students' online data
(5) Data of students' social activities such as participating in competitions, scientific research and extracurricular activities
(6) Data of students participating in physical exercise
(7) Ideological discussion
(8) Use "Baidu Questionnaire Star" to conduct online questionnaire survey.

3.2 Data Collection

The data can be divided into direct data and indirect data according to the type, and the collection methods can be divided into direct query and reading in the database, using Kafka and flume big data collection tools, crawler software acquisition, online questionnaire survey, counselor thought discussion, etc. The collected data information is shown in Table 1:

The data sources of student information are complex and redundant, and the original data collected are incomplete, inconsistent and false, which will affect the accuracy of portrait construction and directly lead to misleading results. Before potential information mining, data cleaning, integration, transformation, specification and other preprocessing are needed. When cleaning, the irrelevant data and duplicate data in the collected original data are eliminated, and the abnormal data are actually investigated and corrected. The

Table 1. Data sources and collection fields

Data source	Acquisition field 1	Acquisition field 2	Acquisition field 3	Acquisition field 4	Acquisition field 5	Acquisition field 6
Student status system	Name	Sex	Student number	Native place	Major	Family background (special)
Educational administration system (course selection)	Name	Sex	Student number	Major	School year	Course selection directory
Educational administration system (score query)	Name	Sex	Student number	Major	School year	Achievement
Educational administration system (competition, scientific research and other data)	Name	Competition	Scientific research	An association	Activity	Award record
All in one card (canteen consumption)	Name	Student number	Card swiping time	Card swiping times	Co existing friends	Credit card amount
All in one card (book borrowing)	Name	Student number	Borrowed bibliography	Borrowing times	Borrowing frequency	Overdue
All in one card (attendance in class)	Name	Student number	Number of classes	Card swiping time	Late times	Length of tardiness

(continued)

Table 1. (*continued*)

Data source	Acquisition field 1	Acquisition field 2	Acquisition field 3	Acquisition field 4	Acquisition field 5	Acquisition field 6
All in one card (physical exercise)	Name	Exercise program	Exercise duration	Exercise frequency	Co existing friends	Match
Statistical analysis of Internet behavior	Account number	Online time	Flow	Forward	Comment	Collection
Mental health test results of Freshmen	Name	Sex	Student number	Major	Test result	Existing problems
Periodic mental health questionnaire	Name	Major	Number of questionnaires	Questionnaire score	Nothing	nothing

expression form of student data from different data sources is inconsistent and heterogeneous, so it is necessary to integrate the data and convert it into a unified standard for collation and storage, so as to prepare for subsequent data mining [15].

3.3 Data Mining and Cleaning

Data mining is the core and key of the process of students' psychological portrait. After cleaning and sorting out, the data also need to be mined and extracted according to the application needs of the portrait, so as to prepare for the follow-up psychological portrait.

(1) Data analysis of all-in-one card canteen: overall consumption ability, consumption preference, personal consumption, meal time and frequency of students in school, analysis of whether students' daily life is regular, whether they can eat on time, and whether they have friends. According to the general meal time in the canteen, breakfast is 6: 30–9: 30, lunch is 11: 30–13: 30, and dinner is 16: 30–19: 30. If the above time period is exceeded, it is considered as "irregular diet". In terms of three meals a day, if the canteen consumption is less than 50 times a month, it is considered as "irregular diet". According to the analysis of the card swiping data in the canteen and the investigation of the actual situation, it is found that if the card swiping time interval is within one minute, in order to eliminate the possibility of accidental collision, if the co-occurrence record of two cards appears more than five times, it is identified as a "co-occurrence friend" with a maximum score of 7 points.

(2) Attendance data analysis of all-in-one card: statistics of students' daily class time, frequency and absence times, and analysis of students' attendance rate. Combined with the curriculum arrangement of the educational administration system, if more than five times a month do not attend classes on time, it will be considered as "poor attendance", with a maximum score of 7 points.

(3) Query and analysis of one-card scores: analyze the examination scores of students in this semester. If they fail more than two subjects, they will be identified as "high academic pressure", with a maximum score of 7 points.

(4) One-card library borrowing data analysis: The catalogue of books borrowed by student libraries, the frequency of borrowing, the length of borrowing, and whether they prefer certain types of books are counted. If they often borrow criminal-related books such as "crime", "murder" and "solving cases" in a short period of time, they are identified as "extreme tendency", with a maximum score of 10 points.

(5) One-card physical exercise data analysis: The students' exercise duration and frequency were analyzed, and the students who exercised less than 10 times per semester and each time for less than 30 min were identified as "lack of exercise" type, with the highest score of 7 points.

(6) One-card scientific research data analysis: analyze the students' participation in competitions, scientific research, associations and collective activities, and identify the students who have never participated in any competitions, scientific research and collective activities as the "lack of public activities" type, with the highest score of 7 points.

(7) Card access control data analysis: based on the school dormitory access control management time, those who do not return to the dormitory within the specified time are identified as "staying out at night", with a maximum score of 8 points.

(8) Statistical analysis of online data: analyze the students' online time and traffic, and dig out the students' network usage and work and rest habits. If the students are still online after 1 am for more than five times in a month, they are identified as "staying up late". At the same time, students' online content is deeply excavated to find students' concerns and analyze students' personality trends. If they often browse extreme websites such as "crime", "murder", "injury" and "love killing" in a short period of time, or buy harmful equipment and drugs, they are identified as "extreme tendency" with a maximum score of 10 points.

(9) Analysis of the family situation of the school roll system: comprehensively analyze the integrity and happiness of the students' families, and identify the students from problem families as the "focus of attention", with the highest score of 7 points.

(10) Analysis of love situation: The counselor shall regularly report the love situation of the students in this class, and list the students who have just broken up recently as the "focus of attention", with the highest score of 7 points.

(11) Mental health test for freshmen: The results of the mental health test for freshmen were analyzed, and the students with psychological problems were identified as the "focus of attention", with the highest score of 8 points.

(12) Periodic mental health questionnaire: The psychological questionnaire was distributed according to the cycle. Through the analysis of the content of the questionnaire, the students with psychological problems were identified as "the focus of attention", with the highest score of 8 points.

In each type of data source, extract the identified students as mentioned above, and assign scores according to their actual situation. The higher the score, the greater the possibility of psychological problems. If the total score exceeds 60 points, an alarm will be issued. Among them, online data and book borrowing are more referential, so once these two items are identified as "extreme tendency", they will be assigned 10 points, and the system will automatically screen out these students and issue an alarm.

4 Portrait Generation

4.1 Establish the Framework of Data Portrait Generation Process

It is necessary to form a complete "data acquisition-data analysis-data feedback" chain for the portrait of students' psychological problems, extract relevant fields from the original database, analyze and sort them out, and finally extract key information to form students' psychological portraits. The tag fields related to students' psychological problems include the types of books borrowed, the contents of Internet browsing, the contents of collections, the results of freshmen's mental health tests, and the results of periodic questionnaires. These contents exist in the complex original database, which can be screened, mined, integrated, extracted, evaluated and generated through the following links. Finally, around the core of "psychological problems", the mental health status of each student is comprehensively evaluated, and the system automatically warns the students who have problems or hidden dangers. The specific process is shown in Fig. 2:

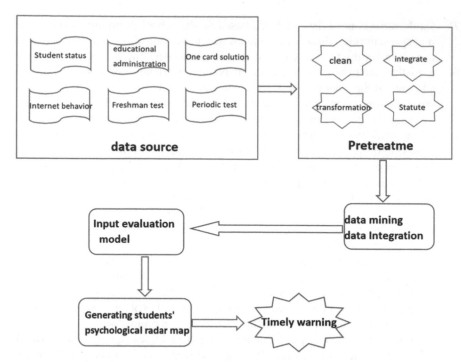

Fig. 2. Generation of college students' psychological portrait

4.2 Realize Data Abstraction by Using Data Transfer Station

Data abstraction refers to abstracting the contents of the selected fields into corresponding mental health status, such as book borrowing data, physical exercise data, Internet data, etc. Through data extraction and data transformation, the data representing the degree of mental health are abstracted, and finally the mental health status of each student is quantified. Taking book borrowing information as an example, the fields related to book borrowing information are extracted from the background of one-card data, the frequency of borrowing, the names of all books borrowed, the length of borrowing of different types of books, whether there is a special preference for a certain type of books, the names and types of bibliographies recommended for purchase, and the names of bibliographies borrowed from other places. The borrowing data of each field is quantified into specific mental health data.

4.3 Define Mental Health Measurement Indicators

Through analysis and evaluation, the selected relevant fields are abstracted into mental health measurement indicators, and the specific mental health measurement indicators include: work and rest, exercise, attendance, performance, online time, types of books of interest, web content of favorite browsing, love, past psychological assessment, random psychological assessment, etc.

This paper refers to the "construction of data portrait evaluation model based on random forest model" [16] proposed by scholar Deng Jiaming in "Research on the Generation of Student Data Portrait in Smart Campus", inputs the selected specific fields into the evaluation model, matches the mental health measurement indicators, and finally outputs the student psychological portrait data.

4.4 Generating the Dimensional Map of Students' Psychological Portraits

The final output of the student's psychological portrait data is displayed in the form of a dimensional map, which clearly shows the student's random psychological assessment results, love situation, regular work and rest, love of exercise, performance, attendance and other information from the beginning of the school to the present. Taking three students from the School of Information as an example, the student's psychological portrait radar map is generated as shown in Fig. 3:

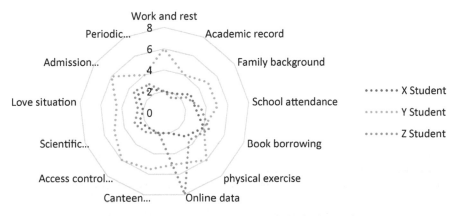

Fig. 3. Radar map of students' psychological portrait

4.5 Timely Warning According to Students' Psychological Portrait Data

According to the data of students' psychological portraits, the students whose total score of the portraits is more than 60 points will be warned by the OA system in time, and the warning information will be directly pushed to the corresponding counselors and head teachers, so as to confirm the actual situation of the students in time, take psychological counseling and avoid tragedies. In this study, 1000 junior students in communication college of a university were collected by big data mining technology, and all kinds of data were cleaned and sorted out, and then the evaluation model was input to match the mental health measurement indicators, and finally the students' psychological portraits were generated. The results showed that the total score of 949 students was less than 60, and they were in good mental health. 51 students scored more than 60 points, of which 5 students scored more than 85 points, and the OA system received early warning information. After communicating with their counselors, roommates and classmates, it is found that the 51 students do have various psychological problems, which is consistent with the actual investigation and has practical research significance.

5   Conclusion

This paper focuses on the analysis of the application of big data technology in college students' mental health, and uses user portrait technology to portray college students' psychology. Big data technology provides data support and technical support for college students' mental health education.Through big data technology, all the characteristics of a student can be effectively obtained and summarized, which is conducive to university educators to obtain key information, so as to make effective adjustments and responses. It is foreseeable that the future of mental health education will rely on big data technology to achieve leaps and bounds. Traditional mental health education, including "ideological and moral" courses, "psychological counselling" and "employment guidance", must be expanded under the impetus of big data, from education mode to self-positioning and personalized service mode. However, the study of college students' mental health factors analysis is not very comprehensive, weight analysis did not carry out more detailed research, in the follow-up study will be concerned about these two aspects of the study.

References

1. Xiaoyi, F., Xiaojiao, Y., Wei, H.: Development of Chinese college students' mental health screening scale. Psychol. Behav. Res. **16**(1), 111–118 (2018)
2. Xiong, Y.: Research on the application of user portrait in personalized recommendation of educational resources. Fujian Comput. **37**(1), 52–53 (2021)
3. Li, J.: Research on precise push of mobile advertising based on user portrait and RBF. Autom. Instrum. **1,** 38–42 (2021)
4. Luo, B., Cui, Y.: J. Comput. Knowl. Technol., **17**(03), 48–49 (2017)
5. Xu, M.: Research on Optimization of Knowledge Recommendation Service of University Library Based on User Portrait. Published by wide angle, vol. 1, 76–78 (2021)
6. Xu, C., Mao, G., Zhou, Y.: Information literacy education of University Library Based on user portrait. J. Univ. Libr. Inf. Sci. **39**(1), 28–31 (2021)
7. Wenxing, S., Shiyun, C.: Improvement of power service level based on user portrait. Electrical age **1,** 28–31 (2021)
8. Jiajia, Y.: Analysis of user portrait system in news field. Fujian Comput. **36**(12), 154–155 (2020)
9. Chunqiu, L.: Research on personalized service of Smart Library Based on user portrait. J. Fuyang Vocat. Tech. Coll. **31**(4), 69–72 (2020)
10. Yu, Y.: Research on precise knowledge service of academic new media based on user portrait. Intell. Explor. **12,** 58–64 (2020)
11. Liu, S.: Which psychological problems are easy to perplex college students. People's Forum **12,** 136–137 (2019)
12. Zeng, D., Hong, Y.: Family background, campus life and college students' health. Dongyue Lun Cong. **41,** 04 (2020)
13. Xiaoyi, F., Yuan Xiaojiao, H., Wei, D.L., Xiuyun, L.: Development of Chinese college students' mental health screening scale. Psychol. Behav. Res. **16**(01), 111–118 (2018)

14. Suhui, G., Quan, W., Chengjie, B.: Research on College Students' behavior early warning and decision-making system based on Hadoop. Comput. Appl. Softw. **38**(1), 6–12 (2021)
15. Ziming, Z., Shouqiang, S.: Research on personalized mobile vision search in smart library based on user portrait. Books Inf. **4**, 84–91 (2020)
16. Jiaming, D.: Research on the generation method of student data portraits in smart campus. Mod. Electron. Technol. **42**(21), 58–62 (2019)

A Multi-stage Event Detection Method

Xiaoshuo Feng[1], Zeyu Lv[2], Wandong Xue[2], Zhengping Sun[2], and Dongqi Wang[2](✉)

[1] Naval Research Institute, Beijing, China
shuoner@163.com
[2] Software College, Northeastern University, Shenyang, China
{2001245,2101254}@stu.neu.edu.cn, szp0928@outlook.com,
wangdq@swc.neu.edu.cn

Abstract. Event detection algorithms are developing rapidly, introducing more and more methods in complex networks. However, the current single-stage event detection method is less effective and far from the expected effect. This paper proposes a multi-stage event detection method by combining multiple techniques, including topic detection (LDA model), complex network community detection method, and machine learning algorithms. Compared to simple topic detection-based solutions. The accuracy and recall are effectively improved on the benchmark event detection data set.

Keywords: Event detection · Community detection · LDA

1 Introduction

In the information age, event detection algorithms have attracted more and more attention. More and more news data flood the network, making it cumbersome and time-consuming to filter practical information to find out what is happening in the real world. As a result, there is a higher and higher demand for automatic mining of news, sorting out and discovering significant events, and assisting personnel in extracting vital information. Researchers have proposed algorithms [1] to solve event detection requirements. However, since there is no universal definition of an event, current event detection algorithms usually cannot satisfy all application requirements. Many algorithms can only generate descriptive words for events and cannot provide more helpful information. In application, descriptive words and events can be well matched when a user pre-realizes an event. Then the users can summarize the event and provides corresponding descriptive vocabulary. While users do not have enough information about the events in advance, they generally have no clear understanding of what is going on by only inspecting the relevant descriptive vocabulary of the event (keywords on the event). As so, event detection solutions that only provide descriptive keywords have limits in applications, making generating more accurate event detection algorithms an urgent requirement.

Existing event detection solutions are mainly single-stage methods, and the detection results are not precise. This paper combines multiple event detection methods to improve the accuracy of event detection. The proposed algorithm will pre-cluster target texts into

© The Author(s), under exclusive license to Springer Nature Switzerland AG 2023
N. Xiong et al. (Eds.): ICNC-FSKD 2022, LNDECT 153, pp. 968–973, 2023.
https://doi.org/10.1007/978-3-031-20738-9_106

groups corresponding to events to which these texts belong. Compared to most existing event detection solutions, the pre-clustering operation will be able to further improve event detection accuracy. Moreover, existing event detection algorithms generally can only describe events by a few words, and the specific event information still needs to be obtained manually. Our proposed solution will accurately recall the text related to each event, making the event detection result more valuable.

This paper has the following contributions:

1. Combines multiple event detection techniques to improve the event detection accuracy.
2. The algorithm can recall the text related to the event while finding the event.
3. The algorithm can identify the category to which the event belongs when finding the event.
4. Compared different community detection algorithms through experiments.

2   Related Work

The event detection algorithm finds the desired events from data such as time signals, text features, GPS features, etc. Early algorithms mainly used only time signals for event detection. With the development of the Internet, text data is gradually enriched, and the time data in the text is sufficient to support event detection algorithms. Therefore, the application of event detection algorithms in news and other texts is becoming more and more extensive. To improve the accuracy of event detection, more and more papers are starting to add more available features that can increase the accuracy of event detection algorithms, such as using GPS information in event detection algorithms.

Event detection was documented as early as 1998 [2]. This method uses only time series data and no other related data such as text. The method expects to find the corresponding event and the location of the event from the changes of the time series data. The method fits the data through a model, and then detects changes in the model parameters to infer whether an event has occurred. Compared with manual interpretation of data curves, this method is more robust in results. Better detection of events in noisy situations.

Algorithms [3] combine textual data and time series to discover events. The algorithm mainly detects events from the frequency of words appearing in news texts over time. The algorithm uses wavelet transform to create a corresponding signal from each word's original signal in the frequency domain. At the same time, trivial words are eliminated through the correlation matrix to improve the accuracy of the algorithm. The algorithm can also output multiple descriptive words for each event, so that users can understand the relevant information of the event. However, the algorithm cannot correlate with the original text data, and cannot obtain complete event information. Algorithms [4] use complex network correlation algorithms to discover events. The algorithm first constructs a graph by treating words as nodes according to the co-occurrence relationship of words in the text, and the edge weights are related to the number of co-occurrences in the text. Instead of using raw word frequency data as the input to the event detection algorithm as in the algorithm [3], the algorithm takes the constructed graph as input. Where the node weight is the sum of the edge weights. The algorithm also detects events from the data in which the node weight changes with time in the graph, and filters out the nodes whose node weight changes drastically, so as to filter out the corresponding key words related to the event. And filter out the text containing these words, and use the topmine algorithm [5] to mine the phrases in the text, so as to supplement the key information to form a better description of the time, and add these phrases to the previous built diagram. The communities in the graph are then partitioned using the Louvain community detection algorithm [6], resulting in all distinct events in that time period. Each community is an event, and the words and noun phrases within the community describe the event. The algorithm also cannot map the event to the original text. Although noun phrases are added to make some supplementary descriptions of the event, the amount of information is still relatively insufficient.

3   Method

Firstly, the proposed method pre-classifies the text and uses latent Dirichlet allocation (LDA) to assign articles to different categories. This way, texts related to other contents are roughly sorted out. We believe that carrying out event detection tasks in specific topics will improve event detection accuracy because the topic assignment will narrow the search scope of event detection tasks. Then, an event detection algorithm for complex networks [4] is employed to generate the corresponding set of representative words (namely vocabulary or event vocabulary) for different events. Then, the vector representations of texts are generated by converting the words into TF-IDF vectors. Finally, the event-related texts will be retrieved using the nearest neighbor calculation with the event vocabulary.

The proposed method combines the Latent Dirichlet Allocation (LDA) model [7], Term Frequency-Inverse Document Frequency (TF-IDF) model [8], complex network community detection methods, and commonly used clustering algorithms for event detection, which improve the event detection accuracy a lot. And after setting the topic category, there is no need to determine the number of events with algorithms such as K-means manually. The event vector representation outputted in step 2 will be used as the cluster center. Furthermore, the implementation of the algorithm does not need to randomly select nodes at the beginning, which reduces the randomness of the algorithm.

Algorithm 1 Event Detection Algorithm

Require: Text T, Number of topics k
Output: Event E
$\{T_i, i \in k\} \leftarrow LDA\ (T, k)$
for i such that $0 \leq i \leq k$ **do**
Output a descriptive vocabulary for each event using a complex network event detection algorithm k_j
Using the TF-IDF algorithm based on descriptive vocabulary k_j to recall corresponding text $\{t_i\}$

---
    end for

The proposed method improves existing algorithms by detecting events with a complex network algorithm. It selects the label propagation algorithm because of its better effect in visual comparison with other methods for complex network community detection. Other than the community detection algorithm in article [9], the proposed method uses a simpler network by deleting the edges with small edge weights in advance. At the same time, since most existing community detection algorithms do not use edge weights for community division, the proposed method is more conducive to the community detection algorithm to divide communities. See Algorithm 1 for an overview of the algorithm.

4 Experiment

The experiment uses the data set provided in the paper [9] and compares the algorithm's accuracy, recall, and F1 value with benchmark event detection algorithms.

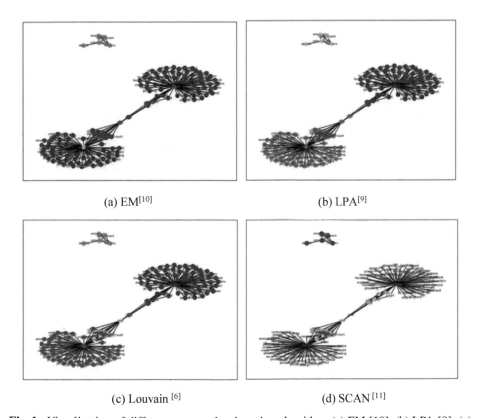

(a) EM[10]

(b) LPA[9]

(c) Louvain [6]

(d) SCAN [11]

Fig. 1. Visualization of different community detection algorithm. (a) EM [10], (b) LPA [9], (c) Louvain [6], (d) SCAN [11]

This data set is constructed by filtering the data set in the original paper [1]. The data set is collected from more than 50 major world news media, covering the most important events within a year. The data set contains 207,722 articles and more than 2000 topics. It covers 4501 events. The news includes sports, technology, politics, economy, art, etc., covering a vast range. The news site with the most articles is dailymail.co.uk. There are also a large number of countries where news organizations are located. It also includes news organizations from China like People's Daily Online and Shanghai Daily. As shown in Fig. 1, representative complex network community detection algorithms are compared. It provides help for algorithm selection and confirms the superiority of the selected community detection algorithm compared with the original algorithm through the numerical test.

Table 1. Experiment result

Algorithm	Accuracy	Recall	F1
RevDet [12]	0.81	0.56	0.66
Ours (Louvain)	**0.94**	0.55	0.69
Ours (LPA)	0.90	0.63	**0.74**
Ours (SCAN)	0.67	**0.66**	0.66

As can be seen from Fig. 1, the community divided by the LPA algorithm is the best, while the community detected by the SCAN algorithm [11] cannot cover most nodes. The critical nodes in the community divided by the SCAN algorithm are missing. The Louvain algorithm is second only to the LPA algorithm. Because the EM algorithm needs to specify the number of communities, it cannot automatically adjust the number of communities, and isn't easy to apply to this algorithm. It can also be seen from Fig. 1 that the EM algorithm cannot separate two distinct communities.

As shown in Table 1, results from Fig. 1 can be further verified. The algorithms using the LPA are the best, the F1 value is the highest, and the accuracy and recall rate also exceeds the baseline algorithm. However, the SCAN algorithm loses many vital nodes, which makes the recalled text of low quality and the most insufficient accuracy.

5   Conclusion

This paper provides an event detection method that combines techniques including LDA, TF-IDF, and the existing community detection-based event detection approach. The experimental result proves that the community divided by the LPA works well. In contrast, the community detected by the SCAN algorithm cannot cover most nodes, which means that compared to the benchmark algorithm, the accuracy and recall of the proposed method on the event detection data set are much superior.

References

1. Hoang, T.A., Vo, K.D., Nejdl, W.: W2E: a worldwide-event benchmark dataset for topic detection and tracking. In: The 27th ACM International Conference ACM (2018)
2. Guralnik, V., Srivastava, J.: Event detection from time series data. Knowl. Discov. Data Min. ACM (1999)
3. Weng, J., Lee, B.S.: Event detection in twitter. In: Fifth International Conference on Weblogs & Social Media DBLP (2011)
4. Moutidis, I., Williams, H.: Complex networks for event detection in heterogeneous high volume news streams. arXiv (2020)
5. El-Kishky, A., et al.: Scalable topical phrase mining from text corpora. Proc. Vldb Endowment **8**(3), 305–316 (2014)
6. Blondel, V.D., et al.: Fast unfolding of community hierarchies in large networks. J. Stat. Mech. abs/0803.0476 (2008)
7. Blei, D.M., Ng, A., Jordan, M.I.: Latent dirichlet allocation. J. Mach. Learn. Res. (2003)
8. Salton, G., Mcgill, M.O., McGill, J.: Introduction to Modern Information Retrieval. McGraw-Hill, New York (1983)
9. Berahmand, K., Bouyer, A.: LP-LPA: a link influence-based label propagation algorithm for discovering community structures in networks. Int. J. Mod. Phys. B, 1850062 (2017)
10. Moon, T.K.: The expectation-maximization algorithm. IEEE Signal Process. Mag. **13**(6), 47–60 (1996)
11. Xu, X., et al.: SCAN: a structural clustering algorithm for networks. ACM, 824–833 (2007)
12. Azeemi, A.H., et al.: RevDet: Robust and Memory Efficient Event Detection and Tracking in Large News Feeds (2021)

Influencing Factor Analysis for Information Technology Training Institutions

Zhenzhen Li[1], Peng Yang[1], Zhenhua Yuan[2], Yan Chen[1], Shaohong Zhang[1($\boxtimes$)], and Deying Liu[3]

[1] School of Computer Science and Cyber Engineering, Guangzhou University, Guangzhou, Guangdong, China
1973163532@qq.com, 770508164@qq.com, 862744285@qq.com, zimzsh@qq.com
[2] School of Mathematics and Information Science, Guangzhou University, Guangzhou, Guangdong, China
2290841141@qq.com
[3] School of Economics and Statistics, Guangzhou University, Guangzhou, Guangdong, China
2112164008@e.gzhu.edu.cn

Abstract. With the continuous development of the information industry and the enormous market demand, the demand for Internet Technology (IT) training is also rapidly increasing. However, the epidemic's sudden outbreak and the implementation of related policies have left training institutions in an existential crisis. To analyze the influencing factors of information technology training institutions, we crawled 20,086 job recruitment information from the website and conducted a questionnaire survey of college students in Guangzhou. The fuzzy comprehensive evaluation method are used to analyze the influencing factors of IT training institutions from the aspects of the institutional foundation, curriculum setting, and publicity mechanism. The research results provide a good reference for the development of IT training institutions.

Keywords: IT training · Data mining · Fuzzy comprehensive evaluation method

1 Introduction

With the rise of the domestic Internet economy, the creation of new industries has created a large demand for IT professionals. According to Ministry of Industry and Information Technology statistics, the domestic IT industry's overall revenue scale increased from 2479.4 billion yuan to 7176.8 billion yuan from 2012 to 2019, with the average growth rate of industry revenue remaining above double digits [1]. Since 2015, in order to promote the rapid development of IT-related industries, the State Council have issued several policies [2]. China's computer talent from 425,000 talent gap to has exceeded one million [3].

© The Author(s), under exclusive license to Springer Nature Switzerland AG 2023
N. Xiong et al. (Eds.): ICNC-FSKD 2022, LNDECT 153, pp. 974–982, 2023.
https://doi.org/10.1007/978-3-031-20738-9_107

With the continuous advancement of IT technology, the demand for IT talent in various industries will rise in the future, and the technical level of IT skills will be required to be higher [4]. From the current supply and demand situation of IT talents, China's higher education graduates can not yet meet the industry's demand for talents [5]. The gap in IT skills provides a broad market space for IT education and training, and this market space will expand rapidly with the rapid development of the IT industry [6]. Menon et al. analyzed the factors influencing students' pursuit of higher education without direct employment in Cyprus. They found that the psychological/individual factor had a significant impact [7], and we will further analyze the motivational factors affecting students' participation in training rather than direct employment.

At the moment, China's IT training industry is highly fragmented, with less than 10% of large-scale training institutions concentrated in the industry. We can see the fierce competition as well as the vast development space in the IT training industry [8]. Therefore, it is necessary to gain more significant development in the market through strategies such as adjusting marketing strategies [9]. In our work, we will consider how IT training institutions can respond to the challenges of the current environment and turn the situation around, and propose improvement suggestions based on the analysis of the influencing factors.

2 Data Collection

2.1 Job Recruitment Data

The recruitment information on the website can intuitively understand the needs of enterprises for talents, to a certain extent, reflects the features of the talent market's demand [10]. We used Python to crawl recruitment information and collected 20,086 job information on recruitment websites such as 51job, Lagou, and Zhilian. A total of 17,120 pieces of recruitment data were retained by cleaning the data, eliminating invalid data, and removing duplicate data [11,12].

2.2 Questionnaire Data

In implementing the questionnaire survey, we took the college students in Guangzhou as the survey object. The questionnaire collected information on the institution's behavioral characteristics, motivation, and evaluation from students who need to improve their IT skills.

The influencing factors were investigated from three aspects: institutional foundation, curriculum, and publicity mechanism. For students of colleges and universities in Guangzhou, The survey used a combination of stratified sampling and multi-stage sampling to obtain samples [13]. Under the premise that the pre-survey ensures that the survey can be implemented, 885 questionnaires were issued in the formal study. The total number of valid questionnaires recovered was 776, and the recovery rate was 87.7%. The obtained data have passed the reliability, validity, normality, and chi-square tests, indicating that the survey data is accurate and reliable.

3   Analysis of the Current State of IT Requirements

3.1   The Demand for IT in Various Professional Positions

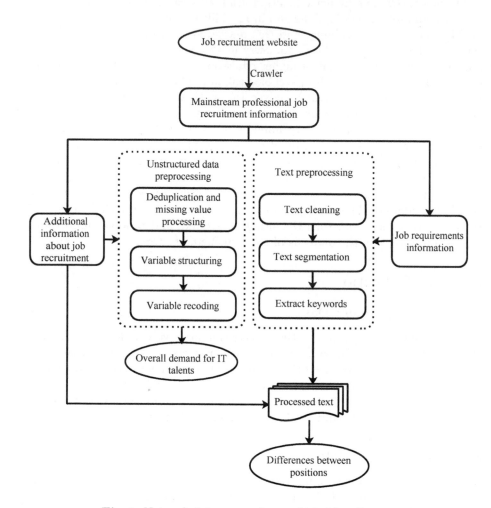

Fig. 1. Network data processing analysis idea diagram

This work analyzes job information crawled by the recruitment website and retains 17,120 pieces of information after data cleaning. The job data processing and analysis ideas are shown in Fig. 1.

Through the analysis of the skill requirements of the position, we found that IT technology has been deeply integrated into various industries. Engineering has a tremendous demand for IT training, accounting for 63%; medical and liberal arts also have a greater demand, reaching more than 30%. In contrast, the demand for agriculture and law majors is lower. There is a sure match between the IT skill requirements of various professional positions and each IT technology. The relevant employment positions in different majors have different requirements for IT technology, as shown in Fig. 2.

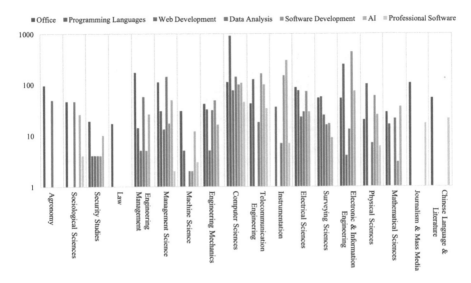

Fig. 2. Map of the distribution of IT technology needs in different specialties

3.2   Questionnaire Analysis of College Students

Based on a questionnaire survey of college students in Guangzhou, we analyzed the information on college students' willingness, motivation, behavioral characteristics, and evaluation of institutional influencing factors. We found that among the surveyed student samples, those who needed to improve IT skill accounted for the majority, 92.14%. It can be seen that the college student market has much room for development.

The results of the analysis of the needs of students in different majors to enhance their IT skills are shown in Fig. 3. Among the students who need to improve their IT skills, the requirements for improving IT skills vary from major to major, which is consistent with the results of job data analysis. According to the questionnaire results, 31.19% of the survey respondents have participated in IT training. More training courses among the survey respondents who have participated in training are office software, programming language, and film editing, followed by software development, web development, and AI technology. It can be seen from Fig. 4 that students of different majors have different needs for IT technology. Students are frequently exposed to training promotion mainly from webpage promotion, but the trust level of students in a webpage is deficient. Students trust the teacher's recommendation the most, while fewer students are exposed to training in this way.

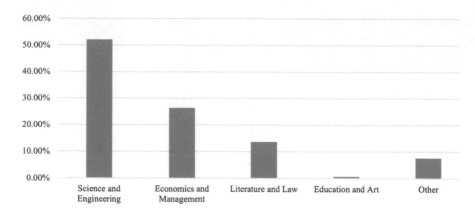

Fig. 3. The proportion of different majors willing to improve IT skills

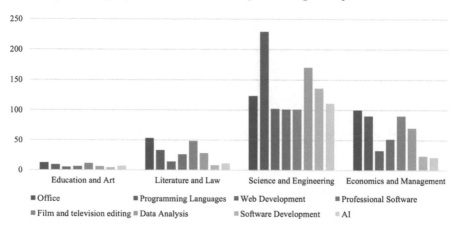

Fig. 4. The proportion of training courses expected by different specialties

4   Influencing Factor Analysis

To analyze the influencing factors of IT training institutions, we collected the scoring of different aspects of institutions. The fuzzy comprehensive evaluation method is used to model student evaluations, and the comprehensive evaluation is analyzed from various perspectives within the institution.

4.1   Fuzzy Comprehensive Evaluation Method Analysis

Given that there are numerous factors influencing IT training participation, and each indicator is not definite. Therefore, the fuzzy comprehensive evaluation method is chosen as a comprehensive evaluation method for the factors influencing IT training. This method can establish a set of index system that reflects the essence of the evaluation object in general, and then combine each index into a comprehensive index that can measure the institution in general [14,15]. Our work roughly divides the indicators into three modules, see Table 1.

Table 1. Table of secondary comprehensive evaluation models

First-level indicators	Second-level indicators
	Teacher resources
	Tuition fees
Evaluation indicators of internet technology institutional	Duration of study
	Institutional environment
	Learning atmosphere
	Project quality
Evaluation indicators of individual scientific research ability	Get more expertise
	Publish high-quality papers
	Increase project experience
Evaluation indicators of individual job search ability	Get internship opportunities
	Obtain professional certificate

Determine a factor set: The factor set of IT training organization evaluation indexes consists of primary and secondary evaluation indexes. First-level indicators: 1–3, and second-level indicators 1–11.

Determine evaluation sets: This paper uses the questionnaire method as an important method of collecting data and uses the more common five-level scale to collect students' evaluation of the institution [16]. The evaluation set of IT training institutions is divided into 5 levels: very satisfied, satisfied, average, dissatisfied, and very dissatisfied.

Determine the weight of each factor: To create an evaluation matrix, different weights are assigned to each factor in both levels. Each factor in the factor set is evaluated individually for the IT training machine.

Weight determination based on analytic hierarchy: In the systematic analysis, the analytic hierarchy method can combine qualitative and quantitative analysis [17]. According to the weight scale method, the relative importance judgment matrix of upper XX, upper YY, upper ZZ corresponding factors follows.

Scale matrix of upper X colon Start 6 By 6 Matrix 1st Row 1st Column 1 2nd Column one third 3rd Column 2 4th Column 3 5th Column 2 6th Column one third 2nd Row 1st Column 3 2nd Column 1 3rd Column one third 4th Column 7 5th Column one third 6th Column 1 3rd Row 1st Column one half 2nd Column 3 3rd Column 1 4th Column 3 5th Column 1 6th Column one third 4th Row 1st Column one third 2nd Column one seventh 3rd Column one third 4th Column 1 5th Column one third 6th Column one seventh 5th Row 1st Column one half 2nd Column 3 3rd Column 1 4th Column 3 5th Column 1 6th Column one third 6th Row 1st Column 3 2nd Column 1 3rd Column 3 4th Column 7 5th Column

3 6th Column 1 EndMatrix$X$: $\begin{pmatrix} 1 & \frac{1}{3} & 2 & 3 & 2 & \frac{1}{3} \\ 3 & 1 & \frac{1}{3} & 7 & \frac{1}{3} & 1 \\ \frac{1}{2} & 3 & 1 & 3 & 1 & \frac{1}{3} \\ \frac{1}{3} & \frac{1}{7} & \frac{1}{3} & 1 & \frac{1}{3} & \frac{1}{7} \\ \frac{1}{2} & 3 & 1 & 3 & 1 & \frac{1}{3} \\ 3 & 1 & 3 & 7 & 3 & 1 \end{pmatrix}$ Scale matrix of upper Y colon

Start 3 By 3 Matrix 1st Row 1st Column 1 2nd Column one third 3rd Column 1 2nd Row 1st Column 3 2nd Column 1 3rd Column 3 3rd Row 1st Column 1 2nd Column one third 3rd Column 1 EndMatrix$Y$: $\begin{pmatrix} 1 & \frac{1}{3} & 1 \\ 3 & 1 & 3 \\ 1 & \frac{1}{3} & 1 \end{pmatrix}$

Scale matrix of upper Z colon Start 2 By 2 Matrix 1st Row 1st Column 1 2nd Column 3 2nd Row 1st Column one third 2nd Column 1 EndMatrix$Z$: $\begin{pmatrix} 1 & 3 \\ \frac{1}{3} & 1 \end{pmatrix}$

Scale matrix of upper U colon Start 3 By 3 Matrix 1st Row 1st Column 1 2nd Column one third 3rd Column one third 2nd Row 1st Column 3 2nd Column 1 3rd Column one third 3rd Row 1st Column 3 2nd Column 3 3rd Column 1 EndMatrix$U$: $\begin{pmatrix} 1 & \frac{1}{3} & \frac{1}{3} \\ 3 & 1 & \frac{1}{3} \\ 3 & 3 & 1 \end{pmatrix}$

The scores of upper X comma upper YX, Y, and upper ZZ are calculated from the weight matrix and the evaluation model, so as to obtain the overall evaluation score of the individual's service to the institution.

Table 2. Overall personal evaluation score of the organization's training services

	Institutional base evaluation	Evaluation of individual scientific research ability	Evaluation of individual job search ability	Total score
Score	72.97	73.22	71.92	72.82

From the evaluation scores in Table 2, it can be seen that students are more satisfied with the evaluation of their individual scientific research ability after training, and the lowest evaluation of their individual job search ability after training, indicating that students attach great importance to improving their job search ability. Training institutions still have shortcomings in improving their job-seeking capacity and need further improvement.

5   Conclusion and Suggestion

In summary, our work analyzes the demand for IT training and the factors influencing it through questionnaire survey and web data mining.

By analyzing the job skill requirements of network data, it can be seen that IT technology has been deeply integrated into various industries. In the sample of students surveyed, the vast majority need to improve IT skills, which shows that institutions have significant room for development in the college student market.

The trust in the promotion method is an essential reference for placing advertisements, and students have the highest trust in teachers' recommendations. Through the fuzzy comprehensive evaluation method obtains a comprehensive evaluation of the institution by the user. Students attach more importance to professional knowledge, and have the lowest evaluation of their job search ability after training. The institution can further optimize according to this feedback.

Based on the analysis of job data, questionnaire data, and research literature, we will make suggestions for IT training institutions.

1. Promoting cooperation with universities and enterprises. Cooperation with colleges and universities can strengthen the teaching team. Further grasp the latest developments in the market, form a competitive advantage for institutions.
2. Updating courses promptly. IT technology changes have accelerated in recent years. Therefore, IT training companies constantly update the content of the lectures so that it is in line with the development of IT industry practice.
3. Adjusting the way of publicity. To successfully create value for customers requires successful delivery of value. The selection and design of marketing channels is the basis for delivering value. IT institutions can increase the promotion of courses and products among college teachers and students.
4. Improving the internal level of the training institution. The quality of the teaching staff directly determines the quality of the students' learning outcomes. IT training institutions should use the industry association evaluation standards to purify the market environment and improve industry quality.

Acknowledgment. The work described in our work was partially supported by grants from the funding of Guangzhou education scientific research project [No. 1201730714], and the Guangdong Basic and Applied Basic Research Foundation [No. 2022A1515011697].

References

1. Tu, X., Wang, M., Sun, K., Zhang, C., Zhang, L.: China internet industry state analysis and prosperity indexes. China Commun. **13**(10), 245–252 (2016)
2. Zhang, C., Chen, Y.: A review of research relevant to the emerging industry trends: industry 4.0, iot, blockchain, and business analytics. J. Indus. Integr. Manag. **5**(01), 165–180 (2020)
3. Spitzer, B., Morel, V., Buvat, J., Kvj, S., Bisht, A., Radhakrishnan, A.: The Digital Talent Gap: Developing Skills for Today's Digital Organizations. Capgemini Consulting (2013)
4. Finnie, R., Mueller, R.E., Sweetman, A.: Information and communication technology talent: the skills we need-framing the issues (2018)
5. Haywood, E., Madden, J.: Computer technology students-what skills do they really need? In: Proceedings of the Australasian conference on Computing education, pp. 139–144 (2000)
6. Avis, J.: Socio-technical imaginary of the fourth industrial revolution and its implications for vocational education and training: A literature review. J. Vocat. Educ. Train. **70**(3), 337–363 (2018)

7. Menon, M.E.: Factors influencing the demand for higher education: The case of cyprus. High. Educ. **35**(3), 251–266 (1998)
8. Palvia, S., Aeron, P., Gupta, P., Mahapatra, D., Parida, R., Rosner, R., Sindhi, S.: Online education: worldwide status, challenges, trends, and implications (2018)
9. Kingsnorth, S.: Digital Marketing Strategy: An Integrated Approach to Online Marketing. Kogan Page Publishers (2022)
10. McCallum, A.: Information extraction: distilling structured data from unstructured text. Queue **3**(9), 48–57 (2005)
11. Jamshed, H., Khan, S.A., Khurram, M., Inayatullah, S., Athar, S.: Data preprocessing: A preliminary step for web data mining. 3c Tecnología: glosas de innovación aplicadas a la pyme **8**(1), 206–221 (2019)
12. Zhang, S., Yang, Z., Xing, X., Gao, Y., Xie, D., Wong, H.S.: Generalized pair-counting similarity measures for clustering and cluster ensembles. IEEE Access **5**, 16904–16918 (2017)
13. Etikan, I., Bala, K.: Sampling and sampling methods. Biometrics Biostatistics Int. J. **5**(6), 00,149 (2017)
14. Zeng, W., Feng, S.: An improved comprehensive evaluation model and its application. Int. J. Comput. Intell. Syst. **7**(4), 706–714 (2014)
15. Zhang, S., Wong, H.S., Shen, Y.: Generalized adjusted rand indices for cluster ensembles. Pattern Recogn. **45**(6), 2214–2226 (2012)
16. Betz, N.E., Hammond, M.S., Multon, K.D.: Reliability and validity of five-level response continua for the career decision self-efficacy scale. J. Career Assess. **13**(2), 131–149 (2005)
17. Vaidya, O.S., Kumar, S.: Analytic hierarchy process: an overview of applications. Eur. J. Oper. Res. **169**(1), 1–29 (2006)

Predicting Impact of Published News Headlines Using Text Mining and Classification Techniques

Parikshit Banerjee, Usha Ananthakumar$^{(\boxtimes)}$, and Shubham Singh

IIT Bombay, Main Gate Rd, IIT Area, Powai, Mumbai, Maharashtra 400076, India
{parikshit.banerjee,shubham.singh}@sjmsom.in,
usha@som.iitb.ac.in

Abstract. The efficient market hypothesis states that security prices reflect all the information relevant to the asset in the public domain. A significant source of such information, which impacts the day-to-day fluctuations of stock prices significantly, is financial news about companies. However, consuming and analyzing the vast amount of textual information published every day about a company in the public requires significant manual effort. This is where we believe text mining, combined with predictive analytics techniques, can significantly enhance the capability of an investor or a financial analyst. The goal of this paper is to create a predictive model that will take published news headlines about Indian companies as input and classify them as Impactful or Not Impactful based on past reactions to similar published news. The experimental results show that the performance of classification algorithms depends on the feature extraction technique that it is paired with and vice-versa. We also demonstrate that machine learning models paired with the right feature extraction technique can, to a reasonable degree, predict market reaction to published news.

Keywords: Text mining · Stock price prediction

1 Introduction

Investors or financial analysts, participating in public stock markets, consume published financial news with the goal of predicting market reaction and identifying investment opportunities. This contradicts the Efficient Market Hypothesis [3], which postulates that all information available at the moment is already reflected in the prices of securities; hence it is impossible to predict stock prices based on available information. However, several individuals and studies have disproved this theory by outperforming the market using available information. We can find an explanation for this by understanding the mechanics behind stock price fluctuations. Like any other commodity, stock prices follow the rules of demand and supply. If the demand for a security exceeds its supply, its price will go up; else, it will fall. Demand is determined by several factors influencing human behavior and rationality, which forms the basis for EMH. But what are the factors that cause the demand for specific security to fluctuate? Published textual information related to the security in question may be considered one of the most significant

N. Xiong et al. (Eds.): ICNC-FSKD 2022, LNDECT 153, pp. 983–992, 2023.
https://doi.org/10.1007/978-3-031-20738-9_108

contributors to such fluctuations. Before we explain why this happens, it is essential to understand the different players who participate in the market and their motivations. Firstly, there are institutional investors, who pool capital on behalf of their organization, other organizations, and individual investors to buy and sell securities in the market. These investors generally hold securities for a medium to long-term time horizon and are not prone to making buying and selling decisions based on daily financial news. On the other hand, retail investors or day-traders buy and sell securities from their personal funds and look for profits on a much shorter time horizon. Hence, retail investors are more likely to react to short-term trends caused by published news.

This study focuses more on the day-to-day fluctuations in stock prices due to published news. It thus captures the behavior of retail investors to a higher degree than institutional investors. As mentioned previously, institutional investors, who control a majority of any financial market globally, do not rely on the hour-to-hour noise generated by published news to make investment decisions on most occasions. Hence, creating an accurate stock price prediction model based solely on published financial news is challenging. Thus, any study of this type is more likely to create a helpful decision support system [10] than a reliable decision-making system. Many investors already leverage published financial news for investment decision support, with our model providing further opportunities to enhance newsreader experience and engagement.

2 Literature Review

Early literature on predicting stock price movements, such as the efficient market hypothesis (EMH) [3] and the random walk theory [14], conclude that stock market prediction is impossible based on available information. However, one of the earliest attempts to analyze textual information to predict stock price movement [19] concluded that a system that analyses textual information to make stock market predictions shows promising accuracy and money-making potential. While their study was limited to analyzing the 400 keywords provided by market experts, their research provided an excellent starting point for further work in this domain. Studies further expanded on this by proposing a system that would monitor a stream of incoming news stories and flag interesting stories that could initiate a significant shift in the security price within a 10 h time horizon from the story [12]. The proposed system achieved a True Positive Rate of between 10 and 40% while limiting False Positives between 0.5 and 15%. Their methodology identified relevant news articles to a particular stock symbol and subsequent alignment of keywords with time-series trends of stock price. This study was significant, as it demonstrated that text-based stock price prediction could assist investors in filtering out useless news articles that would have no bearing on the stock-price fluctuations of security. However, the tools leveraged in the study were limited to Unigram BoW for feature extraction and Naïve Bayes for classification. While the goal of our paper is to create a similar decision support system, we have leveraged many more feature extraction and classification techniques to identify the best combination of techniques.

Now that we have discussed some initial influences of our research, we now move to the specific methodologies followed by other researchers in this domain. The majority of research conducted in this domain relies on a standard 3 step methodology: (i) Dataset

Building, (ii) Feature Extraction, and (iii) Classification. In the following sub-sections, we have discussed each one in detail.

2.1 Dataset Building

This phase involves the collection of raw textual information to be used as predictor space for the research. Financial information for analysis is obtained from three primary sources: Company disclosures/filings, media articles/opinions, and internet-expressed sentiments/opinions [11]. If we look at existing literature in this domain, financial news articles have been predominantly used as predictors, especially in studies trying to estimate short-term fluctuations in security prices. Other than the textual information, this phase also includes collecting relevant stock quotations before and after publishing financial information to understand the relationship between keywords and price fluctuations.

2.2 Feature Extraction

Feature Extraction involves the conversion of the input raw text to machine-readable format for classifier training and prediction. The output of this phase is generally a Document-Term Matrix, with a list of features as column headers and documents as rows. Cell values are numerical values to represent the importance of a particular feature in a document. Bag-of-words (BoW), a flexible and straightforward approach for text representation, is generally considered the de facto standard ([5, 13]) of the financial article. However, comparison of three different textual representations: bag of words, noun phrases, and named entities, established that a Proper Noun scheme performs better than BoW in almost all measures of prediction accuracy [17]. The result can be considered significant as it demonstrated that BoW need not be the standard approach, and improvements in classification accuracy are possible by employing more sophisticated feature selection techniques. One such technique was introduced by performing an empirical comparison of twelve popular feature selection techniques evaluated on a benchmark of 229 prevalent classification problems [4]. They suggested a new feature selection method, named Bi-Normal Separation (BNS), which performed significantly better than the other techniques in almost all evaluation measures.

Hagenau et al. [5] also focused on robust feature selection to improve classification accuracy. They employed Binormal Separation and Chi-Square statistics to provide exogenous feedback for feature selection. They also introduced a 2-word feature representation that employed market feedback to select the most relevant combinations of words. The 2-word feature type is an extension of the much more common 2-g feature selection technique, allowing a distance greater than 0 between 2 words. Combining 2-word combinations with BNS scores, they achieved a predictive accuracy of up to 76% with SVM as the classifier. They did not measure the performance of their classifiers using any other metric, which is problematic considering accuracy is not a good indicator of predictive performance. However, considering the results achieved by Forman [4] and Hagenau et al. [5], we included BNS as a feature selection methodology to compare against BoW.

2.3 Classification Algorithms

While some studies in this domain attempt to predict discrete stock prices based on current stock price and textual information, most studies are classification problems. The target variable could either be predicting the directional change of stock price or predicting if there will be any significant shift at all. One study which falls in the latter category [15] studied the effect of published news on the Brazilian Stock Market. They implemented a binary classification model where the target variable was classified as either "Interesting" or "Not Interesting". As this is one of the few research papers that compares multiple feature selection and classification techniques, we considered it to be significant.

Li et al. [14] added the dimension of sentiments to BoW. They analyzed sentiments in financial news using two different dictionaries: Harvard psychological dictionary and Loughran–McDonald financial sentiment dictionary. They concluded that, at individual stock, sector, and index levels, models with sentiment analysis provide superior performance to bag-of-words models, in both validation and independent testing sets. For the labeling of rows, they used different thresholds of stock-price deviation, and the final threshold was calibrated depending on the results. We also agree that this technique for labeling provides excellent simplicity and flexibility to a study of this kind and allows investors to choose from various thresholds based on their required level of balance between True Positive Rate and False Positive Rate. Hence, we incorporated this technique into our study.

3 Data and Research Methodology

This section discusses the overall approach employed for collecting data, feature extraction and labeling, classification, and classifier performance evaluation.

3.1 Dataset Description

The dataset used for this study consists of 1772 financial news headlines regarding any publicly listed Indian company. While we concede that working with entire news stories may give more features to work with, we have focused on news headlines in our study. In modern media, news headlines play a critical role in attracting viewers, and in many cases, communicate the essence and sentiment of the story that follows. This can further be inferred from the results obtained by Im et al. [8], where they observed that replacing news headlines with the contents of a news article only increased the predictive accuracy of their model by 5–8%. While experienced traders are expected to consume the entire story and make decisions based on their interpretation, our model aims to direct the investor towards the news articles that they can focus on. Hence, we proceeded with a dataset of headlines extracted from news archives for this study. The data was collected from the archives of Hindu Businessline and Moneycontrol.com. The Hindu Businessline is one of India's most widely circulated physical newspapers, while Moneycontrol.com is one of India's foremost digital destinations for financial news.

The criteria to be included in this study was that the news headline must contain the name of any publicly listed Indian company. News headlines published within the 26 months, 1st January 2018 to 29th February 2020, were initially considered for this study. We deliberately removed all news headlines published after the outbreak of COVID-19 since company-specific news articles were no longer sufficient to capture broader market signals. We have also excluded all news headlines mentioning Buy/Sell ratings of these stocks since they may create bias in the model, and it is pretty apparent how these headlines will impact the market. We also deliberately avoided scenarios where multiple articles with different contexts were published about the same company on any particular date. This was done since we are predicting the impact of the news articles on closing price, and there would be no way to separate the impact of multiple articles. Hence, we removed all articles which coincided with other articles about the same company within 24 h.

For historical stock prices, we again referred to Moneycontrol.com. For every news headline included in our dataset, we captured the opening and closing prices on either side of the time when the article was published [5]. For example, if a news article pertinent to our study was published during regular market hours, we collected the opening and closing stock prices for that day. However, if the news article was published before or after market hours, we collected the last closing stock price as the pre-publishing price and the subsequent opening stock price as the post-publishing price.

3.2 Data Labelling

A majority of studies in this domain tend to label texts as either positive or negative based on price movement direction, irrespective of the deviation percentage. We adopted a slightly different labeling strategy since our research aims to produce a model that classifies news headlines as either "Impactful" or "Not Impactful", regardless of the direction of change. One way to do that was to identify an optimal threshold of deviation that may be considered significant enough for a news headline to be considered interesting.

A similar approach was employed by Li et al. [14], who calibrated the thresholds based on the accuracy of predictions for that threshold. We also followed a similar threshold-based strategy for data labeling. Based on the accuracy, they chose a threshold of 50 bps or 0.5% for their final results. In our study, the thresholds required for accurate classification turned out to be higher.

We opted for a labelling strategy for binary classification with each headline labeled as either "Not Impactful" (0 or Negative class) or "Impactful" (1 or Positive class) based on its effect on the price of a security [16]. If the percentage deviation from opening or closing price crossed the threshold, the headline, which caused the mentioned price shift, was labeled as "Impactful". If the price failed to cross the threshold, the headline was labeled as "Not Impactful". In Table 1. Labelling thresholds and distribution of observations, all thresholds used for this study have been listed, with the number of positive and negative occurrences in each threshold.

Table 1 Labelling thresholds and distribution of observations shows that at the thresholds of 0.5 and 3%, positive and negative classes are not represented approximately equally. This scenario is prevalent in real-world examples, where only a tiny percentage of observations are generally considered "abnormal" or "interesting". For example, imbalance on the order of 100 to 1 to 100,000 to 1 is prevalent in several scenarios [16].

Table 1. Labelling thresholds and distribution of observations

Labelling threshold (%)	Positive class observations	Negative class observations	Total
0.50	1511	261	1772
1	1182	590	1772
1.50	918	854	1772
2	704	1068	1772
3	418	1354	1772

3.3 Handling Class Imbalance

To counter the effect of class imbalance in labeling thresholds of 0.5 and 3%, we referred to techniques such as Random Over and Under Sampling, Synthetic Sampling, Cluster-based oversampling (CBO), and Synthetic Minority Over-Sampling Technique (SMOTE) ([2, 6]). For our dataset, SMOTE was the most effective in improving classification performance and hence was included in our study.

In our dataset, the positive class is the majority in the case of the threshold of 0.5%, while the reverse is valid for the threshold of 3%. Hence, we leveraged SMOTE and performed oversampling of negative class observations for 0.5% and positive class observations for 3%.

3.4 Feature Extraction

For textual data, pre-processing of the dataset is essential to ensure that only impactful features are retained and evaluated. Activities in this phase of the research included removing stop words and punctuations, thus allowing our model to focus on the content and keywords that could impact the reader's mindset. We also removed all numbers and percentage values from the headlines during pre-processing. This was done to retain the focus on understanding the contribution of keywords to stock price fluctuations and eliminate confusion about the impact of numerical data.

After the generation of a cleaned corpus of texts, we focused on feature space generation and scoring. Three methods for feature space creation and scoring have been compared in this study.

The first method combined the Bag-of-Words approach for feature creation and the TF-IDF technique for feature scoring. For feature creation, unigrams, bigrams, and trigrams were extracted from the text, but unigrams consistently provided better results; hence the results obtained from using bigrams and trigrams were not considered further. All sparse features occurring in less than 0.5% of texts are removed from the matrix. The output was a Document Term Matrix, of dimension 1772 * 194, with each of the 1772 individual documents represented as a bag-of-words vector with scores against each feature.

The second method also leverages BoW and TF-IDF for feature space creation, but further adds sentiment features and scores. The rationale behind adding sentiment features and scores was to assess the impact of investor sentiments on stock price fluc-tuations. We considered sentiment scores from Henry's finance-specific dictionary and Loughran-McDonald dictionary in our research and interpreted their impact based on improvement in model performance.

The final method uses the BoW method to represent text documents in vector form. However, instead of the TF-IDF technique used in the previous two approaches, we used Binormal Separation (BNS) for feature scoring. BNS gives a higher score to features with a greater True Positive Rate than a False Positive Rate. For example, if a feature is typically found in positive classes, BNS will allot a higher score to that feature than a feature present uniformly across positive and negative classes. BNS generated a Docu-ment Term Matrix of the exact dimensions as the one generated in the first method, with the only difference in feature scores.

3.5 Machine Learning Techniques

Once the initial raw text is converted into a machine-readable format, the next step involves feeding the dataset into a machine learning algorithm for classification. Our study used six classification techniques based on their popularity in contemporary text classification works [1, 9]. The six techniques used are Decision Trees (DT), Random Forests (RF), Support Vector Machines (SVM), K-Nearest Neighbors (KNN), Logistic Regression (LR), and Naive Bayes (NB). Combined with three feature extraction tech-niques and five classification thresholds, this gave us 90 different models to compare across.

For classifier training and performance evaluation, we used the 5-fold cross-validation method. We have used the average performance across the five iterations for model evaluation.

3.6 Model Evaluation Metrics

A majority of studies in this field have used Accuracy, which measures the percentage of observations correctly classified out of all observations. However, existing literature in this domain has already established that Accuracy may not sufficient or appropriate to judge model performance, especially in the case of imbalanced datasets. To counter these limitations, we have used F-Measure, Geometric Mean, and AUC ([10, 18]) to appropriately evaluate our classifier's ability to identify Impactful news headlines and suggest the same to the investors.

4 Results and Discussion

This section presents the summarized results obtained from combining the three short-listed feature extraction and scoring techniques with different classification algorithms and labeling thresholds.

The results obtained provided sufficient evidence towards varying performance of classification algorithms depending on the feature extraction technique used and vice-versa. For all three performance measures mentioned in the previous section, we have considered a performance of 60% or higher to be satisfactory. This is based on our review of previous research in this domain, from which it is evident that a majority of studies failed to achieve a performance of 60% in one or more of these measures.

Compared to contemporary research [7, 16], we can conclude that models performing well in at least one of the above metrics may be considered satisfactory models. However, we were more interested in models performing well across multiple metrics. Hence, we shortlisted models that returned a 60% or higher score on at least two of the three measures for further analysis. This resulted in the selection of the 7 models listed in Table 2. Models performing well across multiple metrics.

Table 2. Models performing well across multiple metrics

Feature selection	Classifier	Threshold (%)	F-measure	Geometric mean	AUC
BoW + TF-IDF	SVM	1.50	0.61	0.64	0.60
BoW + TF-IDF	RF	1.50	0.61	0.59	0.61
BNS	RF	1.50	0.62	0.57	0.59
BoW + TF-IDF + sentiment scores	RF	1.50	0.59	0.60	0.62
BoW + TF-IDF + sentiment scores	RF	2	0.28	0.67	0.60
BNS	KNN	1.50	0.59	0.6	0.62
BNS	KNN	1	0.78	0.42	0.60

As we can see from Table 2. Models performing well across multiple metrics, the threshold of 1.5% offers five models that offered satisfactory results in at least two of the three evaluation metrics. Thus, it can be considered to be the optimal threshold to identify interesting deviations.

Now moving on to classification techniques, we can see that Random Forests (RF) outperformed all other classification techniques, with four of the seven shortlisted models employing RF for classification. SVM (1) and KNN (2) also give satisfactory results when compared to other classification techniques. One interesting point to note here is the absence of Logistic Regression (LR) among top-performing algorithms. LR is very popular and has been employed in several similar kinds of research conducted. However, it seems that for news headlines, its performance may be inferior.

Finally, for feature space creation, BoW combined with TF-IDF scoring gives similar performance to BNS, irrespective of whether it is combined with sentiment scores. Offer similar performance. Adding sentiment scores, with TF-IDF scores of features does not seem to enhance model performance, with BoW and TF-IDF offering very similar results with and without sentiment scores.

Now coming to differences in performance in classification techniques when combined with different feature extraction techniques, we can see an interesting observation when SVM is combined with TF-IDF. For the threshold of 1.5%, SVM and TF-IDF provide the only model which scores at least 60% in all three key metrics. This clearly demonstrates that the performance of models may vary based on the pairing of feature extraction and classification technique. One further example that we observed was that SVM is the superior classifier when TF-IDF is the feature extraction technique, while its performance is average with BNS.

To conclude, we can say that using BNS or TF-IDF with RF, SVM, or KNN should give satisfactory results in a study of this kind.

5 Conclusion

Studies like ours attempting to predict the impact of company-specific financial news articles on stock prices are significant since they provide an additional dimension to support the decision-making of newsreaders. The results achieved by our study indicate that there is significant potential for a stock prediction system incorporating company-specific news as a parameter to outperform other systems relying solely on technical indicators.

Compared to previous research in this domain, our study achieved satisfactory results across three performance metrics. One additional insight generated by our study indicates that the performance of classification algorithms may vary depending on the feature extraction technique that it is paired with.

Future research in this domain should consider pairing advanced classification algorithms, such as neural networks, with the GARCH model to understand the impact of published news while considering volatility.

References

1. Brindha, S., Prabha, K., Sukumaran, S.: A survey on classification techniques for text mining. In: 2016 3rd International Conference on Advanced Computing and Communication Systems (ICACCS), vol. 1, pp. 1–5. IEEE (2016)
2. Chawla, V., Bowyer, K.W., Hall, L.O., Kegelmeyer, W.P.: SMOTE: synthetic minority over-sampling technique. J. Artif. Intell. Res. 16, 321–357 (2002)
3. Fama, E.F.: Efficient capital markets: a review of theory and empirical work. J. Finance 25(2), 383–417 (1970)
4. Forman, G.: An extensive empirical study of feature selection metrics for text classification. J. Mach. Learn. Res. 3, 1289–1305 (2003)
5. Hagenau, M., Liebmann, M., Neumann, D.: Automated news reading: stock price prediction based on financial news using context-capturing features. Decis. Support Syst. 55(3), 685–697 (2013)
6. Haibo, H., Garcia, E.A.: Learning from imbalanced data. IEEE Trans. Knowl. Data Eng. 21(9) (2009)
7. Hájek, P.: Combining bag-of-words and sentiment features of annual reports to predict abnormal stock returns. Neural Comput. Appl. 29(7), 343–358 (2017). https://doi.org/10.1007/s00521-017-3194-2
8. Im, T.L.: Impact of financial news headline and content to market sentiment. Int. J. Mach. Learn. Comput. 4(3) (2014)
9. Jindal, R., Malhotra, R., Jain, A.: Techniques for text classification: literature review and current trends. Webology 12(2) (2015)

10. Junqué De Fortuny, E., de Smedt, T., Martens, D., Daelemans, W.: Evaluating and under-standing text-based stock price-prediction models. Inf. Process. Manag. **50**(2), 426–441 (2014)
11. Kearney, C., Liu, S.: Textual sentiment in finance: a survey of methods and models. Int. Rev. Financ. Anal. **33**, 171–185 (2014)
12. Lavrenko, V., Schmill, M., Lawrie, D., Ogilvie, P., Jensen, D., Allan., J.: Language models for financial news recommendation. In: Proceedings of the 9th International Conference on Information and Knowledge Management (2000)
13. Li, X., Xie, H., Chen, L., Wang, J., Deng, X.: News impact on stock price return via sentiment analysis. Knowl.-Based Syst. **69**, 14–23 (2014)
14. Malkiel, B.G.: A random walk down wall street: the time-tested strategy for successful investing. W.W. Norton, New York (2003)
15. Nizer, P.S.M, Nievola, J.C: Predicting published news effect in the Brazilian stock market. Expert Syst. Appl. **39**, 10674–10680 (2012)
16. Provost, F., Fawcett, T.: Robust classification for imprecise environments. Mach. Learn. **42**, 203–231 (2001)
17. Schumaker, R.P., Chen, H.: Textual analysis of stock market prediction using breaking financial news: the AZFin text system. ACM Trans. Inf. Syst. (TOIS) **27**(2), 1–19 (2009)
18. Sokolova, M., Japkowicz, N., Szpakowicz, S.: Beyond accuracy, F-Score and ROC: a family of discriminant measures for performance evaluation. In: Sattar, A., Kang, B. (eds) AI 2006: Advances in Artificial Intelligence. AI 2006. Lecture Notes in Computer Science, vol 4304. Springer, Berlin, Heidelberg (2006)
19. Wüthrich, B., Permunetilleke, D., Leung, S., Lam, W., Cho, V., Zhang, J.: Daily prediction of major stock indices from textual www data. HKIE Trans. **5**(3), 151–156 (1998)

Public Opinion Analysis for the Covid-19 Pandemic Based on Sina Weibo Data

Feng Wang[(✉)] and Yunpeng Gong

College of Information Science and Engineering, Henan University of Technology,
Zhengzhou 450001, China
wangfeng_haut@haut.edu.cn

Abstract. The outbreak of Covid-19 has been continuously affecting human lives and communities around the world in many ways. In order to effectively prevent and control the Covid-19 pandemic, public opinion is analyzed based on Sina Weibo data in this paper. Firstly the Weibo data was crawled from Sina website to be the experimental dataset. After preprocessing operations of data cleaning, word segmentation and stop words removal, Term Frequency Inverse Document Frequency (TF-IDF) method was used to perform feature extraction and vectorization. Then public opinion for the Covid-19 pandemic was analyzed, which included word cloud analysis based on text visualization, topic mining based on Latent Dirichlet Allocation (LDA) and sentiment analysis based on Naïve Bayes. The experimental results show that public opinion analysis based on Sina Weibo data can provide effective data support for prevention and control of the Covid-19 pandemic.

Keywords: Public opinion · Covid-19 · Pandemic · Sina Weibo

1 Introduction

The novel coronavirus disease (Covid-19) has rapidly outbroken worldwide, leading to a global health emergency. Due to the dangerous nature of high pathogenicity, fast speed of spread, huge number of infected people, and long incubation period, the Covid-19 pandemic is affecting the lives of the citizens and posing significant challenges to all governments globally. Therefore, it has attracted wide attention.

Nowadays, social media websites (such as Twitter, Facebook, Instagram, Reddit, YouTube and discussion forums) have become the primary communication platforms for most people to convey their thoughts. So it is helpful to better understand people's reflections towards a specific event by analyzing the contents from social media. The outbreak of the Covid-19 pandemic has led to millions of discussions and posts on social media every day, thus providing a valuable source of data. Based on social media data, a lot of research works have been carried out. Some Covid-19 datasets of social media have been released to help tracking coronavirus related information [1]. LDA was used to extract the most topical issues about the Covid-19 pandemic [2]. To identify the hidden topics and their temporal evolution from the collected Covid-19 related tweets,

N. Xiong et al. (Eds.): ICNC-FSKD 2022, LNDECT 153, pp. 993–1001, 2023.
https://doi.org/10.1007/978-3-031-20738-9_109

two matrix decomposition methods were developed [3]. A seven-stage context-aware natural language processing (NLP) approach was applied to analyze Covid-19 related comments from six social media platforms [4]. The sentiment and emotion trends of public attention during the pandemic were identified by using different learning approaches [5–7]. To identify situational information and to understand how it was being propagated on social media, the Covid-19 related Weibo data was classified into seven types [8]. By using a large-scale Twitter dataset, aspect-based sentiment analysis was conducted to explore the image of China in the Covid-19 epidemic at different aspect levels [9]. With the development and promotion of vaccines, research works on Covid-19 vaccine related discussions on social media have also been carried out [10]. During the Covid-19 pandemic, vast infodemics that means information of questionable quality have been generated worldwide. Because they are mixed with false/fake or misleading information, different methods have been proposed to fight against such infodemics [11].

In times of crises like the Covid-19 pandemic, public opinion becomes a focal issue. As the biggest Chinese microblogging website, Sina Weibo provides rich data about public opinion and its dynamics. So it is natural to track the evolution of public opinion during the pandemic based on Weibo data. To achieve this goal, methods from NLP and machine learning were applied in this paper to analyze the collected Covid-19 related Weibo posts using feature engineering, cosine similarity measures, word cloud analysis, LDA topic mining, and Naïve Bayes classification. Based on public opinion analysis, more effective measures can be enforced to minimize the effects of the pandemic and confidence can be enhanced to defeat the pandemic.

2   Methodology

2.1   Data Acquisition and Preprocessing

A Python web crawler was developed to collect Weibo users' posts on the Covid-19 pandemic from Sina website. These documents were posted from December 20th 2019 to April 30th 2020 with a total number of 279,775.

To relate social media contents with real-life pandemic situation, the domestic pandemic data was obtained from Wuhan-2019-nCoV that is a GitHub open source project. The world wide pandemic data was obtained from the Covid-19 of Data Packaged Core Datasets that is a GitHub open source dataset.

To prepare the data for feature extraction, the following preprocessing operations are applied on the crawled Weibo data. Then the preprocessed Weibo posts can be converted into a corpus of text.

(1) Data cleaning

The duplicate reviews were firstly removed. Some special characters and symbols, such as punctuation, white space, '@' and 'V', were discarded. Expanding the full text, forwarding Weibo, web links and some other page tags were also eliminated. Compared with text, emojis can also express users' emotional tendencies. Therefore, these non-verbal elements were kept to mine potential information.

(2) Word segmentation

Word segmentation is the premise of all subsequent operations for public opinion analysis. The Jieba tool was used to perform word segmentation on Weibo documents. Then continuous sentences were splitted into a set of separate words.

(3) Stop words removal

On the one hand, traditional stop word list was used to remove those words that frequently occur in the texts without carrying meaningful information, such as conjunctions, prepositions, articles, pronouns, empty words and modal words. On the other hand, a customized stop word table that includes nearly 2,000 words was established to drop stop words that are suitable for Weibo document.

2.2 Feature Extraction

After preprocessing, TF-IDF method is used to perform feature extraction and vectorization. TF-IDF consists of two parts: TF (term frequency) and IDF (inverse document frequency), where TF is a calculation term that calculates the quantity a term appears in a document and IDF is a measure that calculates a word's general importance [7]. The term frequency of term t in document d is calculated by Eq. (1), where $n_{t,d}$ is the count of a term t in document d.

$$f_{tf}(t, d) = \frac{n_{t,d}}{\sum_{t' \in d} n_{t',d}}$$
(1)

IDF denotes how popular or rare a word is in the documents and it is calculated by Eq. (2), where N is the count of the total documents, i.e. $|D|$, and $N_{t,d}$ is the count of the documents that contain term t, as shown in Eq. (3).

$$f_{idf}(t, D) = \log\left(\frac{N}{N_{t,d}}\right)$$
(2)

$$N_{t,d} = |\{d \mid d \in D, t \in d\}|$$
(3)

The value of TF-IDF is calculated by multiplying TF and IDF values, as shown in Eq. (4). And it can be used to filter common words while retaining important words.

$$f_{tf-idf}(t, d, D) = f_{tf}(t, d) \times f_{idf}(t, D)$$
(4)

Suppose the Weibo text corpus consists of N documents, i.e. $D = \{d_1, d_2, ..., d_N\}$, and each document is composed of M terms, i.e. $\{t_1, t_2, ..., t_M\}$. The ith Weibo document can then be denoted by $d_i = \{(t_{i1}, w_{i1}), (t_{i2}, w_{i2}), ..., (t_{iM}, w_{iM})\}$, where w_{ij} is the weight of term t_j in the ith document, which is calculated by Eq. (5).

$$w_{ij} = \frac{f_{tf-idf}(t_j, d_i, D)}{\sqrt{\sum_{d_k \in D} \left[f_{tf-idf}(t_j, d_k, D)\right]^2}}$$
(5)

Obviously, the terms with higher frequency in a specific document and lower frequency in other documents will have higher weights. Then the cosine similarity between two Weibo documents d_i and d_j can be obtained by Eq. (6).

$$S(d_i, d_j) = \frac{\sum_{k=1}^{M} \left(w_{ik} \times w_{jk}\right)}{\sqrt{\sum_{k=1}^{M} \left(w_{ik}\right)^2} \times \sqrt{\sum_{k=1}^{M} \left(w_{jk}\right)^2}} \qquad (6)$$

2.3 Public Opinion Analysis for Covid-19

To understand public opinions about the Covid-19 pandemic, word cloud analysis, topic mining and sentiment analysis are conducted respectively.

(1) Word Cloud Analysis

As an appealing text visualization method for visual description of text keywords, word cloud is widely used for its simplicity and comprehensibility. By comprising the size and visual focusing of words that are weighted by their frequency of occurrence, it can clearly display the important content of the Covid-19 related Weibo posts.

Based on the Jieba word segmentation and TF-IDF features, keywords and their corresponding weights were extracted from the preprocessed Weibo documents. These keywords were sorted in descending order of weight and the top K keywords were selected to depict the word cloud.

(2) Weibo Topic Mining

Given a Weibo corpus D consisting of N documents, each document d is composed of M words. To perform the task of topic mining, the LDA generative process is executed as following steps [10].

Step 1: For each document d ($d \in \{1, 2,..., N\}$), generate a multinomial distribution θ_d whose hyper-parameter α follows the Dirichlet distribution.
Step 2: For each topic t ($t \in \{1, 2,..., K\}$), generate a multinomial distribution φ_t whose hyper-parameter β also follows the Dirichlet distribution.
Step 3: For the nth ($n \in \{1, 2,..., M\}$) position in document d,

(a) Choose a topic $z_{d,n}$ for that position which is generated from $z_{d,n} \sim$ Multinomial(θ_d);
(b) Fill in that position with word $w_{d,n}$ which is generated from the word distribution of the topic picked in the previous step $w_{i,j} \sim$ Multinomial($\phi_{z_{d,n}}$).

The components composed of topics, the per-document topic distribution and the per-word topic distribution are hidden variables that are predicted from the analysis of observed variables, i.e. all words included within N documents.

Therefore, the probability for a Weibo corpus D can be modelled by Eq. (7), where the documents and words are assumed to be independent.

$$P(D|\alpha, \beta) = \prod_{d=1}^{N} \int_{\theta_d} P(\theta_d|\alpha) \times \left(\prod_{n=1}^{M} \sum_{z_{d,n}} P(z_{d,n}|\theta_d)P(w_{d,n}|z_{d,n}, \beta) \right) d\theta_d \quad (7)$$

Based on the above model, words in the set of Weibo posts can be used to generate vocabulary that is then used to discover hidden topics during the Covid-19 pandemic.

(3) Sentiment Analysis

The TF-IDF features extracted from Sect. 2.2 are passed to the Naïve Bayes classifier to perform sentiment classification, which classifies Weibo posts positively, neutrally or negatively. According to the Bayes theorem, the Naïve Bayes classifier can be expressed as Eq. (8), where s is a sentiment and d is a Weibo post.

$$P(s|d) = \frac{P(d|s) \cdot P(s)}{P(d)} \quad (8)$$

Obviously, the posterior class probability for a given Weibo corpus is decided by the conditional density evaluation and class prior probability. And a specific post will be given to the class with the maximum posterior class probability.

3 Experiments

3.1 Word Cloud Analysis

Based on the TF-IDF features extracted from acquired Weibo dataset, the forty most critical keywords were selected to depict the word cloud. From the word cloud, it can be found that 'coronavirus', 'new', 'pneumonia', 'case', 'pandemic' and some other words are very hot, showing people's high attention to the Covid-19 pandemic. In addition, some other words related to the monitoring information on Covid-19 (death, accumulation, patient, discharge, confirmed, cured, suspected cases, new added), the spread of the virus (spread, work, virus, infection, import, China, overseas, America), epidemic prevention (mask, health, prevention, test, isolation), corresponding media message (video, report, Weibo, news) as well as the symptom (asymptomatic) were also being discussed, indicating people's intention to protect themselves and stay safe.

The generated word cloud is interactive. Take the keyword 'pandemic' for example. Its corresponding change curve is shown in Fig. 1. In the initial stage, the pandemic was less talked about. So it was not plotted in the figure. The pandemic was not effectively controlled when it broke out and the confirmed cases increased rapidly. It's obvious that the keyword 'pandemic' was discussed frequently with the rapid spread of virus and people keep high attention to it during the pandemic period.

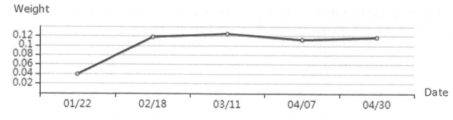

Fig. 1. The change curve of keyword 'pandemic'.

3.2 Weibo Topic Mining

The number of topics in the LDA model was set to be four. Figure 2 shows the top 30 most relevant terms for topic 2, where light blue bar means the overall term frequency while red bar denotes the estimated term frequency within topic 2. As shown in Fig. 2, the most common words under topic 2 include 'mask', 'coronavirus', 'everyone', 'out', 'don't', 'Wuhan', 'prevention', 'new', 'careful', 'virus', 'wash hands frequently', 'prepare', 'protect', 'avoid'. It shows that pandemic prevention gradually became a hot topic with the passage of time. People's sense of self-protection was also strengthen continuously while paying attention to the Covid-19 pandemic. The terms of 'thanks', 'hardship' and 'healthcare workers' reflect high respect for frontline healthcare workers. 'Wish', 'encouragement' and 'safe' embody people's determination to fight the pandemic and confidence in defeating the new coronavirus.

3.3 Sentiment analysis

Figure 3 shows the relationship between the sentimental trends of Weibo posts and the domestic pandemic, where the yellow curve means the total confirmed cases in each day while the green, blue and red curves represent the number of Weibo posts corresponding to neutral, negative and positive sentiments respectively. The neutral posts were found to be in the majority while positive and negative posts were basically the same during the increasing period of the infected cases. It shows that people tend to be neutral and objective about the pandemic. Because the pandemic had been basically under control from February 18th to middle March, the Covid-19 related Weibo posts were also decreased gradually. But there was a sudden increase after middle March. This is due to the sharp increase in the number of confirmed Covid-19 cases worldwide after middle March. It shows that people's attention to the Covid-19 pandemic had been increased rapidly again, due to the rapid outbreak of overseas pandemic. It also shows that people's negative emotions were accumulated seriously because the pandemic continued over and over again.

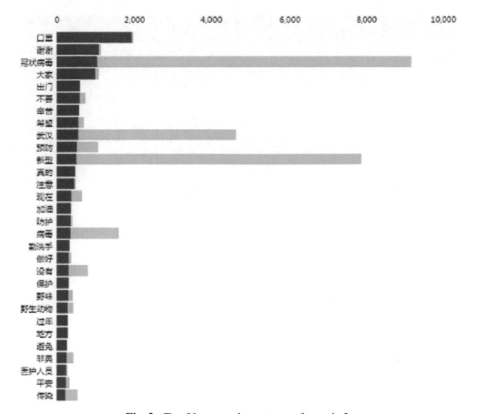

Fig. 2. Top-30 most relevant terms for topic 2.

4 Conclusion

As a global health emergency, the Covid-19 pandemic has disrupted human lives and the social economy development. It not only has led to society division and uncertainty, but also posed great challenges to all governments. In this paper, Weibo data were exploited to analyze public opinions, attitudes, and emotions about the Covid-19 pandemic. The crawled Weibo data were firstly preprocessed, then TF-IDF features were extracted and words were represented as vectors. Visual word cloud analysis, LDA-based topic mining and sentiment analysis based on Naïve Bayes were subsequently conducted respectively. Based on these analyses, it can help authorities to improve the effectiveness of their campaigns about the pandemic response.

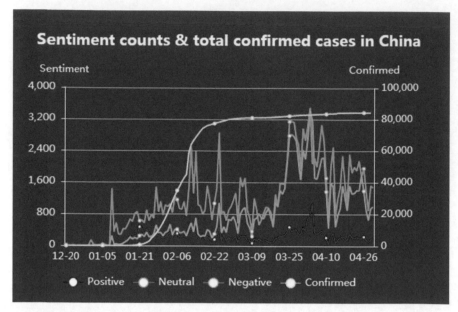

Fig. 3. Sentimental trends illustration of Weibo posts and the domestic pandemic.

Acknowledgment. This work was supported in part by the Research and Practice of Teaching Reform Program of Henan University of Technology, China (Grant No. JXYJ2021008).

References

1. Shuja, J., Alanazi, E., Alasmary, W. et al.: COVID-19 open source data sets: a comprehensive survey. Appl. Intell. **51**, 1296–1325 (2021)
2. Mutanga, M.B., Abayomi, A.: Tweeting on COVID-19 pandemic in South Africa: LDA-based topic modelling approach. Afr. J. Sci. Technol. Innov. Dev. **14**(1), 163–172 (2022)
3. Chang, C.H., Monselise, M., Yang, C.C.: What are people concerned about during the pandemic? Detecting evolving topics about COVID-19 from Twitter. J. Healthc. Inf. Res. **5**, 70–97 (2021)
4. Oyebode, O., Ndulue, C., Mulchandani, D., et al.: COVID-19 pandemic: identifying key issues using social media and natural language processing. J. of Healthc. Inf. Res. **6**(2), 174–207 (2022)
5. Cui, H., Kertész, J.: Attention dynamics on the Chinese social media Sina Weibo during the COVID-19 pandemic. EPJ Data Science **10**(1), 1–16 (2021). https://doi.org/10.1140/epjds/s13688-021-00263-0
6. Zhou, J., Zogan, H., Yang, S., et al.: Detecting community depression dynamics due to covid-19 pandemic in australia. IEEE Trans. on Comput. Soc. Syst. **8**(4), 982–991 (2021)
7. Ahmad, W., Wang, B., Xu, H., et al.: Topics, sentiments, and emotions triggered by COVID-19-related tweets from IRAN and Turkey official news agencies. SN Comput. Sci. **2**(5), 394 (2021)
8. Li, L., Zhang, Q., Wang, X., et al.: Characterizing the propagation of situational information in social media during COVID-19 epidemic: a case study on Weibo. IEEE Trans. Comput. Soc. Syst. **7**(2), 556–562 (2020)

9. Chen, H., Zhu, Z., Qi, F., et al.: Country image in COVID-19 pandemic: a case study of China. IEEE Trans. Big Data **7**(1), 81–92 (2021)
10. Yin, H., Song, X.Y., Yang, S.Q., et al.: Sentiment analysis and topic modeling for COVID-19 vaccine discussions. World Wide Web **25**(3), 1067–1083 (2022)
11. Yang, S., Jiang, J., Pal, A. et al.: Analysis and insights for myths circulating on twitter during the covid-19 pandemic. IEEE Open J. Comput. Soc. **1**, 209–219 (2020)

A Survey on Temporal Knowledge Graphs-Extrapolation and Interpolation Tasks

Sulin Chen[iD] and Jingbin Wang[(✉)] [iD]

Fuzhou University, Fuzhou Fujian 350108, China
chen.s.chen.2020@mumail.ie, wjb@fzu.edu.cn

Abstract. Current research on knowledge graphs focuses mostly on static knowledge graphs while ignoring temporal information. Recently, people have begun to study the temporal knowledge graph, which integrates temporal information into KGC, so that the modeling is constantly changing with the knowledge that evolves over time. In this survey, we summarize the existing temporal knowledge graph research, which is divided into extrapolation tasks and interpolation tasks according to time. The extrapolation task is mainly used to predict future facts and consists of three models: Temporal Point Process, Time Series, and other models. The interpolation task extends the existing KGC models to complement the lack of past temporal information, including five models: Translational Distance, Semantic Matching, Neural, Relational Rotation, and Hyperbolic Geometric models.

Keywords: Temporal knowledge graph · Knowledge embedding · Knowledge completion

1 Introduction

Knowledge graphs (KG) are fundamentally semantic networks or graph data structures. Combining nodes and edges allows for the construction of facts. Nodes can be concepts, attributes, events, or entities, while edges represent relationships between nodes, and edge labels indicate the type of relationship. Facts can be represented as fact triples of the form, for example (*Tiffany*, *is_friend*, *Jessica*) represents Tiffany is Jessica's friend. The process of building a knowledge graph is iteratively updated. When we construct a knowledge graph, it is often accompanied by the loss of entities or linkages, resulting in an incomplete knowledge graph that significantly affects its quality. Knowledge completion aims to infer new knowledge from facts already in the knowledge graph, which might alleviate the issue of link loss; hence, knowledge graph completion is often a job involving link prediction.

Traditional static knowledge graph embedding (KGE), also known as knowledge representation learning (KGL), is one of the crucial ways to approach the issue of knowledge graph completion. It maps entities and relationships into dense low-dimensional real-valued vectors [1]. Through the knowledge representation of the vector, different methods are used for reasoning to predict the relationship between entities to improve the

© The Author(s), under exclusive license to Springer Nature Switzerland AG 2023
N. Xiong et al. (Eds.): ICNC-FSKD 2022, LNDECT 153, pp. 1002–1014, 2023.
https://doi.org/10.1007/978-3-031-20738-9_110

convenience of calculation and facilitate the downstream tasks to operate the knowledge graph, such as question answering tasks [2] and relation extraction [3].

With the development of time, it is found that the traditional KGE method does not consider a time when learning, thus often ignoring important timing information. Recently, an emerging topic on knowledge graphs - temporal knowledge graphs has attracted significant research attention from all walks of life. The temporal knowledge graph is a vector representation that applies time to the knowledge graph, expands the triple (e_i, r, e_j) into a quadruple (e_i, r, e_j, t) indicating the fact happened at time t, and integrates the time of knowledge into the knowledge graph. In the context of learning and knowledge graph completion, the entities and relationships of the KG undergo dynamic modifications over time. Our primary contributions are outlined below.

- **Comprehensive summary**. This paper provides a comprehensive study of the temporal knowledge graph. The main existing temporal models and techniques with the latest trends are presented and summarized. In addition, we compare and summarize with static knowledge graphs.
- **New categories and directions**. We divide the temporal knowledge graph into extrapolation and introversion tasks based on time differences with new categorization. We propose a more detailed classification for the extrapolation task, including Temporal Point Process, Time Series, and other models. We summarize a new temporal embedding model for the extrapolation task based on the traditional static embedding model, including Translational Distance, Semantic Matching, Neural Network, Relational Rotation and Hyperbolic Geometric models.

The remaining of this survey is divided into different sections: In Sect. 2, we provide an overview of temporal knowledge graphs, covering Definitions, Notations, and Relational Work; then, we argue TKG in Sect. 3 from two scopes: Extrapolation with three models and Interpolation with five models; finally, we discuss with a conclusion in Sect. 4 in the end.

2 Overview

2.1 Definitions and Notations

Definitions:

1. People often represent temporal point processes based on the conditional strength function $\lambda(t)$- a stochastic model of the next event time given all previous events,

$$\lambda(t)dt := \mathbf{P}\{eventin[t, t + dt)|T(t)\} = \mathbf{E}[dN(t)|\mathrm{T}(t)], \tag{1}$$

where $\lambda(t)dt$ is the conditional probability of observing an event in a small window $[t, t + dt)$ given the history $T(t) := \{t_\tau < t\}$ up to t, dt is a small window of size which only one event can occur, i.e., $dN(t) \in \{0, 1\}$.

Moreover, the conditional density of the event occurring at moment t is defined as:

$$h(t) = \lambda(t)S(t), \tag{2}$$

where S(t) is the conditional probability that no event occurs.

2. TransE's scoring function:

$$f(e_i, r, e_j) = \|\mathbf{e_i} + \mathbf{r} - \mathbf{e_j}\|_1 \tag{3}$$

3. CompleEx's scoring function:

$$X(e_i, r, e_j) = \mathrm{Re}(< \mathbf{r}, \mathbf{e_i}, \overline{\mathbf{e_j}} >) \tag{4}$$

Table 1. Notations and descriptions.

Notion	Descriptions
Δ	A triple set
(e_i, r, e_j)	A triple of head, relation and tail
(e_i, r, e_j, t)	A quadruple of head, relation, tail and time
$(\mathbf{e_i}, \mathbf{r}, \mathbf{e_j})$	Embedding of head, relation and tail
$(\mathbf{e_i}, \mathbf{r}, \mathbf{e_j}, \mathbf{t})$	Embedding of head, relation, tail and time
τ	The time step corresponding to time t
$\bar{t}$	The most recent time point when entity was involved in an event before time t
$t-$	The time point just before time t
l_1 and l_2	Measure the distance
q	The number of entities
R^d	d dimensional real-valued space
$v \in V$	Vertex in vertical set
$\circ$	The hadamard (elementwise) product
$< \mathbf{e_i}, \mathbf{r}, \mathbf{e_j}, \mathbf{t_\tau} >$	Hermitian dot product
$\otimes_2$	The element-wise geometric product between 2-grade multivector embedding
$[a|b]$	Matrix or vector concatenation
$\mathrm{Re}(\cdot)$	A real vector component

2.2 Related Surveys

Surveys have focused on static knowledge graphs in the past, such as knowledge inference [4], knowledge fusion [5], or KGE [6]. Temporal knowledge graphs, an emerging field in recent years, have also been investigated by some scholars. Ji et al. [7] classified temporal knowledge graphs into four research areas, including temporal embedding, entity dynamics, temporal relational dependencies, and temporal logical reasoning. Mo et al. [8] classified knowledge graphs into static and dynamic models based on the presence or absence of model time but did not classify them for differences in time. Cai et al. [9] summarized the existing temporal knowledge graph completion (TKGC) approaches based on the differences in capturing temporal dynamics using accurate time stamps. Instead, we classified temporal knowledge graphs into two main categories – extrapolation and interpolation – based on differences in past and future time and then further subdivided each category. For example, the extrapolation task is divided into three sub-categories Temporal Point Process, Time Series, and other models; the interpolation task is divided into five sub-categories Translational Distance, Semantic Matching, Neural Network, Relational Rotation, and Hyperbolic Geometric models (Table 1).

3   Temporal Knowledge Graph

Interpolation and extrapolation are standard tools in data processing and function table preparation in modern times, which estimate the assumed value of a variable in terms of the general trend seen from the information [10, 11]. Furthermore, the difference is that interpolation uses a function to predict a dependent value of an independent variable in the data. In contrast, extrapolation indicates a conditional value of an independent variable outside the data range. Recently, this idea has been applied to the temporal knowledge graphs: the extrapolation task predicts the facts in the future, while the goal of interpolation is to compensate for the fact that previous time points are missing.

Also, in statistical theory, interpolation is more accurate because extrapolation assumes that the model will work for values outside the range of the data, which may not be the case. Therefore, the interpolation method can obtain a more efficient estimate. In the temporal knowledge graph, the interpolation model can use the information before and after the predicted time point to complete the missing facts, while the extrapolation model can only use the information before the predicted time point to predict future facts, which is more affected by uncertain factors.

3.1   Extrapolation Task

Extrapolation tasks generally focus more on predicting future facts, and extrapolation models usually predict future facts in terms of previous dynamic points. With the popularity of temporal knowledge graphs, more and more people pay attention to extrapolation tasks. We summarize and categorize the extrapolation task into three modules: Temporal Point Process (TPP) model in 3.1.1, Time Series model in 3.1.2, and other models in 3.1.3.

3.1.1   Temporal Point Process (TPP) Model

As a classical mathematical tool for modeling in the continuous-time domain, temporal point processes have become an essential solution for dealing with various complex temporal behaviors in networks. They are often applied to future prediction or quasi-causal relationship findings [12–14]. A TTP is a stochastic process consisting of a series of events over a while [15].

In developing the temporal point processes, we can divide them into two directions: the traditional statistical point process and the new depth point process [14]. Traditional parameterization methods of statistical point processes generally require explicit intensity function forms. The success of modeling often depends on the choice of the functional model, and a good model can often show solid predictive ability. Recently, a new research depth point process has demonstrated powerful capabilities, which integrates temporal point processes and deep networks, also known as neural point process. For example, Know-Evolve [16] modulates the intensity function $\lambda_r^{e^i, e^j}(t|\bar{t})$ (1–2) with a bilinear score function and incorporates a deep RNN framework to update node embeddings.

Conditional intensity function:

$$\lambda_r^{e^i, e^j}(t|\bar{t}) = h(f(e^i, r, e^j, \bar{t})) \times (t - \bar{t}) \tag{5}$$

Bilinear score function:

$$f(e^i, r, e^j, t) = \mathbf{v}^{e^i}(t-)^{\mathrm{T}} \times \mathbf{R_r} \times \mathbf{v}^{e^j}(t-) \tag{6}$$

where $\mathbf{v}^{e^i}(t-)$, $\mathbf{v}^{e^j}(t-)$ represent the most recently updated vector embeddings of the head and tail entities respectively.

DeRep [17] obtains the node representation of related nodes based on the bilinear depth time point process parameterization.

Bilinear score function:

$$f(e^i, r, e^j, \bar{t}) = w_r^T \times [\mathbf{z}^{e^i}(\bar{t}); \mathbf{z}^{e^j}(\bar{t})] \tag{7}$$

where [;] denotes concatenation and $w_r \in R^{2d}$ is model parameter for learning timescale-specific compatibility. $\mathbf{z}^{e^i}(\bar{t})$, $\mathbf{z}^{e^j}(\bar{t})$ are the node representations learned through the network. Moreover, DeRep divides the node representation into three parts ((i) Self-Propagation (ii) Exogenous Drive (iii) Localized Embedding Propagation) to update local embeddings.

Based on DeRep, Knyazev et al. [18] combines the NRI neural relational inference model and adopts bilinear relations to represent more complex relations between nodes.

3.1.2 Time Series Model

In a temporal knowledge graph, the graph structure is no longer static. Adding a temporal dimension to the data allows entities and their factual links to evolve over time to obtain the most time-sensitive and critical information. The traditional modeling time information method is generally combined with the time series diagram to enhance the performance of the model in these current time series models. There are many traditional methods for modeling time information, such as Recurrent Neural Network (RNN), Gate Recurrent Unit(GRU), and Gate. Formally, the knowledge graph is divided into several snapshots (or subgraphs) in chronological order, and then the temporal evolution between snapshots is modeled.

A. In the time series knowledge graph, historical information is very important, and we often learn historical information to predict future facts. RE-Net [19] builds an autoregressive model on the time series, combines the RNN model on the encoder to encode historical information, and uses the Relational Graph Convolutional Networks(RGCN) aggregator [20] to learn local from the neighborhood Structural dependencies. And MLP decoder defines the joint probability of a current event.

The global representation H_t is expected to preserve the global information about all the graphs up to time t.

$$H_t = RNN(g(G_t), H_{t-1}) \tag{8}$$

where H_{t-1} summarizes the global information from all the past graphs until time $t-1$, g is an aggregate function and $g(G_t)$ is an aggregation over all the events G_t at time t.

In historical learning information, repeated and redundant facts often appear. CyGNet [21] proposed a new time-aware copy-generation model responding to this problem. The mechanism combines two modes: Copy mode and Generation mode, enabling it to learn repetitive facts, identify them, and select future facts from known facts that have emerged in the past. Based on historical facts, HIPNET [22] perceives the underlying patterns behind temporal changes and proposes a HIP network from the perspectives of time, structure, and repetition to transmit historical information to predict future events accurately. Like a query $(e_i, r, ?, t)$, we compute the repetitive history score as follows:

$$f(e_i, r, e_j, t) = softmax(\mathbf{W}[h_{e_i,t}; h_{r,t}] + \mathbf{V}_t^{(e_i,r)})_{e_j}, \tag{9}$$

$$\mathbf{V}_t^{(e_i,r)} = \mathbf{V}_1^{(e_i,r)} + \mathbf{V}_2^{(e_i,r)} + \dots + \mathbf{V}_{t-1}^{(e_i,r)}. \tag{10}$$

where $\mathbf{W} \in \mathbf{R}^{q \times 2\mathbf{d}}$, $h_{e_i,t}, h_{r,t} \in \mathbf{R}^d$ model historical preferences independently of time and structure and are representations between entities and relationships. $\mathbf{V}_t^{(e_i,r)}$ is an q-dimensional multihot indicator vector between head entities e_i and relation r.

Li et al. [23] believes that people often search for and use historical information from their memory. CluSTeR divides the model into two parts: clue search and temporal reasoning. Through the RL (reinforcement learning) system [24, 25] effectively search for historical information, then form it into an input subgraph, use Graph Convolutional Networks (GCN) and GRU for temporal reasoning, and finally get the prediction result.

B. As a powerful tool for sequence modeling, self-attention is often used to extract time-aware representations from entity neighborhoods. In dynamic graphs, the time-changing neighborhood structure affects interactions with nodes, allowing attention itself to become a variable quantity over time. DeRep provides an attention mechanism that determines the attention coefficient and aggregation quantity of node edges depending on the time point process employing time information. DySAT [26] discovers latent node representations in dynamic graphs by integrating self-attention in neighborhood structure and temporal dynamics dimensions. DySAT leverages temporal attention to historical weight information to capture the graph evolution of asynchronous time. The temporal self-attention function is defined as:

$$Z_n = \beta_n(X_n W_v), \beta_n^{ij} = \frac{\exp(e_n^{ij})}{\sum\limits_{k=1}^{T} (e_n^{ik})}, e_n^{ij} = (\frac{((X_n W_q)(X_n W_k)^T)_{ij}}{\sqrt{F'}} + M_{ij}) \tag{11}$$

where n represents any node, $\beta_n \in \mathbf{R}^{T \times T}$ is the attention weight matrix, X_n is input of n and $W_q \in \mathbf{R}^{D' \times F'}, W_k \in \mathbf{R}^{D' \times F'}, W_v \in \mathbf{R}^{D' \times F'}$ are linear projection matrices through which queries, keys and values can be first converted to a different space. $M_{ij} =$

$$\begin{cases} 0, & i \leq j \\ -\infty, & otherwise. \end{cases}$$

Similarly, TemporalGAT [27] uses a temporal self-attention mechanism to learn historical knowledge through TCN (Temporal Convolutional Network) and GAT(Graph Attention Networks) layers.

Compared with static graphs, the evolution of dynamic graphs may be highly non-stationary, resulting in a scarcity of graph data. To address this problem, FTAG [28] proposes a one-shot learning framework by only giving a support Entity pair, and you can effectively infer the actual entity pair of a relationship. A neighborhood encoder with a self-attention mechanism is employed to gather valuable temporal neighborhood information to generate the support set and query representation. The similarity network is then utilized to determine the similarity of the query instances of the support set.

Currently, most models operate in a black-box fashion. In order to explain which historical information improves the accuracy of prediction information, 2021 xERTE [29] proposes a visual link prediction framework by combining temporal dynamics and structural dependencies: using TRGA mechanism (Temporal Relational Graph Attention Mechanism) and the reverse representation updating method to forecast the temporal dynamic graph.

3.1.3   Other Model

To further study software engineering (SE) problems, Ahrabian et al. [30] introduces a new relative time KG embedding in combination with transformer style [31–32], and combines SE problem and temporal knowledge graph embedding to provide a new direction for temporal prediction query models.

Different from most current temporal graph embedding graphs that utilize time series, TGAT [33] utilizes the classical Bochner's theorem from harmonic analysis [34], discusses the time-feature interaction, and proposes a new functional time coding technique.

3.2   Interpolation Task

Compared with extrapolation tasks, interpolation tasks have been studied more deeply. The interpolation task focuses on completing the missing parts of the existing knowledge graph, and its objective is to anticipate the validity of a fact at a specific moment. Most interpolation tasks are based on the original static model to add time to constrain the model so that the model has accuracy and consistency in time. Combined with the classification of the static knowledge graph completion model, interpolation tasks are divided into the following categories: Translational Distance Model in 3.2.1, Semantic Matching Model in 3.2.2, Neural Networks Model in 3.2.3, Relational Rotation Model in 3.2.4 and Hyperbolic Geometric Model in 3.2.5.

3.2.1 Translational Distance Model

Traditional translational distance models map entities and relations into vector spaces and consider relationships to be offsets between entities. The time-based translation distance model inserts time information and expands triples (e_i, r, e_j) into quads (e_i, r, e_j, t) to provide entity features at any point in time. For example, t-TransE [35] expands TransE [36] model, combining temporal order score function:

$$g(r_i, r_j) = \|r_i M - r_j\|_1, \tag{12}$$

where $M \in R^{n \times n}$ is a transition matrix between pair-wise temporal ordering relation pair (r_i, r_j), and TransE's scoring function (3) to get the joint score function:

$$L = \sum_{m^+ \in \Delta} [\ \sum_{m^- \in \Delta'} [\gamma_1 + f(m^+) - f(m^-)]_+$$
$$+ \lambda \sum_{n^+ \in \Omega'_{e_i}, n^- \in \Omega'_{e_i}} [\gamma_2 + g(n^+) - g(n^-)]_+] \tag{13}$$

where $m^+ = (e_i, r_i, e_j) \in \Delta, m^- = (e'_i, r_i, e'_j) \in \Delta, n^+ = (r_i, r_j) \in \Omega_{e_i}, n^- = (r_i, r_j) \in \Omega'_{e_i}, \Omega_{e_i} = \{(r_i, r_j) | (e_i, e_i, e_j, t_m) \in \Delta_\tau, (e_i, e_i, e_j, t_n) \in \Delta_\tau, t_m < t_n\}.$

While TTransE [37] also extends TransE's score function (3) to Vector- based TtransE:

$$f(e_i, r, e_j, t) = -\|e_i + r + e_j - t\|_{l_{1/2}}. \tag{14}$$

Based on TransH [38], HyTE [39] correlates each time stamp with a corresponding hyperplane w_t, so that time can be explicitly integrated into the entity-relationship space and so that having the exact distribution representation has different results at different points in time. There is a mapping function at t time: $P_t(e_i) + P_t(r) \approx P_t(e_j)$ within $P_t(e_i) = e_i - (w_t^T e_i)w_t, P_t(r) = r - (w_t^T r)w_t, P_t(e_j) = e_j - (w_t^T e_j)w_t$ and the scoring function is defined as:

$$f(e_i, r, e_j, t) = -\|P_t(e_i) + P_t(r) - P_t(e_j)\|_{l_{1/2}}. \tag{15}$$

where $\|\cdot\|_{l_{1/2}}$ is the l_1 or l_2-norm of the difference vector.

3.2.2 Semantic Matching Model

Semantic Matching Models are commonly used to capture the underlying semantics between entities and vectors. Traditional semantic matching models RESCAL [40] and Complex [41] represent the model as an order 3 tensor and then extract the hidden information of data through tensor decomposition. The temporal knowledge graph expands the order 3 tensor into the order 4 tensor, TComplex [42] adds timestamp embedding t_τ to modulate multi-linear dot products based on Complex's scoring function (4):

$$X(e_i, r, e_j, t_\tau) = \mathrm{Re}(< r, e_i, \overline{e_j}, t_\tau >). \tag{16}$$

where t_τ can be used interchangeably to modulate entities and relationships in order to obtain a time-dependent representation:

$$< r, e_i, \overline{e_j}, t_\tau > = < r \circ t_\tau, e_i, \overline{e_j} > = < r, e_i \circ t_\tau, \overline{e_j} > = < r, e_i, \overline{e_j} \circ t_\tau >. \tag{17}$$

TIMEPLEX [43] also discovers the embedding of complex-valued entities, relations, and time to execute tensor decomposition of time. Moreover, TeLM [44] goes beyond the complex value representation in the order 4 tensor decomposition. TeLM embeds each entity, relation, and time segment as a K-dimensional 2-grade multi-vector embedding M with multi-vector components and specifies the scoring function as:

$$f(e_i, r, e_j, t) = < Sc(M_{e_i} \otimes_2 M_{r\tau} \otimes_2 \overline{M_{e_j}}, 1 > \tag{18}$$

where $Sc(\cdot)$ represents the real valued vector of the scalar component of a multivector embedding, 1 represents a $k \times 1$ vector having all k elements equal to one, M represents the element-wise conjugation of multivectors.

3.2.3 Neural Networks Model

Many neural network-based inference models have been presented in latest years as neural networks have gained popularity. The traditional neural network models represent triples as vectors as input and are trained through multi-layer neural networks to finally output scores. In temporal Knowledge Graph, in order to achieve time consistency, Liu et al. [45] adopts a deep neural network to encode Context to capture time relations. While TeMP [46] is a frequently-based gating and data antagonism technology, which combines a graph neural network and time dynamic model to complete the temporal knowledge graph, and proposes a solution to the problems of time sparsity and variability.

3.2.4 Relational Rotation Model

The general translational distance and semantic matching models cannot handle the multi-relation model, so the relationship rotation models are proposed. Relational rotation models are usually modeled in complex space and the relationships are regarded as the rotations between head and tail entities to solve multi-type relationships i.e., RotatE [47], QuatE [48]. In the temporal knowledge graph, TeRo [49] specifies the temporal development of entity embedding e_i, e_j as the rotation between the start and end times. And for facts involving time intervals, every relation is represented as an embedding of two double complex numbers with a scoring function of:

$$f(e_i, r, e_j, t) = -\|\mathbf{e_{i,t}} + r + \overline{\mathbf{e_{j,t}}}\|, \tag{19}$$

where $\mathbf{e_{i,t}}, \mathbf{e_{j,t}}$ is the time-specific entity embedding,
And the mapping function is defined as follows:

$$\mathbf{e_{i,t}} = \mathbf{e_i} \circ \tau, \mathbf{e_{j,t}} = \mathbf{e_j} \circ \tau. \tag{20}$$

Similar to the above method, the scoring function of ChronoR [50], which is also embedded through time rotation, is:

$$f(e_i, r, e_j, t) = \langle \mathbf{e_i} \circ Q_{r,t} \circ \mathbf{r_2}, \mathbf{e_j} \rangle, \tag{21}$$

where $Q_{r,t} = [r|t]$ represents the linear operator in k-dimensional space, $r \in \mathbf{R}^{n_r \times k}$, $t \in \mathbf{R}^{n_r \times k}$, $\mathbf{r_2} \in \mathbf{R}^{n_r \times k}$.

3.2.5 Hyperbolic Geometric Model

The traditional hyperbolic geometric model assumes the responsible network as a random geometric graph contained in the hyperbolic space, and leverages the intrinsic properties of the hyperbolic space to expose the characteristics of the complex network [51–52]. DyERNIE [53] uses a product of Manifold's to explore multiple non-Euclidean structures that simultaneously exist in sequential knowledge graphs.

And HERCULES [54] refers to Givens transformations and hyperbolic attention [55] to model different relational models. And define the Curvature of the manifold as the product of relation parameter and time parameter.

4 Conclusion

With the development of time, the temporal knowledge graph has attracted an increasingly popular concern. The paper conducts a board survey on the following two tasks: extrapolation and interpolation tasks. The research on interpolation tasks is generally more comprehensive than on extrapolation tasks. However, most of the current interpolation tasks are still based on the nested traditional KGC model, which still retains the shortcomings of the original model. We summarize the extrapolation task through three models. We believe that extrapolation is usually used to predict that the development law of an object is a gradual rather than a jump change to find a suitable model to reflect the changing trend of the predicted object. Through this survey, we have summarized the current representative research findings and trends, which we hope will serve as an inspiration for future research.

References

1. Wang, Q., Mao, Z., Wang, B., Guo, L.: Knowledge graph embedding: a survey of approaches and applications. In: IEEE Transactions on Knowledge and Data Engineering, vol. 29, pp. 2724–2743 (2017)
2. Bordes, A., Weston, J., Usunier, N.: Open Question Answering with Weakly Supervised Embedding Models. In: Calders, T., Esposito, F., Hüllermeier, E., Meo, R. (eds.) ECML PKDD 2014. LNCS (LNAI), vol. 8724, pp. 165–180. Springer, Heidelberg (2014). https://doi.org/10.1007/978-3-662-44848-9_11
3. Daiber, J., Jakob, M., Hokamp, C., Mendes, P.N.: Improving efficiency and accuracy in multilingual entity extraction. In: Proceedings of the 9th International Conference on Semantic Systems, vol.13, pp. 121–124. ACM Press (2013)
4. Chen, X., Jia, S., Xiang, Y.: A review: knowledge reasoning over knowledge graph. Expert Syst. Appl. **141**, 112948 (2020)
5. Wang, Q., Mao, Z., Wang, B., Guo, L.: Knowledge graph embedding: a survey of approaches and applications. In: IEEE Trans. Knowl. Data Eng. **29**, 2724–2743 (2017)
6. Smirnov, A., Levashova, T.: Knowledge fusion patterns: a survey. Inf. Fusion **52**, 31–40 (2019)
7. Ji, S., Pan, S., Cambria, E., Marttinen, P., Yu, P., S.: A Survey on knowledge graphs: representation, acquisition, and applications. In: IEEE Trans. Neural Netw. Learn. Syst. **33**, 494–514 (2022)
8. Mo, C., Wang, Y., Jia, Y., Liao, Q.: Survey on temporal knowledge graph. In: 2021 IEEE Sixth International Conference on Data Science in Cyberspace (DSC) 294–300. IEEE (2021)

9. Cai, B., Xiang, Y., Gao, L., Zhang, H., Li, Y., Li, J.: Temporal knowledge graph completion: a survey. In: arXiv:2201.08236 (2022)
10. Barnard, E., Wessels, L.F.A.: Extrapolation and interpolation in neural network classifiers. In: IEEE Control Syst. **12**, 50–53 (1992)
11. Taylor, C.: The difference between extrapolation and interpolation. In: ThoughtCo. https://www.thoughtco.com/extrapolation-and-interpolation-difference-3126301. Aaccessed 2022/5/23
12. Yang, S.-H., Zha, H.: Mixture of mutually exciting processes for viral diffusion. In: Proceedings of the 30th International Conference on Machine Learning, pp. 1–9. PMLR (2013)
13. Liu, S., Li, L.: Learning general temporal point processes based on dynamic weight generation. Appl. Intell. **52**, 3678–3690 (2022)
14. Yan, J.: Recent advance in temporal point process: from machine learning perspective. In: IJCAI 2019, p. 7 (2019)
15. Cox, D.R., Lewis, P.A.W: Multivariate point processes. Selected statistical papers of Sir David cox: volume 1. Des. Invest. Stat. Methods Appl. **1**(159) (2006)
16. Trivedi, R., Dai, H., Wang, Y., Song, L.: Know-evolve: deep temporal reasoning for dynamic knowledge graphs. In: Proceedings of the 34th International Conference on Machine Learning, vol. 70, pp. 3462–3471 (2017)
17. Trivedi, R., Farajtabar, M., Biswal, P., Zha, H.: Dyrep: learning representations over dynamic graphs. In: ICLR, p. 25 (2019)
18. Knyazev, B., Augusta, C., Taylor, G. W.: Learning temporal attention in dynamic graphs with bilinear interactions. In: PLoS ONE, vol. 16, pp. e0247936 (2021)
19. Jin, W., Qu, M., Jin, X., Ren, X.: Recurrent event network: autoregressive structure inference over temporal knowledge graphs. In: arxiv:1904.05530v3 (2019)
20. Schlichtkrull, M., Kipf, T.N., Bloem, P., van den Berg, R., Titov, I., Welling, M.: Modeling Relational Data with Graph Convolutional Networks. In: Gangemi, A., et al. (eds.) ESWC 2018. LNCS, vol. 10843, pp. 593–607. Springer, Cham (2018). https://doi.org/10.1007/978-3-319-93417-4_38
21. Zhu, C., Chen, M., Fan, C., Cheng, G., Zhan, Y.: Learning from history: modeling temporal knowledge graphs with sequential copy-generation networks. In: AAAI (2021)
22. He, Y., Zhang, P., Liu, L., Liang, Q., Zhang, W., Zhang, C.: HIP network: historical information passing network for extrapolation reasoning on temporal knowledge graph. In: Proceedings of the Thirtieth International Joint Conference on Artificial Intelligence, pp. 1915–1921 (2021)
23. Teo, T.W., Choy, B.H.: in. In: Tan, O.S., Low, E.L., Tay, E.G., Yan, Y.K. (eds.) Singapore Math and Science Education Innovation. ETLPPSIP, vol. 1, pp. 43–59. Springer, Singapore (2021). https://doi.org/10.1007/978-981-16-1357-9_3
24. Bahdanau, D., Brakel, P., Xu, K., Goyal, A., Lowe, R., Pineau, J., Courville, A.C., Bengio, Y.: An actor-critic algorithm for sequence prediction. In: 5th International Conference on Learning Representations (2017)
25. Feng, J., Huang, M., Zhao, L., Yang, Y., Zhu, X.: Reinforcement learning for relation classification from noisy data. In: Proceedings of the Thirty-Second AAAI Conference on Artificial Intelligence and Thirtieth Innovative Applications of Artificial Intelligence Conference and Eighth AAAI Symposium on Educational Advances in Artificial Intelligence, pp. 5779–5786 (2018)
26. Sankar, A., Wu, Y., Gou, L., Zhang, W., Yang, H.: DySAT: Deep neural representation learning on dynamic graphs via self-attention networks. In: Proceedings of the 13th International Conference on Web Search and Data Mining, pp. 519–527 (2020)
27. Fathy, A., Li, K.: TemporalGAT: Attention-Based Dynamic Graph Representation Learning. In: Lauw, H.W., Wong, R.-W., Ntoulas, A., Lim, E.-P., Ng, S.-K., Pan, S.J. (eds.) PAKDD

2020. LNCS (LNAI), vol. 12084, pp. 413–423. Springer, Cham (2020). https://doi.org/10.1007/978-3-030-47426-3_32

28. Mirtaheri, M., Rostami, M., Ren, X., Morstatter, F., Galstyan, A.: One-shot learning for temporal knowledge graphs. In: arXiv:2010.12144 (2020)

29. Han, Z., Chen, P., Ma, Y., Tresp, V.: xERTE: Explainable reasoning on temporal knowledge graphs for forecasting future links. In: arXiv:2012.15537 (2020)

30. Ahrabian, K., Tarlow, D., Cheng, H., Guo, J.L.C.: Software engineering event modeling using relative time in temporal knowledge graphs. In: arXiv:2007.01231 (2020)

31. Huang, C.-Z.A., Vaswani, A., Uszkoreit, J., Shazeer, N., Simon, I., Hawthorne, C., Dai, A.M., Hoffman, M.D., Dinculescu, M., Eck, D.: Music Transformer. In: arXiv:1809.04281 (2018)

32. Dai, Z., Yang, Z., Yang, Y., Carbonell, J., Le, Q.V., Salakhutdinov, R.: Transformer-XL: attentive language models beyond a fixed-length context. In: arXiv:1901.02860 (2019)

33. Xu, D., Ruan, C., Korpeoglu, E., Kumar, S., Achan, K.: Inductive representation learning on temporal graphs. In: arXiv:2002.07962 (2020)

34. Loomis,L., H.: Introduction to abstract harmonic analysis. Courier Corporation (2013)

35. Jiang, T., Liu, T., Ge, T., Sha, L., Li, S., Chang, B., Sui, Z.: Encoding Temporal information for time-aware link prediction. In: in Proceedings of the 2016 Conference on Empirical Methods in Natural Language Processing, pp. 2350–2354 (2016)

36. Bordes, A., Usunier, N., Garcia-Duran, A., Weston, J., Yakhnenko, O.: Translating embeddings for modeling multi-relational data. In: in Neural Information Processing Systems (NIPS), pp. 1–9 (2013)

37. Leblay, J., Chekol, M. W.: Deriving validity time in knowledge graph. In: Companion of the The Web Conference 2018, pp. 1771–1776. ACM Press (2018)

38. Wang, Z., Zhang, J., Feng, J., Chen, Z.: Knowledge graph embedding by translating on hyperplanes. In: Twenty-Eighth AAAI Conference on Artificial Intelligence (2014)

39. Dasgupta, S.S., Ray, S.N., Talukdar, P.: HyTE: hyperplane-based temporally aware knowledge graph embedding. In: in Proceedings of the 2018 Conference on Empirical Methods in Natural Language Processing, pp. 2001–2011 (2018)

40. Nickel, M., Tresp, V., Kriegel, H.-P.: A three-way model for collective learning on multi-relational data. In: ICML, pp. 809–816 (2011)

41. Trouillon, T., Welbl, J., Riedel, S., Gaussier, E., Bouchard, G.: Complex embeddings for simple link prediction. In: ICML, pp. 2071–2080(2016)

42. Lacroix, T., Obozinski, G., Usunier, N.: Tensor Decompositions for temporal knowledge base completion. In: arXiv:2004.04926 (2020)

43. Jain, P., Rathi, S., Mausam, Chakrabarti, S.: Temporal knowledge base completion: new algorithms and evaluation protocols. In: Proceedings of the 2020 Conference on Empirical Methods in Natural Language Processing (EMNLP), pp. 3733–3747 (2020)

44. Xu, C., Chen, Y.-Y., Nayyeri, M., Lehmann, J.: Temporal knowledge graph completion using a linear temporal regularizer and multivector embeddings. In: Proceedings of the 2021 Conference of the North American Chapter of the Association for Computational Linguistics: Human Language Technologies, pp. 2569–2578 (2021)

45. Liu, Y., Hua, W., Xin, K., Zhou, X.: Context-Aware Temporal Knowledge Graph Embedding. In: Cheng, R., Mamoulis, N., Sun, Y., Huang, X. (eds.) WISE 2020. LNCS, vol. 11881, pp. 583–598. Springer, Cham (2019). https://doi.org/10.1007/978-3-030-34223-4_37

46. Wu, J., Cao, M., Cheung, J.C.K., Hamilton, W.L.: TeMP: temporal message passing for temporal knowledge graph completion. In: Proceedings of the 2020 Conference on Empirical Methods in Natural Language Processing (EMNLP), pp. 5730–5746 (2020)

47. Sun, Z., Deng, Z.-H., Nie, J.-Y., Tang, J.: RotatE: knowledge graph embedding by relational rotation in complex space. In: ICLR (2019)

48. Zhang, S., Tay, Y., Yao, L., Liu, Q.: Quaternion knowledge graph embeddings. In: arXiv:1904.10281 (2019)

49. Xu, C., Nayyeri, M., Alkhoury, F., Yazdi, H.S., Lehmann, J.: TeRo: a time-aware knowledge graph embedding via temporal rotation. In: arXiv:2010.01029 (2020)
50. Sadeghian, A., Armandpour, M., Colas, A., Wang, D.Z.: ChronoR: rotation based temporal knowledge graph embedding. In: arXiv:2103.10379 (2021)
51. Nickel, M., Kiela, D.: Poincaré embeddings for learning hierarchical representations. In: arXiv:1705.08039 (2017)
52. Chamberlain, B.P., Clough, J., Deisenroth, M.P.: Neural embeddings of graphs in hyperbolic space. In: arXiv:1705.10359 (2017)
53. Han, Z., Ma, Y., Chen, P., Tresp, V.: DyERNIE: Dynamic evolution of Riemannian manifold embeddings for temporal knowledge graph completion. In: arXiv:2011.03984 (2020)
54. Montella, S., Rojas-Barahona, L., Heinecke, J.: Hyperbolic temporal knowledge graph embeddings with relational and time curvatures. In: arXiv:2106.04311 (2021)
55. Chami, I., Wolf, A., Juan, D.-C., Sala, F., Ravi, S., Ré, C.: Low-dimensional hyperbolic knowledge graph embeddings. In: Proceedings of the 58th Annual Meeting of the Association for Computational Linguistics, pp. 6901–6914 (2020)

Analysis of Abnormal User Behavior on the Internet Based on Time Series Feature Fusion

Long Zhao[1,2], Yanyan Wang[2(✉)], Wenbin Fan[2], DeXuan Wang[2], Yuan Zhou[2], Yekun Fang[2], and Enhong Chen[1]

[1] School of Computer Science and Technology, University of Science and Technology of China, Hefei 230022, China
lzhao@ustcinfo.com, cheneh@ustc.edu.cn
[2] GuoChuang Cloud Technology LTd., Shenzhen 230031, China
{wang.yanyan, fan.wenbin, wang.dexuan, zhou.yuan, fang.yekun}@ustcinfo.com

Abstract. The rapid increase of data has made data security a major problem that enterprises need to explore urgently. At present, data leakage is mainly divided into external attacks and internal leakage. This paper is focused on internal data leakage. Aiming at the quantifiable problem of abnormal behaviors, we propose a model based on data mining algorithms and time series to discover a criterion of normal behaviors. Via those criteria (called baseline), calculating anomaly scores and sorting the importance of those abnormal behaviors are possible. The purpose of our works is to discover possible data security risks efficiently and quickly and intuitively display their priority by quantifying their anomaly scores.

Keywords: User behavior analysis · Anomaly analysis · UEBA

1 Introduction

In the Internet age, the amount of data generated by enterprises increased dramatically, and these data are used in various fields including industrial services, marketing support, risk control, business operations and other scenarios. Therefore, the importance of data is increased gradually and progressively. However, the losses caused by data breaches are increasing year by year, and the frequency of data breaches is also increasing [1]. In addition, due to the large increase in the amount of data that are all unlabeled, the cost of manual labeling is increasing day by day, and it has become a difficult problem to use traditional supervised algorithms with manual labeling to identify anomalies [2].

To solve the related problems, this paper proposes a feature fusion model based on time series and builds a scoring model for the abnormal rate of user behavior with indicators such as user visit rate and website visit rate, to achieve anomaly detection in big data scenarios.

© The Author(s), under exclusive license to Springer Nature Switzerland AG 2023
N. Xiong et al. (Eds.): ICNC-FSKD 2022, LNDECT 153, pp. 1015–1023, 2023.
https://doi.org/10.1007/978-3-031-20738-9_111

2   Related Works

The model proposed in this paper is based on User Entity Behavior Analytics (UEBA) [3], which is specialized in identifying potential threats. This technology focuses on the user's normal behavior and training a model for describing that behavior by using statistical learning, reinforcement learning, deep learning, and other machine learning algorithms, then based on defined anonymous detecting rules, the anomaly rank can be performed [3, 4]. The model trained is called baseline in this paper.

UEBA currently has a high usage rate in preventing data leakage within enterprises. The model constructs employee models by reviewing the online records of enterprise employees and calculates the abnormal score of future behavior based on the baseline. Then the model warns of possible abnormal access behaviors and updates the baseline if those behavior are proved normally. This technical field is an indispensable part of intelligent data prevention and intelligent anomaly recognition [3, 4].

Ideally, the use of classical or neural network-based supervised classification algorithms is the primary measure to solve the problem of user abnormality detection. However, the data collected at the scene of industrialization are mainly unlabeled. The cost of the manual label is unbearable for enterprises, a possible solution is sampling, but sampling leads to over-fitting or under-fitting, which debases the model. Therefore, an unsupervised model is used in this scenario.

The classical unsupervised anonymous detection algorithm explores the high dimensional mapping of data and calculates hyperplanes to cluster the data with a confidence range [5–7]. This paper makes a comparative experiment on the KNN algorithm based on Euclidean distance and the isolated forest anomaly detection algorithm; the result is shown in the experiment section.

3   Model Construction

Aiming at the analysis of user abnormal behavior, this paper proposes a time series feature fusion model based on the different baselines, including group abnormal behavior detection, employee abnormal behavior detection, and website abnormal access detection. The final abnormal score of one employee is calculated in a weighted sum of these three scores, as shown in Fig. 1.

Group behavior anomaly detection considers employees belonging to the group. The main purpose is to build a sub-model of different groups, that is, train a model to describe the cyclical changes of the online records. The purpose of this model is to describe the average online records of the employees from the group. Employee behavior abnormal detection constructs an independent baseline for each employee. This model is not calculating the average level of the online access frequency, but the employee owns. Abnormal behavior recognition prediction is to identify individuals visiting websites, calculate the probability of individuals visiting websites through certain statistical means, and generate a corresponding individual network access preference matrix. The core purpose is to distinguish between high-frequency and low-frequency websites and to give them a corresponding quantitative outlier.

3.1 Feature Engineering

For easily understanding, we defined the variables used later as time (t), time zone (h) and website access (x). We defined the time in a day as time zone, and the length of time from the earliest data as time, each access of one website as website access.

3.2 Prediction of Group Abnormal Behavior

Group behavior anomaly prediction is used to identify the network access periodicity of different departments, and to predict the network access frequency trend of the group. This part mainly uses the time series model [7–9] to calculate the previous access records of the members in the group and calculate the access tendency of the whole group in a certain period in the future. Then, calculates the confidence deviation according to the weighted sum of the predicted values in a single period and the predicted values in the front and rear neighborhood, and calculates the corresponding anomaly function according to the confidence deviation.

The median value is selected to represent the deviation within each hour. The specific construction method is as follows:

(1) Calculate the total number of times each group t accesses the network in each period.
(2) Build a time series model for the number of visits to each group.
(3) Predict the number of visits in a certain period T in the future.
(4) Calculate the average standard deviation $s(h)$ of each time zone h using the neighborhood standard deviation:

$$s(h) = k * (b_1 * \sigma(h) + b_2 * \sigma(h-1) + b_3 * \sigma(h+1)), \tag{1}$$

where k represents a constant, which corresponds to the confidence region width index in the time series model, which is set to 1 here. In this paper, the neighborhood idea is used, and the fluctuation influence between adjacent time zones is considered. Therefore, the standard deviation of three time zones is used, as $\sigma(h)$, $\sigma(h-1)$, $\sigma(h+1)$. They are the standard deviation of time zone h, the standard deviation of time zone $h-1$ and the standard deviation of time zone $h+1$. b_1, b_2, b_3 are the standard deviation weights respectively. See the experimental part for the specific assignment. The reason why h is used is that different access frequencies will occur in the same period of each day. A credible range is fitted by calculating the standard deviation.

(5) Construct an abnormal score function according to the predicted value $y(t)$, the actual value $r(t)$ of the period t and the time zone h corresponding to the period, and calculate the group abnormal score:

$$g(t) = E\left(\left| \frac{y(t) - r(t)}{s(h)} \right| \right), \tag{2}$$

$$E(m) = \begin{cases} 1, m >= 1 \\ m, m < 1 \end{cases}, \tag{3}$$

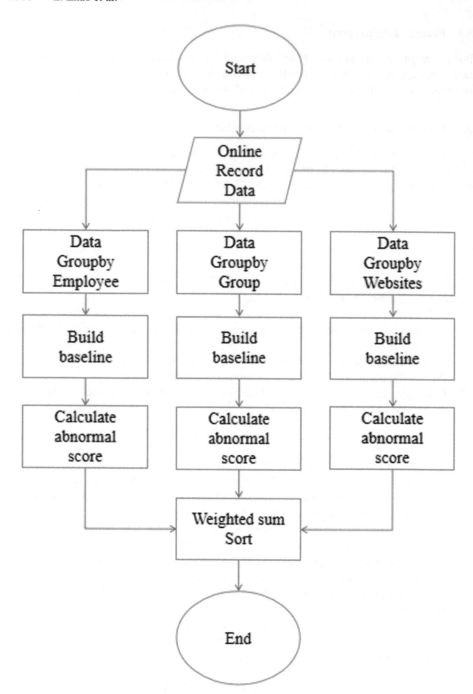

Fig. 1. Flow chart of the algorithm

where $E(m)$ is the piecewise function, and m is used to normalize the outlier score of the group, and the purpose is to scale its value range to $[0,1]$, $y(t)$ is the predict num of the model, and $r(t)$ is the real one, $s(h)$ is the function calculated in 4). Finally, we get the group abnormal score $g(t)$ in a particular time t.

3.3 Prediction of Employee Abnormal Behavior

The modeling method of individual behavior anomaly prediction is similar to that of group behavior anomaly prediction, but there are differences in specific steps 1 and 2. The specific steps of individual behavior anomaly prediction are as follows:

(1) Calculate the total number of times each individual T accesses the network in each period.
(2) Using the time series model for everyone.
(3) The processing of time series in the follow-up part is the same as that in the group behavior anomaly prediction part. Through the follow-up, the personal behavior anomaly score $P(t)$ under the period t is calculated.

3.4 Website Abnormal Behavior Detection

Website abnormal behavior detection detects whether the website visited by an employee has ever appeared in the record, and judges whether the websites are the user's preferred website. Preference websites are defined as websites whose user visit frequency is greater than a threshold. The specific steps are as follows:

(1) For everyone, count the number of visits to all websites they have visited.
(2) Use the number of visits divided by the total number of visits to obtain the website access probability $p(x)$.
(3) According to the website visit probability $p(x)$, calculate the website preference degree $L(x)$ according to formula (4).

$$l(x) = \begin{cases} 1, p(x) > 0.1 \\ 0, p(x) \leq 0.1 \end{cases}. \tag{4}$$

(4) According to the website preference degree $L(x)$, calculate the abnormal behavior score $a(x)$ according to formula (5):

$$a(x) = \begin{cases} 0, l(x) = 0 \\ 1, l(x) = 1 \end{cases}. \tag{5}$$

It is used to quantify the abnormal scores under different preference labels.

3.5 Abnormal Score F(X)

In this paper, the final abnormal score is obtained by weighted fusion of the above three aspects:

$$f(x) = c_1 * g(x) + c_2 * p(x) + c_3 * a(x). \tag{6}$$

where c_1, c_2, and c_3 are the weights of group behavior anomaly prediction, individual behavior anomaly prediction, and abnormal behavior recognition prediction. The specific weights are set in the experimental part.

4 Experiments and Analysis

This paper makes the corresponding abnormal behavior identification strategy in terms of the employee external access record data provided by an enterprise. The purpose is to sort the identified abnormal behaviors, and the data provided include the quantified abnormal behavior score.

4.1 Experimental Setup

The average absolute error [10] mainly reflects the average of the absolute value of the error between the observed value and the real value, expressed as:

$$MAE = \frac{1}{n} \sum_{i=1}^{n} |y^i - \hat{y}^i|, \tag{7}$$

where n represents the number of employees in the dataset, and $\hat{y}^i, y^i$ represent the real score and predicted score of the employee respectively.

Compared with MAE, the square root difference has two advantages. It can be derived and be used in the neural network of deep learning and is more sensitive to outliers.

$$RMSE = \sqrt{\frac{1}{n} \sum_{i=1}^{n} (y^i - \hat{y}^i)^2}. \tag{8}$$

In addition, the sum of the difference among samples are the same, the MAE is the same, but RMSE has a stronger ability to express the deviation range of the data. In this project, the model needs a certain detection ability for some data with large deviation, hence RMSE is chosen as the experimental criterion of this project.

Finally, according to the experimental indicators, after repeated experiments, coefficients of (b_1, b_2, b_3) is determined as $(0.5, 0.25, 0.25)$ and (c_1, c_2, c_3) as $(0.2, 0.3, 0.5)$.

4.2 Comparative Experimental Results and Analysis

To obtain a model with better experimental indicators, we used KNN [11], isolated forest [12], random number [13], and our model of this paper to conduct comparative experiments. KNN and isolated forest are traditional anomaly detection algorithms where KNN calculates the Euler distance of data points in hyperspace and identifies points with abnormal distance. The isolated forest uses some random hyperplanes to cut the original space into several subspaces, and forms data into a leaf in a split tree based on the number of sunspaces uses if a data point is isolated in a subspace. The lower the number of leaf node layers where the data point is located, the more isolated it is. The random number algorithm is the standard 50% of machine learning algorithms. Those experimental results are shown in Table 1:

Table 1. Compared with traditional abnormal detection

Algorithm	RMSE
Our model	**0.24**
KNN anomaly detection	0.32
Isolation Forest	0.315
Random number in the range of [0.5,1]	0.305

Table 1 shows that the method in this paper has achieved the best performance, which can extract user abnormal behavior features better and fit the sample data better. In the data set, the abnormal scores of all access behaviors in some periods are higher than 0.9, and the overall abnormal behavior accounts for up to 50%, which does not meet the definition of abnormality, so it seems that the data set may be highly biased sampling data. In this case, the traditional algorithm, KNN, and isolated forest are not effective for anomaly detection.

4.3 Analysis of User Anomalous Behavior

Firstly, in this paper, we depict the time series of the data, which has strong periodic characteristics. Figure 2 is constructed following the time aggregation method mentioned in Sect. 3.1 from the original data, we use the subscript to call different groups, such as A, B, and C respectively. We have multiple points on each abscissa by merging data in the same time zone to one abscissa for showing the dispersion for the data and it is not high. First, in the model we are building, each group is independent and does not interfere with the other. It can be seen from the figure that different groups have almost the same general trend, that is, around hours 0–3, there is a downward trend, and the number is very low, and at hours 10–16, the overall trend is upward and reaches a peak. After hour 16, it shows a different trend and finally returned to a similar hour 0. Furthermore, groups A and B are still increasing in frequency at hour 23, while group C is decreasing. It seems that they are similar but different, so we construct baselines for them separately (Fig. 3).

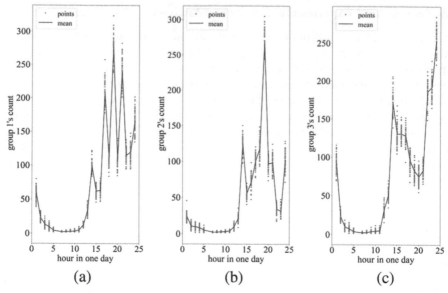

Fig. 2. A figure of the sampled groups' total online access counts to the websites. The following subscript will be used following for referring to each group, as A, B, and C.

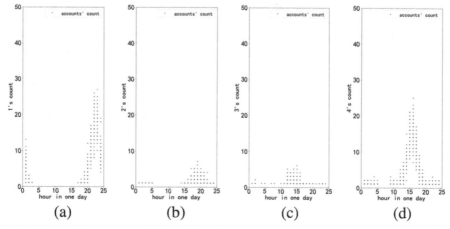

Fig. 3. A figure shows the sampled people's online access counts to the websites, and the following subscript will be used for referring each person, as (a), (b), (c), and (d).

5   Conclusion

This paper proposes an unsupervised anomaly detection model to quantify the anomaly score of website access, group detection, employee detection, and warn of possible data leakage. For the anomaly detection model proposed in this paper, the next optimization direction is to add IP information and extract IP features through NLP. In addition, combined with distributed, the time series prediction model can be distributed to different devices for training through the distributed system, which greatly improves efficiency.

References

1. Shashanka, M., Shen, M.Y., Wang, J.: User and entity behavior analytics for enterprise security. In: IEEE International Conference on Big Data (Big Data). IEEE, (2016)
2. Turcotte, M., Moore et al.: User behavior analytics (2017)
3. Barlow, H.B.: Unsupervised learning. Neural Comput. 1(3), 295–311 (1989)
4. Eskin, E., Arnold, A., Prerau, M., et al.: A geometric framework for unsupervised anomaly detection. Springer, US (2002)
5. Ahmad, S., Lavin, A., Purdy, S. et al.: Unsupervised real-time anomaly detection for streaming data. Neurocomputing (2017)
6. Litan, A.: Forecast snapshot: user and entity behavior analytics. Worldwide (2017)
7. Taylor, S.J., Letham, B.: Forecasting at scale. https://doi.org/10.7287/peerj.preprints.3190v2 (2017)
8. Mehanian, C., Cazzanti, L., Penzotti, J. et al.: Time series-based entity behavior classification: US, US20140074614 A1 (2014)
9. Kalpakis, K., Gada, D., Puttagunta, V.: Distance measures for effective clustering of ARIMA time-series. In: Proceedings 2001 IEEE International Conference on Data Mining. IEEE (2002)
10. Yang, C., Sun, B.: Modeling, optimization, and control of zinc hydrometallurgical purification process, Chap. 4, pp. 63–82 (2021)
11. Gongde, G. et al.: KNN model-based approach in classification. In: OTM Confederated International Conferences. On the Move to Meaningful Internet Systems. Springer, Berlin, Heidelberg (2003)
12. Liu., K., Ming, T., Zhou, Z.: Isolation Forest. In: 2008 Eighth IEEE international conference on data mining. IEEE (2008)
13. Ecuyer, P.L.: Random number generation, pp. 35–71. Handbook of computational statistics. Springer, Berlin, Heidelberg (2012)

A Product Design Scheme Based on Knowledge Graph of Material Properties

Pengcheng Ding[1], Hongfei Zhan[1(✉)], Yingjun Lin[2], Junhe Yu[1], and Rui Wang[1]

[1] Ningbo University, Ningbo 315000, China
{2111081091,zhanhongfei,yujunhe,wangrui}@nbu.edu.cn
[2] Zhongyin (Ningbo) Battery Co., Ltd, Ningbo 315040, China
yjlin0819@163.com

Abstract. Materials are the material basis of product design. In addition to considering the function and structure of the product, the product design of contemporary enterprises should also fully consider the material characteristics. Material characteristics will not only restrict the shape of the product, but also affect the product structure, and even more, it will affect the quality of the product. Therefore, how to rationally use materials for product design has become a new research. On this basis, this paper proposes a product design scheme based on the knowledge graph of material properties. First, the knowledge in the material properties is defined as subject-function-structure through the triple method; secondly, the Bi-directional Long Short-Term Memory (BiLSTM) method is used to first extract entities and then extract relationships to construct a knowledge map; then combine the product function expression model to generate a product design scheme; Finally, a comprehensive evaluation of the program is carried out.

Keywords: Data mining · Knowledge graph construction · Functional expression model · Scheme design · Comprehensive evaluation

1 Introduction

With the rapid economic development and increasingly fierce product competition, the material field has developed rapidly under the impetus of a new round of manufacturing technology revolution. At the end of the 20th century, data mining technology has been widely used in the research of materials science and achieved good results [1]. In the field of materials, machine learning and deep learning are used to study material properties, which has attracted great interest from many scholars at home and abroad [2]. China also seized the opportunity and launched a corresponding special plan in 2015. After years of development, it has also achieved good results in this field [3].

The "material genetic engineering" project provides a new development direction for the material database, which makes the construction of the material database develop rapidly. Therefore, it is very important to develop various data sharing platforms and computing tools. Materials science researchers around the world are actively building the materials database [4].At present, the more famous foreign material information databases are Materials Project (MP) open database [5], computing materials open database (AFLOW) [6], etc. Compared with foreign countries, the domestic construction of material science database is relatively late. But in the "13th five-year" national key research and development plan special support, China's material science database platform gradually built, such as material database query system [7], Chinese Academy of Sciences institute of Metal development, national material science data sharing network [8], material genetic engineering special database (MGED), the national advanced materials network and information center material information network, these are high quality material research platform, this material knowledge graph material data is from the national material science data sharing network.

As an important part of product design, materials affect product design with their own properties. Rational use of material properties can not only design the appearance of products that meet the needs of the public, but also make use of the material's own characteristics to meet the requirements of product performance. Material and product design are inseparable, so how to use material properties in product design is very important. On this basis, this paper proposes a new method for product design based on material attribute knowledge graph.

2   Overall Framework

The overall framework of product design based on material specific knowledge graph is shown in Fig. 1, which is mainly divided into two parts: construct the knowledge graph based on material properties; product design scheme generation.

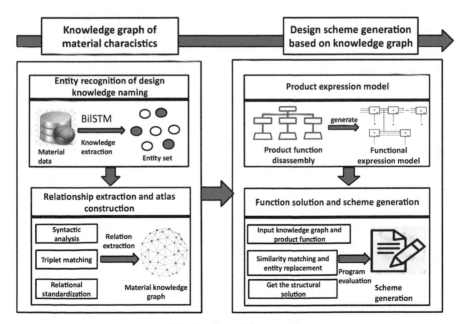

Fig. 1. Overall framework

Section 3 of this paper focuses on building a knowledge graph based on material properties. The first step is to collect relevant information on material properties in the material database, extract entities from the relevant data, and then establish the relationship between entities in the material knowledge graph through performance relationship extraction to form corresponding triples (entity-attribute-object). In the process of establishing the knowledge graph based on material properties, the Subject-Action-Object (SAO) structure is used to describe the knowledge concept related to the scheme design, and it is used as the basic unit of the material knowledge graph. The overall architecture of the material knowledge graph mainly includes the following processes: data acquisition, entity extraction, entity recognition, relation extraction and graph model establishment.

Section 4 of this paper believes that the generation process of product design scheme based on knowledge graph of material properties is described. First, according to the function-structure mapping model, the products to be divided are divided according to the corresponding functions, and then the knowledge map based on the material properties is used to solve the corresponding solutions, and finally the comprehensive evaluation of the scheme is carried out to form the final scheme.

3   Knowledge Graph Construction

The knowledge graph based on material properties (hereinafter referred to as the material knowledge graph) is a knowledge graph constructed by extracting and analyzing the design knowledge of materials in the material database. The graph is helpful for knowledge mining, analysis, and establishment of design knowledge connections, and can provide support for product design. The construction process of the material knowledge graph is shown in Fig. 2.

3.1   Entity Recognition Process

Considering the factors of completeness, accuracy and simplicity of product design process knowledge, this paper chooses to use the SAO [9, 10] structure to build the required material knowledge graph. In addition, utilizing the Bi-directional Long Short-Term Memory (BiLSTM) method can avoid the problem of requiring a large number of people to specify a series of rules, which can improve the efficiency of the named entity process and is especially suitable for large-scale named entity recognition. Recognition for design knowledge naming aims to extract subjects and objects in triplet structures from material data files. The functional carrier of the designed product is the main body, which corresponds to the part structure of each part of the product. What kind of functional goals the product wants to achieve is represented by objects, and the identification process of named entities is shown in Fig. 3.

Through the above steps, the nodes required to build the material knowledge graph can be obtained, but the construction of the material knowledge graph also requires the edges (relationships) used to connect entities and entities, so functional relationship extraction is also required.

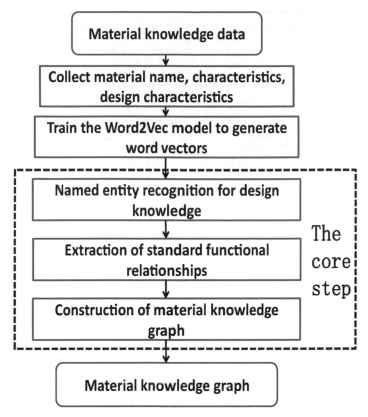

Fig. 2. Graph construction steps

3.2 Extraction of Relationships

An essential link in the triple composition of the material knowledge graph is the extraction of relationships between entities. In this paper, a weakly supervised entity relation extraction method is adopted, which combines semantic information and syntactic template extraction technology. This technology can solve the problem of non-standardized relations in information materials, and is helpful for the construction of material knowledge graph.

Through the above process, the edges (relationships) required for the construction of the material knowledge graph can be obtained by combining the entities obtained by the entity naming and recognition in the previous section. In this case, triples can be established to complete the construction of the material knowledge graph.

4   Scheme Generation

The material knowledge map constructed above provides a reliable and quantifiable design knowledge base for product design. On this basis, the product innovation scheme design can make the design process more scientific and effective, and can make the

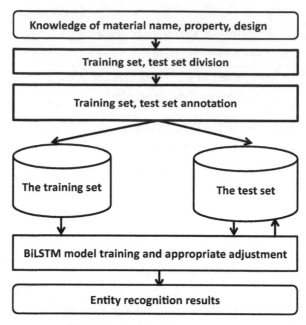

Fig. 3. Entity identification process

product design more reasonable. In the process of generating a product design scheme based on the knowledge graph of material properties, the first step is to establish a product function expression model to obtain the module functions of the existing product, and then use the correlation matrix to establish the relationship between functions and material properties; then, the similarity matching method of material knowledge graph is used to search the structural solution corresponding to the function, and the structural solution set is formed to generate the initial scheme set. Finally, the Analytic Hierarchy Process and the Entropy Weight Method (AHP-EWM) method is used to evaluate the scheme set [11], and the best design scheme is selected. The specific process is shown in Fig. 4.

4.1 Product Function Expression Model

In the generation process of product design scheme based on material knowledge graph, the first step is to disassemble product functions, and use the product function expression model to establish the relationships between various functions, and use this as the input for product function solution.

Product Function Disassembly
Product function disassembly is based on the Function-Behavior-Structure (FBS) [12] model, and the function-behavior-structure mapping model can be used to effectively disassemble product functions. In this mapping model, the reason of product design and product purpose are expressed by function; the composition of product design objects

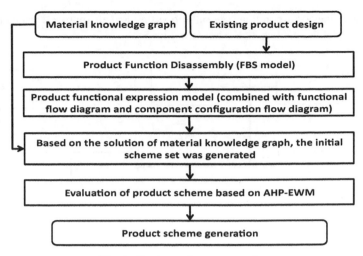

Fig. 4. Product scheme generation

is represented by structure; how the structure realizes product functions is described by behavior, as shown in Fig. 5.

Product Function Expression Model

The most important part of the product design scheme is the product function and expression form represented by the scheme, and then the product function flow diagram and the component configuration flow diagram are combined to represent the functions and forms in the design scheme respectively. This method proposed by Kurtoglu is to use graphical language for modeling and analysis of product design. The relationship between the various functions of the product can be described by the function flow diagram; the structural solution corresponding to the product function can be obtained through the component configuration flow diagram. The fusion of the above two flow diagrams is the product function expression model in this paper. The product function expression model consists of three important parts: function, structure and information. In the process of product function disassembly, all functions represented by leaf nodes have the same granularity, and a function is a set of functions with the same granularity. Taking these functions as the main body of the product scheme expression model, the black-box method is used to model each function, and the structure of each function is determined in combination with product function disassembly. Finally, according to the connection of material, energy and information flow, a product function expression model is formed.

4.2   Product Function Expression Improvement Model

Through the constructed product functions and material properties mapping model, the relationships between product function and material property is established. Based on this, the functional attribute modules in the product function expression model are connected with the corresponding material properties, and an improved product function expression model is established, as shown in Fig. 5.

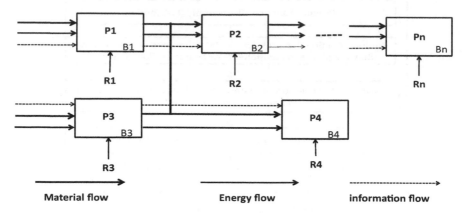

Fig. 5. Product improvement model

Establishment of Product Function and Material Properties Mapping Mode

In the process of establishing the product functions and material properties mapping model, the first step is to refine and disassemble the product function until it cannot be disassembled, and obtain the sub-function set $F = \{f_1, f_2, f_3, \cdots, f_n\}$; the second step is to determine the product manufacturing unit corresponding to each sub function f_i and obtain the process requirements of the manufacturing unit; the third step is to establish the mapping relationships between the sub-function f_i and material properties through the process manufacturing requirements of the manufacturing unit; finally, the mapping model of product functions and material properties is established in turn, as shown in Fig. 6.

4.3 Solution Function

The most important step in the generation of product design scheme is to solve the product function, which is the core of the product design. The functional solving process of the material knowledge graph can be transformed into a process of searching for specific relations and tail entities, and extracting the corresponding head entities. The set of head entities that meet the requirements is the candidate solution for the function solution, as shown in Fig. 7. Specific process:

(1) Input the material knowledge graph and the product function expression. $G = [V, N]$ is used to describe all functions of the product, and the generated function set is $B = \{G_1, G_2, \cdots, G_n\}$. The "action" of a function is represented by V, and the "object" of a function is represented by N.

(2) Functional solution $G = [V_1, N_1]$. The candidate triplet T is generated by searching edges with similar semantic function relation to V_1 from the material knowledge graph and extracting the corresponding triplet of these edges.

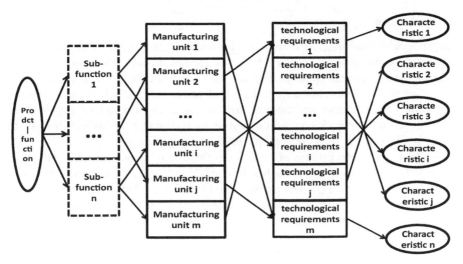

Fig. 6. Mapping of the model

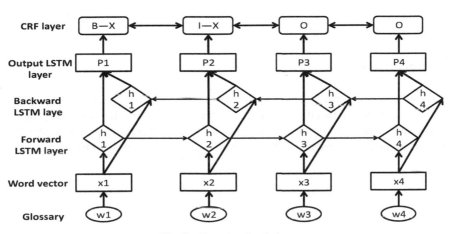

Fig. 7. Functional solution

(3) Each triplet "tail entity" in T generates a word vector, as well as the word vector that gets N_1. Cosine similarity is used to calculate the similarity between "tail entity" and N_1. If the similarity is less than 0.9, this triplet will be deleted from candidate triplet T.

(4) The "head entity" of the remaining triples is extracted from T to form the candidate solution set F_1 of function G_1. If F_1 is an empty set, the structural solution of function is added to F_1.

(5) Repeat steps 2 – 4 until all functions in set B have completed the function solution.

(6) The initial scheme set H is generated according to these solutions.

4.4 Evaluation

Solve the above functions, then combine the functions to form an initial scheme set, and finally use the product design scheme evaluation method based on combination weighting and Euclidean norm to evaluate and select the best scheme. The process of selecting the optimal scheme from the initial design scheme collection will involve many evaluation indicators, and these evaluation indicators are weighted, supplemented by data based on customer satisfaction, and using reasonable methods to evaluate the scheme is a key step to select the optimal product design scheme.

The first step in the process of program evaluation is to use the AHP and consult with experts to assign weights to the first level of the entire two-level metric. The use of second-level indicator assignment is considered that the AHP method is too subjective and lacks objectivity, which will affect the evaluation accuracy. In the second step, the rating evaluation is used as the basis, supplemented by a questionnaire to obtain the customer recognition of each design solution under the second level of indicators, and the indicator data obtained by this process can be assigned with the EWM. The third step is to obtain the comprehensive weight by analyzing and calculating the first and second weights. Finally, referring to the definition of membership in fuzzy theory, Euclidean norm is used to describe the gap between the actual scheme and the ideal scheme, and the final scheme is determined on this basis.

5 Conclusion

Based on the existing research, this paper proposes a new product design method, that is, product scheme design based on material knowledge graph. This method mainly includes two key steps: construction of material knowledge graph; product proposal generation based on material knowledge graph. The main conclusions drawn from this study are as follows:

(1) In the process of establishing the material knowledge graph for the product design proposed in this paper, the knowledge required for design is combined with the map in the form of SAO structure, which improves the integrity of the design structure and gives full play to the advantages of knowledge graph in knowledge management and analysis. The entity recognition of design knowledge, standard relation extraction can make the establishment of graph more accurate.

(2) In this paper, the product functions are disassembled using the FBS model structure in the process of design solution generation. Then, the product functional representation model is established using functional flow diagram and component configuration flow diagram. Next, the functional solution based on material knowledge graph is performed to generate the initial solution set. The combination uses comprehensive weighting and Euclidean norm to evaluate the solutions comprehensively and select the best solution, which can improve the logic of the product design solution generation process, scientific.

The method proposed in this paper has been improved on the basis of existing product design, but product design is a highly complex and highly integrated process. In many fields, the reliability of this design method has yet to be verified due to the imperfect material database. It should be emphasized that the focus of this design method is to improve the functional structure of existing products, and there are certain limitations in the design and development of unknown products. In future research, we will consider applying this theoretical method to various practical cases to verify its reliability, and through further research, we will strengthen its effectiveness in the field of unknown product development and design, and generate more complete and reasonable solutions.

Acknowledgment. I would like to extend my gratitude to all those who have offered support in writing this thesis from National Key R&D Program of China (2019YFB1707101, 2019YFB1707103), the Zhejiang Provincial Public Welfare Technology Application Research Project(LGG20E050010, LGG18E050002) and the National Natural Science Foundation of China(71671097).

References

1. Merayo, D., Rodriguez-Prieto, A., Camacho, A.M.: Prediction of mechanical properties by artificial neural networks to characterize the plastic behavior of aluminium alloys. Materials **13**(22), 5227 (2020)
2. Niu, C.C., Li, S.B., Hu, J.J., et al.: A review of machine learning applications in materials informatics. Materi. Rev. **34**(23), 23100–23108 (2020)
3. Su, Y.J., Fu, H.D., Bai, Y., et al.: Research progress on genetic engineering of materials in China. Acta Metall. Sin. **56**(10), 1313–1323 (2020)
4. Li, Z.X., Zhang, N., Xiong, B., et al.: Application and prospect of materials science database in materials development. Front. Data Comput. **2**(2), 78–90 (2020)
5. Jain, A., Ong, S.P., Hautier, G., et al.: Commentary: the materials project: a materials genome approach to accelerating materials innovation. APL Mater. **1**, 011002 (2013)
6. Curtarolo, S., Setyawan, W., Hart, G.L.W., et al.: AFLOW: an automatic framework for high through put materials discovery. Comput. Mater. Sci. **58**, 218–226 (2012)
7. Yang, L., Su, H., Chai, F., et al.: Current status of application of material database and data mining techniques. Mater. Prog. China **38**(7), 672–681 (2019)
8. Wu, M.M., Liu, L.M., Han, Y.F.: Material design, simulation and top-level design of the database. Mod. Sci. **10**, 53–58 (2018)
9. Park, H., Yoon, J., Kim, K.: Identifying patent infringement using SAO based semantic technological similarities. Scientometrics **90**(2), 515–529 (2012)
10. Franzosi, R.: From words to numbers: a set theory framework for the collection, organization, and analysis of narrative data. Sociol. Methodol. **24**(15), 105 (1994)
11. Huang, J.W., Deng, A.Z., Li, S.B., Luo, S., Sun, J.H.: Decision based on AHP and EWM. Ship Electron. Eng. **41**(11), 18–22 (2021)
12. Qian, L., Gero J.S.: Function-behaviour-structure paths and their role in analogy-based design. AIEDAM **10**(4), 289–312 (1996)

Blockchain Application in Emergency Materials Distribution Under Disaster: An Architectural Design and Investigation

Hongmei Shan[1], Yu Liu[1(✉)], Jing Shi[2], and Yun Zhang[1]

[1] Xi'an University of Posts and Telecommunications, 710061 Xi'an, Shaan Xi, China
shmxiyou@xupt.edu.cn, liuyu2020@stu.xupt.edu.cn, yun@stu.xupt.edu.cn
[2] University of Cincinnati, 45221 Cincinnati, OH, USA
jing.shi@uc.edu

Abstract. In recent years, natural disasters have been raging around the world, among which COVID-19, a public health pandemic, is still spreading all over the world. Emergency material distribution (EMD) in rescue work is facing serious challenges in the term of coordination, efficiency, traceability, identification, credibility, and security, privacy, and transparency. To address the issues, this paper develops a framework based on blockchain technology so that the operation information of EMD can be safely and promptly transmitted to the blockchain system. The scheme could achieve to query, track and record the operation status of emergency material collection, vehicle scheduling, inventory management, relief material allocation and others in real time. By using this framework, a case study on medical rescue materials distribution for COVID-19 is provided to illustrate how blockchain technology and Internet of Things (IoT) technology can be effective. The research provides insights on using the disruptive blockchain technology to effectively coordinate the deployment of emergency supplies, vehicles and personnel, and improve the transparency and credibility of rescue work to minimize the losses caused by disasters.

Keywords: Blockchain · Natural disasters · Emergency materials distribution · IoT · COVID-19

The first and the second author recognizes that the research was supported by the National Social Science Fundation of China (Grant No. 21BJY216); and the Graduate Innovation Fundation of Xi'an University of Posts and Telecommunications.

1   Introduction

Human beings are often faced with sudden and frequent occurrence of disasters, which pose major threats, and cause tremendous damages, casualties, property losses, social insecurity, and public disorder. According to the data of the World Health Organization (WHO), Severe Acute Respiratory Syndrome (SARS) killed at least 774 people and infected 8096 people in the world in 2003, and caused economic losses of over 140 billion USD in mainland China. Currently, COVID-19 is still spreading around the globe, and its severity has far exceeded the global impact of SARS in 2003. The direct losses in retails, catering, and tourism markets alone is immeasurable.

In the event of major disasters such as epidemic, earthquake, flood, and hurricane, emergency materials distribution (EMD) is an essential portion in the rescue work, which directly affects the lives of people and rescue personnel. Most existing EMD research focuses on emergency material delivery and its path optimization, which aim to reduce response time, increase carrier capability, and decrease no-load number. For instance, Lin and Batta [1] propose a multi-cycle vehicle path problem by considering time constraints. Wang et al. [2] propose a two-dimensional multi-target model for the delivery of emergency supplies to achieve the shortest transportation time and the lowest transportation cost, and verify the rationality to solve emergent supply delivery by simulation experiment. Ruan et al. [3] present a method to determine the quantity of rescue medical material allocated to each medical aid point. Nevertheless, making decision for EMD has never been easy. This is because this operation process dramatically differs from the normal logistics distribution thanks to the high level of uncertainty, unpredictability, urgency and harmfulness, and weak economy. Blockchain technology is a disruptive technology that possesses enormous potentials for a variety of applications. In this paper, we aim to apply blockchain technology to facilitate EMD operation. A blockchain-enabled framework is proposed to innovate and broaden managerial concepts of EMD. It is believed that the adoption of blockchain technology could significantly alleviate the barriers presently impeding the EMD operations.

2   Literature Review

2.1   Blockchain Technology

Blockchain technology is primarily a distributed ledger system that completes data transactions in a verifiable and perpetual way instead of relying on a central organization. As such, the vulnerability to attacks, frauds, crashes, damages, or hacks can be reduced [4]. Trust between distributed nodes is ensured in blockchain systems [5]. This facilitates the creation and movement of transaction records and achieve permanent storage of digital assets. In a blockchain, each transaction that occurs over a certain period is recorded on a ledger, forming a "block." Each block is connected to the one before it and the one after it. Mathematically, hashing function is designed to make these blocks chained' together

[6]. Once These blocks will become immutable, or unable to be destroyed or altered by a single entity, once they are linked together as a chain. Instead, automation and common protocols are used to verify and control these blocks [7]. Blockchain embraces smart contracts which reserve negotiated items and supervise the outcomes in accordance with the agreed conditions. The system will automatically execute transactions between nodes once certain conditions are met.

2.2 Blockchain Applications in Supply Chain and Logistics

Countless applications of blockchain technology have been proposed and implemented, while financial related applications are by far the most studied. In the areas related to EMD, scholars have started to study the impacts of blockchain on supply chain management, including supply chain coordination [8]. Westerkamp et al. [9] expand a product traceability in the supply chain by developing a scheme which embodies manufactured goods and their constituents with a blockchain-based composition. Traditionally, building the necessary trust within the multi-tier supply chains is difficult because of lack of transparency and visibility. Within blockchains, however, trust and faith are programmed and implanted into the technological platform. Weber et al. [10] construct a decentralized model of mutual trust and cooperation in supply chains by means of intelligent contract mechanism in blockchain. The model enhances the flexibility of supply chain governance while promoting the interaction and sharing of information among the various entities of supply chain system.Based on the analysis above, it is clear that scholars have paid heavy attention to address the coordination, integrity, efficient, security, transparency, trust, traceability in supply chain management by using blockchain technology. Meanwhile, there are also a handful of studies focusing on logistics. Francisconi [11] studies an information system based on blockchain technology in port logistics and conducts empirical research to evaluate the benefits obtained from the application of blockchain technology. Chang and Park [12] discuss that in the event of COVID-19 outbreak, blockchain technology can help people by providing a transparent process to learn the flow of rescue materials, donated money or in-kind goods. Thus, donors can be confident to make more donations. However, how to integrate blockchain with EMD system is not addressed in study.

3 Emergency Materials Distribution System Based on Blockchain Technology

3.1 Targeting Role of Blockchain on Emergency Materials Distribution

The technical advantages of blockchain technology targets the proposed EMD issues (shows as Fig. 1). The advantages of blockchain, such as decentralization, permanent storage, security, tamper-proof, anti counterfeiting, authenticity and credibility, traceability, verifiability and automation, could pave the way for resolving the challenges faced with EMD. Figure 1 fully interprets the rationality and feasibility of using blockchain to improve management performance of EMD, namely, through blockchain, enhancing the transparency of the EMD process, realizing the traceability and anti counterfeiting of relief materials, simplifying the operation and reducing the consumption of human and materials, ensuring efficiency, safety and reliability in delivering the materials.

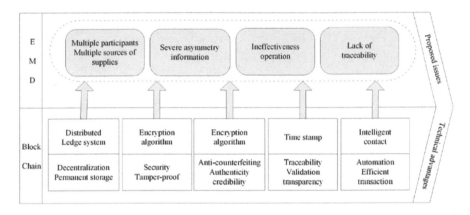

Fig. 1. Coupling of blockchain and EMD

3.2 Proposed Framework

On the basis of the above coupling concept, an overall design of emergency material distribution management system on blockchain based is proposed, as shown in Fig. 2. It can be seen that the framework includes five layers: perception layer, communication channels, data layer, core blockchain and applications levels. The perception layer is basically used to gather the multisource data by various Internet of Things (IoT) devices in a timely, accurate and manner. With the IoT devices, the conventional objects can be transformed into "smart objects". The collected multisource information can then be processed, transmitted, and stored, which provides help and support of decision–making for management.

In the layer of communication channel, the sensed data obtained are transmitted by multiple communication channels including Bluetooth, ZigBee, 4G and 5G. All in all, this layer is responsible for managing diverse communication methods in a way that is easy to control and easy to use, which alternate and cooperate with each other to ensure the data transmission stability. The data layer is responsible for analyzing, summarizing and storing the data collected and transmitted by the first two layers. It is also an aggregation of underlying technologies for blockchain, which mainly includes hashing algorithm, Merkle tree, chained data structure, asymmetric encryption algorithm, timestamp, P2P

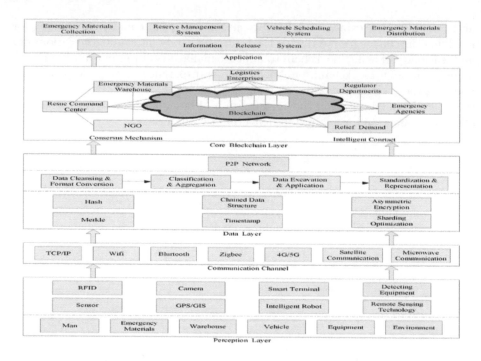

Fig. 2. Framework of blockchain-based emergency material distribution system

network services, LevelDB databases, and so on. The core blockchain layer can be built in multiple forms of chain structures which involve the distributed multiple participants. For EMD, the member nodes include rescue command center, nongovernmental organizations (NGOs), emergency material warehouse, logistics enterprises, relevant regulator departments, relief agencies, victims and their governmental organizations. Whenever a transaction activity is generated or carried out, the credit records, transaction data and other information of any participants will be encapsulated in the individual block. The top application layer provides interface and information services for participants according to their needs. There are many application procedures to support the management of operation process, personnel, charitable fund or vehicle and making-decision, mainly including emergency supplies collection, reserve management system, vehicle scheduling system, emergency materials allocation, information release system.

3.3   Main Functions

Emergency Supply Collection. According to the forecast and demand information released from disaster areas, this function allows the selection of the best time and methods to raise the donated funds to purchase the relief goods, or (and) collect the rescue materials from the society. More importantly, relying

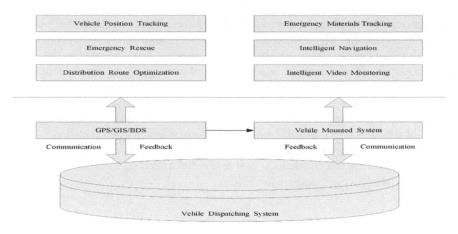

Fig. 3. Vehicle scheduling system

on the blockchain technology, the system connects supply and demand information in real time to maximize the best collection effect of materials. Transaction information, such as rescue fund source, goods purchase, transportation, and storage management, will be written into the distributed ledger in real time, so as to realize the two-way traceability function of material supply source and distribution direction, ensure safety and reliability of rescue materials.

Emergency Material Reserve System. The emergency material reserve system is the integration of IoT technologies, M2M (Machine to Machine) technologies, multi–sensors, advanced logistics equipment, and other technologies to carry out the collection, query, backup, printing and statistical analysis of material information, and then transmit the collected information to the intelligent warehousing management system. The system provides feedback on the calculated results to improve the management efficiency, realize the visualization, mechanization, automation and intelligence of operation activities. At the same time, the multi-sensors system can auto-monitor the temperature and humidity of warehouse environment in real time to ensure the safety of materials.

Vehicle Scheduling System. Vehicle scheduling system, combined with IoT and other technologies, can realize vehicle position tracking, distribution route optimization, electronic map information system, emergency materials tracking, intelligent video monitoring, emergency rescue functions and so on (see Fig 3.). These technologies support and complement each other to achieve the vehicle scheduling service function. Moreover, the reader on the vehicle can also obtain the information of loaded materials in real time and send feedback to the dispatch center through the communication channels, so as to visually ensure the dynamic tracking management of emergency materials.

Emergency Materials Allocation. In the emergency, the relief materials allocation can be mainly managed by government, also can be entrusted to professional social organizations or third-party logistics firms to reduce the burden on the government and realize the separation of emergency command from materials allocation based on blockchain. According to the existing material supplies and actual demands, these organizations reasonably allocate the relief materials to affected people and organizations in disaster areas, and the receiving information of rescue materials will be accurately recorded in the blockchain system. Finally, the final direction will be included in the ledger, and all relevant participating nodes in the blockchain can trace the relief materials destination to make the whole process perfect for EMD.

Information Release System. Information release system mainly includes three parts: central control system, terminal display system and network platform, integration with the support of modern network streaming media and other information technology. Generally, various rescue activities are expected to be carried out in the shortest time. Thus, it is indispensable to construct an information promulgation system which synchronously and instantaneously provides the authenticity information for the rescue command center, relief units, social welfare organizations and public by timely collection, classification, aggregation and dynamic tracking of the data.

4   Case Study-Medical Emergency Materials Distribution

4.1   Background

Considering that prevention and control requirement of the global outbreak of COVID–19, we present a case study on blockchain based medical emergency materials distribution (MEMD) to ponder over the adoption opportunities of blockchain for EMD. The medical supplies urgently for epidemic prevention in public health events include various PPE items such as N95 masks, eye protectors, shoe covers, caps and disinfectant. In the COVID–19 scenario, the sufficient quantity to meet the demand is a key, in which the help and support from all sectors of society are essential. The multiple–participants are involved in the process: (1) the donors (individuals, companies, charities or others) (2) logistics firms; (3) rescue units which receive donations of medical supplies and related staff; (4) Red Cross organizations and related entities. The current problems related to medical supplies distribution of public donations are as follows:

1. The supply and demand of medical emergency materials are asymmetric, and the information is seriously disjointed. The insufficient information sharing among the designated hospitals, Red Cross organization, society donation institutions or individuals negatively affects the performance of medical material supply assurance system for epidemic rescue work.

2. There is lack of traceable logistics records throughout the transportation distribution process. In the early control stage of COVID–19 outbreak in Wuhan, the process of material collection, shipment, delivery, and allocation was chaotic.

3. The typical operation of MEMD is not good coordinated along the supply chain. Again, in the early stage of COVID-19 outbreak in Wuhan China, a large number of funds and materials were donated and collected from the charities and donors in upstream, while the donated fund and materials were piled up in the warehouse of Red Cross and were not distributed in time in downstream.

4.2   Medical Emergency Materials Distribution Solution

In view of the close relationship between various nodes in the process of MEMD, while some participant nodes have certain limitation authorities. The blockchain can link the multiple organizations, designate several nodes for consensus authentication, and other nodes can publish and broadcast transaction activities, but they do not have authentication authority. Other nodes can access the blockchain data with access authority under the authorization of the system. So as to realize the sharing and matching of data information of donated materials from all walks of life. The main procedures are as follows:

1. Relevant technologies such as blockchain smart contract are applied to realize automatic real–time tracking and response of emergencies by automatic contract execution. In the event of a disaster, blockchain-enabled systems use comprehensive real-time information as a data source to dynamically capture event progress information. Then, according to the conditions, the automatic response is made in real time, and the corresponding emergency logistics plan is launched.

2. The recipient hospitals or individuals publish accurate medical and relief material needs including the quantity and standard requirements through the blockchain information release platform. The donors obtain the relief material needs at real time, and then make appropriate donations of relief funds or supplies in platform. The Red Cross organization can gather the medical materials, and collect the rescue funds to purchase, mobilize the volunteers for disaster relief. Logistics firms actively respond to the transportation demand between Red Cross, hospitals and donors.

3. To verify and enhance the social credibility of each participant, the system can generate the identity authentication information of each node (e.g., hospitals, donors, logistics firms), and collect and store the transaction information on the blockchain. To avoid suffering from information disclosure delay, data transmission distortion and other issues caused by the crisis of trust, the participants could make proactive and timely disclosure of process data, and engage effective self-certification.

4. For transparent and traceable medical material distribution, all nodes in the decentralization system are peer–to–peer nodes, which receive information equally and timely, and each participant could observe the entire behavior of other nodes in the system within the scope of authorization.

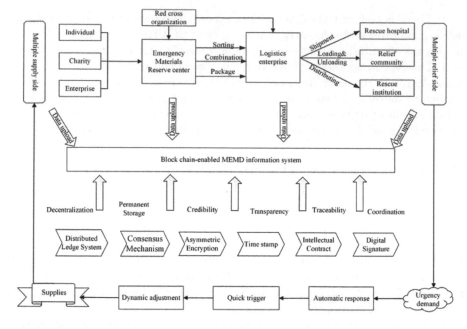

Fig. 4. Medical emergency materials distribution based on block chain technology

The information process of MEMD embedded on blockchain is shown in Fig. 4. To sum up, the blockchain-based material donation platform builds a charity bridge between donors, Red Cross, logistics firms and hospital or community recipients, to maximize the coordination, efficiency, traceability, identification, credibility, and intelligence, security, privacy, transparency in the process of MEMD.

5   Discussion

The application scheme of blockchain technology in EMD is to support and optimize the management of EMD, which aims at improving the response to relief materials demand and enhancing the efficiency and speed of operation. A further objective is to achieve the transparency and traceability of whole process by creating the distributed ledger to record the data of each smart subject including materials, vehicles, equipment and organizations. However, challenges also exist for implementing the proposed framework. The major issues of privacy security, consensus agreement and flux performance are discussed in the following.

5.1   Privacy and Security

The massive data information carried by various kinds of IoT and communication devices causes concerns about privacy and security. Transaction data is openly and transparently stored on the blockchain. Although the relationship between address and user's real identity is, it still brings serious security risks, especially in applications of sensitive data. How to ensure privacy and data security is an important task.

5.2 Consensus Agreement for Internet of Things

Current consensus agreements such as POW (Proof of Work), POS (Proof of Stake) and PoET (Proof of Elapsed Time) generally focus on financial value transfer. A common problem with these consensus protocols is that the consensus process does not end with permanently committed blocks. Instead, the blocks are finalized over a period of time based on the blocks of the next few extended chains, then the longest chains are identified as valid chains. Thus, these consensus agreements are prone to bifurcations, which ultimately lead to delay in the confirmation of transactions.

5.3 Pressure of Flux Performance

The improvement of throughput and performance is a hot issue in the field of blockchain. The current blockchain flux is limited-bitcoin can only handle 7 transactions per second on average, and Ethereum can only handle 15 transactions per second on average. In particular, the embedded devices in the IoT system have increased significantly, and thus the traditional communication models based on centralized data center are hardly able to handle the high–speed data stream from billions of devices. Although the consensus algorithm with higher efficiency can improve the number of transactions per second (TPS) to a certain extent, it is still difficult to meet the business needs.

6 Conclusive Remarks

EMD in rescue work has been facing with many challenges, such as multiple participants, severe asymmetrical information, ineffective operations and lack of traceability. In this paper, a technical framework of EMD management based on blockchain technology combined with IoT is proposed to improve the quick response to the demand for relief materials and achieve the efficiency, transparency and traceability in EMD. The system architecture includes five main layers: perception layer, communication channels, data layer, core blockchain and applications levels. The major functions of system are also defined. This research is a groundbreaking work in exploring the application of blockchain in EMD to improve the capacity of quick response, and achieve the efficiency, transparency and traceability of operation. Nevertheless, the system architecture for addressing EMD issues needs to be realized and pilot tested. This will be a major task for future research. Thereafter, implementation and empirical study could be the next desirable research direction.

References

1. Lin, Y.H., Batta, R., Rogerson, P.A., et al.: A logistics model for emergency supply of critical items in the aftermath of a disaster. Socio-Econ. Plann. Sci. **45**(4), 132–145 (2011)
2. Wang, H., Xu, R., Zijie, X., et al.: Research on the optimized dispatch and transportation scheme for emergency logistics. Procedia Comput. Sci. **129**, 208–214 (2018)
3. Ruan, J.H., Wang, X.P., Shi, Y.: Scenario-based allocating of relief medical supplies for large-scale disasters. ICIC Express Lett. **7**(2), 471–478 (2013)
4. Tapscott, D., Tapscott, A.: How blockchain will change organizations. MIT Sloan Manag. Rev. **58**(2), 10–13 (2017)
5. Wang, Y., Singgih, M., Wang, J., Rit, M.: Making sense of blockchain technology: How will it transform supply chains? Int. J. Prod. Econ. **211**(5), 221–236 (2019)
6. Laurence, T.: Blockchain for Dummies. Whiley, New York (2017)
7. Swan, M.: Blockchain Blueprint for a new economy. O'Reilly, New York (2015)
8. Li, Y., Wang, B., Yang, D.: Research on supply chain coordination based on block chain technology and customer random demand. Discrete Dyn. Nat. Soc. **2019**, 1–10 (2019)
9. Westerkamp, M., Victor, F., Kupper, A.: Tracing manufacturing processes using blockchain-based token compositions. Digit. Commun. Netw **6**(2), 167–176 (2020)
10. Weber, I., Xu, X., Riveret, R., Governatori, G., Ponomarev, A., Mendling, J.: Untrusted business process monitoring and execution using blockchain. Springer International Publishing **8**, 329–347 (2016)
11. Francisconi, M.: An explorative study on blockchain technology in application to port logistics. Delft University of Technology, pp. 67–89 (2017)
12. Chang, M.C., Park, D.: How can blockchain help people in the event of pandemics such as the COVID-19? J. Med. Syst. **44**(5), 102 (2020)

Design and Implementation of Virtual Power Plant System Based on Equipment-Level Power and Load Forecasting

Xu Zhenan[1(✉)], Liu Zesan[1], Meng Hongmin[1], Huang Shu[1], Wen Aijun[1], Li Shan[1], Jin Siyu[2], and Cui Wei[1]

[1] State Grid Information and Communication Industry Group Co., Beijing 100085, China
{xuzhenan,liuzesan,menghongmin,huangshu,wenaijun,lishan,
cuiwei}@sgitg.sgcc.com.cn
[2] State Grid Information and Telecommunication Co., Ltd., Beijing China-Power Information Technology Co., Ltd., Beijing 100085, China
jinsiyu@sgitg.sgcc.com.cn

Abstract. In order to ensure the effective participation of virtual power plants in grid interaction under the novel power system. This paper design and implement a virtual power plant system based on equipment-level power forecasting and load forecasting technology. In order to shield equipment differences, a unified measurement standard for physical equipment of different energy systems is established by constructing virtual machine groups (including equipment attribute information, equipment collection information and equipment evaluation information); using equipment-level neural network power forecasting and multiple load forecasting engines to ensure equipment-level forecast accuracy and flexible aggregation capabilities; building model self-learning capabilities by monitoring equipment historical data to drive model iterative optimization; the cloud edge collaboration solution is used to realize the vertical aggregation of resources, reduce the cloud load and improve the command processing speed. Through deployment and verification, this system can improve prediction accuracy and dynamically aggregate multiple types of resources to participate in grid interaction.

Keywords: Virtual power plant · Virtual machine groups · Power forecasting · Load forecasting · Cloud edge collaboration

1 Introduction

In recent years, with the large-scale development and use of fossil energy, the reserves of coal and oil have been greatly reduced, and the greenhouse effect has also been intensified [1, 2]. In order to alleviate these problems, renewable energy sources such as wind power and photovoltaics have been developed. However, due to the intermittency and randomness of new energy generation such as photovoltaic wind turbines, the challenges of power [3], which leads to the low enthusiasm of the power grid for new energy dispatching, and the phenomenon of abandoning wind and light is serious. As

N. Xiong et al. (Eds.): ICNC-FSKD 2022, LNDECT 153, pp. 1045–1055, 2023.
https://doi.org/10.1007/978-3-031-20738-9_114

a distributed resource aggregation and management technology, virtual power plant has the characteristics of high efficiency, flexibility and friendliness. By aggregating new energy, flexible loads, energy storage units and other adjustable resources as a whole to participate in grid scheduling and electricity market transactions [4], to achieve grid flexibility resource aggregation and coordinated optimal scheduling [5, 6].

At the beginning of the Twenty-first century, virtual power plants appeared in the United States and other countries [7]. The United States mainly focuses on the demand response business. At present, the cumulative resources on the load side in the United States are as high as 30,000 MW, occupying 4% of the peak period of electricity. 19 research institutes in 8 European countries including the United Kingdom, Spain and France have implemented the Flexible Electric network to integrate the expected energy solution project [8], which integrates large-scale virtual power plants and distributed energy to maximize the contribution of distributed energy sources to the power system. Australian company AGL is building a virtual power plant by connecting 1000 energy storage battery packs in southern Australia. This project can store 7 MWh of electricity, which is equivalent to a 5 MW solar peaking power station. China's research on virtual power plants started relatively late, but it has developed rapidly in recent years. In 2019, the State Grid Jibei Company's virtual power plant was put into operation. The project has second-level perception, computing, and storage capabilities. The first phase access control distributed photovoltaic, energy storage, flexible load and other 11 types of 19 kinds of ubiquitous adjustable resources, it's capacity about 160,000 kW. By 2020, 10% of the summer air-conditioning load of Hebei Power Grid can respond in real time response through VPP [9]. In 2021, the virtual power plant in Pinghu County, Zhejiang Province will be put into practical application. It aggregates 6 categories and 18 sub-categories of source-load-storage resources, including 5G base stations, energy storage power stations, and distributed new energy. It is expected to save 300 million yuan in investment costs for power grid construction each year, increase clean energy consumption by 96,000 MWh, and reduce carbon emissions by about 76,000 tons.

The above-mentioned project practice has achieved remarkable results in reducing the curtailment of wind and solar power, and enhancing the interaction between the source, network and load-storage, but the similar-day algorithm is mostly used for the prediction of participating resource power [10]. This algorithm uses historical data as a map and obtains forecast data by calculating historical approximate days. In the case of complex and changeable actual weather conditions, the algorithm's prediction will be inaccurate. Therefore, the existing data on how to improve the prediction accuracy and ensure the validity of the regulation and control instructions of virtual power plants are rarely involved. Therefore, the existing data on how to improve the prediction accuracy and ensure the validity of the regulation and control instructions of virtual power plants are rarely involved. Therefore, how to improve the prediction accuracy and ensure the effectiveness of virtual power plant regulation instructions remains to be studied. Aiming at this problem, this paper uses the neural network model to design and implement a virtual power plant system with self-learning capability for equipment-level power forecasting and load forecasting [11–13]. Using equipment virtualization technology and equipment-level power forecasting and multi-load forecasting engines, equipment-level dynamic aggregation and accurate forecasting are realized; achieving model self-learning by monitoring historical data; Using the cloud-edge collaboration solution, optimize the hierarchical scheduling of regulation and ensure the feasibility of regulation instructions. Through deployment verification, this system can dynamically aggregate multiple types of resources to participate in grid interaction.

2 Functional Design of Cloud Edge Collaborative Vpp

In this solution, the virtual power plant consists of two parts: Cloud operation platform and edge control platform.

2.1 Cloud Operating Platform

The cloud platform connects upwards to the electricity market trading platform and participates in market transactions; it connects downwards to the edge control platform and virtual machine groups for optimal scheduling. It provides functions including forecast configuration, operational monitoring, autonomous operation, ancillary services, equipment regulation, potential analysis, resource management and market transactions. The specific function view is shown in Fig. 1.

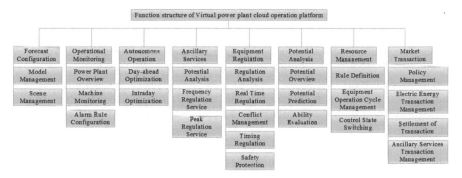

Fig. 1. Function structure of virtual power plant cloud operation platform.

The forecast configuration module provides general parameter configuration functions such as the dimensions of the neural network model required by participating devices, including model management, scene management. Operational Monitoring uses WebSocket to provide real-time monitoring of key indicators for virtual power plant. It includes power plant overview, machine monitoring and alarm rule configuration. Autonomous operation includes two parts: day-ahead optimization and intraday optimization. Ancillary Services include potential analysis, frequency regulation service and peak regulation service. With the gradual maturity of the supporting policies for electricity spot trading, it is necessary to design virtual power plant participate in the market trading business. This module mainly includes policy management, electric energy transaction management and transaction settlement.

2.2 Edge Control Platform

As a lightweight edge node of the cloud platform, the edge platform provides regional resource autonomy and control instruction decomposition functions. It connects upward to the cloud platform and accepts cloud platform scheduling; the downward aggregation of participating resources such as new energy units and flexible loads has the functions of forecast configuration, operational monitoring, potential analysis, ancillary services, equipment regulation and resource management. The specific function view is shown in Fig. 2.

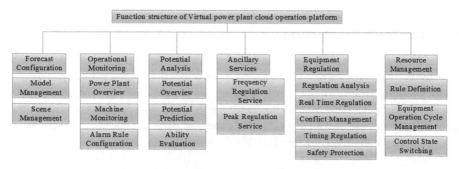

Fig. 2. Function structure of virtual power plant edge control platform.

Resource management is connected to the IoT management platform and combined with the unified device model, it encapsulates the access devices, providing conditions for the dynamic grouping of subsequent devices. The module provides functions such as rule definition, equipment operation cycle management and control state switching. Ancillary services consistent with the auxiliary services of cloud operation platform, including peak regulation service and frequency regulation service. Potential analysis is for timely feedback of device adjustment potential. It mainly includes potential overview, potential prediction and ability evaluation. Equipment regulation is the scheduling center of virtual power plants, including regulation analysis, real time regulation, conflict management, timing regulation and safety protection.

3 Research on Equipment-Level Virtual Power Plant Prediction Technology with Self-learning Ability

3.1 Virtual Device Definition

In order to achieve device-level prediction, it is necessary to regulate the aggregated devices of the virtual power plant. The traditional solution is to rely on the device file module of the IoT management platform to provide device attribute information, and use device control module to acquire real-time data and issue instructions. As for the scheduling capability of equipment, the rated power, real-time power, and regional prediction data are often used for simple calculation. Although this method can complete the decomposition of scheduling instructions, it lacks flexibility and the overall prediction is difficult to reflect the operating conditions of a single equipment. Relying on the idea of unified device model, this paper realizes the virtual device encapsulation technology based on ontology and semantics (Fig. 3).

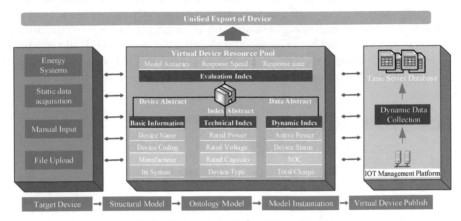

Fig. 3. Package principle of virtual device.

First, the system obtains the basic information of the equipment (such as manufacturer, system, etc.), technical indicators (such as rated power, rated voltage, etc.), dynamic data (such as active power, remaining capacity, etc.) through the IoT management platform, and establishes the index mapping relationship and further abstracted into a device model independent of the IoT management platform.

Secondly, the device model is associated with the IoT device indicators according to the unified model standard. The static data of the device can be synchronized by manual entry, etc., and the dynamic data can be integrated with the time series database to ensure the real-time acquisition of the measurement indicators of the model entity.

Finally, the unified evaluation index of the equipment model is added. The implementation of this technology can effectively shield the physical differences of devices, and lay the foundation for complex business logic such as dynamic grouping of subsequent devices.

3.2 Equipment-Level New Energy Power and Load Forecasting Engine

In electricity market, accurate and fast prediction can improve the accuracy of baseline power reporting of virtual power plants, reduce additional dispatch costs and penalty costs caused by power deviations; in terms of optimal dispatch and system simulation, ultra-short-term prediction accuracy determines the load decomposition. Reasonable, if the accuracy is low, it will often lead to a large deviation of the scheduling effect.

At present, the new energy generation power prediction and load prediction of virtual power plant systems on the market often use similar daily algorithms for prediction. Although this type of method can guarantee a certain accuracy, the accuracy is low, and after the development is completed, self-learning optimization cannot be carried out in the later stage. At the same time, there are also a small number of virtual power plant systems that use neural networks as prediction algorithms, but the model training is usually completed in the deployment phase, and the training set is mostly selected from regions or data from multiple devices of the same type. This method regards the device set as a whole. Although it can provide more accurate prediction when the device is inherently unchanged, it is difficult to ensure accuracy in the face of device partition prediction, dynamic aggregation grouping. This platform uses a long-term and short-term neural network model based on machine learning for power and load forecasting [14], and provides equipment-level autonomous learning and judgment capabilities to ensure the accuracy of forecast data. The technical workflow is shown below.

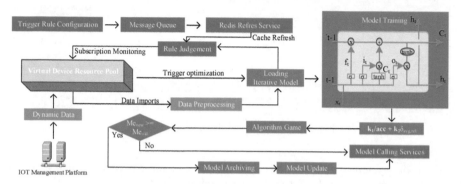

Fig. 4. Update process of forecast engine.

First, the forecasting engine obtains the virtual devices that have been constructed in the virtual device resource pool, reads the properties of the virtual devices, and divides the virtual devices into model scenarios and reads the model dimension information through the prediction configuration function (Fig. 4).

Second, after the first-generation model is generated, the model forecasting engine monitors the virtual device resource pool and dynamic data. When the device is updated or the dynamic data is updated to a certain level, the forecasting engine self-optimization function will be triggered. The forecasting engine will call the preset model and use the historical data to iterative training.

Third, after the iterative model training is completed, forecasting engine performs model evaluation on the iterative model, including the model accuracy as shown in formula 1 and the average relative error as shown in formula 2. After the evaluation index is obtained, the weighted comparison formula of the evaluation index is shown in 3. If the accuracy of the iterative model improves, the equipment model in the forecasting engine will be replaced. If the accuracy of the iterative model is less than or equal to the current equipment model, the current model will be retained.

$$acc = \frac{P_t}{P} \tag{1}$$

$$\delta_{avg,rel} = \frac{1}{P} \sum_{i=1}^{P} \left| \frac{F_i - F_i'}{F_i} \right| \tag{2}$$

$$\text{Assess} = k_1 \big/ \text{acc} + k_2 \delta_{avg,rel} \tag{3}$$

In the formula, acc represents the accuracy rate, which can reflect the proportion of accurate prediction points in the prediction model. The higher the proportion, the more accurate the model prediction. Among them, P_t predicts the number of correct points, and the predicted value here is defined as correct within $\pm 15\%$ of the actual value, P represents the number of predicted points. $\delta_{avg,\ rel}$ represents the average relative error [15], which can reflect the size of the error of the prediction result and can also be used as an accuracy index, F_i represents the actual load value of the ith point, and $F_i{}'$ represents the predicted load of the ith point.

Last, after the current iteration is completed, the forecasting engine provides day ahead forecast and ultra-short-term forecast business call interfaces. At the same time, if the server resources are limited, the user can set the day ahead rolling forecast of the model to reduce the pressure of real-time call.

4   Implementation of Virtual Power Plant System

4.1   Design Overall Architecture

The platform is designed and developed by micro service architecture [16]. At the same time, in order to ensure the flexibility of the front-end presentation layer and the stability of the back-end services, backends for frontends (BFF) are introduced for node combination. From top to bottom, the architecture of the virtual power plant is divided into six layers: Presentation Layer, Backend for Frontend Layer, Service Management Layer, Converged Service Layer, Business Service Layer and Basic Service Layer. (see Fig. 5).

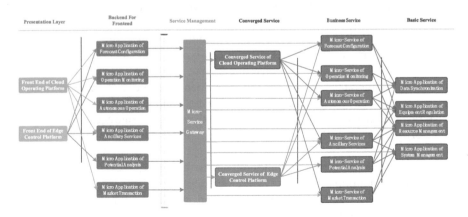

Fig. 5. Overall structure of virtual power plant.

4.2   System Implementation

Both the cloud and the edge of the virtual power plant platform adopt B/S architecture. The back-end development language is Java, JDK version is JDK1.8 and the forecasting engine is the TensorFlow1.13 framework, which is developed in Python. The technical architecture is shown in the figure (Fig. 6).

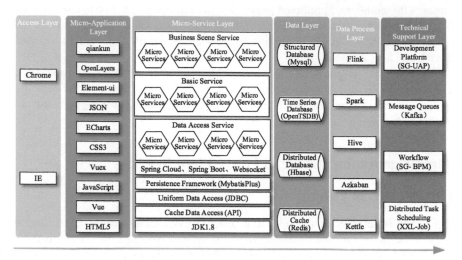

Fig. 6. Technical architecture of virtual power plant.

The platform is developed using SG-UAP, which mainly consists of technical support layer, data process layer, data layer, micro-service layer, micro-application layer and access layer. The technical support layer uses SG-BPM for workflow configuration and XXL-Job for distributed task scheduling; Data process layer provides data processing capabilities such as data processing and streaming data processing. Data layer uses databases such as mysql, hbase, opentsdb, etc. Mysql stores the relational data such as business configuration and equipment nameplate information. Hbase and opentsdb stores strong timing and large amount of data such as equipment measurement; the micro-service layer is implemented based on SpringCloud and SpringBoot; the micro-application layer includes Vue, ECharts and other technologies; the access layer uses IE, Chrome browsers or mobile access. The screenshot of some functions of the platform is shown in Fig. 7.

In order to verify the functions of the platform in this paper, an experimental environment is built for platform verification. The cloud platform is connected to photovoltaic, wind turbines, energy storage units and a side-end management and control platform, using 3 application servers, 1 GPU server 1, and 2 database servers 1 for data service deployment. The side-end management and control platform simulates side-end deployment, connects to photovoltaic, wind turbines, energy storage and other equipment, and uses 1 application server, 1 GPU server 2, and 1 database server 2 for deployment. System deployment resources are shown in Table 1. After the system is deployed, first build

Fig. 7. Screenshot of the virtual power plant system.

a virtual device library, and add device model evaluation indicators for device prediction and subsequent evaluation.

Table 1. System resources

Server type	Server configuration	Number
Application server	*64-bit Centos7.11; CPU: 16C; Memory: 32G; Hard Disk: 500G*	4
GPU server 1	*64-bit Centos7.11; CPU: 16C; Memory: 64G; Hard Disk: 500G; GPU:2 * NVIDIA V100*	1
GPU server 2	*64-bit Centos7.11; CPU: 16C; Memory: 64G; Hard Disk: 500G; GPU:2 * NVIDIA T4*	1
Database server 1	*64-bit Centos7.11; CPU: 8C; Memory: 64G; Hard Disk: 2T*	2
Database server 2	*64-bit Centos7.11; CPU: 8C; Memory: 32G; Hard Disk: 1T*	1

After the construction of the virtual device library, the forecasting engine monitors device dynamic data, and triggers self-optimization when the update volume reaches 40,320. The forecasting engine calls the preset model to iteration train the model. This paper selects a photovoltaic generator and manually pushes the data for verification. After the model iterative training is completed, the calculation results of *acc*, $\delta_{avg,\,rel}$, and *Assess* are shown in Table 2. The iterative model evaluation index is 0.8082, which is less than the current model evaluation index 0.8952. It can be seen from the foregoing description that the iterative model is judged to be better than the current model, and the replacement operation is performed. In order to verify the above rationality, the iterative model and the current model are input to the same validation set for comparison, the results are shown in Fig. 8. The average absolute error percentage of the iterative model prediction curve and the current model prediction curve are calculated respectively. The iterative model result is 35.12 and the current model result is 37.94. The average absolute error of the iterative model is smaller than the current model, which is the same as the forecasting engine conclusion.

Table 2. System resources.

Model	P	P_t	acc (%)	$\delta_{avg, rel}$	Assess
current model	552	439	79.53	0.6537	0.8952
iterative model	552	458	82.97	0.5435	0.8082

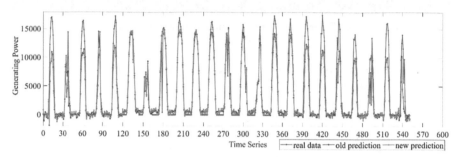

Fig. 8. Output model comparison.

5 Conclusion

To ensure the virtual power plant can effectively participate in the grid interaction. This paper designs and implements a virtual power plant system based on equipment-level power forecasting and load forecasting. Using virtual equipment packaging technology to establish a unified equipment measurement standard; using equipment virtualization technology and equipment-level power generation forecasting and multi-load forecasting engines to ensure forecast accuracy and flexible resource aggregation capabilities; using cloud-edge collaboration solutions to achieve optimal regulation and hierarchical scheduling, to ensure the real-time and effective control instructions. Through deployment and verification, the system can dynamically aggregate multiple types of resources, provide more accurate baseline load forecasting capabilities, and provide guarantees for grid interaction.

References

1. Lin Ze-wei, F., Wang Peng, S., Ren Song-yan, T.: The economic evaluation of energy-intensive industries transition due to the carbon emission peaks: evidence from Shaanxi province. Ecol. Econ. **38**(06), 13–21 (2022)
2. Zhang Zhi-gang, F., Kang Chong-qing, S.: Challenges and prospects for constructing the new-type power system towards a carbon neutrality future. Proceedings of the CSEE **42**(08), 2806–2819 (2022)
3. Li M., F., Yu, Z., Xu, T.: Study of complex oscillation caused by renewable energy integration and its solution. Power Syst. Technol. **41**((4):1035–1042 (2017)
4. Chen Kai-ling, F., Gu Wen, S., Wang Hai-qun, T.: Mode of virtual power plant operation and dispatching in energy blockchain network. J. Syst. Manag. **31**(01), 143–149 (2022)

5. Xu Jia-yin, F., Wang Tao, S., Wang Jia-qing, T.: Distribution network planning method considering flexibility resource attribute of microgrids. Electric Power Constr. **43**(06), 84–92 (2022)
6. Zhang Jing, F., Lin Yu-jun, S., Qi Xiao-guang, T.: Low-carbon economic dispatching method for power system considering the substitution effect of carbon tax and carbon trading. Electric Power Constr. **43**(06), 1–11 (2022)
7. Koraki, F., Strunz, S.: Wind and solar power integration in electricity markets and distribution networks through service-centric virtual power plants. IEEE Trans. Power Syst. (0885–8950) **33**(1), 473–485 (2018)
8. Zhang Kai-jie, F., Ding Guo-feng, S., Wen Ming, T.: Review of optimal dispatching technology and market mechanism design for virtual power plants. Integr. Intell. Energy **44**(02), 60–72 (2022)
9. Wei, X., Yang, D., Ye, B.: Operation mode of virtual power plant in energy Internet and its enlightenment. Electric Power Constr. **37**(4), 1–9 (2016)
10. Sun Le-ping, F., Han Shuai, S., Wu Wan-lu, T.: Coordinated optimal scheduling of multiple virtual power plants in multiple time scales based on economic model predictive control. Energy Storage Sci. Technol. **10**(05), 1845–1853 (2021). https://doi.org/10.19799/j.cnki.2095-4239.2021.0195
11. Hao, X, Song, J., Pei, T.: Application of FCM and neural network in power load data correction. Transducer Microsyst. Technol. (2020)
12. Guan Shu-huai, F., Shen Yan-xia, S.: Short-Term power load forecasting based on fuzzy C-means combined neural network. Transducer Microsyst. Technol. **40**(5), 128–131 (2021)
13. Zhang Lin, F., Lai Xiang-ping, S., Zhong Shu-yong, T.: Electricity load forecasting method based on orthogonal wavelet and long short-term memory. Neural Netw. **39**(01), 72–79 (2022). https://doi.org/10.19725/j.cnki.1007-2322.2021.0070
14. Chen Liang, F., Wang Zhen, S., Wang Gang, T.: Application of LSTM networks in short-term power load forecasting under the framework deep learning framework. Electric Power Inf. Commun. Technol. **15**(5), 8–11 (2017)
15. Wang Gan-jun, F., Xu Yan-hui, S.: Research on accuracy assessment method for load forecast. Guangdong Electric Power **25**(11), 39–42 (2013)
16. Zhang, J., Yu, H, Zhou, Z.: Design and implementation of smart contract micro-service architecture for load aggregator. Autom. Electric Power Syst. 1–14 (2022)

Intrusion Detection Method Based on Minkowski Distance Negative Selection

Yang Lei[(✉)] and Wenxuan Bie

Cryptology Engineering Institute, Engineering University of People's Armed Police Force, Xi'an 710086, China
surina526@163.com

Abstract. Negative selection algorithm is considered as the dominant and important method to realize the intrusion detection because of its excellent adaptive protection mechanism. However, the traditional algorithm does not match the characteristics of the "self" sequence, which often causes the failure to detect the "non-self" pattern sequence, resulting in the "black hole" phenomenon. Therefore, aiming at the defects of too many "black holes" in the negative selection algorithm, this paper proposes an intrusion detection method based on Minkowski distance negative selection. The technique first calculates the Minkowski distance between the detector sequences, and measures the same number of bits between the detector and the self-set. The trained new detectors are input into the mature detector set for integration. Simulation results show that compared with the traditional negative selection algorithm, this method greatly reduces the black holes, the false alarm rate, and improves the detection efficiency.

Keywords: Intrusion detection · Negative selection algorithm · Minkowski distance · Sequential detector

1 Introduction

The computer technology has greatly changed the mode of human production and life, but the following computer security problems have also deeply troubled the majority of users [1]. In recent years, the economic losses caused by computer crimes to various industries all over the world are increasing year by year. In this context, computer information security has become a hot topic in related fields, and has produced a series of technical models and methods [2–5]. Many models based on the artificial immune system have been born, such as clonal selection [6], danger theory [7], negative selection algorithm (NSA) [8–12] and immune network [13], and have been widely employed in the field of computer intrusion detection. Among them, negative selection algorithm is considered to be the dominant and important method to solve intrusion detection problems because of its excellent adaptive protection mechanism.

The negative selection algorithm [14] proposed by Forrest puts forward a new concept and conceptual model for the self-maintenance of computer network system from the perspective of biological autoimmunity. In the whole NSA system, the detector

N. Xiong et al. (Eds.): ICNC-FSKD 2022, LNDECT 153, pp. 1056–1063, 2023.
https://doi.org/10.1007/978-3-031-20738-9_115

generation and training mechanism has always been the key to the performance of the algorithm. Therefore, in recent years, scholars mainly focus on the improvement of detector generation mechanism in classical NSA. NSA successfully simulates the maturation mechanism of T lymphocytes to effectively screen the detector, and determines the illegality of the antigen by checking whether the detector matches the antigen, so as to achieve the purpose of identifying abnormal or invasive information. In addition, the advantage that negative selection algorithm does not need prior knowledge makes it quickly become a kind of algorithm model widely used in intrusion detection, fault diagnosis, computer security and other research fields. On this basis, various improved negative selection algorithms can be developed to adapt to different application environments or solve some practical defects hidden in different traditional algorithms.

In view of this, aiming at the defects of a large number of black holes in the traditional negative selection algorithm, this paper presents a negative selection intrusion detection method based on Minkowski distance. The algorithm calculates the Minkowski distance between the detector sequences and measures the same number of bits between the detector and the self-set. Then the trained new detectors are input into the mature detector set for integration, so as to effectively decrease the number of black holes. Finally, the performance of the proposed method and the traditional NSA intrusion detection model is compared and analyzed through simulation experiments.

2 Negative Selection Algorithm

Negative selection algorithm is an auto/non-auto recognition technique in the immune system. Its mechanism is based on the maturation process of T cells in the thymus. T cells will be eliminated once they recognize auto elements. In the biological immune system, T lymphocytes (T cells for short) produced by bone marrow play a role in coordinating the functions of various parts of the immune system when resisting foreign attacks. T cells develop and differentiate into two types of T cells in the thymus, another central immune organ. One type of unqualified T cells is eliminated due to autoimmunity, while the other type of mature T cells learns to resist a specific invasion. Therefore, the process of selecting mature T cells that can face the invasion from all immature T cells is the negative selection mechanism.

The immune system model proposed by Forrest et al. is very similar to the intrusion detection system model in computer networks. In essence, the model boils down the security problems in computer networks to self and non-self, and intrusion detection seeks a classification mechanism in these two forms to separate normal behavior from intrusion behavior as much as possible to settle the problem of "false alarm" and "false alarm" widely existing in traditional intrusion detection methods. The specific algorithm of traditional negative selection is described as follows [15].

Step 1: Define the self-set. A single binary bit string is set as a set of substrings with equal growth. The set of strings not included in S becomes a non-self N. S and N are both from the space U, that is, $S \cup N = U, S \cap N = \varphi$.
Step 2: Generate a detector. A set of valid detection elements R is generated, in which each element cannot match any element in the self-set. If any detector matches a self pattern, the pattern cannot be a detector and needs to be removed.

Step 3: Monitor the occurrence of abnormal persons. Monitor with the generated valid detection elements. Since all detection elements matching the self pattern are discarded. If the detected pattern matches a valid detection element, the detection element will be activated, which indicates that an abnormal event has occurred.

Step 4: Identify whether there are exceptions by matching rules. In the matching rule of consecutive r bits, it is necessary to see whether the two strings match according to the number of consecutive matching bits. When the number of consecutive matching digits is not less than the R value, they match. If not, they don't match.

3   Intrusion Detection Model Based on Negative Selection

The generation algorithm of anomaly detector based on negative selection is as follows. (1) Define a sequence set M with length l, which is the self-set. (2) Randomly generate a sequence a whose length is also l. (3) The sequence a is matched with the sequence in the self-set M according to the matching rules. (4) If a matches a sequence in M, a will not be the detector, and it will be removed immediately. The step jumps to (3). (5) If a does not match the sequence in M, a becomes a detector mode and is stored in a mature detector (trained to become a mature detector).

In the negative selection intrusion detection model, the black hole phenomenon is often caused by the mismatched self sequence of the negative selection algorithm. However, the black hole cannot detect the non-self pattern sequence, which leads to the phenomenon of false alarm. Therefore, the intrusion detection system should decrease the number of black holes. Define the non-self set as N based on the self-set M. A black hole can be represented as a sequence b, $b \in N$ that exists in the non-self set N and matches a sequence in the self-set M. Intrusion detection system should distinguish self from non-self as far as possible. The more similar the self and non-self in a detection model, the more black holes in the detection system. Therefore, this negative selection algorithm will not only produce too many false positives, but also lead to the increase of time complexity, and ultimately affect the detection efficiency.

4   Intrusion Detection Method Based on Minkowski Distance Negative Selection

Minkowski distance is the most commonly used distance formula between real number vectors and can also calculate the distance between strings. It is a general expression of Manhattan, Euclidean and other distance formulas. The core idea of the negative selection algorithm based on Minkowski distance is to optimize the coverage effect of the detector by adding the parameter r to the string based matching rule, so as to decrease the number of black holes generated in the traditional negative selection algorithm and improve the detection efficiency. The process of the algorithm includes five steps. (1) Randomly generate a detector sequence d with length l. (2) The sequence d is matched with the sequence in the self-set M according to the matching rules. (3) If d matches a sequence in M, it will not become a detector, i.e. delete d immediately, and the step jumps to (2). (4) If d does not match the sequence in M, d is a new detector and is stored in detector set **M**, that is, $R \cup \{d\} \to R$. (5) When **R** changes to the threshold value preset, it ends, otherwise skip to step (1).

The affinity calculation is regarded as the matching rule. In the immune system model, the similarity between antibody and antigen is described, that is, the Minkowski distance between strings is calculated. In the string matching rules-based algorithm, the detection performance of each system is compared by testing its sequence matching efficiency and detection failure rate. Here, it is defined that the sequence matching efficiency EM and the detection failure rate EF are expressed by Eqs. (1) and (2) respectively.

$$E_m = \sum_{i=r}^{1} C_l^i \cdot \left(\frac{1}{2}\right)^i \cdot \left(\frac{1}{2}\right)^{l-i} = \frac{1}{2^l}\left(\sum_{i=r}^{1} C_l^i\right) \tag{1}$$

$$E_f = (1 - E_m)^{N_R} \tag{2}$$

In Eqs. (1)–(2), E_m and E_f are the sequence matching efficiency and detection failure rate respectively.

The following gives the detailed steps of the Minkowski distance based negative selection algorithm.

Step 1: **M** is trained to be an effective detector set, and the threshold R is set, $C = l-r$, and $m = r-1$.
Step 2: A detector d is generate, and the uncertain sequence bits of d is $i = r-1$.
Step 3: If $l > N_\varphi$, d is added to **M** until the number of **M** equals to R;
Step 4: If the Minkowski distance between two detection sequences is $l \le N_\varphi$, the detector d and string are calculated φ_r the same consecutive digits j at the corresponding position. If $j = r$, delete the detector d and jump to Step 2. If $j = r-1$, use all the uncertain sequence bits of the detector d and φ_r inverse phase of R shall be replaced, and the parameter m shall be set to 0 and turned to Step 3. If $j < r-1$ and $j + m \le r-1$, the number of bits i of the determined sequence of the detector d remains unchanged and turns to Step 3. If $j < r-1$ and $j + m > r-1$, the detector d and φ_r randomly generates a new detector sequence with a balance factor $C = 1-(r-1-j)$ such that $t = d$.
Step 5: The sequence matching efficiency E_m and the detection failure rate E_f can be obtained. The updated calculation formulas are expressed by Eqs. (3)–(4), respectively.

$$E_m = \sum_{i=r-t}^{1-t} C_{l-t}^i \cdot \left(\frac{1}{2}\right)^i \cdot \left(\frac{1}{2}\right)^{l-i} \tag{3}$$

$$E_f = (1 - E_m)^{N_R} \tag{4}$$

A good detection algorithm needs to decrease black holes and consider the performance consumption of detection processing. Considering these two aspects comprehensively, we can find a compromise and effective detection algorithm. The algorithm proposed in this paper absorbs the advantages of string based matching rules to increase the parameter r value. When the parameter r value increases, it can enhance the filtering times, so that more and more random detectors will be added to **M**, which greatly enhances the coverage and detection ability of the effective detector set m for non-self sets, and finally can effectively reduce the number and area of black holes. Compared with the traditional negative selection algorithm, the non-self set coverage and detection ability of the effective detector set **M** of this algorithm are greatly improved compared with the traditional selection algorithm, which is bound to decrease the black holes number and improve the detection efficiency.

5 Intrusion Detection Experiment and Analysis

Based on the mentioned above, for the parameters including same detection sequence d, the detection sequence length l, r, C and R are selected to obtain different sequence matching efficiency E_m and the detection failure rate E_f, that is, the above parameters can influence the final performance. However, the detection sequence length l and the parameter r in the negative selection algorithm model are very important. Therefore, different sequence length l and the parameter r can be selected for comprehensive evaluation during the simulation experiments, so as to further obtain the actual performance evaluation results.

5.1 Algorithm Testing and Analysis

Comparison experiment of sequence matching efficiency E_m of two algorithms under different sequence length l and the parameter r.

Select 20 groups of different (l, r) parameter values, and the experimental results of sequence matching efficiency E_m obtained according to NSA and the proposed Minkowski distance negative selection algorithm (MNSA) are shown in Fig. 1.

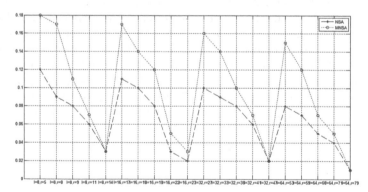

Fig. 1. Comparison of sequence matching efficiency E_M results of the two algorithms

It can be easily found in Fig. 1 that.

(a) If l remains constant, the sequence matching efficiency of NSA and MNSA decreases with the gradual increase of the parameter r, because the value of R involves the filtering times. When R becomes larger, the filtering times will increase exponentially, so the sequence matching efficiency E_m will also decrease with it.

When the R value is small, E_m of MNSA is significantly better than that of the classical NSA. However, When R becomes larger, that is, the filtering times, E_m of MNSA will eventually be close to classical NSA, for example as shown in the pairs ($l = 8, r = 14$).

(b) When l becomes larger, the length of detection sequence also increases. Therefore, the value of detection efficiency E_m will also decrease. However, no matter how the pair (l, r) is taken, the performance of MNSA is better.

Comparative experiment on the number of black holes of two algorithms under different sequence length l and the parameter r.

Select 20 groups of different (l, r) parameter values, and the experimental results of the number of black holes obtained according to NSA and MNSA are shown in Fig. 2.

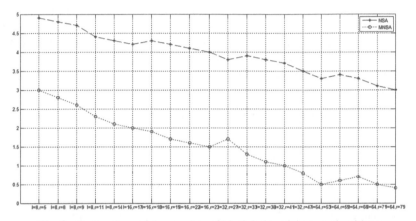

Fig. 2. Comparison of the number of black holes of the two algorithms

It can be easily found in Fig. 2 that.

(a) When the value of l remains unchanged and the value of R gradually increases, the filtering times of the two algorithms are enhanced. Thus, many random detectors are added to **M**, which increases the detection ability of the effective detector set **M** for non-self sets, resulting in the gradual reduction of the number and area of black holes, thus forming a downward trend on black holes of NSA and MNSA in Fig. 2.

(b) When the pair (l, r) are increasing, although the sequence matching efficiency will be reduced, the filtering times of MNSA algorithm as a whole have been increased, and the conditions for random detectors to enter the effective detector set **M** are becoming more and more strict, which will inevitably lead to the improvement of the non-self set coverage and detection ability of the effective detector set **M**, and the black holes number will inevitably decrease.

(c) NSA only filters once. Therefore, although the black holes number obtained by NSA also decreases with the increase of the parameters l and r, the downward trend is slow compared with MNSA.

(d) The sharp decrease in the number of black holes means that many non-self set spaces are detected, which indicates that the non-self set false negative rate in the detection process shows an obvious downward trend, reflecting the advantage of MNSA algorithm proposed in this paper in reducing the false negative rate.

(e) Although the number of black holes in the figure shows a downward trend as a whole, there are still some fluctuations when some parameters are taken, which is caused by the random distribution of self set and non-self set distribution.

5.2 Performance Experiment and Analysis of Intrusion Detection Methods

This section is based on Lincoln Laboratory (http://www.ll.mit.edu). The performance of three intrusion detection models, the traditional negative selection algorithm, the Hamming distance based NSA and MNSA, are verified and compared. In addition, there are two attack modes of intrusion, namely DoS attack and probe attack. The implementation environment of the three algorithms is tested under Windows 10 (Professional), the platform is visual studio 2019, the CPU is Intel i7-12700h 3.0 GHz, and the memory is 128 GB. The performance index results of the three intrusion detection models obtained from the experiment are reported in Table 1.

Table 1. Performance comparison of three methods

	Number of detected attacks	False alarms	Missing alarm rate (%)	Detection rate (%)
Traditional NSA	66	16	21.43	60.12
NSA based on hamming distance	74	13	7.65	79.89
Proposed	81	11	7.71	90.21

As observed in Table 1, the intrusion detection model of NSA is the highest on the missing alarm rate index, and the index values of the latter two models are very close. However, in the detection rate index, the detection rate index value of MNSA is the best, and the value reaches 90.21%, which is far better than NSA and NSA based on Hamming distance. Therefore, it is not difficult to see that the performance of the intrusion detection system model based on the proposed algorithm is the best, which improves the first two models to a great extent, and has good experimental results and feasibility.

6 Conclusions

In this paper, the problem that NSA is prone to produce black holes leading to false positives is studied, and the improved negative selection algorithm is effectively applied to intrusion detection system. By calculating the Minkowski distance between the detector sequences, the algorithm measures the continuous same bits between the detector and the self set string, so as to improve the coverage of the detector and reduce the number of black holes. Simulation experiments are carried out to compare and analyze the sequence matching efficiency, the number of black holes, the false alarm rate and detection rate of the traditional NSA and the method proposed in this paper. The experimental results show that the proposed method greatly reduces the black holes number, the false alarm rate, and improves the detection efficiency.

References

1. Yang, J.H., Woolbright, D.: Correlating TCP/IP packet contexts to detect stepping-stone intrusion. Comput. Secur. **30**(6–7), 538–546 (2011)
2. Yang, J.H., Huang, S.H.S.: Mining TCP/IP packets to detect stepping-stone intrusion. Comput. Secur. **26**(7–8), 479–484 (2007)
3. Wang, Y., Kim, I., Mbateng, G., et al.: A latent class modeling approach to detect network intrusion. Comput. Commun. **30**(1), 93–100 (2006)
4. Pan, Q.Q., Wu, J., Bashir, A.K., et al.: Joint protection of energy security and information privacy for energy harvesting: an incentive federated learning approach. IEEE Trans. Industr. Inf. **18**(5), 3473–3483 (2022)
5. Xie, N., Chen, J.J., Chen, Y.C., et al.: Detection of information hiding at anti-copying 2D barcodes. IEEE Trans. Circuits Syst. Video Technol. **32**(1), 437–450 (2022)
6. Burnet, F.: The clonal selection theory of acquired immunity. Vanderbilt University Press, Nashville (1959)
7. Matzinger, P.: The danger model: a renewed sense of self. Science **296**(5566), 301–305 (2002)
8. Forrest, S., Perelson, A.S., Allen, L. et al.: Self_nonself discrimination in a computer. In: Proceedings of IEEE Symposium on Research in Security and Privacy, pp. 202–212. IEEE Press, Oakland, CA (1994)
9. Chai, Z.Y., Wang, X.R., Wang, L.: Real valued negative selection algorithm for anomaly detection. J. Jilin Univ. (Eng. Ed.) **42**(1), 176–181 (2012)
10. Chai, Z.Y., Wang, X.R., Wang, L.: Real-value negative selection algorithm for anomaly detection. J. Jilin Univ. (Eng. Technol. Ed.) **42**(1), 176–181 (2012)
11. Chen, W., Li, T., Liu, X.J., et al.: A negative selection algorithm based on hierarchical clustering of self. Sci. China (Inf. Sci.) **43**(5), 611–625 (2013)
12. Wang, D.W., Zhang, F.B.: Real-valued negative selection algorithm with boundary detectors. Comput. Sci. **36**(8), 79–81 (2009)
13. Zhang, F.B., Wang, T.B.: Negative selection algorithm for real valued n-dimensional chaotic mapping. Comput. Res. Develop. **50**(7), 1387–1398 (2013)
14. Zhang, X.M., Yi, Z.X., Song, J.S., et al.: Research on negative selection algorithm based on matrix form. J. Electron. Inf. **32**(11), 2701–2706 (2010)
15. Artificial immune system. http://www.dca.fee.unicamp.br/~lnunes/immune.html

Machine Learning and Data Science (19)

Multi-scale Prototypical Network
for Few-shot Anomaly Detection

Jingkai Wu[1], Weijie Jiang[1], Zhiyong Huang[2], Qifeng Lin[1], Qinghai Zheng[1], Yi Liang[3], and Yuanlong Yu[1(✉)]

[1] College of Computer and Data Science, Fuzhou University, Fuzhou, China
`2003200800fzu.edu.cn`, `n1803100005fzu.edu.cn`, `linqf0fzu.edu.cn`,
`zhengqinghai0fzu.edu.cn`, `yu.yuanlong0fzu.edu.cn`
[2] Intelligent Robot Research Center, Zhejiang Lab, Hangzhou, China
`huangzy0zhejianglab.com`
[3] FuJian YiRong Information Technology Co.Ltd, Fuzhou, China
`liangyi0sgitg.sgcc.com.cn`

Abstract. Few-shot anomaly detection needs to solve the problem of anomaly detection when the training samples are scarce. The previous anomaly detection methods showed incompatibility in the case of lack of samples, and the current few-shot anomaly detection methods were not satisfactory. So, we hope to solve this problem from the perspective of few-shot learning. We utilize a pre-trained model for feature extraction and construct multiple sub-prototype networks in multi-scale features to compute anomaly maps corresponding to each scale. The final anomaly map can be used for anomaly detection. Our method does not need to be trained for each category and can be plug-and-play when there are a small number of normal class samples as the support set. Experiments show that our method achieves excellent performance on MNIST, CIFAR10, and MVTecAD datasets.

Keywords: Few-shot · Pre-trained model · Multi-scale features

1 Introduction

Anomaly detection aims to identify data that deviates significantly from other observations to raise suspicion that it is produced by a different mechanism. It has been an indispensable part of computer vision. As we all know, humans can judge whether an image is abnormal and even further locate the abnormality after learning just a few samples. However, because data collection is often expensive, not all tasks can provide an extensive collection of training data. This research on few-shot anomaly detection can make the anomaly detection method based on deep learning closer to the actual application scene and more intelligent. It can also solve the existing problems such as the shortage of training samples, especially labeled samples, and the uneven distribution of training samples. Moreover, following the previous anomaly location task settings, we also need to locate the abnormal areas of high-resolution images in the case of a few training samples. Therefore, this task remains challenging.

Not long ago, Shelly et al. [1] proposed HTDGM when they posed the problem of few-shot anomaly detection. HTDGM is a hierarchical generative model that captures the data distribution of each training image while the discriminator is asked to distinguish between normal or abnormal samples. However, there are many deficiencies, such as complex structure, difficulties in training, and poor adaptability to the number of training samples. Copying the previous anomaly detection method directly cannot adapt well to the few-shot settings. In addition, since the anomaly detection task usually contains many subtasks. Previous works need to train a single model for each class to accomplish the task. In other words, the more categories in the task, the more models and the more training time we need. We hope to design a method that can detect all categories of anomalies, this method does not require a lot of training data and time, but can achieve impressive results.

Prototypical network [2] believes that there is a class center for each class, and can be obtained by calculating the mean value of the feature vector of each class in support set. The class of the test sample is the same as the class of the closest class prototype. By this premise, we use a pre-trained model as the feature extractor for all categories. Unlike original prototype networks, considering that the information in the middle and bottom layers is also essential in anomaly detection, especially defect detection, our prototype network consists of a couple of sub-prototype networks constructed at each feature layer. The corresponding anomaly map of the sub-network is obtained via metric function, and these anomaly maps are added in a particular proportion to get the final anomaly map. This anomaly map can be directly used for anomaly detection and localization.

We make the following contributions in this paper. We use only one model to complete all categories of anomaly detection and even localization tasks well, even without training process and massive data. Moreover, we propose a new prototypical network construction method with the help of multi-scale features to make it suitable for anomaly detection tasks, which significantly improves the representation ability of normal class prototypes. The experimental results on three datasets show that our method significantly improves the performance of few-shot anomaly detection and has apparent advantages in deployment and efficiency.

2   Related Work

2.1   Anomaly Detection

Deep learning methods have long replaced the traditional methods such as frequency domain analysis and have become the mainstream in image anomaly detection. Reconstruction-based methods use the discrepancy of reconstruction ability between normal and abnormal images (reconstruction error) to detect anomalies. Autoencoder based method and GAN based method are the most commonly used during training. The main idea of the classification-based method is to convert single class normal samples into multi-class samples to train the classifier. In this way, the classification surface is constructed in the image space to classify the normal samples and potential abnormal samples. Ruff et al. [3] proposed a deep learning variant of SVDD, DeepSVDD. PatchSVDD [4] extended DeepSVDD to a patch-wise detection method. GEOM [5] applies different transformations to normal images and trains a classifier to classify the transformation applied. GOAD [6] modifies the anomaly score used by GEOM. DROCC [7] trains a classifier to classify the original training samples and the samples resulting from adversarial training. There are also some other neat methods. Salehi et al. [8] proposed multi-resolution distillation to detect anomalies by using the difference in the representation ability of the teacher network and the student network on the test image.

2.2 Few-shot Learning

Deep learning methods are relatively mature with large amounts of training data, but there is still room for further research when training data is insufficient. It is one of the most critical gaps between machines and humans. Few-shot learning aims to solve this problem. Some works adopt augmentation, an intuitive way, to enhance data representation. The meta-learning approach represented by MAML [9] learns a new task according to experience. They mainly study many similar tasks to obtain a good model initialization. When facing new tasks with only a few samples, they can adapt quickly by fine-tuning. Metric learning compares the similarity between samples through learned measurement methods. Sung et al. [10] propose Relation Network to learn and calculate the similarity between unlabeled and labeled samples. Few-shot learning work is generally applied to classification tasks and cannot be applied to anomaly detection directly.

3   Method

Our approach contains two steps. First, we compute the normal prototype, which contains several sub-prototypes by multi-scale features. Then we calculate the anomaly score and anomaly map according to each sub-prototype network.

3.1   Construct Prototype Based on Multi-scale Features

The training dataset D_{train} only consists of normal samples in an anomaly detection task. Few-shot learning uses limited supervision in tasks. In this case, we randomly select a subset of D_{train} and denote it as $D'_{train} = \{x_1, \cdots, x_k\}$. At the same time, the testing dataset D_{test} remains the same (Fig. 1).

Multi-scale features of support set are the raw material for constructing a prototype. The ability of the feature extractor plays a crucial role in prototype construction. The most intuitive approach is self-supervised training from scratch based on training samples. However, in few-shot learning setting, a small number

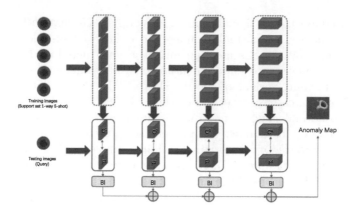

Fig. 1. An overview of our method. The multi-scale features of input can be obtained by the intermediate layer output of the pre-trained feature extractor. Each sub-prototype C^l is produced by the corresponding layer's output. F^l stands for the output of the corresponding layer of the test image. Bilinear interpolation (BI) is used to calculate results.

of training samples usually cannot train a powerful feature extractor to our satisfaction for the subsequent similarity metric calculation. Therefore, we use the Resnet18 [11] pre-trained on ImageNet. This is a very successful pre-trained model with high-level semantic features suitable for the original classification task and very mature mid-level and low-level semantic features that will be useful for our anomaly detection task. Correspondingly, we also innovated prototype construction, making it composed of multiple sub-prototypes. We choose the output of the four blocks of Resnet18.

As mentioned above, the feature extractor will output a series of feature maps in different layers for an input image. We can compute each sub-prototype via the output of the corresponding feature layer on the support set:

$$C^l = \frac{\sum_{i=1}^{k} F^l (x_i)}{k} \qquad (1)$$

where $F^l (x_i) \in R^{h_l \times w_l \times c_l}$ is the feature map in the l-th layer, h_l is the height, w_l is the width, and c_l is the number of channels in that layer. The prototype of a normal class is composed of these sub-prototypes together. That is to say, the prototype network we use for anomaly detection is comprised of various sub-prototypical networks. Following the viewpoint of solving the anomaly detection task from a one-classification perspective, we can also view this problem as a one-way few-shot problem.

3.2 Compute Anomaly Map via Prototype Network

This phase aims to compute the anomaly map for anomaly localization and detection. Same as Sect. 3.1, the multi-scale features of a test sample $X_t \in$

$R^{h \times w \times l}$ will be obtained after forwarding it into the feature extractor. In each sub-prototypical network, we follow the idea of STPM [12] to construct a metric function at position (i, j):

$$M_{ij}^l (x_t) = d \left(\hat{C}_{ij}^l, \hat{F}^l (x_t)_{ij} \right) = \frac{1}{2} \left\| \hat{F}^l (x_t)_{ij} - \hat{C}_{ij}^l \right\|_2^2 \qquad (2)$$

where $C_{ij}^l \in R^{d_l}$ and $F^l (x_t)_{ij} \in R^{d_l}$ are feature vectors at position (i, j), $\hat{C}_{ij}^l$ and $\hat{F}^l (x_t)_{ij}$ are their L2-normalization results.

In each prototypical network, we can get an anomaly map at that scale. To make them form the final anomaly map, we bilinear interpolate them separately so that their sizes match the original image. The final anomaly map(denoted as $M(x_t)$) is the weighted average of the map at different feature scales:

$$M (x_t) = \frac{\sum_{l=1}^{L} \alpha_L \cdot BI \left(M^l (x_t) \right)}{\sum_{l=1}^{L} \alpha_l} \qquad (3)$$

where α_l depicts the importance of the l-th feature scale, A total of L layer features are selected, and the choice of weights will also directly affect the quality of prototype construction. The effect of different layers will be fully verified in Sect. 4.3. BI stands for Bilinear Interpolation Operation. In defect localization, this anomaly image has given pixel-level defect localization. Considering that noise may affect the judgment of defects, we introduce Gaussian filtering to smooth the anomaly map.

3.3   Scoring Function for Anomaly Detection

Finally, we need to give the test image an image-level anomaly score to measure the degree of anomaly. The higher the score, the higher the probability that the image is anomalous. Generally speaking, anomaly refers to the semantic anomaly, a distribution that deviates entirely from the normal distribution. So we use the mean of the anomaly map as anomaly score(denoted as s) to measure the anomaly of the whole image:

$$s = \frac{1}{h \times w} \sum_{i=1}^{h} \sum_{j=1}^{w} M (X_t)_{ij} \qquad (4)$$

Defect detection is a particular case of anomaly detection, and anomalous samples should not be discriminated as anomalies in terms of overall semantics. Only a small part of the region is abnormal when an abnormality occurs, and other regions are normal. So we choose the maximum value instead of the mean as the anomaly score for defect detection.

4   Experiments

We conduct substantial experiments in this section to confirm the validity of our work. We use the best feature layer selection for comparison in the comparative experiments of anomaly detection and defect detection. In the end, we conduct a further discussion on the choice of feature layers in ablation experiments. We evaluate our method on three datasets: CIFAR10 [13], MNIST [14], and MVTecAD [15].

4.1   Anomaly Detection

We perform anomaly detection on CIFAR10 and MNIST. Both cifar10 and mnist contain 50,000 training samples and 10,000 training samples in 10 categories, where the samples of cifar10 are 32*32 color images and the samples of mnist are 28*28 grayscale images. In each scenario, one class is selected as the normal class, and the other classes are used as the abnormal class. The samples in the training set are used as the support set for few-shot learning. Following previous anomaly detection work, AUROC is adopted to measure overall performance.

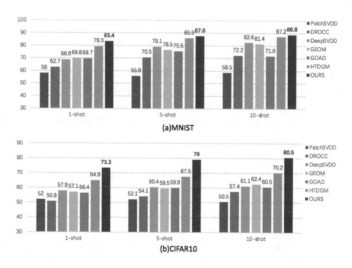

Fig. 2. Average AUROC in % for anomaly detection on (a) MNIST and (b) CIFAR10 for 1-shot,5-shot, and 10-shot settings.

We demonstrate the validity of our method in comparison with a range of methods. HTDGM [1] is most relevant to our mission. At the same time, excellent methods for anomaly detection such as PatchSVDD [4], DROCC [7], DeepSVDD [3], GEOM [5] and GOAD [6] are also used to test the adaptability in few-shot settings. As shown in Fig. 2, these methods offer different adaptability, none of them are inferior to HTDGM, and our method outperforms the above methods in 1-shot, 5-shot, and 10-shot settings on both CIFAR10 and MNIST datasets.

4.2   Defect Detection

Anomaly detection has higher requirements in many applications, such as industrial defect detection. Bergmann et al. [15] proposed MVTecAD, a dataset that contains 5354 samples in 15 categories. Specifically, there are 3639 normal samples for training and 1725 samples for testing. This dataset provides not only image-level labels for test samples, but also pixel-level labels for abnormal regions. In this task, image-level AUROC is used to measure the discriminative performance, while pixel-level AUROC is used to measure the performance of defect localization.

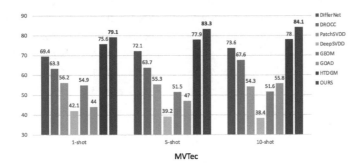

Fig. 3. Average AUROC in % for defect detection on MVTecAD, for 1-shot, 5-shot, and 10-shot settings.

In addition to the comparison methods mentioned in Sect. 4.1, we also compared with DifferNet [16] proposed for the defect detection task. In Fig. 3, we can see that our method outperforms all methods in all cases. In particular, compared to HTDGM, the average AUROC value of our method is 3.5%, 5.4% , and 6.1% higher in the 1-shot, 5-shot, and 10-shot settings, respectively.

In defect detection, the more important goal is to give pixel-level predictions of abnormal regions, which is also relatively demanding, especially in the case of insufficient training samples, so there is currently no good way to locate defects in few-shot settings. However, in the case of sufficient samples, many researchers have carried out intensive exploration, and methods such as AE_{SSIM} [17], AnoGAN [18], VAE-grad [19] and MKD [8] have achieved good results. Table 1 shows that our method in 5-shot setting already slightly (0.9%) outperforms the MKD method with sufficient samples, and the advantage is more pronounced (1.7%) in the 10-shot setting, indicating that our method has little dependence on training samples for defect localization.

4.3   Ablation Study

Our normal class prototype consists of multiple sub-prototypes constructed from intermediate features, and this section explores the impact of the choice of feature layers. We fixed the experiments in this section under the 5-shot setting. Next, we discuss anomaly detection and defect detection cases, respectively.

Anomaly detection experiments are performed on CIFAR10. We experimented with various combinations of feature layers, including single and multi-layer. The results in Table 2 show that high-level features play a more significant

Table 1. Pixel-level AUROC in % for defect localization on MVTecAD.

Method	AE$_{SSIM}$	AnoGAN	VAE-grad	MKD	OURS1	OURS2	OURS3
Bottle	93	86	92.2	96.3	96.9	97.4	97.4
Hazelnut	97	87	97.6	94.6	90.6	93.9	94.1
Capsule	94	84	91.7	95.9	94.5	95.3	95.4
Meta_lNut	89	76	90.7	86.4	67.7	71.9	74.2
Leather	78	64	92.5	98.1	98.8	98.9	98.9
Pill	91	87	93	89.6	84.2	87.8	88.5
Wood	73	62	83.8	84.8	91.5	91.6	92.1
Carpet	87	54	73.5	95.6	98.7	98.9	98.9
Tile	59	50	65.4	82.8	90.3	91.4	92.0
Grid	94	58	96.1	91.8	79	84.4	88.2
Cable	82	78	91	82.4	87.7	90.6	91.2
Transistor	90	80	91.9	76.4	88.9	94.5	94.6
Toothbrush	92	90	98.5	96.1	97.7	98.2	98.2
Screw	96	80	94.5	95.9	77.6	83.8	86.3
Zipper	88	78	86.9	93.9	91.7	95.2	96.1
Mean	87	74	89.3	90.7	83.12	91.6	92.4

OURS1, OURS2, and OURS3 correspond to the results of our method in the 1-shot, 5-shot, and 10-shot settings, respectively. Other baselines are in the case of sufficient training samples

Table 2. Ablation studies for feature selection on CIFAR10.

	1	2	3	4	[1,2]	[2,3]	[3,4]	[1,2,3]	[2,3,4]	[1,2,3,4]
AR$_I$	61.8	63.8	78.1	78.1	65.9	73.4	**79.0**	75.0	74.8	78.2

The performance is measured by the average image-level AUROC (AR_I). The best results are highlighted in bold-face

role in anomaly detection, which we believe is because anomaly detection needs to detect deviations from categories and semantics. Multi-layer features can provide more robust representation capabilities than single-layer features. But it is not necessary to select too many feature layers. For example, the first two layers are not helpful for the anomaly detection task. We choose the output of the highest two blocks to provide the best results.

We then test the effect of feature layer selection on defect detection on MVTecAD dataset. Similarly, we try various options for the selection of feature layers. From the results in Table 3, we can find that the features of the middle and low layers play a more important role in defect detection, both at the image level and the pixel level. Selecting multiple layers of features can also lead to better results. Our analysis is because defects only generally appear in small regions of the image, and most of them are low-level semantic information

Table 3. Ablation studies for feature selection on MVTecAD.

	1	2	3	4	[1,2]	[2,3]	[3,4]	[1,2,3]	[2,3,4]	[1,2,3,4]
AR_I	73.9	81.0	82.1	77.9	81.1	83.1	81.1	**83.3**	81.9	82.0
AR_P	84.7	89.9	90.8	82.1	90.1	91.5	86.2	**91.6**	87.3	87.2

The performance is measured by the average image-level AUROC(AR_I) and pixel-level (AR_P)). The best results are highlighted in bold-face

such as texture and scratches. Lower layer features are also more suitable for anomaly localization. The features of the first three blocks are the best choice.

5 Conclusions

We propose a prototype network containing sub-prototype networks for few-shot anomaly detection. The pre-trained ResNet18 is used as the feature extractor, and the sub-prototype networks are constructed separately using the multi-scale features. Each sub-prototype network can get an anomaly map, and they jointly contribute to the final abnormal map. The whole process uses only one model and requires no training, which is very efficient. We validate our method on CIFAR10, MNIST and MVTecAD with excellent performance in anomaly detection and localization even with small samples. Our innovative transformation of the prototype network is also beneficial for other works on few-shot learning.

Acknowledgements. This work was supported by National Natural Science Foundation of China (NSFC) under grant 61873067, University-Industry Cooperation Project of Fujian Provincial Department of Science and Technology under grant 2020H6101, NSF Project of Zhejiang Province No. LQ22F030023, and Zhejiang Postdoctoral Project No. 2021NB3UB15.

References

1. Sheynin, S., Benaim, S., Wolf, L.: A hierarchical transformation-discriminating generative model for few shot anomaly detection. In: Proceedings of the IEEE/CVF International Conference on Computer Vision, pp. 8495–8504 (2021)
2. Snell, J., Swersky, K., Zemel, R.: Prototypical networks for few-shot learning. Adv. Neural Inf. Process. Syst., **30** (2017)
3. Ruff, L., Vandermeulen, R., Goernitz, N., Deecke, L., Siddiqui, S.A., Binder, A., Müller, E., Kloft, M.: Deep one-class classification. In: International Conference on Machine Learning, pp. 4393–4402. PMLR (2018)
4. Yi, J., Yoon, S.: Patch svdd: Patch-level svdd for anomaly detection and segmentation. In: Proceedings of the Asian Conference on Computer Vision (2020)
5. Golan, I., El-Yaniv, R.: Deep anomaly detection using geometric transformations. Adv. Neural Inf. Process. Syst., **31** (2018)
6. Bergman, L., Hoshen, Y.: Classification-based anomaly detection for general data. arXiv preprint arXiv:2005.02359 (2020)

7. Goyal, S., Raghunathan, A., Jain, M., Simhadri, H.V., Jain, P.: Drocc: Deep robust one-class classification. In: International Conference on Machine Learning, pp. 3711–3721. PMLR (2020)
8. Salehi, M., Sadjadi, N., Baselizadeh, S., Rohban, M.H., Rabiee, H.R.: Multiresolution knowledge distillation for anomaly detection. In: Proceedings of the IEEE/CVF Conference on Computer Vision and Pattern Recognition, pp. 14,902–14,912 (2021)
9. Finn, C., Abbeel, P., Levine, S.: Model-agnostic meta-learning for fast adaptation of deep networks. In: International Conference on Machine Learning, pp. 1126–1135. PMLR (2017)
10. Sung, F., Yang, Y., Zhang, L., Xiang, T., Torr, P.H., Hospedales, T.M.: Learning to compare: relation network for few-shot learning. In: Proceedings of the IEEE Conference on Computer Vision and Pattern Recognition, pp. 1199–1208 (2018)
11. He, K., Zhang, X., Ren, S., Sun, J.: Deep residual learning for image recognition. In: Proceedings of the IEEE Conference on Computer Vision and Pattern Recognition, pp. 770–778 (2016)
12. Wang, G., Han, S., Ding, E., Huang, D.: Student-teacher feature pyramid matching for anomaly detection. arXiv preprint arXiv:2103.04257 (2021)
13. Krizhevsky, A., Hinton, G., et al.: Learning multiple layers of features from tiny images (2009)
14. LeCun, Y.: The mnist database of handwritten digits. http://yannlecun.com/exdb/mnist/ (1998)
15. Bergmann, P., Fauser, M., Sattlegger, D., Steger, C.: Mvtec ad–a comprehensive real-world dataset for unsupervised anomaly detection. In: Proceedings of the IEEE/CVF Conference on Computer Vision and Pattern Recognition, pp. 9592–9600 (2019)
16. Rudolph, M., Wandt, B., Rosenhahn, B.: Same same but differnet: Semi-supervised defect detection with normalizing flows. In: Proceedings of the IEEE/CVF Winter Conference on Applications of Computer Vision, pp. 1907–1916 (2021)
17. Bergmann, P., Löwe, S., Fauser, M., Sattlegger, D., Steger, C.: Improving unsupervised defect segmentation by applying structural similarity to autoencoders. arXiv preprint arXiv:1807.02011 (2018)
18. Schlegl, T., Seeböck, P., Waldstein, S.M., Schmidt-Erfurth, U., Langs, G.: Unsupervised anomaly detection with generative adversarial networks to guide marker discovery. In: International Conference on Information Processing in Medical Imaging, pp. 146–157. Springer (2017)
19. Dehaene, D., Frigo, O., Combrexelle, S., Eline, P.: Iterative energy-based projection on a normal data manifold for anomaly localization. arXiv preprint arXiv:2002.03734 (2020)

Distributed Nash Equilibrium Seeking with Preserved Network Connectivity

Qingyue Wu[✉]

Beijing SunWise Space Technology Ltd., No. 16, South Third Street, Haidian District, Beijing, China
wuqycast@163.com

Abstract. This paper studies constrained aggregative games over multi-agent systems with network connectivity preservation, where each player's communication capability has a limited range. To solve this problem, a distributed continuous-time algorithm is designed to minimize each player's payoff function and search for the generalized Nash equilibrium. Based on the bounded derivatives and sign function, an improved average tracking dynamic is designed to estimate the average of all players' strategies. In addition, the network connectivity is maintained under the proposed algorithm if the communication network at the initial time is connected. Finally, numerical examples validates the performance of proposed algorithm in heating ventilation air conditioning (HVAC) systems.

Keywords: Aggregative games · Connectivity-preserving · Nash equilibrium seeking · Multiagent systems

1 Introduction

Recently, network connectivity has become a research hot. Formation control in multi-agent systems [1] and distributed optimization [2] are typical examples in urgent need of connectivity preservation. In order to ensure that agents efficiently complete tasks through information exchange, it is important to form and maintain the network connectivity between agents. However, most of agents exchange information through wireless networks, which leads to the limited communication capability. If distances between agents are too large, information may be delayed or lost, even interrupted, resulting in failure to complete tasks.

Aggregative games accommodate competitive cases where multiple players adjust strategies to maximize profits. For example, each user adjusts its strategy to minimize the cost of electricity consumption in the electricity market in [3]. Multiple companies maximize profits and achieve Nash Equilibrium by adjusting production and prices in Nash-Cournot game in [4]. In addition, [5] designs an algorithm over time-varying networks and [6] extends this work to coupled constraints.

© The Author(s), under exclusive license to Springer Nature Switzerland AG 2023
N. Xiong et al. (Eds.): ICNC-FSKD 2022, LNDECT 153, pp. 1077–1086, 2023.
https://doi.org/10.1007/978-3-031-20738-9_117

Although remarkable works in distributed Nash equilibrium seeking are available [3–6], there are still some interesting problems to be studied, such as pre-served network connectivity. In above works, the communication networks are time-dependent. In practical applications, the communication network is affected by many factors such as network attacks and communication range limitation. This kind of state-dependent network is considered in distributed optimization problem [2], and consensus problems [7]. The optimal resource allocation problem over state-dependent communication network is considered in [2]. Based on the penalty method and Lagrangian multiplier method, a potential-function is designed to solve the coupling of supply-demand constraint. Also [7] considers the fixed-time average tracking problem where all existing communication links are preserved afterwards. However, there are few results on the distributed Nash equilibrium seeking under the state-dependent communication networks.

In this paper, we aim to solve the Nash equilibrium seeking problem under a state-dependent communication network. The main contributions are high-lighted as follows:

1. A novel distributed algorithm is proposed to solve aggregative games. Specifically, the algorithm incorporates with two parts: a gradient play dynamic for Nash equilibrium seeking and the remaining part to estimate global aggregate strategies.
2. By designing a class of discontinuous function, the network connectivity preservation is achieved. Furthermore, numerical examples show that our algorithm is effective and applicable to energy consumption in HVAC systems.

In the rest of this paper, Sect. 2 introduces preliminaries and formulates aggregative games. The algorithm design and convergence analysis in Sect. 3. Section 4 provides simulation results and Sect. 5 gives conclusions.

Notations: Denote R and R^m as the set of real numbers and m-dimensional real column vectors. $\mathbf{1_n}$ and $\mathbf{0_n}$ stand for the n-dimensional column vectors where all the elements are 1 and 0, respectively. Also $||\cdot||$ denotes the Euclidean norm. $\text{sgn}(x) = \begin{cases} x/||x||_2, ||x||_2 \neq 0 \\ 0, ||x||_2 = 0 \end{cases}$ is the signum function. Define $x^{[p]} = ||x||^p \text{sgn}(x)$ and ∇f is the gradient of function f.

2   Preliminaries and Problem Formulation

In this section, we give introduce the graph theory. Then problem formulation and some common assumptions are introduced.

2.1   Graph Theory

Define $\mathcal{G}(\mathcal{V}, \mathcal{E})$ as an undirected graph with the nodes set $\mathcal{V} = \{1, 2, ..., n\}$ and the edges set $\mathcal{E} \subset \mathcal{V} \times \mathcal{V}$. $A = [a_{ij}]$ is the adjacency matrix with non-negative elements. In addition, $a_{ij} = 1$ if $(i, j) \in \mathcal{E}$ for all $i, j \in \mathcal{V}$, otherwise $a_{ij} = 0$. if there exists a path between any two agents of an undirected graph $\mathcal{G}$, $\mathcal{G}$ is connected. Define L as the Laplacian matrix of the graph $\mathcal{G}$ with $L = L^T \geq 0$. The eigenvalues of matrix L are listed from small to large as $0 < \lambda_2 \leq ... \leq \lambda_n$.

Lemma 1 ([8]). *For an undirected graph $\mathcal{G}$, the Laplacian matrix L satisfies the following properties.*

1. $\lambda_2 = \min_{||x||\neq 0, 1_n^T x = 0}(x^T Lx/||x||^2)$;
2. $x^T Lx \geq \lambda_2 x^T x$ *if* $\sum_{i=1}^n x_i = 0$;
3. *For any* $x = [x_1, \ldots, x_n]^T$, $x^T Lx = (1/2)\sum_{i=1}^n \sum_{j=1}^n a_{ij}(x_j - x_i)^2$.

2.2   Fixed-Time Convergence

Consider the following dynamic system

$$\dot{x}(t) = f(t, x(t)), \ x(0) = x_0, \ t \geq 0 \tag{1}$$

where $x(t)$ is a state vector and f is a nonlinear function. In addition, assume the equilibrium point of system (1) is the origin.

Lemma 2 ([9]). *For a continuous radially unbounded function V,*

1. $V(x(t)) = 0$ *iff* $x(t) = 0$;
2. *if there exist some constants $\alpha, \beta > 0$, $p \in (0,1)$ and $q > 1$ such that the solution of (1) satisfies $V(x(t)) \leq -\alpha V^p(x(t)) - \beta V^q(x(t))$*

the origin is globally fixed-time T stable, where $T \leq 1/(\alpha(1-p)) + 1/(\beta(q-1))$.

2.3   Problem Formulation

We focus on solving aggregative games under the multi-agent system framework. In aggregative games, the player $i \in [N] = \{1, 2, ..., n\}$ has a cost function $J_i(x_i, \bar{x})$ which is only known by oneself and player i aims to take its feasible strategy x_i from local strategy set $\Omega_i \in R^m$ to minimize its cost function:

$$\min \ J_i(x_i, \bar{x}) \quad \text{s.t.} \quad x_i \in \Omega_i \tag{2}$$

where $J_i : R^m \to R$ is the cost function of player i and the global aggregate strategy is $\bar{x} = (1/n)\sum_{i=1}^n x_i \in \bar{\Omega}$ with $\bar{\Omega} = \{\sum_{i=1}^n x_i/n | x_i \in \Omega_i\} \in R^m$. There is no loss of generality in assuming $m = 1$. Furthermore, $x = [x_1, ..., x_n]^T \in \Omega$ represent the strategy profile of game (2), where $\Omega = \prod_{i=1}^n \Omega_i$. The vector of strategy $x^* = [x_1^*, ..., x_n^*]^T$ is a Nash Equilibrium of game (2) if for any $i \in [N]$,

$$J_i(x_i^*, \bar{x}^*) \leq J_i(x_i, \frac{1}{n}x_i + \frac{1}{n}\sum_{j\neq i}^n x_j^*), \ x_i \in \Omega_i \tag{3}$$

2.4   Some Assumptions and Lemmas

Next some assumptions are provided.

Assumption 1 *For any player $i \in [N]$, the strategy set Ω_i is closed, nonempty and convex. Each cost function $J_i(x_i, y)$ is continuously differentiable with respect to (x_i, y) over the set $\Omega_i \times \bar{\Omega}$, where $y = [y_1, \cdots, y_n]^T$. In addition, the function $J_i(x_i, \bar{x})$ is convex for $x_i \in \Omega_i$.*

Under Assumption 1, according to [10], it is easy to get the Nash Equilibrium x^* of game (1) by solving the variational inequality $(x - x^*)^T h(x^*) \geq 0$, $x^* \in \Omega$, where $h(x)$ is the pseudo-gradient map defined as $h(x) = [\nabla_{x_1} J_1(x_1, \bar{x}), ..., \nabla_{x_n} J_n(x_n, \bar{x})]^T$. In order to explicitly show the coordinate of the pseudo-gradient, define a map g as $g(x, y)[g_1(x_1, y_1), ..., g_n(x_n, y_n)]^T$, where $g_i(x_i, y_i) = \nabla_{x_i} J_i(x_i, y_i) + \nabla_{y_i} J_i(x_i, y_i)/n$. It is obvious that $h(x) = g(x, 1_n \bar{x})$.

Assumption 2 *The map $h(x)$ is μ-strongly monotone over Ω such that for all $x, \hat{x} \in \Omega$, there exists a constant $\mu > 0$ satisfying $(h(x) - h(\hat{x}))^T (x - \hat{x}) \geq \mu \|x - \hat{x}\|_2^2$*

Assumption 3 *The map $g(x, y)$ is l_1-Lipschitz continuous on $\Omega \times \bar{\Omega}$ such that for all $x, \hat{x} \in \Omega$ and $y, \hat{y} \in \bar{\Omega}$, $\|g(x, y) - g(\hat{x}, \hat{y})\|_2 \leq l_1(\|x - \hat{x}\|_2 + \|y - \hat{y}\|_2)$.*

Furthermore, we introduce some supporting lemmas which will be used in the subsequent analysis.

Lemma 3 ([10]). *Under Assumptions 1-3, there is a unique Nash equilibrium x^* of game (1).*

Lemma 4. *Let $P_\Omega(x) = \arg\min_{w \in \Omega} \|x - w\|$ denote as the projection of $x \in R^m$ on a closed convex set $\Omega \in R^m$. For any $x, w \in R^n$, $\|P_\Omega(x) - P_\Omega(w)\|_2 \leq \|x - w\|_2$ holds.*

Considering the communication range limitation, our goal is to design a distributed Nash equilibrium seeking algorithm s over state-dependent communication networks.

3   Main Results

A distributed Nash equilibrium dynamic with network preserving connectivity is designed in this section and then analyse its convergence.

3.1   Algorithm Design

If the state of player i and j satisfy $\|y_i(t) - y_j(t)\| < R$ at time t, then they can send information with each other. Assume that all players have the same sensing range R. The distributed control protocol for player $i \in [N]$ as follows:

$$\begin{aligned}
\dot{x}_i &= P_{\Omega_i}(-\nabla g_i(x_i, y_i)) \\
\dot{y}_i &= \theta_i + \dot{x}_i \\
\theta_i &= \sum_{j=1}^{n} \alpha(\|y_i - y_j\|)(\beta \text{sgn}(y_j - y_i) + (y_j - y_i)^{[\gamma]})
\end{aligned} \tag{4}$$

where $\gamma > 1$ and $\beta > 0$ is control gain to be given later. In addition, the initial states are $x_i(0) \in \Omega_i$ and $y_i(0) = x_i(0)$. The function $\alpha(.)$ is

$$\alpha(\|y_i - y_j\|) = \begin{cases} \alpha_1(\|y_i - y_j\|), \|y_i - y_j\| < R \\ \alpha_2(\|y_i - y_j\|), \quad otherwise \end{cases} \tag{5}$$

where

$$\alpha_1(||y_i - y_j||) = \begin{cases} \frac{R}{(R-||y_i-y_j||)^2}, & ||y_i - y_j|| < R \\ 0, & otherwise \end{cases} \tag{6}$$

$$\alpha_2(||y_i - y_j||) = \begin{cases} 1, & ||y_i - y_j|| < R \\ 0, & otherwise \end{cases} \tag{7}$$

The right hand side of (4) is discontinuous, where the solution is defined in the Filippov sense.

Lemma 5 ([7]). *For an undirected and connected network $\mathcal{G}$ at $t = 0$ with $||y_i(0) - y_j(0)|| < R$, if $||y_i(t) - y_j(t)|| < R$, then the followings hold*

1. *$L_1(t) - L_2(t)$ is symmetric positive semi-definite;*
2. *$\lambda_2(L_1(t)) \geq \lambda_2(L_2(t))$*

where $L_1(t)$ and $L_2(t)$ are the Laplacian matrix of the communication network at time $t \neq 0$ and $t = 0$, respectively.

3.2 Convergence Analysis

The following lemma shows the graph is always connected over time under (4).

Lemma 6. *For the distributed control protocol (4), $\mathcal{G}(t)$ is undirected and connected for all $t \geq 0$ if the undirected graph $\mathcal{G}(0)$ is connected.*

Proof. Select the following Lyapunov function

$$V_1(y(t)) = \frac{1}{2} \sum_{i=1}^{n} \sum_{j=1}^{n} \int_0^{||y_i-y_j||} \alpha(s)(\beta + s^\mu) ds$$

and its derivative along (9) is

$$\dot{V}_1 = \frac{1}{2} \sum_{i=1}^{n} \sum_{j \in N_{i(t)}} \alpha_{ij}(t)(\beta \mathrm{sgn}(y_j(t) - y_i(t)) + (y_j(t) - y_i(t))^{[\gamma]})^T (\dot{y}_i - \dot{y}_j)$$

$$= \sum_{i=1}^{n} \sum_{j \in N_{i(t)}} (\beta \mathrm{sgn}(y_j(t) - y_i(t)) + (y_j(t) - y_i(t))^{[\gamma]})^T \dot{y}_i$$

$$= \sum_{i=1}^{n} -(\theta_i(t) + \dot{x}_i)^T \theta_i(t)$$

Next we prove this lemma holds in virtue of the contradiction method. First, suppose that t_1 is the first time such that $\lim_{t \to t_1^-} ||y_i(t) - y_j(t)|| = R$ for $(i, j) \in \mathcal{E}(0)$. Denote $I_R = \{i| \lim_{t \to t_1^-} ||y_i(t) - y_j(t)|| = R, (i, j) \in \mathcal{E}(0)\}$. There exist $i_0 \in I_R$ and $w_0 \in \{1, \cdots, n\}$ satisfying $y_{i_0 w_0}(t_1) \geq y_{j w_0}(t_1)$. According to (5), we obtain

$$\alpha(\|y_i(0) - y_j(0)\|) = \frac{R}{(R - \|y_{i0} - y_j\|)^2} \to \infty$$

which implies $\lim_{t \to t_1^-} \theta_i(t) \to +\infty$. Since $\dot{x}_i(t)$ is bounded, then it is obvious to get that $\lim_{t \to t_1^-} \dot{V}_1(t) \to -\infty$. On the other hand, since $\lim_{t \to t_1^-} R/(R - \|y_{i0} - y_j\|)^2 \to +\infty$, we have $\lim_{t \to t_1^-} V_1(t) \to +\infty$, which contradicts with $\lim_{t \to t_1^-} \dot{V}_1(t) \to -\infty$. Thus the initial interaction network is preserved.

Since Ω_i is a closed convex set and the map g is l_1-Lipschitz continuous, $\dot{x}_i(t)$ is bounded. Define the upper bound $\eta = \sup_{i \in [N]} \{\dot{x}_i(t)\}$. The following lemma shows $y_i \to \bar{x}$ in fixed-time, where $\bar{x} = (1/n) \sum_{i=1}^n x_i$.

Lemma 7. *Suppose that Assumptions 1–3 hold and the undirected graph $\mathcal{G}(0)$ is connected. If $\beta = \tilde{\beta} + \eta n^{1/2}/\lambda_2^{1/2}$ with $\tilde{\beta} > 0$, then $y_i \to \bar{x}$ under algorithm (4) in fixed-time $T = 2/(\tilde{\beta}\lambda_2(L_2)^{1/2}) + 2/(2^\gamma(\gamma - 1)n^{1-\gamma}\lambda_2(L_2)^{(1+\gamma)/2})$.*

Proof. Firstly, we show that $y_i = \cdots = y_j$ in fixed time. Define $\tilde{y}_i(t) = y_i(t) - (1/n) \sum_{i=1}^n y_i$ and $\tilde{y} = [\tilde{y}_1^T, \cdots, \tilde{y}_n^T]^T$. According to (9), we have

$$\dot{\tilde{y}}_i(t) = \theta_i(t) + \dot{x}_i(t) - \sum_{j=1}^n \dot{x}_j(t)/n$$

Select a Lyapunov function $V_2(\tilde{y}(t)) = \frac{1}{2} \sum_{i=1}^n \tilde{y}_i^T(t)\tilde{y}_i(t)$. Noting that $\sum_{i=1}^n \tilde{y}_i(t) = 0_n$ and $\sum_{i=1}^n \tilde{y}_i(t) \sum_{i=1}^n \dot{x}_j(t) = 0$, then the derivative of V_2 is

$$\dot{V}_2(\tilde{y}(t))$$

$$= \sum_{i=1}^n \tilde{y}_i^T(\theta_i(t) + \dot{x}_i(t))$$

$$= \frac{1}{2} \sum_{i=1}^n \sum_{j=1}^n (\tilde{y}_i - \tilde{y}_j)^T (\alpha(\|y_i - y_j\|)(\beta \operatorname{sgn}(y_j(t) - y_i(t))$$

$$+ (y_j(t) - y_i(t))^{[\gamma]})) + \sum_{i=1}^n \tilde{y}_i^T \dot{x}_i(t)$$

$$= \sum_{i=1}^N \tilde{y}_i^T \dot{x}_i(t) - \frac{\beta}{2} \sum_{i=1}^n \sum_{j=1}^n (\alpha(\|y_i - y_j\|)\|\tilde{y}_i - \tilde{y}_j\|$$

$$- \frac{1}{2} \sum_{i=1}^n \sum_{j=1}^n (\alpha(\|y_i - y_j\|)\|\tilde{y}_i - \tilde{y}_j\|^{[1+\gamma]}$$

$$\leq \eta n^{\frac{1}{2}} (\sum_{i=1}^n \|\tilde{y}_i\|^2)^{\frac{1}{2}} - \frac{\beta}{2} (\sum_{i=1}^n \sum_{j=1}^n (\alpha(\|y_i - y_j\|)^2 \|\tilde{y}_i - \tilde{y}_j\|^2)^{\frac{1}{2}}$$

$$- \frac{1}{2} n^{1-\gamma} (\sum_{i=1}^n \sum_{j=1}^n (\alpha(\|y_i - y_j\|)^{\frac{2}{1+\gamma}} \|\tilde{y}_i - \tilde{y}_j\|^2)^{\frac{1+\gamma}{2}} \tag{8}$$

where $(\sum_{i=1}^{n} u_i^l)^{1/l} \leq (\sum_{i=1}^{n} u_i^l)^{1/r} \leq n^{(1/r-r/l)}(\sum_{i=1}^{n} u_i^l)^{1/l}$ in [11] with $l > r \geq 1$ is used. Based on Lemma 1, it is easy to get

$$\sum_{i=1}^{n}\sum_{j=1}^{n}(\alpha(||y_i - y_j||)^2||\tilde{y}_i - \tilde{y}_j||^2 \geq 4\lambda_2(L_0)V_2$$

$$\sum_{i=1}^{n}\sum_{j=1}^{n}(\alpha(||y_i - y_j||)^{2/(1+\gamma)}||\tilde{y}_i - \tilde{y}_j||^2 \geq 4\lambda_2(L_\gamma)V_2 \tag{9}$$

Based on Lemma 5 and (9), (8) is converte'd to

$$\dot{V}_2 \leq \eta n^{\frac{1}{2}}V_2^{\frac{1}{2}} - \beta\lambda_2^{\frac{1}{2}}(L_2)V_2^{\frac{1}{2}} - 2^\gamma n^{1-\gamma}\lambda_2^{\frac{1+\gamma}{2}}(L_2)V_2^{\frac{1+\gamma}{2}}$$
$$\leq -\tilde{\beta}\lambda_2^{\frac{1}{2}}(L_2)V_2^{\frac{1}{2}} - 2^\gamma n^{1-\gamma}\lambda_2^{\frac{1+\gamma}{2}}(L_2)V_2^{\frac{1+\gamma}{2}} \tag{10}$$

Integrating on both sides of (10) from t to $(t + \tau)$ yields

$$\int_t^{t+\tau} \dot{V}_2(s)ds$$
$$\leq \tau(-\tilde{\beta}\lambda_2^{\frac{1}{2}}(L_2) \min_{s\in[t,t+\tau]} V_2(s)^{\frac{1}{2}} - 2^\gamma n^{1-\gamma}\lambda_2^{\frac{1+\gamma}{2}}(L_2) \min_{s\in[t,t+\tau]} V_2(s)^{\frac{1+\gamma}{2}})$$

According to Lemma 2, it is obvious that variable y_i achieves consensus after T.

Next, we show that $y_i \to \bar{x}$. Recalling algorithm (4) and $y_i(0) = x_i(0)$, we get $\sum_{i=1}^{n}\dot{x}_i = \sum_{i=1}^{n}\dot{y}_i$, which implies $\sum_{i=1}^{n}x_i(t) = \sum_{i=1}^{n}y_i(t)$ for all $t \geq 0$. Thus $y_i \to \bar{x}$ after T.

Theorem 1. *Suppose that Assumptions 1–3 hold and the undirected graph $\mathcal{G}(0)$ is connected. The trajectory x along algorithm (4) asymptotically converges to the Nash equilibrium x^*.*

Proof. Select a Lyapunov function $V(t) = \frac{1}{2}||x - x^*||^2$ and its derivative is

$$\dot{V} = (x - x^*)^T P_\Omega(g(x,y))$$
$$\leq -(x - x^*)^T(g(x,y) - g(x^*, 1_n\bar{x}^*)))$$
$$\leq -(x - x^*)^T(g(x, 1_n\bar{x}) - g(x^*, 1_n\bar{x}^*))$$
$$\quad -(x - x^*)^T(g(x,y) - g(x, 1_n\bar{x}))$$
$$\leq -\mu||x - x^*||^2 - l_1(x - x^*)^T||y - 1_n\bar{x}|| \tag{11}$$

According to Lemma 7, the variable y_i achieve consensus as $y \to 1_n\bar{x}$ for $t \geq T$. Since $\dot{x}(t)$ and $\dot{y}(t)$ are bounded, which implies $x(t)$ and $y(t)$ are bounded in finite time T. Define $e(t) = ||x - x^*|| \, ||y - 1_n\bar{x}||$ and integrates on both sides of $e(t)$ from T to t gives rise to

$$\int_T^t ||e(s)||ds \leq \tilde{e} \int_T^t ||\bar{y}(s) - 1_n \bar{x}(s)||ds < +\infty \tag{12}$$

where $\tilde{e} = \sup ||x(t) - x^*||$. Combining with (12), (11) is converted to

$$\dot{V} \leq -\mu ||x - x^*||^2 + l_1 e(t) \tag{13}$$

Integrating on both sides of (13) from T to ∞ is

$$0 \leq \int_T^{+\infty} \mu ||x(s) - x^*||^2 ds \leq V(0) - \lim_{t \to +\infty} V(t) + \lim_{t \to +\infty} \int_T^t ||a(\tau)||d\tau < +\infty \tag{14}$$

which implies $x(t) \to x^*$ asymptotically. The proof is completed.

4   Numerical Examples

This section considers energy consumption in HVAC systems with $n = 5$ electricity users in [3]. The cost function of electricity user $i \in [N]$ is

$$J_i(x_i, \bar{x}) = (x_i - \zeta_i)^2 + p(\bar{x})x_i, \quad x_i \in [\tilde{x}_i, \hat{x}_i]$$

where the pricing function $p(\bar{x}) = 0.04 \times n\bar{x} + 5$. In addition, x_i denotes the actual energy consumption and ζ_i represents nominal value of energy consumption. Relevant parameters are consistent with those in [3].

Next, we validate the effectiveness of our algorithm. The initial undirected graph with 5 users shown in Fig. 1(a). Also the undirected graph at $t = 10s$ shown in Fig. 1(b), where dashed lines indicate the new communication links. The nominal values and constrained sets are shown in Table 1. The initial state $x_i(0) = \zeta_i$, $y_i(0) = \zeta_i$, $\beta = 100$ and the communication rang $R = 5$. x_i generated by algorithm (4) converges to the Nash equilibrium shown in Fig. 2, where dotted lines denote the optimal energy consumption $x^* = [41.5, 46.4, 51.3, 56.2, 61.1]^T$. Numerical examples above supports theoretical results.

Table 1. Parameters setting

	user 1	user 2	user 3	user 4	user 5
η_i	50	55	60	65	70
$\tilde{x}_i$	40	44	48	52	56
$\hat{x}_i$	60	66	72	78	84

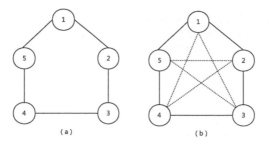

Fig. 1. (a) The initial undirected graph; (b)The undirected graph at $t = 10$

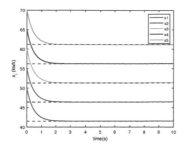

Fig. 2. Users' energy consumptions generated by algorithm (4)

5   Conclusion

This paper has investigated aggregative games with local constrained strategy sets in a distributed way over state-dependent communication networks. By designing auxiliary dynamics related to the communication range, the network connectivity is preserved over time. Using the Lyapunov stability theory, the proposed algorithm achieves fixed-time estimation and converges to Nash equilibrium. The validity of our results has been further illustrated by simulations.

References

1. Lin, Z., Wang, L., Han, Z., et al.: Distributed formation control of multi-agent systems using complex Laplacian. IEEE Trans. Autom. Control **59**(7), 1765–1777 (2014)
2. Wang, X., Yang, S., Guo, Z., et al.: A distributed dynamical system for optimal resource allocation over state-dependent networks. IEEE Trans. Netw. Sci. Eng. **9**(4), 2940–2951 (2022)
3. Ye, M., Hu, G.: Game design and analysis for price-based demand response: An aggregate game approach. IEEE Trans. Cybern. **47**(3), 720–730 (2016)
4. Zhu, R., Zhang, J., You, K., et al.: Asynchronous networked aggregative games. Automatica **136**, 110054 (2022)
5. Koshal, J., Nedic, A., Shanbhag, U.V.: Distributed algorithms for aggregative games on graphs. Oper. Res. **64**(3), 680–704 (2016)

6. Belgioioso, G., Nedic, A., Grammatico, S.: Distributed generalized Nash equilibrium seeking in aggregative games on time-varying networks. IEEE Trans. Autom. Control **66**(5), 2061–2075 (2020)
7. Hong, H., Yu, W., Yu, X., et al.: Fixed-time connectivity-preserving distributed average tracking for multiagent systems. IEEE Trans. Circuits Syst. II: Express Briefs **64**(10), 1192–1196 (2017)
8. Godsil, C., Royle, G.F.: Algebraic Graph Theory, 2nd edn. Springer, New York, USA (2001)
9. Polyakov, A.: Nonlinear feedback design for fixed-time stabilization of linear control systems. IEEE Trans. Autom. Control **57**(8), 2106–2110 (2011)
10. Facchinei, Pang, J.: Finite-dimensional variational inequalities and complementarity problems. 2nd edn. Springer, New York, USA (2003)
11. Goldberg, M.: Equivalence constants for l_p norms of matrices. Linear Multilinear Algebra **21**(2), 173–179 (1987)

Heterogeneous Network Representation Learning Guided by Community Information

Hanlin Sun[1,2,3(✉)], Shuiquan Yuan[1,2], Wei Jie[4], Zhongmin Wang[1,2,3], and Sugang Ma[5]

[1] School of Computer Science and Technology, Xi'an University of Posts and Telecommunications, Xi'an, China
{sunhanlin,zmwang}@xupt.edu.cn, 2003210125@stu.xupt.edu.cn
[2] Shaanxi Key Laboratory of Network Data Analysis and Intelligent Processing, University of Posts and Telecommunications, Xi'anXi'an, China
[3] Xi'an Key Laboratory of Big Data and Intelligent Computing, University of Posts and Telecommunications, Xi'an, China
[4] School of Computing and Engineering, University of West London, London, UK
wei.jie@uwl.ac.uk
[5] School of Electronic Engineering, Xi'an University of Posts and Telecommunications, Xi'an, China
msg@xupt.edu.cn

Abstract. Network representation learning usually aims to learn low-dimensional vector representations for nodes in a network. However, most existing methods often ignore community information of networks. Community structure is an important topology feature in complex networks. Nodes belonging to a community are more densely connected and tend to share more common attributes. Preserving community structure of network during network representation learning has positive effects on learning results. This paper proposes a community-enhanced heterogeneous network representation learning algorithm. It introduces the community information of a heterogeneous network into its node representation learning, so that the learned results can maintain both the properties of the micro-structure and the community structure. The experiment results show that our algorithm can greatly improve the quality of heterogeneous network representation learning.

Keywords: Heterogeneous network learning · Network representation learning · Community structure · Random walk

1 Introduction

In our daily life, many real world relationships can be represented as network structure data, where objects and their relationships are represented by nodes and edges, respectively. Examples include transportation networks, Social networks, citation networks, and so on. Most of these networks have multiple types of nodes and edges and are referred as heterogeneous networks. For example, in the scholar network Aminer, there are three types of nodes, *paper*, *author*, and *conference*, and two types of linking relationships,

N. Xiong et al. (Eds.): ICNC-FSKD 2022, LNDECT 153, pp. 1087–1094, 2023.
https://doi.org/10.1007/978-3-031-20738-9_118

writing (author-paper) and *publish* (paper-Conference). Heterogeneous networks have richer semantic and structural information than homogeneous networks, which contain only one type of node and edge, and are more in line with real-life environments.

Network representation learning (NRL), also known as graph embedding or network embedding, is an emerging network analysis method, especially for large-scale networks. Generally, The purpose of NRL is to learn real-valued, low-dimensional and dense vector representations for nodes in the a network. The learned vectors can be further applied in subsequent network analysis tasks, such as node classification [1, 2], link prediction [3, 4], recommendation system [5, 6], and so forth. However, representation learning of heterogeneous networks is harder than that of homogeneous networks, not only to consider the heterogeneity of nodes and edges but also to consider various potential semantic relationships.

One basic requirement of NRL is to preserve network structural properties in learned low-dimensional representations. Community structure, which is prevalent in many real-world networks, is the most obvious one. A community is a subgroup of nodes of a network that have denser connections among them, compared with the connections between them and the rest nodes. In addition, the member nodes of a community usually share similar properties. A heterogeneous network usually has different community structures under different semantemes. However, traditional heterogeneous network representation learning algorithms often ignore community characteristics of networks.

In this paper, we propose a heterogeneous network representation learning algorithm enhanced by community information, which is named as CHNRL (Community enhanced Heterogeneous Network Representation Learning). It incorporates community information of a heterogeneous network extracting under a specific semanteme into its node representation learning process and thus can maintain community properties in learned node representations. As a result, the quality of NRL can be improved.

2   Related Works

In this section, we review the works mostly related to ours in this paper.

2.1   Network Representation Learning

There exist a number of methods for network representation learning. For homogeneous networks, Deepwalk, Node2vec, and LINE are popular ones. Deepwalk [7] bases on the idea of Word2vec [8, 9] and captures properties of a network by random walk sequences. Node2vec [10] improves Deepwalk by introducing two walking control parameters, one for exploring local neighbors and the other for far neighbors. LINE [11] considers the first-order and the second-order similarities between neighbor nodes while learning embedding of nodes. For heterogeneous networks, metapath2vec [12] is similar as Deepwalk but collects random walks under a given meth-path, and thus learns representations for different type nodes. Esim [13] learns the embedded representation of nodes by sampling positive and negative meta-path instances from the network. HIN2vec [14] conducts multiple training tasks to learn representation of nodes in heterogeneous networks. HAN [15] first converts a heterogeneous network to several meta-path based homogeneous networks and then learns node embeddings on each network. Finally, it uses a graph attention network to aggregate embeddings of a node from different homogeneous network and obtains a total representation for the node. However, these methods ignore the influence of network community structure on representation learning.

2.2 Community Preserving in NRL

Realizing the impacts of community structure on NRL, several works started to consider incorporating community properties. M-NFM [16] combines two non-negative matrix factorization(NMF) models for node representation learning and community structure detection, and jointly finds a disjoint community structure for the network and learns its node representations that can retain community properties at the same time. CARE [17] provides a new way to integrate community features into node representation learning. It first uses Louvain ' s community discovery method to detect the community structure of the network and uses the community guidelines obtained for random walks, which is used for node representation learning as in Deepwalk. However, such algorithms are developed for homogeneous networks.

3 Methodology

In this section, we describe the details of our CHNRL. It consists of three steps: (1) extracting community structure information of a heterogeneous network under a given meta-path, (2) generating random walk sequences, and (3) learning node representations, as shown in Algorihtm1. The second step is the same as our early work [18, 19]. That is, Community information extracted from the network is used to guide the generation of node sequences. Here the community information refers to whether a node and its neighbor stay in at least one same community. While selecting the next walk for the current node, a priority is given to neighboring nodes within the same community. As a result, the collected walks can incorporate the community structure properties of the network. The third step is same as in DeepWalk, that uses the local information obtained by truncated random walk to learn potential representations by treating a walk as an equivalent of a sentence. Here we focus on the first step.

Meta-path plays an important role in heterogeneous network analysis. A meta-path is defined as a sequence of relations connecting two nodes in a heterogeneous network. And stands for a specific semanteme. In CHNRL, we extract the community structure information of a heterogeneous network under an appointed meta-path. Specifically, we first create a homogeneous network from a heterogeneous network by including the nodes with type at the end of the given meta path and connecting two such nodes if there is a meta-path instance between them. Then, we extract the community structure information of the created homogeneous network through a community detection algorithm. Currently, we use the OSLOM [20], which is reported to have good performance in overlapping community detection. By using the random walk model in [18, 19], the community structure property will be implicitly preserved in the collected walk sequences and finally in the learned node representations.

Algorithm 1: **CHNRL**

Input: heterogeneous network: G
meta-path: p
number of random walks per node: μ
random walk max length: L
walk priority: α
representation size: d
window length: k
Output:
node embeddings: $X \in R^{v \times |d|}$
1.   G' =ConvertNetwork(G, p)
2.   com_info = ExtractCommunityInfo(G')
3.   walk_list =RandomWalk_Com(G', com_info, μ, L, α)
4.   X = SkipGram(walk_list, d, k)

---

4   Experiments

In this section, we evaluate the performance of CHNRL. We compare our algorithm against Node2vec [10], LINE [11], Metapath2vec [12] and GCN [21] on multi-label classification and visualization task.

4.1   Datasets

We use two heterogeneous network datasets in our experiments, the AMiner Computer Science (CS) [22] and the Database and Information System (DBIS) [23]. There are three types of nodes, including *author*, *paper*, and *conference*, in them. Details are shown in Table 1.

Table 1. Statistics of datasets.

Datasets	Node#	Edge#	Label#
DBIS	Author, 60447 Paper,72902 Conference,463	P-A,192421 P-C,72902	8
Aminer	Author,1693531 Paper,3194405 Conference,3883	P-A,9323739 P-C,3194405	8

In our experiments, we use the networks converted from the two heterogeneous networks under the meta-path APA. That is two author nodes in the converted network are connected if there is at least one instance of meta-path APA between them. However, the topologies of the two converted networks are not consistent with their node labels.

Therefore, we first revise the labels of the converted networks, namely: (1) if a node has a label, but it has no edges with such nodes having the same label, the label is dropped by the node; and (2) if a node has no the label that its connecting nodes possess and the number of such nodes holding the label is at least equal to half of the number of connecting nodes, the label is added by the node.

On the two converted networks, OSLOM detects 4732 communities for DBIS and 84333 communities for Aminer.

4.2 Parameter Settings

In the experiment, we set the node embedding dimension of all algorithms to 128. For the random walk based algorithms, the walk sequence length is 100, the number of walks per node is 20, and the learning window size is 5. In Node2vec, the two parameters p and q are 0.25 and 4, respectively. The walking priority to sharing community neighbors of CHNRL is 0.9 by testing. The learning rate of GCN is 0.01.

4.3 Multi-label Classification

In multi-label classification experiments, node representations are learned by CHNRL or compared algorithms. LINE and Node2vec run on the original networks, but ignore the different types of nodes, i.e. treat heterogeneous networks as homogeneous networks. Metapath2vec collects random walks following the meta-path APA. GCN uses the same networks as in our CHNRL. Then, the representations are divided into two parts, training and testing. The training node representations and their associated labels are used to train a logistic regression classifier, and the testing node representations are fed to the classifier to predict their labels. The prediction results are measured by Macro-F1 and Micro-F1 scores. Tables 2 and 3 show the scores of the converted DBIS and AMiner networks, respectively. We set the ratio of training nodes as 50–90% of all nodes, increasing by 10%.

As can be seen from the two tables, Our CHNRL is significantly superior to other algorithms according to the two metrics. The mathpath2vec is the second best since it takes into account meta-path information of the two networks. As comparing math-path2vec against CHNRL, the latter further considers community structure information and thus has better performance.

4.4 Visualization

To intuitively show the performance of different algorithms, we also visualize the learned node representations of DBIS network. For clarity, we randomly select the nodes of nine communities found by OSLOM and project their 128-dimension representations to a 2-dimensional space using LargeVis [24]. The results are shown in Fig. 1, where different colors indicate different communities. It is clear that node2vec and LINE cannot preserve the community structure, while metapath2vec and CHNRL can. Intuitively, CHNRL is better than metapath2vec. The reason lies in that it obviously takes account of community structure information of networks.

Table 2. Experiment results on the DBIS datasets for the node classification task.

Metric	Method	50%	60%	70%	80%	90%
Macro-F1	LINE	0.1128	0.1131	0.1132	0.1132	0.1129
	Node2vec	0.1211	0.1229	0.1226	0.1237	0.1248
	Metapath2vec	0.2929	0.2982	0.3003	0.3029	0.3090
	GCN	0.1827	0.1840	0.1924	0.1992	0.2085
	CHNRL	**0.4806**	**0.4843**	**0.4847**	**0.4877**	**04885**
Micro-F1	LINE	0.8219	0.8217	0.8225	0.8252	0.8297
	Node2vec	0.8199	0.8227	0.8244	0.8261	0.8290
	Metapath2vec	0.8316	0.8311	0.8316	0.8340	0.8392
	GCN	0.8288	0.8285	0.8288	0.8307	0.8319
	CHNRL	**0.8695**	**0.8706**	**0.8736**	**0.8757**	**0.8796**

Table 3. Experiment results on the Aminer datasets for the node classification task.

Metric	Method	50%	60%	70%	80%	90%
Macro-F1	LINE	0.0878	0.0879	0.0880	0.0882	0.0889
	Node2vec	0.0991	0.0992	0.0993	0.0999	0.1001
	Metapath2vec	0.3300	0.3301	0.3307	0.3318	0.3327
	GCN	0.1820	0.1841	0.1879	0.1932	0.2062
	CHNRL	**0.5973**	**0.5971**	**0.5975**	**0.5977**	**0.5982**
Micro-F1	LINE	0.5417	0.5417	0.5419	0.5425	0.5427
	Node2vec	0.5428	0.5427	0.5428	0.5436	0.5438
	Metapath2vec	0.6011	0.6012	0.6015	0.6014	0.6019
	GCN	0.5645	0.5650	0.5660	0.5675	0.5711
	CHNRL	**0.7378**	**0.7378**	**0.7379**	**0.7381**	**0.7387**

It should be noted that Fig. 1 shows some communities of the DBIS author nodes with different colors, but not node labels. The detected communities are finer topology structures than label structures. Members of a community tend to have same labels, while a label may be held by a number of communities.

5   Conclusion

In this paper, we propose a new representation learning algorithm, CHNRL, for heterogeneous networks. It takes the community structure information under a specific meta-path semanteme of a heterogeneous network into consideration during its node representation learning, and thus can preserve network topology properties well in the learned low dimensional representations. Experiments of multi-label classification and visualization both show the topology preserving capability of CHNRL. In fact, a heterogeneous network usually has several different meta paths, that stand for different semanteme. We will explore the impacts of community structures under different meta-paths on heterogeneous network representation learning and the integration of such impacts in our future work.

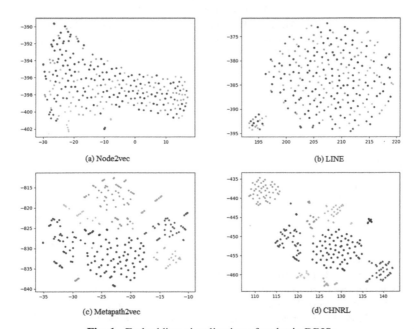

Fig. 1. Embedding visualization of nodes in DBIS.

Acknowledgement. This work is partially supported by the International Science and Technology Cooperation Program of the Science and Technology Department of Shaanxi Province, China (Grant No. 2019KW-008), the Natural Science Basic Research Program of Shaanxi Province, China (Grant No. 2022JM-342), and the Project Supported by Science and Technology Project in Shaanxi Province, China (Grant No. 2019ZDLGY07–08).

References

1. Bhagat, S., Cormode, G., Muthukrishnan, S.: Node classification in social networks. In: Social Network Data Analytics, pp. 115–148. Springer, Boston, MA (2011)
2. Koller, D., Friedman, N., Džeroski, S.: Introduction to statistical relational learning. MIT Press (2007)
3. Liben-Nowell, D., Kleinberg, J.: The link-prediction problem for social networks. J. Am. Soc. Inform. Sci. Technol. **58**(7), 1019–1031 (2007)

4. Yang, S.H., Long, B., Smola, A.: Like like alike: joint friendship and interest propagation in social networks. In: Proceedings of the 20th International Conference on World Wide Web, pp. 537–546 (2011)
5. Zhang, J., Shi, X., Zhao, S.: Star-gcn: Stacked and reconstructed graph convolutional networks for recommender systems. arXiv preprint arXiv:1905.13129 (2019)
6. Backstrom, L., Leskovec, J.: Supervised random walks: predicting and recommending links in social networks. In: Proceedings of the Fourth ACM International Conference on Web Search and Data Mining, pp. 635–644 (2011)
7. Perozzi, B., Al-Rfou, R., Skiena, S.: Deepwalk: Online learning of social representations. In: Proceedings of the 20th ACM SIGKDD International Conference on Knowledge Discovery and Data Mining, pp. 701–710 (2014)
8. Mikolov, T., Chen, K., Corrado, G.: Efficient estimation of word representations in vector space. arXiv preprint arXiv:1301.3781 (2013)
9. Mikolov, T., Sutskever, I., Chen, K.: Distributed representations of words and phrases and their compositionality. Adv. Neural Inf. Process. Syst., **26** (2013)
10. Grover, A., Leskovec, J.: node2vec: Scalable feature learning for networks. In: Proceedings of the 22nd ACM SIGKDD International Conference on Knowledge Discovery and Data Mining, pp. 855–864 (2016)
11. Tang, J., Qu, M., Wang, M.: Line: Large-scale information network embedding. In: Proceedings of the 24th International Conference on World Wide Web, pp. 1067–1077 (2015)
12. Dong, Y., Chawla, N. V., Swami, A.: metapath2vec: Scalable representation learning for heterogeneous networks. In: Proceedings of the 23rd ACM SIGKDD International Conference on Knowledge Discovery and Data Mining, pp. 135–144 (2017)
13. Shang, J., Qu, M., Liu, J.: Meta-path guided embedding for similarity search in large-scale heterogeneous information networks. arXiv preprint arXiv:1610.09769 (2016)
14. Fu, T., Lee, W. C., Lei, Z.: Hin2vec: Explore meta-paths in heterogeneous information networks for representation learning. In: Proceedings of the 2017 ACM on Conference on Information and Knowledge Management, pp. 1797–1806 (2017)
15. Wang, X., Ji, H., Shi, C.: Heterogeneous graph attention network. In: The world Wide Web Conference, pp. 2022–2032 (2019)
16. Wang, X., Cui, P., Wang, J.: Community preserving network embedding. In: Thirty-first AAAI Conference on Artificial Intelligence. pp. 203–209 (2017)
17. Keikha, M.M., Rahgozar, M., Asadpour, M.: Community aware random walk for network embedding. Knowl.-Based Syst. **148**, 47–54 (2018)
18. Sun, H., Jie, W., Wang, Z., Wang, H., Ma, S.: Network representation learning guided by partial community structure. IEEE Access **8**, 46665–46681 (2020)
19. Sun, H., Jie, W., Loo, J.: Network representation learning enhanced by partial community information that is found using game theory. Information **12**(5), 186 (2021)
20. Lancichinetti, A., Radicchi, F., Ramasco, J.J.: Finding statistically significant communities in networks. PLoS ONE **6**(4), e18961 (2011)
21. Kipf, T.N., Welling, M.: Semi-supervised classification with graph convolutional networks. arXiv preprint arXiv:1609.02907 (2016)
22. Tang, J., Zhang, J., Yao, L.: Arnetminer: extraction and mining of academic social networks. In: Proceedings of the 14th ACM SIGKDD International Conference on Knowledge Discovery and Data Mining, pp. 990–998 (2008)
23. Sun, Y., Han, J., Yan, X.: Pathsim: Meta path-based top-k similarity search in heterogeneous information networks. Proc. VLDB Endowment **4**(11), 992–1003 (2011)
24. Tang, J., Liu, J., Zhang, M.: Visualizing large-scale and high-dimensional data. In: Proceedings of the 25th international conference on world wide web, pp. 287–297 (2016)

Prediction and Allocation of Water Resources in Lanzhou

Lei Tang$^{(\boxtimes)}$, Simeng Lin, Zhenheng Wang, and Kanglin Liu

School of Traffic and Transportation, Beijing Jiaotong University, Beijing 100044, China
18271077@bjtu.edu.cn, 18251145@bjtu.edu.cn, 18251155@bjtu.edu.cn,
klliu@bjtu.edu.cn

Abstract. Due to the scarcity as well as the unbalance between supply and demand of water resources, it is of great importance to improve the accuracy of water demand prediction. Lanzhou, a typical northwestern city in China, is facing acute issues on the sustainable development of water resources. In this paper, we forecast the water demand in Lanzhou based on back propagation neural network, and then propose an optimization model to better allocate the limited water resources based on the prediction results. Comprehensive numerical experiments validate the feasibility of the proposed model, which shed light on several managerial insights associated with the sustainable water planning strategies in Lanzhou.

Keywords: Prediction · Bp neural · System dynamics · Water source

1 Introduction

Water is one of the most precious resources, but its supply is limited. According to WHO&UNICEF(2019) [1], 2.2 billion people worldwide do not have access to safe drinking water. Lanzhou, a typical northwestern city in China, has a varied topography, including mountains, hills, platforms, river valleys, basins, and etc. The per capita available water resources in Lanzhou is only 720 m^3, equivalent to only 33% of the national constituting per capita (2150 m^3). Making accurate water resources prediction can help the government to catch the current development situation, optimize the allocation plan, and make sustainable decisions. The majority of existing prediction papers use a single method for prediction. In this paper, we creatively propose a joint prediction scheme so that the adverse impacts caused by a single method can be reduced. Meanwhile, the scheme can also be adjust different areas (such as agriculture, industry, ecology and domestic), which has more flexibility.

Nowadays, back propagation (BP) neural network has become an emerging prediction method. The method is praised for the self-learning, nonlinear mapping, and robust predicting abilities. One challenge that affects the prediction accuracy is the selection of input properties. To overcome this difficulty, we first list all the factors that affect the water resources in Lanzhou by system dynamics, and then figure out important factors to improve prediction based on the principal component analysis (PCA). Meanwhile, according to the different characteristics of these important factors, Holt linear trend

N. Xiong et al. (Eds.): ICNC-FSKD 2022, LNDECT 153, pp. 1095–1104, 2023.
https://doi.org/10.1007/978-3-031-20738-9_119

method and exponential equations are employed to predict the changes of these factors in the next 15 years. Finally, panel and time series data are gathered as input features. The contributions of this paper can be summarized as follows.

- We develop a tailored BP neural network by selecting appropriate input features of Lanzhou, where system dynamic, PCA, Holt linear trend method and exponential equations are incorporated.
- An optimization model is proposed to improve the allocation strategies of limited water resources in Lanzhou based on the aforementioned predicted results.
- Several managerial insights are put forward and validates the feasibility of the proposed method. For example, improving sewage reuse rate and adjusting the current water resource policy will contribute significantly to the sustainable development of water resources in Lanzhou.

2   Literature Review

Since 1960s,s, scholars began to focus on the research of urban water demand prediction. Jain et al. [2] constructed an artificial neural network model for urban hourly water demand prediction. Xu [3] proposed a virtual water trade way to optimize water resource allocation.

Based on the specific national conditions of China, scholars have devoted much effort on the prediction and sustainable development of water resources. Pioneers in this area often employ empirical method, i.e., making rough estimations based on the experience of experts according to the historical data. In recent years, the empirical method is gradually replaced by more scientific and systematic demand prediction models, including univariate/multivariate regression analysis, grey system prediction, time series prediction and neural network [4]. System dynamics has been widely applied to water-related research due to its advantages in holistic understanding and strategic decision making of complex systems [5]. However, there is little research on water resources prediction for Chinese northwestern cities, where water shortage has become an acute issue, and our paper bridges this gap by investigating the specific properties of Lanzhou.

Although the current research on water resources has made some progress, most of the existing papers study the prediction and optimization of water resources in a separate way [6], the joint prediction-optimization scheme is relatively rare [7]. In the past decade, intelligent prediction models such as artificial neural network and fuzzy programming, have attracted much attention [8]. However, in the area of water prediction, existing papers only employ time-irrelevant features which might lead to inaccurate results. In this paper, both panel and time series data are involved to make long predictions.

3   Problem Statement and Formulation

This section is composed of two sections, i.e., prediction and allocation of water resources. For prediction, we first analyze the structure of the water system using system dynamics; and then, determine the time-dependent parameters by Holt linear trend and exponential equations; next, select 12 important input variables by PCA; finally, make predictions on the water resources based on BP neural network. As for allocation, we first determine the benefits by analytic hierarchy process (AHP), and then use a linear program to determine the best allocation policy.

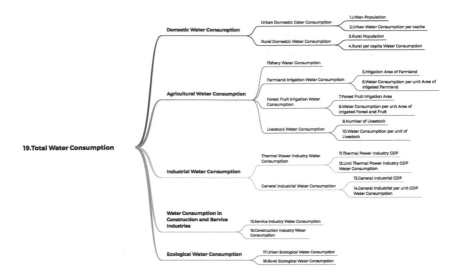

Fig. 1. Water demand system

3.1   Prediction

System dynamics a combination of qualitative and quantitative methods. The water demand system mainly includes life, agriculture, industry, construction and tertiary ecology subsystems. Subsystems are connected into a complex one through different feedback channels, see Fig. 1.

We take Holt linear trend and exponential equation to predict model parameters for the next 15 years based on historical data. In addition, the historical water use data of the study is a one-dimensional random non-stationary process. Due to the influence of random factors, anomalous points will be generated in the water use data sequence. These anomalous points will increase the overall noise of the historical data, thereby increasing the unpredictability of its changing law.

We use BP neural network to predict the demand of water resources, see Fig. 2. Note that, the input layer number can be found in 1,and they are the important variables after PCA in 4.2.The neural network algorithm adopts the gradient descent to improve the convergence speed with normalized input data, use historical data with important parameters as input layer and historical water consumption data as output layer.

3.2   Allocation

Due to the differences in the socio-economic structure of different regions, water resources have different benefits in various places, so we use analytic hierarchy process (AHP) to determine the benefit weights of several different types of water-using. Then, we obtain the $L \times m$ matrix of the ratings of the m indicators at L experts, set to w.

The subjective weights obtained from AHP are then corrected according to the entropy value method. For all $i = 1, 2, \cdots, L, k = 1, 2, \cdots, m$, let $E_i = (e_{i1}, e_{i2}, \cdots, e_{im})$ be the vector of each expert's rating, then

$$e_{ik} = 1 - (x_{ik} - \overline{x_{ik}}) / \max x_{ik} \tag{1}$$

denotes the level of the L_i expert's evaluation result of the target $B(B_1, B_2, \cdots, B_m)$. The expert evaluation model based on information entropy can be as follow.

Let the entropy value $H_i = \sum_{k=1}^{m} h_{ik}$ be the degree of uncertainty, where

$$h_{ik} = \begin{cases} -e_{ik} \ln e_{ik}, & 1/e \leq e_{ik} \leq 1 \\ 2/e - e_{ik} \left| \ln e_{ik} \right|, & 0 < e_{ik} < 1/e \end{cases}. \tag{2}$$

The smaller the value of H_i, the higher the level of decision and reliability of the expert L_i. Conversely, the higher the entropy value, the lower the reliability of the evaluation conclusion given by the expert, and less scientific. $C_i = \frac{1/H_i}{\sum_{i=1}^{m} 1/H_i}$ indicates the weight of each expert's evaluation, and the larger the value of C_i, the greater the weight of i expert's opinion in the evaluation. Next, we fuse the expert evaluation water use benefit matrix w with the expert evaluation weight matrix C_i. The weight matrices for different types of water-using are $W_i = w * C_i$. Finally, with the objective of maximizing the comprehensive benefits of regional water resources, the objective function of optimal water resources allocation is established, and the model is solved by linear program-

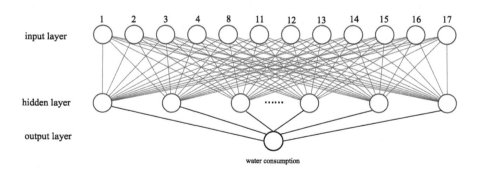

Fig. 2. Layers of BP neural network

ming method with the constraint of water supply demand.

$$\max E = \max \sum_{i=1}^{n} W_i Q_i \tag{3a}$$

$$\text{s.t.} \quad \sum_{i=1}^{n} Q_i \leq Q_{total}, \tag{3b}$$

$$Q_{imin} \leq Q_i Q_{imax} \quad i = 1, 2, 3, 4, \tag{3c}$$

$$0 \leq Q_i \quad i = 1, 2, 3, 4, \tag{3d}$$

where, (3a) indicates the objective function of maximizing water resources efficiency, w_j indicates the weights of different types of water using, and Q_i indicates different types of water using; (3b) indicates that Q_{total} in the above equation indicates the total amount of water available in the current year, and the water supply from 2021–2025 predicted in this paper is used here as the calculation data; in equation (3c), Q_{imin} and Q_{imax} denote the maximum and minimum values of water consumption in the current year, respectively; (3d) is a non-negative constraint. Solving the linear programming above, the water allocation results for each water use sector can be obtained.

4   Case Study

4.1   Instance Generation

The industrial GDP, surface water resources utilization, urban population and other data used in the model are all inquired from "Gansu Water Resources Bulletin" [9] and other official reports"Thirteenth Five-Year Development Plan of Lanzhou City", "Urban Water Conservation Plan of Lanzhou city" or obtained through mathematical or statistical methods. According to survey [10], the treatment conversion rate of domestic sewage is 60%, and the conversion rate of industrial sewage is 40%. We use this benchmark as the predicted value of waste water conversion rate in the next 15 years.

The water demand system is composed of five subsystems including agricultural water, industrial water, domestic water, with total of 18 variables, and we adopt the neural network model to select important input parameters for neural network. The model variables above are number 1–18 in Fig. 1.

4.2   Prediction

We use SPSS to accomplish PCA, which successfully reduce the dimensions from 18 to 12.

Above are the 18 variables that have an impact on water consumption constructed by the system dynamics model. Some of the variables above may have a tiny impact on water consumption. In order to ensure the accuracy of the neural network model, we use PCA in SPSS to eliminate variables with small influencing factors. For comparison purposes, we added the factor of water consumption to the PCA for analysis.

Table 1. Component matrix

	Component				Component				Component				Component		
Field	1	2	3	Field	1	2	3	Field	1	2	3	Field	1	2	3
19	0.957			1	−0.900			2	0.759	0.399		9	0.579	−0.769	
12	0.948			4	0.895			15	−0.727	−0.471	−0.398	5	0.422	0.749	
14	0.947			8	−0.867	−0.460		17	−0.708		0.534	18	0.550	0.582	
13	−0.945			11	−0.851			7	0.369	0.884		10			0.789
16	0.921	−0.313		3	0.804	−0.391		6	−0.550	0.779					

Note the numbers can be found in Fig. 1

In Table 1, the total water consumption field has the highest degree of interpretation for the first component, reaching 0.957. We can conclude that the first principal component explains the change in water consumption.

The variables with smaller correlation coefficients are eliminated, among the variables that affect the first principal component, here we set variables with correlation coefficients less than 0.7 will be eliminated.

The variables finally entered into the backpropagation neural network model are Lanzhou City's water consumption per unit of general industrial GDP, unit thermal power industry GDP water consumption, general industrial GDP, construction industry water consumption, urban population, rural unit population water consumption, thermal power industry GDP, unit water consumption for irrigating forest and fruit area, urban per capita water consumption, rural population, service industry water consumption, urban ecological water consumption.

For input parameters that have a significantly close to linear trend, the Holt linear trend method is suitable for prediction. For example, the urban population in the following figure has small fluctuations between 1999 and 2018 and has a clear linear upward trend, so this variable is suitable for Holt linear trend model to predict. Other parameters applicable to this method are: water consumption per unit of forest and fruit irrigation area, thermal power industry GDP, general industrial GDP, rural population, service industry water consumption, urban ecological water consumption.

On the other side, although some parameters that have an impact on water consumption may show a downward trend, the derivative of the downward trend will slow down over time. It can't be well fitted if use the Holt linear trend method. In terms of daily life, per capita water consumption will have a limit maximum value, and the closer to this limit value, the slower the rate of decrease in per capita water consumption will be. Therefore, this parameter is more suitable for prediction by exponential model. Other parameters applicable to this method are: water consumption per unit of thermal power industry GDP, water consumption per unit of general industrial GDP, urban per capita water consumption, and construction industry water consumption.

Model Validation We use the date of 1999–2015 as the training set and the date of 2016–2018 as the prediction set, before using neural network model and system dynamics model to simulate and predict the development of the next 15 years. The prediction data of 2016–2018 are obtained through the above two methods (holt linear trend and exponential equations). The model must meet the accuracy test to ensure that the model can reflect the reality.

In the neural network model's prediction set, the predicted values of the parameters for 2016–2018 are Table 2.

Table 2. 2016–2018 parameters prediction

Year	2016	2017	2018	Year	2016	2017	2018
$8(10^4 mu/10^9 m3)$	0.020	0.019	0.017	$2(10^9 m^3/10^4$ people)	0.0031	0.0031	0.0030
$12(m^3/10^4$ CNY)	415	385	355	$3(10^4$ people)	64.8	61.7	58.6
$11(10^9$CNY)	6.0	5.0	4.1	$4(m^3/10^4$ people)	0.0014	0.0013	0.0013
$16(10^9 m^3)$	0.0080	0.0080	0.0080	$1(10^4$ CNY)	317.5	326.4	335.3
$15(10^9 m^3)$	0.44	0.45	0.47	$14(m^3/10^4$ CNY)	54.9	45.7	36.5
$17(10^9 m^3)$	0.65	0.70	0.74	$13(10^9$ CNY)	537.4	546.7	555.9

Note the numbers can be found in Fig. 1

The accuracy calculation results of water consumption based on the above predicted values are shown in the Table 3.Also Holt exponential smoothing (Holt), linear regression (LR), and ARIMA were compared together, with historical data for 1999–2015 and predicted results for 2016–2018. The model error we use is smaller than other methods, so the model is suitable for used.

Table 3. 2016–2018 water consumption $(10^9 \text{m}^3/\text{a})$

Year	Actual value	BP Prediction	BP Error	LR Prediction	LR error	Holt Prediction	Holt Error	ARIMA	ARIMA Error
2016	12.64691	12.60897	0.00301	13.4944168	0.062804	11.83106	−0.06896	12.7386	0.007198
2017	12.52075	12.61889	−0.00778	13.2510611	0.055113	11.33158	−0.10494	12.7986	0.021709
2018	12.4251	12.73218	−0.02412	13.0077054	0.044789	10.8321	−0.14706	12.7749	0.027382

And for water supply model, since the historical date of Lanzhou's water supply and water consumption are exactly the same, it is speculated that the date of Lanzhou's water supply does not reflect the local water supply capacity of Lanzhou. Combining multiple references, we build the water supply model which the parameters in it almost involved all the influencing factors of the water supply, so we use this model for prediction directly.The model variables are:Surface water resources, Ground water resources and Reservoir capacity.

Forecast Water Supply and Demand in the Next 15 Years Set historical data from 1999–2018 as training set of the neural network, and set the prediction results of the important parameters selected by PCA in the next 15 years as the prediction input, output the following water demand development.

We can find from the trends in the Fig. 4b that the water consumption of Lanzhou City will be on the rise in the next 15 years, and the rate of increase will change gradually. The lowest value of water consumption is 1.25 billion m^3 in 2021, and the highest value is 1.57 billion m^3 in 2033.

We can find that although the water supply volume in the next 15 years will fluctuate slightly, the total amount of change will be small. The minimum value is 0.409 billion m^3, and the maximum value is 0.416 billion m^3, the change is 0.0686 million m^3.

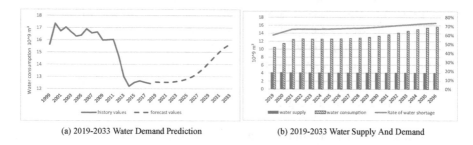

(a) 2019-2033 Water Demand Prediction   (b) 2019-2033 Water Supply And Demand

Fig. 3. Perdiction Result

It can be seen from the Figure (3b) that Lanzhou's local water resources will be in a state of long-term scarcity if there's no intervention plan being taken. The water shortage rate would be risen from 67% in 2019 to 74% in 2033. Predictable water shortage minimum value would be 836.7 million m^3 in 2020, and the maximum value is 1166.1 million m^3 in 2033.

4.3 Allocation

In the optimal allocation of water resources, we first used AHP for scoring by experts and the scoring matrix is summarized in Table 4. After that, the entropy value method is used to calculate the weights of the experts' scores, and the weights of each expert are obtained as Table 5.

Table 4. AHP matrix

Water-use Type	Weight				
	Expert 1	Expert 2	Expert 3	Expert 4	Expert 5
Industrial	0.061	0.110	0.067	0.304	0.062
Ecological	0.150	0.055	0.273	0.429	0.334
Agricultural	0.193	0.541	0.533	0.206	0.206
Domestic	0.597	0.294	0.128	0.061	0.397

Table 5. Entropy and weight matrix

Expert number	Entropy	Weight
1	0.95068	0.171162
2	0.662118	0.245758
3	0.742313	0.219208
4	1.195002	0.136168
5	0.714589	0.227712

With the fusion of weights, the weight shares of industrial, agricultural, domestic and ecological water are 0.10763, 0.23359, 0.35763 and 0.30116 respectively. The final objective equation is derived as

$$\max E = 0.10763Q_1 + 0.23359Q_2 + 0.35763Q_3 + 0.30116Q_4 \tag{4}$$

The objective equation can be solved by using the constraints in Sect. 3.2, where the constraints for the maximum and minimum values of water consumption are obtained by using Holt linear trend and Exponential equation, as shown in Table 6.

Table 6. Maximum and minimum water use

Year	Industrial		Agricultural		Domestic		Ecological	
	Min	Max	Min	Max	Min	Max	Min	Max
2021	1.93	4.33	5.20	6.42	0.66	1.56	0.05	2.76
2022	1.74	4.40	5.15	6.38	0.60	1.62	0.02	3.17
2023	1.57	4.44	5.10	6.33	0.54	1.68	0.00	3.57
2024	1.43	4.46	5.05	6.29	0.49	1.74	0.00	3.95
2025	1.30	4.47	5.00	6.24	0.44	1.79	0.00	4.32

Table 7. Water allocation

Year	Industrial	Agricultural	Domestic	Ecological	Benefit coeff.
2021	1.9338	6.2684	1.5588	2.759	3.0607
2022	1.7361	5.9851	1.6248	3.174	3.1219
2023	1.5701	5.6984	1.6835	3.568	3.1767
2024	1.4274	5.4078	1.737	3.9478	3.227
2025	1.3029	5.1134	1.7863	4.3172	3.2737

By solving the above linear programming, the results of water resources allocation for each water department can be obtained, as shown in Table 7. The above scheme is of positive significance to the sustainable development of water ecological civilization construction in Lanzhou, and can provide technical support for the decision-making of water resources allocation in Lanzhou.

5   Conclusion

This paper analyzes the water demand prediction and allocation strategies in Lanzhou. The prediction method is based on a tailored BP neural network by incorporating panel and time-dependent properties. Moreover, we propose a linear programming model to better allocate the water resources and contribute to the sustainable development of Lanzhou. According to numerical results, we find the proposed solution approach results in an prediction accuracy error within 3%, and observe several valuable managerial insights for decision makers. Future research directions include: (1) applying the prediction and allocation method to other cities; (2) optimizing the water allocation model by fully balancing the trade-off between multi objectives, such as efficiency, equity and sustainable development; (3) incorporating demand and supply uncertainties to the joint prediction-optimization scheme by stochastic programming or robust optimization.

Acknowledgement. This research is partly supported by the Fundamental Research Funds for the Central Universities (grant number 2021RC202), the National Natural Science Foundation of China (grant numbers 72101021), and the fellowship of China Postdoctoral Science Foundation (grant number 2021M690009)

References

1. WHO&UNICEF: Progress on drinking water, sanitation and hygiene: 2000–2017: special focus on inequalities, https://data.unicef.org/resources/progress-drinking-water-sanitation-hygiene-2019/
2. Jain, A., Varshney, A.K., Joshi, U.C.: Short-term water demand forecast modelling at t kanpur using artificial neural networks. Water Resour. Manage. **15**(5), 299–321 (2001)
3. Xu, Z., Yao, L., Zhang, Q., Kiyoshi, D., Long, Y.: Inequality of water allocation and policy response considering virtual water trade: a case study of lanzhou city, china. J. Clean. Prod. **269**(4), 122326 (2020)
4. Liu, C., Zeng, Z., Pang, Y., Hongfei, L.U., Bai, F., Gao, F.: Comparison of urban water demand forecasting methods. Water Resour. Prot. **31**(6), 179–183 (2015)
5. Winz, I., Gary, Brierley, Trowsdale, S.: The use of system dynamics simulation in water resources management. Water Resour. Manage., (2008)
6. Zhao, G.L., Mu, E., Wen, X., Rayburg, Z., Tian, P.S.: Changing trends and regime shift of streamflow in the yellow river basin. Stochastic Environ. Res. Risk Assess. **29**(5), 1331–1343 (2015)
7. Yaping, J., Xin, Z., Yan, L.: Combined forecasting of urban water demand based on grey neural network and maldives chain. Northwest China J. Univ. Agricult. Forestry (NATURAL SCIENCE EDITION). **39**(7), 229–234 (2011)
8. Liu, H., Zhang, H., Tian, L.: Artificial neural network method for forecasting hourly water consumption. China Water Wastewater., **18**
9. Lanzhou city government: Gansu Water Resour. Bull., 2000–2018
10. Shuang, Z.: Based on the sustainable use of water resources in the perspective of lanzhou optimiztaion and adjustment of industrial structure. Ph.D. dissertation, Lanzhou University (2015)

A Deterministic Effective Method for Emergency Material Distribution Under Dynamic Disasters

Yingfei Zhang[1], Xiangzhi Meng[2], Ruixin Wang[2], and Xiaobing Hu[1(✉)]

[1] College of Safety Science and Engineering, Civil Aviation University of China, No. 2898, Jinbei Road, Dongli District, Tianjin, China
{2021095013,xbhu}@cauc.edu.cn
[2] Sino-European Institute of Aviation Engineering, Civil Aviation University of China, No. 2898, Jinbei Road, Dongli District, Tianjin, China
{2018122047,rxwang}@cauc.edu.cn

Abstract. It has great practical significance in solving the problem of emergency material distribution in a dynamic disaster environment, and there are usually four considerations: the corresponding optimal relationship between emergency material supply storage and demand nodes, the shortest paths to deliver emergency materials, the timeliness of path optimization and the success rate of optimization solutions. At present, in solving the distribution path problem between the multiple emergency material supply storage and multiple demand nodes, it is difficult to the static path optimization and dynamic path optimization that ensure the optimality of distribution path planning in a dynamic disaster environment, and they even fail to complete the distribution task. In this study, by improving co-evolutionary path optimization and ripple-spreading algorithm, the emergency material distribution problem can be solved with an optimality guarantee, high timeliness, and success rate, and simulation experiments verify the advantages of the reported method.

Keywords: Emergency material distribution · Dynamic disaster · Ripple-spreading algorithm · Timeliness

1 Introduction

Disaster mitigation and emergency management have become more challenging for governments [1]. Optimal and fast delivery is the primary goal of emergency material distribution path planning. The path from storage to demand nodes is the premise of an effective emergency rescue process. However, achieving optimal emergence material distribution for existing methods in a dynamic disaster environment. Because the corresponding optimal relationship between emergence material supply storage and demand nodes is likely different in dynamic disaster events from the static plan, suppose we cannot find the best corresponding relationship and distribution path at the beginning; the delivery truck could waste time and eventually fail to get to demand nodes because the dynamic disaster gradually cuts off all roads to demand nodes.

In a dynamic routing environment, it is always challenging to consider four considerations: a) the corresponding optimal relationship between emergency material supply

© The Author(s), under exclusive license to Springer Nature Switzerland AG 2023
N. Xiong et al. (Eds.): ICNC-FSKD 2022, LNDECT 153, pp. 1105–1116, 2023.
https://doi.org/10.1007/978-3-031-20738-9_120

storage and demand nodes, b) the shortest paths to deliver emergency materials, c) the timeliness of path optimization, and d) the success rate of optimization solutions. There are many ways to effectively distribute emergence materials between storage and demand nodes [2–4]. Some methods complete the material distribution by setting up central processing points [5–7]. Kemball-Cook was first proposed in 1984 to distribute emergency materials needed for target management [8]. Linet developed a mathematical model with the minimum required time delay to deal with the order of emergency dispatch [9]. Chang established a temporary distribution center for flood emergency supplies based on geographic information and constructed a material distribution model for rescue teams [10]. Liang built a routing optimization model considering the timeliness of emergency material distribution based on a genetic algorithm (GA) [11]. Fox introduced a dynamic road network into the traveling salesman problem, which has good guiding significance for emergency logistics scheduling [12].

To get the optimal path of emergency material distribution, we must pay attention to the second and the third considerations in the path optimization algorithm. Traditional path optimization algorithms usually implement the nearest principle, static path optimization (SPO), and dynamic path optimization (DPO). SPO means each emergency material supply storage has its area of responsibility. SPO does not consider the accessibility of paths under a dynamic disaster. Considering dynamic disasters usually have a particular impact on the SPO, it does not meet the actual condition very well. Therefore, DPO is often used.

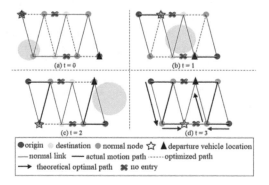

Fig. 1. Traditional DPO path planning diagram

Emergency material supply storage and demand point are usually multiple. This involves the many-to-many path optimization problem (POP), and the fourth consideration should be included. There are three main categories of algorithms to solve POP: deterministic algorithm, evolutionary algorithm, and potential field algorithm. The traditional DPO based on these algorithms usually repeats the calculation of one-to-one POP to solve a many-to-many POP [13]. Its computational efficiency is low, and the origin node and destination node must be paired correctly in advance. Once a link is destroyed by disaster, it will cause all paths through this link to fail. As Fig. 1 shows, the path of the traditional DPO is not theoretically optimal. This method leads to returning behavior and wastes time. This paper solves the emergency material distribution problem using the co-evolutionary path optimization (CEPO) method. Then we use a simulation experiment to verify its feasibility.

2 Problem Description and Mathematical Model

In a dynamic disaster environment, we need to consider the optimal relationship between emergency material supply storage and demand nodes, the shortest paths to deliver emergency materials, the timeliness of path optimization, and the success rate of optimization solutions. Firstly, we describe the mathematical models of SPO, DPO, and CEPO by taking a single pair of emergency material supply storage and a single demand point under a dynamic road network as an example.

2.1 Problem Description

As shown in Fig. 2, there is one origin and one destination. The emergency material distribution path planned by SPO is fixed. Once the dynamic disaster impacts the road network, this method will lose its practicability. The path planned by the traditional DPO seems feasible at the initial time. But when the disaster moves, the vehicle could have to go back to the origin and replan the path. This returning behavior is inefficient in the calculation and wastes time. Differently, as the blue line, CEPO can predict the area of dynamic disaster influence, and it can just avoid the obstacle in the period. So, the vehicle can reach its destination within the shortest distance.

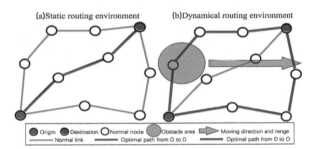

Fig. 2. Material distribution case in different methods

2.2 Mathematical Model

The above description expresses the optimal path based on SPO as follows.

$$P_{SPO} = \sum_{1 \le i \le N_D} P_{t_0}^i \tag{1}$$

In Eq. (1), $P_{t_0}^i$ is the optimized path in SPO for the i-th origin node and destination node pair according to the routing environment $R(t0)$ at the initial time t_0. N_D is the number of destination nodes which is the demand nodes. On the other hand, the mathematical model of DPO is as follows.

$$P_{DPO} = \sum_{\substack{n \in N \\ 1 \leq i \leq N_D}} P_t^i \Big|_{t+nT}^{t+(n+1)T} \tag{2}$$

In Eq. (2), $P_t^i \big|_{t+nT}^{t+(n+1)T}$ is the optimized path for the i-th origin node and destination node pair according to the road network environment $R(t)$ from time $t + nT$ to $t + (n + 1)T$. And this formula indicates that after a while T, the optimized path P_t^i will be updated. According to the new road network environment $R(t + T)$, the new optimized path P_{t+T}^i is formed and implemented immediately. Until the final optimized path is found, the iteration will not stop.

Different from the calculation procedures of these two methods, CEPO can combine the dynamic disaster model with the calculation procedure for the optimal path of emergency material distribution so that we calculate only once and obtain the optimal paths from multiple emergency material supplies storage to multiple demand nodes. Thus, we can describe the mathematical model of CEPO as follows.

$$P_{CEPO} = \min_{P \in \Omega_P} \sum_{\substack{1 \leq i \leq N_O \\ 1 \leq n \leq N_D}} f_T(P_{i,n}, S(P_{i,n})) \tag{3}$$

Suppose there are N_O origin nodes and ND destination nodes in the road network R. $P_{i,n}$ is the path from origin node i to destination node n. ΩP is the set of all possible paths from the origin node to the destination node. $S(Pi,n)$ is the number of all nodes in $P_{i,n}$. For example, $P_{i,n}(1)$ is the origin node and $Pi,n(S(Pi,n))$ is the destination node. $f_T(P,m)$, $1 \leq m \leq S(P)$-1, is the time function that calculates the time for traveling to node m along path P. Furthermore, CEPO also includes the following formulas.

$$A_{k|0}(P_{i,n}, P_{i,n}(m+1)) = 1 \tag{4}$$

At time $t = 0$, Eq. (4) means that it predicts node $Pi,n(m)$ is connected to node $Pi,n(m + 1)$ at the future time k, $k = km, \dots, km + 1$.

$$
\begin{aligned}
f_T(P_{i,n}, \ m+1) &= \max(f_T(P_{i,n}, \ m), \ k_m|0) + T_{k_m|0}(P_{i,n}(m), P_{i,n}(m+1)) \\
&\quad + W_{k_m|0}(P_{i,n}(m), P_{i,n}(m+1))
\end{aligned} \tag{5}
$$

From Eq. (5), $f_T(P_{i,n}, 1)=0$. At time $t = 0$, $T_{k_m|0}(P_{i,n}(m), P_{i,n}(m+1))$ and $W_{k_m|0}(P_{i,n}(m), P_{i,n}(m+1))$ respectively mean the predicted traveling time and the predicted waiting time if traveling from node $Pi,n(m)$ to node $Pi,n(m + 1)$ at the time k. Equation (5) gives the time of reaching node $Pi,n(m + 1)$ along path Pi,n. The value of traveling time $T_{k_m|0}$ and waiting time $W_{k_m|0}$ are connected with a dynamic road network. The f_{RY} is the time-varying function of a dynamic road network. It shows that:

$$[A_{k_m+1|0}, \ T_{k_m+1|0}, \ W_{k_m+1|0}] = f_{RY}(A_{k_m|0}, \ T_{k_m|0}, \ W_{k_m|0}) \tag{6}$$

3   The Improvement of the Algorithm

IT is necessary that to enable the traditional methods to compare with CEPO, these algorithms need to be improved. At first, the improvement of methods is given. They are used to be compared with RSA in the following experiments.

3.1   The Improvement of Traditional Methods

To solve the many-to-many POP, SPO fails in planning path in the dynamic disaster environment because the algorithm considered the responsibility area limitation, that is CSPO. So, scholars have introduced waiting behavior, that is WSPO. The vehicle might need to wait when a planned path is impossible due to dynamic disaster. Until the road is restored, the vehicle will continue to travel along the planned path. The improved SPO wastes too much time in the waiting process. So, scholars put forward the DPO.

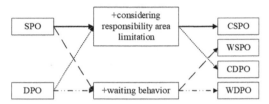

Fig. 3. Four improvement traditional methods

The improved DPO must try all origin and destination node combinations to find the optimal path. Essentially, it still is a static path optimization method, it calculates at every period. Secondly, when the road network becomes complex, the method calculation's efficiency is low. In addition, the improved DPO is limited to calculating the optimal path in the road network at the current time. Theoretically, the optimality of the results will be lost in the future due to changes in the road network. To compare the differences of SPO, DPO, and CEPO algorithms, DPO is also improved similarly. Considering responsibility area limitation, the DPO becomes CDPO. Adding the waiting behavior, the DPO becomes WDPO. They are shown in Fig. 3.

3.2   The Improvement of CEPO

The ripple spreading algorithm (RSA) is the core algorithm of CEPO [14]. It is necessary to improve RSA to solve the emergency material distribution problem so that RSA can solve the many-to-many POP.

The Principle of One-to-One RSA
RSA is a path search algorithm inspired by the natural ripple-spreading phenomena. In a dynamic road network, there is one origin node and one destination node, the ripple of origin node is first activated and then it will trigger the next node ripple at a future time, with a certain radius and speed. As shown in Fig. 4, these ripples are likened to a relay

race, triggering another new ripple, until the destination node ripple has activated. A ripple excited at a node spread to its adjacent nodes. Once all adjacent nodes are excited, the ripple will disappear. In a dynamic road network, a link could become impassable due to dynamic disasters. In such a case, the ripple will wait until the link becomes normal and continues to spread. Finally, we can get the optimal path by backtracking from the destination node to the origin node.

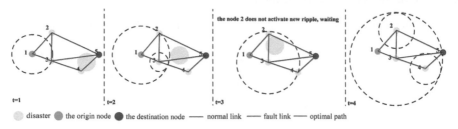

Fig. 4. The principle of one-to-one RSA in finding the optimal path

In the emergency material distribution problem, the origin node of RSA is emergency material supply storage, and the destination node of RSA is the material demand point. This way, in the road network affected by dynamic disasters, the path can be searched from emergency material supply storage to the demand point. Integrating disaster dynamics into ripple relay race can ensure the path will avoid disaster influence in the right place at the right time. Thus, the first and the second considerations can be solved. This calculation method can also prevent the returning behavior in those traditional path search algorithms and improve the computational efficiency of path searching procedures. Thus, the third and fourth considerations can be solved.

The Improvement of RSA
For basic RSA, a new ripple is activated by the last ripple, and the algorithm requires each node is activated to produce one ripple at most. This rule cannot solve many-to-many POP. The RSA is improved to solve the problem. The algorithm needs to change and meet the following conditions: a) Increase the number of origin nodes and destination nodes in the algorithm, and set a rule that every origin node has not a destination node at the initial stage. b) Change the properties of nodes in the algorithm, so that each node can be activated for multiple times, but the upper limit of this number cannot exceed the number of origin nodes. c) Follow-up ripples are activated by the origin nodes' ripples as long as one ripple is triggered at a destination node, the path searching procedure for the associated origin node will stop. d) After all destination nodes' ripples are triggered, the ripple relay race stops.

As shown in Fig. 5, the road network has N_N nodes, and there are N_O origin nodes and N_D destination nodes in it. N_R is the number of current ripples. $S_R(r)$ represents the state of ripple r. $S_R(r) = 0/1/2$ represents the state of ripple r is inactivated/waiting/activated. $R(r)$ is the radius of ripple r. $CL,z|0(D(r), n)$ is the L-th link of node i and node j at the time z. Thus, each node can be activated to generate N_O new ripples at most. And one origin ripple can only activate a new ripple at a linked node. If there is a destination

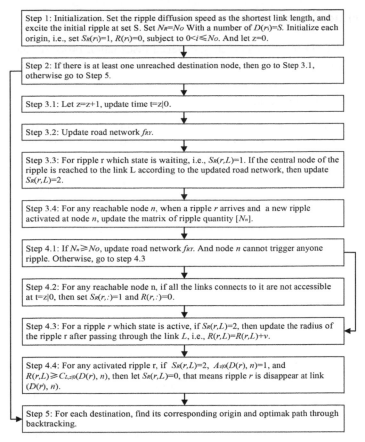

Fig. 5. The flowchart of the improved RSA

node that has not been activated, then the ripple relay race needs to continue. In this way, RSA can ensure the optimal corresponding relationship between emergency material supply storage and demand nodes, the shortest paths to deliver emergency materials, the timeliness of path optimization, and the success rate of optimization solutions in the dynamic disaster road network environment.

Thus, the emergency material distribution problem can be solved by CEPO. Firstly, the ripples of emergency material supply storage are activated. When these ripples reach adjacent nodes, the ripples of adjacent nodes are activated. Secondly, if the path around this node has been destroyed by dynamic disaster, it must wait for the path recovery, and continue to spread. Other ripples can spread normally. Thirdly, only when the ripples of all the adjacent nodes are activated, will the ripples of this node die. Finally, when the demand point (destination node) is activated, the optimal distribution path can be obtained by backtracking from the demand point to the emergency material supply storage. After all the distribution paths are found, the ripple relay race is ended.

4   Experiments and Results

To verify the CEPO for the emergency material distribution problem under dynamic disasters, we take 3 scale experiments to compare the results of SPO, DPO, and CEPO.

4.1   The Experimental Setup

These experiments are based on 50×50, 70×70, and 100×100 road networks. The coordinate range of them is $[-1000\ 1000\ -1000\ 1000]$, and 3 dynamic disaster areas are initially located in $[-300\ 500]$, $[-1500\ 300]$, and $[1500\ -700]$. The radius of the dynamic disaster area is 100, and the obstacle moves at a constant speed of 20. Delivery vehicles travel at a constant speed of 50. These details are shown in Table 1.

Table 1. Origin nodes and destination nodes number of path networks with different scale

Experiment number	Scale of road networks	Origin node number	Destination node number
1	2500	1	1120
		50	1131
		2451	1320
		2500	1331
2	4900	1	2200
		70	2211
		4831	2690
		4900	2701
3	10000	7	4040
		100	4060
		9901	5840
		10000	5860

Assuming disasters are moving in three directions, emergency material supply storage is located in four corners, and the demand nodes are located in the center of the road network. After the obstacle moves, the path can still access. Suppose the supply storage has a specified responsibility area. To show the difference between these methods, the comparison is designed as follows: Considering the responsibility area limitation of SPO(CSPO), adding the waiting behavior to SPO(WSPO), considering the responsibility area of DPO(CDPO), and adding the waiting behavior to DPO(WDPO), and the improved CEPO. The difference is mainly reflected by the length of the shortest path (PL), the shortest transportation time (TT), and the time of calculating the path (CT). Suppose A is the number of responsibility area's demand nodes and B is the total number of demand nodes, the success of emergency material distribution of A-B can be represented by DDSR (demand-demand success rate). If all the emergency material supply storage corresponds to their responsibility area, the corresponding rate of this method is 100%. Otherwise, the corresponding number is CR (corresponding rate).

4.2 The Results

There are many similarities in the results, and we take the results of the 70×70 road network as an example, shown in Fig. 6.

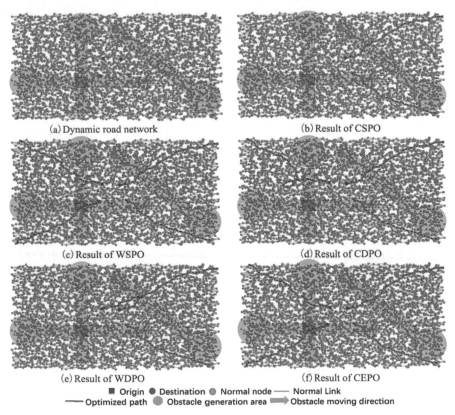

(a) Dynamic road network

(b) Result of CSPO

(c) Result of WSPO

(d) Result of CDPO

(e) Result of WDPO

(f) Result of CEPO

■ Origin ● Destination ● Normal node —— Normal Link
—— Optimized path ● Obstacle generation area ⇒ Obstacle moving direction

Fig. 6. The comparison results of 4900 nodes of different methods

Four red boxes represent emergency material supply storage, and four blue circles represent demand nodes. Figure 6(a) is a schematic diagram of the dynamic disaster road network structure, which indicates the direction of disaster movement. The DDSR of CEPO and WSPO is 100%, and the DDSR of CSPO is the lowest, only 50%. Only CEPO and WSPO planned the path to deliver emergency materials. According to the results, we can gain the following conclusions.

Firstly, the primary consideration is the success rate of emergency material distribution in the dynamic disaster environment. From Table 2, only WSPO and CEPO in all road network environments have successfully planned the emergency material distribution paths. The worst performance is CSPO, followed by CDPO and WDPO. The success rate of CSPO and CDPO is low because the responsibility area influences the path planning of these two methods. The success rate of WSPO is 100%, but the success rate of WDPO decreases with the complexity of the road network. This result reflects the low calculation efficiency and success rate in complex road network environments. Among them, ' >' represents the failure of emergency material distribution path planning, and we will no longer compare those failure methods with others. CEPO break the limitation of fixed responsibility area and successfully plan paths for emergency material distribution problem in each case.

Secondly, TT, PL, and CT are also the key indicators of our concern. It can provide ample time in the preparatory phase of the emergency evacuation. As shown in Table 2, the shortest paths are generated by WSPO in the 50 × 50 road network. The long waiting time takes 12.19 s longer than CDPO, WDPO, and CEPO in terms of transportation time on average. Although CDPO, WDPO, and CEPO have the same PL and TT, the CT of the CEPO is the shortest. The same observation also appears in the 100 × 100 road network. To effectively avoid returning behavior, the path planning time and vehicle waiting time under CEPO can be reduced.

Table 2. Comparison of all results of different methods

Scale	Indexes	CSPO	WSPO	CDPO	WDPO	CEPO
2500	DDSR	75%	**100%**	100%	100%	**100%**
	CR	100%	100%	100%	100%	100%
	PL(m)	>3882.09	5098.85	5109.82	5109.82	**5109.82**
	TT(s)	>279.51	380.10	367.91	367.91	**367.91**
	CT(s)	0.48	0.73	0.66	2.80	**0.14**
4900	DDSR	50%	**100%**	75%	75%	**100%**
	CR	100%	100%	100%	66.67%	100%
	PL(m)	>2619.86	**5253.04**	>4513.21	>4160.13	5258.31
	TT(s)	>188.63	410.46	>324.95	>299.53	**378.60**
	CT(s)	3.52	8.59	6.38	32.22	**0.28**
10000	DDSR	25%	**100%**	25%	25%	**100%**
	CR	100%	100%	100%	100%	75%
	PL(m)	>1264.20	5074.33	>1264.21	>1264.21	5370.93
	TT(s)	>91.02	394.20	>91.02	>91.02	**386.71**
	CT(s)	3.44	16.86	4.75	29.23	**0.77**

5 Conclusion

This paper uses CEPO to solve the problem of many-to-many POP emergency material distribution under dynamic disasters. It is compared with different methods in the same simulation experiment. We find that CEPO can break the limitation of the fixed responsibility area and obtain the corresponding optimal relationship between emergency material supply storage and demand nodes. Meanwhile, it can get the shortest emergency material distribution path, lower transportation time consumption, and ensure a 100% success rate. CSPO and CDPO are limited by the fixed responsibility area, so their success rates are low. Although WSPO introduces waiting behavior to obtain a 100% success rate, the transportation time is increased. It makes the delivery vehicles expose for too long in the disaster environment. Thereby the potential risk is increased. WDPO ensures a 100% success rate in a simple road network, but the planning process takes a long time and cannot guarantee the optimality of the results. The road network is more complex, the efficiency of WDPO is lower, and the success rate will decrease significantly. In summary, CEPO has four advantages in solving the emergency material distribution problem within dynamic disasters: the optimal relationship between emergency material supply storage and demand nodes, the shortest paths to deliver emergency materials, the timeliness of path optimization, and the success rate of optimization solutions.

References

1. Bimantoro, M.F., Solamat, L.A.: Shortcut evacuation road concept as a solution for disaster mitigation to reduce impacts of earthquake and tsunami at Gunung Kidul Coast, pp. 49–53, Yogyakarta, Indonesia (2016)
2. Smith, S.F., Gallagher, A., Zimmerman, T.L.: Distributed management of flexible times schedules. In: Proceedings of the 6th International Joint Conference on Autonomous Agents and Multiagent Systems, pp.74:1–74:8, Honolulu, Hawaii (2007)
3. Musliner, D.J., Durfee, E.H., Wu, J., Dolgov, D.A., Boddy, M.S.: Coordinated plan management using multiagent MDPs. In: Distributed Plan and Schedule Management, pp. 73–80, Stanford, CA (2006)
4. Chapman, A.C., Anna, M.R., Ramachandra, K., Jennings, N.R.: Decentralized dynamic task allocation using overlapping potential games. Comput J 9(53), 1462–1477 (2010)
5. Ramchurn, S.D., Polukarov, M., Farinelli, A., Truong, C., Jennings, N.R.: Coalition formation with spatial and temporal constraints. In: Proceedings of the 9th International Conference on Autonomous Agents and Multiagent Systems, pp. 1181–1188, Toronto, Canada (2010)
6. Koes, M., Nourbakhsh, I.R., Sycara, K.P.: Heterogeneous multirobot coordination with spatial and temporal constraints. In: Proceedings 20th National Conference on Artificial Intelligence, pp. 1292–1297. Pittsburgh, PA (2005)
7. Barbulescu, L., Rubinstein, Z.B., Smith, S.F., Zimmerman, T.L.: Distributed coordination of mobile agent teams: the advantage of planning ahead. In: Proceedings of 9th International Conference on Autonomous Agents and Multiagent Systems, pp. 1331–1338 (2010)
8. Kemball-Cook, D., Stephenson, R.: Lesson in logistics from Somalia. Disaster 8(1), 57–66 (1984)
9. Zdamar, L., Ekinci, E., KüÜkyazici, B.: Emergency logistics planning in natural disasters. Ann. Oper. Res. 129(7), 217–245 (2004)
10. Chang, M.S., Tseng, Y.L., Chen, J.W.: A scenario planning approach for the flood emergency logistics preparation problem under uncertainty. Transp. Res. Part E: Logist. Transp. Rev. 43(6), 737–754 (2007)
11. Liang, Y.M., Fang Z.M., Huang, J.H.: Emergency material distribution route optimization method considering timeliness. Logist. Technol. 44(6), 38–41 (2021)

12. Fox, K.R., Gavish, B., Graves, S.C.: An n-constraint formulation of the time-dependent traveling salesman problem. Oper. Res. **28**(4), 1018–1021 (1980)
13. Hu, X.B., Wang, M., Leeson, M.S., Di Paolo, E.A., Liu, H.: Deterministic agent-based path optimization by mimicking the spreading of ripples. Evol. Comput. **24**(2), 319–346 (2016)
14. Zhang, M.K., Hu X.B., Wang, J.A.: Research on the evacuation path of high-rise buildings considering the dynamic fire diffusion process. Chin. Saf. Sci. J. **29**(3), 32–38 (2019)

Unsupervised Concept Drift Detectors: A Survey

Pei Shen[1], Yongjie Ming[1], Hongpeng Li[1], Jingyu Gao[2(✉)], and Wanpeng Zhang[2]

[1] HBIS Digital Technology Co., Ltd. Hebei, Shijiazhuang 050022, China
{shenpei,hbmingyongjie,hblihongpeng}@hbisco.com

[2] School of Computer Science and Technology, Xidian University, Xi'an, Shaanxi, China
21031211570@stu.xidian.edu.cn

Abstract. Concept drift mainly refers to the change of the current data distribution in the data streams due to the dynamic evolution of the external environment, which leads to the failure of machine learning or data mining models to show the desired effect. As a result, the forecast or decision model needs to be constantly updated. To find the appropriate time for model updating, a large number of studies have proposed methods for detecting concept drift, which can be divided into supervised detection methods and unsupervised detection methods. Because an unsupervised concept drift detector does not make strong assumptions about the availability of data annotations for real application scenarios, it has more robust availability and generality, and there is less summary work related to unsupervised detectors. Therefore, in this paper, we will summarize all the existing unsupervised concept drift detection methods and give a new classification basis to classify the unsupervised concept drift detection methods.

Keywords: Concept drift · Data streams · Unsupervised detector · Survey

1 Introduction

With the continuous development of machine learning, more and more machine learning systems are applied to other fields to perform predictive classification tasks. As the machine learning system constantly interacts with the outside world for a long time, the distribution of data input into the system will also change with time, which is challenging to observe, which is named the concept drift problem [1].

To solve the impact of concept drift, a large number of research and implementation of concept drift detection methods. Gama et al. Proposed a DDM method to detect concept drift [2]. The alarm signal will be triggered when the error rate rises above the warning threshold. When the error rate increases to the drift threshold, the new model will be directly used to replace the old model. A similar method is EDDM [3], which uses the distance between two correct classification results to judge the concept drift, improving the method's sensitivity to a certain extent. There are many drift detection methods, such as ADWIN [4], ECDD [5].

Although these drift detection methods have achieved good results in the research field, most are supervised drift detection methods. These methods assume that their application scenarios can get the actual annotation of the input data in a short time, which is

N. Xiong et al. (Eds.): ICNC-FSKD 2022, LNDECT 153, pp. 1117–1124, 2023.
https://doi.org/10.1007/978-3-031-20738-9_121

not guaranteed in many practical application scenarios. The unsupervised drift detection method only detects the input data and does not need the actual results corresponding to the data, so it has more robust applicability. Therefore, we have decided to summarize all the unsupervised drift detection methods and give appropriate classification methods.

The main contribution of this paper is to make a detailed summary of all current unsupervised drift detection methods. At the same time, we give our reasonable classification basis, unsupervised drift detection methods based on data distribution differences and model performance, and describe the current unsupervised drift detection methods.

The organizational structure of the rest of this paper is as follows. Section 2 introduces some existing work on the classification of drift detection and gives the classification basis of the unsupervised drift detection method in this paper. Section 3 describes and summarizes the unsupervised drift detection method based on data distribution differences. Section 4 analyzes and describes the unsupervised drift detection method based on the model's performance. Finally, Sect. 5 provides a summary of this paper and future work.

2   Related Work

2.1   Existing Taxonomy

Lu et al. [6] summarized the field of the whole concept drift detection method. The document summarized the entire process of the drift detection method into four stages and divided the drift detection method into three categories according to the method used in each stage of the method. The first group is the drift detection method based on the error rate, which will test the error rate of the prediction results of the prediction model in real-time. The second is the drift detection method based on data distribution. This method directly starts from the data distribution and calculates the distribution difference between the current and the reference data. The third type is the drift detection method based on multiple hypothesis tests, which mainly improves the first two types of methods. This kind of method can effectively reduce the false positive rate of the drift detection method.

Wares et al. [7] Classified and summarized supervised drift detection methods, which were divided into four categories: statistical methods-based, sliding window-based, integration detector-based on the data block, and incremental integration detectors. On this basis, Roberto et al. [8] compared the performance of mainstream supervised drift detection methods. In this document, 14 supervised drift detection methods were selected and tested on data sets with different drift types and data sizes, in which the measured indicators include accuracy and recall.

Gemaque et al. [9] summarizes and divides the unsupervised drift detection methods into two categories. The first is the drift detection method based on batch processing. This method will accumulate the data to form a specific scale of data block first and then trigger the detection method to use some metrics to distinguish the differences between the two data blocks. The second kind of method is the online drift detection method, which is different from the batch detection method. Each new data instance of this method will trigger the drift detector for detection.

2.2 Taxonomy

This section will give our classification of unsupervised drift detection methods and the characteristics and differences of various detection methods. Referring to the classification form of the unsupervised drift detection method in [9], we divide the unsupervised drift detection method into two categories and divide the method in each category into two subcategories.

The first kind is an unsupervised drift detection method based on data distribution differences. This method directly measures the distribution difference between new data samples and reference data. The different measurement methods used can be further divided into detection methods based on regional density and detection methods based on statistical tests. Among them, the density-based detection method inputs the sample density of data in different regions in the sample space. To express the distribution differences between data sets, the other is to directly use the two samples statistical test method or distance index to indicate the gap between sample data explicitly.

The second kind of detection method is the unsupervised drift detection method based on the model's performance. This method will detect concept drift through the decline of some model performance indicators. However, because the application scenario of the unsupervised drift detection method cannot obtain the actual label, additional label-independent indicators need to be designed to reflect whether the model's performance is declining. The drift detection method can be divided into the detection method based on the classifier model and the detection method based on other models according to the specific situation and different detection models.

Different from the classification method in [9], the classification method in this paper mainly distinguishes the unsupervised drift detection method from the basis of judging the occurrence of drift detection. In contrast, [9] primarily determines the unsupervised drift detection method from the accumulation and use of data. In addition, we have added more unsupervised drift detection methods based on [9].

3 Data Distribution Difference-Based Methods

3.1 Regional Density-Based Methods

The regional density is usually obtained by the proportion of the number of samples in the region to the total sample size of the dataset. Therefore, the data distribution difference between the two datasets can be judged by comparing whether the regional density of all subregions between the two datasets is similar. An effective method is to use kdq-Tree [10] to divide the restricted space. The reference data and new data are divided into different spaces through kdq-Tree, and the KL distance is calculated for the data in each corresponding subspace. Because the method divides the spatial region, it can detect the local region where the concept drift occurs.

Similarly, the nearest neighbor-based density variation identification (NN-DVI) [12] uses the most relative neighbor division method to divide the space, which is divided into three steps. The first is data modeling. NN-DVI uses the KNN method to divide spatial regions. At the same time, the diversity function is introduced to ensure that the samples in each subspace are composed of different small data areas as much as possible. The second step of the method is mainly to calculate the change in regional density. NN-DVI quantifies the difference between the two data sets by accumulating the number of instance particle differences.

Statistical change detection (SCD) [11] proposed a density test method without spatial division. For the given baseline data and the data to be tested, the density test method will first evenly divide the baseline data into two parts, which can be approximately considered that the data distribution of the two parts is similar. Then, a kernel density estimator is trained by using one part of the baseline data. The kernel density estimator is used to calculate the kennel density on the other part of the baseline data and the data to be tested. The distribution difference is determined as the logarithmic probability density difference of the nuclear density estimator on the two parts of the data. Equal density estimation (EDE) [13] proposed an unsupervised drift detection method based on the equal density region, which needs to ensure that the number of data samples in the two data sets is the same. At the same time, EDE defines a measurement index called Density-Scale. Theoretical proof shows that the larger the value of this index is, the higher the density difference between the two data sets in the area of such density. EDE effectively solves the instability and low efficiency of traditional regional division methods.

To solve the enormous computational overhead and time consumption caused by regional density estimation for multidimensional data, PCA-based change detection (PCA-CD) [14] introduced the method of principal component analysis (PCA) to detect the changes in the multidimensional data flow. Other density-based drift detection methods include LDD-Density Synchronized Drift Adaptation (LDD-DSDA) [15] etc.

3.2 Statistical Test-Based Methods

The drift detection based on the statistical test does not need to divide the feature space of the data and directly compares the two data sets as a whole. This kind of unsupervised drift detection method usually has a fixed detection process.

Detecting change in data streams (DCDS) [16] is an earlier method that applies the statistical test method to drift detection. DCDS introduces the KS test and designs a data distribution measurement index called A-distance. For each time window, a triple is constructed as the input of distance function d, in which the triple includes the threshold of drift alarm, the start point and the end point of time window. The KS test used by DCDS is a very effective test method. Still, because the non-parametric test cannot summarize the parameters of the data distribution, all the data need to be involved in the calculation during the test, which will make the calculation of the whole test process become very large. Incremental KS test-based drift detector (IKS-BDD) [18] proposed an incremental KS test method, which improved the K-S test in terms of time complexity from the original O (NLGN) to O (LGN). Thus, a fast-unsupervised online concept drift detection is realized.

Hellinger distance-based drift detection (HDDDM) [17] uses Hellinger distance to calculate the difference between the distributions of the two data sets. HDDDM detects the concept drift in the form of batch processing, takes the training set of the prediction model as the initial reference data, and then accumulates the input data of the model to calculate the Hellinger distance between the two data blocks, then calculate the difference between the distance and the Hellinger distance between the previous data block and the reference data block. If it exceeds the set threshold, it is considered that concept drift has occurred.

Fast and accurate analog detection (FAAD) [19] uses information calculation and minimum spanning tree clusters to reduce the redundant dimension of sequence data. In addition, random sampling and subsequence partitioning based on the index probabilistic suffix tree is proposed to speed up the model construction and ensure the detection rate of sequence data flow.

4 Model Performance Representation-Based Methods

4.1 Classifier-Based Methods

Margin density drift detection (MD3) [20] introduces the concept of edge density. When the classifier classifies the data, the model will be difficult to deal with the data falling in the area near the decision boundary, and the number of samples falling in the area is the edge density. The greater the edge density, the greater the amount of data that the current classifier is difficult to handle, which indicates that the current classifier has suffered from performance degradation and cannot adapt to the current data distribution, that is, concept drift.

The online modified version of the page-Hinkley test (OMV-PHT) [21] method is based on the coincidence degree of the deterministic distribution of the classifier output. The classifier will give the distribution of the category it belongs to for the input. The part where the probability distributions of the two types coincide indicates that the classifier cannot classify this part of the data well. Therefore, the larger the area of this part is, the lower the classifier's performance will appear.

Semi-supervised adaptive novel class detection (SAND) [22] mainly detects whether there is concept drift based on the confidence of multiple classifiers, which is realized by maintaining multiple classifiers in a model set. SAND first detects the abnormal value of the data and puts the odd value into the buffer for accumulation. When the abnormal value accumulates to a certain amount, it detects the new category. If a new class is found, it creates a new one. Otherwise, it predicts the known class according to the model. Finally, the confidence degree of the new class and known classes' confidence degree is calculated. If there is a significant difference in the confidence degree, it is proved that there is a drift.

Plover [23] introduced the concept of uniform stability to detect concept drift. Uniform stability means that when an instance in the training set is replaced and the replacement object follows the same distribution, the difference in the loss function of the model on the two data sets will be less than a certain threshold. First, calculate the measurement function of the current data window, then calculate the difference value between the previous and current measurements, and finally calculate the divergence according to all the current difference values. If it exceeds the threshold, it is considered that concept drift has occurred.

4.2 Other Model-Based Methods

There are a small number of unsupervised drift detection methods based on other models. Among them, the more representative is concept drift detection via competency models (CM) [24]. This method mainly judges whether concept drift occurs based on the change of case representation ability in the case-based reason model. CM builds a capability model for the instance and measures the distribution change of the two example sets by using the difference in the example representation capability of the two instance sets on the whole group.

Another method is introducing the calculation of Shapely value in interpretability into the drift detection method. Label-less concept drift detection and explanation (L-Code) [25] belongs to this method. Shapely value can measure the importance of an input feature to the prediction result, which implies the correlation between the input feature and the outcome. Therefore, if the importance distribution between the input features changes, the previous prediction model cannot show the original performance on the new data because the old model will continue to regard the features with the currently reduced importance as essential features, resulting in the deviation of the prediction results. L-Code mainly includes two parts: an interpretation model and a detector. The interpretation model will receive input data, calculate the Shapely value of each feature, and input the Shapely value distribution into the detector. The detector will perform a statistical test according to the Shapely value distribution of the data in the reference window and the current window. Since L-Code only calculates the distribution of Shapely values for input data to judge the degradation of model performance, the drift detection method is model-independent.

5 Conclusion

In this paper, we analyze and summarize the existing unsupervised drift detection methods and divide the current ones into two categories: Based on data distribution differences and model performance. The primary classification basis is the method to judge the method of concept drift. On the one hand, the unsupervised drift detection method based on the difference in data distribution starts with the data directly. When it detects the change in data distribution, it is considered that the concept drift occurs. On the other hand, the unsupervised drift detection method based on the performance of the model starts from the perspective of the model and indirectly proves the decline of the model performance through label-independent indicators to detect the concept drift.

We also analyze ten methods based on data distribution differences and distinguish them according to their different calculation methods of data distribution changes. Most of them calculate data distribution changes based on regional density. We also analyze six performance methods based on the model and classify them into subclasses according to the different models used, most of which are based on classifier detection methods. In future work, we will experiment with the unsupervised detection method under the same conditions and select the one with the best performance under different conditions to make up for the vacancy in this part of the work.

Acknowledgments. This work is supported by the National Natural Science Foundation of China under Grant No. 62172316 and the Key R&D Program of Hebei under Grant No. 20310102D. This work is also supported by the Key R&D Program of Shaanxi under Grant No. 2019ZDLGY13-03-02, and the Natural Science Foundation of Shaanxi Province un-der Grant No. 2019JM-368.

References

1. Widmer, G., Kubat, M.: Learning in the presence of concept drift and hidden contexts. Mach. Learn. **23**(1), 69–101 (1996)
2. Gama, J., Medas, P., Castillo, G., Rodrigues, P.: Learning with drift detection. In: Brazilian Symposium on Artificial Intelligence, pp. 286–295. Springer, Berlin, Heidelberg (2004)
3. Baena-García, M., del Campo-Ávila, J., Fidalgo, R., Bifet, A., Gavalda, R., Morales-Bueno, R.: Early drift detection method. In: Fourth International Workshop on Knowledge Discovery from Data Streams, vol. 6, pp. 77–86 (2006)
4. Ross, G.J., Adams, N.M., Tasoulis, D.K., Hand, D.J.: Exponentially weighted moving average charts for detecting concept drift. Pattern Recogn. Lett. **33**(2), 191–198 (2012)
5. Bifet, A., Gavalda, R.: Learning from time-changing data with adaptive windowing. In: Proceedings of the 2007 SIAM International Conference on Data Mining, pp. 443–448. Society for Industrial and Applied Mathematics (2007)
6. Lu, J., Liu, A., Dong, F., Feng, G., Gama, J., Zhang, G.: Learning under concept drift: a review. IEEE Trans. Knowl. Data Eng. **31**(12), 2346–2363 (2018)
7. Wares, S., Isaacs, J., Elyan, E.: Data stream mining: methods and challenges for handling concept drift. SN Appl. Sci. **1**(11), 1–19 (2019). https://doi.org/10.1007/s42452-019-1433-0
8. Barros, R.S.M., Carvalho Santos, S.G.T.: A large-scale comparison of concept drift detectors. Inf. Sci. **451**, 348–370 (2018)
9. Gemaque, R.N., Costa, A.F.J., Giusti, R., Dos Santos, E.M.: An overview of unsupervised drift detection methods. Wiley Interdiscip. Rev. Data Min. Knowl. Disc. **10**(6), e1381 (2020)
10. Dasu, T., Krishnan, S., Venkatasubramanian, S., Yi, K.: An information-theoretic approach to detecting changes in multi-dimensional data streams. In: Proceedings of Symposium on the Interface of Statistics, Computing Science, and Applications (2006)
11. Song, X., Wu, M., Jermaine, C., Ranka, R.: Statistical change detection for multi-dimensional data. In: Proceedings of the 13th ACM SIGKDD International Conference on Knowledge Discovery and Data Mining, pp. 667–676 (2007)
12. Liu, A., Jie, L., Liu, F., Zhang, G.: Accumulating regional density dissimilarity for concept drift detection in data streams. Pattern Recogn. **76**, 256–272 (2018)
13. Gu, F., Zhang, G., Lu, J., Lin, C.-T.: Concept drift detection based on equal density estimation. In: 2016 International Joint Conference on Neural Networks (IJCNN), pp. 24–30. IEEE (2016)
14. Qahtan, A.A., Alharbi, B., Wang, S., Zhang, X.: A PCA-based change detection framework for multidimensional data streams: Change detection in multidimensional data streams. In: Proceedings of the 21th ACM SIGKDD International Conference on Knowledge Discovery and Data Mining, pp. 935–944 (2015)
15. Liu, A., Song, Y., Zhang, G., Lu, J.: Regional concept drift detection and density synchronized drift adaptation. In: IJCAI International Joint Conference on Artificial Intelligence (2017)
16. Kifer, D., Ben-David, S., Gehrke, J.: Detecting change in data streams. VLDB **4**, 180–191 (2004)
17. Ditzler, G., Polikar, R.: Hellinger distance based drift detection for nonstationary environments. In: 2011 IEEE Symposium on Computational Intelligence in Dynamic and Uncertain Environments (CIDUE), pp. 41–48. IEEE (2011)

18. dos Reis, D.M., Flach, P., Matwin, S., Batista, G.: Fast unsupervised online drift detection using incremental Kolmogorov-Smirnov test. In: Proceedings of the 22nd ACM SIGKDD International Conference on Knowledge Discovery and Data Mining, pp. 1545–1554 (2016)
19. Li, B., Wang, Y.-J., Yang, D.-S., Li, Y.-M., Ma, X.-K.: FAAD: an unsupervised fast and accurate anomaly detection method for a multi-dimensional sequence over data stream. Front. Inf. Technol. Electron. Eng. **20**(3), 388–404 (2019). https://doi.org/10.1631/FITEE.1800038
20. Sethi, T.S., Kantardzic, M.: On the reliable detection of concept drift from streaming unlabeled data. Expert Syst. Appl. **82,** 77–99 (2017)
21. Lughofer, E., Weigl, E., Heidl, W., Eitzinger, C., Radauer, T.: Recognizing input space and target concept drifts in data streams with scarcely labeled and unlabelled instances. Inf. Sci. **355**, 127–151 (2016)
22. Haque, A., Khan, L., Baron, M.: Sand: semi-supervised adaptive novel class detection and classification over data stream. In Proceedings of the AAAI Conference on Artificial Intelligence, vol. 30, no. 1. 2016
23. de Mello, R.F., Vaz, Y., Grossi, C.H., Bifet, A.: On learning guarantees to unsupervised concept drift detection on data streams. Expert Syst. Appl. **117**, 90–102 (2019)
24. Lu, N., Zhang, G., Jie, L.: Concept drift detection via competence models. Artif. Intell. **209**, 11–28 (2014)
25. Zheng, S., van der Zon, S.B., Pechenizkiy, M., de Campos, C.P., van Ipenburg, W., de Harder, H., Nederland, R.: Labelless concept drift detection and explanation. In: NeurIPS 2019 Workshop on Robust AI in Financial Services: Data, Fairness, Explainability, Trustworthiness, and Privacy (2019)

Music Video Search System Based on Comment Data and Lyrics

Daichi Kawahara[✉], Kazuyuki Matsumoto, Minoru Yoshida, and Kenji Kita

Tokushima University, 2-1, Minamijosanjima-Cho, Tokushima 770-8506, Japan
daikawahara4@gmail.com, st_gakmuk@tokushima-u.ac.jp

Abstract. Many people currently use video-sharing services such as YouTube. Keyword search is prevalent in these services. It can be hard to find a video that matches the users' interests using keyword search unless appropriate words are used. In this study, we propose a method for retrieving music videos with similar impressions by analyzing comment data from YouTube viewers and music lyrics. The proposed method converts comments and lyrics into vectors using Word2Vec, and music videos with similar impressions are retrieved using fuzzy c-means clustering. According to the mean reciprocal rank (MRR) scores, it was clear that the output of the music videos had the same impression within the top three songs.

Keywords: Fuzzy c-means · Word2Vec · YouTube

1 Introduction

According to a survey conducted by the Ministry of Internal Affairs and Communications in Japan, the average time spent per weekday by all age groups by Internet usage category was 38.7 min for "watching video posting/sharing services," and teens and 20-somethings both spent more than 100 min per day off. This indicates that video-sharing services are a part of the lives of young people. Among these, YouTube is the most popular. There are more than two billion users worldwide, with more than one billion hours of listening time and billions of views per day. These changes affect how people experience music. According to a survey by the Recording Industry Association of Japan, 54.9% of the respondents stated that they use YouTube when they want to listen to music.

YouTube users watch a variety of videos from a variety of videos. However, it is difficult to find videos suited to one's interests by using keyword searches alone. Therefore, there is a need for a system that allows users to easily search for videos that are similar in content to the videos they are interested in or that may contain the desired information. Keyword search entails associating keywords to find unwatched videos, channels, and artists that are likely to interest the users. Therefore, a keyword-independent video-search system is required.

YouTube allows viewers to post comments on the content of each video, and we thought it would be possible to estimate the impression of a video by reading the emotions from multiple comments. This research targets music videos uploaded to YouTube and uses features extracted from comment data as queries to build a system that searches for similar videos based on likely user impressions. In addition, because comment data alone contain information unrelated to the music itself, which may adversely affect search performance, lyrics information, which is considered to be directly related to the impression of the music, is also used.

Comments express the viewer's impressions of the video, descriptions of the video's content, and reflections. However, existing keyword search functions that use the name of the artist or character of the music included in the comments as a keyword only yield search results that are directly related to the keyword or videos in which the artist appears. In this case, it is anticipated that users will not be able to search for videos they have not yet watched. The goal of this research is to produce songs that are similar in terms of the universal impressions that viewers are likely to have of the videos and the content they are looking for, rather than to determine the similarity of videos based on whether specific keywords are included. As a result, in the analysis of comment data, we exclude proper nouns such as people and characters and capture the characteristics of words such as "cool" and "impressed."

2   Related Research

This section describes previous studies using lyrics and comments as features for music estimation.

2.1   Lyric Features in Music Impression Estimation

Hu [1] compared the classification accuracy of song lyrics with that of acoustic features such as acoustic signals. The proposed method is based on the TF-IDF method, which weighs the features obtained by dividing the lyrics into word sets. A total of 2829 songs were used in the evaluation experiment, and the results showed that the method using only lyrics slightly outperformed the classification method using acoustic features, and that the combination of lyrics and acoustic features further improved the classification accuracy.

The problem with this study is that the accuracy drops when the lyrics do not match the impression of the music very well. This study used viewer comments as information other than lyrics to cope with this situation.

2.2   Video Classification Using Comments

Ren et al. [2] classified videos posted on Bilibili using time-synchronized comments (TSC). They combined a deep neural network with a multilayer neural network. The method focuses on the TSC of each video and the TSC of each user and learns based on video features, user features, and comment time.

Konishi [3] employed the Louvain method [4] to cluster videos posted on Nico Nico Douga (NND)[1] based on the distribution of comments in the TSC by calculating the importance of users who commented on the features of the comment distribution.

---

[1] https://www.nicovideo.jp/.

Sabu [5] used Vader and Text-Blob to extract emotions from movie trailer comments, which he then combined with historical and upcoming movie data to generate a list of recommendations for upcoming movies.

Melody [6] collected comments on trending videos over two years and used BERT to predict the categories of videos attached to the comments. The results indicated a classification accuracy of approximately 53%.

2.3 Music Estimation Using Comments

Nawaz et al. [7] and Phakhawat et al. [8] conducted research on emotion estimation from YouTube comments and achieved up to 84% accuracy.

Yamamoto et al. [9] attempted to classify music videos based on user-submitted time-synchronized NND comments based on impressions such as "cute" and "cheer up." In this study, by restricting the comments used to adjectives and adjectival verbs and using comments in the chorus section of the song, it was possible to classify the impression of the song with high accuracy, with a macro mean F value of 0.659. Tsuchiya et al. [10] performed impression estimation based on comments from a chorus section. He observed that it is easy to estimate impressions that easily express emotions, such as "cute," from comments, whereas it is difficult to estimate impressions such as "eccentric" from comments.

In this study, by adding new information such as lyrics and comments, it is possible to capture not only similarities in video content but also similarities in impressions of song content. Furthermore, this study uses comment data from NND but differs in that it also uses comments from YouTube.

3 Methodology

3.1 Overview of the Proposed Method

The method proceeds as follows. The flow is depicted in Fig. 1.

- Selection of music videos for comment collection.
- Collection of comment data and lyrics of songs.
- Extracting features from collected data and assigning information.
- Search for similar music videos based on the information given.

There is a method for music retrieval based on impressions received from a music piece. A song impression is a viewer's subjective impression of a song, such as "cheer up" or "sad." If we can read the emotions that viewers may feel from the lyrics in addition to the comments, which directly reflect their emotions, we can determine the impression of a song and the similarity of the impression with other songs. The similar video retrieval system proposed in this study can retrieve videos of songs with similar impressions to a specific song by using the comments and lyrics of a song video as a search query, but it does not directly estimate a song's impression. Section 3.2 and subsequent sections provide more information.

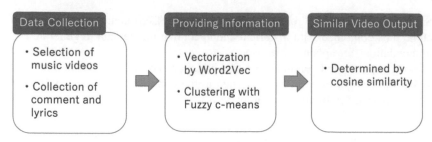

Fig. 1. Flow of the method

3.2   Collection of Comment Data and Lyrics

In this study, we collected comment data using Google's YouTube Data API v3[2]. We selected eight categories of impressions of Japanese songs and collected comments by referring to Oricon Music Store[3] and other sites. Table 1 lists the categories selected.

Table 1. Target impression categories

Songs to cheer up	Songs of heartbreak
Songs of gratitude	Songs that make you want to meet
Songs of friendship	Songs to get you excited
Songs of mutual love	Songs for when you're feeling down

The lyrics were retrieved using the lyrics search service site Utanet[4].

3.3   Vectorization of Comment Data and Lyrics

Vectors were assigned to the created comment and lyrics datasets using a model (Wikipedia entity vector) from Word2Vec [11, 12], which was pre-trained using Japanese Wikipedia sentences. Features were extracted for each word by referring to the Word2Vec model using the comment sentences and lyrics as input, and a list of word vectors was generated. The generated list of word vectors is averaged and considered a single vector. The output vector has 200 dimensions.

3.4   Clustering of Comment Data and Lyrics

In this study, the set of comments and lyric vectors was clustered, the degree of belonging to each cluster was calculated for each vector, and the average degree of belonging was calculated for each video to create a new vector. Clustering was used to calculate the

---

[2] https://developers.google.com/youtube/v3.

[3] https://music.oricon.co.jp/php/special/Special.php?pcd=sp147.

[4] https://www.uta-net.com/.

features of each video. Figure 2 shows the clustering flow. Fuzzy c-means [13–15] was used for clustering. The distance measure was the Euclidean distance.

While non-hierarchical clustering, such as k-means, can only determine whether an individual belongs to a cluster as 0–1, fuzzy clustering can be expressed in the range of 0 to 1, making it possible to belong to multiple clusters, which is a characteristic of fuzzy clustering. Comments are expressed in various forms, such as short sentences or words to express the viewer's impression or explanation of the video or in the style of dialogue. Therefore, the use of words is not uniform, and considering the diversity of their meanings, it is difficult to imagine a situation in which a word vector belongs to a single cluster. In addition, the growth of music in recent years has made the elements of music more complex, and it is not always possible to categorize music into only one category, therefore, so fuzzy clustering was used to allow for multiple affiliations.

The comment and lyrics vectors of each video were used as input, the degree of belonging of the comments and lyrics to each cluster was output, the average of the degree of belonging to each cluster was calculated, and a vector of the number of cluster dimensions representing the average degree of belonging of each cluster was created for each set of comments and lyrics of a song. Because the value of affiliation is also close when the cluster distribution of comments and lyrics is close, we believe that the affiliation value can be used to compare whether the impressions of the music are similar. In other words, we attempt to express the characteristics of the comments and lyrics of the music videos in vectors created by clustering.

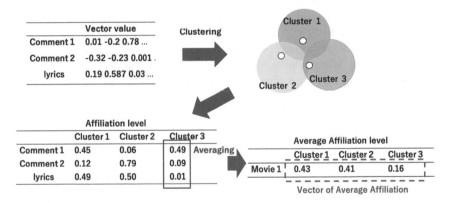

Fig. 2. Clustering process flow

3.5 Similar Music Output

Using the average affiliation vector for each video created in the previous section, we determine the cosine similarity between the vectors for each video for all videos, where A and B are the two mean affiliation vectors. Then, the cosine similarity can be expressed by Eq. (1). Figure 3 shows an example of actual search results. The 0th video of the output video was identical to that of the input video.

$$cosine(A, B) = \frac{A \cdot B}{\|A\| \|B\|} \tag{1}$$

	title	similarity	category
0	Iruka / Nagori Yuki	1.000000	4
1	RADWIMPS / me me she	0.940787	4
2	Nisino Kana / Darling	0.906276	3
3	miwa / Kimi Ni Deaeta Kara	0.902627	2
4	AAA / Sayonara No Maeni	0.894657	4

Fig. 3. Example of output results

4   Experimental Result

In this section, we confirm the performance of the proposed music video retrieval system. In the performance evaluation, we considered the retrieval performance to be higher when the impression category of the input video matched the impression category of the top music video in the output music video list.

4.1   Experiment Preparation

Data used in the experiment. Comment data for the eight categories of music videos listed in Table 2 were obtained in this experiment, as described in Sect. 3.2. Videos were obtained from official music videos of artists and music company channels as of December 2021, if they existed, or from videos with more than 500,000 views and the most comments, if no official videos existed. Each video received one hundred comments, beginning with the highest viewer ratings.

4.2   Experiment

Experimental contents. Vector values were assigned to the acquired comment and lyrics data using Word2Vec. Clustering was used to determine the degree of belonging of the comments and lyrics to each cluster, and the average value of each cluster was expressed as a cluster average degree of belonging vector for all videos. Clustering was performed with the number of clusters set to 5, 10, 20, 30, 40, 50, and 60, and the parameter M set to 1.2–2.0 for each cluster to see if the performance differed depending on the clustering settings.

Calculate the cosine similarity between the average affiliation vectors for all videos. For each input video (search query), five songs were selected as search results in the order of similarity among all videos, excluding the search query.

Search evaluation by mean reciprocal rank. For a quantitative similarity music video retrieval system, the mean reciprocal rank (MRR) was used as an evaluation index. MRR is calculated based on the rank of the first input music category number and the same music output from the top of the output list, and the reciprocal of the rank is the score. U is assumed to be the set of all songs, U is each song, and k_u is the rank of the

Table 2. Number of experimental data

Category No.	Category name (abbreviation)	Number of videos	Number of comments
0	Songs to cheer up (cheer)	40	4000
1	Songs of gratitude (gratitude)	28	2800
2	Songs of friendship (friendship)	23	2300
3	Songs of mutual love (mutual)	29	2900
4	Songs of heartbreak (heartbreak)	38	3800
5	Songs that make you want to meet (meet)	20	2000
6	Songs to get you excited (excited)	20	2000
7	Songs for when you're feeling down (down)	20	2000
Total		218	21,800

first occurrence of a song of the same category as U in the search list of song U, the calculation of MRR can be expressed as in Eq. (2).

$$MRR = \frac{1}{|U|} \sum_{u \in U} \frac{1}{k_u} \qquad (2)$$

The MRR was calculated using all videos as input, and the average of the obtained values was defined as the score of a similar music video retrieval system. Ten attempts were made to calculate the MRR. And the average of each score was obtained because the initial array values for clustering by fuzzy c-means were random. Some of these scores are presented in Table 3.

Scores were measured for various numbers of clusters and parameters and uniformly resulted in an average score between 0.33 and 0.4. This score indicates that on average, the output of similar videos in the same category as the input videos is within the top three tracks.

Table 4 shows the MRR and percentage of correct responses for each category. The correctness rate is the percentage of songs in the same category as the input song among the top 10 songs retrieved, as indicated by the average score of 10 trials. The results are shown in Table 3 when the number of clusters with the highest average score was 60 and the parameter m = 1.2 was used.

As shown Table 4, the values varied significantly depending on the impression category. Some categories had scores approaching 0.5, where as others had scores below 0.33, the overall average score, suggesting that the number of videos per category had no small effect.

Performance comparison with simple data set. The experiments in the previous section demonstrated that the constructed system can achieve a certain degree of accuracy. This section shows that we compare the performance of this system with a retrieval

Table 3. Score by MRR

Number of clusters	Parameter	Average score	Maximum score	Minimum score
50	1.2	0.354	0.396	0.305
50	1.3	0.339	0.415	0.300
50	1.4	0.333	0.369	0.297
50	1.5	0.332	0.351	0.309
50	1.6	0.342	0.381	0.311
10	1.2	0.342	0.360	0.318
20	1.2	0.338	0.363	0.320
30	1.2	0.345	0.358	0.322
40	1.2	0.356	0.369	0.335
60	1.2	0.359	0.375	0.344

Table 4. Scores per impression category

Category	Average score	Maximum score	Minimum score	Percentage of correct (%)
Cheer	0.419	0.451	0.386	33.6
Gratitude	0.301	0.331	0.292	20.2
Friendship	0.297	0.333	0.288	15.8
Mutual	0.342	0.374	0.315	16.2
Heartbreak	0.445	0.480	0.386	28.6
Meet	0.284	0.320	0.253	12.5
Excited	0.278	0.315	0.256	11.8
Down	0.298	0.325	0.266	12.1

system using only comments and a retrieval system using only lyrics to demonstrate the usefulness of the proposed system by combining comment data with lyrics information. As the feature for both, the cluster average affiliation vector from fuzzy c-means was used, and the average MRR score was obtained after 10 trials for each number of clusters and parameters. Table 5 displays the average score comparison results.

Although there were some areas where the scores were higher for comments alone depending on the number of clusters and parameters, overall, the scores for both comments and lyrics were slightly higher.

5 Discussion

Overall, search accuracy was achieved; however, accuracy decreased in some categories. Table 6 displays the search results for these categories.

Table 5. Score comparison by MRR

Number of clusters	Parameter	Comments only	Lyrics only	Comments and lyrics
50	1.2	0.350	0.329	0.354
50	1.3	0.334	0.330	0.339
50	1.4	0.344	0.339	0.333
50	1.5	0.332	0.326	0.332
50	1.6	0.340	0.319	0.342
10	1.2	0.334	0.309	0.342
20	1.2	0.338	0.334	0.338
30	1.2	0.347	0.321	0.345
40	1.2	0.356	0.315	0.356
60	1.2	0.354	0.344	0.359
Average		0.339	0.326	0.344

Table 6. Example of Failures

Input/output	Song titles	Category
Input	Hirai Ken / LIFE is…	Down
Output 1st	Official HIGE DANdism/ FIRE GROUND	Excited
Output 2nd	Little Glee Monster / My Best Friend	Friendship
Output 3rd	Yuzu / Tomo ∼Tabidachi No Toki∼	Friendship
Output 4th	The Drifters / Ii Yu Dana	Cheer
Output 5th	UVERworld / Hi! Mondaisaku	Cheer

Table 6 shows an example in which a song with the same impression as the input did not appear in the top five search results. The data from the six songs' comments were examined, and it was confirmed that the word "deru" was common and frequently appeared in the comments. In the "Down" category, there were many comments such as "Namida ga deru (I am crying)," but in the "Excited" and "Cheer" categories, the word "deru" was often used with a different meaning from the input music, such as "Genki ga deru (I feel energetic)". This case can be considered a failure caused by Word2Vec's inability to capture differences in word meanings based on context.

6 Conclusions

In this study, we proposed a system to retrieve similar music videos based on the impression that the viewers of music videos tend to have, using comment data and lyrics, and after assigning vectors to comments using a pre-trained word variance representation model by Word2Vec, we proposed a system to retrieve similar music videos using Fuzzy c-means. Clustering and similar video retrieval were accomplished by calculating cosine similarity between videos based on average affiliation. The mean vector cosine similarity between all videos was calculated in the experiment using the comment data and lyrics set of 218 collected music videos as input. In addition, we used MRR as a quantitative system evaluation index to assess the retrieval performance of music videos with the same impression. The results show that the system can achieve high accuracy as a retrieval system to a certain extent. However, there were cases in which words whose meanings changed depending on context were not well captured.

In this experiment, the number of targeted music videos and comments was small, and the search targets were biased, making it difficult in some cases to determine whether the videos were similar or not. We believe that increasing the number and categories of target music videos and comments will create more diversity in the features included in the comments and have a greater impact on the average degree of affiliation by clustering. On the other hand, increasing the amount of target data may result in noisy data, and some songs may not be considered to belong to only one impression category, so it is necessary to improve the data set collection method and search performance evaluation method. Furthermore, it was not possible to capture word impressions influenced by context for some words. As a result, since Word2Vec, which returns one vector per word, is insufficient, we would like to look into BERT or its extended version, Sentence BERT, which takes an entire sentence as input and is good at understanding context.

Acknowledgements. This work was supported by the 2022 SCAT Research Grant and JSPS KAKENHI Grant Number JP20K12027, JP21K12141.

References

1. Hu, X., Downie, J., Ehmann, A.: Lyric text mining in music mood classification. In: 10th International Society for Music Information Retrieval Conference (2009)
2. Ren, H., Wang, D.: TRRS: temporal recurrent recommender system based on time sync comments. In: Proceedings of International Conference on Machine Learning and Soft Computing, ICMLSC, pp. 123–127 (2019)
3. Konishi, A., Hosobe, H.: Clustering Nico Nico Douga videos by using the distribution of time-synchronized comments. In: Proceedings of the 2020 IEEE/WIC/ACM International Joint Conference on Web Intelligence and Intelligent Agent Technology, WI-IAT, pp. 520–525 (2020)
4. Blondel, V.D., Guillaume, j.-L., Lambiotte, R., Lefebvre, E.: Fast unfolding of communities in large networks. J. Statist. Mech. **2008**(10008), 1–12 (2008)
5. Sahu, S., Kumar, R., MohdShafi, P., Shafi, J., Kim, S., Ijaz, M.F.: A hybrid recommendation system of upcoming movies using sentiment analysis of YouTube trailer reviews. Mathematics. https://doi.org/10.3390/math10091568. Last accessed 1 June 2022
6. Melody, L.: Classifying comments on YouTube via pre-training of deep bidirectional transformers for language understanding. A Master's paper for the M.S. in I.S degree, 37 p (2021)
7. Nawaz, S., Rizwan, M., Rafiq, M.: Recommendation of effectiveness of YouTube video contents by qualitative sentiment analysis of its comments and replies. Pak. J. Sci. **71**(4) (2019)

8. Phakhawat, S., Thanaruk, T., Choochart, H.: Classifying emotion in Thai YouTube comments. In: IC-ICTES (2015)
9. Yamamoto, H., Nakamura, S.: Using viewers' time-synchronized comments for mood classification of music video clips. Information Processing Society of Japan, Transactions on Databases TOD **6**(3), 61–72 (2013)
10. Tsuchiya, S., Ono, N., Nakamura, S., Yamamoto, H.: A proposed method for estimating music video impressions from social comments. DEIM Forum (2016)
11. Mikolov, T., Chen, K., Corrado, G., Dean, J.: Efficient estimation of word representations in vector space. In: International Conference on Learning Representations, ICLR (2013)
12. Mikolov, T., Sutskever, I., Chen, K., Corrado, G., Dean, J.: Distributed representations of words and phrases and their compositionality. In: Twenty-Seventh Conference on Neural Information Processing Systems, NIPS (2013)
13. Dunn, J.: A fuzzy relative of the ISODATA process and its use in detecting compact well-separated clusters. J. Cyber. **3**(3) (1973)
14. Ross, T.J.: Fuzzy Logic with Engineering Applications, 3rd edn. Wiley (2010)
15. Winkler, R., Klawonn, F., Kruse, R.: Fuzzy c-means in high dimensional spaces. Int. J. Fuzzy Syst. Appl. **1**(1) (2011)

Data Analysis of Nobel Prizes in Science Using Temporal Soft Sets

Feng Feng[1](✉), Jing Luo[2], Jianke Zhang[1], and Qian Wang[3]

[1] Department of Applied Mathematics, School of Science, Xi'an University of Posts and Telecommunications, Xi'an 710121, China
fengf@xupt.edu.cn, jiankezhang@xupt.edu.cn
[2] School of Communications and Information Engineering, Xi'an University of Posts and Telecommunications, Xi'an 710121, China
luoj@stu.xupt.edu.cn
[3] School of Economics and Management, Beijing University of Posts and Telecommunications, Beijing 100876, China
chandler@bupt.edu.cn

Abstract. In many real-life applications, some item sets are only frequent during certain periods of time. However, traditional association analysis techniques may fail to find interesting connections between item sets in this scenario, owing to the ignorance of temporal information. To overcome this difficulty, we try to enhance classical rule extraction by means of temporal soft sets in this study. The concept of temporal granulation mappings is used to produce the granular structure associated with a given set consisting of temporal transactions. By virtue of temporal granulation mappings, temporal soft sets and Q-clip soft sets are defined. On this basis, we construct a temporal soft set based framework for describing and mining temporal association rules. We also develop a temporal association rules mining algorithm, which extends the Apriori method in the temporal soft set setting. Moreover, a case study regarding Nobel Prizes in science is conducted on a real-life data set to validate the efficacy of the presented approach. Experimental results verify the efficacy of the proposed method, and comparative analysis further reveals that our approach can extract some strong and promotive rules that are ignored by traditional methods.

Keywords: Data mining · Temporal soft set · Q-clip soft set · Association rule · Fuzzy set

1 Introduction

In the process of data mining, there exists a lot of uncertain information in the massive data, which affects our accurate perception and prediction of real problems in the cognitive process. Molodtsov put forward the notion of soft sets for the first time [1], providing a new research direction for handling uncertainty.

© The Author(s), under exclusive license to Springer Nature Switzerland AG 2023
N. Xiong et al. (Eds.): ICNC-FSKD 2022, LNDECT 153, pp. 1136–1145, 2023.
https://doi.org/10.1007/978-3-031-20738-9_123

Soft set theory describes uncertainty based on the perspective of parameterization. With the parameter space associated with the universe of discourse, it can achieve more abundant information description and processing.

As a prominent branch of data mining, association rule mining aims at finding potentially meaningful connections between item sets. Association rule mining techniques can help discover business values hidden in transaction data sets (TDSs). Agrawal et al. [2] proposed the concept of association rules for the first time, and put forward the description of association rules mining problems and Apriori algorithm. It is a representative algorithm of candidate item sets generation. Candidate patterns are generated by permutation and combination for several times to increase the length of patterns. Then obtain the frequent patterns by successive screening, so as to generate the association rules. Time feature is an inherent attribute characteristic of a database. The appearance of temporal data makes us have to consider the influence of time factor in data mining analysis. Temporal association rule mining focuses on discovering valuable association rules during different periods of time. Ye et al. [3] proposed an association rules mining approach that can be directly applied to temporal data sets. Ou-Yang et al. [4] put forward the time interval extension and merging technology and a new temporal association rules discovery method, which further popularized the application of association rules.

In 2011, Herawan and Deris [5] opened up a new direction of mining association rules with the soft set theory. Feng et al. [6] modified some important concepts in literature [5] and further improved and clarified the nature of association rules mining based on soft sets. Subsequently, Feng et al. [7] developed a novel research route, using soft sets theory and soft sets logic formulas as the main tools to describe and mine association rules. In 2020, Feng et al. [8] provided unified mathematical characterizations of some core concepts used in maximal and regular association rule mining. Besides, three algorithms were developed to extract maximal association rules using logical formulas over soft sets. However, none of the above methods take the temporal factor into consideration. In real-world scenario, it may happen that some item sets are frequent only in a specific period, but not in the time span of the whole data set. The above-mentioned soft set based methods all ignore the temporal association between some item sets and cannot identify some useful association rules. To fill this gap, it is meaningful and necessary to consider how to extract the association rules which are strong during certain periods of time.

In this paper, an enhanced association rule mining method with temporal soft sets is developed. We introduce temporal soft sets and their Q-clip soft sets by virtue of temporal granulation mappings, and design the Apriori-based strong and promotive temporal association rule mining (Apriori-SPTARM) approach for extracting temporal association rules with temporal soft sets. Moreover, we employ the Apriori-SPTARM method to analyze a data set regarding Nobel Prizes in science and compare our approach with the Apriori method in terms of the quantity of strong and promotive rules, so as to validate the effectiveness and superiority of our approach.

2   Preliminaries

Let U be the universe of discourse. All parameters associate with the objects in U constitute the parameter space E. In what follows, $P(U)$ represents the class of all subsets of U.

Definition 1. [1] We call an ordered pair $\Omega = (G, B)$ as a soft set over U, in which $B \subseteq E$ is a parameter set and $G : B \to P(U)$ is referred to as the approximate function of the soft set Ω.

The subset $G(b)$ determined by the approximation function is referred to as the b-approximation set of the soft set Ω, where $b \in B$ is any parameter.

Definition 2. [6] Let $\Omega = (G, B)$ be a soft set over U and $u \in U$. Then we call $Co_\Omega(u) = \{b \in B : u \in G(b)\}$ as the parameter coset of U in Ω.

Let $I = \{i_1, i_2, \cdots, i_{|I|}\}$ be an item domain, which is the set of all items in the TDS. Every transaction is a nonempty subset of I and corresponds to a unique transaction identifier. A TDS $D = \{t_1, t_2, \cdots, t_{|D|}\}$ contains all transactions.

Suppose that D is a TDS, $t \in D$ is a transaction and $\mathscr{P} = \{p_1, p_2, \cdots, p_{|\mathscr{P}|}\}$ is a set composed of pairwise disjoint periods of time. If $p_t = \gamma(t) \in \mathscr{P}$ (meaning that t appears during the period p_t), then we call p_t as the period marker of t, and we refer to $W = (D, \gamma, \mathscr{P})$ as a temporal TDS.

3   Association Rule Mining Using Temporal Soft Sets

3.1   Temporal Soft Sets

Definition 3. [9] Assume that $\mathscr{P}$ is a set of pairwise disjoint time periods and U is an universe of discourse. Then we call $\gamma : U \to \mathscr{P}$ as a temporal granulation mapping.

Definition 4. [9] Let $\mathfrak{W} = (G, B, \gamma, \mathscr{P})$ be a quadruple such that
(1) (G, B) is a soft set;
(2) $\gamma : U \to \mathscr{P}$ is a temporal granulation mapping.
Then we refer to $\mathfrak{W}$ as a temporal soft set (TSS).

As an abstract data representation, the TSS $\mathfrak{W}$ can describe temporal information that cannot be depicted by its underlying soft set (G, B).

Definition 5. [9] Suppose that $\mathfrak{W} = (G, B, \gamma, \mathscr{P})$ is a TSS over U, and $Q \subseteq \mathscr{P}$. The soft set $\mathfrak{W}_Q = (H, B)$ over $\gamma^{-1}(Q) = \{u \in U | \gamma(u) \in Q\}$ is referred to as the Q-clip of $\mathfrak{W}$, where $H(b) = G(b) \cap \gamma^{-1}(Q)$ for all $b \in B$.

3.2 Temporal Association Rules Characterized by Soft Sets

Definition 6. [9] Suppose that $Y \subseteq B$ is a nonempty item set and $\mathfrak{W} = (G, B, \gamma, \mathscr{P})$ is a TSS over U, where $\Omega = (G, B)$ is its underlying soft set. Then we refer to

$$\triangle_{\mathfrak{W}}^{Q}(Y) = \{u \in \gamma^{-1}(Q) : Y \subseteq Co_{\Omega}(u)\}$$

as the Q-realization of Y.

We call the cardinality of the set $\triangle_{\mathfrak{W}}^{Q}(Y)$ as Q-support of Y in $\mathfrak{W}$, denoted by $S_{\mathfrak{W}}^{Q}(Y)$. The expression $Y \Rightarrow^{Q} Z$ is called a temporal association rule (TAR) in $\mathfrak{W}$, where $Y, Z \subseteq B$ are two disjoint nonempty item sets. And we say that Y is the antecedent of the TAR, and Z is the consequent.

Definition 7. [9] The temporal realization of the TAR $Y \Rightarrow^{Q} Z$ in $\mathfrak{W}$ is given by:

$$\triangle_{\mathfrak{W}}^{Q}(Y \Rightarrow^{Q} Z) = \triangle_{\mathfrak{W}}^{Q}(Y \cup Z). \tag{1}$$

We say that

$$S_{\mathfrak{W}}^{Q}(Y \Rightarrow^{Q} Z) = S_{\mathfrak{W}}^{Q}(Y \cup Z) = \mathrm{card}(\triangle_{\mathfrak{W}}^{Q}(Y \Rightarrow^{Q} Z)) \tag{2}$$

is the temporal support of $Y \Rightarrow^{Q} Z$ in $\mathfrak{W}$.

Definition 8. [9] The temporal confidence of $Y \Rightarrow^{Q} Z$ in $\mathfrak{W}$ is given by:

$$C_{\mathfrak{W}}^{Q}(Y \Rightarrow^{Q} Z) = \frac{S_{\mathfrak{W}}^{Q}(Y \Rightarrow^{Q} Z)}{S_{\mathfrak{W}}^{Q}(Y)} = \frac{S_{\mathfrak{W}}^{Q}(Y \cup Z)}{S_{\mathfrak{W}}^{Q}(Y)}. \tag{3}$$

In addition, $C_{\mathfrak{W}}^{Q}(Y \Rightarrow^{Q} Z) = 0$ if $S_{\mathfrak{W}}^{Q}(Y) = 0$.

Definition 9. The temporal lift of $Y \Rightarrow^{Q} Z$ in $\mathfrak{W}$ is given by:

$$L_{\mathfrak{W}}^{Q}(Y \Rightarrow^{Q} Z) = \frac{C_{\mathfrak{W}}^{Q}(Y \Rightarrow^{Q} Z)}{S_{\mathfrak{W}}^{Q}(Z)/|\gamma^{-1}(Q)|} = \frac{S_{\mathfrak{W}}^{Q}(Y \cup Z) \cdot |\gamma^{-1}(Q)|}{S_{\mathfrak{W}}^{Q}(Y) \cdot S_{\mathfrak{W}}^{Q}(Z)}. \tag{4}$$

In addition, $L_{\mathfrak{W}}^{Q}(Y \Rightarrow^{Q} Z) = 0$ if $S_{\mathfrak{W}}^{Q}(Y) \cdot S_{\mathfrak{W}}^{Q}(Z) = 0$.

If $L_{\mathfrak{W}}^{Q}(Y \Rightarrow^{Q} Z) > 1$, the appearance of Y promotes the occurrence of Z. That is, there exists a positive correlation between Y and Z. In this case, the TAR $Y \Rightarrow^{Q} Z$ in $\mathfrak{W}$ is said to be a promotive rule during a period in Q. In order to extract meaningful rules in the TDSs, we should set the minimum temporal support (min-Tsupp) and the minimum temporal confidence (min-Tconf). The item set Y is a temporal frequent item set during a period in Q, if $S_{\mathfrak{W}}^{Q}(Y) \geq$ min-Tsupp. When $S_{\mathfrak{W}}^{Q}(Y \Rightarrow^{Q} Z) \geq$ min-Tsupp and $C_{\mathfrak{W}}^{Q}(Y \Rightarrow^{Q} Z) \geq$ min-Tconf, the TAR $Y \Rightarrow^{Q} Z$ in $\mathfrak{W}$ is deemed as a strong rule during a period in Q.

3.3   The Proposed Algorithm

To extract promotive temporal association rules (PTARs) which are strong, we present a mining method combining the Apriori with TSS based association rule mining. In the sequel, our method will be abbreviated to Apriori-SPTARM. The pseudo-code description of our method is given below.

Algorithm   Apriori-SPTARM method

Input: $W = (D, \gamma, \mathscr{P})$, $Q \subseteq \mathscr{P}$, min-Tsupp: $\alpha \in N^*$, min-Tconf: $\beta \in (0, 1]$.
Output: The set $\text{SPTAR}(W, Q, \alpha, \beta)$ containing all temporal association rules which are strong and promotive during a period in Q.
1: Establish a temporal soft set $\mathfrak{W} = (G, I, \gamma, \mathscr{P})$
2: Compute $\gamma^{-1}(Q)$ and establish the Q-clip soft set $\mathfrak{W}_Q = (C, I)$
3: Scan the data set divided by $\mathfrak{W}_Q = (C, I)$ to generate the candidate 1-item sets, and retain frequent 1-item sets
4: Generate candidate $(k + 1)$-item sets with connecting and pruning, and retain the frequent $(k + 1)$-item sets
5: Construct the set $\text{FoS}(\mathfrak{W}_Q, \alpha_Q)$ containing all temporal frequent item sets
6: **for** $Y \in \text{FoS}(\mathfrak{W}_Q, \alpha_Q)$ **do**
7:     **for** $Z \in \text{FoS}(\mathfrak{W}_Q, \alpha_Q)$ **do**
8:         **if** $Y \cap Z = \emptyset$ **then**
9:             Calculate $S_{\mathfrak{W}}^Q(Y \Rightarrow^Q Z)$
10:        **end if**
11:        **if** $S_{\mathfrak{W}}^Q(Y \Rightarrow^Q Z) \geq \alpha$ **then**
12:            Calculate $C_{\mathfrak{W}}^Q(Y \Rightarrow^Q Z)$ and $L_{\mathfrak{W}}^Q(Y \Rightarrow^Q Z)$
13:            **if** $C_{\mathfrak{W}}^Q(Y \Rightarrow^Q Z) \geq \beta$ and $L_{\mathfrak{W}}^Q(Y \Rightarrow^Q Z) > 1$ **then**
14:                Put the rule $(Y \Rightarrow^Q Z)$ into the set $\text{SPTAR}(W, Q, \alpha, \beta)$
15:            **end if**
16:        **end if**
17:    **end for**
18: **end for**
19: **return** $\text{SPTAR}(W, Q, \alpha, \beta)$

---

4   Data Analysis of the Nobel Prizes in Science

The data set contains the information of the Nobel Laureates between the year from 1901 to 2020. Only the data related to the Nobel Prizes in Chemistry (NPC), Physics (NPP) and Physiology or Medicine (NPPM) are analyzed here. We finally obtained 337 pieces of simplified data satisfying the conditions by preprocessing the original data set. And it only contains three attributes of award year, category and affiliation country.

The initial domain U contains all the Nobel Prizes in science from 1901 to 2020, and the attributes related to the Nobel Prizes constitute the corresponding parameter space E. Suppose that there is a parameter set $B = C_1 \cup C_2$, where $C_1 = \{c_1, c_2, \cdots, c_{30}\}$ is the set about affiliation countries, and the set $C_2 = \{n_1, n_2, n_3\}$ represents the category, where n_1, n_2, n_3 are NPC, NPP, and

NPPM respectively. Assume that $\mathscr{P} = \{p_1, p_2, \cdots, p_6\}$ is a set of pairwise disjoint periods, where $p_1 = [1901, 1920], \cdots, p_6 = [2001, 2020]$. The temporal granulation mapping can induce the following divisions: $\{\gamma^{-1}(p_k)|p_k \in \mathscr{P}\} = \{\{u_1, \cdots, u_{52}\}, \cdots, \{u_{278}, \cdots, u_{337}\}\}$, where $k = 1, 2, \cdots, 6$. Therefore, we can establish a TSS $\mathfrak{W} = (G, B, \gamma, \mathscr{P})$ over U.

Table 1. Supports of parameter sets of eight major countries in $\mathfrak{W}$

Parameter set	Supp	p_1-Supp	p_2-Supp	p_3-Supp	p_4-Supp	p_5-Supp	p_6-Supp
{USA}	179	3	10	28	41	49	48
{UK}	76	8	13	13	19	6	17
{Germany}	42	19	16	1	0	0	6
{France}	27	9	3	0	4	4	7
{FRG}	19	0	0	4	8	7	0
{Switzerland}	19	2	2	3	2	7	3
{Japan}	16	0	0	1	1	2	12
{Sweden}	15	3	3	3	3	3	0

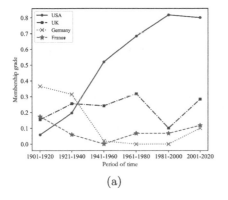 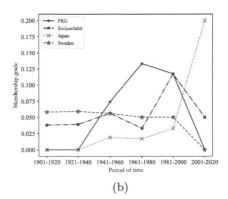

(a)   (b)

Fig. 1. Membership grade of eight countries during each period

The country $c_j (j = 1, 2, \cdots, 8)$ whose support of parameter set is not less than 12 in $\mathfrak{W}$ is selected as the main country. We can construct a fuzzy set $T^{p_k}(c_j) = \frac{S_{\mathfrak{W}}^{p_k}(\{c_j\})}{|\gamma^{-1}(p_k)|}$ in C_1 with the supports of parameter sets in Table 1, where $k = 1, 2, \cdots, 6$, and $j = 1, 2, \cdots, 8$. The fuzzy set T can describe the overall scientific influence of the eight countries in each period. According to the results shown in Fig. 1, Germany's scientific influence has gradually declined contrast to the USA, while the UK has remained steady. From 1941 to 2000, the strength about scientific and technological innovation of Japan improved smoothly, and developed rapidly after the 21st century, only second to the UK and the USA.

It can also study the scientific influence of the USA and the UK about different fields during six periods. Define a fuzzy set $F_{n_l}^{p_k}(c_j) = \frac{S_{\mathfrak{M}}^{p_k}(\{c_j\})}{|\gamma_{n_l}^{-1}(p_k)|}$ in C_1, where $\gamma_{n_l}^{-1}(p_k) = \gamma^{-1}(p_k) \cap G(n_l)$, $k = 1, 2, \cdots, 6$, $l = 1, 2, 3$, and $j = 1, 2$ (representing the USA and the UK respectively). Then calculate the membership grade of these two countries in each period. As depicted in Fig. 2, it is not hard to know that the impact of the USA increased year by year, and its impact in the three fields were monopolized after 1981. From 1901 to 1980, continuous progress was made in the field of chemistry in the UK and it eventually reached the highest level. In the physics domain, Britain's strength was strongest from 1921 to 1940, but declined over the next 60 years.

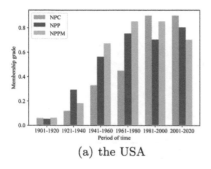

(a) the USA

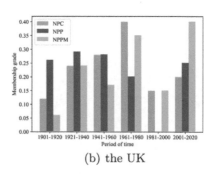

(b) the UK

Fig. 2. Membership grade of two countries in different fields during each period

Based on the constructed TSS, the Apriori-SPTARM method was used to extract the PTARs which are strong the six periods of time. The min-Tsupp and the min-Tconf in each period were set to 2 and 70% respectively.

Consider the PTAR {NPP} $\Rightarrow^{p_4}$ {USA}, we have $S_{\mathfrak{M}}^{p_4}(\{NPP\} \Rightarrow^{p_4} \{USA\}) = 15$ and $C_{\mathfrak{M}}^{p_4}(\{NPP\} \Rightarrow^{p_4} \{USA\}) = 75\%$, and p_4-lift of the rule is $L_{\mathfrak{M}}^{p_4}(\{NPP\} \Rightarrow^{p_4} \{USA\}) = 1.10$. This rule indicates that 75% of the Nobel Prizes in Physics were honored with the United States between 1961 and 1980. For the PTAR {NPPM} $\Rightarrow^{p_5}$ {USA}, whose p_5-support is 17, p_5-confidence is 85%, and p_5-lift is 1.04, representing that from 1981 to 2000, the USA won 85% of the NPPM.

Assume that $Q = \mathscr{P} = \{p_1, p_2, \cdots, p_6\}$, set the min-supp is 12, and min-conf is 70% during a period in Q. Consider the rule {NPPM} $\Rightarrow^Q$ {USA}, its Q-support is $S_{\mathfrak{M}}^Q(\{NPPM\} \Rightarrow^Q \{USA\}) = 64 > 12$, Q-confidence is $C_{\mathfrak{M}}^Q(\{NPPM\} \Rightarrow^Q \{USA\}) = 57.66\% < 70\%$, and Q-lift is $L_{\mathfrak{M}}^Q(\{NPPM\} \Rightarrow^Q \{USA\}) = 1.09 > 1$. Obviously, in the whole time span from 1901 to 2020, the rule {NPPM} $\Rightarrow^Q$ {USA} is not a strong association rule. According to the above analysis, the rule {NPPM} $\Rightarrow^{p_5}$ {USA} is identified as a PTAR which is strong during the period p_5.

Consider the rule $\{NPC,UK\} \Rightarrow^Q \{USA\}$, we have $S_{\mathfrak{W}}^Q(\{NPC,UK\} \Rightarrow^Q \{USA\}) = 7 < 12$ and $C_{\mathfrak{W}}^Q(\{NPC,UK\} \Rightarrow^Q \{USA\}) = 26.92\% < 70\%$. The Q-lift of this rule is $L_{\mathfrak{W}}^Q(\{NPC,UK\} \Rightarrow^Q \{USA\}) = 0.51 < 1$. Thus, this rule is not frequent during a period in Q, and can not be identified as a strong rule or even a promotive rule. For the rule $\{NPC,UK\} \Rightarrow^{p_6} \{USA\}$, with p_6-support is 4, p_6-confidence is 100%, and p_6-lift is 1.25. So it is a PTAR during the period p_6. And it can be interpreted as, all of the NPC awarded to Britain were cooperated with America between the year from 2001 to 2020.

Finally, we will compare our new method (Apriori-SPTARM) with the Apriori approach according to the quantity of strong and promotive rules extracted from the data set regarding Nobel Prizes in science.

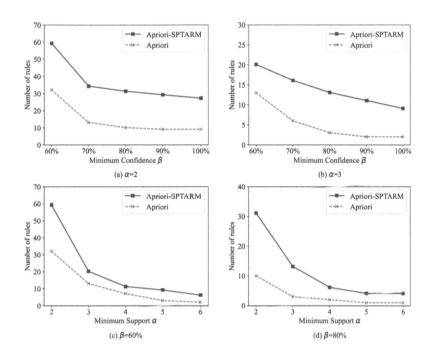

Fig. 3. Comparison between the quantity of rules obtained by two approaches.

More specifically, the quantity comparison is illustrated in Fig. 3. In the light of the results in Fig. 3 (a) and (b), when the min-supp $\alpha = 2$ or $\alpha = 3$, and the min-conf β is set to 60%, 70%, 80%, 90% and 100%, the Apriori-SPTARM method can extract more strong and promotive rules by considering the influence of time factor. In addition, the Fig. 3 (c) and (d) also demonstrate that the quantity of strong and promotive rules obtained by our new approach is greater than the classical Apriori method when $\beta = 60\%$ or $\beta = 80\%$, and α is assigned as 2, 3, 4, 5, and 6, respectively. More importantly, it can be seen that several item sets are only frequent during some specific periods of time. When employing the Apriori method, the confidence of some rules may not exceed the minimum threshold, and so these rules cannot be identified as strong rules. On the contrary, it is likely that the temporal confidence of these rules can exceed the minimum threshold during certain periods of time. Thus, such rules can be recognized as strong rules during certain periods after dividing the data set according to temporal granularity. In conclusion, the Apriori-SPTARM method can disclose more useful details hidden in the collected data which might be ignored by the traditional method.

5   Conclusions

This study has demonstrated how to use temporal soft sets to enhance classical association rule mining and overcome the difficulty owing to the ignorance of temporal information in some traditional methods. In order to produce the granular construction of a temporal transaction data, we defined temporal granulation mappings. TSSs and their Q-clip soft sets were introduced, which can be used to construct a TSS based framework for extracting TARs. By integrating the Apriori method into the mining framework based on TSSs, we proposed the Apriori-SPTARM method for extracting PTARs which are strong. A case study concerning Nobel Prizes in science has been done to confirm the feasibility of the Apriori-SPTARM method. In addition, comparative analysis regarding the quantity of strong and promotive rules revealed that our newly method can produce better mining results than the Apriori approach. Numerical experiments indicated that the presented approach can extract some strong and promotive rules missed by the Apriori method. It overcomes the limitation of classical mining methods ignoring temporal message.

Acknowledgments. The authors are thankful to the anonymous referees for their helpful suggestions. This work was partially supported by the Key Research and Development Projects of Shaanxi Province (Grant No. 2021SF-480).

References

1. Molodtsov, D.: Soft set theory-first results. Comput. Math. Appl. **37**(4–5), 19–31 (1999)
2. Agrawal, R., Srikant, R.: Mining sequential patterns. In: Proceedings of the 11th International Conference on Data Engineering, Taipei, China, pp. 3–14 (1995)
3. Ye, X.F., Keane, J.A.: Mining association rules in temporal databases. In: 1998 IEEE International Conference on Systems, Man and Cybernetics, San Diego, USA, pp. 2803–2808 (1998)
4. Ou-Yang, W., Cai, Q.: Discovery of association rules with temporal constraint in databases. J. Softw. **10**(5), 80–85 (1999)
5. Herawan, T., Deris, M.M.: A soft set approach for association rules mining. Knowl.-Based Syst. **24**(1), 186–195 (2011)

6. Feng, F., Cho, J., Pedrycz, W., Fujita, H., Herawan, T.: Soft set based association rule mining. Knowl.-Based Syst. **111**, 268–282 (2016)
7. Feng, F., Zhang, L., Zhang, Q.: Description and mining method of maximal association rules based on logical formulas on soft set. J. Jilin Univ. (Nat. Sci.) **56**(4), 901–908 (2018)
8. Feng, F., Wang, Q., Yager, R.R., Alcantud, J.C.R., Zhang, L.: Maximal association analysis using logical formulas over soft sets. Expert Syst. Appl. **159**, 113557 (2020)
9. Liu, X., Feng, F., Wang, Q., Yager, R.R., Fujita, H., Alcantud, J.C.R.: Mining temporal association rules with temporal soft sets. J. Math., 7303720 (2021)

Few-Shot Learning for Aspect-Based Sentiment Analysis

Heng Ruan[1,2,3]([✉]), Xiaoge Li[1,2,3], Xianliang Li[4], Huikai Jiang[1,2,3],
and Yingchao Li[1,2,3]

[1] Computer Science and Technology, Xi'an University of Post and Telecommunications,
Xi'an 710121, Shaanxi, China
{317892305,hollis.jiang,lyc0114}@stu.xupt.edu.cn,
lixg@xupt.edu.cn
[2] Computer Science and Technology, Shaanxi Key Laboratory of Network Data Analysis and
Intelligent Processing, Xi'an 710121, Shaanxi, China
[3] Computer Science and Technology, Xi'an Key Laboratory of Big Data and Intelligent
Computing, Xi'an 710121, Shaanxi, China
[4] Computer Science and Technology, Inspur Software Technology Co., Ltd., Davis, USA
li-xl@inspur.com

Abstract. Aspect-based sentiment analysis (ABSA) is a fine-grained sentiment analysis task. However, the lack of annotated sequence data in many domains hinders the effectiveness of monitoring methods. In order to address this issue, we explored the cross-domain ABSA tasks with only a few labeled support sentences. In our work, we exploited three benchmark datasets from different domains (Laptop, Restaurant, and Service), and used one of them as a source domain for training, used other domains as the target domain for testing. Our system uses a supervised sequence labeling model as a feature extractor, and construct a prototype representation of each tag, and then we calculate the distance between the tokens of the target domain and the prototype representation to determine the label type. We further use the Viterbi decoder to determine dependencies of the labels. We show that our method achieves state-of-the-art performance on the cross-domain ABSA task, the F1 scores of the latest model is 57.03 and 30.34 under the 5-shot and 1-shot settings, respectively, which are 9.88 and 1.43 higher than the strongest baseline under the same settings.

Keywords: Aspect-based sentiment analysis · Few-shot learning · Prototypical networks · Transfer learning

1 Introduction

Aspect-based sentiment analysis (ABSA) is a fine-grained natural language processing task, and can provide more detailed information than general sentiment analysis. As shown in the Fig. 1, in the sentence "The sweet shrimp was excellent but the okra was lousy", aspect term extraction aims to identify the aspect terms "sweet shrimp" and "okra", and opinion term extraction aims to determine opinion terms "excellent" and

N. Xiong et al. (Eds.): ICNC-FSKD 2022, LNDECT 153, pp. 1146–1157, 2023.
https://doi.org/10.1007/978-3-031-20738-9_124

"lousy", and aspect sentiment classification assign a sentiment polarity of each opinion: "excellent (positive)"and "okra (negative)".

Typically, those three tasks are formulated as the sequence labeling problem [3, 18]. However, solving the ABSA as a sequence labeling task involves processing unstructured text and categorizing each word, most of the existing works usually rely heavily on the scale of labeled data [1, 2]. At the same time, the joint extraction of aspect terms and opinion terms face rapid changes in domains, but for new domains with fewer samples, labeled data is usually scarce.

Fig. 1. Illustration of ABSA tasks, the same color indicates the aspect term and its corresponding opinion term

That makes Few-shot learning [8–10] get people's attention. It uses prior knowledge to identify new classes with limited supervised information, and learn prior experience from the original field and quickly adapt to and classify/label the new field when there are only a few labeled examples. A lot of work has been proposed for few-shot learning [5, 10, 17], among these methods, the prototype network [10] is a promising one. It is simple and effective, and followed meta learning paradigm, build N-way K-shot meta-task set. The goal of a meta-task is to infer the query-set with the learning of a few tagged support-set. In our work, we formulate the aspect term extraction, opinion term extraction and aspect term extraction cross-domain ABSA tasks as a few-shot sequence labeling problem, and use the unified tagging scheme to label the samples [11, 18]. And then, in the few-shot learning situation, we use a prototype network to train a label classifier, and use the Viterbi algorithm to determine the dependencies between labels, use a progressive optimization strategy to optimize the model training process, then, the training of the model is completed. We tested a metric learning technique in a task that often occurs in the real world, identifying instances with few examples of tags, and our work shows that the method achieves state-of-the-art performance on the cross-domain ABSA task.

2 Related Works

2.1 Few-Shot Learning

Few-shot learning is called the closest to real artificial intelligence, which learns the prior knowledge and obtains the ability to recognize new knowledge fast with limited supervised information. Previous few-shot learning method are highly dependent on hand-made features [12], and with research, various types of few-shot methods have been proposed. However, few-shot learning for ABSA is less investigated [7, 16]. Used the attention mechanism to mitigate the noise on the support set and query set, and then use the network to achieve the aspect sentiment classification. Compared with it, our method implements the three subtasks of few-shot ABSA in one unified model.

2.2 Aspect-Based Sentiment Analysis

Previous works for the ABSA tasks are mainly divided into unsupervised methods and supervised methods. Unsupervised methods extract aspects by mining semantic associations or co-occurrence frequencies [6]. These methods require large corpora to mine aspect knowledge but have limited performance. Supervised methods automatically learn useful representations through handcrafted features [4]. For real applications, a typical approach is to concatenate these subtasks together, but this approach becomes ineffective due to accumulated errors between tasks. In our work, we formulate the three sub-tasks as a uniform few-shot sequence labeling problem, and adopt a prototype-based few-shot learning method to solve this problem.

3   Problem Statement and Setup

3.1   Sequence Labeling

ABSA involves three sub-tasks, those tasks can be formulated as a unified sequence labeling problem [11, 18].Specially, give a sentence $X = (x_1, x_2 \cdots, x_n)$ as a sequence of words, aiming to assign each token x_i a label $y_i \in Y$ over the unified tags, where $Y =$ {B-POS, I-POS, B-NEG, I-NEG, B-NEU, I-NEU, B_OPI, I_OPI, O}. Tagging schemes such as bio is used to indicate whether the token is at the beginning or inside of the aspect term and opinion term. The "POS, NEG, NEU" represent the positive, the negative or the neutral sentiment of the aspect term, respectively.

3.2   Few-Shot Method

Typically, few-shot learning tasks are trained on a source domain and then work on other target domains without fine-tuning. The target contains only few labeled samples, called support-set $S = \{x_i, y_i\}^N$, , where N is the number of training data. There are only K examples for each of N labels (N-way). According to the training, assigned the token x tag y* corresponding to the most similar token in the support set:

$$y^* \in Y = (y_1, y_2 \cdots, y_n = \arg\max_y P(y|x, s)) \tag{1}$$

Few-shot learning usually has two sampling strategies, which are greedy-based sample and random sample strategies, and the greedy-base sampling strategies is always used in previous works [7], because greedy-based sampling is more targeted, and all categories can be trained with samples in each original training process. They used N-way K-shot greedy-based sample strategies to iterative judge whether an example could be added into support-set, but this method will become gradually strict with the sampling. It is not suitable for the ABSA sequence labeling tasks because of the dense comment data tags. In this paper we use a N-way K ~ 2K-shot setting for the support-set construction, as shown in Algorithm 1, and the primary purpose is to ensure that each class in Support-set contains K ~ 2K examples, which method effectively alleviates the limitation of sampling.

Algorithm 1 Greedy N-way K ~ 2K-shot Sampling algorithm
   Input: Datasets X, Label set Y, N, K

```
1   S ← ∅.
2   for i=1 to N do
3       Count[i]=0;
4   repeat
        {
5   Randomly sample (x, y) ∈ X.
6   Compute |Count| and Countᵢ after update;
7       if |Count[i]| > N or ∃Count[i] > 2K.
8           then
9               Continue;
10          else
            {
11      S = S ∪ (x, y);
12      update Countᵢ ;
            }
        }
13  Until Countᵢ ≥ K for i = 1 to N;
```

4 Model

In this section, we preset our few-shot cross-domain ABSA algorithm based on progressive optimization prototype learning. Our method uses a sequence labeling model trained on the source domain as a token embedder to generate the contextual representation for all tokens. At inference, these static representations are simply used for token classification.

4.1 The Unified Tagging Scheme

In this paper, we use the unified tagging scheme to label the samples. The three subtasks can be more easily modeled and implemented using a unified labeling scheme. Table 1 shows an example of the tagging scheme. By this way, the joint modeling of three sub-tasks is easier to implement.

Table 1. Tagging schemes used in the work

Sentence	The	Sweet	Shrimp	Was	Excellent	,	But	The	Okra	Was	Lousy	
Joint	O	B_POS	I_POS	O	O	O	O	O	B_NEG	O	O	O
	O	O	O	O	B_OPI	O	O	O	O	O	B_OPI	O
Unified	O	B_POS	I_POS	O	B_OPI	O	O	O	B_NEG	O	B_OPI	O

4.2 Pre-trained Sequence Labeling Models as Token Embedders

We consider two popular neural architectures for our sequence labeling model: a BiL-STM and a BERT-based model. Sequence labeling models usually consist of a token embedder $f_\theta(\cdot)$ and a linear classifier $W \in D^{M \times L}$, where M is the token embedding size, and L is the number of labels. We follow the setting from their original papers to train these models on the source domain. At inference time, our ProABSA uses the Bi-LSTM and Transformer encoders just before the final linear classification layers as token embedders. Given a sequence $X = (x_1, x_2 \cdots , x_n)$, for each token x_i, the encoder generates the contextual representation of each token is::

$$z = [z_1, z_2 \cdots , z_n] = f_{\theta 0}([x_1, x_2 \cdots , x_n]) \tag{2}$$

These models are trained to minimize the cross-entropy loss on the training data in the source domain. The cross-entropy is:

$$L(x, y) = \sum_{(X,Y) \in D^L} \sum_{i=1}^{T} kl(y_i || y_i x_i) \tag{3}$$

where the kl divergence between two distribution is $kl(p||q) = E_p \log(p/q)$.

4.3 Prototype-Based Classification for Sequence Labeling of ABSA

We build our work based on prototype network [19]. In concretely, it is representing every token as vectors in the same representation space, and then take the average representation of all tokens belonging to the n-th class in the support set as the prototype Cn for that class:

$$C_n = \frac{1}{|S_n|} \sum_{x \in S_n} f_{\theta 0}(x) \tag{4}$$

where S_n is the tokens set of the n-th type in S, and $f_{\theta 0}$ is defined in (2). And then calculates the distance between the embedded query instance and all prototypes.

Query set:The hotel is very clean.

Distance   SoftMax

Support set:The seafood was excellent .
The owners are extremely friendly.

Fig. 2. Prototype-method

We use the l_2 distance as the metric function $d(f_\theta(x), C_i) = ||f_\theta(x) - C_i||_2^2$ in our work. The predicted distribution is computed by computing the distances between x and all other prototypes, then using a soft-max function of the distances between x and all term prototypes. For example, the prediction probability for the n-th prototype is:

$$q(y = \mathrm{II}_m x) = \frac{\exp(-d(f_{\theta 0}(x), c_m))}{\sum_{m'} \exp(-d(f_{\theta 0}(x), c_{m'}))} \tag{5}$$

We provide a simple example to illustrate the prototype method in Fig. 2.

The prototype set is constructed by averaging all labeled features in the support set belonging to a given entity type. (e.g. The prototype for B-POS is an average of the tokens: seafood, owners).

At each training iteration, a new episode is sampled and the model parameter θ_0 is updated by inserting (5) into (4). And in the test step, the prediction is the label of the prototype closest to x. That is, for the support set S_y of type y, the query condition is x, and the prediction process is as follows:

$$y^* = \arg \min_{y \in Y} d_y(x), \tag{6}$$

$$d_y(x) = d(f_\theta(x), C_n) \tag{7}$$

5 Experiment

We evaluate few-shot ABSA on three standard fine-grained sentiment analysis data sets.

5.1 Datasets

Table 2 shows the dataset that contains reviews from the different domain in SemEval ABSA challenge 2014 [13–15].

Table 2. Data statistics. Corresponding datasets sizes in different domains.

Datasets	Domain	Sentences	Tokens	Training	Testing
L	Laptop	1752	29.2k	1223	529
R	Restaurant	3900	56.8k	2166	1734
S	Service	1469	31.5k	960	509

5.2 Evaluation Tasks

Domain transfer In our work, in order to test the robustness of our framework, we construct 9 transfer pairs like L→S with the above three different domains to cross-validate the models. Each time, we use one of the three domains as the source domain for training, and chose one of others target domain for testing separately. The support-sets are sampled from the corresponding training set of the three corpora.

5.3   Baselines

We consider two popular neural architectures for our sequence labeling model: a BiL-STM model and a BERT-based model to embedders. Our work has verified the effectiveness of our method by comparing with the following popular approaches. Except SimBERT, all others are domain transfer models, that means we pre-train it on a source domain and test on a target domain.

Matching Network is a metric-based few-shot learning method, and the distance is measured by cosine similarity. We employ the matching network [20] with BERT embedding for classification in our work.

SimBERT is a model based on a pre-trained BERT encoder, and predicts labels based on the cosine similarity of non-fine-tuned BERT word embeddings. For each word x_j, SimBERT finds its most similar word x_k in the support set, and predicts the label of x_j as the label of x_k.

TransferBERT is similar with SimBERT. The only difference is that we pre-train it on source domains and test on a target domain.

Transfer Selective Adversarial Learning is a novel selective adversarial learning method for jointly extracting of aspects and sentiments.

NNShot&STRUCTSHOT is a few-shot sequence labeling system based on nearest neighbor learning. The basic structure of StructShot is the same as NNShot, but with an additional Viterbi decoder used in the inference phase.

5.4   Results and Analysis

We report the experimental results of various methods under different settings in Tables 3, 4, 5 and 6 respectively. The first column represents the migration target (e.g. The L->R indicates that the source domain is Laptop and the target domain is Restaurant), and other columns show the F1 score with one domain as the target domain and another domain as the source domains. The best score for each metric is marked in bold. The experimental results demonstrate the effectiveness of the method.

   Overall Performance Our ProABSA achieves good performance in transfer across domains, Tables 3 and 4 show the 1-shot and 5-shot ProABSA results. Which achieves an average of 30.34 and 57.03 F1 scores on the 1-shot and 5-shot label set tasks, respectively. Our ProABSA surpassing the strongest few-shot learning baseline model, and our model significantly outperforms TransferE2E indicating that it is difficult to train an effective fine-grained attribute word and sentiment word extraction model using only a small amount of labeled data using traditional machine learning and transfer learning methods. In addition, the performance of Simbert shows that the metric learning method is more advantageous than the traditional machine learning model in the case of fewer shot. More specifically, models refined on the support set such as Transfer-bert mainly tend to predict with random labels. These systems can only handle tokens that are easily generalized.

Table 3. F1 scores on 1-shot cross-domain ABSA tasks. ProABSA (1-shot) are our methods, which achieve the best performance. Ave. shows the average scores. The best results are **bold.**

Transfer Pair	BiLSTM + CRF	Structshot	Transfer-bert	TransE2E	ProABSA (1-shot)
L->L	10.92	27.34	**38.73**	7.84	32.40
L->R	13.38	26.53	**36.22**	11.22	29.80
L->S	18.10	21.30	24.54	8.96	**26.50**
R->R	15.46	35.41	**36.25**	14.23	29.00
R->L	16.42	25.43	30.02	15.06	**31.30**
R->S	21.54	**30.37**	17.57	12.39	22.80
S->S	18.37	27.89	36.44	9.20	**36.50**
S->L	19.30	22.57	18.72	13.04	**40.10**
S->R	21.35	17.48	22.24	12.17	**24.70**
Ave	17.20	26.04	28.97	11.57	**30.34**

Table 4. F1 scores on 5-shot cross-domain ABSA tasks. ProABSA (5-shot) are our methods, which achieve the best performance. Ave. shows the average scores. The best results are **bold**.

Transfer pair	BiLSTM + CRF	Structshot	Transfer-bert	TransE2E	ProABSA (5-shot)
L->L	25.37	42.45	55.36	53.12	**58.24**
L->R	20.59	46.18	48.93	52.65	**61.37**
L->S	32.74	53.26	28.54	48.55	**54.52**
R->R	32.52	37.58	57.67	45.18	**58.83**
R->L	27.36	41.73	22.46	43.75	**53.56**
R->S	24.81	39.85	28.59	42.23	**61.32**
S->S	31.23	35.25	**58.40**	51.29	55.35
S->L	23.56	29.76	39.74	**49.38**	48.79
S->R	24.62	38.42	42.36	38.26	**61.27**
Ave	26.97	40.49	42.45	47.15	**57.03**

However, for the text data without fixed types of attribute words and emotion words in the sentiment analysis data set, the traditional transfer learning or machine learning methods for model training will inevitably lead to poor generalization ability of the classifier and over-fitting phenomenon. Therefore, the metric-based methods are better to deal with such problems.

BiLSTM token embedder vs. BERT token embedder
Comparing the results of BiLSTM-based and BERT-based, it can be found that the F1-score of the BERT-based system significantly outperforms the BiLSTM-based system

on the 1-shot and 5-shot labels, exceeding 7.0 and 10.37, respectively. Therefore, we believe that pre-training language models are crucial for the encoding of low-resource natural language processing including few-show learning process. Table 5 and 6 show the F1 score of 1-shot and 5-shot cross-domain ABSA task under BiLSTM-based and BERT-based.

6 Conclusion

Due to the scarcity of data, the effectiveness of ABSA's supervised methods is limited. Our work is to quickly adapt and classify tasks by using few-shot learning to term prototypes from the original domain, and with fewer labeled samples in the new domain. Large numbers of experiments have proved the effectiveness of this method. The proposed ProABSA method may be extended to other domain adaptation methods.

Table 5. F1 score of 1-shot cross-domain ABSA task under BiLSTM-based and BERT-based. ProABSA (1-shot) is our method Ave. Shows the average score. The best results are in **bold**.

Transfer pair	BiLSTM					Bert				
	Transe2e	NN	Struct	Match	ProABSA (1-shot)	Simbert	Struct	Match	Transferbert	ProABSA (1-shot)
L->L	7.84	26.94	30.42	13.76	31.23	19.76	27.34	16.55	**38.73**	32.40
L->R	11.22	22.42	23.72	10.98	19.54	–	26.53	17.47	**36.22**	29.80
L->S	8.96	15.77	18.94	12.53	20.16	–	21.30	15.32	24.54	**26.50**
R->R	14.23	19.24	26.12	11.44	21.74	19.43	35.41	24.53	36.25	29.00
R->L	15.06	15.38	21.33	8.92	23.23	–	25.43	22.33	30.02	**31.30**
R->S	12.39	16.43	23.34	9.34	26.91	–	30.37	19.61	**27.57**	22.80
S->S	9.20	17.09	19.57	14.07	25.26	16.34	27.89	20.03	36.44	**36.50**
S->L	13.04	21.22	26.45	6.80	23.73	–	22.57	18.25	18.72	40.10
S->R	12.17	19.65	20.63	11.34	18.66	–	17.48	19.66	22.24	**24.70**
Ave	11.57	19.35	23.33	10.98	23.34	–	25.99	19.27	30.04	**30.34**

Table 6. F1 score of 5-shot cross-domain ABSA task under BiLSTM-based and BERT-based. ProABSA is our method Ave. Shows the average score. The best results are in **bold**.

Transfer pair	BiLstm					Bert				
	Transe2e	NN	Struct	Match	ProABSA (5-shot)	Simbert	Struct	Match	Transferbert	ProABSA (5-shot)
L->L	23.45	19.54	22.52	33.70	44.94	44.32	47.37	37.22	55.36	**58.24**
L->R	31.53	22.93	31.53	42.36	51.11	–	38.26	52.95	48.93	**61.37**
L->S	37.41	31.24	26.54	39.64	39.47	–	**57.05**	36.74	28.54	54.52
R->R	19.30	39.21	19.18	42.34	41.82	49.72	52.64	44.53	57.67	**58.83**
R->L	31.20	31.93	27.45	38.78	52.33	–	53.44	39.82	22.46	**53.56**
R->S	22.52	25.36	31.25	42.64	57.60	–	51.52	60.26	28.59	**61.32**
S->S	27.43	29.87	25.80	45.73	47.95	35.58	45.32	52.71	58.40	**55.35**
S->L	31.97	33.38	19.82	44.27	45.20	–	39.67	46.43	39.74	**48.79**
S->R	27.81	29.30	24.51	24.63	39.56	–	50.00	38.95	42.36	**61.27**
Ave	28.07	29.20	25.40	39.34	46.66	–	48.36	45.51	42.45	**57.03**

References

1. Peng, H., Xu, L., Bing, L., et al.: Knowing what, how and why: A near complete solution for aspect-based sentiment analysis. In: Proceedings of the AAAI Conference on Artificial Intelligence, vol. 34, no. 05, pp. 8600–8607 (2020)
2. Yan, H., Dai, J., Ji, T., et al.: A Unified Generative Framework for Aspect-Based Sentiment Analysis (2021)
3. Li, Z., Li, X., Wei, Y., et al.: Transferable End-to-End Aspect-Based Sentiment Analysis with Selective Adversarial Learning (2019)
4. Kiritchenko, S., Zhu, X., Cherry, C., et al.: Nrc-canada-2014: detecting aspects and sentiment in customer reviews. In: Proceedings of the 8th International Workshop on Semantic Evaluation (SemEval 2014), pp. 437–442 (2014)
5. Finn, C., Abbeel, P., Levine, S.: Model-agnostic meta-learning for fast adaptation of deep networks. In: International Conference on Machine Learning (PMLR), pp. 1126–1135 (2017)
6. Schouten, K., Van Der Weijde, O., Frasincar, F., et al.: Supervised and unsupervised aspect category detection for sentiment analysis with co-occurrence data. IEEE Trans. Cybern. **48**(4), 1263–1275 (2017)
7. Hu, M., Zhao, S., Guo, H., et al.: Multi-label few-shot learning for aspect category detection. arXiv preprint arXiv:2105.14174 (2021)
8. Miller, E.G., Matsakis, N.E., Viola, P.A.: Learning from one example through shared densities on transforms. In: Proceedings IEEE Conference on Computer Vision and Pattern Recognition (CVPR 2000) (Cat. No. PR00662), vol. 1, pp. 464–471. IEEE (2000)
9. Toprak, C., Jakob, N., Gurevych, I.: Sentence and expression level annotation of opinions in user-generated discourse. In: Proceedings of the 48th Annual Meeting of the Association for Computational Linguistics, pp. 575–584 (2010)
10. Snell, J., Swersky, K., Zemel, R.: Prototypical networks for few-shot learning. In: Advances in Neural Information Processing Systems, p. 30 (2017)
11. Li, X., Bing, L., Li, P., et al.: A unified model for opinion target extraction and target sentiment prediction. In: Proceedings of the AAAI Conference on Artificial Intelligence, vol. 33, no. 01, pp. 6714–6721 (2019)
12. Fink, M.: Object classification from a single example utilizing class relevance metrics. In: Advances in Neural Information processing Systems, p. 17 (2004)
13. Pontiki, M., Galanis, D., Papageorgiou, H., et al.: Semeval-2016 task 5: aspect based sentiment analysis. In: International Workshop on Semantic Evaluation, pp. 19–30 (2016)
14. Pontiki, M., Galanis, D., Papageorgiou, H., et al.: Semeval-2015 task 12: aspect based sentiment analysis. In: Proceedings of the 9th International Workshop on Semantic Evaluation (SemEval 2015), pp. 486–495 (2015)
15. Manandhar, S.: Semeval-2014 task 4: sentiment analysis. In: Proceedings of the 8th International Workshop on Semantic Evaluation (SemEval 2014) (2014)
16. Fritzler, A., Logacheva, V., Kretov, M.: Few-shot classification in named entity recognition task. In: Proceedings of the 34th ACM/SIGAPP Symposium on Applied Computing, pp. 993–1000 (2019)
17. Hou, Y., Zhou, Z., Liu, Y., et al.: Few-shot sequence labeling with label dependency transfer and pair-wise embedding. arXiv preprint arXiv:1906.08711 (2019)
18. Mitchell, M., et al.: Open domain targeted sentiment (2013)
19. Fei-Fei, L., Fergus, R., Perona, P.: One-shot learning of object categories. IEEE Trans. Pattern Anal. Mach. Intell. **28**(4), 594–611 (2006)
20. Vinyals, O., Blundell, C., Lillicrap, T., et al.: Matching networks for one shot learning. In: Advances in Neural Information Processing Systems, p. 29 (2016)

Multimodal Incremental Learning via Knowledge Distillation in License Classification Tasks

Jian Qiu[1,2($\boxtimes$)], Zongrui Zhang[1,2], and Hao He[1,2]

[1] State Key Lab of Advanced Optical Communication System and Network, School of Electronic Information and Electrical Engineering, Shanghai Jiao Tong University, Shanghai, China
`zzr965586083@sjtu.edu.cn, hehao@sjtu.edu.cn`
[2] China Institute for Smart Court, Shanghai Jiao Tong University, Shanghai, China
`qiujian1232@sjtu.edu.cn`

Abstract. Most current license classification algorithms are based on handcrafted rules or single visual feature for image classification, which cannot make full use of the multi-modal attributes existed in the licenses, resulting in limited accuracy. Moreover, the update and abolishment of licenses have put forward requirements for the rapid response and learning ability of license classification models, so that the model can learn new knowledge incrementally, and finding a balance between stability and plasticity has always been very important. However, most of the existing incremental learning algorithms are designed for a single modality. In this paper, we propose a incremental multimodal classification algorithm to make full use of the multimodal properties of licenses, and integrate multimodality and incremental learning algorithm based on knowledge distillation. Compared to single visual feature solution, the proposed algorithm has better performance, which is 11% higher than that of fine-tuning, and about 7% higher than that of single-modal incremental learning.

Keywords: Incremental learning · Multimodal fusion · License classification · Knowledge distillation

1 Introduction

Within a certain period of time in social development, the types and styles of licenses are relatively fixed, but with the continuous development of society and relevant policy changes and adjustments, old licenses will be abolished and new licenses will appear. The rapid response learning ability of the model puts forward requirements, and it also makes incremental learning practical.

N. Xiong et al. (Eds.): ICNC-FSKD 2022, LNDECT 153, pp. 1158–1166, 2023.
https://doi.org/10.1007/978-3-031-20738-9_125

Licenses naturally have a variety of modal features, including visual features, text features, etc. Existing license classification algorithms are mostly based on handcrafted rules or single visual features to classify images, but ignore the multimodal attributes of the license itself, resulting in limited accuracy. However, most of the existing incremental learning algorithms are incremental learning for a single modality, which further leads to the ineffectiveness of incremental learning in the problem of license classification.

Multimodal license classification with visual and textual product descriptions is one of the most fundamental problems in any government system. The traditional way of handling license classification involves using rules for classification, or using a single modality (visual features only or text information only) for classification, lack of scalability and generality. In the face of new tasks, similar fine-tuning to incrementally learn new knowledge often encounters catastrophic forgetting [3], after learning new tasks, the performance of the model on the old task catastrophically degrades. The use of multimodal features to classify licenses is in line with the multimodal feature attributes of licenses themselves, but it also puts forward new requirements for incremental learning.

Our proposed algorithm uses both visual features and textual features for license classification, and then integrates the incremental learning algorithm based on knowledge distillation according to the multi-modal characteristics to train the model on the new task without the need for old samples, and obtain a A new model that maintains a certain performance on the old task, while having a good performance on the new task. In the design of our loss function, we learn from the way humans learn pictures and text at the same time. When training old models on new tasks, we focus not only on visual features and text features, but also on the relationship between the two features. The new features after multimodal fusion. It is hoped that in this way, the impact of the new task on the performance of the old task is minimized, while the performance of the new task is guaranteed.

2   Related Works

In recent years, deep learning models are no longer limited to a single modality, but are increasingly multimodal, such as text modalities, audio modalities, and image modalities. The reason is that due to the limitations of single-modal information, the information cannot be fully expressed, but the complementarity and redundancy between multi-modalities should make up for the limitations of single-modal information. The goal of the multimodal classification task is to build a model that can handle multimodal data, making use of unimodal features while leveraging the interaction of multimodal features to achieve better classification performance and make the system more robust [1,2].

Incremental learning has been researched for nearly 30 years, and incremental learning is roughly equivalent to the concepts of Continuous Learning [4,5] and Lifelong Learning [6–8]. They can all be seen as training a model in a continuous stream of data, and over time, more new Data becomes available gradually, including new datasets from different tasks [9–12] or homogeneous data from similar datasets [13–19], while older data may become unavailable due to reasons such as storage constraints or privacy protection.

In order to overcome catastrophic forgetting while maintaining performance on new tasks, we hope that models must exhibit the ability to integrate new knowledge and refine existing knowledge from new data on the one hand (Plasticity), and on the other hand must prevent new inputs from affecting existing knowledge. Significant disturbance (Stability) of knowledge. These two conflicting needs constitute the so-called stability-plasticity dilemma.

According to the information of how to store and use the previous task in the subsequent learning in incremental learning, it can be divided into the following three paradigms: incremental learning algorithm based on regularization, incremental learning algorithm based on data replay and incremental learning algorithms based on Parameter Isolation.

Regularization-based incremental learning algorithms provide a way to alleviate catastrophic forgetting under certain conditions by introducing additional losses to correct gradients and preserve old knowledge learned by the model. However, although the current deep learning models are over-parameterized, the model capacity is ultimately limited, and we usually need to make tradeoffs between the performance of old tasks and new tasks. Most of the current research preserves the retrieval performance on old datasets through the incremental learning algorithm of Knowledge Distillation (KD). Hinton [20] proposed knowledge distillation technology in 2015, which simplifies the training process of deep network by extracting the student network from the teacher network. This algorithm transfers the useful information in the teacher network to the student network for training. The common probability distribution between the network and the teacher network outputs the difference index, that is, the KL divergence between the two as the objective function, so that the incremental model retains the key parameters in the old model. At present, most of the incremental learning algorithms focus on single-modal features. In this paper, multi-modal features are integrated, and the incremental learning algorithm based on knowledge distillation is used to improve the model performance.

Compared with the method proposed by Yan et al. [24] to freeze the original learning representation and expand its parameters in each incremental learning, our method consumes less resources, and the performance after each incremental learning is more stable.

3   Method

In order to study the influence of single modality and multimodality on the results of incremental learning, this paper takes the license classification problem as the reference object. The algorithm is integrated into the model to realize the multimodal incremental learning of the license classification problem. Figure 1 is the structure diagram of the model.

In this paper, ResNet50 [21] is used as the pre-training classification network, which is pre-trained on the old data set and used as the teacher network of the

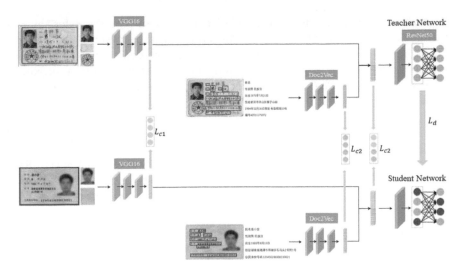

Fig. 1. Multimodal incremental learning model via knowledge distillation

multi-modal license classification model. By constraining the visual features and text features in the teacher network to retain the consistency of their output probability distribution to guide students network for learning. Among them, the visual modality and the text modality both extract features of the same dimension, and the knowledge distillation loss guides the respective incremental network learning on the two modalities, so as to retain the performance of the old model. The output of the student network is used as a classification loss constraint. Considering the inline relationship between modalities, the mutual relationship between different modalities is also used as a classification loss constraint, thereby further enhancing the performance on new samples. In summary, the objective function of the overall network architecture consists of three loss functions:

$$\min_{\theta_x, \theta_y} = L_d + L_c + L_p \tag{1}$$

where θ_x is the image channel parameter of the incremental model, and θ_y is the text channel parameter.

L_c is the class error cross entropy loss, which enables the model to learn new classes of visual and textual information from new samples.

$$L_c = \frac{1}{M} \sum (c_i(-\lg p_i(X_i) - \lg p_i(Y_i))) \tag{2}$$

p_i represents the probability distribution of different modal feature categories, c_i represents the category label of the certificate sample in the current new data set, X_i and Y_i represent the visual feature and text feature of the i sample, respectively, M is the newly added training batch sample set size.

We adopt the multimodal knowledge distillation loss L_d to transfer the classification performance of the original multimodal license classification model on the old sample data set to the new model. KL divergence can help us measure the amount of information lost when using one distribution to approximate another distribution. The greater the difference between the two distributions, the greater the KL divergence, which is defined as follows:

$$D_{KL}(p|q) = \sum p \lg \frac{p}{q} \tag{3}$$

where $p(u)$ is the target distribution of the student network, $q(u)$ is the distribution of the teacher network to match, if the two distributions match exactly, then the KL divergence $D_{KL} = 0$. So our multimodal-based knowledge distillation loss L_d is designed as follows:

$$L_d = D_{KL}(p_x|q_x) + D_{KL}(p_y|q_y) \tag{4}$$

where

$$q_i = -\langle \frac{c_i^{* \frac{1}{\theta_i}}}{\sum c_i^{* \frac{1}{\theta_i}}}, \frac{c_i^{\frac{1}{\theta_i}}}{\sum c_i^{\frac{1}{\theta_i}}} \rangle \tag{5}$$

and $*$ represents the model parameters obtained by the teacher network using the old samples.

The role of L_p lies in the generation of multimodal cross-correlation codes, which are used to guide the classification loss constraints, so that the performance of the new model on new samples can be guaranteed.

$$L_p = \frac{1}{M} \sum (c_i(-\lg p_i((X, Y)_i)) \tag{6}$$

where (X, Y) means the concatenation of the two modal feature vectors.

So the whole algorithm flow is as follows:

Input: Add image data set x, add text data set y, label set C
Output: The student network parameters θ
1: Initialize the student network parameters θ_0
2: Disassemble the old and new label items for statistical analysis to obtain the co-occurrence probability matrix of the old and new labels
3: Calculate L_c according to formula (2)
4: Calculate L_d according to formula (4)
5: Calculate L_p according to formula (6)
6: Update parameters θ_x, θ_x by stochastic gradient descent according to formula (1)
7: Repeat $3-6$
8: **return** θ

---

4   Experiment

4.1   Dataset

The license data set is a collection of 3968 images from the network, covering 20 types of licenses such as the first generation of resident identification cards, the second generation of resident identification cards, Road transport operation permit and medical device trading permit. The main method of collecting data is to use crawlers to parse web pages and download pictures, and perform data desensitization processing according to the specified format. The annotation description of the image contains all the text and category labels in the corresponding image. In order to achieve incremental sample iteration, this paper divides the 20 class labeled original certificate set into 16 class old labeled data set and 4 class new labeled data set, and will include all samples of any of these 4 class new labeled data sets As the new sample set, the rest are used as the old sample set, and the ratio of old and new samples after sorting is 801:3167.

4.2   Experimental Details

In the experiment, the pre-trained VGG16 [22] neural network model is used to extract visual features from the modal data of the license image, including 13 convolutional layers, 5 pooling layers, 3 fully connected layers, and the fully connected layer is used as the image feature output layer. For text modality data in license images, a pre-trained doc2vec [23] model is employed to extract sentence-level text features. Different modal channels solve the subsequent expansion and multi-modal fusion problems by adding a multi-layer perceptron network to generate the same dimensional code. Image classification adopts ResNet as a pre-trained classification network.

In order to compare the advantages of the multimodal incremental learning algorithm, we also constructed a simple incremental learning network based on visual features, still using the training VGG16 neural network model to extract visual features from the modal data of the license image, and using ResNet as the pre-training classification network. In experiments, we use the average accuracy for evaluation.

4.3   Result

In our experiments, we use accuracy as a measure, which is the ratio of the number of samples correctly classified by the classifier to the total number of samples on the specified test dataset.

As mentioned above, Table 1 shows the comparison of the two algorithms on our self-made license classification data set. From Table 1, it can be found that the algorithm of incremental learning using multimodal features proposed in this paper has a smoother forgetting trend. Compared with incremental learning using only a single visual feature, the average accuracy is higher, compared

Table 1. After adding 2 and 4 categories respectively, the accuracy (%) of the model is compared

Method	Incremental 2 class	Incremental 4 class
Fine-tune model	50.72	48.23
Model based on visual feature	54.65	50.67
DER	**64.88**	58.31
Our model	61.81	**59.45**

(a) The first generation

(b) The second generation

Fig. 2. The resident identification cards

with directly using new data samples for model training, it has better anti-forgetting performance. Compared with DER, it performs better in the case of incremental learning of more new categories. As a typical image with multimodal information, it is not enough to use the visual information in the license image for classification. The resident identification cards (Fig. 2) classification results are shown in Table 2.

Table 2. The accuracy (%) of the model (after incremental 2 class) on the resident identification cards

Model	The first generation of resident identification cards	The second generation of resident identification cards
Fine-tune model	47.19	50.23
Model based on visual feature	53.24	56.31
DER	**61.41**	62.89
Our model	61.11	**63.35**

It can be seen from the results in Table 1 that the multimodal incremental learning algorithm can make good use of visual features and text features, which makes the model more advantageous through the multimodal incremental learning algorithm during the training and learning process.

5   Conclusion

This paper proposes a multimodal incremental learning algorithm for license classification problems, and constructs a Chinese license classification dataset. This algorithm only uses new class samples for model expansion and builds a multi-modal knowledge distillation network. The purpose is to minimize the impact of new tasks on the performance of old tasks, while ensuring the performance of new tasks. The experimental results show that the algorithm in this paper has better performance in incremental expansion than the single mode.

In the next step, we will consider comparing other incremental learning algorithms based on data playback and parameter isolation, as well as fine-grained feature representation for paired modal data, through effective fine-grained pairwise feature fusion, thereby improving the model performance.

References

1. Tao, H., Hou, C., Yi, D., Zhu, J.: Multiview classification with cohesion and diversity. IEEE Trans. Cybern. **50**(5), 2124–2137 (2018)
2. Liang, X., Qian, Y., Guo, Q., Cheng, H., Liang, J.: AF: An association-based fusion method for multi-modal classification. IEEE Trans. Pattern Anal. Mach. Intell. (2021)
3. McCloskey, M., Cohen, N.J.: Catastrophic interference in connectionist networks: the sequential learning problem. In: Psychology of Learning and Motivation, vol. 24, pp. 109–165. Academic Press (1989)
4. Delange, M., Aljundi, R., Masana, M., Parisot, S., Jia, X., Leonardis, A., Tuytelaars, T.: A continual learning survey: defying forgetting in classification tasks. IEEE Trans. Pattern Anal. Mach. Intell. (2021)
5. Lopez-Paz, D., Ranzato, M. A.: Gradient episodic memory for continual learning. In: Advances in Neural Information Processing Systems (2017)
6. Aljundi, R., Chakravarty, P., Tuytelaars, T.: Expert gate: lifelong learning with a network of experts. In: Proceedings of the IEEE Conference on Computer Vision and Pattern Recognition, pp. 3366–3375 (2017)
7. Chen, Z., Liu, B.: Lifelong machine learning. Synth. Lect. Artif. Intell. Mach. Learn. **12**(3), 1–207 (2018)
8. Li, Y., Chen, X., Li, N.: Online optimal control with linear dynamics and predictions: algorithms and regret analysis. In: Advances in Neural Information Processing Systems (2019)
9. Chaudhry, A., Dokania, P.K., Ajanthan, T., Torr, P.H.: Riemannian walk for incremental learning: understanding forgetting and intransigence. In: Proceedings of the European Conference on Computer Vision (ECCV), pp. 532–547 (2018)
10. Davidson, G., Mozer, M.C.: Sequential mastery of multiple visual tasks: Networks naturally learn to learn and forget to forget. In: Proceedings of the IEEE/CVF Conference on Computer Vision and Pattern Recognition, pp. 9282–9293 (2020)
11. Hu, W., Lin, Z., Liu, B., Tao, C., Tao, Z., Ma, J., Yan, R.: Overcoming catastrophic forgetting for continual learning via model adaptation. In: International Conference on Learning Representations (2018)
12. Zhao, B., Xiao, X., Gan, G., Zhang, B., Xia, S.T.: Maintaining discrimination and fairness in class incremental learning. In: Proceedings of the IEEE/CVF Conference on Computer Vision and Pattern Recognition, pp. 13208–13217 (2020)

13. Castro, F.M., Marín-Jiménez, M.J., Guil, N., Schmid, C., Alahari, K.: End-to-end incremental learning. In: Proceedings of the European Conference on Computer Vision (ECCV), pp. 233–248 (2018)
14. Hou, S., Pan, X., Loy, C.C., Wang, Z., Lin, D.: Learning a unified classifier incrementally via rebalancing. In: Proceedings of the IEEE/CVF Conference on Computer Vision and Pattern Recognition, pp. 831–839 (2019)
15. Hu, X., Tang, K., Miao, C., Hua, X.S., Zhang, H.: Distilling causal effect of data in class-incremental learning. In: Proceedings of the IEEE/CVF Conference on Computer Vision and Pattern Recognition, pp. 3957–3966 (2021)
16. Liu, Y., Su, Y., Liu, A.A., Schiele, B., Sun, Q.: Mnemonics training: multi-class incremental learning without forgetting. In: Proceedings of the IEEE/CVF Conference on Computer Vision and Pattern Recognition, pp. 12245–12254 (2020)
17. Rebuffi, S.A., Kolesnikov, A., Sperl, G., Lampert, C.H.: ICARL: incremental classifier and representation learning. In: Proceedings of the IEEE Conference on Computer Vision and Pattern Recognition, pp. 2001–2010 (2017)
18. Wu, Y., Chen, Y., Wang, L., Ye, Y., Liu, Z., Guo, Y., Fu, Y.: Large scale incremental learning. In: Proceedings of the IEEE/CVF Conference on Computer Vision and Pattern Recognition, pp. 374–382 (2019)
19. Tao, X., Hong, X., Chang, X., Dong, S., Wei, X., Gong, Y.: Few-shot class-incremental learning. In: Proceedings of the IEEE/CVF Conference on Computer Vision and Pattern Recognition, pp. 12183–12192 (2020)
20. Hinton, G., Vinyals, O., Dean, J.: Distilling the knowledge in a neural network. arXiv preprint arXiv:1503.02531 (2015)
21. He, K., Zhang, X., Ren, S., Sun, J.: Deep residual learning for image recognition. In: Proceedings of the IEEE Conference on Computer Vision and Pattern Recognition, pp. 770–778 (2016)
22. Simonyan, K., Zisserman, A.: Very deep convolutional networks for large-scale image recognition. arXiv preprint arXiv:1409.1556 (2014)
23. Le, Q., Mikolov, T.: Distributed representations of sentences and documents. In: International Conference on Machine Learning, pp. 1188–1196. PMLR (2014)
24. Yan, S., Xie, J., He, X: DER: dynamically expandable representation for class incremental learning. In: Proceedings of the IEEE/CVF Conference on Computer Vision and Pattern Recognition, pp. 3014–3023 (2021)

FERTNet: Automatic Sleep Stage Scoring Method Based on Frame Level and Epoch Level

Xuebin Xu, Chen Chen$^{(\boxtimes)}$, Kan Meng, Xiaorui Cheng, and Haichao Fan

School of Computer Science and Technology, Xi'an University of Posts & Telecommunications, Xi'an, Shaanxi 710121, China
xuxuebin@xupt.edu.cn, {2103210107,mengkan,cxxxxxr, fhc7911}@stu.xupt.edu.cn

Abstract. Sleep disorders can cause a series of physical problems, such as cardiovascular and cerebrovascular diseases, lack of energy, and so on. This has a great impact on people's daily life. Sleep staging is the basis for solving this problem. However, traditional staging requires manual completion, which is time-consuming and labor-intensive. Based on this, we propose a deep learning architecture, FERTNet, to learn the temporal context between frame-level and epoch-level to complete sleep staging. Specifically, frame-level features are extracted using the improved ResNet, and temporal convolutional neural networks learn their temporal context dependencies at different levels from the extracted features. This model is validated on the public sleep dataset sleep-edf. The 20-fold cross-validation achieved an accuracy of 85.8%, a macro f1 score of 79.0, and a Cohen's κ of 0.804. In addition, the used TCN structure makes the network training time relatively short, which is 5.1 h. This model has the advantages of convenience and efficiency and is suitable for clinical use.

Keywords: Deep learning · Automatic sleep stage classification · Long time series

1 Introduction

Sleep is a physiological process that is vital to humans. Adequate sleep can effectively restore the body's mechanisms. However, due to the influence of unhealthy lifestyles, many people have problems with sleep disorders. Prolonged sleep disturbance can lead to a series of diseases such as lack of energy, cardiovascular and cerebrovascular diseases [1]. Therefore, early diagnosis of sleep disorders is of practical significance.

Sleep stage diagnosis is an effective way to detect sleep disturbances. In actual clinical practice, first use polysomnography (PSG) to obtain biological signals such as EEG, OMG, EMG, and pulse, and then professional physicians conduct sleep staging according to the internationally accepted sleep staging standard [2]. Data were manually divided into sleep periods. However, manual division takes a lot of time and is very inconvenient. Therefore, the development of automatic sleep staging is urgently needed.

In recent years, automatic sleep staging methods are mainly based on two methods, machine learning, and deep learning. Automatic sleep staging methods based on

© The Author(s), under exclusive license to Springer Nature Switzerland AG 2023
N. Xiong et al. (Eds.): ICNC-FSKD 2022, LNDECT 153, pp. 1167–1175, 2023.
https://doi.org/10.1007/978-3-031-20738-9_126

machine learning usually include two steps: feature extraction and sleep stage classification. First, sleep features are extracted by the time domain analysis method, frequency domain analysis method, etc., and then sleep stages are classified by classifiers such as random forest and support vector machine. The automatic sleep stage classification method based on deep learning mainly relies on the characteristics of a deep neural network to automatically extract useful features, which can avoid a lot of manpower and time-consuming. Deep learning models can improve the feature expression ability of the network by deepening the network structure and stacking different functional network layers. Compared with machine learning, deep learning models have advantages in automatic feature extraction and model optimization. The most common framework for automatic sleep staging is the CNN-RNN framework. For example, time and frequency domain features are extracted by one-dimensional CNN, multi-view features are learned by channel self-attention mechanism and finally sent to Bi-LSTM for sequence learning [3]. Using a deep belief network (DBN) to extract multi-level hierarchical features, long short-term memory (LSTM) determines the final category prediction label [4]. Combining Convolutional Neural Networks and Recurrent Neural Networks to extract sleep-specific subjective variable features from RF signals and capture the time course of sleep [5].

Although the current automatic sleep staging methods have achieved good results, there are still several prominent problems: (1) Inaccurate feature recognition caused by too deep-stacked networks; (2) The temporal correlation of long-term series cannot be captured. This paper proposes an automatic sleep staging model (FERTNet). The main contributions are summarized as follows (1) An improved ResNet network is proposed to extract frame-level time-invariant features. ResNet's residual structure can solve the problem of gradient disappearance caused by too deep network [6], reduce information loss and enhance network flexibility. (2) This paper utilizes Temporal Convolutional Networks (TCN) to capture the temporal correlations of long sequence features. We conduct experiments on the public dataset sleep-edf and achieve 86.2% accuracy. In Sect. 2, related work will be introduced. The third section introduces the model framework and improvements. Section 4 introduces the experimental part. Finally, our work will be summarized.

2   Related Work

2.1   ResNet Network

CNN can learn the representative characteristics of the sleep stage through convolutional layers, but as the number of network layers continues to increase, the network convergence speed will decrease, and the recognition rate will also decrease, and the ResNet network is proposed to solve the problem of network degradation [7]. ResNet networks have a unique residual module that enables stacked network layers to accurately adapt to residual mapping. In addition, it has feed-forward neural networks that can connect quickly, capable of reducing network parameters. However, the original ResNet did not solve the problem of network degradation. The ResStage group we propose replaces the original ResBlocks and contains three stages that can effectively enhance network performance capabilities.

2.2 TCN Network

The dilated convolution of TCN has a certain effect on the recognition of long-range temporal patterns and can map sequences of arbitrary length to output sequences of the same length. TCN is simpler and more accurate than canonical recurrent networks such as LSTM and GRU. We use a TCN network with two hidden layers, which can not only capture the causal relationship of time series but also effectively expand the receptive field of TCN and alleviate the drawbacks of feature loss.

3 Model Overview

3.1 FERTNet Network Model

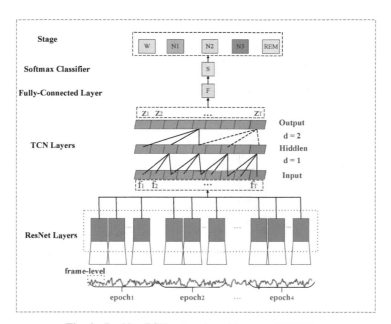

Fig. 1. ResNet-TCN network architecture (FERTNet)

Figure 1 shows two steps. The first is to use the improved ResNet to extract the frame-level invariant features from the original single-channel EEG signal, and then the feature sequence extracted in the first step is to mine the time-varying features of the sequence level through TCN and then complete the sleep staging. Take the 30s as an epoch, each epoch is divided into n frame sequences, and each frame sequence contains partial overlapping features of the previous frame sequence. To model the long sequence, we entered 4 epochs, frame-level invariant features extracted by the improved ResNet:

$$F = \{f_1, \ldots, f_T\} \tag{1}$$

Time-varying features of feature sequences can be extracted through the TCN network:

$$Z = \{z_1, \ldots, z_T\} \tag{2}$$

The extracted time-varying features are imported into the fully connected layer to output the target epoch $p(y|\{z_1, \ldots, z_T\})$. Finally, the output by softmax belongs to the sleep phase epoch.

3.2 Improved ResBlock Group

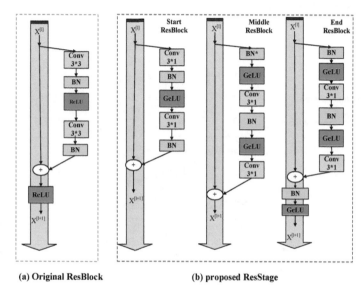

(a) Original ResBlock   (b) proposed ResStage

Fig. 2. **a** Original ResStage [7]; **b** proposed ResStage

As shown in Fig. 2, the ResStage we propose contains a starting ResBlock, and an ending ResBlock, and the rest are intermediate ResBlocks. Compared to the original ResBlock, to compensate for the shortcomings of the ReLU function and reduce the effect of negative signals, we replaced all linear rectification functions (ReLU) with GeLU functions:

$$GeLU(x) = 0.5x\left(1 + \tanh\left(\sqrt{2/\pi}\left(x + 0.044715x^3\right)\right)\right) \tag{3}$$

this change can also effectively improve the training speed.

In addition, we redistributed the position of bn and GeLU functions, and in Start Res-Block, the batch normalization (BN) layer is after the second convolutional layer, which makes it more convenient to add elements using projection shortcuts. End Resblock's main propagation path ends up with bn layers and GeLU functions, making information propagation faster and preparing for the next stage.

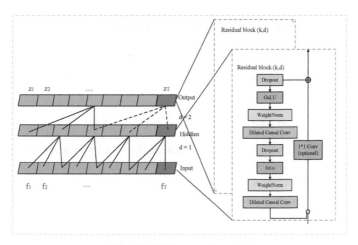

Fig. 3. TCN network model

3.3 TCN Network Structure

As shown in Fig. 3, TCN can satisfy mappings such as $f : f^{T+1} \rightarrow z^{T+1}$, and due to causal limitations, $\{z_T\}$ depends only on $\{f_1, \ldots, f_T\}$, , in addition, the expanded convolution of TCN can effectively expand the receptive field of the network, and the expanded convolution is defined as (d represents the expansion factor, j represents the filter size, and $k - d \cdot i$ is the past direction)

$$H(k) = \sum_{i=0}^{j-1} f(i) * X_{k-d*i} \tag{4}$$

when $d = 1$, it is a regular convolution, and the receptive field of TCN can be increased by increasing the expansion factor d or filter size k.

4 Experiment and Result Analysis

4.1 Datasets

The sleep-edf dataset consisted of two groups of subjects, one for SC records of 20 healthy subjects without sleep-related disorders, and the other for ST records of 22 subjects studying the effects of temazepam on sleep. This study uses Fpz-Cz channel data from SC. Table 1 lists the sample distributions of the sleep-edf dataset. The EEG sampling rate of this database is 100Hz, and the data is sliced at the 30s. According to the R&K standard, it is divided into W, N1, N2, N3, N4, REM, MOVEMENT, UNKNOWN. In this experiment, the EEG signal of the Fpz-Cz channel was used. According to the AASM judgment rule, N3 and N4 were combined as the N3 period, and the sleep data from before going to bed to 30 min after waking up were intercepted.

Table 1. Number of epochs by class in the dataset

Dataset	W	N1	N2	N3	REM	Total
Sleep-edf	8285	2804	17,799	5703	7717	42,308

4.2 Evaluation Metrics and ResNet Network Parameters

To evaluate model effectiveness, there are the following metrics: Accuracy per Class (PR), Per Class Recall (RE), Per Class F1 Score (F1), Overall Accuracy (Acc), Macro Average F1 Score (MF1), and Cohen Kappa Coefficient (κ).

$$PR_i = \frac{e_{ii}}{\sum_{j=1}^{N} e_{ij}} \tag{5}$$

$$PR_i = \frac{e_{ii}}{\sum_{j=1}^{N} e_{ji}} \tag{6}$$

$$F1_i = \frac{2PR_i RE_i}{2PR_i + RE_i} \tag{7}$$

$$ACC = \frac{\sum_{j=1}^{N} e_{ij}}{\sum_{i=1}^{N} \sum_{j=1}^{N} e_{ij}} \tag{8}$$

$$\kappa = \frac{p_o + p_e}{1 - p_e} \tag{9}$$

where PR represents the accuracy with which the model distinguishes sleep stages from other stages. RE represents the accuracy of the model for predicting sleep stages, e_{ij} are the element in the ith row and jth column of the confusion matrix, N is the number of sleep stages, p_o is the actual agreement rate, and p_e is the theoretical agreement rate.

Table 2 is the parameter setting of ResNet. To better process the timing information, we use one-dimensional processing.

4.3 Datasets and Evaluation Metrics

To verify the effectiveness of our improved network, we compare the experimental results of the original ResNet and the improved ResNet. The confusion matrix for 20-fold cross-validation on sleep-edf is shown in Fig. 4. The classification performance of the improved network in N1, N3, and REM stages is improved, and the REM stage is the most significant, with an improvement of 5.4%. The results show that the improvements made can improve the network's ability to identify features and make a more accurate classification of sleep stages.

4.4 Comparison with the Forward Approach

Table 3 shows the comparison results of MultitaskCNN, AttnSleep, and FERTNet networks under the same experimental conditions. It can be seen that the accuracy rate,

Table 2. Comparison of original ResNet and improved ResNet parameters

Layer	Resnet-34		Improved Resnet-34	
	Blocks	Output	Blocks	Output
conv1	7 × 7, 64, stride2	112 × 112	7 × 1, 64, stride2	1500 × 1
conv2_x	3 × 3maxpool, stride2	56 × 56	3 × 1maxpool, stride2	750 × 1
	$\begin{bmatrix} 3 \times 3,\ 64 \\ 3 \times 3,\ 64 \end{bmatrix} \times 3$		$\begin{bmatrix} 3 \times 1,\ 16 \\ 3 \times 1,\ 16 \end{bmatrix} \times 3$	
conv3_x	$\begin{bmatrix} 3 \times 3,\ 128 \\ 3 \times 3,\ 128 \end{bmatrix} \times 4$	28 × 28	$\begin{bmatrix} 3 \times 1,\ 32 \\ 3 \times 1,\ 32 \end{bmatrix} \times 4$	375 × 1
conv4_x	$\begin{bmatrix} 3 \times 3,\ 256 \\ 3 \times 3,\ 256 \end{bmatrix} \times 6$	14 × 14	3 × 1maxpool, stride2	94 × 1
			$\begin{bmatrix} 3 \times 1,\ 64 \\ 3 \times 1,\ 64 \end{bmatrix} \times 6$	
conv5_x	$\begin{bmatrix} 3 \times 3,\ 512 \\ 3 \times 3,\ 512 \end{bmatrix} \times 3$	7 × 7	$\begin{bmatrix} 3 \times 1,\ 128 \\ 3 \times 1,\ 128 \end{bmatrix} \times 3$	47 × 1
out	average pool, 1000-d fc, softmax	1 × 1	dropout(p = 0.5)	47 × 1

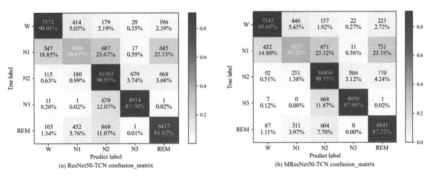

Fig. 4. Confusion matrix of ResNet-TCN and improved ResNet-TCN, **a** Resnet50-TCN confusion **b** MResnet50-TCN confusion

macro f1 score, and Cohen's κ of the FERTNet network are higher than those of the other two networks, indicating that it can effectively extract deep features and make the classification more accurate. Due to the small number of N1 samples, it has the problem that the characteristics are not obvious and the classification effect is poor. Compared with other methods, the method in this paper has certain advantages in dealing with long-term sequences, and can effectively learn the conversion rules between various stages and improve the classification performance of each category. In addition, the

TCN network used in this paper can effectively shorten the network training time and save resources.

Table 3. Comparison of FERTNet with other methods

Method	Overall performance			Per-class performance					Avg traing times
	ACC	MF1	K	W	N1	N2	N3	REM	
MultitaskCNN [8]	84.6	78.0	0.79	90.9	39.7	83.2	76.6	73.5	2.6 h
AttnSleep [9]	84.4	78.1	0.79	89.7	42.6	88.8	90.2	79.0	21 min
FERTNet	85.8	79.8	0.808	92.0	45.5	89.4	88.6	83.5	15 min

5   Conclusion

In this paper, a sleep staging model based on single-channel raw EEG signal FERTNet is proposed. FERTNet can extract both invariant features of EEG signals at the frame level and time-varying features at the epoch level. Experimental results show that this model can improve the degradation of deep network models, the extraction of useful features is small, and the dependencies of long-term series are difficult to capture. In addition, our models increase network training time and save machine consumption. It can be applied to the medical and health field.

References

1. Luyster, F.S., Strollo, P.J., Zee, P.C., Walsh, J.: K. Sleep: a health imperative. Sleep **35**(6), 727–734 (2012)
2. Berry, R.B., Budhiraja, R., Gottlieb, D.J., Gozal, D., Iber, C., Kapur, V.K., Tangredi, M.M., et al.: Rules for scoring respiratory events in sleep: update of the 2007 AASM manual for the scoring of sleep and associated events: deliberations of the sleep apnea definitions task force of the American Academy of Sleep Medicine. J. Clin. Sleep Med. **8**(5), 597–619 (2012)
3. Yuan, Y., et al.: A hybrid self-attention deep learning framework for multivariate sleep stage classification. BMC Bioinformatics **20**(16), 1–10 (2019)
4. Giri, E.P., Fanany, M.I., Arymurthy, A.M.: Combining generative and discriminative neural networks for sleep stages classification. arXiv preprint arXiv:1610.01741 (2016)
5. Zhao, M., Yue, S., Katabi, D., Jaakkola, T.S., Bianchi, M.T.: Learning sleep stages from radio signals: a conditional adversarial architecture. In: International Conference on Machine Learning (pp. 4100–4109). PMLR (2017)
6. Konovalenko, I., Maruschak, P., Brevus, V. Steel surface defect detection using an ensemble of deep residual neural networks. J. Comput. Inf. Sci. Eng., **22**(1) (2022)
7. He, K., Zhang, X., Ren, S., Sun, J.: Deep residual learning for image recognition. In: Proceedings of the IEEE Conference on Computer Vision and Pattern Recognition (pp. 770–778) (2016)

8. Phan, H., Andreotti, F., Cooray, N., Chén, O.Y., De Vos, M.: Joint classification and prediction CNN framework for automatic sleep stage classification. IEEE Trans. Biomed. Eng. **66**(5), 1285–1296 (2018)
9. Eldele, E., et al.: An attention-based deep learning approach for sleep stage classification with single-channel eeg. IEEE Trans. Neural Syst. Rehabil. Eng. **29**, 809–818 (2021)

Container Load Prediction Based on Extended Berkeley Packet Filter

Zhe Qiao$^{(\boxtimes)}$, LiJun Chen, and YuXuan Bai

Xi'an University of Posts and Telecommunications, Xi'an 710121, China
{qiaozhe,baiyuxuan}@stu.xupt.edu.cn, cljun@xupt.edu.cn

Abstract. The widespread use of container cloud puts forward higher requirements for the accuracy and real-time prediction of container resources. Aiming
at the problem of low efficiency of existing container load prediction ability, a
model based on the combination of eBPF and LSTM and container resource prediction is proposed. The model uses the cgroup mechanism in the Linux kernel
as the medium, uses eBPF technology to extract the load data parameters during
container operation, and combines the LSTM prediction algorithm to predict the
historical load change trend of the container resource usage over a period of time
in the future, and then compared to raw test data. Experimental results show that
the combination of eBPF and LSTM has good predictive power in container load
prediction, and can accurately predict the load data of container load in the future
period of time It has high practical application value.

Keywords: Containers · LSTM · eBPF · Cgroup · Linux

1 Introduction

As a new generation of virtualization technology, containers have quickly become the
virtualization technology of choice for enterprise deployment services with this high
flexibility, high availability, and high efficiency. With the large number of containers
deployed in the production environment, the number of services on a single host will
rise sharply, and in the process of providing services to users, various resource shortages
will inevitably occur in the process of competing for host resources between containers
when the business is busy. Thus affecting the quality of service of the business. However,
the business on some containers is empty at a certain time, and there are a large number of
idle resources, resulting in waste of resources. There-fore, how to get the container load
information in a timely and accurate manner and predict the future resource occupation
to ensure the safe and reliable operation of the container on the host machine has become
one of the hot issues in the current research.

At present, a large number of domestic and foreign scientific researchers, cloud
service providers and Internet companies have made a lot of contributions in this regard.
Xie et al. [1]. Present the Autoregressive Integrated Moving Average Model (ARIMA)
hybrid model and a triple exponential smoothed real-time prediction of docker container
resource load models designed to solve the problem of predicting linear and non-linear

relationships in container workloads in container environments, thereby improving the accuracy of container load prediction. Kumar et al. [2]. Proposed a hybrid model of Long short-term Memory (LSTM) and Bidirectional Long short-term Memory (BLSTM) to predict the load of the container over time in the future from a multivariate perspective, which obtained better prediction results than other algorithmic models. Meng et al. [3]. Designed a container load prediction model for time series analysis. In practical applications, it not only has high prediction accuracy, but also has good scalability of resources. However, when making container load predictions, they spend most of their energy on the optimization of prediction algorithms, and have done less research on data acquisition. According to the 2022 Cloud Native Security and Usage Report released by sysdig, 23% of enterprise containers have more than 1000 containers [4], and the scale of data generated by these containers in operation is bound to pose greater challenges to data acquisition systems. The original acquisition system was difficult to meet due to its poor real-time and accuracy.

Extended Berkeley Packet Filter (eBPF) is Alexei Starovoitov on the basis of the original Berkeley Packet Filter (BPF), expanded the number of registers and register width, and deeply optimized for hardware, so that eBPF instruction execution speed is greatly improved [5]. It can extract kernel running data without modifying the Linux kernel source code, with low system overhead, fine data granularity, and high accuracy. As the top subsystem of the Linux kernel, eBPF has become an ideal tool for Linux kernel observability, networking, and kernel security [6]. LSTM is a variant of Recurrent Neural Network (RNN) proposed by Hochreiter and Schmidhuber that solves the problem of exponential anomalies in RNN training in long-term sequence training. Over the years, LSTMs have excelled in solving long-series data problems and are widely used in a variety of data prediction problems [7].

This paper combines the advantages of eBPF and LSTM, proposes a model for predicting container load, the model has two parts, the data acquisition module is based on the observability of eBPF to extract cgroup data in container resource utilization and limitation, because the data extracted by eBPF itself has a time attribute, so the difficulty of data extraction is reduced. The forecast module is implemented using LSTM to ensure the accuracy of the prediction results.

2   Related Work

2.1   Gets the Container Load Data

At present, most of the container load information collection uses monitoring tools provided by third parties, or outputs data in the form of log printing at specific functions [8]. However, these methods have drawbacks, although the former can facilitate the output of monitoring data, but the tool itself will occupy a large system resource, not suitable for application on resource-constrained devices. The latter has no additional performance overhead, but requires kernel modification and recompilation, which is time-consuming, laborious, and inflexible. The eBPF technology used in this article allows the insertion of a Verifier-verified eBPF instruction in the kernel space, when the corresponding type event in the kernel is triggered, this instruction will be executed, and the kernel function information is saved in the eBPF Maps, the user-state program can obtain the core current state information only through Maps, not only has a high degree of accuracy in data extraction, but also the system execution of eBPF instructions has very low hardware resource overhead [9]. The eBPF works as shown in Fig. 1. BPF Compiler Collection (BCC) is a set of front-end frameworks that support high-level language compilation, which use user-state Python programs to inject C into the kernel for probing, simplifying eBPF programming. BCC not only provides a safe and reliable environment for C injection, but also provides a large number of Python interfaces to the user state, simplifying Python programming.

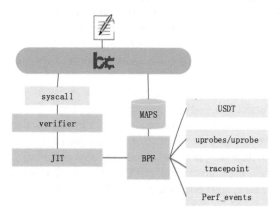

Fig. 1. How eBPF works

cgroup is a mechanism provided by the Linux kernel to record, allocate, and restrict the use of physical resources by process groups, and it is the cornerstone of implementing container technology[10]. Cgroup uses sub-file systems to manage different hardware resources of the system, and adopts a hierarchical structure for resource limits of different processes, and a process can be added to cgroups belonging to different levels, which in turn are attached to different resource control subsystems. This allows you to arbitrarily limit the use of process resources. When the container is successfully created by the kernel, the corresponding folder is generated in the cgroup file system, and the container resources can be managed through different files under the folder. The relationship between cgroup and process is shown in Fig. 2.

The resource statistics control information of the cgroup subsystem has specific data structure management [11]. Such as memory resources are managed by mem_cgroup data structures, and provide hook functions for operating the memory subsystem, but mem_cgroup data structures are not all exposed, only the cgroup_subsys_state data structures are exposed, and the rest of the fields are maintained and managed by the memory subsystem. The relationship between these two data structures is shown in Fig. 3. As can be seen from the figure, the cgroup_subsys_state data structure is in the first field position of the mem_cgroup data structure, so when we get the cgroup_subsys_state data structure, we only need to cast the data structure type to get all the fields of the mem_cgroup data structure.

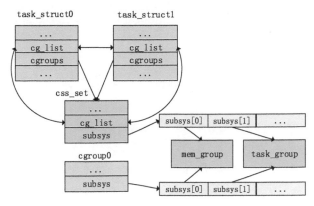

Fig. 2. Process-cgroup relationship

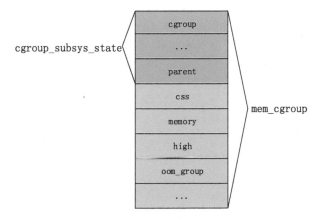

Fig. 3. The relationship between the cgroup_subsys_state and the mem_cgroup structure

Second, in order to accurately obtain container resource information, it is also necessary to distinguish between container processes and ordinary processes. The pid namespace structure is a means for the Linux kernel to implement process isolation, which can renumber the process pid, allowing the same pid number to exist under different namespaces, so that the kernel determines the process according to the pid namespace and id number, thus achieving isolation between container processes. There is a level field in the pid namespace to record the depth of the pid_ns, because the container process and the normal process belong to different levels, and the child processes created by the ordinary process will be added to the system root namespace by default, so the container process can be determined by the leave field. The pid namespace hierarchy relationship is shown in Fig. 4:

2.2 Long Short Term Memory Predictive Model

At first, scholars at home and abroad mostly used the RNN model when using neural networks to process time series data12. It consists of input stage, concealment stage and

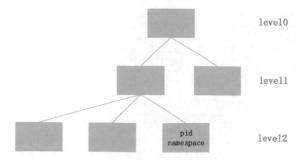

Fig. 4. The pid namespace topology

output stage. As shown in Fig. 5, compared to neural networks, recurrent neural networks are mainly reflected in all time steps, and the weights and biases of cyclic neurons in the hidden layer are the same13. Since the RNN output depends on the previous N step state, the RNN will have difficulty linking up to the previous relevant information when solving long-distance data. Therefore, it does not perform well when dealing with long-time series problems.

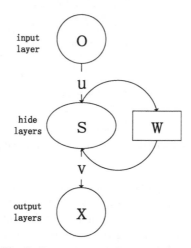

Fig. 5. Recurrent neural network model

LSTM is a special RNN model that has the same chain structure as RNN. But in the loop module, LSTM has four layers of neural networks, three "gates" for processing state information. Resolve long-distance dependencies by selectively deleting and adding information through selective decisions of forgetting gates and input gates. The LSTM structure is shown in Fig. 6. Where h_{t-1} is the state value output at the previous moment, x_t is the input value of the current moment, c_{t-1} is the state value at the previous moment, c_t is the updated value of the current state output, f_t is the for-gotten gate, it is the input gate, o_t is the output gate, and σ and tanh represent the activation function [14].

The LSTM cycle module processing is mainly divided into three stages:

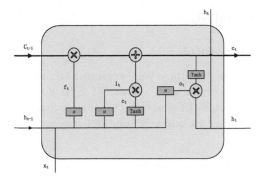

Fig. 6. LSTM hidden layer structure

Mainly based on the input data and the state data output of the previous stage, it is judged by the vector value output by the sigmoid function. Where 1 represents the retention state and 0 represents the deletion state. The calculation formula for this stage is shown in (1).

$$f_t = \sigma\big(w_f \times [h_{t-1}, x_t] + b_f\big) \tag{1}$$

Enter the stage. This stage mainly selectively adds input information x_t to c_{t-1} to avoid recording information with little relevance. There are two steps in total, x_t and h_{t-1} through the sigmoid activation function to get the input gate i_t, and the input gate control is obtained by h_{t-1} and x_t through the tanh activation function $\tilde{c}_t$. is used to update c_{t-1}. This part of the calculation formula is shown in (2), (3) and (4).
Enter the gate calculation formula: i_t

$$i_t = \sigma\big(w_i \times [h_{t-1}, x_t] + b_i\big) \tag{2}$$

Feature state calculation formula: $\tilde{c}_t \tilde{c}_t$

$$\tilde{c}_t = \tanh\big(w_c \times [h_{t-1}, x_t] + c_t\big) \tag{3}$$

Output status ct calculated formula:

$$c_t = f_t \times c_{t-1} + i_t \times \tilde{c}_t \tag{4}$$

Output stage. This stage is controlled by output gate o_t how much information in c_t is output as the current state. The output gate calculation formula is shown in (5), and the output state h_t calculation formula is shown in (6).

$$o\_t = \sigma\big(w\_o[h\_(t-1), x_t] + b_o\big) \tag{5}$$

$$h_t = o_t \times \tanh(c_t) \tag{6}$$

3 Experimental Analysis

3.1 Experimental Environment

The experiments covered in this article were performed on the Ubuntu20.04 operating system, Linux 5.15 kernel version, CPU version E5 2666 V3, docker version 20.10.12. The eBPF program is written by BCC to extract container load parameters. The specific configuration of the hardware and software used in the experiment is shown in Table 1.

Table 1. Hardware and software configuration information

Name	Version information
BCC	0.24
Memory	32GB DDR4
CPU	E5 2666 V3
Docker	20.10.12
Linux	Linux-5.15
LSTM	Cyclic memory network

3.2 Experimental Steps

First, create a container on the Ubuntu system and run the service to simulate the business load. Then run the eBPF program to collect the container's memory and CPU utilization data, respectively. Because there is a data discontinuity in the use of eBPF to obtain the system timestamp, the system timestamp is converted to a numeric serial number process. The data is then saved in a data file for model training and testing.

Next, export the collected data and standardize the preprocessing of the dataset using the mean and standard deviation. Considering that the data collected from adjacent time points in the dataset is not very different, it is necessary to define a sliding window to selectively extract data features, and then divide the dataset into training dataset, validation dataset and test dataset, use the training dataset for net-work model training and save the data model for future data prediction.

Finally, the predictions of the model are evaluated according to the test data set and the accuracy of the model is judged.

3.3 Analysis of Experimental Results

Figure 7 shows a dotted plot of the predicted and original values of the container CPU utilization. Figure 8 shows a dotted plot of container memory utilization and raw data. The two models iterated a total of 50 times, and the last 2% of the data set was selected as the original detection data set. From the figure, we can conclude that the results obtained by the LSTM prediction model are very close to the original data.

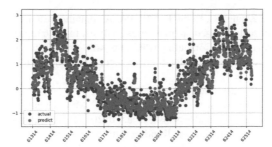

Fig. 7. Container CPU utilization forecast data compared with actual data

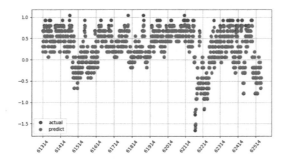

Fig. 8. Container memory utilization prediction data compared with actual data

4   Conclusion

In this paper, eBPF technology is combined with LSTM prediction model to analyze the load change of the container in the future. In order to solve the problem of ob-taining accurate load information for containers, this paper analyzes the cgroup mechanism in the Linux kernel, including the correspondence between processes and cgroup data structures, and focuses on how to extract data information in the memory subsystem using eBPF technology. In the model verification stage, in order to illustrate the predictive power of the model, we will take a part of the collected data set as a test set, and compare the prediction results of the model with the test set. Experimental results show that the difference between the prediction results of the model and the test set is very small, which shows that the model can well predict the change of container load in the future period of time.

References

1. Xie, Y., et al.: Real-time prediction of docker container resource load based on a hybrid model of arima and triple exponential smoothing. IEEE Trans. Cloud Comput. **10**(2), 1386–1401 (2022)
2. Kumar, S., Muthiyan, N., Gupta, S., Dileep, A.D., Nigam, A.: Association learning based hybrid model for cloud workload prediction. In 2018 International Joint Conference on Neural Networks (IJCNN), pp. 1–8. IEEE (2018)

3. Meng, Y., Rao, R., Zhang, X., Hong, P.: CRUPA: A container resource utilization predic-tion algorithm for auto-scaling based on time series analysis. In 2016 International Conference on Progress in Informatics and Computing (PIC), pp. 468–472. IEEE (2016)
4. Sysdig, https://sysdig.com/2022-cloud-native-security-and-usage-report. Last accessed 03 Jun 2022
5. eBPF, https://ebpf.io. Last accessed 03 Jun 2022
6. Miano, S., Bertrone, M., Risso, F., Tumolo, M., Bernal, M.V.: Creating complex net-work services with eBPF: experience and lessons learned. In 2018 IEEE 19th International Conference on High Performance Switching and Routing (HPSR), pp. 1–8. IEEE (2018)
7. Zhu, Y., Liu, J., Guo, C., Song, P., Zhang, J., Zhu, J.: Prediction of battlefield target tra-jectory based on LSTM. In 2020 IEEE 16th International Conference on Control and Automa-tion (ICCA), pp. 725–730. IEEE (2020)
8. Chen, L., Liu, J., Xian, M., Wang, H.: Docker container log collection and analysis system based on ELK. In 2020 International Conference on Computer Information and Big Data Applications (CIBDA), pp. 317–320. IEEE (2020)
9. Weng, T., Yang, W., Yu, G., Chen, P., Cui, J., Zhang, C.: Kmon: An In-kernel transparent monitoring system for microservice systems with eBPF. In 2021 IEEE/ACM International Workshop on Cloud Intelligence (CloudIntelligence), pp. 25–30. IEEE (2021)
10. Zhuang, Z., Tran, C., Weng, J., Ramachandra, H., Sridharan, B.: Taming memory related performance pitfalls in linux Cgroups. In 2017 International Conference on Computing, Networking and Communications (ICNC), pp. 531–535. IEEE (2017)
11. Stan, I.-M., Rosner, D., Ciocîrlan, Ş.-D.: Enforce a global security policy for user ac-cess to clustered container systems via user namespace sharing. In 2020 19th RoEduNet Conference: Networking in Education and Research (RoEduNet), pp. 1–6. IEEE (2020)
12. Park, N., Ahn, H.K.: Multi-layer RNN-based short-term photovoltaic power forecast-ing using IoT dataset. In 2019 AEIT International Annual Conference (AEIT), pp. 1–5. IEEE (2019)
13. Ye, Q., Yang, X., Chen, C., Wang, J.: River water quality parameters prediction method based on LSTM-RNN model. In 2019 Chinese Control And Decision Conference (CCDC), pp. 3024–3028. IEEE (2019)
14. Zhang, X., Liu, Z., Bai, J.: Linux network situation prediction model based on eBPF and LSTM. In 2021 16th International Conference on Intelligent Systems and Knowledge En-gineering (ISKE), pp. 551–556. IEEE (2021)

A Review on Pre-processing Methods for Fairness in Machine Learning

Zhe Zhang[1(✉)], Shenhang Wang[2], and Gong Meng[2]

[1] School of Computer Science and Technology, Xidian University, Xi'an, Shaanxi, China
21031211522@stu.xidian.edu.cn
[2] Beijing Aerospace Automatic Control Institute, Beijing, China
{wangshh,mengg}@casc.com.cn

Abstract. With the development of artificial intelligence (AI) technology, the application of machine learning (ML) algorithms has become more extensive, and AI algorithms have begun to make decisions in some important fields (finance, law, and medical health). However, studies have shown that due to social, historical, and other factors, the data for training machine learning algorithms already contain human biases, so machine learning algorithms will learn or even amplify these biases, resulting in unfair decision-making. There have been many studies on fairness in machine learning, including how to define and measure fairness and enhance fairness in ML. The existing means of lightening bias in ML can be classified into three types which are pre-processing, in-processing, and post-processing, according to the life cycle of ML. In this paper, we survey the pre-processing techniques and summarize them according to different categories. At the same time, we also introduce commonly used fairness measures to study fairness.

Keywords: Machine learning · Fairness in machine learning · Pre-processing · Algorithmic bias

1 Introduction

Due to the rapid development of AI technology and the wider application of machine learning, machine learning is no longer only used in some simple scenarios such as spam classification. Today, ML algorithms are starting to make important decisions in a growing number of domains, such as deciding whether to loan applicants, whether to grant parole to criminals or whether to recruit candidates applying for jobs, and more.

Previously, it was widely believed that machine learning algorithms would produce more objective or fair decisions than human decisions, but this was not the case, because the datasets on which machine learning algorithms depended could contain bias or discrimination, while data-driven algorithms inadvertently encode existing human biases and introduce new biases. For example, the U.S. justice system uses a piece of software called COMPAS to predict the probability of a criminal committing another crime, thereby helping judges decide whether to grant parole to offenders. However, a research [1] found that COMPAS would be more inclined to misjudge African Americans as re-offending and African-Americans were twice as likely to commit crimes again as whites in its predictions. Therefore, it is crucial to explore the fairness of machine learning in a context where people's lives can be significantly affected by the decisions made by ML algorithms.

The causes of unfairness in machine learning algorithms include bias in the training data collected by humans, bias in the machine learning algorithm itself, and biased data generated after the algorithm interacts with the user. As shown in Fig. 1, this process can be described by a ring diagram, and the most direct impact on machine learning fairness is the bias in the data. In the previous machine learning fairness review articles [18–20], the approaches to enhance fairness are generally divided into three categories according to the lifecycle of ML, as shown in Fig. 2. The first method is pre-processing, which corresponds to the data preprocessing stage in ML. The second one is in-processing, which works in the model training stage. The last one is post-processing method, which mainly focuses on using the prediction results to lighten the bias. The pre-processing method focuses on reducing or eliminating bias in the dataset by preprocessing it. Since the bias contained in datasets is a major cause of unfairness in machine learning algorithms, the importance of pre-processing methods also becomes self-evident.

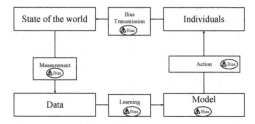

Fig. 1. Bias placed in the data, model, and user interaction feedback loop

This paper introduces fairness in machine learning and pre-processing methods to enhance fairness. Unlike other survey works in this field, this paper focuses on investigating fairness-enhancing ways based on pre-processing ideas and categorizes them. The fairness of machine learning discussed in this paper mainly involves classification tasks because most fairness problems are binary classification (i.e., whether to get a loan). The rest of this article is structured as follows. Section 2 presents a few most used measures for fairness in ML algorithms. Section 3 reviews existing pre-processing methods for enhancing fairness. Section 4 provides concluding remarks.

2 Measures of ML Fairness

To reduce or eliminate bias in machine learning algorithms, we should define a measure of fairness first. However, there is no universally applicable fairness measure. Therefore, this paper introduces some commonly used measures. They can be divided into group fairness and individual fairness according to different objects of achieving equity.

To facilitate subsequent discussions, we need to introduce some notations that will be used later. Let D be a binary classification dataset, p is the binary protected attribute

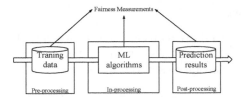

Fig. 2. An overview of ML fairness intervention methods

in D (gender, race, age). $p = 0$ means the protected group (female, black), and $p = 1$ means the unprotected (male, white) group. Let y' be the predictions of the classifier. $y' = 1$ indicates a positive prediction, such as granting the applicant loan eligibility, $y' = 0$ indicates a negative prediction, and y indicates a true result.

2.1 Group Fairness

The group-based fairness measure judges the fairness between two or more groups, often divided by the values of sensitive attributes.

Disparate Impact. The concept of disparate impact [14] was originally proposed in the legal field, and the formula for this measure is described below.

$$\frac{\Pr(y' = 1|p = 0)}{\Pr(y' = 1|p = 1)} \tag{1}$$

Intuitively, this measure requires that the selection rate predicted to be positive in different groups should be similar. For example, if a positive represents a loan to an applicant, the probability of getting a loan should be identical between blacks and whites. Generally speaking, the measure uses the 80% rule to judge the algorithm's fairness. The algorithm is relatively fair if the ratio is higher than 80%. Otherwise, it means that there are different impacts.

Demographic Parity. The basic idea of demographic parity [21] (statistical parity) and disparate impact is the same, requiring similar selection rates in different groups. But the way demographic parity measures similarity is the difference rather than the ratio. the formula for this measure is as follows:

$$\Pr(y' = 1|p = 0) = \Pr(y' = 1|p = 1) \tag{2}$$

Equalized Odds. This measure [22] means that the algorithm is fair enough if the False Positive Rate (FPR) and True Positive Rate (TPR) are the same between the protected group and the unprotected group.

$$\Pr\big(y' = 1|p = 0, y = 0\big) = \Pr\big(y' = 1|p = 1, y = 0\big) \tag{3}$$

$$\Pr\big(y' = 1|p = 0, y = 1\big) = \Pr\big(y' = 1|p = 1, y = 1\big) \tag{4}$$

2.2 Individual Fairness

This measure indicates that if two individuals are similar enough, then their predictions should also be similar. Different tasks can have different similarity measures [23]. A common description of individual fairness is as follows

$$y_i' = y_j' \quad ; \quad if \; s(i,j) = 1 \tag{5}$$

where $s(i,j)$ represents the similarity metrics between individual i and individual j.

3 Pre-processing Approach

Bias usually comes from the data itself, and pre-processing methods recognize the problem and try to change the sample distribution or perform specific transformations on the data to eliminate discrimination. The benefits of doing so include: 1. Training a model with a processed dataset can naturally reduce discrimination in machine learning algorithms. 2. Similar to feature engineering, the pre-processing method is more general because it does not require what model to use later. This paper divides the existing pre-processing methods into five types: Relabeling, Resampling, Reweighing, Fair representation, and Adversarial learning-based. Next, we will introduce them one by one.

3.1 Relabeling

The most intuitive idea of modifying the dataset is to directly change the dataset's labels to satisfy a certain fairness measure. Luong et al. [24] were inspired by KNN and defined a distance metric between two instances. Assuming an instance belongs to the protected group, if there is a significant difference in the treatment between its neighbors belonging to the protected group and the unprotected group, then mark it as discriminated against and change its label. Kamiran et al. [25] proposed a method of massaging the dataset. They train a classifier ranker to output the probability that the instance is positive. According to the score of the ranker, then select the same number of instances closest to the decision boundary separately from the two groups and flip their labels. Hajian et al. [26] proposed an approach based on rule protection and rule generalization, which can handle both direct and indirect discrimination uniformly. Relabeling and counterfactual studies are also very relevant since we need to worry about whether flipping the dependent variable will affect the classification results. Wang et al. [27] avoid discrimination against protected groups by identifying proxy attributes for sensitive attributes and building a preprocessor to use counterfactual distributions.

3.2 Resampling

Resampling methods refer to selecting representative but unbiased instances from the original dataset for training. Kamiran et al. [25] sample according to the weights of instances and propose the concept of preferentially sampled. Instances close to the decision boundary are preferentially sampled because they are most likely to be discriminated against or favored. Celis et al. [28] proposed a generalized P-DPP of (determinantal point process) k-DPP. After extracting subsets from the dataset, it can satisfy both combinatorial diversity and geometric diversity while ensuring fairness. Chouldechova et al. [30] implemented a model comparison framework that automatically identifies the subgroups that differ most between models. Iosifidis et al. [29] address the underrepresentation and class imbalance of the protected group by generating samples. Ustun et al. [10] used sensitive attributes to train decoupled classifiers, i.e., using different classifiers for different groups, instead of using a pooled model that ignores group membership. Oneto et al. [9] used multi-task learning (MTL) augmented by fairness constraints to learn unbiased classifiers suitable for different groups jointly.

3.3   Reweighing

Reweighing refers to giving weights to instances without changing the dataset. Kamiran et al. [25] assign weights to instances according to their types so that the sensitive attributes and predictions of the dataset are independent of each other, thereby improving the fairness of the dataset. Krasanakis et al. [11] train a classifier initially for learning the instances' weights, and use these weights to update the classifier iteratively. Jiang et al. [12] took a similar approach, where they first identify sensitive instances and then gain weights for them to optimize the selected measure.

3.4   Fair Representation

This approach refers to learning an intermediate representation for the dataset. The new representation satisfies the need for fairness while preserving helpful information in the dataset. Zemel [13] et al. offer a supervised representation learning method optimized for the three objectives of accuracy, individual fairness, and demographic parity. Feldman et al. [14] modified the dataset's features to make it indistinguishable between protected and non-protected groups, thus ensuring that the algorithm did not have a preference for a specific group. They also use a parameter to balance fairness and accuracy. Louizos et al. [15] learn an intermediate representation of the data based on a deep variational autoencoder to make sensitive attributes and potential variation factors as independent as possible while retaining as much residual information as possible. [16] introduced a new probabilistic formulation for data preprocessing to reduce discrimination. A convex optimization approach is proposed to learn data transformations with three objectives: controlling discrimination, limiting distortion of individual data instances, and preserving utility. Samadi et al. [17] studied PCA dimensionality reduction techniques that preserve the similarity of different groups. Backurs [4] et al. applied fair representation learning to fair clustering. Lahoti et al. [6] employed an unsupervised representation learning approach to optimize for two objectives: individual fairness based on k-nearest neighbor graphs and individual fairness based on fairness graphs.

3.5   Adversarial Learning-Based

Generative Adversarial Networks (GANs) [5] are often used to generate data samples. Fair adversarial learning has recently attracted increasing attention for fair classification and representation generation. Xu et al. [8] proposed a fairness-aware generative adversarial network (FairGAN), whose generator can generate data with fairness as the goal and maintain good utility. Then they [7] built a causal fairness-aware generative adversarial network (CFGAN), which can learn tight distributions from a given dataset while maintaining causal fairness according to a given causal graph. Madras et al. [3] relate common group fairness measures to adversarial learning and learn a fair representation of the data by providing an appropriate adversarial objective function for each fairness measure. Feng et al. [2] developed a minimax adversarial network that uses a generator to generate latent representations of the data and a discriminator to make the distributions similar between different protected groups.

4   Conclusion

In this survey, we introduce unfairness issues in machine learning algorithms. We then review some commonly used fairness measures. Finally, we review pre-processing methods for enhancing machine learning fairness, categorizing them into five types. We note that a recent trend in pre-processing methods is to transform data into fair representations where we can take GANs into consideration.

At present, machine learning fairness still faces great challenges. For example, most works of literature only focus on binary classification, and diversification is urgently needed; we also need more real-world datasets; preprocessing methods may introduce new biases, etc. Therefore, we need more efforts to understand the root causes of bias in ML and collect more unbiased data to achieve fairness in ML algorithms.

Acknowledgment. This work is supported by the National Natural Science Foundation of China under Grant No. 62172316 and the Key R&D Program of Hebei under Grant No. 20310102D. This work is also supported by the Key R&D Program of Shaanxi under Grant No. 2019ZDLGY13–03-02, and the Natural Science Foundation of Shaanxi Province under Grant No. 2019JM-368.

References

1. https://www.propublica.org/article/machine-bias-risk-assessments-in-criminal-sentencing
2. Feng, R., et al.: Learning fair representations via an adversarial framework. arXiv preprint arXiv:1904.13341 (2019)
3. Madras, D., et al.: Learning adversarially fair and transferable representations. International Conference on Machine Learning. PMLR (2018)
4. Backurs, A., et al.: Scalable fair clustering. International Conference on Machine Learning. PMLR (2019)
5. Goodfellow, I., et al.: Generative adversarial nets. Adv. Neural Inf. Process. Syst., **27** (2014)
6. Lahoti, P., Gummadi, K., Weikum, G.: Operationalizing individual fairness with pairwise fair representations. Proc. VLDB Endowment **13**(4), 506–518 (2019)
7. Xu, D., et al.: Achieving causal fairness through generative adversarial networks. In Proceedings of the Twenty-Eighth International Joint Conference on Artificial Intelligence (2019)
8. Xu, D., et al.: Fairgan: fairness-aware generative adversarial networks. In 2018 IEEE International Conference on Big Data (Big Data). IEEE (2018)

9. Oneto, L., et al.: Taking advantage of multitask learning for fair classification. In Proceedings of the 2019 AAAI/ACM Conference on AI, Ethics, and Society (2019)
10. Ustun, B., Liu, Y., Parkes, D.: Fairness without harm: decoupled classifiers with preference guarantees. In International Conference on Machine Learning. PMLR (2019)
11. Krasanakis, E., et al.: Adaptive sensitive reweighting to mitigate bias in fairness-aware classification. In Proceedings of the 2018 world wide web conference (2018)
12. Jiang, H., Nachum, O.: Identifying and correcting label bias in machine learning. In International Conference on Artificial Intelligence and Statistics. PMLR (2020)
13. Zemel, R., et al.: Learning fair representations. In International Conference on Machine Learning. PMLR (2013)
14. Feldman, M., et al.: Certifying and removing disparate impact. In Proceedings of the 21th ACM SIGKDD International Conference on Knowledge Discovery and Data Mining (2015)
15. Louizos, C., et al.: The variational fair autoencoder. arXiv preprint arXiv:1511.00830 (2015)
16. Calmon, F., et al.: Optimized pre-processing for discrimination prevention. Adv. Neural Inf. Process. Syst., **30** (2017)
17. Samadi, S., et al.: The price of fair pca: One extra dimension Adv. Neural Inf. Process. Syst., **31** (2018)
18. Mehrabi, N., et al.: A survey on bias and fairness in machine learning. ACM Comput. Surveys (CSUR) **54**(6), 1–35 (2021)
19. Caton, S., Haas, C.: Fairness in machine learning: a survey. arXiv preprint arXiv:2010.04053 (2020)
20. Pessach, D., Shmueli, E.: A review on fairness in machine learning. ACM Comput. Surveys (CSUR) **55**(3), 1–44 (2022)
21. Calders, T., Verwer, S.: Three naive Bayes approaches for discrimination-free classification. Data Min. Knowl. Disc. **21**(2), 277–292 (2010)
22. Hardt, M., Price, E., Srebro, N.: Equality of opportunity in supervised learning. Adv. Neural Inf. Process. Syst., **29** (2016)
23. Dwork, C., et al.: Fairness through awareness. In Proceedings of the 3rd Innovations in Theoretical Computer Science Conference (2012)
24. Luong, B.T., Ruggieri, S., Turini, F.: k-NN as an implementation of situation testing for discrimination discovery and prevention. In Proceedings of the 17th ACM SIGKDD International Conference on Knowledge Discovery and Data Mining, pp. 502–510 (2011)
25. Kamiran, F., Calders, T.: Data preprocessing techniques for classification without discrimination. Knowl. Inf. Syst. **33**(1), 1–33 (2012)
26. Hajian, S., Domingo-Ferrer, J.: A methodology for direct and indirect discrimination prevention in data mining. IEEE Trans. Knowl. Data Eng. **25**(7), 1445–1459 (2012)
27. Wang, H., et al.: Avoiding disparate impact with counterfactual distributions. NeurIPS Workshop on Ethical, Social and Governance Issues in AI (2018)
28. Celis, L.E., Deshpande, A., Kathuria, T., Vishnoi, N.K.: How to be fair and diverse?. arXiv preprint arXiv:1610.07183 (2016)
29. Iosifidis, V., Fetahu, B., Ntoutsi, E.: Fae: A fairness-aware ensemble framework. In 2019 IEEE International Conference on Big Data (Big Data), pp. 1375–1380. IEEE (2019)
30. Chouldechova, A., G'Sell, M.: Fairer and more accurate, but for whom?. arXiv preprint arXiv: 1707.00046 (2017)

Analysis of Test Scores of Insurance Salesman Based on Improved K-means Algorithm

Wei Bai$^{(\boxtimes)}$ and Jianhua Liu

Xi'an University of Posts and Telecommunications, Xi'an 710061, China
weibaih@isoftstone.com, xytx04@xupt.edu.cn

Abstract. In order to study the problems existing in the training and examination of marketing staff in insurance companies, it is necessary to conduct k-means clustering analysis on the learning status of marketing staff, so as to adopt different training programs for different groups and improve the theoretical level and actual productivity of insurance marketing staff. However, the traditional K-means algorithm has the problems of sensitive selection of initial points and unstable clustering results. This paper proposes an improved K-means algorithm. The algorithm combined the advantages of K-means++ and dichotomized K-means, weakened the dependence of clustering results on the initial clustering center, and evaluated the iterative clustering effect by contour coefficient. The experimental results prove the practicability of the improved K-means algorithm in the analysis of the test scores of insurance salesmen, and the clustering results can provide scientific guidance for the development of training programs of insurance salesmen and the improvement of their academic performance. Moreover, the clustering result of the algorithm is more stable than that of the traditional K-means algorithm, and the clustering effect is better.

Keywords: Cluster analysis · K-means algorithm · Data mining · Insurance marketer · Performance analysis

1 Introduction

In recent years, China's insurance industry has developed rapidly and its status has been improved unprecedeningly. However, for insurance companies, field marketing staff is the first impression of customers on insurance companies. In order to establish a good image for customers, every insurance company will choose to start from the construction of marketing staff. In insurance companies, effective training can improve the comprehensive quality of marketing staff and increase productivity. Training examination is the most direct and efficient means to improve the level of marketing staff. But most insurance company stays only to the examination result of sale member register, store wait for the surface job, lack the analysis of latent information behind examination result of sale member, cause the waste of training resource thereby [1]. In recent years. Data mining technology has been widely used, which can mine valuable information and knowledge from a large amount of data, among which K-means clustering algorithm is

© The Author(s), under exclusive license to Springer Nature Switzerland AG 2023
N. Xiong et al. (Eds.): ICNC-FSKD 2022, LNDECT 153, pp. 1192–1201, 2023.
https://doi.org/10.1007/978-3-031-20738-9_129

widely used in data analysis, image analysis, market research and other fields [2]. The application of K-means clustering algorithm can discover the distribution characteristics of scores, find the relationship between scores, better understand the learning ability and learning style of marketing staff, and provide scientific decision-making guidance for the capacity building of marketing staff [3].

However, in the traditional K-means algorithm, the initial cluster number K value is selected by subjective experience in the analysis of test scores of marketing staff [4]. If the K value is improperly selected, the final clustering result will not be the global optimal solution, but the local optimal solution, and the clustering effect will be not ideal. In addition, the traditional K-means algorithm randomly selects the initial clustering center, so the clustering result obtained by the K-means algorithm may be different each time it is executed, leading to the problem of unstable results. These are the defects of K-means algorithm. In order to overcome these defects, many scholars put forward a series of solutions. Based on the principle of minimum spanning tree, the initial clustering center was obtained by pruning in literature [5]. In literature [6], the maximum distance product method was used to select the high-density sample point with the maximum distance product among all the initialized cluster centers as the initial cluster center. Literature [7] proposed that variance was taken as the evaluation index for selecting cluster centers, and sample points in different regions with the smallest variance in the sample set were taken as the initial cluster centers. Literature [8] turned the sample set into K-DIST difference graph, and selected the sample points corresponding to smaller values in the K-DIST difference graph as the initial clustering center. In this paper, dichotomy K-means is combined with K-means++, and the contour coefficient is used to evaluate the effect of each clustering. The clustering result with the best clustering effect is selected as the final classification result. This improvement is more consistent with the actual situation that the initial training and learning status of insurance salesmen is unknown. Moreover, the improved K-means algorithm is easier to be embedded into the internal systems of insurance companies related to marketing staff.

2   K-means Algorithm

2.1   Traditional K-means Algorithm

The traditional K-means algorithm is an iterative relocation method. Its basic idea is to use the Euclidean distance between sample points to represent the similarity between sample points. The smaller the Euclidean distance is, the higher the similarity is. The larger the Euclidean distance, the lower the similarity. In all the sample points in the first place, selected K sample points as the initial clustering center, each remaining sample points are calculated respectively to the clustering center K Euclidean distance, which the minimum Euclidean distance, which clustering sample points and the most similar, will be assigned to the most similar categories, wait until after each sample points assigned to a category, Calculate the new clustering center of each category, and mark the category of each sample point again according to the new clustering center. Then, the process is iterated until the specified number of times is reached or the cluster center no longer changes.

Basic Concepts and Formulas. With clustering sample set $X = \{x_i | x_i \in \mathbb{R}^\vartheta, i = 1, 2, \cdots, n\}$, n is the number of samples, ϑ is dimension. The Euclidean distance between samples can be calculated by:

$$D(x_i, x_j) = \sqrt{(x_i - x_j)^T (x_i - x_j)} \tag{1}$$

The calculation formula of cluster center of C_i is:

$$c_i = \frac{\sum_{x \in C_i} x}{|C_i|} \tag{2}$$

Advantages and disadvantages of traditional K-means algorithm. The traditional K-means algorithm is simple to implement and widely used, and it is the basis of other optimization algorithms. However, it also has obvious defects:(1) it needs to randomly select the initial cluster center; (2) More sensitive to outliers; (3) The number of categories needs to be specified, and the clustering result is easy to fall into local optimum.

2.2 K-means++ Algorithm

The K-means++ algorithm improves the traditional K-means algorithm by optimizing the selection of the initial cluster center point. The distance between the selected initial cluster centers should be as far as possible. All other steps are the same.

Related Basic Concepts and Formulas. Suppose m cluster centers $C = \{c_1, c_2, \cdots, c_m\}$ have been selected. The distance between sample x and the nearest clustering center is:

$$D(x) = \min_{1 \le i \le m} D(x, c_i) \tag{3}$$

The probability of sample x being selected as the next clustering center is:

$$P(x) = \frac{D(x)^2}{\sum_{x \in X} D(x)^2} \tag{4}$$

Advantages and Disadvantages of K-means++ Algorithm. This algorithm does not randomly select K initial clustering centers like the traditional K-means algorithm, but selects them based on the principle of making the distance between them as far as possible, which reduces the number of iterations and speeds up the convergence speed. However, it still does not overcome the disadvantage of artificially specifying the number of clusters K, so the K-means++ algorithm also has certain limitations in use.

2.3 Binary K-means Algorithm

The binary clustering is carried out on the sample set every time to weaken the influence of the random selection of the initial clustering center. In fact, every dichotomous clustering of the sample set is a traditional K-means clustering with a specified number of clusters of 2.

Related Basic Concepts and Formulas. The sum of squares of errors within the cluster is:

$$sse_i = \sum_{x \in C_i} D(x, c_i)^2 \tag{5}$$

sse_i shows the degree of aggregation of cluster C_i, c_i is the center of cluster C_i. The larger sse value is, the higher degree of internal sample aggregation of cluster C_i.

The sum of the squares of the total errors is:

$$SSE = \sum_{i=0}^{K} sse_i \tag{6}$$

SSE is the result of calculating the sum of squares of errors within each cluster C_i in the final clustering result and then summing it, which can be used to evaluate the clustering effect.

Advantages and Disadvantages of BiK-means Algorithm. BiK-means algorithm has greatly reduced the impact of the traditional K-means algorithm on randomly selecting the initial clustering center. The clustering effect of each cluster is evaluated by introducing the sum of squares of errors within the cluster, and the cluster with the worst clustering effect is selected as the data set for the next K-means division. In this way, the clustering divided each time is more aggregation than the last one. However, the clustering number K of BiK-means algorithm also needs to be manually set, so there is still room for optimization of Bik-means.

According to the advantages and disadvantages of K-means++ algorithm and BiK-means algorithm, this paper combines the two algorithms, optimizes the defects of traditional K-means, and proposes a BiK-means++ algorithm, which will use this algorithm to conduct cluster analysis on the test scores of insurance company salesmen.

2.4 Improved Algorithm BiK-Means++

With the previous explanation of K-means++ and BiK-means algorithm, the improved BiK-means++ algorithm is easy to understand. It effectively combines K-means++ and BiK-means algorithm. First, when BiK-means algorithm is used for binary classification, the traditional K-means algorithm is no longer used, and K-means++ algorithm is used for binary classification, so that the clustering speed can be improved again and the initial value sensitivity problem introduced by randomly selecting two initial clustering centers can be overcome. Second, the clustering effect is evaluated when the number of clusters increases, and the classification result of each iteration is saved. Finally, the classification result with the largest contour coefficient is selected as the final clustering result. In this way, the number of clusters can be dynamically selected to get the best cluster.

Related Basic Concepts and Formulas. The sample set data is divided into K clusters, and the clustering contour coefficient is:

$$S(i) = \frac{b(i) - a(i)}{\max(a(i), b(i))} \tag{7}$$

i represents one of the samples, and $a(i)$ represents the average sum of the distances between sample i and all the other samples in its cluster. Define the average sum of the distances from sample i to a cluster that doesn't contain it as the distance from sample i to that cluster. Calculate the distance between sample i and each cluster that does not contain it, where the smallest distance is $b(i)$. As shown in Fig. 1.

It can be seen that the range of contour coefficient value is $[-1,1]$, and the closer it is to 1, it indicates that the degree of aggregation and separation of clustering results are relatively better, and the clustering effect is better.

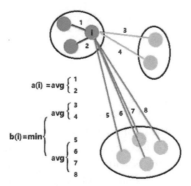

Fig. 1. Contour coefficient

2.5   Improved Algorithm BiK-Means ++ Basic Steps

- Step 1. Select the cluster number range $[K_{min}, K_{max}]$, general $K_{min} = 2$, $K_{max} = \sqrt{n}$, where n is the number of cluster samples;
- Step 2. Take all sample points as a cluster, and then use K-means++ algorithm to divide the cluster into two;
- Step 3. Calculate the number m of existing clusters and the SSE of each cluster. Calculate and save the contour coefficient when the current number of clusters is m. Save the cluster center to which each sample belongs in the current cluster;
- Step 4. If m is less than K_{max}, indicating that the partition has not reached the maximum number of clusters, and then select the cluster with the largest current SSE for K-means++ binary clustering;
- Step 5. Repeat Step3 and Step4 until m equals K_{max};
- Step 6. The number of clusters corresponding to the maximum contour coefficient is selected as the final cluster number, so as to obtain the cluster center of each sample, that is, the final cluster result.The flow chart of BiK-means++ algorithm is shown in Fig. 2.

2.6   Characteristics of Improved Algorithm BiK-Means++

Absorbing the advantages of K-means++ and BiK-means++ algorithms, the contour coefficient is introduced to evaluate the clustering effect. (1) The influence of the initial clustering center on the clustering results is minimized; (2) Fast convergence speed and easy to obtain global optimization; (3) Reduce the probability of selecting abnormal points; (4) According to the contour coefficient, the number of clusters and clustering results with the best clustering effect are dynamically selected.

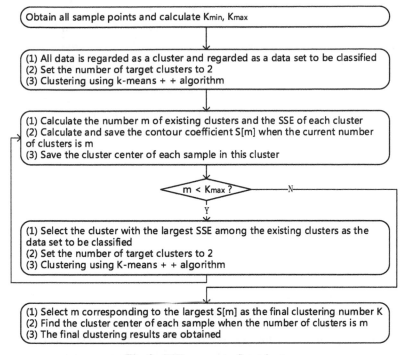

Fig. 2. BiK-means++ flow chart

3   Experiment and Analysis of Bik-Means++ algorithm

3.1   Experimental Data

The experimental data is the test results of all marketers of an insurance company in a city. The number of marketers is 218, and the courses can be divided into four categories. Table 1 shows some of the test scores of marketers.

3.2   Data Preprocessing

In order to better perform cluster analysis, we have standardized the data. The data after standardization are in the same order of magnitude. The range of standardized values is [0,1], and the data can be compared and analyzed. In this paper, the min max standardization method (formula 8) is used to standardize the original data. The standardization results are shown in Table 2.

Table 1. Test results of marketers

Job number	Corporate culture	Insurance Law	Marketing management	Insurance products
XIAN1237	90	80	70	100
XIAN3578	100	90	80	100
XIAN6871	95	90	80	90
XIAS7975	100	85	85	100
XIASC740	95	90	85	95

Table 2. Results after min max standardization

Job number	Corporate culture	Insurance Law	Marketing management	Insurance products
XIAN1237	0.900000	0.777778	0.700000	1.000000
XIAN3578	1.000000	0.888889	0.800000	1.000000
XIAN6871	0.950000	0.888889	0.800000	0.800000
XIAS7975	1.000000	0.833333	0.850000	1.000000
XIASC740	0.950000	0.888889	0.850000	0.900000

3.3 Experimental Results and Analysis

This paper uses pycharm2019 integrated environment and python3.6 as programming language to program the improved algorithm BiK-means++. Because the number of clusters is unknown in advance, K is set $K_{min} = 2$, $K_{max} = 14$, cluster the standardized data and get the clustering results in Table 3.

Table 3. Clustering results

Job number	Course name			
	Corporate culture	Insurance Law	marketing management	Insurance products
1(37)	0.8–1.0	0.8–1.0	0.8–1.0	0.9–1.0
2(79)	0.7–0.9	0.7–1.0	0.6–0.8	0.8–1.0
3(79)	0.7–1.0	0.6–0.8	0.8–1.0	0.7–0.9
4(20)	0.5–9.0	0.5–0.7	0.4–0.6	0.3–0.7
5(3)	0.1–0.4	0.0–0.3	0.0–0.2	0.0–0.4

It can be seen from Table 3 that the class 1 marketers have achieved relatively good results in all courses, indicating that they have a correct learning attitude and work hard during the training period, and can serve as an example for others to learn and teach some excellent learning experience. The second type of marketers' learning is medium, accounting for 36% of the total. The scores of the two courses of corporate culture and marketing management are not high. After a little review, you can enter the class 1 group. The learning advice given to them is to strengthen appropriately study of corporate culture and marketing management courses. The learning of type 3 marketers is similar to that of type 2, and also belongs to medium level. It also accounts for 36% of the total number. Properly strengthening the study of insurance law and insurance products can also be promoted to class 1 group. For the marketers of the second and third groups, the lecturers of the marketing training department can do some targeted training on weak courses, which can improve the training efficiency and improve their performance. Type 4 marketers have relatively poor academic performance, incorrect learning attitude, laziness and lack of self-discipline and initiative. Most of the course scores failed, so it is necessary to reorganize the training and examination for these groups to meet the basic job requirements. Class 5 marketers failed all courses and need to meet separately to learn about the situation. Deal with it accordingly.

Among these five groups, the first, second and third groups account for 89% of the total number. This figure shows the success or failure of this marketing training on the whole. On the one hand, it shows that the lecturers' training skills are excellent. On the other hand, it shows that most of the marketers are learning well and have mastered the knowledge and skills. The effect of this training and examination for marketers is ideal. If this figure is relatively low, it is likely to be caused by the poor level of the training instructors or the unreasonable design of the training program, so the training team needs to make corresponding adjustments and improvements.

According to the characteristics of each cluster, instructors can dynamically adjust the course focus in the training, formulate different training and guidance strategies, implement some personalized training contents, and carry out team collaborative learning.In order to verify the stability of the BiK-means++ algorithm proposed in this paper, we use the traditional K-means algorithm and BiK-means++ algorithm to cluster the test scores of marketers for five times, and calculate the contour coefficient of each clustering result. The results are shown in Table 4.

As can be seen from Table 4, the experimental results of running the traditional K-means algorithm with the specified number of clusters of 5 for 5 times show that the contour coefficient values of each clustering result are different, while the contour coefficient values of BiK-means++ algorithm are the same, which shows that the BiK-means++ algorithm is more stable than the traditional K-means algorithm when clustering the test scores of marketers. In addition, we can also see that the contour coefficient value of BiK-means++ algorithm is higher than that of any experiment of traditional K-means algorithm, which shows that the clustering effect of BiK-means++ algorithm is better than that of traditional K-means algorithm.

Table 4. Comparison of BiK-means ++ and K-means results

Number of experiments	Algorithm	Contour coefficient
1	K-means	0.522538
	BiK-means++	0.771253
2	K-means	0.468886
	BiK-means++	0.771253
3	K-means	0.478526
	BiK-means++	0.771253
4	K-means	0.459041
	BiK-means++	0.771253
5	K-means	0.503113
	BiK-means++	0.771253

4 Conclusion

Effective use of data mining technology to analyze the test results of marketers can obtain valuable information in a deeper level and provide scientific guidance for the training of marketers. This paper proposes an improved algorithm BiK-means++, which overcomes the shortcomings of the traditional K-means algorithm and applies it to the test results of insurance marketers for experimental analysis. The experimental results show that BiK-means++ can cluster the learning situation of marketers, and adopt personalized training programs according to the characteristics of each group, so as to effectively improve the training efficiency of instructors and the learning performance of marketers. It is also proved that BiK-means++ is more stable and effective than traditional K-means algorithm.

References

1. Chou, C.H., Hsieh, S.C., Qiu, C.J.: Hybrid genetic algorithm and fuzzy clustering for bankruptcy prediction. Appl. Soft Comput **56**, 298–316 (2017)
2. Liu, B., Wan, C., Wang, L.P.: An efficient semi-unsupervised gene selection method via spectral biclustering. IEEE Trans. Nanobiosci. **5**(2), 110–114 (2006)
3. Fu, X.J., Wang, L.P.: Data dimensionality reduction with application to simplifying RBF network structure and improving classification performance. IEEE Trans. Syst. Man Cybernetics Part B: Cybernetics **33**(3), 399–409 (2003)
4. Wu, J., Wang, R.Z.: An empirical study on the relationship between training and productivity of insurance marketers. Guangdong Sci. Technol. **23**(1), 10–11 (2014)
5. Feng, B., Hao, W.N., Cheng, G.: Optimization of initial cluster center selection for K-means algorithm. Comput. Eng. Appl. **49**(14), 182–192 (2013)
6. Zheng, D., Wang, Q.P.: K-means initial clustering center selection algorithm. Comput. Appl. **32**(8), 2186–2192 (2012)

7. Zhou, W.B., Shi, Y.X.: Density based optimization algorithm for k-means clustering center selection. Comput. Appl. Res. **29**(5), 1726–1728 (2012)
8. Xiong, Z.Y., Chen, R.T., Zhang, Y.F.: An efficient K-means clustering center initialization method. Comput. Appl. Res. **28**(11), 4188–4190 (2011)

The Difference Between the Impact of Intellectual Capital on Corporate Market Value and Book Value in the Taiwan Electronics Industry GMM Method

Li-Wei Lin[1], Xuan-Ze Zhao[2(✉)], Shu-Zhen Chen[3], and Kuo-Liang Lu[1]

[1] College of Business Administration, Fujian Jiangxia University, Fuzhou 350108, China
[2] School of Information, Zhejiang University of Finance and Economics Dongfang College, Zhejiang, China
zhaoxuanze@126.com
[3] Zhejiang University of Finance and Economics, Hangzhou, Zhejiang, China
chen.sz@zufe.edu.cn

Abstract. This paper uses the econometric model and GMM estimation method from the two dimensions of industry and region to interpret the difference between the book value and market value generated by intelligent capital in the electronics industry in Taiwan. Because in the constant changes of the environment, the company's intelligent capital and the market value and value difference of the enterprise show a complex change relationship, but currently there is no relevant quantitative analysis.

Keywords: Intellectual capital · GMM estimation method · Tobin's Q

1 Introduction

As early as Roos et al., the study of intelligent capital is not only limited to measuring intangible assets, but should also include the management of intangible assets [1]. Johanson et al., observed 11 Swedish companies that successfully measured and managed intangible assets and found that they have evolved management procedures to ensure that pointer measurement results can be translated into necessary actions, demonstrating the importance of managing intellectual capital [2]. Bontis pointed out that managers should shift their future thinking direction from short-term product strategy focus to long-term strategic focus on manpower, structure, and customer capital [3].

2 Literature Review

2.1 Definition and Measurement of Intelligent Capital

Resource-based theory argues that the internal resources an organization possesses are the basis of shaping the competitive advantage for the organization, and the organization's competitive advantage (increasing customer value or reducing the value of cost creation) drives superior organizational performance [4, 5]. At present, intellectual capital plays a significantly more important role in the economy based on knowledge, therefore, organizations prioritize intellectual capital to produce competitive advantages and achieve better performance [6].

2.2 Literature Discussion on Intelligent Capital to Market Value and Book Value

A combination can create value for the corporate. Manuel et al., analyzed the data of 309 Mexican manufacturing firms, and found that three dimensions of intellectual capital (including human capital, structural capital and relational capital) have a positive and significant influence on organizational performance [6]. Antonio et al., explored the relationship between intellectual capital disclosure quality and firm value based on a sample of 110 companies [7]. The findings suggest a significantly positive relationship between all three components of IC (structural, human, social and relationship) and corporate value.

3   Research Design

3.1 Data Analysis

This paper selects the listed companies in the electronics industry in Taiwan as a sample of research, and uses 2006–2017 as a sample interval to study the impact of smart capital on the difference between corporate net worth and value.

After the above steps were performed on the original samples, a total of 7776 samples of 648 companies in 12 years were obtained.

3.2 Model Setting and Variable Selection

We have designed a new research model as follows:

(1) Testing the dynamic relationship

We have done the relationship between the cause of the cause of the regression square model and the depent variable. The lag order of the dependent variable is determined by the regression equation of the 3.1 type.

$$TobiQ_{it} = \alpha_0 + \sum \alpha_p * TobiQ_{i,t-p} + Controls + \varepsilon_{it} \tag{3.1}$$

In Eq. 3.1, $TobiQ_{it}$ represents the difference between the company's net worth and value in the current period In the 3.1 formula, the intelligent capital variable is not added yet, and the model estimates use three different methods: random effect, OLS, and fixed effect.

(2) **Testing endogenous problems**

We use the regression equation to explain the influence of the explained variables on the explanatory variables and determine the endogenous problems in the model. In the process of studying the difference between intellectual capital and enterprise value and value, we are endogenous. Generally speaking, the root of endogeneity mainly includes the following three aspects.

But these unobservable factors usually reflect some characteristics of the individual. To capture these unobservable individual characteristics, the model is set to 3.2.

$$TobiQ_{it} = \alpha + \beta TobiQ_{i,t-1} + \gamma X_{it} + \lambda Z_{it} + \eta_i + \varepsilon_{it} \qquad (3.2)$$

4 Empirical Results and Analysis

4.1 Descriptive Statistics

In order to reduce the deviation of the abnormal value from the estimation result, before the empirical study, this paper firstly performs 1% tailing treatment on all variables of the sample data. Descriptive statistics for each of the main variables are shown in Table 1.

Table 1. Descriptive statistics

Variable	Average value	Standard deviation	Minimum	Maximum
TobiQ	1.124	0.668	0.370	4.720
Hu	5.577	1.201	2.758	9.035
In	0.050	0.062	0.000	0.385
Pr	6.575	1.249	2.063	9.582
Cu	0.057	0.046	0.000	0.293
Scale	15.170	1.293	12.58	19.72
Age	3.011	0.438	1.601	3.815
Lev	0.396	0.159	0.065	0.804
Investment	0.038	0.397	−0.753	3.662
DS	0.210	0.120	0.045	0.605

Note Descriptive statistical analysis among various variables

The sample data includes more than 7000 observations of Taiwan's electronics industry listed companies in 2006–2017. Table 1 shows that the average value of TobiQ is 1.124, which means that the average value created by the enterprise is greater than the cost of the invested assets.

4.2 Determine the Lag Order of the Interpreted Variable

We estimate the regression equation of 3.1, and then select the estimation results of the lag one, two and three lag dependent variables, respectively, see Table 2 columns (1), (2), and (3) of 4.3.

Table 2. Determine the lag order (OLS) of the dependent variable TOBIQ

	(1)	(2)	(3)
L.TobiQ	0.661***	0.584***	0.615***
	(0.009)	(0.013)	(0.015)
L2.TobiQ	–	0.128***	0.103***
	–	(0.012)	(0.015)
L3.TobiQ	–	–	0.074***
	–	–	(0.012)
Scale	−0.002	−0.007*	−0.011**
	(0.004)	(0.004)	(0.005)
Age	−0.060***	−0.042***	−0.028*
	(0.013)	(0.014)	(0.015)
Lev	−0.283***	−0.226***	−0.223***
	(0.037)	(0.038)	(0.039)
Investment	0.027**	0.022	0.027*
	(0.013)	(0.014)	(0.015)
DS	0.016	−0.020	−0.036
	(0.043)	(0.043)	(0.045)
Constant	0.731***	0.053	1.085***
	(0.080)	(0.082)	(0.085)
Year	Y	Y	Y
Industry	Y	Y	Y
Observations	6108	5337	4645
R-squared	0.588	0.602	0.617

Note ***, **, * and Y correspond to the significance levels of 10%, 5%, 1% and "Yes"

See Table 3.

4.3 Identify Endogenous Problems

In this paper, estimating the empirical results of 4, respectively, the results of the first-phase dependent variable on human capital, innovation capital, process capital, customer capital are shown in Table 4. Columns (1), (2), (3), and (4) of 5and 6.

Table 3. Determine the lag order of the dependent variable (fixed effect)

	(1)	(2)	(3)
L.TobiQ	0.357***	0.339***	0.365***
	(0.012)	(0.014)	(0.016)
L2.TobiQ	–	−0.002	0.015
	–	(0.013)	(0.015)
L3.TobiQ	–	–	−0.006
	–	–	(0.013)
Scale	−0.027	0.001	−0.054**
	(0.020)	(0.022)	(0.025)
Age	−0.319***	−0.223**	−0.345***
	(0.082)	(0.098)	(0.119)
Lev	−0.244***	−0.309***	−0.368***
	(0.068)	(0.072)	(0.079)
Investment	0.045***	0.032**	0.034**
	(0.013)	(0.014)	(0.015)
DS	0.064	0.010	0.082
	(0.102)	(0.108)	(0.119)
Constant	2.207***	1.000**	3.083***
	(0.366)	(0.424)	(0.510)
Year	Y	Y	Y
Industry	Y	Y	Y
Observations	6108	5337	4645
R-squared	0.336	0.327	0.298

Note ***, **, * and Y correspond to the significance levels of 10%, 5%, 1% and "Yes"

Table 4. Endogenous test (OLS)

	(1)	(2)	(3)	(4)
L.TobiQ	0.078***	0.010***	0.436***	0.001
	(0.017)	(0.001)	(0.027)	(0.001)
Scale	0.720***	−0.006***	0.326***	−0.008***
	(0.009)	(0.001)	(0.015)	(0.000)
Age	−0.048*	−0.011***	−0.035	0.004***

(*continued*)

Table 4. (*continued*)

	(1)	(2)	(3)	(4)
	(0.025)	(0.002)	(0.043)	(0.001)
Lev	−0.577***	−0.089***	−0.330**	−0.040***
	(0.070)	(0.005)	(0.129)	(0.004)
Investment	0.058**	−0.003**	−0.002	0.000
	(0.025)	(0.002)	(0.045)	(0.001)
DS	0.263***	−0.041***	−0.179	−0.006
	(0.081)	(0.005)	(0.139)	(0.004)
Constant	−4.881***	0.238***	1.227***	0.159***
	(0.151)	(0.010)	(0.262)	(0.008)
Year	Y	Y	Y	Y
Industry	Y	Y	Y	Y
Observations	6135	6163	4730	6148
R-squared	0.584	0.324	0.195	0.179

Note ***, **, * and Y correspond to the significance levels of 10%, 5%, 1% and "Yes"

See Table 5.
See Table 6.

Table 5. Endogeneity test (fixed effect)

	(1)	(2)	(3)	(4)
L.TobiQ	0.031***	−0.004***	0.206***	−0.002***
	(0.009)	(0.001)	(0.030)	(0.001)
Scale	0.462***	−0.013***	1.019***	−0.011***
	(0.015)	(0.001)	(0.057)	(0.001)
Age	0.093	0.004	−0.996***	0.005
	(0.062)	(0.004)	(0.201)	(0.004)
Lev	−0.098*	−0.009**	−1.343***	0.008**
	(0.052)	(0.004)	(0.197)	(0.003)
Investment	0.028***	0.001	0.065*	0.001*
	(0.010)	(0.001)	(0.034)	(0.001)
DS	−0.149*	−0.007	0.685**	−0.005
	(0.078)	(0.005)	(0.283)	(0.005)

(*continued*)

Table 5. (*continued*)

	(1)	(2)	(3)	(4)
Constant	−1.550***	0.242***	−5.675***	0.206***
	(0.275)	(0.019)	(0.978)	(0.018)
Year	Y	Y	Y	Y
Industry	Y	Y	Y	Y
Observations	6135	6163	4730	6148
R-squared	0.175	0.092	0.135	0.058

Note ***, **, * and Y correspond to the significance levels of 10%, 5%, 1% and "Yes"

Table 6. Endogeneity test (random effect)

	(1)	(2)	(3)	(4)
L.TobiQ	0.027***	−0.004***	0.299***	−0.002***
	(0.009)	(0.001)	(0.028)	(0.001)
Scale	0.542***	−0.012***	0.473***	−0.010***
	(0.013)	(0.001)	(0.027)	(0.001)
Age	0.021	−0.005*	−0.194**	0.002
	(0.045)	(0.003)	(0.078)	(0.003)
Lev	−0.172***	−0.017***	−0.710***	0.002
	(0.051)	(0.004)	(0.162)	(0.003)
Investment	0.027***	0.001	0.048	0.001*
	(0.010)	(0.001)	(0.034)	(0.001)
DS	−0.093	−0.013***	0.356*	−0.006
	(0.075)	(0.005)	(0.206)	(0.005)
Constant	−2.339***	0.313***	−0.250	0.196***
	(0.224)	(0.015)	(0.453)	(0.013)
Year	Y	Y	Y	Y
Industry	Y	Y	Y	Y
Observations	6135	6163	4730	6148
R-squared				

Note ***, **, * and Y correspond to the significance levels of 10%, 5%, 1% and "Yes"

5   Conclusions and Recommendations

The research in this chapter finds that the improvement of human capital level will significantly increase the current TobiQ of the enterprise, but the human capital level has a significant negative impact on the next period of TobiQ; the improvement of the innovation capital level significantly inhibits the current TobiQ of the enterprise, but The next issue of TobiQ has a significant promotion effect, and there is a significant lag effect on innovation capital; the increase of process capital level will significantly increase the current TobiQ of the enterprise, but it will not significantly promote the next period of TobiQ; the improvement of customer capital level Significantly reduced the company's TobiQ.

References

1. Roos, J., Roos, G., Dragonetti, N. C., Edvinsson, L., Intellectual Capital[J]
2. Johanson, U., Martensson, M., Skoog, M.: Mobilizing change through the management control of intangibles. Acc. Organ. Soc. **26**(7–8), 715–733 (2001)
3. Bontis, N.: Intellectual capital; an exploratory study that develops measures and models. Manag. Decis. **36**, 63–76 (1998). https://doi.org/10.1108/00251749810204142
4. Barney, J.: Firm resources and sustained competitive advantage. Acad. Manag. Exec. **9**(4), 49–61 (1995)
5. Porter, M.E.: Competitive Advantage[M]. New York, Free Press (1996)
6. Ibarra Cisneros, M. A., HernÃ¡ndez Perlines, F., MarÃa RodrÃguez GarcÃa.: Intellectual capital, organisational performance and competitive advantage. [J]. Euro. J. Int. Manage., **14**(6) (2020)
7. Vitiello, A.: The potential role of Gliptins to fight COVID-19. Authorea (2020). https://doi.org/10.22541/au.159493007.742298

A Synchronous Secondary Index Framework Based on Elasticsearch for HBase

Xiaohui Lin[1,2,3], Wenzhong Guo[1,2,3(✉)], and Kun Guo[1,2,3]

[1] College of Computer and Data Science, Fuzhou University, Fuzhou 350116, China
{200327064,guowenzhong,gukn}@fzu.edu.cn
[2] Fujian Key Laboratory of Network Computing and Intelligent Information Processing (Fuzhou University), Fuzhou 350108, China
[3] Key Laboratory of Spatial Data Mining and Information Sharing, Ministry of Education, Fuzhou 350108, China

Abstract. In the era of big data, HBase has been widely used in big data applications due to its fast write throughput and fast rowkey lookup. However, HBase only optimizes the index for rowkeys, and the query performance of non-rowkey columns is low. To solve this problem, we propose a synchronous secondary index framework based on Elasticsearch (SSIFE). In the framework, Elasticsearch is used to build a secondary index of HBase's non-rowkey columns, using the efficient and multi-condition retrieval function of Elasticsearch to achieve multi-conditional queries on the vast amount of data stored in HBase. Furthermore, an index maintenance strategy based on HBase replication is designed to reduce the cost of index data synchronization. Finally, a parallel query mechanism is designed to achieve higher query speed. Experimental results on the real-world power dataset show that the performance of SSIFE is better than other secondary index schemes and index maintenance strategies.

Keywords: Secondary index · Index maintenance · HBase · Elasticsearch

1 Introduction

In the era of big data, traditional data storage and management methods have been unable to cope with the system scalability and performance problems caused by the sharp increase in data scale. To overcome the above issues, a class of storage systems with flexible storage, low cost, high scalability, and common write/query latency has emerged in the industry. These systems are collectively referred to as NoSQL [1] databases. HBase is one of the mature systems that is modeled after Google's Bigtable [2] in the Hadoop [3] ecosystem. Elasticsearch [4] is a distributed, highly scalable, high real-time search and data analysis engine. It can easily make large amounts of data have the ability to search, analyze and explore, so it is widely used to retrieve large amounts of data.

© The Author(s), under exclusive license to Springer Nature Switzerland AG 2023
N. Xiong et al. (Eds.): ICNC-FSKD 2022, LNDECT 153, pp. 1210–1218, 2023.
https://doi.org/10.1007/978-3-031-20738-9_131

HBase creates a B+tree [5]-like index on the rowkey, which can support fast queries based on the rowkey. For queries on non-rowkey columns, a table needs to be scanned due to the lack of indexes. The query delay of a full table scan under massive data is unacceptable, significantly affecting HBase non-rowkey columns' query performance. In recent years, various secondary index technologies have been proposed, providing solutions for HBase fast query. However, most of these schemes build index tables based on HBase or directly embed index data into original data tables and use coprocessors of HBase to maintain index consistency when writing data. The coprocessor is similar to triggers in relational databases that trigger specified operations before and after written data. This can significantly slow down writes, affecting the system's ability to handle high write throughput.

In order to solve the above problems, we propose an efficient synchronous secondary index framework based on Elastisearch. In the framework, we use Elasticsearch to store the secondary indexes of non-rowkey columns in HBase data tables to achieve fast multi-conditional queries on the massive data stored in HBase. Meanwhile, an index maintenance strategy based on HBase replication is designed to alleviate the problem of reducing HBase write throughput caused by frequent index synchronization operations. A parallel query mechanism is designed in the return table query stage to improve the data query's performance. The contributions of this paper are summarized as follows.

(1) The designed secondary index and an index maintenance strategy improve the query performance of non-rowkey columns and reduce the cost of index data synchronization.
(2) The designed parallel query mechanism disassembles the returned rowkey set into multiple query subtasks, which improves the query performance in the return table query stage.
(3) Query and write performance experiments are performed on our proposed framework using real-world power dataset. The experimental results show that SSIFE has better performance than other index schemes and index maintenance strategies.

2   Related Work

To improve the query performance of non-rowkey columns, the solution typically adds a secondary index to the non-rowkey columns that need to be queried. According to the storage location of the index data, the current secondary index strategy can be classified into local and global indexes [6]. At present, most secondary index schemes belong to global indexes.

Hindex [7] is Huawei's open-source HBase secondary index framework. It builds a local index in each shard of the HBase data table. When querying data, the coprocessor on each region processes the query request and returns the queried data from each region after aggregating.

CCIndex [8] and LCIndex [9] are two typical global indexes. By creating an index table to speed up the query on non-primary key columns, the data is

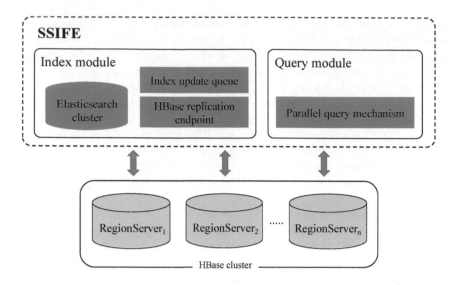

Fig. 1. The architecture of SSIFE

rearranged in the order of the specified index columns. HiBase [10] uses Redis [11] to cache hot data in the index table, which effectively improves the query efficiency of the index table. Diff-index [12] focuses on research on index maintenance strategies ,it provides a series of synchronous and asynchronous index maintenance schemes, which are sync-full, sync-insert, and async-simple. DELI [13] decomposes the index maintenance task into several subtasks, and only executes cheap tasks synchronously, while delaying expensive ones. Apache Phoenix [14] is an relational database engine for HBase. It implements local and global indexes and supports multiple consistency models.

These index schemes build index tables based on HBase or directly embed index data into data tables and use the coprocessor of HBase to achieve index data synchronization, which will lead to a series of problems such as degraded system write performance and excessive index space consumption.

3   SSIFE

The architecture of SSIFE is shown in Fig. 1, which is mainly composed of an index module and a query module. The Index module consists of an HBase replication endpoint, an Elasticsearch cluster, and an index update queue. Among them, the HBase replication endpoint is used to synchronize index data, the Elasticsearch cluster stores index data, and the index update queue is an in-memory queue used to cache index data. The query module is mainly used to process the user's query request. The following sections will introduce the index module and query module separately.

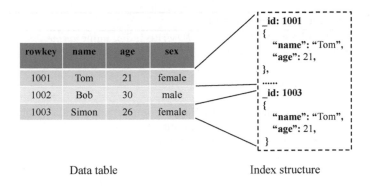

Data table   Index structure

Fig. 2. Data table and its corresponding index structure

3.1 Index Module

To avoid a full table scan when querying HBase non-rowkey columns and provide fast non-rowkey columns query capability, Elasticsearch is used to build indexes for non-rowkey columns in HBase user tables. Each index in Elasticsearch is used to manage all non-rowkey columns in the data table that can be used for queries. The HBase data table and the corresponding index structure are shown in Fig. 2. On the left is a table of HBase, and on the right is the index corresponding to the table in Elasticsearch. Each row in the table corresponds to a document in the index, associated with a row key and an id. A document stores the fields that can be used for querying. When a user initiates a query request, they can specify the value of one or more fields in the document. Elasticsearch will return the document's id that meets the query conditions. Then, the id collection can be used to obtain complete data in batches in HBase.

Before HBase completes the data write operation, it first records the write operation in the write-ahead-log (WAL) [15] to avoid memory loss caused by server failure. HBase provides a replication mechanism to obtain data write operations recorded in WAL in real-time and can customize the processing logic of these obtained data. The index maintenance strategy is implemented based on this feature, we customize the implementation logic of replication and embed it into each node in the HBase cluster. In each replication, we obtain the written data on the node in real-time and then filter out the index fields that has been established. Field updates are written to the index update queue and synchronized to Elasticsearch. The index update queue is an in-memory queue whose primary function is to avoid the write delay caused by directly writing data to the Elasticsearch cluster and reduce the efficiency of index synchronization. The details of the index synchronization process are described in Algorithm 1. Function is Indexed(c) determines whether column c is indexed.

Algorithm 1 Index synchronization

Input: A row set to be written to HBase L, write index name $name$
Output: Number of sync indexes cnt

1: $cnt \leftarrow 0$
2: // iterate over each modified row
3: **for** $p \in L$ **do**
4:      $dict \leftarrow \{\}$ // empty dictionary with key=columns name, value=columns value
5:      $columns \leftarrow$ p.columns // get the columns of row p
6:      $rowkey \leftarrow$ p.rowkey // get the rowkey of row p
7:      **for** $c \in columns$ **do**
8:          **if** isIndexed(c) **then**
9:              dict.put($c.name$, $c.value$) // put ($c.name$, $c.value$) to dict
10:          **end if**
11:      **end for**
12:      $cnt = cnt+$ saveIndexData($rowkey$, $dict$, $name$) // save index to Elasticsearch
13: **end for**
14: **return** cnt // return number of sync indexes

3.2   Query Module

The query module mainly provides external query services. There are two main types of queries: rowkey-based queries and multi-condition queries. The system will automatically determine the query type. The data will be obtained directly from HBase if the query is based on rowkeys. If it is a multi-condition query, get the rowkey set through Elasticsearch and quickly obtain complete data from HBase through the parallel query mechanism in the return table query stage. The details of the parallel query mechanism are described in Algorithm 2.

Algorithm 2 Parallel query

Input: Rowkey set $rowkeys$, HBase table Name $name$, batch size $batch$.
Output: Query result set R_s.

1: n $\leftarrow$ rowkeys.size
2: **if** rowkeys.size $mod\ batch == 0$ **then**
3:      $loop\_time \leftarrow n\ /\ batch$ // get the concurrency
4: **else**
5:      $loop\_time \leftarrow n\ /\ batch + 1$
6: **end if**
7: // parallel for
8: **for** $loop = 1$ to $loop\_time$ **do**
9:      $end \leftarrow \min((loop + 1) \times batch, n)$ // boundary of rowkey subset
10:      $part \leftarrow$ rowkeys.subList($loop \times batch, end$) // get part of rowkey set
11:      $r \leftarrow$ getDataFromHBase($part, name$) // get data from HBase based on rowkey
12:      R_s.add(r) // add subresult set to final result set
13: **end for**
14: **return** R_s // return the final result set

---

4 Experiments

We first introduced the experimental environment and the datasets used in this section. Then there are experimental results and analysis, including query performance experiments and write performance experiments.

4.1 Experimental Environment

The experiment deployed an HBase cluster with five nodes. The hardware of each machine in the cluster is a dual-core CPU 2.60 GHz, 8 GB memory and 500 GB hard disk. The operating system is Ubuntu 18.04.5. For comparative experiments, this article also implemented HiBase and installed Apache Phoenix. Software installed on the version is given in Table 1.

Table 1. Cluster software

SoftwareName	Version
HBase	1.3.1
Elasticsearch	7.6.2
Phoenix	4.16.1
Redis	6.0.2

4.2 Experimental DataSet

The data in this experiment comes from a power company in Fujian Province. The total data size is 11.6 GB, with about 50 million lines. Part of the power consumption data and some fields are shown in Table 2. Among them, ppq_mon represents the monthly power supply, accu_ppq represents the cumulative power supply, type represents the support type, and ext_provincial_flag represents the data source province.

Table 2. Power consumption data.

id	ppq_mon	accu_ppq	accu_type	ext_provincial_flag
1001	21080704	6123275	08	1
1002	11164155	7427562	21	2

4.3   Query Performance Experiment

This experiment evaluates the query performance of SSIFE, HiBase, and Apache Phoenix. We designed different query conditions to return result sets of different sizes and count the query latency of different result sets. As shown in Fig. 3, the time consumed by the query increases gradually as the amount of returned data increases. However, SSIFE has more obvious advantages over other solutions as response data increases. SSIFE has lower query latency than HiBase and Apache Phoenix, mainly because SSIFE uses multiple threads to fetch data in parallel from HBase in batches based on rowkeys.

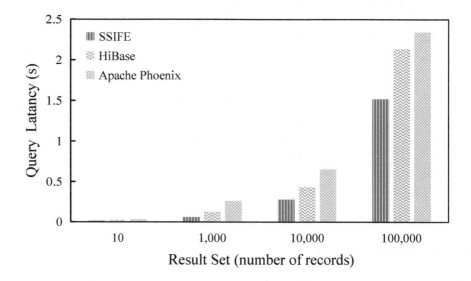

Fig. 3. Query performance of the frameworks

4.4   Write Performance Experiment

In writing performance experiments, we compare the performance of three different index maintenance strategies. Coprocessor-async and coprocessor-sync are asynchronous and synchronous index maintenance strategies based on the HBase coprocessor, and REP is the index maintenance strategy based on HBase replication proposed in this paper. Compare the average update latency corresponding to each scheme by setting different system throughput. As shown in Fig. 4, the update latency for each index maintenance scheme increases as the system throughput increases. REP has a lower write response time than coprocessor-async and coprocessor-sync because coprocessor-sync needs to complete the index synchronization operation during data writing. Although coprocessor-async does not need to immediately meet all index synchronization operations, it also needs to write all operations to an asynchronous update queue. All operations of Rep are done asynchronously, and there is no need to perform any index synchronization-related operations before and after writing, so it takes the least amount of time.

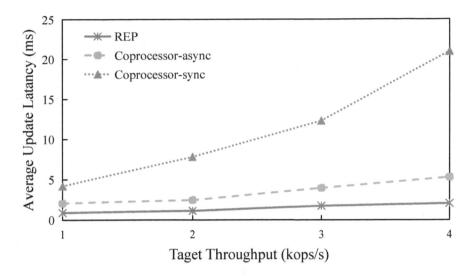

Fig. 4. Write performance of three index maintenance strategies

5   Conclusions

We propose an efficient synchronous secondary index framework based on Elastisearch. In the framework, Elasticsearch is uscd to build secondary indexes for non-rowkey columns in HBase. Using the efficient and multi-condition retrieval function of Elasticsearch to achieve multi-conditional queries on the massive amount of data stored in HBase. Furthermore, implements a lightweight index maintenance strategy based on HBase replication. And a parallel query mechanism is designed to improve the query speed further. Experiments demonstrate the effectiveness of SSIFE. In the future, we plan to integrate a fault-tolerant mechanism of index synchronization to avoid data inconsistency caused by index data loss, so as to enhance the fault tolerance of the framework.

Acknowledgements. This work was supported by the National Natural Science Foundation of China under Grant No. 62002063 and No. U21A20472, the National Key Research and Development Plan of China under Grant No. 2021YFB3600503, the Fujian Collaborative Innovation Center for Big Data Applications in Governments, the Fujian Industry-Academy Cooperation Project under Grant No. 2017H6008 and No. 2018H6010, the Natural Science Foundation of Fujian Province under Grant

No. 2020J05112, the Fujian Provincial Department of Education under Grant No. JAT190026, the Major Science and Technology Project of Fujian Province under Grant No. 2021HZ022007 and Haixi Government Big Data Application Cooperative Innovation Center.

References

1. Scherzinger, S., Sidortschuck, S.: An empirical study on the design and evolution of NoSQL database schemas. In: International Conference on Conceptual Modeling, pp. 441–455. Springer, Cham (2020)
2. Chang, F., Dean, J., Ghemawat, S., Hsieh, W.C., Wallach, D.A., Burrows, M., et al.: Bigtable: a distributed storage system for structured data. ACM Trans. Comput. Syst. (TOCS) **26**(2), 1–26 (2008)
3. Husain, B.H., Zeebaree, S.R.: Improvised distributions framework of Hadoop: a review. Int. J. Sci. Bus. **5**(2), 31–41 (2021)
4. Du, Y.: Massive semi-structured data platform based on Elasticsearch and MongoDB. In: Signal and Information Processing. Networking and Computers, pp. 877–888. Springer, Singapore (2021)
5. Wang, Q., Lu, Y., Shu, J.: Sherman: a write-optimized distributed B+ tree index on disaggregated memory. In: Proceedings of the 2022 International Conference on Management of Data, pp. 1033–1048 (2022)
6. Qader, M.A., Cheng, S., Hristidis, V.: A comparative study of secondary indexing techniques in LSM-based NoSQL databases. In: Proceedings of the 2018 International Conference on Management of Data, pp. 551–566 (2018)
7. Huawei Hindex: https://github.com/Huawei-Hadoop/hindex
8. Zou, Y., Liu, J., Wang, S., Zha, L., Xu, Z.: CCIndex: a complemental clustering index on distributed ordered tables for multi-dimensional range queries. In: IFIP International Conference on Network and Parallel Computing, pp. 247–261. Springer, Berlin, Heidelberg (2010)
9. Feng, C., Yang, X., Liang, F., Sun, X.H., Xu, Z.: Lcindex: a local and clustering index on distributed ordered tables for flexible multi-dimensional range queries. In: 2015 44th International Conference on Parallel Processing, pp. 719–728. IEEE, Beijing, China (2015)
10. Wei, G., et al.: HiBase: a hierarchical indexing mechanism and system for efficient HBase query. Chin. J. Comput. **39**(01), 140–153 (2016)
11. Gade, A.N., Larsen, T.S., Nissen, S.B., Jensen, R.L.: REDIS: a value-based decision support tool for renovation of building portfolios. Build. Environ. **142**, 107–118 (2018)
12. Tan, W., Tata, S., Tang, Y., Fong, L.L.: Diff-index: differentiated index in distributed log-structured data stores. In: EDBT, pp. 700–711 (2015)
13. Tang, Y., Iyengar, A., Tan, W., Fong, L., Liu, L., Palanisamy, B.: Deferred lightweight indexing for log-structured key-value stores. In: 2015 15th IEEE/ACM International Symposium on Cluster, Cloud and Grid Computing, pp. 11–20. IEEE, Shenzhen, China (2015)
14. Akhtar, S., Magham, R.: Using phoenix. In: Pro Apache Phoenix, pp. 15–35. Apress, Berkeley, CA (2017)
15. Qi, H., Chang, X., Liu, X., Zha, L.: The consistency analysis of secondary index on distributed ordered tables. In: 2017 IEEE International Parallel and Distributed Processing Symposium Workshops (IPDPSW), pp. 1058–1067. IEEE, Lake Buena Vista, FL, USA (2017)

MOOC Resources Recommendation Based on Heterogeneous Information Network

Shuyan Wang, Wei Wu$^{(\boxtimes)}$, and Yanyan Zhang

Xi'an University of Posts and Telecommunications, Xi'an 710121, Shaanxi, China
wsyly@xupt.edu.cn, {ww,18893722140}@stu.xupt.edu.cn

Abstract. Aiming at the problem that the existing MOOC recommendation mechanism cannot meet the dynamic and diversified learning needs of different individuals, a MOOC resource recommendation model based on heterogeneous information network is proposed. First by capturing MOOC platform of the heterogeneity between multiple entities in building its corresponding heterogeneous information network, and then through the node level attention and meta-path level fusion of attention, will learn to the user and the knowledge incorporated into the extended matrix factorization framework, to predict user preferences for knowledge, to carry on the personalized recommendation service. Experimental results show that this model has better recommendation performance than other commonly used models, and effectively solves the problem of personalized recommendation for learners.

Keywords: Heterogeneous information network · Attentional mechanism · Recommendation system

1 Introduction

After entering the twenty-first century, with the rapid iteration of information technology, MOOCs ushered in a period of rapid development. Tens of thousands of well-designed online courses meet the learning needs of millions of students and are widely favored by students. However, with the continuous development and expansion of MOOC platform, other difficulties have arisen [1].

First of all, MOOC does not have a training plan similar to that of regular colleges and universities, cannot organize courses systematically and scientifically, and cannot provide learners with effective learning guidance, resulting in information overload and low course completion rate. Secondly, direct course recommendation ignores learners' learning needs for specific knowledge points. Therefore, the recommendation system needs to quickly find the courses and knowledge points that users are interested in from the mass data, so as to promote the accurate dissemination of knowledge.

This paper proposes a MOOC resource recommendation model based on heterogeneous information network [2]. In order to fully explore the relationship between learners and course resources, the complex relationship between entity classes is modeled. The optimal combination of nodes and meta-paths is obtained by hierarchical attention fusion so that the aggregated node embedding can better capture the sophisticated

© The Author(s), under exclusive license to Springer Nature Switzerland AG 2023
N. Xiong et al. (Eds.): ICNC-FSKD 2022, LNDECT 153, pp. 1219–1227, 2023.
https://doi.org/10.1007/978-3-031-20738-9_132

semantic information in heterogeneous graph. To summarize, the main contributions of this paper are as follows:

- This paper collects and preprocesses MOOC data, constructs heterogeneous information network based on the complex interrelation of multiple objects, and designs different meta-paths to capture its context information from the perspective of users and knowledge points.
- This paper proposes a model HGATRec (**H**eterogeneous **G**raph **A**ttention Network **Rec**ommendation) for MOOC resources. In this model, the optimal combination of neighborhood and multi-meta-paths is obtained by means of hierarchical attention. Finally, the matrix decomposition framework is extended to predict the user's preference score for specific knowledge points.
- In this paper, several experiments are conducted on MOOC data sets, and the results show that the HGATRec model is real and effective.

2 Related Work

In recent years, most of the course recommendation researches are carried out by constructing user rating matrix. For example, matrix decomposition method SVD++ [3] and factor decomposition method LibFM. In addition, Jing et al. [4] combined the relationship between user interest and course prerequisite, and made course recommendation through collaborative filtering. Jiang et al. [5] proposed a goal-based curriculum recommendation algorithm to recommend appropriate series of relevant courses for users based on the knowledge model they have learned. Huang et al. [6] proposed a resource recommendation model based on DBN, which fully considers user learning characteristics and can improve user learning efficiency.

Relevant research shows that the recommendation research algorithms of adaptive learning mainly include collaborative filtering algorithm, clustering algorithm and content-based recommendation algorithm. However, how to study resource recommendation from the perspective of heterogeneous information network is still in the exploratory stage. This paper will make full use of heterogeneous information to improve the performance of recommendation system.

3 Proposed Model

The Framework of HGATRec model is shown in Fig. 1. The framework consists of four parts. The first part is feature extraction, preprocessing MOOC data sets and analyzing the relationship between different types of entities. The second part is the embedding of meta-path, building a structured HIN to model the relationship between different types of entities, and describing the correlation of knowledge points through different meta-paths. The third part is graph attention network. Firstly, node-level attention is used to learn the weights of neighbors based on the meta-path and aggregate them to obtain node embedding specific to the meta-path. Then, HGATRec differentiates differences through meta-path-level attention, and obtains the weight of the meta-path under the corresponding node, so as to obtain the optimal weighted combination embedded in a specific node. The fourth part is the scoring prediction. After generating the representation of users and knowledge points, the method based on extended matrix factorization is used to predict users' preferences.

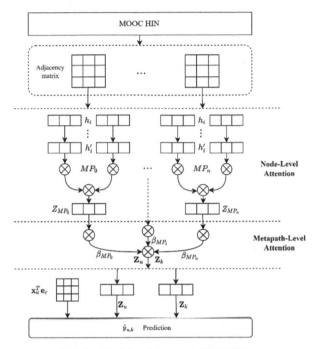

Fig. 1. Model architecture of HGATRec

3.1 Feature Extraction

Generally speaking, each course contains different knowledge points, and the course content also contains rich semantic information. In this paper, word embedding and user entity feature embedding of knowledge points are generated from the course information through Word2vector [7]. In addition, there is rich contextual information between different entity types, which can be used to establish corresponding adjacency matrix.

3.2 Meta-path Based Relationship

In this paper, four entity type nodes are selected: users, courses, knowledge points and teachers. The relationship is shown in Fig. 2. This paper considers user-user and knowledge point-knowledge point relationships via different meta-paths. The results are shown in Table 1, where U represents the user, C represents the course, K represents the knowledge point, and T represents the teacher. For example, $U \rightarrow K \xrightarrow{-1} U$ indicates that users have learned the same knowledge point, $U \rightarrow C \xrightarrow{-1} U$ indicates that users have learned the same course, and so on.

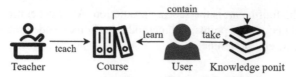

Fig. 2. MOOC entity relationship diagram

Table 1. The establishment of meta-paths.

Type	Meta-path
User	$U \rightarrow K \xrightarrow{-1} U$
	$U \rightarrow C \xrightarrow{-1} U$
	$U \rightarrow C \rightarrow T \xrightarrow{-1} C \xrightarrow{-1} U$
	$\ldots$
Knowledge point	$K \rightarrow U \xrightarrow{-1} K$
	$K \rightarrow C \xrightarrow{-1} K$
	$\ldots$

3.3 GATs Network

In view of the diversity of HIN node features, a new recommendation model is designed by using attention network. The attention network considers the importance of node attributes and edges with node-level attention and meta-path-level attention to obtain the optimal weighted combination.

Node-level Attention

Since different types of nodes have different feature spaces. Therefore, for each type of node, a specific transformation matrix $\mathbf{M}_{MP_i}$ is designed to project the features of different types of nodes into the same feature space. The projection is shown as follows:

$$h'_i = \mathbf{M}_{MP_i} \cdot h_i \tag{1}$$

where h_i and h'_i are the initial and projected features of node i respectively, and MP_i represents the meta-path of node i.

Then, given a node pair (i, j) connected by meta-path MP, the importance of node i to node j, namely, e_{ij}^{MP}, can be obtained through the deep neural network of node-level attention represented by att_{node}. The specific calculation method is as follows:

$$e_{ij}^{MP} = att_{node}\left(h'_i, h'_j; MP\right) \tag{2}$$

Importance e_{ij}^{MP} of node $j \in \mathcal{N}_i^{MP}$ is then calculated, where $\mathcal{N}_i^{MP}$ represents the meta-path-based neighbor of node i. By splicing the projection features of node i and node j, the weight of node j to node i is obtained. After obtaining the importance between node pairs, they are normalized by softmax function to get the weight coefficient α_{ij}^{MP}:

$$\alpha_{ij}^{MP} = softmax_j\left(e_{ij}^{MP}\right) = \frac{\exp\left(\sigma\left(\mathbf{a}_{MP}^{\mathrm{T}} \cdot \left[h_i' \| h_j'\right]\right)\right)}{\sum_{k \in \mathcal{N}_i^{MP}} \exp\left(\sigma\left(\mathbf{a}_{MP}^{\mathrm{T}} \cdot \left[h_i' \| h_j'\right]\right)\right)} \tag{3}$$

where σ represents the activation function, $\|$ represents the join operation, and $\mathbf{a}_{MP}$ is the node-level attention vector of meta-path MP.

Then, the meta-path-based embedding of node i is aggregated with the projected features of its neighbors, and the corresponding coefficients are as follows:

$$Z_i^{MP} = \sigma\left(\sum_{j \in \mathcal{N}_i^{MP}} \alpha_{ij}^{MP} \cdot h_j'\right) \tag{4}$$

where Z_i^{MP} is the learning embedding of node i of meta-path MP. Given user-user meta-path set $\{MP_1, \ldots, MP_n\}$, after node features are input into node-level attention, n groups of semantic-specific node embeddings are obtained, represented as $\{Z_{MP_1}, \ldots, Z_{MP_n}\}$.

Meta-path-Level Attention

In heterogeneous graph, each meta-path contains several types of semantic information. In order to learn more complex node embedding, these different semantic information need to be fused. Through the deep neural network of meta-path-level attention, the importance of each meta-path can be obtained, as shown in Eq. (5):

$$\left(\beta_{MP_1}, \ldots, \beta_{MP_n}\right) = att_{sem}\left(Z_{MP_1}, \ldots, Z_{MP_n}\right) \tag{5}$$

Then, the importance of all meta-paths is normalized by softmax function to obtain the weight β_{MP_i} of corresponding meta-path MP_i, as shown in Eq. (6):

$$\beta_{MP_i} = \frac{\exp\left(\omega_{MP_i}\right)}{\sum_{i=1}^{n} \exp\left(\omega_{MP_i}\right)} \tag{6}$$

The higher β_{MP_i} is, the more important meta-path MP_i is. The learned weights are used as coefficients and fused with semantic-specific embedding to obtain the final user representation $\mathbf{Z}_u$, as shown below:

$$\mathbf{Z}_u = \sum_{i=1}^{n} \beta_{MP_i} \cdot \mathbf{Z}_{MP_i} \tag{7}$$

3.4 Prediction

After the representation of user and knowledge point is obtained, the preference score of user for knowledge point is obtained through the extended matrix decomposition framework, which is calculated as follows. Where $\hat{y}_{u,k}$ is the preference score, $\mathbf{x}_u$ and $\mathbf{e}_c$ are the potential features of matrix factorization, and b_c is the bias term. In addition, $\mathbf{M}$ is a trainable matrix so that $\mathbf{Z}_u$ and $\mathbf{Z}_k$ are in the same space, and γ is a trainable parameter.

$$\hat{y}_{u,k} = \mathbf{x}_u^T \mathbf{e}_c + \gamma \cdot \mathbf{Z}_u^T \cdot \mathbf{M} \cdot \mathbf{Z}_k + b_c \tag{8}$$

4 Experiments

4.1 Datasets

In addition to MOOC data set, this experiment also adopted a classical data set named DBLP to verify the effectiveness and wide availability of HGATRec (Table 2).

Table 2. Datasets details.

Dataset	Entities	Relations(A–B)	Meta-path
DBLP	User	Paper-Author	APA, APCPA, APTPA…
	Author	Paper-Conf	
	Conf term	Paper-Term	
MOOC	See Table 1		

MOOC: The MOOC dataset contains hundreds of thousands of course selections from 1738 teachers, 706 real online courses, 104,863 knowledge points, and 189,156 users.

DBLP: DBLP is an integrated database system for computer English literature. In this paper, 20 conferences, 14,328 papers, 4057 authors and 8789 keywords were extracted from DBLP to conduct experiments.

4.2 Baselines

PMF [8]: A classical probability matrix decomposition model, which only makes recommendations by explicitly decomposing the scoring matrix into two low-dimensional matrices.

NeuMF [9]: Based on deep learning recommendation algorithm, which combines traditional matrix decomposition and multi-layer perceptron, it can simultaneously extract low and high dimensional features for recommendation.

Metapath2vec [10]: This model builds different neighbors of nodes in the network based on the selection of the original path, and finally uses the heterogeneous Skip-Gram model to embed the heterogeneous nodes.

SemRec [11]: It is a recommendation model based on weighted meta-path, which calculates user similarity matrix through the weighted meta-path and makes score prediction by combining the results obtained from different meta-paths.

4.3 Results

In the experiment of this paper, the whole data set is randomly divided into training set, verification set and test set according to the ratio of 6:2:2. According to the performance of the model in the verification set, the corresponding adjustment of hyperparameters to achieve the optimal model performance, and the model performance is verified on the test set.

To evaluate the performance of different models, adapt three widely used evaluation metrics: Hit Ratio of top-K items (HR@K), Mean Reciprocal Rank(MRR@K) and Normalized Discounted Cumulative Gain (NDCG@K), and set K to 10.

Comparison of Recommended Model Performance
Table 3 shows the performance of HGATRec model and five baseline methods in three data sets. It can be seen from the table:

(1) In MOOC data sets, HGATRec improved by 0.0243, 0.0108 and 0.0207 respectively in HR, MRR and NDCG indicators compared with SemRec model, and the recommendation performance was always superior to all comparison models, as well as other data sets. This is because HGATRec model adopts the hierarchical attention mechanism to embed the specific node obtained by node-level attention and aggregate the weight of the meta-path under the corresponding node obtained by meta-path-level attention, so as to improve the recommendation performance.
(2) The recommendation performance of HIN-based Metapath2vec, SemRec and HGATRec models is superior to that of traditional PMF and NeuMF models, which verifies the effectiveness of heterogeneous information for improving the performance of the recommendation model.
(3) Compared with the HIN-based Metapath2vec and SemRec models, HGATRec model adopts a more effective method to improve the recommendation system and makes full use of heterogeneous information.

Table 3. Overall performance comparison.

Model	MOOC (@10)			DBLP(@10)		
	HR	MRR	NDCG	HR	MRR	NDCG
PMF	0.3381	0.2468	0.2567	0.3496	0.1997	0.2386
NeuMF	0.3505	0.2483	0.2518	0.3545	0.2076	0.2385
Metapath2vec	0.3997	0.2818	0.2787	0.3864	0.2196	0.2660
SemRec	0.4059	0.2930	0.2820	0.3990	0.2278	0.2693
HGATrec	**0.4302**	**0.3038**	**0.3027**	**0.4080**	**0.2480**	**0.2951**

The Influence of Different Meta-paths on the Model

In order to further analyze the influence of different meta-paths on recommendation performance, these meta-paths were gradually merged into the proposed recommendation model HGATRec in this experiment, and the performance changes of the recommendation model were examined. The results are shown in Fig. 3. As can be seen from the figure, the recommendation performance of HGATRec model improves with the addition of meta-paths in general, and the more meta-paths there are, the better the recommendation performance. Meanwhile, different meta-paths have different effects on recommendation performance. In addition, with only a small amount of metadata merged, the model can quickly achieve relatively good score prediction performance, suggesting that a small number of high-quality meta-paths can lead to a large performance improvement. Therefore, the complexity of the model can be controlled by choosing some efficient meta-paths.

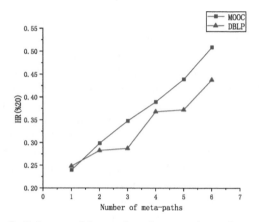

Fig. 3. Influence of the number of meta-paths on the model

5   Conclusions and Future Work

In this paper, HGATRec, a recommendation model based on heterogeneous graph attention network, is proposed to predict the knowledge points that users may be interested in, and the auxiliary information in MOOC is effectively utilized for recommendation. By comprehensively considering node attributes and meta-path weight, the two are weighted and fused with the attention mechanism. Experimental results show that HGATRec model has better prediction and recommendation performance than other common recommendation models. The meta-path selection in this article is manually designed, and in the future, a more efficient way to automatically select the optimal meta-path in HIN will be considered.

Acknowledgement. This research was supported by the Key R&D Project of Shaanxi Province (Grant No.2020GY-010), the key project of Shaanxi Education reform (Grant No. 21BG038).

References

1. Goopio, J., Cheung, C.: The MOOC dropout phenomenon and retention strategies. J. Teach. Travel Tour. **21**(2), 177–197 (2021)
2. Peng, H., Li, J., Song, Y., et al.: Streaming social event detection and evolution discovery in heterogeneous information networks. ACM Trans. Knowl. Discov. Data **15**(5), 1–33 (2021)
3. Anwar, T., Uma, V., Srivastava, G.: Rec-CFSVD++: implementing recommendation sys-tem using collaborative filtering and singular value decomposition (SVD)++. Int. J. Inf. Technol. Decision Making **20**(04), 1075–1093 (2021)
4. Jing, X., Tang, J.: Guess you like: course recommendation in MOOCs. In: Proceedings of the International Conference on Web Intelligence, pp. 783–789 (2017)
5. Jiang, W., Pardos, Z. A., Wei, Q.: Goal-based course recommendation. In: Proceedings of the 9th International Conference on Learning Analytics and Knowledge, pp. 36–45 (2019)
6. Zhang, H., Huang, T., Lv, Z., et al.: MOOCRC: a highly accurate resource recommendation model for use in MOOC environments. Mobile Netw. Appl. **24**(1), 34–46 (2019)
7. Mikolov, T., Chen, K., Corrado, G., et al.: Efficient estimation of word representations in vector space. arXiv preprint arXiv:1301.3781 (2013)
8. Mnih, A., Salakhutdinov, R. R.: Probabilistic matrix factorization. In Proceedings of the 20th International Conference on Neural Information Processing Systems, pp: 1257–1264 (2007)
9. He, X., Liao, L., Zhang, H., et al.: Neural collaborative filtering. In: Proceedings of the 26th International Conference on World Wide Web, pp. 173–182 (2017)
10. Dong, Y., Chawla, N. V., Swami, A.: metapath2vec: scalable representation learning for heterogeneous networks. In: Proceedings of the 23rd ACM SIGKDD international conference on knowledge discovery and data mining, pp: 135–144 (2017)
11. Shi, C., Zhang, Z., Ji, Y., Wang, W., Yu, P.S., Shi, Z.: SemRec: a personalized semantic recommendation method based on weighted heterogeneous information networks. World Wide Web **22**(1), 153–184 (2018). https://doi.org/10.1007/s11280-018-0553-6

Model Inversion-Based Incremental Learning

Dianbin Wu[1], Weijie Jiang[1], Zhiyong Huang[2], Qinghai Zheng[1],
Xiaodong Chen[1], WangQiu lin[3], and Yuanlong Yu[1(✉)]

[1] College of Computer and Data Science, Fuzhou University, Fuzhou, China
{200327090,n180310005,yu.yuanlong}@fzu.edu.cn
[2] Intelligent Robot Research Center, Zhejiang Lab, Hangzhou, China
huangzy@zhejianglab.com
[3] FuJian YiRong Information Technology Co. Ltd, FuZhou, China
wangqiulin@sgitg.sgcc.com

Abstract. As developments in the field of computer vision continue to be achieved, there is a need for more flexible strategies to cope with the large-scale and dynamic properties of real-world object categorization situations. However, regarding most existing traditional incremental learning methods, they ignore the rich information of the previous tasks embedded in the trained model during the continuous learning process. By innovatively combining model inversion and generative adversarial networks, this paper proposes a model inversion-based generation technique, which makes the information contained in the images generated by the generator more informative. To be specific, the information in the model, which has been trained by the previous task, can be inverted into an image, which can be added to the training process of the generative network. The experimental results show that the proposed method alleviates the catastrophic forgetting problem in incremental learning and outperforms other traditional methods.

Keywords: Incremental learning · Deep learning · Catastrophic forgetting

1 Introduction

The visual system of living in real world is essentially gradual in its development, new visual information is continuously absorbed, while existing knowledge should be retained at the same time [1]. However, to address the multi-classification tasks, most existing deep neural network learning models rely on a large-number of labeled training samples. For most existing learning methods, they assume that all samples are available during the training phase, and there-fore, when a new data set needs to be learned, the network parameters of the entire dataset need to be retrained to adapt to changes in the data distribution. For the training task of sequential tasks without access to data from previous tasks, the performance of traditional neural network models, which learned on previously tasks, decreases significantly when facing new tasks. To alleviate this unpleasant phenomenon, namely catastrophic forgetting, DGR [2] proposes a scholar model uses generators to generate pseudo-samples of previous tasks, which enabling access to the sample distribution of previous tasks However, the problem of aforementioned approaches that the quality of samples generated by the generator decreases continuously as the number of tasks increases, Consequently, it is still challenging to improve the quality of generated samples without accessing to samples from previous tasks.

To address the aforementioned limitation in this paper, we creatively apply the way of model inversion into the training process of the generator, and subsequently propose a novel network training method based on the model inversion. To be clear, the proposed method can invert images from the already trained network and add them to the training process of the generator, which effectively alleviates the forgetting problem caused by the inaccessibility of previous samples in the past incremental learning process. It has been experimentally demonstrated that our proposed method outperforms DGR's [2] generator-only approach and DEEP's [3] model-inversion-only approach in terms of accuracy of the dataset.

2   Related Work

2.1   Parameter-Based Isolation Approach

This approach uses different model parameters for each task to prevent any possible forgetting. When no constraints are applied to the neural network size, new branches can be added for new tasks while freezing the previous task parameters [4,5], or a copy of the model can be specified for each task [6]. Alternatively, the architecture is kept static and a fixed fraction is assigned to each task. During the training of a new task, previous task sections are masked out or imposed at the parameter level [7,8], or unit level [9]. These methods usually require a task sequence number to be known in advance and the corresponding mask or task branch to be activated during the prediction. As a result, they are restricted to a multi-header setup and cannot handle tasks between a shared header. The expert gate [6] avoids this problem by learning an automatic coding gate. In summary this approach is limited to task increment settings and is more suitable for learning long sequence tasks when the model capacity is not constrained and optimality is given priority.

2.2   Regularization-Based Methods

This type of approach chooses not to store the original input, prioritizing data privacy, with the aim of reducing memory requirements. Instead, an additional regularization term is introduced into the loss function to consolidate the previously learned knowledge when learning new data. Silver et al. [10] first proposed to use the output of the previous task model given the input image of the new task, mainly to improve the performance of the new task. The approach has been reintroduced by LwF [11] to alleviate forgetting and transfer knowledge using the previous model output as a soft label for previous tasks. However, it has been shown that this strategy performs poorly on the existence of domain transfers between tasks. To overcome this problem, Triki et al. [12] promoted incremental integration of shallow autoencoders to constrain task features in their corresponding learned low-dimensional spaces. Changes to important parameters are penalized in the training of subsequent tasks. Elastic weight consolidation (EWC) [9] was the first to establish such an approach. Variational continuous learning (VCL) introduced a variational framework for this approach,resulting in a series of Bayesian-based works. Zenke et al. estimated importance weights online during task training.

2.3   Model Inversion

Generative adversarial networks [13] have been at the forefront of generative image modeling, producing high-fidelity images, e.g., using BigGAN [14]. Although good at fitting image distributions, the generator that trains the GAN needs access to the original data. Another area of work in security focuses on synthesizing images from a single CNN. Fredrikson et al. [15] proposed a model inversion strategy to obtain class images from the network by performing gradient descent on the input. Subsequent work has improved or extended methods to cope with new scenarios. In vision, researchers visualize neural networks to understand their properties, and Mahendran et al. explore inversion, activation maximization, and caricature to synthesize "natural preimages" from trained networks. Nguyenet et al. [16] used a trained GAN generator as a preprocessor and then transformed the trained CNN into an image, and their subsequent work Play further improved the diversity and quality of the image by preprocessing potential codes. Bhardwajet et al. used training data clusters prime to improve the inversion. These methods still rely on auxiliary dataset information or additional pretrained networks. Of particular relevance to this type of approach is Mordvintsev et al.'s DeepDream [17], which is able to "dream" on natural images given a pretrained CNN. Despite the significant progress, synthesizing high-fidelity and high-resolution natural images from deep networks remains challenging.

3   Method

3.1   Background

We first define a continuous task sequence learning process. In our continual learning framework,we define the sequence of tasks to be solved as a task sequence $T = (T_1, T_2, T_3, ..., T_n)$ of N tasks.

Next, we define the model at t task as M_t. For the proposed method, it is able to learn a new task while the knowledge learned in the previous tasks from 1 to $t-1$ is retained as much as possible.

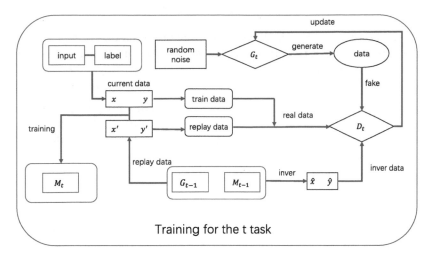

Fig. 1. Illustration of a generator training process based on model inversion. (x, y) is a sample of the current task, (x', y') is the sample generated by the trained generator in the previous task, $(\hat{x}, \hat{y})$ is the sample of the inverse performance of the trained model in the previous task.

3.2 Inversion-Based Replay

At beginning, we consider an incremental training task on the model, meanwhile, it also trains the model with the current task data. Since there is no access to samples from previous tasks, we need trains a generator, which is capable to generate samples for the previous task. When the model is trained, we need to continue training the generator with the current task samples and the generated samples, so that we can generate all the task samples up to the current task. In addition, we include the inverse image of the model to the generator's training process in order to introduce prior information from the previously trained model. Specifically, we describe our training process in Fig. 1.

3.3 Loss Function

Retain knowledge learned from previous tasks involves two separate processes, i.e., training the generator and the model. First,the new generator receives current task input x and replayed inputs and inversion inputs from previous tasks and models. The loss function of the model has two parts, the current task sample and the true label, and the generated sample and the generated label, and the specific formula is:

$$\mathcal{L} = -y_t \cdot \log \hat{y}_t + -g_t \cdot \log \hat{g}_t \tag{1}$$

where y_t is the true label, $\hat{y}_t$ is the output value of the model for the real sample, $\hat{g}_t$ is the output value of the model for the generated samples, and g_t is the label of the generated sample.

Discriminators_Loss : Since the inverse image is added to the training process of the generator, different from the traditional loss function, We add the inverse images to the training process of the discriminator, so that the discriminator can discriminate the images with the previous task information, and construct the following loss function:

$$L_D = l(net(D(x), Tlabel)) + l(net(D(fake), Flabel)) + l(net(D(x'), Ilabel)) \tag{2}$$

where x is the true sample, fake is the sample generated by the generator, x' is the image obtained by inversion of the model. In addition, $Ilabel$ was experimentally tested and we chose to mark it as 0.

Inversion_Loss : For the model inversion process, we use the inversion method [20] of the paper. A random noise image is input and the mean and variance are calculated for each batch normalization layer in the already trained model, as a constraint on the training process of noisy images. Therefore, the trained inverse image is rich in prior information from the previous task. The specific loss function is as follows:

$$\mathcal{R}_{\text{feature}}(\hat{x}) = \sum_l \|\mu_l(\hat{x}) - E\left(\mu_l(x) \mid \mathcal{X}\right)\|_2 + \sum_l \|\sigma_l^2(\hat{x}) - E\left(\sigma_l^2(x) \mid \mathcal{X}\right)\|_2 \tag{3}$$

where $\mu_l(\hat{x})$ and $\sigma_l^2(\hat{x})$ are the batch-wise mean and variance estimates of feature maps corresponding to the l_{th} convolutional layer. The $E(\cdot)$ and $\|\cdot\|_2$ operators denote the expected value and l_2 norm calculations, respectively.

For the expected value $E(\mu_l(x)|\chi)$ and $E(\sigma_l^2(x)|\chi)$, we employ the running average statistics stored in the widely used BatchNorm (BN) layer for estimation, and we have:

$$E\left(\mu_l(x) \mid \mathcal{X}\right) \simeq \text{BN}_l(running\_mean) \tag{4}$$

$$E\left(\sigma_l^2(x) \mid \mathcal{X}\right) \simeq \text{BN}_l(running\_variance) \tag{5}$$

4   Experiments

To verify that our proposed algorithm can alleviate the forgetting problem effectively, this paper employs the following two widely used datasets, i.e., mnist and cifar10. For each dataset, we divided into five tasks, each of which contains two categories. Additionally, during the training phase, only the data set of the current task is accessible, while during the testing phase, the data of the trained tasks between are tested. We compare our method with the following two state-of-the-art methods, namely DGR and DEEP.

For our method, We used RESNET18, which is fast to train, as the network structure of the model. For the network structure of the generator, we choose the classical DCGAN.

4.1   Example of a Sample Generated by the Generator

The results of the samples generated by the generator trained with 50 epochs are shown in Fig. 2. As can be seen from Fig. 2, the trained generator is able to generate samples of past tasks without touching the data.

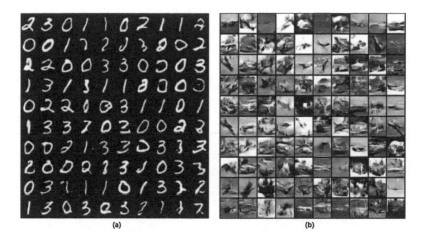

Fig. 2. (a) is the result of the generator trained with the samples of mnist. (b) is the result obtained by the generator after training with the sample of cifar10.

4.2   Illustration of Model Inversion

The dream picture obtained by inversion of the model that ha s been trained is shown in Fig. 3. We can clearly see that the image out of the inversion has information from the previous task.

4.3   Experimental Results

On the mnist dataset we choose to compare with DGR, which also uses a generator to preserve the knowledge of the old task, but our approach uses model inversion added to the training of the generator. In addition, we compared it

1234   D. Wu et al.

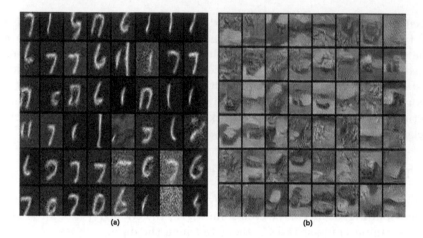

(a)   (b)

Fig. 3. (a) is the image of the inverse performance in the model trained on the mnist dataset. (b) is the image of the inverse performance of the trained model on the cifar10 dataset.

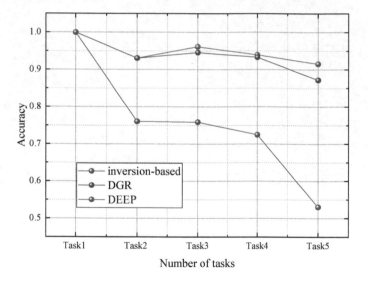

Fig. 4. Experimental results for the mnist dataset.

with DEEP, which uses only model inversion. In the experimental setup, the generators all choose DCGAN, epoch, and model size are consistent.

As shown in Fig. 4, our method is 3% higher than DGR and 20% higher than DEEP. The reason for this result is that the inverted images are added to the training of the generator, which improves the discriminator and makes the generator generate images with more a priori information. Experimental results show that the method combining model inversion and generator is superior to DGR and DEEP.

To challenge the experimental effect on more complex datasets, we choose the cifar10 dataset as the task increment. We also compare with DGR and also with DEEP, which uses only model inversion and does not use any other method of preserving knowledge. From the experimental results, our method is 4% more accurate than DGR and 2% more accurate than DEEP at the end, and outperforms other methods at each step of the incremental process. It is again demonstrated that even on complex datasets, pictures with a priori task information enable the generator to preserve more knowledge of previous tasks and finally alleviate the forgetting problem.

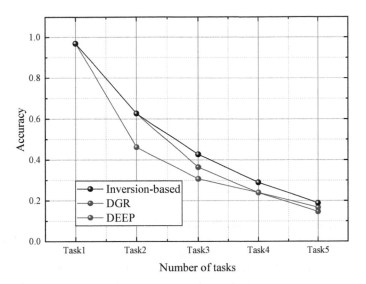

Fig. 5. Experimental results of cifar10.

5 Conclusion

In this paper, we propose a novel approach to train the generator by introducing the rich prior information in the model into the training process of the generator. Based on the comparative experimental results, the proposed training method achieves the better performance in mitigating catastrophic forgetting. In future work, we will focus on the task of improving the network structure, aming to make the generator images more realistic and the inverse images with more a priori information.

Acknowledgements. This work was supported by National Natural Science Foundation of China (NSFC) under grant 61873067, University-Industry Cooperation Project of Fujian Provincial Department of Science and Technology under grant 2020H6101, NSF Project of Zhejiang Province No. LQ22F030023, and Zhejiang Postdoctoral Project No. 2021NB3UB15.

References

1. Chen, Z., Liu, B.: Lifelong machine learning. Synth. Lect. Artif. Intell. Mach. Learn. **12**(3), 1–207 (2018)
2. Shin, H., Lee, J.K., Kim, J., Kim, J.: Continual learning with deep generative replay. Adv. Neural Inf. Process. Syst. **30** (2017)
3. Yin, H., Molchanov, P., Alvarez, J.M., Li, Z., Mallya, A., Hoiem, D., Jha, N.K., Kautz, J.: Dreaming to distill: data-free knowledge transfer via deep inversion. In: Proceedings of the IEEE/CVF Conference on Computer Vision and Pattern Recognition, pp. 8715–8724 (2020)
4. Xu, J., Zhu, Z.: Reinforced continual learning. Adv. Neural Inf. Process. Syst. **31** (2018)
5. Rusu, A.A., Rabinowitz, N.C., Desjardins, G., Soyer, H., Kirkpatrick, J., Kavukcuoglu, K., Pascanu, R., Hadsell, R.: Progressive neural networks. arXiv preprint. arXiv:1606.04671 (2016)
6. Aljundi, R., Chakravarty, P., Tuytelaars, T.: Expert gate: lifelong learning with a network of experts. In: Proceedings of the IEEE Conference on Computer Vision and Pattern Recognition, pp. 3366–3375 (2017)
7. Fernando, C., Banarse, D., Blundell, C., Zwols, Y., Ha, D., Rusu, A.A., Pritzel, A., Wierstra, D.: Pathnet: evolution channels gradient descent in super neural networks. arXiv preprint. arXiv:1701.08734 (2017)
8. Mallya, A., Lazebnik, S.: Packnet: adding multiple tasks to a single network by iterative pruning. In: Proceedings of the IEEE Conference on Computer Vision and Pattern Recognition, pp. 7765–7773 (2018)
9. Kirkpatrick, J., Pascanu, R., Rabinowitz, N., Veness, J., Desjardins, G., Rusu, A.A., Milan, K., Quan, J., Ramalho, T., Grabska-Barwinska, A., et al.: Overcoming catastrophic forgetting in neural networks. Proc. Natl. Acad. Sci. **114**(13), 3521–3526 (2017)
10. Silver, D.L., Mercer, R.E.: The task rehearsal method of life-long learning: overcoming impoverished data. In: Conference of the Canadian Society for Computational Studies of Intelligence, pp. 90–101. Springer (2002)
11. Li, Z., Hoiem, D.: Learning without forgetting. IEEE Trans. Pattern Anal. Mach. Intell. **40**(12), 2935–2947 (2017)
12. Rannen, A., Aljundi, R., Blaschko, M.B., Tuytelaars, T.: Encoder based lifelong learning. In: Proceedings of the IEEE International Conference on Computer Vision, pp. 1320–1328 (2017)
13. Han, S., Mao, H., Dally, W.J.: Deep compression: compressing deep neural networks with pruning, trained quantization and Huffman coding. arXiv preprint. arXiv:1510.00149 (2015)
14. Brock, A., Donahue, J., Simonyan, K.: Large scale Gan training for high fidelity natural image synthesis. arXiv preprint. arXiv:1809.11096 (2018)
15. Furlanello, T., Lipton, Z., Tschannen, M., Itti, L., Anandkumar, A.: Born again neural networks. In: International Conference on Machine Learning, pp. 1607–1616. PMLR (2018)
16. Nguyen, A., Dosovitskiy, A., Yosinski, J., Brox, T., Clune, J.: Synthesizing the preferred inputs for neurons in neural networks via deep generator networks. Adv. Neural Inf. Process. Syst. **29** (2016)
17. Mordvintsev, A., Olah, C., Tyka, M.: Inceptionism: Going Deeper into Neural Networks, 2015. https://research.googleblog.com/2015/06/inceptionism-going-deeper-into-neural.html (2015)

Knowledge Representation by Generic Models for Few-Shot Class-Incremental Learning

Xiaodong Chen[1], Weijie Jiang[1], Zhiyong Huang[2], Jiangwen Su[3],
and Yuanlong Yu[1(✉)]

[1] College of Computer and Data Science, Fuzhou University, Fuzhou, China
200320050@fzu.edu.cn, n180310005@fzu.edu.cn, yu.yuanlong@fzu.edu.cn
[2] Intelligent Robot Research Center, Zhejiang Laboratory, Hangzhou, China
huangzy@zhejianglab.com
[3] FuJian YiRong Information Technology Co. Ltd, FuZhou, China
sujiangwen@sgitg.sgcc.com.cn

Abstract. Few-shot class-incremental learning (*FSCIL*) means identifying new classes in a few samples while not forgetting the old ones. The challenge of this task is that new class has only few supervised information during the learning process. Aiming to boost the performance of FSCIL, we propose a novel method in this paper. To be clear, our method has two contributions as follows. First, we elegantly employ the principal component analysis (*PCA*) and adopt a model with a strong prior for feature extracting, specifically, we decouple the feature extractor from classifier in the incremental learning process. Second, we innovatively introduce the data augmentation during the learning process of FSCIL to enhance the sample diversity and get a more accurate class prototype based on enriched samples. Excellent experimental results on CIFAR-100, miniImageNet, and CUB200 datasets verify the superiority of our method, compared to several existing methods.

Keywords: Few-shot · Class-incremental learning · Pre-trained model · PCA · Data augmentation

1 Introduction

With the development of science and technology, deep learning has made great progress in many areas of computer vision [1,2]. However, a classification model can only make predictions on predefined categories of images when the model is trained by current supervised information. To extend the trained model to new classes, a large amount of training data containing both new and old classes is required. More generally, in real-world scenarios, it is impossible to obtain large

© The Author(s), under exclusive license to Springer Nature Switzerland AG 2023
N. Xiong et al. (Eds.): ICNC-FSKD 2022, LNDECT 153, pp. 1237–1247, 2023.
https://doi.org/10.1007/978-3-031-20738-9_134

amounts of training samples. Unlike deep learning methods, humans can learn new knowledge quickly from a small sample without forgetting old knowledge. This gap between humans and deep learning models lead us to a new scenario, namely few-shot class-incremental learning (*FSCIL*), which aim is to design a deep learning model that can quickly scale a trained model to new classes with only a small number of samples and maintain a certain level of old knowledge. In this paper, we devote into the study of few-shot class-incremental learning tasks and propose a solution namely Knowledge Representation by Generic Models to the aforementioned limitations of FSCIL from the following two aspects.

First, due to the extreme imbalance in sample size between base class and new class in FSCIL scenario [3], we choose to decouple feature extractor and classifier. Specifically, we use a pre-trained model, which is the encode image module in clip [4], as the base framework. Clearly, the feature extractor in base session and new session is the same, and we can update classifier only in new session. As clip's feature extractor is a model trained from a massive dataset, the embeddings may not be fully applicable to current dataset, so there may be a lot of noisy information interfering with the classification task. Therefore, we adopt principal component analysis (*PCA*) with the sufficient samples in base session to obtain covariance matrix and the corresponding eigen vectors with dimensionality reduction. Subsequently, in new session, we also perform dimensionality reduction operation using the same covariance matrix. The underlying idea is that data in new session and base session belong to the same distribution.

Second, since the limitation of few-shot scenario, there are only five training samples for each class in each new session, which will not allow our classifier to clearly describe the classification boundary of the whole class. Therefore, we introduce the data augmentation on training samples, so as to obtain more diverse supervised samples and make our classifier closer to the center of classified class.

We have experimented on CIFAR100, CUB200 and MiniImageNet, the results outperform several existing methods. The following points are the main contributions of this paper. 1) To obtain features that are more robust to the current data set, we propose the operation of feature dimensionality reduction for model trained on a large number of data using principal component analysis. 2) We employ the data augmentation for the pre-trained model to get more accurate category prototypes for new sessions.

2   Related Work

2.1   Class-Incremental Learning

Class incremental learning has become a hot topic in the field of deep learning, and its mainstream approaches can be specifically divided into two main categories based on whether or not samples from older classes are preserved.. The methods without exemplar preservation mainly use regularization terms to limit the output of model or adjust dynamic architecture of model to fit the need of new classes. EWC [5] measures the importance of each parameter through Fisher information matrix and uses regularization terms to make the parameters with high importance change as little as possible on the training of new tasks. In addition to Fisher information matrix, there are other methods to measure the importance of parameters, such as MAS [6] and SI [7]. On the other hand, other methods based on not preserving samples try to meet the needs of the new task by extending or pruning the network architecture [8,9].

Exemplar retention methods aim to preserve representative data from old task and replay it when training new task. iCaRL [10] uses replay of old samples and knowledge distillation to retain knowledge from previous task. BiC [11] uses the retained samples to construct a validation set and optimize an additional rescaling layer. GEM [12] uses retained samples to limit the direction of projection of the gradient to overcome forgetting. Other approaches consider preserving high-dimensional features instead of the original picture [13], or using generative models for data rehearsal [14].

2.2 Few-Shot Class-Incremental Learning

Few-shot class-incremental learning has recently been used to address small and medium sample scenarios for class incremental learning. TOPIC [15] defines a benchmark-setting in FSCIL that uses a neural gas network to preserve the topology between new classes and old class features. Exemplar Relation Graph [16] maintains a graph that can represent the relationships between classes and uses the graph to do knowledge distillation. The idea of FSLL [17] comes from CIL scenario, which picks a small set of parameters to update at each incremental session to prevent overfitting. Semantic-aware knowledge distillation [18] considers the use of auxiliary word embeddings to fuel model updates noting the characteristics of few shot. Self-promoted prototype refinement [19] adapts feature representation and considers a single stage capability through many generated class incremental episodes. CEC [3] decouples the entire training process into two parts, the learning of the feature extractor and the learning of the classifier. It also uses an additional graph model to share semantic information about the context on the classifiers. The existing approach borrows heavily from previous approaches in the field of few-shot and class-incremental learning, which has the advantage of solving the immediate problems in FSCIL, such as catastrophic forgetting and too few samples, but lacks a comprehensive consideration of both problems and does not use richer prior information. Although word embeddings are used as prior knowledge [18], word embeddings do not necessarily describe the relationships between categories completely, and at the same time, the differences between word embeddings and feature representations can lead to poor results. For reasons of appeal, our feature extraction uses a model pre-trained on top of a large number of samples. In addition, feature downscaling and data augmentation are performed for the pre-trained model.

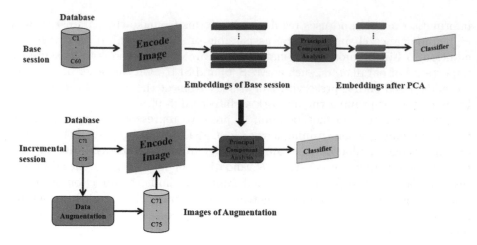

Fig. 1. Our few-shot class-incremental learning task can be divided into two parts: (1) In base session we do PCA dimensionality reduction using enriched samples $C1, ..., C60$ and fix the encode image (decoupling) in the following sessions. (2) In incremental session we perform dimensionality reduction operation using the feature vector matrix obtained in base session and then enhance samples using data augmentation.

3   Method

3.1   Principal Component Analysis

Our model framework can be illustrated in Fig. 1. It can be seen that we perform PCA by a large number of embeddings in base session to achieve the dimensionality reduction. Specifically, we input N images to get the embeddings $X \in R^{N \times D}$, where N is the number of samples and D represents the data dimensionality. Zero-meaning the features of each dimension of X, i.e., subtracting the mean of this feature, the formula can be written as follows:

$$X_{mean} = \frac{1}{D} \sum_{i=1}^{D} X_i \tag{1}$$

$$\tilde{X} = X - X_{mean} \tag{2}$$

The covariance matrix of the data after zero-meaning can be calculated:

$$\Sigma = \frac{1}{N} \tilde{X}^T \tilde{X} \tag{3}$$

Eigenvalue decomposition of the covariance matrix:

$$\Sigma = U^T \lambda U \tag{4}$$

where U and λ denote the set of eigenvectors and eigenvalues of Σ, respectively. Corresponding eigenvectors are arranged according to eigenvalues from smallest to largest, and the first K rows of eigenvectors are selected to form a matrix $U_K = (u_1, u_2, ..., u_k)$, which is then multiplied by the original data X, i.e., projected onto principal component axis:

$$X_{rotate} = XU_K \tag{5}$$

In this paper, we also do a whitening operation on the features after PCA, specifically, let the data on each dimensional feature be divided by the standard deviation of that dimensional feature, and the formula can be written as:

$$X_{PCA,White} = \frac{X_{rotate}}{\sqrt{\lambda}} \tag{6}$$

All the above processes are done in base session, so that we get the reduced dimensional embedding. The eigenvector matrix U_K and eigenvalue matrix λ of PCA operations performed in subsequent incremental sessions are all the same as base session.

3.2 Data Augmentation

Data augmentation methods are widely used in the field of few-shot learning and have achieved good results. In this paper, a model pre-trained on rich samples is used as backbone. On the basis of this, it is possible to make classifier closer to the center of class prototype if the diversity as well as the perturbation of samples is enhanced. To systematically investigate the effect of data augmentation, we tried several operations of data augmentation such as cropping and resizing (with horizontal flipping), cutout, color distortion, etc.

To better understand the approach of each data enhancement method, we investigated effect of using different data augmentation operations tested on the model of this paper, and the specific results are shown in Table 1. We show the accuracy of last incremental session, Without-data-aug means that no data augmentation is used, i.e., only the decoupling method is utilized. From the table, we can see that center croping, resizing with horizontal flipping, and color distortion produce positive effects on top of our model, but the others interfere with the class prototype, based on which we apply a combination of these three operations to our model. At the same time, in order to demonstrate that our data augmentation operations only work for the model in this paper, we perform the same operation on top of CEC [3], which is another FSCIL method, and obtain the results shown in Fig. 2. As before, we only show the accuracy of last session. It can be seen that the method without data enhancement (blue) is significantly higher than the method with data enhancement introduced (red). But on top of our approach, the effect is just the opposite.

4   Experiments

4.1   Seeting

Datasets. Based on the benchmark settings, we evaluated the experimental effects on CIFAR100, CUB200, and MiniImageNet. CIFAR100 and MiniIma-

Table 1. Results of the last session in the FSCIL scenario on the CUB200 after using different data augmentation operations.

Operations	Acc of last session
Center crop	**59**
Crop resize flip	**59.58**
Color drop	48.89
Color jitter	**58.01**
Cutout	54.96
Gaussian noise	56.04
Gaussian blur	56.88
Without data Aug	57.8

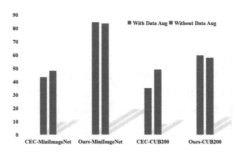

Fig. 2. The effect of CEC (Continuously Evolved Classifier) and Ours method on MiniImageNet as well as CUB200 after using data enhancement.

geNet contain 100 classes, each consisting of 500 training images and 100 test images. CUB200 is a fine-grained dataset containing 200 different bird classes. In detail, CIFAR100 and MiniImageNet are divided into 60 base classes, 40 new classes, and the 40 new classes are in turn sliced into 8 incremental sessions. For each new class there are only five training images, i.e., a 5-way 5-shot task. For CUB200, the 200 classes are divided into 100 base classes and 100 new classes, and the 100 new classes are sliced into 10 incremental sessions, each new class also has only five training samples, i.e., a 10-way 5-shot task.

Implementation Details. We use the encode image in clip as our back bone, which name is ViT-B/32. The feature dimension of all three datasets is 512. The dimensionality of CIFAR100 and MiniImageNet is reduced to 160 after PCA, and CUB200 is reduced to 100 dimensions. We use three data enhancement methods in each incremental learning session: center crop, crop resize flip , and color jitter. We broaden the data to 325 samples (350 for CUB200) which includes 25 (50 for CUB200) original images, with a ration of 1:1:1 on three operations. With the settings as [3], we use Cosine as our classifier and parameterize the classifier weights using data embeddings, where the weight vector for each category is initialized using the average of feature embeddings in the training set.

Table 2. Ablation experiment on CUB200. PCA is a principal component analysis operation.

PCA	DAG	Decoupled	Sessions											Average
			0	1	2	3	4	5	6	7	8	9	10	
		✓	74.1	71.7	68.7	65.0	64.6	62.0	61.3	58.3	57.7	58.0	57.8	63.6
✓		✓	75.6	**73.2**	70.2	65.9	65.9	63.3	61.4	59.7	59.0	59.0	58.8	64.6
	✓	✓	74.1	72.0	69.4	65.8	65.5	63.7	62.7	60.4	59.0	59.7	59.5	64.7
✓	✓	✓	**75.6**	73.1	**70.6**	**66.6**	**66.6**	**64.2**	**62.5**	**60.8**	**59.7**	**60.1**	**60.2**	**65.4**

DAG indicates data augmentation

4.2 Ablation Study

To demonstrate the effectiveness of our proposed method, we have done ablation experiment on CUB200. Our architecture consists of two main components: principal component analysis (PCA) and data augmentation (DAG). To validate the effect of these two components, we show the effect of modeling the addition of these two methods separately, as well as the effect of using them together. As shown in Table 2, the PCA module and the DAG module improve by 1% and 1.7%, respectively, compared to the decoupled methods alone. The combined effect improves by 2.4%.

4.3 Comparison with Existing Methods

In order to better evaluate the effect of our method, we compare it with existing FSCIL methods (CEC, FACT and Limit) and some classical CIL methods (iCaRL and EEIL). Decoupled-Cosine refers to only using decoupling without adding PCA and data augmentation methods. Additionally, we set fine-tuned method as the baseline. For CIFAR100, it can be seen from Table 3 that the accuracy of each session exceeds previous method. Although the accuracy of our base session is about 5% higher than FACT, the effect of our last session exceeds FACT 15%. What's more, our PD is the lowest, which means our model has good anti-forgetting effect. For MiniImageNet, as shown in Fig. 3, our effect has also surpass the existing FSCIL methods and several CIL methods. Even though the accuracy of our base session is comparatively higher than previous methods, it can be seen that our curve is relatively smooth, that is, the drop in accuracy is much smaller under our method than previous methods. For CIFAR100 and MiniImageNet, the addition of PCA and data enhancement modules improved the results by 8.73% and 2.33% respectively.

4.4 Implications of PCA for Class-Incremental Learning

From Table 2, Table 3, and Fig. 3, we can see that our method outperforms Decoupled-Cosine in session 0 on all three datasets, which is due to the introduction of PCA. In order to be more clear about the classification effect of PCA on the categories in incremental session, we verified it on CUB200. In each

Table 3. Experimental results on CIFAR100. PD means degraded precision.

Method	Accuracy in each session(%) ↑									PD ↓
	0	1	2	3	4	5	6	7	8	
Finetune	64.10	39.61	15.37	9.8	6.67	3.8	3.7	3.14	2.65	61.45
iCaRL [10]	64.10	53.28	41.69	34.13	27.93	25.06	20.41	15.48	13.73	50.37
EEIL [20]	64.10	53.11	43.71	35.15	28.96	24.98	21.01	17.26	15.85	48.25
TOPIC [15]	64.10	55.88	47.07	45.16	40.11	36.38	33.96	31.55	29.37	34.73
CEC [3]	73.07	68.88	65.26	61.19	58.09	55.57	53.22	51.34	49.14	23.93
LIMIT [21]	73.81	72.09	67.87	63.89	60.70	57.77	55.67	53.52	51.23	22.58
FACT [22]	74.60	72.09	67.56	63.52	61.38	58.36	56.28	54.24	52.10	22.50
Decoupled-Cosine	72.90	69.97	67.50	64.72	62.79	61.49	61.14	59.74	58.21	14.69
Ours	**79.78**	**77.31**	**75.39**	**72.60**	**71.36**	**70.22**	**70.02**	**68.75**	**66.94**	**12.84**

incremental session, we only tested the accuracy of classes from session 1 to the current session. For example, on CUB200, if the current Session is 3, then we will test the accuracy of 30 classes included from session 1 to 3. The results are shown in Table 4. It can be seen that after adding the PCA method, the classification effect of classes in each incremental session is significantly improved, which shows that PCA has a certain role in resisting catastrophic forgetting in class-incremental learning scenario.

4.5   The Effect of Image Size

In fact, the data enhancement operation in this paper is only performed on CUB200 and MiniImageNet, because the image size of these two datasets is larger than CIFAR100, and the size of CIFAR100 image is only 32×32, this may be the reason for the poor performance of data augmentation operations on CIFAR100. In order to verify this conjecture, we resize the images of CUB200 and MiniImageNet to 32×32, and then perform data enhancement operation, the obtained effect is shown in Fig. 4, where CUB-Normal refers to the use of the original image, and CUB-Small refers to the use of reduced images(as well as MiniImageNet). Obviously, the experimental effect of zooming the image to a small size is far inferior to the effect of using a large image, indicating that the CLIP model is only sensitive to large-sized images, but not to the small images.

Table 4. On CUB200, the accuracy of remaining classes after removing the base classes in each incremental session.

	Accuracy without base classes in each session(%) ↑										Average
Method	1	2	3	4	5	6	7	8	9	10	
Without PCA	51.38	49.91	42.44	45.17	42.88	43.72	41.91	40.81	43.57	45.33	44.71
With PCA	**57.30**	**53.92**	**44.94**	**48.59**	**47.04**	**46.56**	**45.58**	**44.03**	**46.30**	**47.96**	**48.22**

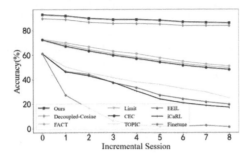

Fig. 3. Results on MiniImgeNet

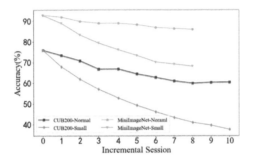

Fig. 4. Results after zooming.

5 Conclusion

In this paper, we propose to apply a pre-trained model to few-shot class-incremental learning scenario to cope with catastrophic forgetting and overfitting. We propose to use PCA to do feature downscaling on the model to remove noise for different datasets, and have obtained features that are more robust to new classes in incremental sessions. Secondly, we propose data augmentation methods for the pre-trained model to obtain more accurate class prototypes using the diversity of samples. We have done extensive experiments on three widely known datasets, CIFAR100, CUB200, MiniImageNet, and achieved significant results. Our method will overfit the base class because our feature extractor will be trained only in base session, which will make the feature space filled by base class, so how to flow out more space for the new class in incremental session will become the future work.

Acknowledgements. This work was supported by National Natural Science Foundation of China (NSFC) under grant 61873067, University-Industry Cooperation Project of Fujian Provincial Department of Science and Technology under grant 2020H6101, NSF Project of Zhejiang Province No. LQ22F030023, and Zhejiang Postdoctoral Project No. 2021NB3UB15.

References

1. Sun, X., Yang, Z., Zhang, C., Ling, K.V., Peng, G.: Conditional Gaussian distribution learning for open set recognition. In: Proceedings of the IEEE/CVF Conference on Computer Vision and Pattern Recognition, pp. 13480–13489 (2020)
2. Zhang, C., Yao, R., Cai, J.: Efficient eye typing with 9-direction gaze estimation. Multimedia Tools Appl. **77**(15), 19679–19696 (2018)
3. Zhang, C., Song, N., Lin, G., Zheng, Y., Pan, P., Xu, Y.: Few-shot incremental learning with continually evolved classifiers. In: Proceedings of the IEEE/CVF Conference on Computer Vision and Pattern Recognition, pp. 12455–12464 (2021)
4. Radford, A., Kim, J.W., Hallacy, C., Ramesh, A., Goh, G., Agarwal, S., Sastry, G., Askell, A., Mishkin, P., Clark, J., et al.: Learning transferable visual models from natural language supervision. In: International Conference on Machine Learning, pp. 8748–8763. PMLR (2021)
5. Kirkpatrick, J., Pascanu, R., Rabinowitz, N., Veness, J., Desjardins, G., Rusu, A.A., Milan, K., Quan, J., Ramalho, T., Grabska-Barwinska, A., et al.: Overcoming catastrophic forgetting in neural networks. Proc. Natl. Acad. Sci. **114**(13), 3521–3526 (2017)
6. Aljundi, R., Babiloni, F., Elhoseiny, M., Rohrbach, M., Tuytelaars, T.: Memory aware synapses: learning what (not) to forget. In: Proceedings of the European Conference on Computer Vision (ECCV), pp. 139–154 (2018)
7. Zenke, F., Poole, B., Ganguli, S.: Continual learning through synaptic intelligence. In: International Conference on Machine Learning, pp. 3987–3995. PMLR (2017)
8. Xu, J., Zhu, Z.: Reinforced continual learning. Adv. Neural Inf. Process. Syst. **31** (2018)
9. Yoon, J., Yang, E., Lee, J., Hwang, S.J.: Lifelong learning with dynamically expandable networks. arXiv preprint. arXiv:1708.01547 (2017)
10. Rebuffi, S.A., Kolesnikov, A., Sperl, G., Lampert, C.H.: iCARL: incremental classifier and representation learning. In: Proceedings of the IEEE conference on Computer Vision and Pattern Recognition, pp. 2001–2010 (2017)
11. Wu, Y., Chen, Y., Wang, L., Ye, Y., Liu, Z., Guo, Y., Fu, Y.: Large scale incremental learning. In: Proceedings of the IEEE/CVF Conference on Computer Vision and Pattern Recognition, pp. 374–382 (2019)
12. Lopez-Paz, D., Ranzato, M.: Gradient episodic memory for continual learning. Adv. Neural Inf. Process. Syst. **30** (2017)
13. Iscen, A., Zhang, J., Lazebnik, S., Schmid, C.: Memory-efficient incremental learning through feature adaptation. In: European Conference on Computer Vision, pp. 699–715. Springer (2020)
14. Xiang, Y., Fu, Y., Ji, P., Huang, H.: Incremental learning using conditional adversarial networks. In: Proceedings of the IEEE/CVF International Conference on Computer Vision, pp. 6619–6628 (2019)
15. Tao, X., Hong, X., Chang, X., Dong, S., Wei, X., Gong, Y.: Few-shot class-incremental learning. In: Proceedings of the IEEE/CVF Conference on Computer Vision and Pattern Recognition, pp. 12183–12192 (2020)
16. Dong, S., Hong, X., Tao, X., Chang, X., Wei, X., Gong, Y.: Few-shot class-incremental learning via relation knowledge distillation. In: Proceedings of the AAAI Conference on Artificial Intelligence, vol. 35, pp. 1255–1263 (2021)
17. Mazumder, P., Singh, P., Rai, P.: Few-shot lifelong learning. arXiv preprint. arXiv:2103.00991 (2021)

18. Cheraghian, A., Rahman, S., Fang, P., Roy, S.K., Petersson, L., Harandi, M.: Semantic-aware knowledge distillation for few-shot class-incremental learning. In: Proceedings of the IEEE/CVF Conference on Computer Vision and Pattern Recognition, pp. 2534–2543 (2021)
19. Zhu, K., Cao, Y., Zhai, W., Cheng, J., Zha, Z.J.: Self-promoted prototype refinement for few-shot class-incremental learning. In: Proceedings of the IEEE/CVF Conference on Computer Vision and Pattern Recognition, pp. 6801–6810 (2021)
20. Castro, F.M., Marín-Jiménez, M.J., Guil, N., Schmid, C., Alahari, K.: End-to-end incremental learning. In: Proceedings of the European Conference on Computer Vision (ECCV), pp. 233–248 (2018)
21. Zhou, D.W., Ye, H.J., Zhan, D.C.: Few-shot class-incremental learning by sampling multi-phase tasks. arXiv preprint. arXiv:2203.17030 (2022)
22. Zhou, D.W., Wang, F.Y., Ye, H.J., Ma, L., Pu, S., Zhan, D.C.: Forward compatible few-shot class-incremental learning. arXiv preprint. arXiv:2203.06953 (2022)

Information Systems (28)

A Medical Privacy Protection Model Based on Threshold Zero-Knowledge Protocol and Vector Space

Yue Yang[1], Rong Jiang[1(✉)], Chenguang Wang[1], Lin Zhang[1], Meng Wang[1], Xuetao Pu[2], and Liang Yang[2]

[1] Yunnan University of Finance and Economics, Kunming 650221, China
{yangyue6,zhanglin2,wangmeng1}@stu.ynufe.edu.cn,
jiang_rong@aliyun.com, wangchenguang@stu.ynufe.com
[2] Kunming University of Science and Technology, Kunming 650093, China
{puxuetao,yangliang}@stu.kust.edu.cn

Abstract. Recent years have witnessed the rapid development of technologies such as the Internet and the Internet of Things, which have driven the development of all walks of life including economics, medicine, and so on. In the healthcare industry, big data has become a core asset in transforming personal care, and public health, and has been officially incorporated into the national development strategy as a national basic strategic resource. However, privacy security and leakage through medical big data become a serious issue in the healthcare industry, resulting in exploiting details specific to a disease, identity theft, destruction of databases, etc. This study proposes a privacy protection model for medical data based on threshold zero-knowledge protocol and vector space. The model uses a threshold zero-knowledge protocol to verify the user's identity from the data source. After the verification, the system will vectorize its historical access records to calculate the correlation between their access requests and historical access records. The experimental results show that the model can issue the access permission accurately and reliably, as well as reduce the misjudgment of the system.

Keywords: Medical big data · Privacy protection · Threshold zero-knowledge protocol · Access control

1 Introduction

Medical privacy is closely related to each of our lives, involving our clothing, food, housing, transportation, birth, aging, illness and death, and other life processes, and is the core asset of medical big data. In the future era of holographic digitalization, the whole world will generate about 1000 megabits of data every year, which makes the industry very promising. However, the privacy and security of medical data also face huge challenges. In 2016, Symantec released the Internet Security Threat Report revealing the top 10 industries with more serious data leakage, and the medical industry topped the list with 116 data leakage incidents and an incident rate of 37.2%; in second place was the retail industry with 34 data leakage incidents and an incident rate of 10.9%, and the third was the education The third industry is education, with 31 data breaches and an incidence rate of 9.9%. We can see that the medical industry has become the biggest victim of data leakage, and its leakage rate is much higher than other industries.

In addition, there is no privacy, and if a higher-ranking doctor wants to access patient information, he or she can access all the patient's medical records by simply logging into the terminal, while a general practitioner can access all the patient information within his or her workstation [1]. Therefore, there is an urgent need to protect medical information that concerns their interests as a way that can effectively protect private information from being leaked, and access control technology will play a role that cannot be ignored. However, in the context of Big Data, traditional techniques are no longer able to meet the requirements of such large-scale data access, which requires us to innovate and improve the traditional access control models to adapt to today's Big Data environment.

2   Related Work

Scholars have studied the current situation of privacy protection and disclosure of medical data. Through the collection of relevant literature, we found that the current academic research in this field mainly adopts technologies such as differential privacy, encryption algorithm, anonymity, and authentication, but there are few studies involving access control [2]. Ji SL [3] summarizes the anonymization and de-anonymization quantization techniques of graphic data in the past decade. However, most anonymous schemes can conditionally retain most graph utility and lose some application utility at the same time. Tsou YT [4] proposed DP based data risk estimation system is proposed, in which DP based noise addition mechanism is used to synthesize data sets. LV CX et al. [5] proposed a random forest classification algorithm based on differential privacy protection that can reduce the impact. For example, Rao et al. [6] introduced a role recommendation model for RBAC systems. It automatically cancels and refreshes user role assignment by observing user access behavior, and is used to provide role assignment as a service in the cloud to optimize the cost of built-in roles; Several experiments were carried out with Amazon access sample data set to verify the proposed model. However, the traditional DAC model is only applicable to small applications and is vulnerable to Trojan attacks and exploit software. Although there is more research on access control, traditional access control does not work in cross-domain environments.

Lin [7] proposed a mutual trust-based access control (MTBAC) model that considers both the behavioral trust of the user and the trustworthiness of the cloud service nodes and establishes a trust relationship between the user and the cloud service nodes through a mutual trust mechanism trust relationship between the user and the cloud service node. Line et al. [8] proposed a possible access control method, and evaluate it according to the standards obtained from the risk analysis of the coordination system of environmental planning of surgical hospitals. It is concluded that the future work should focus on expanding the location-based method through situational awareness, and increasing the support for using pop-up windows or handheld devices to share the most sensitive information. Ashtiani [9] proposed a customizable trust evaluation model based on fuzzy logic, which demonstrated the integration of processes after interaction through business interaction view and reliability adjustment.

To sum up, the main innovations of this paper are as follows: ① This paper uses the method of threshold zero knowledge protocol to verify the identity of doctors and ensure the credibility of the data source. ② The model proposed in this paper improves the security and initiative of the model, and the vectorized access record improves the fitting degree of the model. Therefore, the research of this paper is an important supplement to the research of medical big data access control.

3   A Medical Privacy Protection Model

3.1   Overview of the Model

At present, most hospitals adopt a role-based access control model, which grants different permissions according to different roles of doctors, and doctors in different working areas have different work and access areas. Under normal circumstances, the patient's case will be printed uniformly and sent to the archives for storage after discharge. Older doctors will get higher privileges. In addition to accessing the patient information of their department, they can even access the patient treatment information of the whole hospital. In addition, the amount of information required by doctors to complete the task will vary according to the patient's history, number and other factors. Therefore, it is difficult to adapt to the actual situation of the medical environment by evaluating the amount of information and the sensitivity of data accessed by doctors.

In summary, the basic idea of the model is rough as follows: when a doctor initiates a request to access a patient case, the system first identifies the doctor based on a threshold zero-knowledge protocol; secondly, the similarity between his access case and the access in the historical record is evaluated based on the person coefficient calculation. Finally, we determine whether to grant access rights according to the medical big data access control policy.

3.2   Quantification of Access Metrics

Doctor's identity quantification. This project intends to verify the identity of the doctor before the visit to ensure that the visitor is a user allowed to visit the hospital. The (t, n) threshold zero-knowledge protocol ($t < < n$) is proposed to be used to authenticate the identity of the doctor. The specific steps are shown in Fig. 1:

Step 1:   The doctor initiates the certification application to the certification center.
Step 2:   The certification center asks the doctor questions for verification.
Step 3:   The doctor calculates the problem and draws the result, and feeds the problem result back to the certification center.
Step 4:   The certification question base checks the results obtained by the certification center, and the certification center records whether the results are correct.
Step 5:   If the result is correct, the certification center will issue an identity certificate to the hospital to complete the identity certificate.

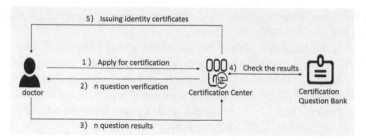

Fig. 1. (t, n) threshold zero-knowledge protocol framework

Physician visit behavior quantification. The basic assumption of the vector space model (VSM): when judging the meaning of an access request, we only need to know some feature items and the frequency of these feature items appearing in the system and do not need to know the position and order of their appearance, that is, the case's Meaning can be represented by the characteristic words that make up the case and their word frequency in the case. When measuring the similarity between cases, the similarity of two cases can be displayed by two common feature words and their weights.

Based on the above content, the basic idea of modeling cases with a vector space model is given. Suppose there are n case documents in the historical record case set L, and each historical record l has m different disease type feature items. These feature items are Independent of each other, there is formula 1, where t_k—the type of disease, such as gastroenteritis, appendicitis, etc.:

$$l = (t_1, t_2, \ldots, t_m) \tag{1}$$

If w_k is used to represent the weight of the feature item t_k in the historical record l, then $(t_1, t_2,\ldots, t_m)$ is the coordinate axis of the m-dimensional space, and $(w_1, w_2,\ldots, w_m)$ is its corresponding coordinate value, then there is formula 2:

$$l = (t_1 w_1, t_2 w_2, t_3 w_3, \ldots, t_n w_n) \tag{2}$$

If the weight of each feature item is used as the component to abstract the history record l, the feature vector can be obtained, then the history record l can be expressed as Eq. 3:

$$\vec{V} = (w_1, \ w_2, \ \ldots, \ w_m) \tag{3}$$

Thereby, the history of visits L can be expressed as Eq. 4:

$$L = \begin{matrix} \vec{V1} \\ \vec{V2} \\ \vdots \\ \vec{Vm} \end{matrix} = (t_1, \ t_2, \ \ldots, \ t_m) = \begin{bmatrix} w11 & \cdots & w1m \\ \vdots & \ddots & \vdots \\ wm1 & \cdots & wmm \end{bmatrix} \tag{4}$$

Physician visit similarity calculation. We can use the Pearson correlation coefficient to calculate the degree of difference between the patient cases visited by the physician and the historical visit records. Suppose the feature vector of physician visits and the feature vector of historical visits are $\vec{Li} = (w1, w2, \ldots, wm)$ and $\vec{Lj} = (w1, w2, \ldots, wm)$, Then the Pearson correlation coefficient is shown in Eq. 5:

$$Ri, L_j = \frac{1}{n} \sum_{j=1}^{n} \frac{m_j \sum_{k=1}^{m_j} w_{1ik} w_{2jk} - \left(\sum_{k=1}^{m} w_{ik}\right)\left(\sum_{k=1}^{m_j} w_{jk}\right)}{\sqrt{m_j \sum_{k=1}^{m_j} w_{1ik} w_{2jk} - \left(\sum_{k=1}^{m} w_{ik}\right)2}\sqrt{m_j \sum_{k=1}^{m_j} w_{1ik} w_{2jk} - \left(\sum_{k=1}^{m} w_{ik}\right)2}} \tag{5}$$

where m_j denotes the number of characteristic terms in the j_{th} patient case, $R(I, L_j)\varepsilon[0,1]$, $+1$ means two diseases are completely positively correlated, and 0 means two places are not correlated. Larger $R(I, L_j)$ indicates a greater correlation and smaller difference between diseases.

For the calculated correlation coefficients, we can use the k-means algorithm to classify them. k-means clustering is commonly used clustering algorithms, which is a distance-based algorithm with high computational efficiency. By the k-means clustering algorithm, the correlation coefficients can be divided into five clusters, and the data in the same clusters maintain a relatively high similarity, while different clustering intervals have low similarity. The clustering algorithm based on Euclidean distance consists of the following three steps.

Step 1: Initialize the k cluster centers: $c_1, c_2, \ldots, c_k$;
Step 2: Calculate the Euclidean distance between each data in the data set and each cluster center, compare the cluster center with the closest Euclidean distance for each data, and assign the data to the cluster to which the closest cluster center belongs.
Step 3: Based on the last clustering result, we calculate the attribute value and the sum of the number of each cluster, and use the quotient of the attribute value and the sum of the number to update the clustering center and get the new clustering center.

4 Simulation Experiment

To more clearly calculate the correlation between user access request cases and historical records in the access control model, and to demonstrate the feasibility and advantages of the dynamic medical privacy protection model based on threshold zero-knowledge protocol and VSM proposed in this paper in applications, simulation experiments are designed in this chapter for verification.

4.1 Experimental Process

In the third chapter, the medical privacy protection model based on threshold zero-knowledge protocol and VSM has been introduced. Before visiting a doctor, it is necessary to vectorize the doctor's historical access records to generate a historical access matrix. Therefore, this paper will conduct simulation experiments from the following three steps.

Step 1: The data used for access behavior is loaded into SPSS software.

Step 2: The correlation coefficient is calculated by processing the user's current access request with the historical access records using the correlation analysis algorithm.

Step 3: Based on the calculated correlation coefficients do scatter plot analysis, compare the parameter values according to the final record, and use the k-means clustering algorithm to cluster the user's access permission.

Calculate the correlation coefficient. After the data is loaded into the work area, the correlation coefficient between the processed data and the historical access records is calculated using the correlation analysis algorithm, and the access rights are granted by extracting the correlation coefficient between the current access request and the historical records.

The correlation coefficient plot shown in Fig. 2 is a scatter plot of the relationship between the correlation coefficients between the current access request and each group of historical records for 60 randomly selected groups (Fig. 3 and Table 1).

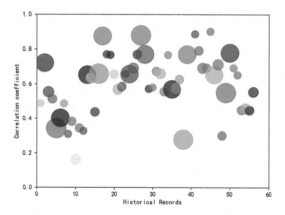

Fig. 2. Scatter plot of a correlation coefficient.

4.2 Experimental Results

Analyzing the above, we can find that after 4 iterations, the final clustering centers are 0.158, 0.334, 0.524, 0.696, and 0.850. By comparing access requests with historical access records, the results demonstrate that the introduction of Pearson correlation coefficients in the vector space model can reduce the misclassification of the system to a certain extent, and validate the effectiveness of the medical privacy protection model based on the threshold zero-knowledge protocol and VSM. In summary, the system design achieves the desired requirements.

5 Conclusion

This paper takes medical big data as the research background, analyzes the problems of patient privacy in the big data environment such as security zone and leakage, and

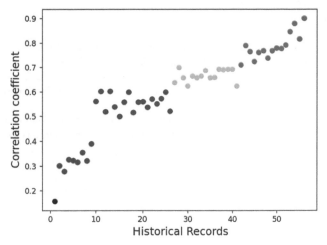

Fig. 3. Clustering results of correlation coefficients.

Table 1. Final clustering center

	Final clustering center				
	1	2	3	4	5
Correlation coefficient	0.158	0.334	0.523	0.695	0.850

points out the necessity of solving this problem from a practical situation. Through the analysis of the current situation of domestic and foreign research, it is found that the current information technology used to solve the privacy leakage research has been done by scholars, but in the method mainly based on differential privacy and anonymity protection technology, in the threshold-based zero-knowledge protocol and VSM access control is still in the preliminary stage of mapping. Therefore, in this paper, the correlation between user access requests and historical access records is systematically and deeply analyzed based on the access control model establishment of VSM, and then on the trusted verification of user identity. In addition, by comparing with the traditional access control model, it is found that this paper can reduce the misjudgment of the system to a certain extent.

The next work will consider adding a reward and punishment mechanism to better encourage users to access normally and develop an incentive mechanism suitable for this scheme.

Acknowledgments. This work was supported by National Natural Science Foundation of China (Nos. 71972165, 61763048), Science and Technology Foundation of Yunnan Province (No. 202001AS070031), Postgraduate Innovation Foundation of Yunnan University of Finance and Economics (2022YUFEYC089).

References

1. Esposito, C., Santis, A.D., Tortora, G., et al.: Blockchain: a panacea for healthcare cloud-based data security and privacy? IEEE Cloud Comput. **5**(1), 31–37 (2018)
2. Habib, M.A., Faisal, C.M.N., Sarwar, S., et al.: Privacy-based medical data protection against internal security threats in heterogeneous Internet of Medical Things. Int. J. Distributed Sensor Netw., **15**(9) (2019)
3. Ji, S.L., Mittal, P., Beyah, R.: Graph data anonymization, de-anonymization attacks, and de-anonymizability quantification: a survey. IEEE Commun. Surveys Tutorials **19**(2), 1305–1326 (2017)
4. Tsou, Y.T., Chen, H.L., Chen, J.Y., et al.: Differential privacy-based data de-identification protection and risk evaluation system. In Proceedings of the International Conference on Information and Communication Technology Convergence (ICTC), South Korea, F Oct 18–20 (2017)
5. Lv, C., Li, Q., Long, H., et al.: A differential privacy random forest method of privacy protection in cloud. In 2019 IEEE International Conference on Computational Science and Engineering (CSE) and IEEE International Conference on Embedded and Ubiquitous Computing (EUC), pp. 470–475. IEEE (2019)
6. Rao, K.R., Nayak, A., Ray, I.G., et al.: Role recommender-RBAC: optimizing user-role assignments in RBAC. Comput. Commun. **166**, 140–153 (2021)
7. Lin, G., Wang, D., Bie, Y., et al.: MTBAC: a mutual trust based access control model in cloud computing. China Communications **11**(4), 154–162 (2014)
8. Line, M.B., Tøndel, I.A., Gjare, E.A.: A risk-based evaluation of group access control approaches in a healthcare setting. In International Conference on Availability, Reliability, and Security, pp. 26–37. Springer, Berlin, Heidelberg (2011)
9. Schmidt, S., Steele, R., Dillon, T.S., et al.: Fuzzy trust evaluation and credibility development in multi-agent systems. Appl. Soft Comput. **7**(2), 492–505 (2007)

Hierarchical Medical Services Data Sharing Scheme Based on Searchable Attribute Encryption

Chenguang Wang[1], Rong Jiang[1](✉), Lin Zhang[1], Meng Wang[1], Yue Yang[1],
Xuetao Pu[2], and Liang Yang[2]

[1] Yunnan University of Finance and Economics, Kunming 650221, China
wangchenguang@stu.ynufe.com, jiangrong@ynufe.edu.cn, {zhanglin2,
wangmeng1,yangyue6}@stu.ynufe.edu.cn
[2] Kunming University of Science and Technology, Kunming 650093, China
{puxuetao,yangliang}@stu.kust.edu.cn

Abstract. In the referral process of hierarchical medical services, it involves the problem of two-way medical data sharing services. In order to solve the problem of data security in this process, the article first uses an attribute proxy re-encryption scheme that supports keyword search to encrypt data, which enables both data sharing and data forwarding services, and also supports flexible access control, so that patients can flexibly control the target hospitals they want to refer to through attribute sets and initiate request services to the target hospitals, and the target hospitals that meet the patient's expectations can decrypt and view these requests and obtain the corresponding patient medical information at the time of treatment. In addition, using the decentralized feature of blockchain, we build a federation chain node in major hospitals and run smart contracts as a trusted authority center, which serves to store key information and perform cryptographic algorithms (e.g., public parameter generation, token generation, and other algorithms) and attribute verification, etc. The innovation of this paper is to create a safe and efficient data sharing model by taking hospital as node and combining searchable encryption and attribute encryption.

Keywords: Blockchain · Attribute re-encryption · Graded care · Data sharing

1 Introduction

The mismatch of resources for medical services is a problem faced by all countries in the world in the process of development. In China, especially in the area of medical services for children, the inability of parents to rationally judge the condition of their children and the lack of trust in the conditions of primary medical services have led to the reduction of resources and the crowding of resources in children's hospitals and even the phenomenon of queuing for hours for emergency services [1]. When it comes to the implementation of graded care services on the ground, the cooperation of various medical institutions faces challenges in terms of information exchange barriers, differences in medical data

© The Author(s), under exclusive license to Springer Nature Switzerland AG 2023
N. Xiong et al. (Eds.): ICNC-FSKD 2022, LNDECT 153, pp. 1259–1266, 2023.
https://doi.org/10.1007/978-3-031-20738-9_136

systems and difficulty in coordinating information [2]. Secondly, the implementation of the two-way referral system is further hindered by the lack of communication among the participants in the hierarchical diagnosis and treatment, the lack of consensus among the participants, and the problems of data leakage and centralized single point of failure of the respective medical data storage. Blockchain and distributed storage technology can solve the problems of trust, security, and centralized single point of failure.

The article is divided into five main parts: the first part is the significance and purpose of the research; the second part provides an overview of the current state of relevant theoretical research; the third part proposes a searchable attribute encryption model for medical data sharing services; the fourth part conducts model analysis; the fifth part concludes this thesis and analyses and outlooks its shortcomings.

2 Related Work

A great deal of research has been done on the sharing and protection of healthcare data services, and scholars at home and abroad have proposed their own solutions. For privacy protection of data, the literature [3] proposes an HGD architecture that allows patients to control their own privacy information and the shared medical data does not involve their own sensitive information, effectively controlling the privacy disclosure of data on the chain. The literature [4] analyzes several defects in the current data sharing model, proposes a decentralized data sharing model based on blockchain, and establishes a domain index to solve the problem of efficient data discovery, and uses differential privacy to ensure the privacy security of the shared data.

The rise of Proxy Re-encryption (PRE), a technology that re-encrypts encrypted data with re-encryption keys to any specified object, has provided a new way of thinking to break down "data silos". [5] and [6] proposed blockchain-based privacy-preserving e-health data sharing protocols that use dual chains to improve remote sharing and treatment efficiency, respectively. The above schemes only consider data access control, without considering that the data itself and the access policy may also reveal sensitive information [7]. Although attribute encryption schemes can solve the fine-grained access control problem, there are still challenges such as inefficiency and access policy failure.

3 Searchable Attribute-Encrypted Medical Data Sharing

3.1 System Models and Entities

As shown in Fig. 1, the project solution has the following main players: the data owner, the IPFS interstellar file system, the hospital information system, the alliance chain, and the data requester.

3.2 Specific Program Process

System initialization phase. This phase is dominated by the alliance blockchain initialization system and key generation.

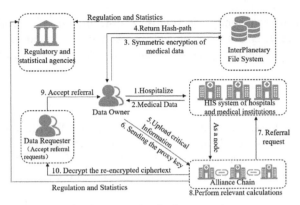

Fig. 1. Example of a figure caption

Initialization procedure. Setup (1^k, u) → (*Params, MSK*): The algorithm is implemented by the alliance chain. The main function of the algorithm is to generate relevant parameters of the encryption system, and to generate keys and call algorithms. Algorithm input is the security parameter 1^k with the full set u of attributes. The output is a bilinear pair e: $G \times G \rightarrow G_T$, where G and G_T are multiplicative cyclic groups of order prime q, g is a generating element in G. Select the random generating element $g_1 \in G$, and choose $a, \alpha \in_R Z_p^*$ (the $\alpha \in_R Z_p^*$ denotes the selection of an element from the set Z_p^* elements from the set uniformly at random α), define the TCR hash function:

$$H_1 : \{0, 1\}^{2k} \rightarrow Z_p^* \ H_2 : G_T \rightarrow \{0, 1\}^{2k} \ H_3 : \{0, 1\}^* \rightarrow G$$

$$H_4 : \{0, 1\}^* \rightarrow G \ H_5 : \{0, 1\}^k \rightarrow Z_p^* \ H_6 : \{0, 1\}^* \rightarrow G$$

The public parameter *Params* and the master key *MSK* are output:

$$Params = \langle e, p, g, g_1, g^\alpha, e(g, g)^\alpha, H_1, H_2, H_3, H_4, H_5, H_6, MF \rangle$$

$$MSK = \langle g^\alpha, a \rangle \tag{1}$$

Finally the public parameter *Params* is uploaded to the Alliance chain as a parameter share for node calls.

Key extraction process. KeyGen(Params, MSK, S) → SK_σ: The algorithm is executed by the alliance chain to generate a private key with attributes. The algorithm requires the input of a master key *MSK*, a set of user attributes $S \subseteq u$ where S has the user identity label u_σ. Select $t \in_R Z_p^*$, output SK_σ private key.

$$SK_\sigma = K = g^\alpha g^{at}, L = g^t, \left\{ K_x = H_3(x)^t \right\}_{\forall x \in S} \tag{2}$$

Finally, the alliance chain adds $\left(u_\sigma, g^{at} \right)$ to the user list L_{DU}.

Search key extraction. SearchKeyGen (MSK, Params) → SK': The algorithm is computed passively by the alliance blockchain, and when the user performs a search for the keyword Kw_i, the user randomly selects $u_i \in_R Z_p^*$, the order $q_u \in g^u$, and the user's own identity tag u_σ and q_u is sent to the alliance blockchain. The alliance chain will verify the user's identity tag and then generate a search key SK_i for the user corresponding to the keyword Kw_i:

$$SK_i = g^{at}, q_u^\alpha \tag{3}$$

Data storage phase. This phase is divided into interstellar file system storage and alliance blockchain storage. The medical data is encrypted and uploaded to IPFS using efficient and secure symmetric encryption. For the hash fingerprint path returned by IPFS and the symmetric key, the key medical information is encrypted and uploaded to the alliance blockchain using searchable attribute proxy re-encryption and a private key and search key for the key medical information is obtained.

Raw data encryption. $E_{filekey}(file) \rightarrow CF$: Input the original ciphertext, the patient or the node of the healthcare facility he/she visits encrypts it using a symmetric key and outputs the ciphertext CF, which is later stored in IPFS and obtains a hash fingerprint of the file storage path.

Data encryption. Encrypt(Params, $(N, \rho), m) \rightarrow CT$: The algorithm in which N is used as $l \times n$ matrix by a non-unary projection function ρ to perform the mapping into the set of attributes, the $m \in \{0, 1\}^k$ As plaintext information, here the symmetric key filekey, the hash returned by IPFS path and the key medical information are encrypted. The specific algorithm is: choose $\beta \in_R \{0, 1\}^k$ and compute $s = H_1(m, \beta)$. Select the random vector used to share the secret index $s{:}v = (s, y_2, \cdots, y_n) \in Z_p^n$, for $i = 1, 2, \cdots, l$, such that $\gamma_i = v \cdot N_i$, and $r_1, r_2, \cdots, r_l \in_R Z_p^*$ Thus the secret text CT is obtained as follows:

$$CT = \begin{cases} A_1 = (m\|\beta) \oplus H_2(e(g, g)^{\alpha s}), A_2 = g^s, \{B_i = g^{a\gamma_i} H_3(\rho(i))^{-r_i}\}_{i \in [l]}, \\ \{C_i = g^{r_i}\}_{i \in [l]}, A_3 = G_1^s, D = \left(H_4(A_1, A_3, (B_i, C_i)_{i=1}^l, (N, \rho))\right)^s \end{cases} \tag{4}$$

Finally, the cipher text is uploaded to the alliance blockchain for storage.
Relevant index token generation and validation.

$$\text{IndexParams, KW} \rightarrow \text{IX,}$$

$$\text{Token}(SK_\sigma, Kwi, S, SK') \rightarrow \text{TK, Verify(TK, CT)} \rightarrow (CT, Q_{CT})$$

Proxy re-encryption phase. When patients need to update their access control policies (e.g. for referral operations), a proxy re-encryption operation is performed to update the attribute set mapping, enabling flexible and secure data sharing.

Re-encryption key generation.Re_KeyGen(Params, $SK_\sigma, (N', \rho'), S) \rightarrow rk$: N' is similar to the data encryption phase in that it is a matrix and consists of ρ' mapped into attributes. The algorithm begins with the user performing an encryption operation on the private key and the desired access control policy.

Setting $\beta', \delta \in_R \{0, 1\}^k, s' = H_1(\delta, \beta')$; choose the random vector used to share the secret index $s{:}v' = \left(s', y_2', \cdots, y_n'\right) \in Z_p^n$, for $i = 1, 2, \cdots, l'$, such that $\gamma_i' =$

$v' \cdot N_i', r_1, r_2, \cdots, r_l \in_R Z_p^*$ Calculation:

$$rk_4 = \left\{ \begin{array}{l} A_1' = (\delta \| \beta') \oplus H_2\left(e(g, g)^{\alpha s'}\right), A_2' = g^{s'}, \left\{ B_i' = g^{a\gamma_i'} H_3\left(\rho'(i)\right)^{-r_i'} \right\}_{i \in [l]}, \\ \left\{ C_i' = g^{r_i'} \right\}_{i \in [l]}, A_3 = G_1^s, D'_i = \left(H_4\left(A_1', A_2', (B_i', C_i')_{i=1}^l, S, (N', \rho')\right)\right) \end{array} \right\}_s$$
(5)

Select $\theta \in_R Z_p^*$ and the re-encryption key is

$$rk = \begin{array}{l} rk_1 = K^{H_5(\delta)} \cdot g_1^{\theta}, \ rk_2 = g^{\theta}, \\ rk_3 = L^{H_5(\delta)}, \ rk_4, \left\{ R_x = K_x^{H_5(\delta)} \right\}_{\forall x \in S} \end{array}$$
(6)

After the re-encryption key rk is generated, the user sends rk to the alliance blockchain node for use by the node for re-encryption.

Data re-encryption.Re_Encrypt(Params, rk, CT) $\rightarrow$ *CT'*: performed by the blockchain alliance blockchain, if the attribute set and access structure are validated by Eq. (7) then the ciphertext is validated by Eq. (8).

$$e\left(A_2', H_6\left(A_1', A_2', (B_i', C_i')_{i=1}^l, S, (N', \rho')\right)\right) = e(g, D')$$
(7)

$$e(A_2, g_1) = e(A_2, g) \cdot \prod_{i \in I} e\left(C_i^{-1}, H_3(\rho(i))^{\omega_i}\right)$$
(8)

Output CT' $= \left(A_1, A_2, A_3, A_4, rk_4, \left(B_i', C_i'\right)_{i=1}^l, D, S, (N, \rho)\right)$.

Data sharing phase. *Data decryption. Decrypt(Params, CT, SK_σ, S)* $\rightarrow m$: A one-to-many decryption algorithm that decrypts any requesting node that satisfies the patient access policy; i.e., for the corresponding S SK_σ, and the CT associated with (N, ρ) associated CT, the algorithm passes the test of Eq. (8) and after passing it calculates:

$$Z = \frac{e(A_2, K)}{Q_{CT}^{u_i}}, \ m\|\beta = H_2(Z) \oplus A_1$$

If $A_3 = g_1^{H_1(m,\beta)}$, then the message m can be decrypted, otherwise output $\perp$.

Decryption of re-encrypted data. Decrypt_{re}(Params, CT', SK_σ', S') $\rightarrow m$: The data requester runs the algorithm to recover the plaintext m by entering the re-encrypted ciphertext as well as the attribute key.

The algorithm runs in the following steps.

$$Validation: e\left(A_2', H_6\left(A_1', A_2', (B_i', C_i')_{i=1}^l, S, (N', \rho')\right)\right) = e(g, D')$$
(9)

If the validation passes, calculate: $Z' = \frac{e(A_2', K)}{\left(Q_{CT'}'\right)^{u_i}}, \delta \| \beta' = H_2(Z') \oplus A_1$.

If $A_3 = g_1^{H_1(m,\beta)}$ and $D = H_4\left(A_1, A_3, (B_i, C_i)_{i=1}^l, (N, \rho)\right)^{H_1(m,\beta)}$ then can will get m, otherwise will return $\perp$.

3.3 Smart Contracts

The smart contract mainly consists of a decryption key request contract and a proxy re-encryption contract. The user requests a key from the Alliance Chain through the decryption key request contract, and verifies the security of the key through the accompanying hash verification and timestamp verification; secondly, when the user's demand changes, i.e. the access policy changes, the proxy re-encryption contract generates the proxy re-encryption key, sends the key to the Alliance Chain for re-encryption, and then updates the attribute matrix to formulate a new access policy for referral operation. The basic smart contract algorithms are shown in Fig. 2.

```
//The requester sets time T and sets the contract condition
//Time T as timeout limit
if(current usage time Tn≤T) Then
//continue if not timed out
Verify the identity and signature of the requested user.
if (validation passes) Then
1)Execution Re_Encrypt(Params,rk,CT)→CT'
2)Send a re-encrypted ciphertext CT to the data requester.
else Return
end
endif
```

```
//The requester sets time T and sets the contract condition
//Time T as timeout limit
if(current usage time Tn≤T) then
//continue if not timed out
if(user authentication passes) then
1) Running KeyGen(Params. MSK. S)→SK_σ
2) Hash the key and add a timestamp etc. to verify the
signature
3) sending the decryption key SK to the user.
else Return⊥
end
endif
```

Fig. 2. Algorithm 1 and Algorithm 2

4 Analysis of Results

4.1 Comparison of Functions

This section analyzes and compares the functional features of data sharing encryption schemes in relevant literature (Table 1). The combined scheme used in this paper has some advantages. The scheme [8] cannot conduct keyword search, which is not conducive to data search and sharing. Schemes [9] are lower in search validation efficiency than the proposed schemes. Proxy re-encryption can meet the flexible data sharing strategy of two-way referral. Scheme [10] cannot carry out proxy re-encryption, so it is not suitable for data forwarding mechanism. Therefore, this paper can not only realize the functions in the table, but also have a good fit effect for medical data sharing and two-way referral.

4.2 Testing of Smart Contracts

Simulation experiment: the author in Tencent cloud server according to Linux system Ubuntu 18.04.1 LTS. FISCO BCOS development tool was used to build blockchain smart contract, and experimental simulation was conducted on the author's scheme.

The gas consumption test can reflect the operation efficiency of a smart contract. The normal gas consumption range is about 20,000 gas. The author programmed the smart contracts in the experiments with Solidity and ran simulated consumption tests using the Remix platform. The test results in Table 2 showed that the smart contract costs were within the normal range and the key initialization contract only ran once during system initialization. According to the time cost test, the average time cost of every 200 smart contracts is about 5 s, also shown in Table 2.

Table 1. Comparison of the functional characteristics of the programmes

Programme	Keyword search	Property encryption	Proxy re-encryption	Decryptable ciphertext
Literature Liang et al. [8]	×	✓	✓	✓
Literature Liang and Suslio [9]	✓	✓	✓	✓
Literature Zheng et al. [10]	✓	✓	×	×
This paper	✓	✓	✓	✓

Table 2. Gas consumption and smart contract time overhead

Smart contract	Specific part of contract	Gas consumption	Average time cost (ms)
Algorithm 1	Initialization procedure	103,412	5.33
	Key extraction process	31,342	4.93
Algorithm 2	Re-encryption key generation	19,756	5.12
	Data re-encryption	20,375	5.19

5   Conclusion

The article firstly points out the current data sharing problems of hierarchical diagnosis and treatment services, and proposes a data sharing scheme suitable for two-way referral services based on real cases: using symmetric encryption to store patient data encrypted in the interstellar file system, re-encrypting the patient data through searchable attribute agents, and encrypting the key information on the chain by blockchain service as a trusted authoritative center. The blockchain service acts as a trusted authority center to run key distribution, re-encryption smart contracts, and send referral service requests to medical institutions that comply with the access control policy. After security verification, functional analysis and efficiency analysis, the solution has good security as well as efficiency and functionality, and can meet the data sharing needs of two-way referral. The article also has some shortcomings: (1) the experimental results lack cryptographic computational cost evaluation (2) the lack of detailed crypto-graphic indistinguishability game for the encryption algorithm. A deeper analysis and comparison of algorithmic schemes can be investigated in the next step.

Acknowledgements. This work was supported by National Natural Science Foundation of China (Nos. 71972165, 61763048), National Statistical Science Research Project (No. 2020LY028), Science and Technology Foundation of Yunnan Province (No. 202001AS070031).

References

1. Pan, Z.H., Yao, N., Qi, J.G.: Research on the current situation of children's medical treatment in China and the problems and Countermeasures of developing graded diagnosis and treatment. Chinese General Practice **21**(10), 1177–1182 (2018)
2. Svensson, A.: Challenges in Using IT Systems for Collaboration in Healthcare Services. International Journal of Environmental Research and Public Health **16**(10), 1773(2019)
3. Yue, X., Wu, H.J., Jin, D.W., et al.: Healthcare data gateways: found healthcare intelligence on blockchain with novel privacy risk control. J. Med. Syst. **40**(10), 1–8 (2016)
4. Dong, X.Q., Guo, B., Shen, Y., et al.: An efficient and secure decentralized data sharing model. J. Comput. **41**(05), 1021–1036 (2018)
5. Pournaghi, S.M., Bayat, M., Farjami, Y.: MedSBA: a novel and secure scheme to share medical data based on blockchain technology and attribute-based encryption. J. Ambient. Intell. Humaniz. Comput. **11**(11), 4613–4641 (2020). https://doi.org/10.1007/s12652-020-01710-y
6. Salman, S., Minahil, Khalid, M., et al.: A secure blockchain-based e-health records storage and sharing scheme. Journal of Information Security and Applications **55**, 102590(2020)
7. Zhang, L.Y., Hu, G.C., Mu, Y., et al.: Hidden Ciphertext Policy Attribute-Based Encryption with Fast Decryption for Personal Health Record System. IEEE Access **7**, 33202–33213 (2019)
8. Liang, K.T., Fang, L.M., Suslio, W., et al.: A Ciphertext-Policy Attribute-Based Proxy Re-encryption with Chosen-Ciphertext Security. in 2013 5th International Conference on Intelligent Networking and Collaborative Systems, 552–559(2013)
9. Liang, K.T., Suslio, W.: Searchable Attribute-Based Mechanism with Efficient Data Sharing for Secure Cloud Storage. IEEE Trans. Inf. Forensics Secur. **10**(9), 1981–1992 (2015)
10. Zheng, Q.J., Xu, S.H., Ateniese, G.: VABKS: Verifiable attribute-based keyword search over outsourced encrypted data. In IEEE INFOCOM 2014 - IEEE Conference on Computer Communications, 522–530 (2014)

Spatial and Temporal Distribution Analysis of Regional Industrial Patents of Injection Molding Machine

Chengfan Ye[1], Hongfei Zhan[1(✉)], Yingjun Lin[2], Junhe Yu[1], and Rui Wang[1]

[1] Ningbo University, Ningbo 315000, China
{2111081063,zhanhongfei,yujunhe,wangrui}@nbu.edu.cn
[2] Zhongyin (Ningbo) Battery Co., Ltd., Ningbo 315040, China

Abstract. The regional injection molding machine patent resources are analyzed based on the time series and the space and industry direction to reflect the development trend of the regional injection molding machine industry and provide a corresponding reference for the study of the regional injection molding machine industry development. This paper uses the patent data of injection molding machines in Ningbo, through data processing and analysis, and uses the data of four time periods to study the distribution of its space and industrial development direction through standard deviation ellipse method, kernel density analysis method and patent text cluster analysis., analyze the direction distribution characteristics, density structure characteristics and industrial development trend of injection molding machine patents in Ningbo area. The results show that the development of Ningbo's injection molding machine industry has shifted from the core urban area to the development along the entire Ningbo area. The highest value of patent density is always in the core urban area and slowly shifted to the east, and the number of patents has increased exponentially at each stage. Spatially northwest of Beilun District and injection molds, hydraulic motors, manipulators and injection molding machines of the industrial direction are the development trends of the injection molding machine industry in Ningbo. This article provides help for the development of injection molding machines in Ningbo from the perspective of space and industry.

Keywords: Development direction · Distribution characteristic · Patent density · Injection molding machine

1 Introduction

With the rapid development of science and technology, we have entered the era of knowledge economy. As an emerging factor of production, patents are playing an increasingly important strategic role in inter-regional market competition, and have become a key factor that affects and even determines the strength of regional competitiveness [1]. In recent years, the number of patent applications has increased significantly, and enterprises and universities have realized the importance of patent development. The spatial

N. Xiong et al. (Eds.): ICNC-FSKD 2022, LNDECT 153, pp. 1267–1273, 2023.
https://doi.org/10.1007/978-3-031-20738-9_137

distribution of patents is not random, and is subject to social, economic and other conditions [2]. Through the analysis of patents, the development direction and trend of regional industrial clusters can be reflected.

2   Related Research Analysis

In recent years, there have been few studies on the temporal and spatial distribution of patents in my country. Looking at the existing studies in my country, most of the quantitative studies on patents are based on patent data in a certain period or region [3]. Zeng Peng et al. analyzed the temporal and spatial differences of patents through three types of patents, and put forward development suggestions for different regions [4]. Part of the research studies the density and regional innovation development of each region based on temporal changes [5–8]. Du Jiangyuan et al. conducted analysis and research on the distribution of patent number and application institutions in combination with the analysis of patent hotspots, the annual distribution of patent transfer status and licensing status, and the conversion rate of patent scientific and technological achievements, and put forward targeted problems and strategies [9]. The above studies are based on the changes in time series or the distribution of patents in various regions to analyze patent resources, and there is no analysis of patents based on industrial content and spatial distribution. This paper analyzes the trend of regional injection molding machine patent resources in space and development direction based on time, and reflects the development trend of injection molding machine industry in this region.

3   Analysis Method

3.1   Data Source

The patents used in this article are the corresponding patent data obtained from patyee patent database through advanced search of keywords and applicant addresses. Each piece of data exported records the title, abstract, its address, and filing date.

3.2   Standard Deviation Ellipse Method

The standard deviation ellipse method can analyze the agglomeration degree and direction characteristics of a certain element in its area, and the center of the ellipse is the center of the average distribution of the element.

The formula is as follows:

Center:

$$SDE_x = \sqrt{\frac{\sum_{i=1}^{n}(x_i - \overline{X})^2}{n}} \tag{1}$$

$$SDE_y = \sqrt{\frac{\sum_{i=1}^{n}(y_i - \overline{Y})^2}{n}} \tag{2}$$

Azimuth angle:

$$\tan\theta = \frac{\left(\sum_{i=1}^{n}\tilde{x}_i^2 - \sum_{i=1}^{n}\tilde{y}_i^2\right) - \sqrt{\left(\sum_{i=1}^{n}\tilde{x}_i^2 - \sum_{i=1}^{n}\tilde{y}_i^2\right)^2 + 4\left(\sum_{i=1}^{n}\tilde{x}_i^2\tilde{y}_i^2\right)^2}}{2\sum_{i=1}^{n}\tilde{x}_i^2\tilde{y}_i^2} \quad (3)$$

In Eqs. (1) and (2), x_i and y_i represents the geographic coordinates of the patent data, and $\overline{X}$ and $\overline{Y}$ represents the average value of the coordinates. In Eq. (3), $\tilde{x}_i$ and $\tilde{y}_i$ represents the coordinate difference relative to the center.

3.3 Nuclear Density Analysis

Kernel density analysis can reflect the spatial density and aggregation characteristics of a certain element. The kernel density analysis of point elements is used to calculate the density around each output grid pixel, and a smooth surface is covered above each point. The surface value is the highest at the point where the point is located, and as the distance from the point increases, the surface value will gradually decrease, and the surface value at the point equal to the search radius of the point is zero. Its expression is:

$$f(x) = \frac{1}{nh}\sum_{i=1}^{n}k\left(\frac{x - x_i}{h}\right) \quad (4)$$

In Eq. (4), x_i is independent patent data samples, $f(x)$ is the kernel density function at the patent sample point, and h is the search radius.

3.4 Patent Text Clustering

Cluster analysis is a statistical analysis method for classifying research objects. This method can divide the data set into different categories. The data in the same category can have a high degree of similarity and can be distinguished from the data in other categories. In the patent data, the industry-related content includes the patent name and abstract, from which the development direction of each stage of the injection molding machine industry patent can be classified through text clustering [10].

4 Research Results and Analysis

4.1 Data Processing

Through patyee patent database keywords 'injection molding machine' and the applicant's address 'Ningbo' search, and then a simple application number merger, screening out a total of 6121 patent data, according to data screening and checking, deleting and modifying the error information, increase the reliability of the data. Four groups of data are obtained by taking every five years of patent passage as a node. Then through the POI of Amap through python code to achieve patent address and latitude and longitude transformation. According to the patent data, they are divided into four groups, the first stage until April 2007, the second stage until April 2012, the third stage until April 2017, and the fourth stage until April 2022.

4.2 Directional Analysis

According to the above four groups of patent data by standard deviation ellipse analysis method, the calculation results are Table 1.

Table 1. Table of parameter changes of standard deviation ellipse method.

Cut-off time	Ellipse circumference	Ellipse area	X center coordinates	Y center coordinates	X-axis length	Y-axis length	Direction angle
2007.4	1.0843	0.0660	121.5847	29.9171	0.2395	0.0878	110.7073
2012.4	1.1075	0.0826	121.6399	29.9042	0.2277	0.1155	102.1589
2017.4	1.6867	0.2224	121.6078	29.8752	0.2964	0.2388	121.1705
2022.4	1.7736	0.2432	121.6007	29.8724	0.3195	0.2424	125.2055

From Table 1, it can be concluded that the perimeter and area of the ellipse in the four time periods increased in multiples. It can be found that the change of the center coordinate of the ellipse is relatively small, and the direction angle of the ellipse changes greatly in the four time periods. From the above results, it can be concluded that patents in Ningbo have increased significantly, increasing exponentially every five years, and the distribution of patents has grown rapidly from the beginning in Ningbo city to other areas, expanding to the entire Ningbo area. The change of the direction and angle of the ellipse is found from the beginning of the ellipse to basically conform to the geographical trend of Ningbo urban area. The reason is that the development of Ningbo urban area is relatively saturated, the development of cost is gradually increasing, and the industry needs to be distributed to the whole city.

4.3 Density Analysis

The four groups of data were formed into a distribution grid based on ArcGIS software, and the kernel density analysis map was formed by appropriate output pixel size and search radius. The classification method was the natural discontinuity point classification method (Fig. 1).

According to the core density analysis results shown in Fig. 1, the patent core density values of the first stage are the highest in Qiuai Town, Yinzhou District, Yufan East Road, Zhenhai District, Daqing North Road, Jiangbei District, and Beilun District Processing Trade Zone, and the density decreases in turn. In the second stage, the highest density of patent cores is located in Yufan East Road, Zhenhai District, Beilun District Xiaogang and the Processing Trade Zone. The density of patents is relatively concentrated in Beilun District. In the third stage and the fourth stage, the nuclear density value of Xiaogang Jiangnan Middle Road in Beilun District is the highest, followed by the processing trade zone in Beilun District. The distribution of patents in Cixi City, Yuyao City, Fenghua District, Ninghai County and Xiangshan County has increased significantly, but the density is far lower than that of Ningbo core city, and most of them are scattered. The reason is that Beilun District has a superior geographical location, abundant land resources, low cost, and is close to Beilun Port, so it has a good development prospect.

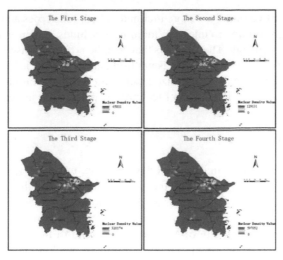

Fig. 1. Analysis chart of patent nuclear density in four stages.

4.4 Text Clustering Analysis

Through word segmentation, weight calculation and vector quantization analysis, the first few keywords of cluster analysis are counted, as shown in Table 2.

Table 2. Cluster analysis table of four stages.

Stage	Key words of patent data
The first stage	Injection molding machine, electromagnetic vibrator, safety door, injection molding machine clamping, plastic injection
The second stage	Injection molding machine, mold hydraulic, injection mold, ductile iron, injection molding machine
The third stage	Injection molding machine, camera module, injection mold, injection mold, hydraulic motor, manipulator
The fourth stage	Injection molding machine, injection mold, hydraulic motor, manipulator, injection molding machine

The order in Table 2 represents the frequency of occurrence of keywords in patent data. In the first stage, injection molding machine, electromagnetic vibrator, safety door, injection molding machine clamping and plastic injection have higher frequency. The second stage is injection molding machine, mold hydraulics, injection mold, ductile iron and injection molding machine injection. The third stage is injection molding machine, camera module, injection mold, hydraulic motor and manipulator. The fourth stage is injection molding machine, injection mold, hydraulic motor and manipulator. It can be found that the first two stages have higher frequency of instrument types and processing types, and the latter two stages are more inclined to hydraulic types and equipment manufacturing types. In addition to injection molding machines, the proportion of injection molds has always been high. The reason is that Ningbo is mainly based on low-end manufacturing, the added value of products is not high, and the profit is low. In recent years, the industrial structure has been optimized, scientific and technological innovation has been emphasized, and the proportion of technology patents has increased.

5   Conclusions

Patent as an important factor in industrial development, many enterprises put the patent in a very important position, so through the patent can also reflect the development of the whole industrial cluster. Through the standard deviation ellipse analysis, the development direction of the injection molding machine industry cluster in Ningbo is consistent with the trend of the central city of Ningbo to the trend of the whole city of Ningbo, reflecting the gradual layout of the injection molding machine industry in the whole Ningbo area. Through the analysis of nuclear density, it is reflected that the industrial distribution density of Ningbo in the core urban area is much higher than that of other counties and cities. The region with the peak intensity gradually moves eastward, and the northwest of Beilun District in the fourth stage is the highest. Through the text clustering analysis of patent resources, the keywords with the highest frequency of patents in each stage are extracted. From the initial processing category to the present hydraulic motor, manipulator and injection molding machine, the core technologies of injection molding machines account for the majority. Throughout the four stages, according to the above analysis of the development trend of Ningbo in the future, the northwest of Beilun District is the center of future development in space. In the direction of development, injection mold, hydraulic motor, manipulator and injection molding machine may become the development trend of injection molding machine in Ningbo.

Acknowledgment. I would like to extend my gratitude to all those who have offered support in writing this thesis from National Key R&D Program of China (2019YFB1707101, 2019YFB1707103), the Zhejiang Provincial Public Welfare Technology Application Research Project (LGG20E050010, LGG18E050002) and the National Natural Science Foundation of China (71671097).

References

1. Sun, W., Sun, Q.L., Chen, Y.: Spatial and temporal evolution characteristics of regional patent density and its distribution—An empirical analysis based on urban data in Northeast China. Technol. Econ. **33**(08), 42–47 (2014)
2. Fan, L.N.: Analysis on the spatial distribution and influencing factors of patents in Mainland China. J. Beijing Norm. Univ. (Soc. Sci. Ed.) (02), 138–145 (2005). Journal (Philos. Soc. Sci. Ed.) (01), 74–80 (2014)
3. Wang, C.F., Xu, H.B., Zhang, G.P.: Research on the difference of urban regional innovation ability—Based on the perspective of patent quality. J. Shandong Univ. (Philos. Soc. Sci. Ed.) **01**, 74–80 (2014)

4. Zeng, P., Zhao, C.: Research on the spatial and temporal differences of three types of patent density distribution in China. Sci. Technol. Prog. Countermeasures **33**(21), 117–125 (2016)
5. Lin, Z., Wang, L.L., Lin, F.: Spatial-temporal evolution analysis of patent resources distribution in various regions of China. Inf. Sci. **36**(10), 164–170 (2018)
6. Wang, J.W., Sun, Y., Lin N.B.: Spatial and temporal evolution characteristics and countermeasures of distribution pattern of urban innovation activities—Taking Hangzhou as an example. Urban Dev. Res. **27**(01), 12–18+29 (2020)
7. Li, W.H., Ye, Z.Y.: Research on technological innovation and development of medical mask industry in China based on patent information. Chin. Inventions Pat. **18**(3), 34–42 (2021)
8. Zhang, Y., Duan, J.H., Shi, L.: Analysis of patent behavior characteristics under innovation-driven development in China. Mod. Manag. Sci. **03**, 3–9 (2022)
9. Du, J.Y., Hao, X.L., Zhao, Y., Bao, W.S.: Research on patent distribution and achievement transformation of quantum science and technology in China. China High-Tech **06**, 144–146 (2022)
10. Wu, Y., Yu, J.H., Zhan, H.F., Xu, B.: Data-driven industrial cluster identification and spatial correlation analysis. Technol. Econ. **33**(04), 51–55 (2020)

Design of Frequency Reconfigurable Antenna Based on Liquid Metal

Jingjing Ren[✉] and Xiaofeng Yang

School of Electronic Engineering, Xi'an University of Posts and Telecommunications,
Xi'an 710121, China
1134353695@stu.xupt.edu.cn, xfyang@mail.xidian.edu.cn

Abstract. This paper presents a new design of liquid metal frequency reconfigurable antenna. The antenna takes the electrically driven patch antenna as the
idea. Based on the microstrip antenna, the slot is cut at the appropriate position
of the microstrip antenna radiation patch to change the resonant length of the
antenna. Using the fluidity and good electrical characteristics of gallium indium
alloy (EGaIn). Injecting liquid metal into a microfluidic channel filled with sodium
hydroxide solution. And the liquid metal flows to the appropriate position under
the driving of voltage, so as to supplement the radiation patch, make the antenna
completely in the path state, and realize the reconfiguration between 2.44 and
2.5 GHz frequency range through the change of radiation patch state. From the
simulation results, we can easily conclude that the antenna has good performance,
has various advantages of microstrip antenna and is simple to manufacture.

Keywords: Liquid metal · Frequency reconfigurable antenna · Microstrip
antenna · Simulation

1 Introduction

With the vigorous development of economy and science and technology today, people's
lifestyle is constantly changing. With the rapid development of wireless communication,
people's life has been completely inseparable from the word communication. Similarly,
modern life has continuously improved the requirements for information transmission
and acquisition, which makes the requirements for communication technology and system realizable functions higher and higher. The communication system that provides a
single service function can no longer meet people's needs [1]. In order to achieve the
goal that the same communication system can realize multiple functions, it is often the
case that one communication platform is equipped with multiple subsystems. This not
only increases the volume, weight, and cost of a system but also leads to electromagnetic
compatibility between subsystems, which destroys the stability of the system. To simplify the system, the concept of the reconfigurable antenna [2] is proposed. Functionally,
reconfigurable antennas can be divided into pattern, frequency and polarization reconfigurable antennas. In terms of reconfigurable methods [3], it is common to change the
structural state of the antenna by means of electronic devices, machinery and changing

materials, so as to affect the magnetic current distribution of the antenna, so as to achieve the purpose of reconfigurable [4].

In recent years, with the emergence of liquid metal, researchers try to combine this liquid metal with the reconfigurable antenna [5]. Liquid metal, a new type of alloy, with its good conductivity and fluidity, can make up for the shortcomings and shortcomings of the traditional solid-state antenna [6] when combined with the antenna, and also provides a new idea for the breakthrough and development of the antenna. Due to the limitation of its material and structural characteristics, the traditional solid-state antenna design is often greatly limited. For example, solid materials have a certain elastic limit and are prone to mechanical fatigue in practical use, resulting in unstable antenna performance; The feed point of a solid-state antenna cannot be closely combined with the radiation patch, resulting in complex electromagnetic interference; Due to the limitation of materials, it is often difficult to modify the size of the mechanism after the design is completed, so it is difficult for the traditional solid-state antenna to realize the reconfigurability of various characteristics. Liquid metal can also receive signals well because of its good conductivity, and it does not have plasticity. Compared with solid-state antenna, liquid metal antenna avoids many limitations [7].

It is found that the current reconfigurable antenna design based on liquid metal is based on its elasticity and fluidity. This paper is also based on the fluidity of liquid metal and the idea of electrically driving the patch antenna to slot the appropriate position of the microstrip antenna radiation patch to change the resonant length of the antenna. The liquid metal is injected into the microfluidic channel filled with sodium hydroxide solution. Driven by voltage, the liquid metal flows to the appropriate position, so as to supplement the radiation patch and make the antenna completely in the path state, so as to achieve the purpose of reconfiguration. Compared with the traditional microstrip slot antenna[8], the antenna does not need to add an RF switch, which avoids the influence of the welding joint of the electronic switch on the performance of the antenna. This paper emphasizes the application of liquid metal instead of an electronic switch to realize reconfigurability, so the antenna model in this paper adopts a simple microstrip slot antenna.

2 Basic Principle of Frequency Reconfigurable Antenna

2.1 Antenna Bandwidth

As a transmitting and receiving device in wireless transmission equipment, the antenna plays the role of energy conversion and directional radiation. Ideally, we hope the antenna can work in any frequency band. However, in practical application, the frequency of the antenna is also changing with the change of its input impedance, gain, and various parameters of the pattern. Therefore, all types of antennas have their specific working frequency range. The bandwidth of the antenna is the frequency range of the antenna that meets the specified standards of various parameters. We can take a range around the center frequency as its bandwidth, and within this bandwidth, the performance of the antenna is within an acceptable range compared with that at the center frequency. The expression of antenna bandwidth has absolute bandwidth, which refers to the actual working frequency range of the antenna. Refers to the difference between high-end frequency and low-end frequency. It can also be expressed by relative bandwidth, which refers to the percentage of absolute bandwidth to center frequency.

2.2 Design Principle

Due to the magnetic current and current on the antenna, the antenna generates radiation. The bandwidth, polarization, gain, and radiation direction of the antenna all depend on the distribution of electromagnetic current on the antenna. Frequency reconfigurable antenna is also based on changing the distribution of antenna electromagnetic flow, so as to realize the real-time change of antenna bandwidth. For the resonant antenna such as dipole and microstrip antenna, there is a specific electromagnetic current distribution during resonance, so it has a specific frequency. Generally speaking, the resonant antenna has a fixed length rather than a physical resonant antenna. For this kind of antenna, frequency reconfiguration is generally completed by changing its electrical size.

The liquid metal mixed with gallium indium in a specific proportion has good conductivity, liquid at room temperature, and good fluidity. Combining this metal with the antenna not only has the electrical characteristics required by the antenna but also has better elasticity than copper. Microstrip antenna has the characteristics of small volume, lightweight, and low profile. In this paper, the design takes a microstrip antenna as the model to realize the purpose of frequency reconfiguration while having the advantages of a microstrip antenna.

3 Antenna Design

This design is based on the coaxial line fed microstrip rectangular microstrip antenna structure. Figure 1 shows the basic mechanism of the antenna. The structural parameters are shown in Table 1. The dielectric substrate adopts an FR4 epoxy resin plate with a thickness of 1.6mm, the antenna adopts 50-Ω coaxial feeding mode, and the radiation patch size is $L_0 \times W_0$ mm^2, the antenna works in the main mode TM$_{10}$. A rectangular groove is opened on the surface of the patch, with the length of Ls and the width of W, and the distance between the gap and the radiation edge is Ps.

Realization of frequency reconfiguration: slot the appropriate position of the antenna radiation patch to change its radiation length, so as to change the antenna frequency. Place the acrylic plate containing the microfluidic channel above the slotted position of the radiation patch, and then inject the liquid metal into the microfluidic channel. Here, the liquid metal should be wrapped in the sodium hydroxide solution to prevent the liquid metal from generating an oxygen layer in the channel. Then, the appropriate voltage is applied to generate the potential difference to drive the liquid metal to flow to the appropriate position to fill the vacancy of the slot of the radiation patch. At this time, the antenna recovers the frequency of the complete radiation patch. At this time, we use the fluidity of the liquid metal and good electrical characteristics to realize the frequency reconfiguration of the antenna.

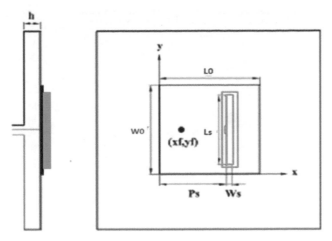

Fig. 1. Antenna model

Table 1. Antenna structural dimension parameters mm

H	L0	W0	L1
1.6	28	37.26	6.6
Length	Ws	Ps	Ls
30	2	15	28

4   Antenna Simulation

Next, the electromagnetic simulation software Ansoft HFSS is used to analyze the characteristics of the liquid metal antenna with liquid metal in different positions. When the liquid metal does not move to the position above the gap, it can be equivalent to that the liquid metal does not exist and the radiation patch has a gap, as shown in Fig. 2a; When the voltage is applied and the liquid metal flows directly above the gap of the radiation patch, the radiation patch is complete, as shown in Fig. 2b. The above model approximates the liquid metal. Here, we mainly focus on the working principle of directional analysis antenna, and the similarity is high, which can well predict the radiation characteristics of the antenna.

Figures 3 and 4 shows the electric field distribution generated in two states when the microstrip patch antenna works in TM_{10} mode, the path length of the surface current is approximately half the dielectric wavelength. Therefore, when the radiation patch is slotted, the resonant frequency of the antenna is at a higher frequency. When the radiation patch is supplemented. The resonant frequency of the antenna goes to one end of low frequency. Figure 5 shows the frequency sweep result analysis of the return loss of the antenna signal port when the liquid metal is in two states and gives the resonant frequency curve of the antenna. It is obvious from the figure that the antenna resonant frequency shifts significantly when the liquid metal is in different states.

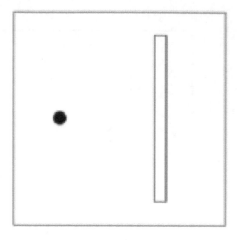

(a) No liquid metal state

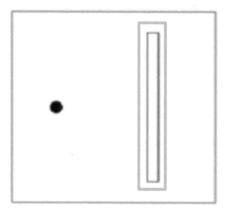

(b) Liquid metal filling gap state

Fig. 2. Equivalent model of slotted microstrip patch antenna in two states of liquid metal

Figure 6 shows the resonant frequency of the liquid metal frequency reconfigurable antenna when the slot length is different. From that figure, it can be seen that when the slot length changes, the change of the resonant frequency of the antenna can be ignored. Figure 7 shows the influence of the slot at different positions on the resonant frequency of the liquid metal antenna. It is not difficult to see from the figure that when the slot is close to the edge of the patch, the higher the radiation frequency of the antenna, and the closer the slot is to the center of the patch, the lower the resonant frequency of the antenna.

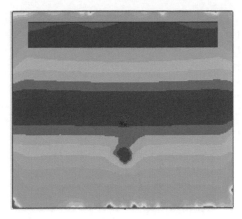

Fig. 3. Surface current distribution of injected liquid metal patch

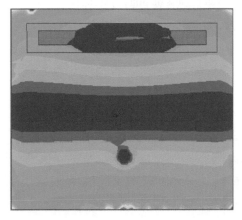

Fig. 4. Current distribution on the surface of uninjected liquid metal patch

5 Conclusion

This paper provides a new idea for making frequency reconfigurable antenna. Using the good electrical characteristics and fluidity of liquid metal, the antenna takes the idea of electrically driving the patch antenna. By placing a microfluidic channel filled with sodium hydroxide solution above the slotted microstrip patch antenna. Then liquid metal, a new material, is injected into the channel. And the liquid metal flow is driven by the voltage to supplement the integrity of the radiation patch, so as to realize the reconfigurability of the antenna in the frequency range of 2.44–2.5 GHz. Simulation with HFSS electromagnetic simulation software shows that the antenna has good impedance matching, bandwidth and radiation characteristics.

The frequency reconfigurable antenna design proposed in this paper not only has various advantages of microstrip patch but also has simple structure, convenient operation, easy adjustment, and good radiation characteristics, which meets the requirements

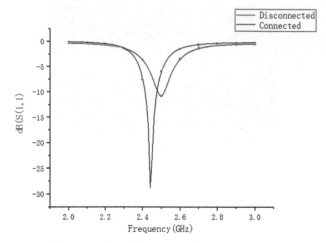

Fig. 5. Returns loss of slotted patch antenna in two states

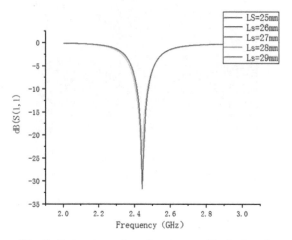

Fig. 6. Return loss of patch antenna with slot length

of various indexes of antenna design. Of course, there are also many problems and deficiencies, which is worthy of consideration and improvement in the future research process.

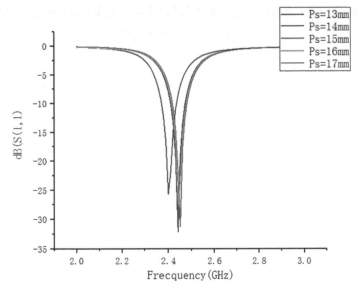

Fig. 7. Return loss of patch antenna with slot position

References

1. Shereen, M.K., Khattak, M.I., Al-Hasan, M.: A hybrid reconfigurability structure for a novel 5G monopole antenna for future mobile communication. Frcqucnz **75**(3–4), 71–82 (2021)
2. Abdulraheem, Y.I., et al.: Design of frequency reconfigurable multiband compact antenna using two PIN diodes for WLAN/WiMAX applications. IET Microwaves Antennas Propag. **11**(8), 1098–1105 (2017)
3. P.B. Saha, R.K. Dash and D. Ghoshal , "A Compact Uplink-Downlink Band Switchable Wide-band Antenna for C-band Satellite Applications." 2020 7th International Conference on Signal Processing and Integrated Networks (SPIN) ,2020,pp. 262–266
4. Kumar, A., Singh, A.P.: Design of Micro-Machined Reconfigurable Patch Antenna using MEMS Switch. IEEE Indian Conference on Antennas and Propogation (InCAP) **2018**, 1–4 (2018)
5. Chang, K.: RFand microwave wireless systems. Wiley, New York (2000)
6. K. L. Wong, Compact and Broadband Microstrip Antennas. John Wiley andSons, Ltd, 2002
7. Tang, S.Y., Tabor, C., Kalantar-Zadeh, K., Dickey, M.D.: Gallium Liquid Metal: The Devil's Elixir. Annu. Rev. Mater. Res. **51**, 381–408 (2021)
8. J.Y. Yang, D. Tang, J.P. Ao , T. Ghosh, T.V. Neumann, D.G. Zhang, Y. Piskarev, T.T. Yu, V.K. Truong, K. Xie, Y.C. Lai, Y. Li, M.D. Dickey, "Ultrasoft Liquid Metal Elastomer Foams with Positive and Negative Piezopermittivity for Tactile Sensing." Advanced Functional Materials,2020,vol.30,no.36

About the Parsing of NMEA–0183 Format Data Streams in GPS

Siyao Dang⊙, Haisheng Huang(✉)⊙, and Xin Li⊙

Xi'an University of Posts and Telecommunications, Xi'an 710121, China
{hhs,lixin}@xupt.edu.cn

Abstract. NMEA-0183 protocol parsing module is the most important software module in navigation, therefore, this paper on the basis of analyzing NMEA 0183 statements, through the C language of the effective parsing and study of the middle field of different statements, wrote its related port communication program, and finally realized NMEA-0183 original data extraction, parsing processing and transmission process, verified by STM32 development board and host computer communication.

Keywords: NMEA-0183 · Parsing · Validation · Navigation · Serial

1 Introduction

GPS system can provide real-time, all-weather, global and high-precision services, widely used in all walks of life, GPS receiver through the antenna unit, through the antenna unit to receive satellite signals, the signal for band-pass filtering, downconversion mixing, AGC amplification A/D conversion and a series of processing, to obtain a digital if the frequency signal, and then the intermediate frequency signal to capture, track and demodule the user's longitude, longitude, time, speed, altitude, latitude and other navigation message, This information is encapsulated via NMEA-0183 and output via serial port. This article mainly analyzes NMEA-0183 packets on the basis of National Marine Electronics Association [1].

National Marine Electronics Association is the standard format developed by the NMEA Electronics Association for asynchronous navigation communication of marine electronic devices, and has become a unified standard protocol for GPS navigation equipment. The first task of NMEA-0183 after receiving it is to parse the position, speed, heading and satellite status information in the statement, and to ensure the availability of navigation data, the data must also be verified. Aiming at the problems and requirements in the practical application of NMEA-0183, this article details the parsing of real-time data streams based on the serial NMEA-0183 format.

2 Introduction and Analysis of NMEA-0183 Format

National Marine Electronics Association is the standard format for electronic devices by the NMEA and is now widely used in data transmission between devices in multiple

N. Xiong et al. (Eds.): ICNC-FSKD 2022, LNDECT 153, pp. 1282–1289, 2023.
https://doi.org/10.1007/978-3-031-20738-9_139

fields. NMEA standard format output using ASCII code, each ASCII data code is 8 bits long, the baud rate of serial communication is 9600 bits per second, the data bit is eight bits, the start bit is one bit, the stop bit is one bit, and there is no parity bit [2].

The NMEA-0183 protocol consists of statements, each of which begins with the character "$t' as the statement starting flag, separated by commas between the data, preceded by a checksum leaf, and finally ended with a checksum value and carriage return, and a newline character [3]. There are 6 most commonly used statement formats: $GPGGA (satellite positioning information), $GPGLL (geolocation information), $GPGSA (satellite PRN data), $GPGSV (visual satellite information), $GPRMC (recommended positioning information), $GPVTG (ground speed information) [4]. The protocol description is shown in Table 1.

Data for the NMEA0183 protocol is sent in sentences, with the structure of each sentence shown in Fig. 1. If a value in the data field is invalid, the value will simply be ignored but will still need to be sent. It should be noted that different statement types have different data frame lengths. The validation of the NMEA-0183 protocol, as known from the figure, * each frame is followed by a checksum of hh, which is obtained by calculating the XOR operation of all characters ASCII code between $ and *[5], and the result will be expressed in ASCII characters as the checksum.

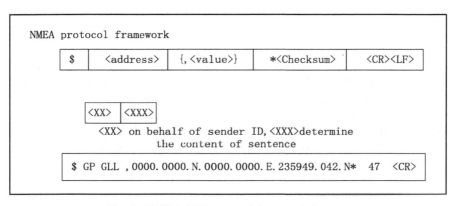

Fig. 1. NMEA-0183 protocol framework diagram

3   Introduction to the Parsing Process

3.1   Overall Design of the System

Take the ATK-S1216F8 Beidou navigation chip as an example to illustrate, the chip in turn output 6 kinds of NMEA information, namely: $GPGGA, $GPGLL, $GPGSA, $GPGSV, $GPRMC, $GPVTC, serial port default settings: baud rate 9600 bps, start bit one bit, data bit eight bit, no parity bit and stop bit one bit, NMEA-0183 protocol parsing system Overall block diagram as shown in Fig. 2, ATK-S1216F8 navigation chip receives satellite signals through the antenna unit, processes a series of processing such as signal filtering, downconversion, etc., to obtain a series of processes such as digital if the signal is filtered, downconversion, analog-to-digital conversion and etc., to obtain a digital if the signal is obtained, and then the if the ifnc signal is captured, tracked, bit synchronization and frame synchronization are performed, the navigation data is demodrated, and finally the user's latitude, longitude, height, speed, time and other information are calculated. Packaged into a data frame in NMEA-0183 protocol format, and then output data through a standard serial port, on the one hand, the serial data emitted by the chip is received through the serial port, and the received signal is processed by the parsing module, and on the other hand, the parsing module extracts the required information according to the data format of the NMEA-0183 protocol.

antenna

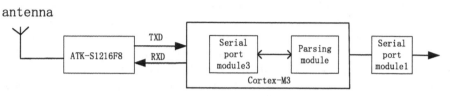

Fig. 2. Overall system block diagram

4 Serial Communication

The device uses RS-232 serial communication, the serial communication can be achieved with the help of a variety of development languages, such as serial communication can be implemented with the help of a variety of development languages, such as Visual C++, JAVA and C language, etc., due to the complexity of different development languages, so the degree of ease and stability of program implementation is not the same, through the programming to find C language, can quickly and easily achieve serial communication, but also quickly and stably achieve data checksum extraction.

In this article, the usart3 serial port will be connected to the analysis module and the Beidou positioning module, and the usart1 module serial port is connected to the computer, of usart1 which usart1 stands many parameters through the serial port, which saved in the FLASH, which is to use easily. Connect the ATK-S1216F8-BD GPS/Beidou module, and according to the serial port UST3, therefore display GPS Beidou information through the LIQUID crystal, and at the same time, through the USMART tool, set the refresh rate and clock pulse width configuration of the Beidou module. The code of the usart3 port mainly includes interrupt, initialization and display. The code for the break is:

```
void UST3_IH(void)
{
A8 resnn;
if(UST_GetITStatus(UST3, UST_IT_RXNE)!=resnET) //Data re-ceived
{
resn=UST3_ReceiveData(UST3);
```

```
if((UST3_RX_STA &(1<<15)==0 //A batch of data that has been received and has
not been processed is not receiving other data
{
if(UST3_RX_STA<UST3_MAX_RECV_LEN) // Data can also be re-ceived
{
TIM_SetCounter(TIM7,0); //The counter is cleared
if(UST3_RX_STA==0) //enables the interrupt of the timer
{
TIM_Cmd(TIM7,ENABLE); //Enable the timer
}
UST3_RX_BUF[UST3_RX_STA++]=resn; //Records the value re-ceived
}else
{
UST3_RX_STA|=1<<15; //Force the mark to complete
}
```

4.1 Acquisition and Analysis

The code can be split in two, the first part was NMEA-0183 record parsing part, the communication part between the serial port and the host computer, and the process of parsing the serial port in the gps.h function is: first read all the data in the cache, and then parse the data.

NMEA-0183 protocol parsing section, this article uses a simple number comma method to parse. This article has introduced the NMEA-0183 data format, it can be seen that the NMEA-0183 protocol is similar to the beginning of $GPGSV, and then there is a fixed output format, regardless of whether there is data output, there are commas, and will be '*' as the end of valid data, this code (writing process) to achieve NMEA-0183 protocol of $GNGGA, $GPGSA, $BDGSV, $GNRMC and $ GNVTG and other 6 types of frame parsing, the NMEA-0183 protocol encapsulated data stored in the array u8 * buf.

Take $GPGSV and $GNGSA as examples to illustrate some of the parsing codes. $GPGSV statement formats Visual satellite status output statement, Its format is usu-ally the first message, the second message, the third message, the fourth message, the fifth message, and then four, five, six, seven to start the cycle. It represents all sentences, Global Security Verification number, total number of satellites, satellite number, satellite elevation and azimuth angle, and signal-to-noise ratio. Statements begin with (1)(2)(3) and (4)(5)(6)(7) statements loop through, representing the send number, elevation angle, and azimuth signal-to-noise ratio cycle, where each GSV statement includes informa-tion from up to four satellites according to the regulations, and information from other satellites will be output in the next $GPGSV statement. $GSGSA differs from $GPGSV in that the data format of the two statements is different.

The partial code for resolving $GPGSV and $GNGSA is:
```
void N_G_A(n_m *gps,A8 *buf1)
{
A8 len,i,j,slx=0;
```

```
A8 PX;
A8 *p,*F1,dx;
p=buf1;
F1=(A8*)strstr((const char *)p,"$GPGSV");
len=F1[7]-'0';// Get the number of GPGSV
PX=NM_Comma_Pos(F1,3); //Parse the information after the third comma to get
the total number of visible satellites
    if(PX!=0XFF)GX->svnum=NM_Str2num (F1+PX,&dx); //Enters the loop from the
fourth comma
    for(i=0; i<len; i++)
    {
F1=(A8*)strstr((const char *)p,"$GPGSV");
for(j=0; j<4; j++)// Each statement contains up to four satellite information, so j<4
{
PX=NM_Comma_Pos (F1,4+j*4); // indicates a satellite number after the fourth
comma
    if(PX!=0XFF)GX->slmsg[slx].num=NM_Str2num(F1+PX,&dx); //Get the satel-
lite number else break;
    PX=NM_Comma_Pos(F1,5+j*4);// From the fifth comma indicates the satellite
elevation angle
    if(PX!=0XFF)GX->slmsg[slx].eledeg=NM_Str2num (F1+PX, &dx);// to get the
satellite elevation angle else break;
    PX=NM_Comma_Pos (F1,6+j*4); // indicates the satellite azimuth from after the
sixth comma if(PX!=0XFF)GX->slmsg[slx].azideg=NM_Str2num (F1+PX,&dx);// to
get the satellite azimuth
    else break;
    PX=NM_Comma_Pos (F1,7+j*4); // indicates the satellite signal-to-noise ratio from
after
the seventh comma if(PX!=0XFF)GX->slmsg[slx].sn=NM_Str2num(F1+PX,&dx); //
Get the satellite signal-to-noise ratio
    else break;
    slx++;
    }
p=F1+1;//Switch to the next message
    }
    }
    void NM_GNGSA_Analysis(n_m *gps,A8 *buf111)
    {
A8 *F1,dx;
A8 PX;
A8 i;
F1=(A8*)strstr((const char *)buf1,"$GNGSA");
PX=NM_Comma_Pos(F1,2);//Get the positioning type
if(PX!=0XFF)GX->fixmode=NM_Str2num(F1+PX,&dx);
for(i=0; i<12; i++) // get the number of the positioning satellite
```

```
{
PX=NM_Comma_Pos(F1,3+i);
if(PX!=0XFF)GX->possl[i]=NM_Str2num(F1+PX,&dx);
else break;
}
PX=NM_Comma_Pos(F1,15);
if(PX!=0XFF)GX->pdop=NM_Str2num(F1+PX,&dx); // The PDOP position accu-
racy factor is obtained
PX=NM_Comma_Pos(F1,16);
if(PX!=0XFF)GX->hdop=NM_Str2num(F1+PX,&dx); // Get the accuracy factor
for the HDOP position
PX=NM_Comma_Pos(F1,17);
if(PX!=0XFF)GX->vdop=NM_Str2num(F1+PX,&dx); //The accuracy factor for
the VDOP position is obtained
}
```

5   Display of GPS Data

Data display is the visual process of parsing out ASCII strings with a specific meaning into intuitive numbers or graphs. Navigation applications only focus on the NMEA-0183 protocol defined in the GGA, GLL, GSA, GSV, RMC and VTG these five statements, in order to facilitate the description, we more commonly used GGA and RMC these two statements as an example to introduce the analysis of NMEA0183 protocol. The Fig. 3 below is the $GNRMC data received by the development board in the case where the antenna receives GPS information.

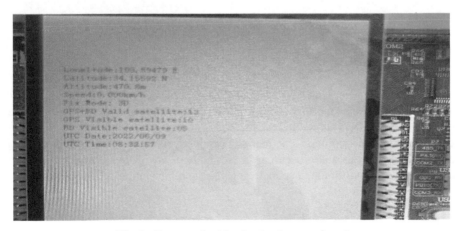

Fig. 3.   Data received by the development board

The next step is to verify the communication function of the port and the host computer, and the data received by the NMEA-0183 protocol is shown in Fig. 5. Communicate with the host computer through the serial port, under ideal circumstances, the

NMEA-0183 input of the serial port can be correctly processed in real time, and the actual time and date, positioning status, longitude and latitude height and other information required should be obtained after its analysis processing, presented in the host computer, the experimental results show that the data sending and receiving ends prepared in this paper can operate normally and effectively, and the normal screenshot of serial communication is shown in Fig. 6.

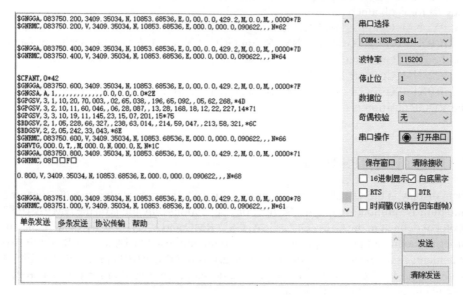

Fig. 5. NMEA-0183 protocol data

6   Conclusion

In view of the characteristics of NMEA-0183 protocol, this paper introduces embedded technology, designs the NMEA0183 protocol analysis method based on embedded technology, and the test shows that the method is stable and reliable, which can ensure the requirements of long-term continuous normal work. It also has certain practical value for data analysis and processing in the fields of navigation equipment working status monitoring, positioning accuracy analysis, and navigation product research and development.

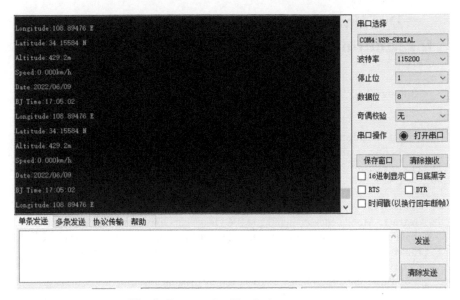

Fig. 6. Data received by the host computer

References

1. Liu, F., Guo, C., Jia, Z.: Implementation of a new NMEA-0183 protocol parsing method. GPS19–20 (2017)
2. Lei, Q., Tan, H., Le, Y.: Design of shipboard integrated display terminal based on NMEA 0183&2000 protocol. Transp. Sci. Technol. 124–128 (2020)
3. Dejun, Q., ZheHu, Z.: NMEA-0183 protocol analysis. Electron. Devices 698–701 (2007)
4. Wei, R., Zhang, K.: Analysis of NMEA 0183 data protocol format commonly used in hydrographic surveying and mapping. In: Zhejiang Water Conservancy Science and Technology (2020)

Dynamic Computing Offloading Strategy for Multi-dimensional Resources Based on MEC

Jihong Zhao[1,2], Zihao Huang[1]([✉]) ⓘ, Xinggang Luo[1], Gaojie Peng[1], and Zhaoyang Zhu[1]

[1] Xi'an University of Posts & Telecommunications, Xi'an 710121, China
{Zihao_huang,XgangL,Penggj,zhuzy}@stu.xupt.edu.cn
[2] Xi'an Jiaotong University, Xi'an 710049, China

Abstract. Mobile edge computing (MEC) is a technology that extends network capabilities from the core network to the edge network, which effectively solves the problems of devices in terms of computing capacity, service energy consumption, and service delay. In MEC networks, conventional models cannot meet the actual needs when computing-intensive services arrive dynamically. In addition, the large energy consumption of computing-intensive services makes User Equipment (UE) energy consumption a vital issue. Meanwhile, UE energy consumption is critical to the system economic benefits of the entire MEC network. Aiming at the above problems, a stochastic optimization model that can dynamically combine queue length, computing offloading, resource allocation, and energy consumption control is designed, and a Multi-Dimensional Resource Dynamic computation Offloading algorithm (MDRDO) based on Lyapunov optimization theory is proposed. The algorithm used Lyapunov optimization theory to solve the closed solution of computation offloading, bandwidth resource and computing resource allocation problem, and solved the above stochastic optimization model at the same time. Under the premise of ensuring the stability of the service queue, the algorithm improved the economic benefits of the overall MEC network system. The simulation results indicate that, compared with other algorithms, the proposed MDRDO could save the energy consumption of UE, and at the same time significantly improved the overall economic benefit of the system.

Keywords: Mobile edge computing · Lyapunov optimization · Computing offloading · System economic benefit

1 Introduction

With the continuous development of mobile communication technology, UE's requirements for Quality of Experience (QoE) are gradually increasing, and the Central Processor Unit (CPU) performance and battery performance of UE are not enough to meet the increasing demands [1]. As a conventional computation offloading method, Mobile Cloud Computing (MCC) cannot meet the requirements of QoE due to the long transmission distance between UE and remote computing center which leads to the large transmission delay between UE and remote computing center [2]. MEC refers to deploying computing and storage resources at the edge of the network to provide IT service environment and cloud computing capability for the network as an emerging technology, so as to provide users with network service solutions with ultra-low latency and high bandwidth [3]. Among them, computation offloading technology is a technology in which UE offload part or all computation tasks to Mobile Edge Computing Serves (MECS) processing, so as to solve the shortcomings of UE in resource storage, computation performance and energy consumption [4]. Aiming at the shortcomings of tasks processing in UE and MCC, MEC offloads computation tasks to MECS, which provides a brand-new solution to the problems of latency sensitivity and energy consumption sensitivity [5] (Table 1).

Table 1. The comparison of MEC and MCC

Technical characteristics	MCC	MEC
Distance to UE	Far away	Closer
Deployment method	Centralized deployment	Centralized deployment& Distributed deployment
Deployment costs	Lower	Higher
Processing delay	Higher	Lower
Computing capacity	Higher	Lower
Storage capacity	Higher	Lower

MEC is becoming a growing concern in the academia and industry, which prompts researchers to put forward different offloading methods. The literature [6] focused on media services such as video and images, proposed that compress the data streaming and partially computation offloads to the MECS, compared with traditional binary offloading method, this method could save energy consumption. However, it is too limited in the types of services and not universal. The literature [7] classified tasks as latency-sensitive and latency-insensitive that processed latency-sensitive tasks in MECS. When the data streaming increased to a threshold, the tasks would be offloaded to the remote server and queued the data streaming in the MECS for the M/M/S queuing model. Nevertheless, it only considered the processing capacity of MECS and MCCS, neither the processing capacity of UEs, nor the key factor of energy consumption performance of UEs were taken into account. The literature [8] proposed the problem of the choice between multiple operators and multiple MECS, which was about the storage resources, bandwidth resources and computation resources of the MECS. The problem of service's placement was implemented by using Lyapunov optimization theory. Compared with the conventional methods, this method was able to consider the overall benefits from multiple perspectives, but ignored the processing power of UE. The literature [9] explored the capabilities of MECS with MCCS together and proposed an algorithm for allocating computation resources in split time slots, which could reallocate computation resources by subdividing the time length so as to improve the utilization of computation resources.

Although there were many studies of offloading methods in MEC researches, the existing methods tend to consider only some of the resources, it is noteworthy that in today's mobile service explosion environment, the energy consumption of UE is becoming more and more serious, thus it is extremely important to focus on the energy consumption resources of UE. The current researches had not fully worked on the coordination of multi-dimensional resources in the overall MEC network, such as the computation resources of MECS, bandwidth resources in transmission links, and energy consumption resources of UE. Because there are blanks in these professional fields how to coordinate the various network resources so as to maximize the economic benefits of the whole system seems arduous in MEC.

The proposed method in this article considered the problem of maximizing the overall economic benefits of multiple network resources in a MEC network that allowed for computational offloading. Firstly, the service in the network is stochastic, therefore dynamic data arrival and data queue stability need to be considered [10]. Secondly, the economic benefits maximization problem is defined as a stochastic optimization problem by jointly optimizing the computation offloading decision, bandwidth resources, MECS computation resources, and local device energy consumption resources. What's more a Multi-Dimensional Resource Dynamic computation Offloading algorithm (MDRDO) for solving stochastic optimization problems is proposed using Lyapunov optimization theory [11]. Specifically, as a dynamic online algorithm, the MDRDO can be adapted to dynamically computation offloading data queue without knowing the incoming traffic in advance. In the MDRDO, the economic benefit maximization problem is firstly modeled as a Mixed Integer Nonlinear Programming (MINLP). Then, the problem is decoupled into computation offloading decision problem, bandwidth resource allocation problem and computing resource allocation problem. By solving the above sub-problems, a closed solution to the problem of maximizing the economic benefit of the system is therefore obtained.

2   MDRDO Algorithm in MEC

2.1   Question Description

This article focuses on the MEC network. As shown in the Fig. 1, a MEC network consists of a Base Station (BS), a MECS and N UEs, where MECS is deployed on BS to provide computation offloading services to UEs. The distance d between the UE and the BS affects whether or not the UEs choose to perform computation offloading. The key factor affecting UE energy consumption is the Path Loss Exponent (PLE), which depends on the surroundings. The more obstacles in the surroundings environment the higher the value of PLE, the higher the value of PLE the serious the attenuation. Typical PLE value is 2 for free space, 3–5 for shaded urban areas and 4–6 for buildings with obstructions [12].

To adapt to dynamic changes, the system divides time into time slots $t \in \{0, 1, 2, \ldots\}$, in which 1 s represents one single time slot. At the start of each time slot, the UE has computing-intensive services to process and updates its resource scheduling policy. There is a decision center in MECS. UE's computing-intensive service is decided by the decision center to be processed in UE or offloaded to MECS for executing. When

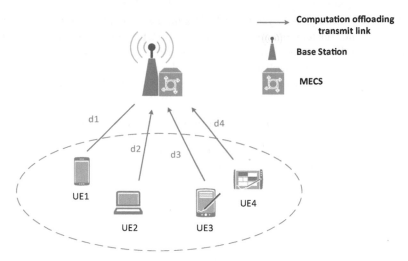

Fig. 1. Architecture of MEC computation offloading

multiple UEs in the same time slot are computation offloaded, the decision center will allocate a certain amount of computation and bandwidth resources to each UE based on the economic benefit of the whole system.

In each time slot, the UEn generates different computing-intensive services $\Lambda_n(t) = \{D_n(t), C_n(t)\}$, $D_n(t)$ represents the input data size (in bits), $C_n(t)$ indicates the number of CPU cycles to accomplish the application. $D_n(t)$ is limited by $0 \leq D_n(t) \leq D_n^{max}$ and $C_n(t)$ is limited by $0 \leq C_n t \leq C_n^{max}$. $x_n(t)$ is the offloading decision of UEn at time slot t, where $x_n(t) = 1$ represents that the application is offloaded to MECS, and $x_n(t) = 0$ represents that the application is processed by UEn locally. If the decision center decides that $\Lambda_n(t)$ is to be executed locally, the execution time is $T_n^{loc}(t) = \frac{C_n(t)}{f_n^{loc}}$, where f_n^{loc} denotes the local compute capability (in CPU cycles/s) of UEn. If the decision center decides that the service is to be executed in MECS, the execution time is divided into two parts, the time required to upload the data $T_n^{trans} = \frac{D_n(t)}{r_n(t)}$ and the time required for the MECS execution. Let the total compute capacity of MECS be F (in CPU cycle/s) and the computation resource allocated to UEn be $f_n(t)$, the computation resource allocation vector within the time slot is $f(t) = \{f_1(t), f_2(t), \ldots, f_N(t)\}$. During the execution of the computation offloading by the UE, the data $D_n(t)$ will be pushed into the data transfer queue $Q_n(t)$ and then be uploaded to the MECS for calculation via wireless transferring, the transmission rate vector is $r(t) = \{r_1(t), r_2(t), \ldots, r_N(t)\}$ and $r_n(t)$ represents the amount of data delivered to the MECS from UEn in slot t, data queues $Q_n(t)$ evolves:

$$Q_n(t+1) = \max\left[Q_n(t) - r_n(t)x_n(t), 0\right] + D_n(t)x_n(t) \tag{1}$$

In most application situations, the MECS feedback to the UE is extremely small, thus the delay required for downlink is not considered. When UEn offloads the service into MECS, the total execution delay is:

$$T_n^{mec}(t) = \frac{C_n(t)}{f_n} + \frac{D_n(t)}{r_n(t)} \tag{2}$$

The process of the whole system is: UEn generates computing-intensive services, the decision center decides whether the services of UEn will be processed locally or offloaded to MECS according to the economic benefit of the overall system, if offloaded to MECS, the decision center will allocate bandwidth resource and computation resource to MECS to ensure the maximum economic benefit of the whole system.

2.2 Multi-dimensional Resource Economic Benefit Problem Model

In MEC system when UEn offloads computing-intensive service to MECS, system rents wireless bandwidth resource to UEn to gain benefit, the price is θ per bps. The economic benefit gained by the system in renting out wireless bandwidth is:

$$\mathcal{E}_n^{\text{brand}}(t) = \theta r_n(t) \tag{3}$$

At the same time, MECS provides computing resource $f_n(t)$ for UEs that choose to offload. For each UEn it has its own local computing resource f_n^{loc}, system only charges for the addition part of computing resource between $f_n(t)$ and f_n^{loc}, λ is the price for each unit addition computation task. The economic benefit of MEC system in the process of computing resource allocation is:

$$\mathcal{E}_n^{compute}(t) = \lambda \left(\frac{f_n(t)}{C_n(t)} - \frac{f_n^{loc}}{C_n(t)} \right) \tag{4}$$

When the offloading decision $x_n(t) = 0$, service is executed by the UEn, and the energy consumed by the UEn is:

$$E_n^{loc} = \frac{C_n(t)}{f_n^{loc}} p_n^{loc} \tag{5}$$

where p_n^{loc} is the power of UEn in computing. When the offloading decision $x_n(t) = 1$, The service is offloaded to the MECS for processing, at which point the energy consumption of the UEn is generated by wireless transmission. Energy consumption is related to transmission power p_n^{trans} and transmission time T_n^{trans}, where the transmission power p_n^{trans} is given by [13]:

$$p_n^{trans} = \left(\frac{4\pi f_c}{c} \right)^2 d^\sigma \frac{P_{rx}}{G_r G_t} \tag{6}$$

where f_c is the carrier frequency, c is the wave velocity, d is the distance between UEn and MECS, σ is the PLE, P_{rx} is the received power, and G_r, G_t are the receiver and transmitter antenna gains respectively.

The energy consumed by UEn at this time slot is:

$$E_n^{trans} = p_n^{trans} T_n^{trans} = \left(\frac{4\pi f_c}{c} \right)^2 d^\sigma \frac{P_{rx}}{G_r G_t} \frac{D_n(t)}{r_n(t)} \tag{7}$$

The price of energy is K per J in the MEC system. In terms of energy resources, the expenditure of the system due to energy consumption is:

$$\mathcal{E}_n^{energy}(t) = K\left\{-[1-x_n(t)]\frac{C_n(t)}{f_n^{loc}}p^{loc} - x_n(t)\left[p_n^{trans}\frac{D_n(t)}{r_n(t)}\right]\right\}$$

$$= K\left\{-\frac{C_n(t)}{f_n^{loc}}p^{loc} + x_n(t)\left[p^{loc}\frac{C_n(t)}{f_n^{loc}} - p_n^{trans}\frac{D_n(t)}{r_n(t)}\right]\right\}$$

$$= K\left\{-\frac{C_n(t)}{f_n^{loc}}p^{loc} + x_n(t)\left[\frac{C_n(t)}{f_n^{loc}}p^{loc} - \left(\frac{4\pi f_c}{c}\right)^2 d^\sigma \frac{P_{rx}}{G_r G_t}\frac{D_n(t)}{r_n(t)}\right]\right\} \quad (8)$$

For UEn, maximizing economic benefits is also minimizing energy consumption expenditure that $\min \mathcal{E}_n^{energy}$.

2.3 Mathematical Model for Maximizing Economic Benefit

The optimization goal is to maximize the economic benefit of the overall system while maintaining the stability of the queue, and according to the above process it is known that maximizing the economic benefit is:

$$\max U(t) = \sum_{n\in\mathcal{N}}\left[\mathcal{E}_n^{brand}(t) + \mathcal{E}_n^{compute}(t) - \mathcal{E}_n^{energy}(t)\right]$$

$$= \sum_{n\in\mathcal{N}}\left\{\left[\theta r_n(t) + \lambda\left(\frac{f_n(t)}{C_n(t)} - \frac{f_n^{loc}}{C_n(t)}\right)\right] - K\left[p_n^{loc}\frac{C_n(t)}{f_n^{loc}} - p_n^{trans}\frac{D_n(t)}{r_n(t)}\right]\right\}x(n)$$

$$+ K\frac{C_n(t)}{f_n^{loc}}p_n^{loc} \quad (9)$$

This article develops a mathematical model for stochastic optimization problems on offloading decisions bandwidth allocation, and computing resource allocation, with the overall average economic benefit maximization as the optimization objective function:

$$\max \overline{U} = \lim_{t\to\infty}\frac{1}{t}\sum_{\tau=0}^{t-1}\mathbb{E}\{U(\tau)\} \quad (10)$$

$$s.t.\begin{cases} \lim_{t\to\infty}\frac{1}{t}\sum_{\tau=0}^{t-1}\mathbb{E}\{Q_n(\tau)\} < \infty \quad \forall n \in \mathcal{N} \\ x(n) \in \{0,1\} \quad \forall n \in \mathcal{N} \\ r_n(t) \geq 0 \quad \forall n \in \mathcal{N} \\ \sum_{n\in\mathcal{N}} f_n(t)x_n(t) \leq F \\ \left(f_n(t) - f_n^{loc}\right)x_n(t) \geq 0 \quad \forall n \in \mathcal{N} \\ 2 \leq \sigma \leq 8 \end{cases}$$

2.4 Transformation of Economic Benefit Maximization Problem

The above optimization equation is a MINLP problem is also NP-hard. With the number of network users and user data increasing, the complexity of the problem rises exponentially and cannot be applied in practice. In this article, a stochastic optimization method based on Lyapunov optimization theory is proposed to solve the optimization equation.

Introduction to Lyapunov Optimization Theory

The Lyapunov optimization method was first applied in cybernetics as a method for solving the stability of dynamic systems [14]. The literature [15] presented a connection between Lyapunov optimization method and communication network theory to model queues in dynamic systems, solved the original objective function and considered the queue length and making the original system reach a steady state as well. In past researches on MEC and computation offloading, researchers have used algorithms such as convex optimization, greedy strategies [16, 17], stochastic gradient algorithms [18, 19]. However, none of these methods can be determined based on the current time slot state. In contrast, the Lyapunov optimization method is able to automatically adapt to the network state in each time slot system change. Unlike other algorithms, it does not require a priori parameter input while can effectively control dynamically changing systems in real time, also ensuring relatively low algorithmic complexity. Using the Lyapunov optimization method, complex long-term optimization problems can be transformed into a series of real-time optimization problems [14].

Construction of a Lyapunov Optimization Model

Let $\mathcal{Q}(t)$ be the queue vector and defines the Lyapunov equation as follows:

$$L(\mathcal{Q}(t)) \triangleq \sum_{n \in \mathcal{N}} \frac{1}{2} Q_n^2(t) \tag{11}$$

The smaller the $L(\mathcal{Q}(t))$ is, the lower the queue density lies. Defines the Lyapunov drift function for a time slot as:

$$\Delta(\mathcal{Q}(t)) \triangleq \mathbb{E}\{L(\mathcal{Q}(t+1)) - L(\mathcal{Q}(t)) | \mathcal{Q}(t)\} \tag{12}$$

where $L(\mathcal{Q}(t+1)) - L(\mathcal{Q}(t)) \leq B + \sum_{n \in \mathcal{N}} Q_n(t)(D_n(t) - r_n(t))x_n(t)$, and the constant B is given by:

$$B = \frac{1}{2} \sum_{n \in \mathcal{N}} \left[(D_n^{\max})^2 + (r_n^{\max})^2 \right] \tag{13}$$

In the Lyapunov optimization method, it is necessary to maximize the Lyapunov drift- plus-penalty function to find the maximum value of the optimization equation, the drift-plus-penalty function is given by:

$$\Delta(\mathcal{Q}(t)) - V\mathbb{E}\{U(t) | \mathcal{Q}(t)\} \tag{14}$$

where V is a non-negative control constant that controls the length of the queue and the overall economic benefit.

$$\Delta(\mathcal{Q}(t)) - V\mathbb{E}\{U(t)|\mathcal{Q}(t)\}$$

$$\leq B + \sum_{n \in \mathcal{N}} \mathbb{E}\{Q_n(t)(D_n(t) - r_n(t))x_n(t)|\mathcal{Q}(t)\} - V\mathbb{E}\{U(t)|\mathcal{Q}(t)\}$$

$$= B + \sum_{n \in \mathcal{N}} \mathbb{E}\ominus x(n)\left\{Q_n(t)[D_n(t) - r_n(t)] - \frac{\lambda V}{C_n(t)}\left(f_n(t) - f_n^{loc}\right)\right.$$

$$\left. - V\theta r_n(t) + KV\left[p_n^{loc}\frac{C_n(t)}{f_n^{loc}} - p_n^{trans}\frac{D_n(t)}{r_n(t)}\right]|\mathcal{Q}(t)\right\} \quad (15)$$

According to the Lyapunov optimization method, the desired optimization can be obtained by minimizing the maximum value of the drift-plus-penalty function. Therefore, the original optimization function is transformed as follows:

$$\min \sum x_n(t)\left\{Q_n(t)(D_n(t) - r_n(t)) - \frac{\lambda V}{C_n(t)}\left(f_n(t) - f_n^{loc}\right)\right.$$

$$\left. -V\theta r_n(t) + KV\left[p_n^{loc}\frac{C_n(t)}{f_n^{loc}} - p_n^{trans}\frac{D_n(t)}{r_n(t)}\right]\right\} \quad (16)$$

3 Problem Solving

The function of Eq. (16) is solved for the offloading decision, bandwidth resource allocation, and computing resource allocation problems.

3.1 Offloading Decision

According to Eq. (16), to simplify the representation notation, define the function as:

$$W_n(t) = Q_n(t)(D_n(t) - r_n(t)) - \frac{\lambda V}{C_n(t)}\left(f_n(t) - f_n^{loc}\right)$$

$$- V\theta r_n(t) + KV\left[p_n^{loc}\frac{C_n(t)}{f_n^{loc}} + p_n^{trans}\frac{D_n(t)}{r_n(t)}\right] \quad (17)$$

When the variable $r_n(t), f_n(t)$ is determined, the offloading decision making subproblem of (16) is transformed into the following binary integer programming problem.

$$\min \sum_{n \in \mathcal{N}} x_n(t)W_n(t) \quad (18)$$

$$s.t.\, x(n) \in \{0, 1\}, \forall n \in \mathcal{N}$$

Each UEn offloads service independently, while decoupling this centralized minimum-sum problem into an offloading decision problem for a single UEn. The offloading decision is:

$$x_n(t) = \begin{cases} 1 \text{ if } W_n < 0 \\ 0 \text{ if } W_n \geq 0 \end{cases}$$

Let the set of UEn offloaded into the MECS be $\mathcal{N}_1$ i.e. $x_n(t) = 1$. For UEn $\in \mathcal{N}_1$, the maximum value problem of optimization equation can be reduced by discarding the constant term as:

$$\max_{n \in \mathcal{N}_1} \sum [Q_n(t) + \theta V] r_n(t) + [K V p_n^{trans} D_n(t)] \frac{1}{r_n(t)} + \frac{\lambda V}{C_n(t)} f_n(t) \qquad (19)$$

decoupling the above problem for bandwidth resource allocation and computing resource allocation.

3.2 Bandwidth Resource Allocation

The bandwidth resource allocation subproblem of (20) is given as:

$$\max_{r_n(t)} \sum_{n \in \mathcal{N}_1} [Q_n(t) + \theta V] r_n(t) + [K V p_n^{trans} D_n(t)] \frac{1}{r_n(t)} \qquad (20)$$

$$s.t. \begin{cases} Q_n(t) + \theta V \geq 0 \\ K V p_n^{trans} D_n(t) \geq 0 \\ 0 \leq r_n(t) \leq r_n^{max} \end{cases}$$

For UEn, the computation offloading services are performed independently, so for each UEn, the bandwidth resources are allocated as:

$$r_n(t) = \begin{cases} r_n^{max} \text{ if } r_n^{max} \geq \sqrt{\frac{K V p_n^{trans} D_n(t)}{Q_n(t) + \theta V}} \\ 0 \qquad \text{otherwise} \end{cases}$$

3.3 Computation Resource Allocation

The subproblem of computation resource allocation is given as:

$$\max_{f_n(t)} \sum_{n \in \mathcal{N}_1} \frac{\lambda V}{C_n(t)} f_n(t) \qquad (21)$$

From which the computation resource allocation could be obtained as (Table 2):

$$f_n(t) = \begin{cases} \left(F - \sum f_n^{loc} \right), \text{ if } n = \text{argmax} \left(\frac{\lambda V}{C_n(t)} \right) \\ f_n^{loc}, \qquad\qquad \text{otherwise} \end{cases} \qquad (22)$$

Table 2. Multi-dimensional dynamic resource unloading algorithm

Input: The queue length $Q_n(t)$
Output: Computation offloading decisions $x_n(t)$, bandwidth resource allocation $r_n(t)$,computation resource allocation $f_n(t)$
1. Get the queue length $Q_n(t)$ in time slot t
2. The computing offloading decision $x_n(t)$ is obtained by solving binary programming problem based on (18)
3. The bandwidth resource allocation $r_n(t)$ is obtained by solving (20)
4. According to (21), the computation resource allocation $f_n(t)$ is obtained
5. Collect computing offloading decisions $x_n(t)$, bandwidth resource allocation decisions $r_n(t)$, computing resource allocation $f_n(t)$ decisions in time slot t
6. $Q_n(t+1)$ is updated according to (1)

4 Simulation Results

In this section, the simulation results are given to evaluate the performance of the proposed algorithm. The following parameters are set to default values unless otherwise stated: there is 1 MECS in the system and the number of UEn N = 20. $D_n(t) = 0.8 \times 10^5 \sim 4 \times 10^5$ bit, $C_n(t) = D_n(t) * 50$ CPU cycles, $f_n^{loc} = 10 \sim 50$ M CPU cycles/s. F = 2 G CPU cycles/s, total radio link bandwidth is 1 Mbps, $\lambda = 2 \times 10^{-2}$/(unit extra computation task), d = 50 ~250 m, $P_{rx} = -33$dBm ~33dBm, $f_c = 2.4$Ghz.

DJORC is an algorithm that considers charging for bandwidth and computation resource in MEC [20]. The Stackelberg-Uniform algorithm can be applied to model the interaction between the edge cloud and users [21]. The no-offloading algorithm means that the UE executes all computing-intensive services locally and does not offload to the MECS. The MDRDO is compared with the above three algorithms.

Figure 2a indicates the economic benefit of the MDRDO growing cumulatively over time, while the DJORC algorithm, which does not take energy resource into account, resulting in lower economic benefit than the MDRDO in the same environment. The economic profit of Stackelberg-Uniform algorithm increases over time, but the growth rate is lower than that of MDRDO and DJORC algorithms. Meanwhile, all the services of no-offloading algorithm are executed locally in UE. As time goes by, the gradual increase of energy consumption leads to a gradual increase in local expenditure and a gradual decline in economic benefit. Figure 2b shows the difference between the economic benefit of proposed algorithm and the three comparison algorithms as the maximum user transmission rate changes. In the no-offloading algorithm, the economic benefit does not change as the maximum transmission rate changes. The economic benefit of proposed algorithm, the DJORC algorithm and the Stackelberg-Uniform algorithm grow as the maximum transmission rate increases, but the DJORC and the Stackelberg-Uniform algorithm have lower growth rate than the proposed algorithm because they do not take energy benefit into account.

Figure 3 shows the average energy consumption of the UEn using the MDRDO or the DJORC or Stackelberg-Uniform or no-offloading algorithms. The energy consumption

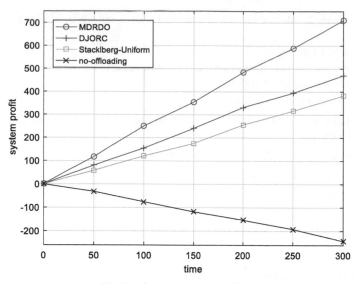

(a) Total economic benefit over time

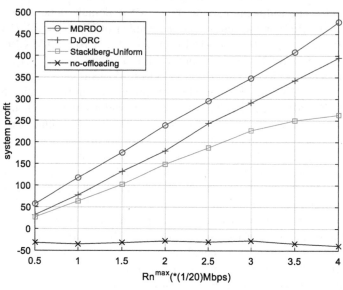

(b) Total economic benefit vs. r_n^{max}

Fig. 2. Total economic benefit

of all four algorithms gradually increases over time, with the trend towards increased energy being most pronounced in the no-offloading algorithm due to the more pronounced consumption of energy resources by local computing. The MDRDO has been optimized for energy consumption as controlling the energy consumption of the equipment more effectively when compared to DJORC and Stackelberg-Uniform algorithm. For UE, where energy storage is extremely limited, the choice of using the MDRDO shows great advantages in terms of energy savings.

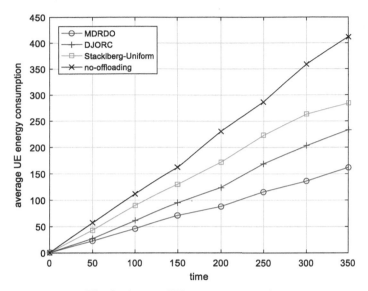

Fig. 3. Average UE energy consumption

As shown in Eq. (14), V in the function is a non-negative control constant to control the length of the queue and the overall economic benefit. Figure 4 indicates that as V grows, the queue length grows and the delay required by the user to complete the computation offloading increases progressively. Figure 5 indicates that the overall economic benefit of the system grows as V increases, with a faster upward trend when V is small and a slower upward trend after V is greater than 2×10^{10}. To balance the issues of queue length and economic benefit, this article defaults to $V = 3 \times 10^{10}$.

5 Conclusion

In this article, MDRDO based on Lyapunov stochastic optimization is proposed for the computation offloading and resource allocation problem in MEC network systems. Using the objective of maximizing the economic benefit of the overall system while ensuring the stability of the overall queue. The original problem is decomposed into computation offloading, bandwidth resource allocation, and computation resource allocation subproblems. By Lyapunov stochastic optimization, the above subproblems are solved to maximize the economic benefits of the system. Simulation results show that the algorithm has the ability to maximize the economic benefit of the system while ensuring the stability of the queue, as well as providing a good improvement in the energy consumption of the UE.

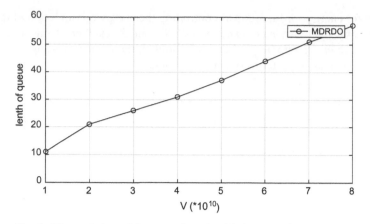

Fig. 4. The variation of the queue length with the control parameter V.

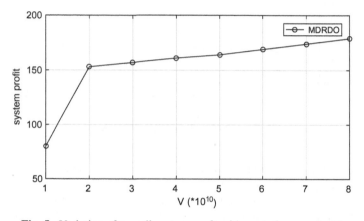

Fig. 5. Variation of overall system profit with control parameter V.

As the design process only considers in single-MECS multi-UE environment. The multi-MECS multi-UE environment in MEC is not taken into account, which means the algorithm cannot control the computation offloading policy and multi-dimensional resource allocation in large-scale MEC environment. Further work will focus on constructing multi-MECS multi-UE stochastic optimization problems to improve the adaptability of controlling strategies for large-scale edge network operation.

References

1. Li, Y., Xia, S., Zheng, M., Cao, B., et al.: Lyapunov optimization-based trade-off policy for mobile cloud offloading in heterogeneous wireless networks. IEEE Trans. Cloud Comput. **10**(1), 491–505 (2022)
2. Mahmoodi, S.E., Uma, R.N., Subbalakshmi, K.P.: Optimal joint scheduling and cloud offloading for mobile applications. IEEE Trans. Cloud Comput. **7**(2), 301–313 (2019)

3. Xie, R., Lian, X., Jia, Q., Huang, T., et al.: Survey on computation offloading in mobile edge computing. J. Commun. **39**(11), 138–155 (2018)
4. Yang, X., Luo, H., Sun, Y., et al.: Coalitional game-based cooperative computation offloading in MEC for reusable tasks. IEEE Internet Things J. **8**(16), 12968–12982 (2021)
5. Yang, X., Luo, H., Sun, Y., et al.: Energy-efficient collaborative offloading for multi-player games with cache-aided MEC. In: ICC2020—2020 IEEE International Conference on Communications (ICC), pp. 1–7, Dublin, Ireland (2020)
6. Chouhan.: Energy optimal partial computation offloading framework for mobile devices in multi-access edge computing. In: 2019 International Conference on Software, Telecommunications and Computer Networks (SoftCOM), pp. 1–6, Split, Croatia (2019)
7. Zhao, T., Zhou, S., Guo, X., et al.: A cooperative scheduling scheme of local cloud and internet cloud for delay-aware mobile cloud computing. In: 2015 IEEE Globecom Workshops (GC Wkshps), pp. 1–6, San Diego, USA (2015)
8. Lei, Z., Xu, H., Huang, L., et al.: Joint service placement and request scheduling for multi-SP mobile edge computing network. In: 2020 IEEE 26th International Conference on Parallel and Distributed Systems (ICPADS), pp. 27–34, Hong Kong, China (2020)
9. Li, Z., Zhang, X.: Resource allocation and offloading decision of edge computing for reducing core network congestion. Comput. Sci. **48**(3), 281–288 (2021)
10. Zhai, D., Sheng, M., Wang, X., et al.: Leakage-aware dynamic resource allocation in hybrid energy powered cellular networks. IEEE T. Commun. **63**(11), 4591–4603 (2015)
11. Gao, J., Chang, R., Yan, Z., et al.: A task offloading algorithm for cloud-edge collaborative system based on Lyapunov optimization. Cluster Comput. (2022)
12. Rappaport, T.S.: Wireless Communications: Principles and Practice, 2nd edn. Prentice Hall, Upper Saddle River, NJ (2002)
13. Proakis, J.G.: Digital Communications, 4th edn. McGrawHill, New York (2001)
14. Tassiulas, L., Ephremides, A.: Stability properties of constrained queueing systems and scheduling policies for maximum throughput in multihop radio networks. IEEE Trans. Autom. Control **37**(12), 1936–1948 (1992)
15. Neely, M.J.: Stochastic network optimization with application to communication and queueing systems. Synth. Lect. Commun. Netw. **1**(1), 1–211 (2010)
16. Zeng, M., Fodor, V.: Energy minimization for delay constrained mobile edge computing with orthogonal and non-orthogonal multiple access. Ad Hoc Netw. **98**, 102060 (2020)
17. Wei, F., Chen, S., Zou, W.: A greedy algorithm for task offloading in mobile edge computing system. China Commun. **15**(11), 149–157 (2018)
18. Jing, Z., Yang, Q., Qin, M., et al.: Momentum-based online cost minimization for task offloading in NOMA-aided MEC networks, In: 2020 IEEE 92nd Vehicular Technology Conference (VTC2020-Fall), pp. 1–6, Victoria, Canada (2020)
19. Zhou, J., Zhang, X., Wang, W.: Joint resource allocation and user association for heterogeneous services in multi-access edge computing networks. IEEE Access **2019**, 12272–12282 (2019)
20. Du, J., Zhao, L., Feng, J., et al.: Economical revenue maximization in cache enhanced mobile edge computing. In: 2018 IEEE International Conference on Communications (ICC), pp. 1–6, Kansas City, USA (2018)
21. Liu, M., Liu, Y.: Price-based distributed offloading for mobile-edge computing with computation capacity constraints. IEEE Wirel. Commun. Lett. **7**(3), 420–423 (2018)

A Novel Ultra Wideband Circularly Polarized Antenna for Millimeter Wave Communication

Zhong Yu, Zhenghui Xin$^{(\boxtimes)}$, Yanli Cui, Li Shi, and Dixuan Liu

Xi'an University of Posts and Telecommunications, Xi'an 710121, China
yz327123@163.com, {1002928675,cc,18709188176,
ldx}@stu.xupt.edu.cn

Abstract. An ultra-wideband circularly polarized microstrip antenna suitable for millimeter-wave communication is proposed. To meet the demand of channel capacity in millimeter-wave communication system, the antenna uses nested circular oscillators with defective ground design to achieve the ultra-wideband characteristics of the antenna, and the dual-frequency circular polarization characteristics of the antenna are achieved by opening X-shaped slots for the internal radiation unit. The antenna has a small size of $4.6 \times 3.5 \times 0.254$ mm^3. The results show that the -10dB bandwidth of the antenna is 53% (26.7GHz–43.8GHz), which can cover N258, N259, N260 and N261, and has circular polarization characteristics in the 28 and 38 Ghz bands. The measured peak gain was 5.7dBi at 28 GHz and 7.1dBi at 38 GHz. Therefore, the antenna proposed in this research has good applications value in millimeter wave communication systems.

Keywords: Ultra wideband (UWB) · Circular polarization (CP) · Millimeter wave

1 Introduction

The widespread use of millimeter wave has put higher requirements on ultra-high rate wireless communication technology, so the antennas in communication systems are gradually evolving towards broadband, high gain and miniaturization [1]. Circular polarization (CP) shows many important advantages in radio wave propagation and has been widely used in such as satellite, RFID, radar and mobile communication systems [2].

Circularly polarized (CP) antennas radiate circularly polarized waves with strong stability to Faraday rotation, which helps to reduce multipath fading and achieve high reliability and data rates [3]. In addition, CP transceiver systems do not suffer from polarization mismatch between the transmitter antenna and receiver antenna and have large spectral efficiency [4]. Currently, there is a high demand for millimeter wave circularly polarized antennas, many researchers have done a large amount of investigations on Circularly polarized millimeter wave antenna. In [5], A slotted circular patch antenna based on SIW with a thin substrate thickness of 0.254 mm is proposed. However, the bandwidth of circular patch is less than 8%, which greatly restricts its broadband application scenarios. A circularly polarized millimeter-wave antenna based on folded C-type substrate integrated waveguide is proposed in [6]. The amplitude difference and phase difference are adjusted through a differential feed network to make the antenna work in the 28 Ghz band with an axial bandwidth of 2 Ghz. However, this kind of antenna is difficult to integrate directly with the millimeter-wave front circuit.Some wideband millimeter-wave antennas [7] with low profile have good circular polarization performance [8], but the single-band antenna cannot meet the current requirements due to its single application scenario.Therefore, the wideband millimeter wave circularly polarized antenna with low throw surface still needs further optimization and improvement.

In this paper, a single-fed, dual resonant point ultra-wideband, CP antenna is designed. The antenna is fed by nested slotted circular patches using microstrip line pairs and covers the frequency band 27–44.5 GHz with circular polarization characteristics in the 28 GHz and 38 GHz bands. The antenna is structurally simple, easy to produce, and has a stable radiation pattern in both bands.

2   Antenna Design

2.1   Antenna Model

Figure 1 shows the antenna structure layout. The antenna composed of two circular slotted patches, a smaller circle nested within another circle, connected by two rectangular microstrip lines. The inner circular patch has an X-shaped slot that increases the axial ratio bandwidth of the antenna.

An X-shaped slot that increases the axial ratio bandwidth of the antenna is underneath the internal circular patch. A recessed circle was dug out of the center of the ground. The antenna is designed on an RT Duroid 5880 substrate with dimensions of $4.6 \times 3.5 \times 0.254$ mm^3 (relative permittivity of 2.2 and loss tangent of 0.002).

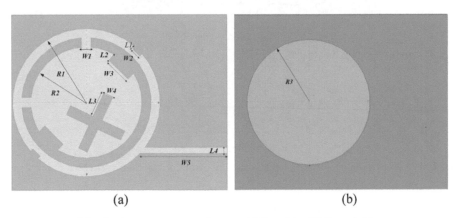

(a)   (b)

Fig. 1. Antenna structure layout: (a) Top side; (b) Dorsal side.

The antenna parameters are optimized and analyzed by simulation software ANSYS HFSS15.0. The dimensions of the proposed antenna is shown in Table 1.

Table 1. The dimensions of the proposed antenna.

Parameter	Value (mm)	Parameter	Value (mm)	Parameter	Value (mm)
L_1	0.1	W_1	0.2	R_1	1.5
L_2	0.2	W_2	0.2	R_2	1.1
L_3	0.55	W_3	0.5	R_3	1.25
L_4	0.12	W_4	0.2	W_5	1.8

2.2 Antenna Analysis

The design of UWB CP millimeter wave antenna structure evolution is shown in Fig. 2.

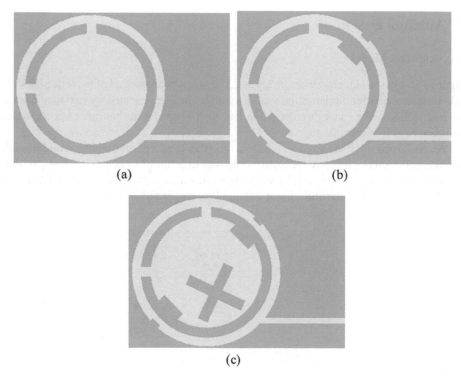

Fig. 2. UWB CP millimeter wave antenna structure evolution: (a) Step 1; (b) Step 2; (c) Step 3.

As shown in Fig. 2a, two rectangular patches are used to make the inner and outer circular radiating units connected to produce two resonant frequencies. Then, the rectangular slots with different lengths and widths are subtracted for the two circulars separately as shown in Fig. 2b. Figure 2c shows the X-shaped slot for the internal circular patch.

Figure 3 shows the results of the antenna design steps. The antenna of Step-1 has linear polarization LP characteristics in both bands and the performance of S11 is poor only −20 dB for almost the whole bandwidth range. The rectangular slotting by the circular patch creates a capacitive component which improves the matching performance of the antenna and gives the antenna a good matching The lower ARBW of step 2 was improved by cutting an X-shaped slot in the internal ring and by cutting a circular slot in the ground at the back of the antenna to improve the S11 performance and ARBW of step 3.

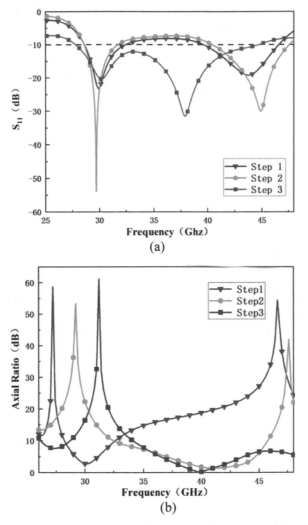

Fig. 3. Results of the antenna design steps: (a) S_{11}; (b) Axial ratio.

From Fig. 3(a), it can be seen that the resonant paths of Step-1 antenna are respective corresponding circular rings, so the resonant points of the Step 1 antenna are 30 GHz and 43 GHz. In order to make the antenna have better impedance matching performance, rectangular radiator (L1, W2) slotting is added on both sides of the external circular radiator. Meanwhile, the wide slot of the internal radiation patch breaks the symmetry of the high frequency current and results in a circular polarization effect. But the range of ARBW is not wide enough. In order to realize circular polarization characteristics at high frequencies and expand the axial ratio bandwidth, an X-shaped slot is opened for the non-center of the internal circular radiator. The high frequency CP character of the antenna is achieved due to the current flowing from the outer loop through the x-slot from the inner loop boundary, creating a 90° phase difference. The circular pattern is adjusted and cut on the ground to reduce the absorption of the antenna by the ground and widen the antenna bandwidth.

3   Parameter Study and Surface Current Distributions

Figure 4 shows the influence of parameters on antenna performance. In order to get the optimal antenna size, the parameters of the slotted band (L2, W3) of the inner ring of the antenna are compared and analyzed, as shown in Fig. 4a.It can be seen that (L2, W3) is the key factor that influence the antenna impedance matching. The parameters of the internal X-shaped slot (L3, W4) of the antenna are compared and analyzed, and the changes of L3 and W4 affect the internal heat sink surface current path and amplitude. The X-shaped slot makes the internal current path grow and reduces the resonant frequency point at high frequency. Figure 4b shows the influence of parameters on antenna performance, and the final impedance bandwidth of the antenna is 48.5% (26.7–43.8GHz).

The current flows into the internal circle through the X-slot to the internal circle boundary producing a 90° phase difference by improving the internal slot size. Therefore, the high-frequency ARBW becomes wider. Figure 5 shows the ARBW and gain of antenna. The antenna has an ARBW of 1.22 GHz (27.54–28.76 GHz) and 3.4 GHz (37.51–40.91 GHz) for AR < 3 dB. In addition, the antenna shows high efficiency in the operating band over 5 dBi.

The working mechanism of the antenna can be explained by analyzing the antenna current distribution. Figure 6 shows the simulated surface current distributions when the antenna works in 28 GHz and 38 GHz bands, respectively.

From Fig. 6a and b, it can be seen that when the antenna works at 28GHz, the current on the outer circular surface of the double circular slot is stronger than that on the inner side, which is mainly involved in the radiation of low frequency band by the outer layer. By observing Fig. 6c and d can be found that the surface current energy near the cross slot on the inner patch increases significantly, at this time mainly by the inner circular surface current involved in the radiation of the high frequency band.

The 2-D radiation patterns of the CP antenna in the xz-plane and yz-plane at 28 GHz and 38 GHz are shown in Fig. 7. It can be seen by Fig. 7a and b that when the antenna operates at 28 GHz is, the main polarization of the antenna in is 60° RHCP in +z direction. Figure 7c and d show that when the antenna is operating at 28 GHz the main polarization of the antenna is LHCP at 60° in the +z direction.

4   Conclusion

In this paper, a single-fed, ultra-wideband, dual-band circularly polarized microstrip antenna for future millimeter-wave communications is proposed. The antenna achieves

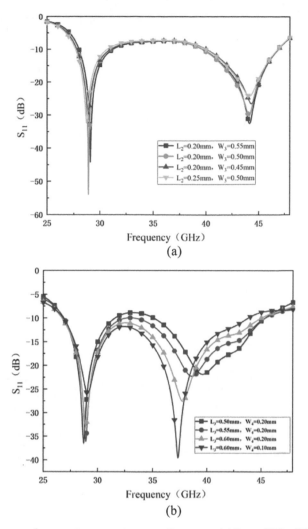

Fig. 4. Influence of parameters on antenna performance: (a) L_2 and W_3; (b) L_3 and W_4

dual-band radiation by using nested slotted circular patches, and changes the current path to achieve ultra-wideband characteristics as well as dual-band CP characteristics by opening an X-shaped slot in the center patch. The final antenna covers 26.7GHz–43.8GHz band, achieving 53% absolute bandwidth with circular polarization characteristics in 28 GHz and 38 GHz bands. The antenna is simple in structure, easy to fabricate, and has stable radiation patterns in both bands. Therefore, the antenna has good application value in millimeter wave communication environment.

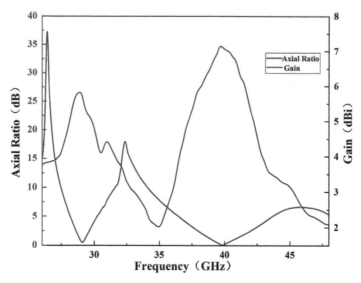

Fig. 5. Axial ratio and gain of antenna

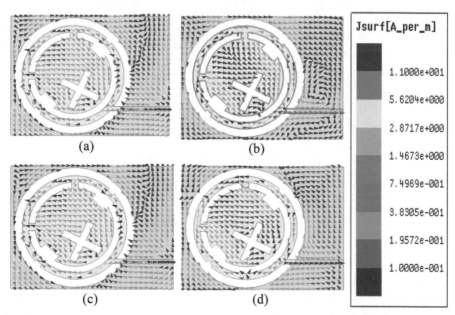

Fig. 6. Simulated surface current distributions: (a) 27 GHz, J = 0°; (b) 27 GHz J = 90°; (c) 38 GHz J = 0°; (d) 38 GHz J = 90°

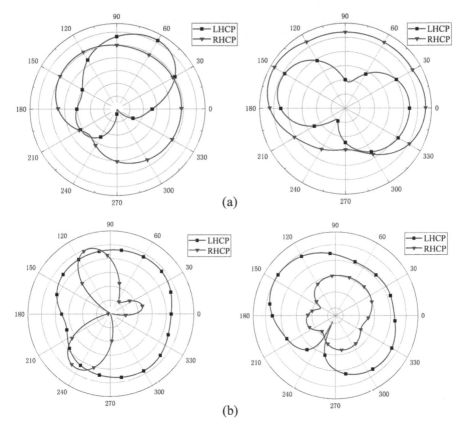

Fig. 7. 2-D radiation patterns of the CP antenna. (a) 28 GHz, (b) 38 GHz

References

1. Phuong, K.-T., Huy Hung, T., Tuan, T.: Circularly polarized MIMO antenna utilizing parasitic elements for simultaneous improvements in isolation, bandwidth and gain. Aeu-Int. J. Electron. Commun. **135**, 21–28 (2021)
2. Liu, X., Zhang, W., Hao, D., Liu, Y.: A 3743 GHz endfire antenna element based on ball grid array packaging for 5g wireless systems. Prog. Electromagnet. Res. Lett. **99**, 135–142 (2021)
3. Fakharian, M.M.: A massive MIMO frame antenna with frequency agility and polarization diversity for LTE and 5G applications. Int. J. RF Microw. Comput. Aided Eng. **31**(10), 152–162 (2021)
4. Biswas, A.K., Chakraborty, U.: A compact single element dielectric resonator MIMO antenna with low mutual coupling. Frequenz **75**(5–6), 201–209 (2021)
5. Xu, J., Hong, W., Jiang, Z.H., Zhang, H.: Low-profile circular patch array fed by slotted substrate integrated waveguide. IEEE Trans. Antennas Propag. **67**(2), 960–970 (2019)
6. Sun, Q., Ban, Y.L., Liu, Y., Yan, F.Q.: Millimeter-wave switchable Quadri-polarization array antenna based on folded C-Type SIW. IEEE Antennas Wirel. Propag. Lett. **20**(6), 1088–1092 (2021)
7. Shu, C., Yao, Y., Alfadhl, Y., Chen, X.: Wideband dual-circular-polarization antenna based on the grooved-wall horn antenna for millimeter-wave wireless communications. In: 2021 14th

UK-Europe-China Workshop on Millimetre-Waves and Terahertz Technologies, pp. 1750–1763. IEEE Electrical Electronics Engineers Inc., Hoes Lane, Anepiscataway (2021)

8. Ying, W.N., Xu, T., Lu, Y., He, G.Q., Yang, X.X.: A broadband linear to circular polarization converter for 5G millimeter wave communication. In: 2021 International Conference on Microwave and Millimeter Wave Technology (ICMMT), pp. 1–3. IEEE Electrical Electronics Engineers Inc., Nanjing, China (2021)

Calculation of Phase Difference and Amplitude Ratio Based on ADRV9009

Jin Wu$^{(\boxtimes)}$, Heng Wen, Xiangyang Shi, and Ling Yang

School of Electronic Engineering, Xi'an University of Posts and Telecommunications, Xi'an 710121, China

lifewujin@xupt.edu.com, {w18710336696,SXY, yangling}@stu.xupt.edu.cn

Abstract. With the rapid development of digital processing and integrated circuit technology, data acquisition and comparison have been applied to many disciplines today, especially in the fields of communication and radar. In some applications where the phase difference and amplitude ratio need to be clearly known, it is very important to be able to obtain the phase difference and amplitude ratio within the error range. In this paper, ADRV9009 and ZYNQ are used to calculate the phase difference and amplitude ratio of two signals of the same frequency. The phase difference can be controlled within 4 degrees, and the amplitude ratio can be controlled within 0.9–1.1.

Keywords: ADRV9009 · JESD204B · Phase difference · Amplitude ratio

1 Introduction

Both the phase and the amplitude of the signal are very important parameters of the signal. The phase difference represents the difference between two signals of the same frequency, and the amplitude represents the magnitude of the signal. Signal phase difference measurement and amplitude measurement are widely used in navigation, ranging, mobile communication, antenna, radar and other fields, and both have extremely important applications and significance. The measurement of signal phase difference and amplitude is also one of the focuses of microwave communication and radio frequency communication. With the development of science and technology, the measurement of phase difference and the measurement of amplitude will have a broader development prospect when combined with other technologies [1, 2].

At present, the most commonly used methods in the measurement of phase difference include correlation method, spectral estimation method, spectral analysis method, Hilbert transform method and so on [3]. In addition, the measurement of signal phase is also divided into two aspects: frequency domain and time domain. At present, the most important measurement method is to use a vector network analyzer. The advantage of this method is high accuracy, but it also has its shortcomings. Because the SMA port needs to be used when measuring the signal phase difference with the vector network analyzer, and in some occasions, this port cannot be provided, so this method is not applicable in

© The Author(s), under exclusive license to Springer Nature Switzerland AG 2023
N. Xiong et al. (Eds.): ICNC-FSKD 2022, LNDECT 153, pp. 1313–1319, 2023.
https://doi.org/10.1007/978-3-031-20738-9_142

some application scenarios [4, 5]. In addition, since most high-frequency vector network analyzers are relatively expensive, the cost of measuring with this method is relatively high. The system designed in this paper uses ADRV9009 and ZYNQ, which is not only simple in design and low in cost, but also small in size, easy to carry, and simple to operate. It can be applied to various occasions and meets the accuracy requirements of most scenarios.

2   System Framework Design

The traditional method of calculating the phase difference and amplitude of the two-channel signal is to use multiple ADC chips to set different frequency points to receive analog signals of different frequencies. After the analog signal is sampled by the ADC chip, it becomes a digital signal and is sent to the FPGA, and then the digital signal is digitally down-converted in the FPGA, and finally the phase difference and amplitude are calculated. This scheme is more complicated and has many steps. In addition, the high-speed interface used by most high-speed ADCs to communicate with the outside world is LVDS. Although this interface has a small delay, the number of pins required is much more than the JESD204B interface, so using the LVDS interface requires more board area. And the LVDS interface requires that the data to be routed to the FPGA interface must be routed at the same length, so the layout is more complicated. In addition, it is necessary to strictly ensure that the reference clocks of each ADC have the same phase, equal amplitude, and equal slope, and also ensure the complete consistency of the front-end RF link of each ADC.

Based on the above analysis, the front-end receiving channel of this system adopts the agile transceiver ADRV9009 of ADI Company [6]. The ZYNQ chip of Xilinx Company is used to calculate the phase difference and amplitude of the two signals. Compared with the traditional FPGA + ARM architecture, using this chip, the hardware design is simpler, the board area required for the PCB board is reduced, and the layout and wiring are simpler and easier. ZYNQ chip has built-in dual A9 ARM processors, which can be very convenient to configure peripheral devices, especially ADRV9009, which can reduce a lot of work and shorten development time.

AS shown in Fig. 1, is The system block diagram. As the core processing unit and control chip of the whole system, ZYNQ is not only responsible for the calculation of the phase difference and amplitude of the two-channel signal, but also for configuring the ADRV9009 to work at the required frequency and bandwidth. Data communication between the ADRV9009 and the ZYNQ is carried using the JESD204B high-speed interface. The process of digital baseband signal processing is carried out at the PL end of ZYNQ. The PL side is equivalent to an FPGA, which is very suitable for processing digital signals. Then send the processed result to the PS side of ZYNQ. The PS side is an ARM processor. The calculation results are sent to the upper computer for further data analysis through the Ethernet port on the PS side. The DDR on the PS side of ZYNQ is used as the running memory of ARM. The clock management unit provides clocks to each circuit module of the entire system, and the PS terminal of ZYNQ configures the clock management unit so that each circuit module can obtain the required clock frequency.

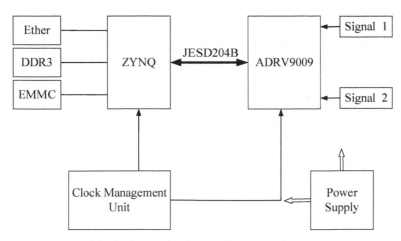

Fig. 1. System implementation block diagram

3   Function and Structure of ADRV9009

ADRV9009 provides dual transmitter and receiver, integrated synthesizer and digital signal processing functions, is a highly integrated RF agile transceiver. This chip has the most advanced dynamic range, the receiving end is composed of two independent wide bandwidth conversion receivers. This chip supports observation path receivers for TDD applications. The complete receiver subsystem of this chip has quadrature error correction, digital filtering, DC offset correction, automatic attenuation control, manual attenuation control, thus eliminating the need for these functions in the digital baseband. In addition, the Digital-to-Analog Converter (DAC), Analog-to-Digital Converter (ADC) are integrated in ADRV9009. The chip's external gain control mode is extremely flexible, allowing significant flexibility in dynamically setting system-level gains.

The structure diagram of ADRV9009 is shown in Fig. 2. This chip integrates two independent 16 bit ADCs, which can complete the conversion of two analog signals to digital signals. Each receiving channel has a tunable input from 75 to 6 GHz, and its bandwidth can reach a maximum of 200 MHz. In addition, it also integrates two 14 bit DACs, which can complete the conversion of two digital signals to analog signals. Each transmit channel has a tunable output from 75MHz to 6 GHz, and its transmit synthesis bandwidth can reach a maximum of 450 MHz.

4   JESD204B High Speed Interface

JESD204B is a high-speed serial interface protocol [7], which is often used for data transmission between ADC and DAC digital conversion chips and FPGA or ASIC chips. Compared with LVDS and CMOS interfaces, JESD204B has more advantages in cost, size, speed, power consumption, etc.

The JESD204B interface block diagram is shown in Fig. 3. JESD204B supports multiple converters and multi-channel data transfer, increasing deterministic latency

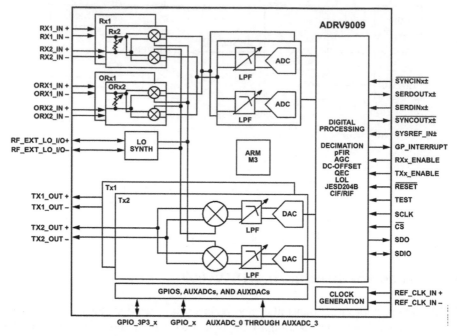

Fig. 2. Schematic diagram of ADRV9009 structure

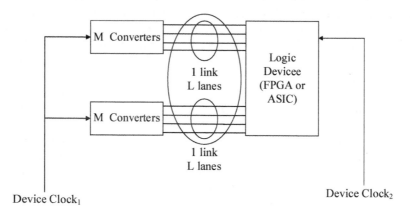

Fig. 3. JESD204B interface block diagram

and increasing transfer speeds to 12.5 Gbit/s. The JESD204B protocol divides the interface into three layers: transport layer, link layer, and physical layer. The transport layer mainly completes the mapping of converted samples to frames; The link layer mainly completes 8B/10B encoding and decoding, determines the sending and receiving rules, link operation and other functions. The link operation includes code group synchronization, SYNC ~ signal combination, initial frame synchronization, frame sequence detection and calibration[8]; The physical layer completes the serial-to-parallel conversion or parallel-to-serial conversion of the data, and provides the underlying path for the transmission and reception of high-speed serial signals [9].

The flow chart of code group synchronization is shown in Fig. 4. The receiver sends a synchronization request through the synchronization interface, that is, sends a SYNC ~ signal; After the transmitter receives the SYNC ~ signal, it transmits the symbol stream of /K28.5/; When the receiver measurement continuously receives at least four valid /K28.5/ symbols, it is regarded as synchronization, and the SYNC ~ signal is invalid at the same time; When the transmitter detects that all receivers have stopped their synchronization requests, the transmitter continues to transmit /K28.5/ symbols until the arrival of the next LMFC boundary;

The JESD204B interface protocol specification defines three subclasses. The ADRV9009 chip uses subclass 1, so subclass 1 is also used in this system. Subclass 1 uses the deterministic latency of the system reference clock (SYSREF). The signals required to establish a data link are Device clock, Lanes (the number of channels of 204B), and SYNC ~ (synchronous clock). In this system design, two receiving channels of ADRV9009 need to be used, each receiving channel has dual ADC, the number of bits is 16bit, and its sampling rate is 245.76 MHz. The rate calculation formula of the JESD204B interface is: Lane rate = (M × N' × [10/8] × Fs)/L; Where M is the number of ADCs, N' is the number of bits of ADC, Fs is the clock used, 10/8 is the link overhead for 8b/10b encoding, and L is the number of channels. Finally, the rate of each channel can be calculated to be 9830.4 MHz. Since the maximum rate of the GTH interface of ZYNQ is 12.5 GHz. Therefore, the transfer rate between ZYNQ and ADRV9009 is satisfied.

5 Measurement Principle

ADRV9009 outputs IQ quadrature signal after sampling the input analog signal. In real life, signals are real signals, but in digital signal processing, complex numbers are suitable for processing and computing signals. The real signal can be represented by a complex vector. In the quadrature processing of data, the spectrum of the data is divided into real part and imaginary part, the in-phase component is the real part of the spectrum, and the quadrature component is the imaginary part of the spectrum. In the time domain, the signal is a real signal, and the real signal has positive spectral components and negative spectral components. The positive spectral components and negative frequency components of the in-phase spectrum of the real signal are symmetrical around the zero frequency point.

The block diagram of quadrature sampling is shown in Fig. 4. After the input signal passes through the quadrature oscillator, low-pass filter, and ADC successively, the digital baseband IQ signal is finally obtained [10].

Suppose two signals to be tested can be expressed as:

$$F_1(\phi_1) = I_1(\phi_1) + jQ_1(\phi_1) = A_1 e^{-j\phi_1} \tag{1}$$

$$F_2(\phi_2) = I_2(\phi_2) + jQ_2(\phi_2) = A_2 e^{-j\phi_2} \tag{2}$$

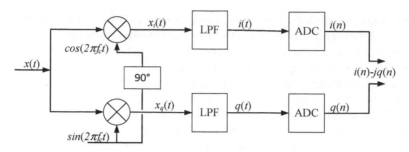

Fig. 4. Quadrature sampling block diagram

where, F_1 and F_2 respectively represent the two signals to be measured, I_1 and I_2 respectively represent the in-phase component of the two signals to be measured, Q_1 and Q_2 respectively represent the quadrature components of the two signals to be measured, ϕ_1 and ϕ_2 respectively represent the two channels the phase of the signal to be measured, A_1 and A_2 respectively represent the amplitude of the two signals to be measured. Multiplying $F_1(\phi_1)$ by its conjugate with $F_2(\phi_2)$ gives:

$$F_1 \cdot \overline{F}_2 = (I_1 I_2 + Q_1 Q_2) + j(Q_1 I_2 - I_1 Q_2) = A_1 A_2 e^{-j(\phi_1 - \phi_2)} \tag{3}$$

It can be seen from formula (3) that $\phi_1 - \phi_2$ is the result of the phase difference between F_1 and F_2. Taking signal $F_1(\phi_1)$ as an example, multiplying $F_1(\phi_1)$ by its conjugate with $F_1(\phi_1)$ gives:

$$F_1(\phi_1) \cdot \overline{F}_1(\phi_1) = I_1^2(\phi_1) + Q_1^2(\phi_1) = A_1^2 \tag{4}$$

Therefore, the amplitude of the signal is:

$$A = \sqrt{I^2(\phi) + Q^2(\phi)} \tag{5}$$

6   Conclusion

This paper introduces the design and implementation of a portable signal phase difference and amplitude calculator based on ADRV9009 and ZYNQ, and briefly expounds JESD204B, the most widely used high-speed data interface in the industry. The signal phase difference and amplitude calculator realized by this scheme has the advantages of small size, portability, wide bandwidth and simple circuit structure. It is convenient for users to measure and use it anytime and anywhere outdoors, and can be used in a variety of scenarios and occasions. Since the ADRV9009 directly converts the analog signal into a digital baseband IQ signal, the algorithm used in this paper directly multiplies and adds the IQ signal to obtain the phase difference result and the signal amplitude result, making full use of the advantages of ADRV9009 and ZYNQ, There is no need to do other redundant processing and calculation of the signals sent by ADRV9009 to ZYNQ, which makes the algorithm implementation simpler and more convenient, shortens the development time, and increases the real-time performance of the calculated results.

Acknowledgments. Supported by Shaanxi Province Key Research and Development Project (2021GY-280); Shaanxi Province Natural Science Basic Research Program Project (2021JM-459).

References

1. Zhong, H., Ruan, H., Sun, B.: Passive locating method using phase difference rate algorithm based on least squares fitting, In: 2019 IEEE 3rd Information Technology, Networking, Electronic and Automation Control Conference (ITNEC), pp. 1504–1508 (2019)
2. Li, M., Wan, P., Xiao, W., Xu, W., Luo, K.: Performance analysis of phase difference estimation method in IOT based on correlation theory. In: 2020 4th International Symposium on Computer Science and Intelligent Control (ISCSIC), pp. 1–6 (2020)
3. Pérez-Díaz, B., Araña-Pulido, V., Cabrera-Almeida, F., Dorta-Naranjo, B. P.: Phase shift and amplitude array measurement system based on 360° switched dual multiplier phase detector. IEEE Trans. Instrum. Meas. **68**(12), 1–11 (2021)
4. Choi, U.-G., Kim, H.-Y., Han, S.-T., Yang, J.-R.: Measurement method of amplitude ratios and phase differences based on power detection among multiple ports. IEEE Trans. Instrum. Meas. **68**(12), 4615–4617 (2019)
5. Shao, Y., Yao, Y., Liu, H., Lv, R., Zan, P.: Power harmonic detection method based on dual HSMW window FFT/apFFT comprehensive phase difference. In: 2021 40th Chinese Control Conference (CCC), pp. 6784–6787 (2021)
6. Yang, Z., Xiong, W., Zhao, Y.: Software defined radio hardware design on ZYNQ for signal processing system. In: 2019 8th International Symposium on Next Generation Electronics (ISNE), pp. 1–3 (2019)
7. Yin, P., Zeng, X., Tang, F.: Built-in self-test circuits for high-speed JESD204B transceiver controller. In: 2021 12th International Symposium on Advanced Topics in Electrical Engineering (ATEE), pp. 1–5 (2021)
8. Li, X., Liu, Y.: Efficient implementation of the data link layer at the receiver of JESD204B. In: 2019 International Conference on Intelligent Computing, Automation and Systems (ICICAS), pp. 918–922 (2019)
9. Liu, B., Wang, Z., Zhang, T., Zhang, L.: A 12.5 Gbps novel SerDes transmitter for JESD204B physical layer. In: 2021 5th International Conference on Communication and Information Systems (ICCIS), pp. 6–12 (2021)
10. Deng, J.H., Prasad, P.V.S., Tseng, I.-F., Huang, W.-C.: Joint I/Q imbalance and nonlinear compensation design in direct downconversion receiver. In: 2021 44th International Conference on Telecommunications and Signal Processing (TSP), pp. 23–26 (2021)

Cross-Polarized Directional Antenna with Y-Shaped Coupling Feed Structure for Ultra-wideband

Zhong Yu, Li Shi$^{(\boxtimes)}$, and ZhengH ui Xin

Xi'an University of Posts and Telecommunications, Xi'an 710121, China
yz327123@163.com, {18709188176,1002928675}@stu.xupt.edu.cn

Abstract. This paper presents a unique high-gain, low-profile, two-element multilateral ultra-wideband (UWB) directional antenna with a height of only $0.19\lambda_0$ and a width of $0.76\lambda_0$. The antenna consists of a square plane reflector and resonant unit, the top of which is fed by two mutually orthogonal Y-shaped structures, and the bottom antenna is composed of two cross-polarized dual-element multilateral antenna radiation units. The antenna impedance bandwidth is extended by two sets of multilateral gradient radiation units and Y-shaped feed structure, and fed by coaxial line through the hole, and finally a nylon column is added at the bottom of the radiation unit to connect the reflector to realize directional radiation. Good agreement between simulation and measurement proves that the antenna can achieve 75% (2322–5111 MHz) of the relative value of impedance bandwidth, the return loss of both input ports is less than 10 dB, in the operating band, 3 dB flap width 76°, in-band gain 8.5–10.5 dBi, peak gain can reach 10.5 dBi, radiation efficiency is about 68%. The antenna proposed in this paper has low profile, highly symmetric structure, and good radiation performance, completely covers all N7/N38/N41/N77/N78/N79 bands of 5G NR below Sub-6GHz as well as LTE, WiMAX, and WLAN wireless communication systems, as a high gain UWB directional antenna for micro-base station, which is a very promising solution.

Keywords: Coupling feed · Cross-polarized · Directional antenna · Dual-polarized · High gain · Ultra-wideband

1 Introduction

The fifth-generation mobile communication technology has ten times faster transmission rate, lower network latency, longer communication distance and wider coverage than 4G. As a key receiving and transmitting device, the antenna is an indispensable component to achieve this leap forward, which should not only meet the far-field directional map characteristics of directional fouling, but also meet many requirements such as wide band, low profile, directional map stability, etc. Among them, the dual-polarized antenna with wide working band and beamwidth is more likely to meet the requirements of 5G, and the antenna with miniaturized features is more convenient for application in practical engineering [1–3]. Recently, dual-polarized antennas have gained a large amount of interest from researchers in the field of wireless communications, because dual-polarized antennas can provide two non-interfering channels on a single antenna to improve spectral efficiency and solve multipath fading. Meanwhile, directional antennas can improve the spatial reuse rate and reduce the energy consumption of wireless networks [4]. In order to achieve the fastest 5G services, the designed antenna should include multiple resonant modes to cover multiple 5G-NR bands, so designing an ultrawideband (UWB) directional antenna is the best choice. In order to design an efficient wideband antenna, the antenna size needs to be increased, which contradicts the need for miniaturization of antennas, so the design methods of miniaturization, wideband/multiband, and high radiation efficiency of antennas need to be investigated. The common dual-polarized horn antennas [5], crossed dipoles [6, 7], independently controllable [8], waveguide antennas [9], Vivaldi antennas [10], quasi-Yagi [11], and horns [12], which have good operating bandwidth and directional radiation characteristics, generally suffer from the drawback of excessive size and their applications are greatly limited. Therefore, dual-polarized patch directional antennas are widely studied and applied because of the advantages of miniaturization.

Many studies [5, 6, 8, 13] on broadband antennas have been reported in many literatures. In the literature [5], a log-periodic array antenna |S11| ≤10 dB with an operating band of 3–6 GHz was fabricated using a coplanar waveguide fed structure. In the literature [6], a wideband back-cavity bowtie antenna was proposed. A standing wave ratio (VSWR) bandwidth of 91.4% was achieved by using a triangular butterfly dipole. In addition, a large cavity and annular structure are adopted to boost the directional radiation pattern. The antenna uses a new transition structure with microstrip slotted lines and operates in the band of 3.6–11.6 GHz with a gain of 4.3–6.8 dBi. The broadband unidirectional quadratic antenna is proposed in the literature [13]. The antenna has 83.7% impedance bandwidth for VSWR with integrated balun and ripple boundary radiation unit. Yet, in the up-operation band, the front-to-back ratio (FBR) is below 10 dB. The antenna suggested in this paper is an evolving of the classical quadratic antenna [13], consisting of four identical patches, a coaxial cable grid and a ground plane with an working bandwidth of 50%. In the literature [14] several technologies are available to extend its operating range. Unfortunately, the radiation in the working band becomes unacceptable. In the literature [15, 16], by adding a tuning plate under the patch and modifying four patches, the antenna operates in the band of 800–2200 MHz. A good unidirectional pattern was obtained at low frequencies, but a depression in the +z direction was observed in the absence of high frequencies.

In this paper, we proposed a dual element multilateral compact and broadband directional antenna based on the existing antenna miniaturization, broadband and high gain technologies. The designed dual-polarized broadband directional antenna uses ±45° cross-polarization to achieve S11 ≤10 dB at 2.32–5.11 GHz, and the mutual coupling between the two antenna units of the antenna system is below −20 dB in the required operating band, which completely covers all N7/N38/N41/N77/N78/N79 bands of 5G NR below Sub-6 GHz and LTE, WiMAX, and WLAN wireless communication systems.

2   Research Method

Based on the above analysis, a dual element multilateral directional antenna is designed to realize small broadband. A cross-fed structure is adopted to improve the isolation between antennas, and the broadband of the antenna is realized by gradually increasing the transverse current path length in the form of differentiation through a gradual multi-lateral structure, so that the radiation structure covers many operating frequency points. The antenna simulation model was constructed and simulated using ANSYS Electronics Desktop 2019 R2-HFSS (solution type Terminal).

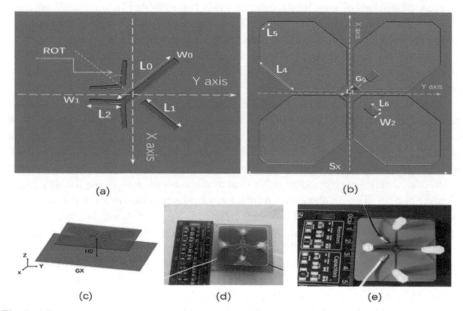

Fig. 1. View of the proposed broadband low-profile dual element multilateral directional antenna: (a) the 2D top-view (b) the 2D bottom-view (c) the 3D schematic of the designed antenna (d) and (e) Antenna production and processing after the samples

Figure 1 shows the structure diagram of the proposed broadband low-profile dual element multilateral directional antenna, with four multilateral radiating units (MRUs) structures uniformly distributed in the horizontal plane, a group of two, and a Y-shaped orthogonal feed Structure. The antenna is fed to the microstrip line from the bottom of the dielectric board by using IPEX-K wire. The antenna is printed on FR-4 ($\varepsilon_r = 4.4$, $\tan\delta = 0.02$) and thickness of 1mm, the dimension of dielectric plate is $50 \times 50 \times 1$ mm^3 reflective surface size is $60 \times 60 \times 1$ mm^3 and profile height is 15mm, which realizes the miniaturization and broadband of antenna and has high gain characteristics.

Figure 1a shows the 2D top-view of the designed antenna. The antenna adopts the crossed Y-shape structure of the upper and lower feed points to achieve the broadband operation of the antenna, adjust the microstrip feed line length L0, L1, and width W0 to achieve the accurate matching of the microstrip feed line. At the same time, the antenna is loaded with the microstrip feed line structure equivalent to the distributed capacitance, such loading is due to its own height introduces inductive component to further regulate the impedance matching. In order that the Y-shaped feed structure can be heavily coupled to the lower MRUS, the two branch lengths L2 and width W1 of the Y-shaped structure, and the angle variable ROT are uniquely designed to reduce the antenna reflection coefficient while finding the maximum value that ensures the bandwidth. The maximum value of the antenna is found by reducing the reflection coefficient. Figure 1b shows the 2D bottom-view of the designed antenna, which is the schematic diagram of the MRU structure, on this structure the multilateral length L4 and the multilateral resonant unit vertex length L5 on the MRU are continuously optimized to broaden the low frequency bandwidth. By changing the spacing of the two MRUs and the feed position, the isolation between the unit structure is improved and the bandwidth is increased at the same time. Figure 1c shows the 3D schematic of the designed antenna. At a certain frequency, the passband BW is inversely related to the quality factor Q. In order to avoid excessive Q value and sidelobe caused by too close distance between reflector and radiating element, thus reducing the antenna passband, the distance H0 is continuously optimized in HFSS, and the Q value is reduced to realize UWB. The final optimized dimensions of the antenna are shown in Table 1. Based on the previous discussion, the proposed antenna was made and assembled to verify its performance, as shown in Fig. 1d and e.

Table 1. The final optimized dimensions of the proposed antenna.

Parameter	Value (mm)	Parameter	Value (mm)	Parameter	Value (mm/deg)
L0	11.00	S_X	50.00	W0	1.00
L1	7.00	H0	15.00	W1	0.50
L2	4.40	GX	60.00	W2	2.00
L3	4.00	L5	3.50	G0	2.80
L4	11.30	L6	3.00	ROT	47.00deg

3   Antenna Performance

3.1   Conductivity Measurements

The simulation data are shown in Fig. 2. Due to the asymmetry of the antenna system structure coupling feed, it must lead to the difference of impedance matching, so S22 is shifted 200 MHz to low frequency compared with S11. There is a resonance point at 2.62 GHz, and the image is basically the same (maximum error 3.1 dB), S12 and S21 are also basically the same (maximum error 3.3 dB). From Fig. 2a, the reflection coefficient of the two antenna units of the simulated experimental antenna system is lower than -10 dB in the desired operating band (S11 is 2.32–5.11 GHz). By optimizing the lengths L4 and L5 on a single multilateral radiating unit, the optimal variant is obtained when

L4 = 11.3 mm and L5 = 2.8 mm. From Fig. 2b, the mutual coupling between two antenna units of the antenna system is roughly below −20 dB in the required working band, and the average isolation degree is 25 dB in the whole working band. Although there are some deteriorations of isolation degree at high frequency (4.7–5.2 GHz), but basically the influence of mutual coupling between two sets of multilateral radiation units can be avoided. From the comparison between simulated and measured results in Fig. 2c, the reflection coefficient of this multilateral directional antenna Port1 in the required operating band 1.81–6.00 GHz is below −10dB, and the return loss is below −5dB in the 0.6–6 GHz band, which basically covers the Sub-6G band. There are some experimental errors between the simulation and the measured data, which is due to the error of processing accuracy in the actual process, and the fact that the four nylon pillars are not considered in the simulation model to have a certain impact on the radiation unit, so there is a difference in the working frequency band. The above measured data were measured in Agilent E5063A vector network analyzer, as shown in Fig. 2d.

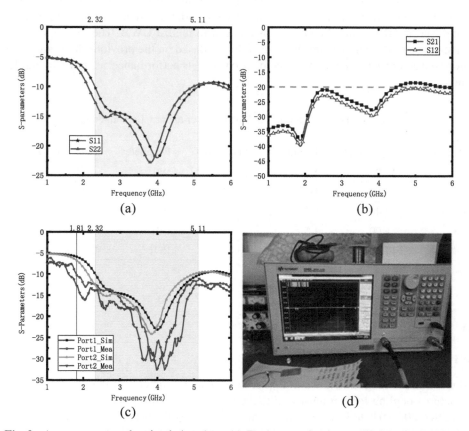

Fig. 2. Antenna system the simulation data: (a) Dual-port reflection coefficient (b) Dual-port transmission coefficient (c) Comparison of simulation and measurement of reflection coefficient (d) Agilent E5036A vector network analyzer

Figure 3 shows the surface current distribution of MRUs, according to the arm length of the symmetrical oscillator more than $0.65\lambda0$ ($\lambda0$ free space wavelength), there will be a side flap, so that the gain decreases, which is due to the reverse current on the arm. To improve this antenna, the shape of the oscillator is changed from straight to polygon. Such as Y-shaped oscillator and Gaussian curve oscillator, the idea is to compensate the phase difference of the current by using the difference of the wave range of the ray between the current elements at each point on the line. The adjustable length and angle of the Y-shaped oscillator is the most common choice, that is, by adjusting the length and angle of the oscillator can obtain the best radiation characteristics. The Y-shaped feed structure can be coupled to the radiation unit at all four operating frequencies, but the coupling effect of single Y-shaped feed line needs to be improved when the antenna works at 4.00 GHz. Overall, the current strength and coupling effect are excellent and can operate in the target frequency band.

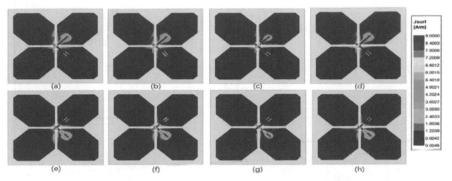

Fig.3. Simulation results of surface current distribution of MRUs. Front side: (a) 2.5GHz (b) 3.0 GHz (c) 4.0GHz (d) 5.0GHz Back side: €2.5 GHz (f) 3.0GHz (g) 4.0GHz (h) 5.0GHz

3.2   Radiation Measurements

The radiation test environment of the directional antenna is Satimo-SG64 Anechoic Chamber system, as shown in Fig. 4a. As shown in Fig. 4b, the 3dB wave width (phi = 0°) is 73° and (phi = 90°) is 76° at the center frequency point (3.65GHz) of the dual element multilateral directional antenna, and the front-to-back wave rejection ratio is 14.6 dB to the requirements of micro base station antenna.

As shown in Fig. 4c, the simulated and measured gain comparison, both have good agreement, the gain in the test band (2.32–5.00 GHz) is 7.5–10.1 dBi, the peak gain is up to 10.1 dBi, the average gain is more than 8.5 dBi after the antenna is larger than 3.5 GHz, which meets the gain requirement of directional micro-base station antenna design. The average gain is more than 8.5 dBi after the antenna is larger than 3 GHz, which meets the gain requirement of directional micro-base station antenna design. As shown in Fig. 4d, the efficiency of the dual multilateral antenna is more than 60% in the high frequency band (3.06–4.67 GHz), 55% in the low frequency band (2.00–3.00 GHz), and 85% in the peak efficiency at 4670 MHz.

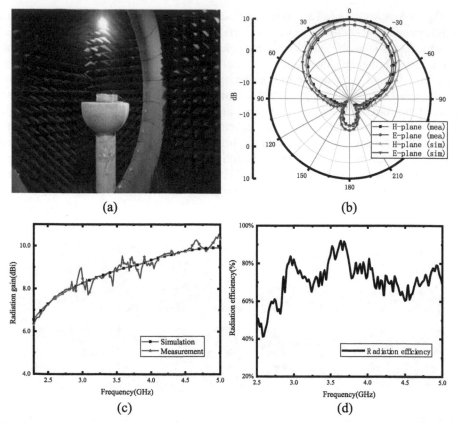

Fig. 4. Antenna system parameters: (a) Radiation measurements environment (b) Far-field radiation direction figure of the antenna system (c) Comparison of Simulated and measured gain (d) Antenna measured radiation efficiency

4 Conclusion

In this paper, a unique Y-shaped feed structure is proposed, and the multi-sided patch radiating element is designed by using the gradual electrical length path method and cross polarization technology, and its advantages and disadvantages are explained. On this basis, based on the unique feed structure, a dual-polarized directional antenna operating in 2.32–5.11 GHz frequency band is designed. The antenna can work in N7/N38/N41/N77/N78/N79 frequency bands and ultra-wideband wireless communication systems such as LTE, WiMax, WLAN, etc. Through conduction and radiation tests, it has good radiation performance. Finally, we compare it with the previous work as shown in Table 2, which shows its advantages in technical application and provides a choice for the selection of micro base station antenna.

Table 2. Comparison of the proposed antenna with prior work

References	Frequency bands (MHz)		Total size (mm)	Mea-gain (dBi)	Peak gain (dBi)	Efficiency(%)	Ave-Efficiency (%)
	Lower band	Higher band					
[5]	2300–2750	5090–6160	43.0 * 43.0 * 0.9	1.93–2.28 3.07–3.94	2.28 3.94	>80.00 >73.00	83.60 74.23
[6]	2260–2680	3280–4070 4750–6040	33.0 * 33.0 * 1.0	/	1.67 2.75 5.50	/	/
[8]	4000–5780	6830–8220	20.0 * 23.0 * 1.6	1.20–3.00	− 2.66 (5.3G) 4.83 (9.9G)	>90.00	
[13]	704–1095	1640–2615	15.0 * 25.0 * 4.0	−0.60 to 0.50	0.50	40.0–53.0 41.0–70.0	/
[14]	824–975	1450–2825	15.0 * 29.0 * 1.6	−0.90 to 0.80 0.80–4.20	/	43.0–57.0 44.0–70.0	49.62 54.74
[15]	2400–3940		100 * 100 * 30	6.72–9.04	8.50	/	/
[16]	3280–3840	4950–6020	56 * 45	2.00–4.50	4.50	75.0–87.0 72.0–84.0	70.00
Our antenna	2322–5111		60 * 60 * 15	7.50–10.10	10.10	42.0–85.0	68.00

References

1. An, W., Li, Y., Fu, H., Ma, J., Chen, W., Feng, B.: Low-profile and wideband microstrip antenna with stable gain for 5G wireless applications. IEEE Antennas Wireless Propag. Lett. **17**(4), 621–624 (2018)
2. Li, Y., Sim, C.-Y.-D., Luo, Y., Yang, G.: High-Isolation 3.5 GHz Eight-Antenna MIMO array using balanced open-slot antenna element for 5G smartphones. IEEE Trans. Antennas Propag. **67**(6), 3820–3830 (2019)
3. Feng, B., Lai, J., Sim, C.-Y.-D: A building block assembly dualband dual-polarized antenna with dual wide beamwidths for 5G microcell applications. IEEE Access **8**, 123359–123368 (2020)
4. Yang, L., Weng, Z., Zhang, C., Du, R.: A dual-wideband dual-polarized directional magneto-electric dipole antenna. Microw. Opt. Technol. Lett. **59**(5), 1128–1133 (2017)
5. Li, H., Kang, L., Xu, Y., Yin, Y.-Z.: Planar dual-band wlan mimo antenna with high isolation. Appl. Comput. Electromagnetics Soc. J. **31**(12), 1410–1415 (2016)
6. Moeikham, P., Akkaraekthalin, P.: A compact printed slot antenna with high out-of-band rejection for WLAN/WiMAX applications. Radioengineering **25**(4), 672–679 (2016)
7. Chen, H.-D., Tsai, Y.-C., Sim, C.-Y.-D. Kuo, C.: Broadband eight-antenna array design for Sub-6 GHz 5G NR bands metal-frame smartphone applications. IEEE Antennas Wireless Propag. Lett. **19**(7), 1078–1082 (2020)
8. Awan, W.A., Zaidi, A., Hussain, N., Iqbal, A., Baghdad, A.: Stub loaded, low profile UWB antenna with independently controllable notch-bands. Microwave Opt. Technol. Lett. **61**(11), 2447–2454 (2019)
9. Aram, M.G., Aliakbarian, H., Trefna, H.D.: A phased array applicator based on open ridged-waveguide antenna for microwave hyperthermia. Microwave Opt. Technol. Lett. **63**(12), 3086–3091 (2021)
10. Li, D.-H., Zhang, F.-S., Xie, G.-J., Zhang, H., Zhao, Y.: Design of a miniaturized UWB MIMO Vivaldi antenna with dual band-rejected performance. IEICE Electron. Express **17**(16) (2020)
11. Nella, A., Bhowmick, A., Rajagopal, M.: A novel offset feed flared monopole quasi-Yagi high directional UWB antenna. Int. J. RF Microwave Comput. Aided Eng. **31**(6) (2021)
12. Yadav, S. V., Chittora, A.: A compactultra-wideband transverse electromagnetic mode horn antenna for high power microwave applications. Microwave Opt. Technol. Lett. **63**(1), 264–270 (2021)
13. Yong-Ling, B., Cheng-Li, L., Zhi, C., Li, J. L.-W., Kai, K.: Small-size multiresonant octaband antenna for LTE/WWAN smartphone applications. IEEE Antennas Wireless Propag. Lett. **13**, 619–622 (2014)
14. Ban, Y.-L., Liu, C.-L., Li, J. L.-W., Li, R.: Small-size wideband monopole with distributed inductive strip for seven-band WWAN/LTE Mobile Phone. IEEE Antennas Wireless Prop. Lett. **12**, 7–10 (2013)
15. Jin, G., Li, L., Wang, W., Liao, S.: Broadband polarisation reconfigurable antenna based on crossed dipole and parasitic elements for LTE/sub-6GHz 5G and WLAN applications. IET Microwaves, Antennas Propag. **14**(12), 1469–1475 (2020)
16. Kundu, S.: A compact printed ultra-wideband filtenna with low dispersion for WiMAX and WLAN interference cancellation. Sādhanā **45**(1), 1–7 (2020). https://doi.org/10.1007/s12046-020-01495-y

Scattering Properties of High Order Bessel Vortex Beam by Perfect Electrical Conductor Objects

Zhong Yu, Yanli Cui$^{(\boxtimes)}$, and Li Shi

Xi'an University of Posts and Telecommunications, Xi'an 710121, China
yz327123@163.com, {cc,18709188176}@stu.xupt.edu.cn

Abstract. This paper investigates the scattering properties of perfect electrical conductor (PEC) objects irradiated by a high-order Bessel vortex wave. The scattering field is modelled by the generalized Lorentz-Mie theory (GLMT) and the calculations of the radar cross section (RCS) are obtained. In addition, the scattered field of the object with the vortex wave is established by numerical method. The scattering properties, which includes phase characteristics and variations, are analyzed. The OAM beams provide more information for target detection and identification in a specific propagation direction compared with a plane wave. New prospects for radar detection and identification could be offered.

Keywords: Orbital angular momentum (OAM) · Scattering field · Radar cross section (RCS)

1 Introduction

The scattering characteristics of radar targets play an important role in detection techniques [1, 2]. Vortex electromagnetic waves (EM) carrying orbital angular momentum (OAM) have been used for target detection in recent years. In contrast to plane waves, a distinctive feature of the OAM waves is the spiral pattern of their phase distribution along the direction of propagation [3, 4], which produces rich scattering phenomena and carries a wealth of phase information when interacting with illuminated objects. Many articles have already researched the scattering properties of OAM beams. In [5], by applying Generalized-Lorenz–Mie theory (GLMT), the author studied the scatter properties of a Bessel vortex beam and a charged sphere. With the GLMT method, the electromagnetic scattering of a Laguerre–Gaussian vortex beam by two anisotropic homogeneous spheres is analyzed using the expressions of the analytical method. In [6], the backscattering property of the EM wave and the radar cross section (RCS) with different OAM modes is studied and evaluated. Wu et al. use numerical calculations and simulations to calculate scattering field, and the results effectively lay the foundations for the study of the properties of vortex fields [7]. Liu et al. study the backward scattering interaction between OAM beams and perfect electrical conductor (PEC) sphere and PEC cube at microwave frequencies [8]. From the above literatures, the interaction

© The Author(s), under exclusive license to Springer Nature Switzerland AG 2023
N. Xiong et al. (Eds.): ICNC-FSKD 2022, LNDECT 153, pp. 1329–1337, 2023.
https://doi.org/10.1007/978-3-031-20738-9_144

of particles with diffraction-free Bessel beams has been adequately studied. However, adequate researches of the target scattering characteristics of the PEC sphere lighted by high-order vortex Bessel beams is still in need.

In this work, the interaction of high-order Bessel beams and PEC objects is investigated by GLMT. In addition, the scattering characteristics and RCS are analyzed. The paper is outlined as follows. Section 2 presents the basic formulations of OAM wave scattering theory. In Sect. 3, the discussion of the numerical results and simulation of the scattering of OAM waves are presented. Finally, Sect. 4 concludes the paper.

2   Methodology

Bessel beams are the solution to the Helmholtz equation and have great interest to researches in recent years [9]. High-order Bessel beams have zero intensity on the axis and are surrounded by concentric light rings. The diameter of the dominant intensity ring is minimal and increases with the topological charges under the same conditions, as can be seen in Fig. 1. Now considering a PEC sphere lighted by a high-order Bessel beam, the beam coordinate system is presented in Fig. 2. The electric field distribution of the lth-order Bessel vortex beam can be expressed by Bessel function, and the time-dependent form $\exp(-i\omega t)$ is omitted in the description that follows.

$$E(r, \phi, z) = E_0 e^{ik_z z} J_l(k_R R) e^{\pm il\phi} \tag{1}$$

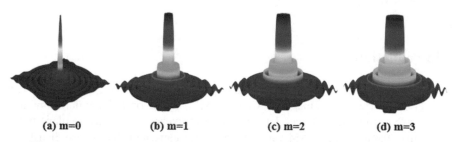

| (a) m=0 | (b) m=1 | (c) m=2 | (d) m=3 |

Fig. 1. Schematic of the intensity distribution of Bessel beams for different orders

where E_0 is the electric field amplitude. The radial distance and azimuthal angle in the column coordinate system are $R = \sqrt{x^2 + y^2}$ and $\phi = \tan^{-1}(y/x)$, respectively. The radial and axial wave numbers are $k_R = k \sin \alpha$ and $k_z = k \cos \alpha$, respectively. k signifies the wave number and α denotes the half-cone angle. The function $J_l(\cdot)$ is the lth-order Bessel function of first kind.

In the coordinate system $Oxyz$, derived from the Maxwell vector equations and the Lorenz gauge condition, each component of the Bessel vortex beam electromagnetic field can be expressed as [10, 11].

$$E_x = \frac{1}{2} E_0 \left\{ \exp[i(k_z z + l\phi)] \times \left[\left(1 + \frac{k_z}{k} - \frac{k_R^2 x^2}{k^2 R^2} + \frac{l(l-1)(x-iy)^2}{k^2 R^4} \right) J_l(kR) \right. \right.$$

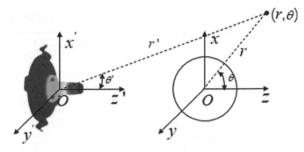

Fig. 2. Schematic diagram for Bessel waves scattered by a PEC sphere.

$$
-\frac{k_R\left(y^2 - x^2 - 2ilxy\right)}{k^2 R^3}\, J_{i+1}(k_R R)\Bigg]\Bigg\} \tag{2}
$$

$$
E_y = \frac{1}{2}E_0 xy\left\{\exp[i(k_z z + l\phi]\left[\begin{array}{c}\dfrac{l(l-1)\left(2 + i\frac{x^2-y^2}{xy}\right) - k_R^2 R^2}{k^2 R^4}\, J_l(kR)\\[3mm] +\dfrac{k_R\left(2 + il\frac{y^2-x^2}{xy}\right)}{k^2 R^3}\, J_{l+1}(k_R R)\end{array}\right]\right\} \tag{3}
$$

$$
E_z = \frac{1}{2}iE_0\frac{x}{kR}\left(1 + \frac{k_z}{k}\right)\left\{\begin{array}{c}\exp[i(k_z z + l\phi]\times\\[2mm]\left[\dfrac{l\left(1 - i\frac{y}{x}\right)}{R} J_l(k_R R) - k_R J_{l+1}(k_R R)\right]\end{array}\right\} \tag{4}
$$

$$
H_x = E_y\sqrt{\varepsilon_0/\mu_0} \tag{5}
$$

$$
H_y = \frac{1}{2}E_0\sqrt{\varepsilon_0/\mu_0}\left\{\exp\left[i(k_z z + l\phi]\times\left[\left(1 + \frac{k_z}{k} - \frac{k_R^2 y^2}{k^2 R^2} + \frac{l(l-1)(y+ix)^2}{k^2 R^4}\right)J_i(k_R R)\right.\right.\right.
$$
$$
\left.\left.\left. -\frac{k_R\left(x^2 - y^2 + 2ilxy\right)}{k^2 R^3}J_{l+1}(k_R R)\right]\right]\right\} \tag{6}
$$

$$
H_z = \frac{1}{2}iE_0\sqrt{\varepsilon_0/\mu_0}\frac{y}{kR}\left(1 + \frac{k_z}{k}\right)\left\{\begin{array}{c}\exp[i(k_z z + l\phi]\times\\[2mm]\left[\left(l\left(1 + i\frac{x}{y}\right)\right)J_l(k_R R) - k_R J_{l+1}(k_R R)\right]\end{array}\right\} \tag{7}
$$

The expressions of the electric and magnetic field components strengths in a coordinate system are converted into expressions for the components in a spherical coordinate system. And using the integral localized approximation, which change kr to $(n + 1/2)$ and θ to $\pi/2$. Where in this case the radial component of the electric field E_r and the magnetic field H_r can be obtained as follows.

$$
E_r = \frac{1}{2}E_0 \cdot \exp[i(k_z z + l\varphi]
$$

$$\cdot \left[\cos\varphi \left(1 + \cos\alpha - \sin^2 a\right) J_t(\rho) + \frac{\frac{l(l-1)(\cos\varphi - i\sin\varphi)}{(n+0.5)^2} J_l(\rho) +}{\frac{\sin\alpha(\cos\varphi + il\sin\varphi)}{(n+0.5)} J_{l+1}(\rho)} \right] \tag{8}$$

$$H_r = \frac{1}{2} H_0 \cdot \exp[i(k_z z + l\varphi]$$

$$\cdot \left[\sin\varphi \left(1 + \cos\alpha - \sin^2\alpha\right) J_l(\rho) + \frac{l(l-1)(\sin\varphi + i\cos\varphi)}{(n+0.5)^2} J_l(\rho) + \frac{\sin a(\sin\varphi - il\cos\varphi)}{(n+0.5)} J_{l+1}(\rho) \right] \tag{9}$$

with

$$\rho = (n + 1/2) \sin\alpha \tag{10}$$

using integral localized approximation, the beam factor can be calculated as

$$\begin{bmatrix} g_{n,\text{TM}}^m \\ g_{n,\text{TE}}^m \end{bmatrix} = \frac{Z_n^m}{2\pi} \int_0^{2\pi} \begin{bmatrix} E_r^{\text{Loc}}(r,\theta,\varphi)/E_0 \\ H_r^{\text{Loc}}(r,\theta,\varphi)/H_0 \end{bmatrix} \exp(-im\varphi) d\varphi \tag{11}$$

in which

$$Z_n^m = \begin{cases} \frac{2n(n+1)i}{2n+1}, & m = 0 \\ \left(\frac{-2i}{2n+1}\right)^{|m|-1}, & m \neq 0 \end{cases} \tag{12}$$

$g_{n,\text{TM}}^m$ and $g_{n,\text{TE}}^m$ are the TM wave beam factor and TE wave beam factor, respectively. And Z_n^m is the transmission factor. After taking the radial component of the electromagnetic field intensity of the Bessel vortex beam and using the orthogonality of the exponential and trigonometric functions, the expression for the beam factor of the axially incident high-order Bessel vortex beam with sphere scattering is calculated as

$$\begin{cases} g_{n\cdot\text{TM}}^m = \frac{Z_n^m}{4} \exp(ik_z z) \left[\begin{array}{l} \left(1 + \cos\alpha - \sin^2\alpha\right) \cdot J_l(\rho) \\ + \frac{(l+1)\sin\alpha}{n+1/2} J_{l+1}(\rho) \end{array} \right], & m = l+1 \\ g_{n,\text{TE}}^m = -i g_{n\cdot\text{TM}}^m \end{cases} \tag{13}$$

$$\begin{cases} g_{n,\text{TM}}^m = \frac{Z_n^m}{4} \exp(ik_z z) \left[\begin{array}{l} \left(1 + \cos\alpha - \sin^2\alpha\right) \\ \cdot J_l(\rho) + \frac{2l(l-1)}{(n+1/2)^2} J_l(\rho) \\ - \frac{(l-1)\sin\alpha}{n+1/2} J_{l+1}(\rho) \end{array} \right], & m = l-1 \\ g_{n,\text{TE}}^m = i g_{n,\text{TM}}^m \end{cases} \tag{14}$$

The scattered electric field strength of θ and φ components can be expressed as

$$
\begin{bmatrix} E_\theta^{sca} \\ E_\varphi^{sca} \end{bmatrix} = \frac{E_0 \exp(-ikr)}{kr} \begin{bmatrix} i\sum_{n=1}^{\infty}\sum_{m=-n}^{+n} \frac{2n+1}{n(n+1)}\Big[a_n g_{n,TM}^m \tau_n^{|m|}(\cos\theta)+ \\ imb_n g_{n,TE}^m \pi_n^{|m|}(\cos\theta)\Big]\exp(im\varphi) \\ -\sum_{n=1}^{\infty}\sum_{m=-n}^{+n}\frac{2n+1}{n(n+1)}\Big[ma_n g_{n,TM}^m \pi_n^{|m|}(\cos\theta)+ \\ ib_n g_{n,TE}^m \tau_n^{|m|}(\cos\theta)\Big]\exp(im\varphi) \end{bmatrix}
\tag{15}
$$

where $P_n^m(\cos\theta)$ is the associated Legendre polynomials and n represents degree and m represents order, respectively. a_n, b_n are Mie scattering coefficients.

The scattering characteristics of a sphere to high-order vector Bessel vortex beams is analyzed by numerical calculations based on the above theoretical basis, The scatter field E^{sca} is obtained. Then, the RCS σ is measured by

$$
\sigma = \lim_{r\to\infty} 4\pi r^2 \frac{|E^{sca}|^2}{|E_0|^2}
\tag{16}
$$

3 Results

In this section, the characteristics of the scattering field are analyzed, and the RCS values of the PEC sphere and PEC cube lighted by a high order Bessel beam are calculated, respectively. To prove the validity of the method, the RPIM (radial point interpolation method) [12, 13] algorithm is used to verify the correctness. The RPIM, as a numerical method for calculating electromagnetic scattering, offers a higher computational speed and accuracy than other algorithms, and using node distribution instead of establishing a grid. In addition, the components of the scattered field are also analyzed by the RPIM algorithm.

In this paper, we consider 10 GHz as resonance frequency and a sphere of radius = 50mm. The sphere is located long the z-axis. As shown in Fig. 3, the results obtained by both the RPIM and GLMT method are generally consistent, indicating the correctness of the method we proposed. Besides spheres, the interaction of PEC cubes with Bessel beams has also been studied, and the diagram for cube scattering is shown in Fig. 4, which the side length of the cube is set to 50 mm. Figure 5 compares the RCS values for Bessel beams of different orders, including plane wave ($l = 0$) and vortex waves. The RCS curves vary equally for two modes in both Phi and Theta directions, and the maximum value occurs at Phi = 180°, and for Theta direction, the minimum value occurs at 180°. Compared with the plane wave ($l = 0$), Bessel beams with mode $l = 1$ obtain smaller RCS, and vortex electromagnetic waves can detect low RCS targets more effectively, which is of value for the detection of low RCS targets. This shows the importance of OAM waves in modern radar technology.

Similarly, the phase distribution of the scattered field of the PEC cube is also a vortex field, showing a complete first-order mode scattered field, as can be seen in

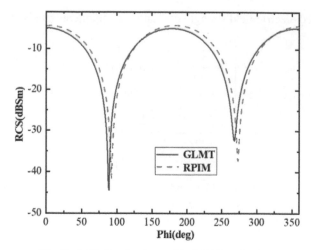

Fig. 3. RCS distributions with GLMT and RPIM

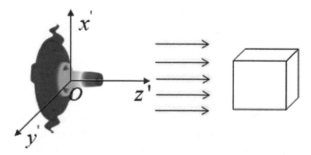

Fig. 4. Illustration for the scattering of PEC cube

Fig. 7. Compared with the PEC sphere, the phase distribution of the first-order scattered field is not very different. That is because the two objects are closer in size and the vortex waves are less able to recognize them.

Now the sphere is displaced by 50 mm along the direction perpendicular to the Bessel vortex wave axis. It can be predicted that the scattering phase diagram of the sphere is no longer in a complete spiral gradient at this point. The degree of phase distortion becomes larger, as can be seen in Fig. 7. Similarly, the cube is shifted 50 mm perpendicular to the axis of the vortex wave, as can be seen in Fig. 6. It can be seen from Fig. 7 that the scattered phase of the cube on the first-order Bessel is also not a complete first-order spiral distribution. Moreover, the vortex waves can be seen to have a higher discrimination of the cube. The simulation data shown in Fig. 7 verifies it. The degree of offset can be inferred from the degree of phase distortion of the object.

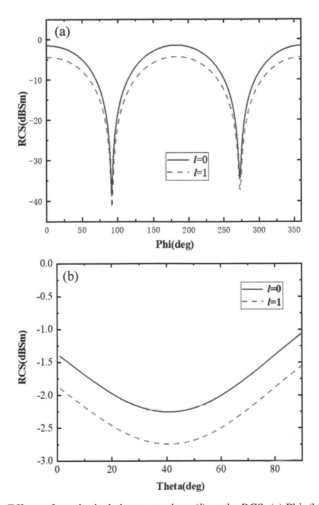

Fig. 5. Effects of topological charge numbers (*l*) on the RCS. (a) Phi, (b) Theta

4   Conclusion

In this paper, a mathematical description of the interaction between a high-order Bessel vortex beam and a target is established using an analytical method. It is also verified by numerical methods. The RCS values of the higher-order Bessel show that its scattering properties are different from those of plane waves (*l* = 0), which could offer new prospects for radar detection and identification. It is also shown that the scattered fields of different targets irradiated by OAM waves exhibit different phase distributions. In addition, more conclusions can be drawn from the phase distribution of the scattered echo signals of different objects, providing a new technical approach to target detection and identification. In future work, the scattering of OAM beams by complex structural objects will be investigated.

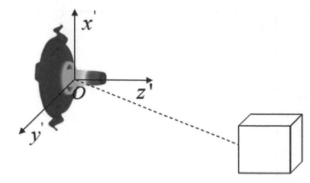

Fig. 6. The cube shifted perpendicular to the axis of the OAM wave

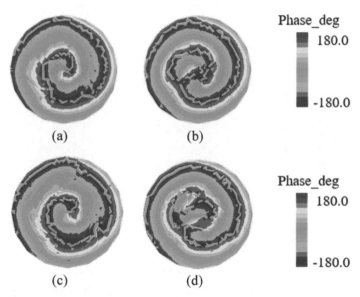

Fig. 7. Scattering phase by the Bessel beam with $l = 1$. (a) PEC sphere, (b) PEC sphere shifted (c) PEC cube and (d) PEC cube shifted.

References

1. Liu, K., Gao, Y., Li, X. et al.: Target scattering characteristics for OAM-based radar. AIP Advances **8** (2018)
2. Chen, H.-T., Zhang, Z.-Q., Yu, J.: Near-field scattering of typical targets illuminated by vortex electromagnetic waves. Appl. Comput. Electromagnet. Soc. J. (ACES) **129–134** (2020)
3. Barnett, S.M., Allen, L., Cameron, R.P. et al.: On the natures of the spin and orbital parts of optical angular momentum. J. Opt. **18** (2016)
4. Yang, Y., Nie, Z., Feng, Y. et al.: Internal and near-surface fields for a charged sphere irradiated by a vector Bessel beam. J Quant. Spectrosc. Radiat. Transfer **240** (2020)
5. Gong, S., Li, R., Liu, Y., et al.: Scattering of a vector Bessel vortex beam by a charged sphere. J. Quant. Spectrosc. Radiat. Transfer **215**, 13–24 (2018)

6. Zhang, C., Chen, D., Jiang, X.: RCS diversity of electromagnetic wave carrying orbital angular momentum. Sci. Rep. **7**, 15412, Nov 13 (2017)
7. Wu, Z., Qu, T., Wu, J. et al.: Scattering of electromagnetic waves with orbital angular momentum on metallic sphere. IEEE Antennas Wirel. Propag. Lett. **19**, 1365–1369, Aug (2020)
8. Liu, K., Liu, H., Sha, W.E.I., et al.: Backward scattering of electrically large standard objects illuminated by OAM beams. IEEE Antennas Wirel. Propag. Lett. **19**, 1167–1171 (2020)
9. Volke-Sepulveda, K., Garcés-Chávez, V., Chávez-Cerda, S., et al.: Orbital angular momentum of a high-order Bessel light beam. J. Opt. B: Quantum Semiclassical Opt. **4**, S82–S89 (2002)
10. Qu, T., Li, H., Wu, Z., Shang, Q., Wu, J., Kong, W.: Scattering of aerosol by a high-order Bessel vortex beam for multimedia information transmission in atmosphere. Multimedia Tools Appl. **79**(45–46), 34159–34171 (2020). https://doi.org/10.1007/s11042-020-08773-1
11. Mitri, F.G.: Electromagnetic wave scattering of a high-order Bessel vortex beam by a dielectric sphere. IEEE Trans. Antennas Propag. **59**, 4375–4379 (2011)
12. Luo, K., Duan, Y.T., Yi, Y. et al.: Implementation of PEC boundary condition for Laguerre-RPIM meshless method with novel uniform nodal distribution. Int. J. RF Microwave Comput.Aided Eng. **27** (2017)
13. Yiqiang, Y., Zhizhang, C.: A 3-D radial point interpolation method for meshless time-domain modeling. IEEE Trans. Microw. Theory Tech. **57**, 2015–2020 (2009)

A Zero Trust and Attribute-Based Encryption Scheme for Dynamic Access Control in Power IoT Environments

Wenhua Huang[1], Xuemin Xie[1(✉)], Ziying Wang[2], and Jingyu Feng[1]

[1] Xi'an University of Posts and Telecommunications, Xi'an 710121, China
{huangwh,fengjy}@xupt.edu.cn, xxm08042@163.com
[2] Electric Power Research Institute, State Grid Jiangsu Electric Power Co, Ltd, Nanjing 211103, China
wangzy16@js.sgcc.com.cn

Abstract. With the rapid development of information technologies in the power industry, many power devices are connected to the Internet, thus expanding the exposure. Attackers could control some devices with weak security capabilities as compromised devices to steal sensitive data inside the power Internet of Things (IoT). Traditional access control schemes are suitable for ordinary devices, but they cannot provide enough data protection against compromised devices. In this paper, we propose a zero trust and attribute-based encryption scheme for dynamic access control in Power IoT environments. In order to protect the privacy information, we hide part of the access policy to ensure that the data owner verifies the attribute set of access entities without knowing the complete access policy structure. Meanwhile, we continuously monitor the network behavior of the access entities and calculate their trust value in real-time, which can avoid access entities with unauthorized attribute sets and abnormal network behavior to gain access permissions. The simulation results show that our scheme can increase the interception rate of malicious access entities.

Keywords: Zero trust · Access control · Compromised devices · Power IoT · Trust evaluation

1 Introduction

With the in-depth development and widespread use of Internet of things (IoT) applications, more and more industries are driven by innovation [1]. Power IoT is a concrete manifestation of the application of IoT and emerging technologies in the power environments industry. It connects users, enterprises, and devices, and shares the data generated to achieve the interconnection between the various parts of the power environments [2, 3]. The services and functions provided also tend to be more diversified and precise. However, while the construction scale of the power IoT continues to expand, its complexity is also increasing. The types and numbers of smart devices in the access system have increased significantly. How to ensure the confidentiality of sensitive data such as

© The Author(s), under exclusive license to Springer Nature Switzerland AG 2023
N. Xiong et al. (Eds.): ICNC-FSKD 2022, LNDECT 153, pp. 1338–1345, 2023.
https://doi.org/10.1007/978-3-031-20738-9_145

business information and user electricity consumption information stored in the intranet has become an urgent problem to be solved.

To protect the data security and privacy information in power IoT environments. Chim et al. [4] proposed a signature authentication scheme based on hash message verification codes. This signature scheme has low computation and transmission overhead. The BLS short signature authentication used by Lu et al. [5] is a bilinear cryptographic system, which sacrifices a certain amount of calculation and communication overhead while ensuring security. Ni et al. [6] used a combination of trapdoor hash function and Paillier encryption technology to ensure the confidentiality and integrity of data, and used homomorphic authenticators to resist attacks from malicious gateways. Yang et al. [7] proposed a blockchain-based power IoT data security sharing mechanism, which controls attribute access and uses blockchain to record authorization data. The abovementioned schemes with centralized digital authentication, which are difficult to satisfy the requirements for efficient interconnection between smart devices in power IoT environments. Devices with weak identity authentication are easily targeted by attackers. Once a malicious attacker enters the intranet under the cloak of a legal identity, it is likely to perform ultra vires operations and steal confidential data, causing immeasurable losses. In a word, the core problem facing the security protection of the power IoT is that the boundary between the network and the application is being broken [8], and the entire network security construction needs to be coordinated from the architectural level. It is necessary to establish the same protection model at the macro level, and satisfy the requirements of individualization and dynamics at the micro level [9]. But most of them are based on systems with strong security protection capabilities and do not consider security access authentication in the case of compromised devices. The security boundary of the power IoT is gradually blurred, and some devices with weak security capabilities may be attacked and become compromised devices which can be controlled to steal sensitive data inside Power IoT environments.

In this paper, we proposed a zero trust and attribute-based encryption scheme for dynamic access control in Power IoT environments. Construct the dynamic authorization to guarantee security during the transmission of sensitive data.

2   Zero Trust Architecture and Compromised Devices

2.1   Zero Trust Architecture

Zero trust is an architecture that is different from the traditional one with only pays attention to border protection security, but pays more attention to data protection. The core principle of zero trust is "never trust, always verify" [10]. Google's "BeyondCorp" model [11] is a case of implementing access control under a zero trust architecture. Its access policy is based on device attribute information, status, and corresponding users, and pays more attention to the correlation analysis between user behaviors and device status. In 2019, the National Institute of Standards and Technology (NIST) standardized the zero trust principle, the logical components in the architecture, and the deployment scenarios of the architecture [12]. Currently, the application of zero trust architecture in the industry is relatively mature, but it is still in its infancy in the academic world. The zero trust architecture is mainly composed of three modules as follows:

Control plane: It consists of two parts: dynamic trust evaluation and security access control. The control plane is responsible for continuous monitoring and evaluation of the access entities, which can ensure the security and controllability of the access process and provide an execution basis for the data plane.

Data plane: The trusted access agent is considered as the first gateway to secure access and the middleware that the access entities must pass through. The implemented policy is provided by the identity security infrastructure and the control plane.

Identity security infrastructure: The key support component is used to realize the ability of the zero trust system. Complete identity management and authority management, and provide access policy for trust access agents.

2.2 Compromised Devices

With the wide application of emerging technologies such as 5G, cloud computing, and blockchain in power IoT environments, more and more smart devices are being accessed and used [13], and their functions are not limited to simple data collection, but more diversified functions such as data analysis and monitoring have been added. Once these sensitive data involving user and system private information are leaked, it will cause immeasurable economic losses.

The growing number and type of smart devices are blurring the security boundaries of power IoT, and thus increasing the risk of sensitive data leakage. The current network defense system featuring border isolation is inadequate to deal with such attacks [14]. Attackers can look for smart devices with vulnerabilities on a large scale, control them as compromised devices in order to bypass border defenses under the guise of "legitimate identities." Then attackers can use compromised devices as the springboard to penetrate inside the power IoT environments, and thus control the high authority devices through APT attacks, supply chain attacks and other attacks to steal sensitive data.

3 Zero Trust Access Control Scheme

3.1 System Architecture

As shown in Fig. 1, our scheme is consisted of data owner (DO), access entity (AE), power IoT data center (PDC), zero trust access control center (ZTAC), and continuous perception analysis center (CPAC).

DO: A DO is responsible for the encryption and upload of sensitive data, and formulating access policies. When receiving a data access request, send part of the access tree structure of the requested data to ZTAC.

AE: As the data access party, it can be a user operating device or a smart device of the power IoT. If it can pass the authentication of ZTAC (both the attribute set and the trust value of AE must satisfy the conditions), then the decryption key of the required data can be obtained.

PDC: Sensitive data in the power IoT are stored in terms of ciphertext. PDC also provides a list of IP addresses for trusted access entities.

ZTAC: The ZTAC is responsible for the authentication, identification, evaluation of AE, and decryption of the requested sensitive data.

CPAC: The CPAC will continuously monitor and analyze the behavior of the AE, calculate the current trust value of the AE, and provide a basis for the dynamic policy-making of ZTAC.

The overview procedure of our scheme can be divided into three phases:

Encryption and Policy Setting Phase: In this phase, the system parameters will be initialized, such as bilinear group and *ElGamal* signature system. AE sets its own attribute set. DO builds an access policy corresponding to sensitive data. Then, DO uses the AES-256 encryption algorithm to encrypt the data and uploads it to PDC, and the key of AES-256 will be encrypted by CP-ABE algorithm.

Access Control Phase: AE initiates a data access request to ZTAC. ZTAC requests an access policy from DO and verifies whether the attribute set of AE satisfies the policy. Then request the trust value BT_k of AE current cycle from CPAC, and compare it with the trust threshold of the requested data TS_{data}.

Data Access Phase: After successfully passing ZTAC authentication, AE can obtain the AES-256 e-key and decrypt it with SK to obtain the key for decrypting the requested data.

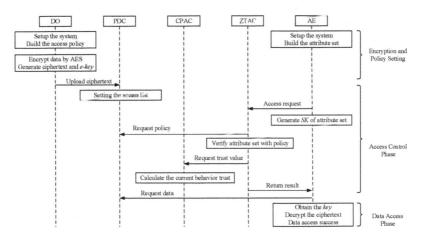

Fig. 1. The overview procedure of the scheme

3.2 Dynamic Access Control

In our zero trust access control system, four phases are involved to protect sensitive data, including authentication, identification, evaluation, and decryption.

Authentication. AE initiates a data access request to the zero trust gateway (ZTAG). After ZTAG receives the access request, it will determine whether it is a legitimate request. The access request is forwarded to PDC through ZTAG, and the IP of AE is inquired whether it is in the list of accessible devices. If it does not exist, ZTAG will directly reject the request. If it exists, the *ElGamal* signature system is used to sign the data packet, and the signed access request data packet is forwarded to the zero trust policy engine (ZTAE) for identification.

Identification. After ZTAG completes the authentication of AE. It verifies the signed message to ensure that the message has not been tampered with, and to secure the integrity and correctness of the message. After the verification is completed, ZTAP initiates the access policy tree structure corresponding to the requested data to DO. After DO receives the request, it sends the access tree with the end leaf nodes removed to ZTAP. After that, ZTAP calls the recursive algorithm *DecryptNode* of CP-ABE algorithm, and substitutes all the values of AE into the computation to verify whether the attribute set of AE satisfies the policy P_{data}. Finally, AE will return the result to the zero trust access engine (ZTAE).

Evaluation. If AE successfully passes the authentication and identification, and ZTAE receives the result returned by the authentication algorithm as True, it will sign the IP information and timestamp information *tp* of AE accessed this time and send them to CPAC. Once receiving this information, CPAC calculates the trust value of AE in real-time based on the historical data set of AE, and compares with the trust threshold TSdata of the requested data. If it is greater than the threshold, AE obtains the access right of the requested data, otherwise the ruling result returns False.

We divide the behavioral features of AE into two parts: network behavior and threat behavior. The network behavior records the statistical data of AE log, and the threat behavior records the illegal access behavior of AE and whether there are possible attacks, such as illegal links and unauthorized operations. The specific behaviors are shown in Table 1.

Table 1. AE behavior features.

Types of behavior	Represent	Specific behavior
Network behavior *NT*	qt_1	Number of suspicious packets per unit time
	qt_2	Number of abnormal throughput rate per unit time
	qt_3	Access time per unit time
Threat behavior *PT*	pt_1	Access request port address exception times per unit time
	pt_2	Number of ultra vires operations per unit time

The trust evaluation value of each dimension is determined by the deviation value between the current behavior data and the historical behavior data. The duality of access behavior can evaluate the trust value. Cqt_i and Hqt_i respectively represent the historical value and current period value of the AE network behavior, which are calculated as follows.

$$Hqt_i = \frac{1}{n-1} \sum_{k=1}^{n-1} qt_{ik} \qquad (1)$$

Among them, n represents the access total number of AE, and qt_{ik} represents the value of various behavior of AE for the kth time. As access number continues to increase, the value of Hqt_i has also been in continuous dynamic changes.

We calculate the trust value of the network behavior of AE_k through the function $Bate(\alpha, \beta)$. Tae_k and Fae_k respectively represent the number of normal and abnormal trust behavior, when $Cqt_i < Hqt_i$, $\alpha = Tae_k + 1$, otherwise $\beta = Fae_k + 1$. The network trust value of AE_k can be obtained as follows.

$$NT = Beat(Tae_k + 1, Fae_k + 1) \qquad (2)$$

Considering $\varphi(x) = (x - 1)$, when x is an integer, the expectation of the function is as follows.

$$E[Beta(\alpha, \beta)] = \frac{\alpha}{\alpha + \beta} \qquad (3)$$

Then we can obtain NT as follows.

$$NT = \frac{Tae_k + 1}{Tae_k + Fae_k + 2} \qquad (4)$$

For the sake of stealing data, the malicious attackers will conduct a large number of unauthorized link behaviors after entering the system. It is necessary to introduce a threat attenuation factor to improve the basic trust value. The threat factor PT is calculated as follows.

$$PT = \frac{1}{\lambda} \sum_{i=1}^{n} (pt_i \cdot w_{pt_i}) \qquad (5)$$

$\lambda = \sum_{i=1}^{n} w_{pt_i}$ is the normalized coefficient, which ensures the value range of PT is $[0,1]$. w_{pt_i} represents the proportion of different behaviors of threat behavior in this trust value, which is defined as follows.

$$w_{pt_i} = \left| \sin(\frac{pt_i}{2 * S} * \pi) \right| \qquad (6)$$

In the above formula, S represents the sum of the values of all behaviors. After introducing threat factors, the dynamic trust value BT_k of AE_k is calculated as follows.

$$BT_k = \begin{cases} \frac{Tae_k+1}{Tae_k+Fae_k+2}, & PT = 0 \\ \frac{Tae_k+1}{Tae_k+Fae_k+2} - \frac{1}{\lambda} \sum_{i=1}^{n} (pt_i \cdot w_{pt_i}), & PT > 0 \end{cases} \qquad (7)$$

Decryption. If the ruling result of AE access request at this time is True, the zero trust computing engine (ZTAU) verifies the accuracy of the signed data packet. After the verification is passed, the decryption algorithm Decrypt of the CP-ABE algorithm is called to decrypt e-key, and sent key to AE. After obtaining the key, AE requests the data ciphertext from PDC, and uses key to decrypt C_{data}.

4 Performance Analysis

We perform simulation and performance testing on our scheme, and analyze the results with 64-bit PC with Intel Core i5-7200 CPU (2.50 GHz) and 4 GB of RAM. Then, we conduct simulation experiments on the VAST Challenge dataset [15] to observe the effect of our scheme against compromised devices. We analyze the ability of our scheme to intercept malicious AEs and compare it with the access control scheme based on CP-ABE [16]. The interception rate is defined as the ratio of the number of malicious AEs actually intercepted to the total number of malicious entities. In our experiment, we set the proportions of malicious entities of different sizes, and calculate the interception rates of the two schemes under different proportions. The results obtained are shown in Fig. 2a. We can see that as the proportion of malicious entities, the interception rate of the access control scheme based on CP-ABE gradually decreases, but the interception rate of our scheme has always been maintained at about 90%.

We also analyze the precision and recall rates under different AE densities With a view of analyzing the comprehensive evaluation effect of our scheme, we compare our scheme with the BTRES method [17]. The misjudgment situation is set as follows. A normal AE whose trust value is higher than 0.6 but not satisfying the access tree structure is false positive. The trust value is in the interval [0.5, 0.7] but the malicious Aes satisfying the access tree structure is false negative. To avoid the randomness of the experimental data, each experiment was performed 10 times, and the average of the statistical results was taken as the final data. Figure 2b. Shows the simulation results under different access entity densities when the proportion of malicious nodes is 30%. It can be seen from Fig. 2c. That the precision and recall rate of our program is always higher than BTRES.

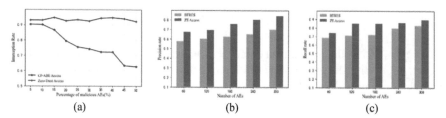

| (a) | (b) | (c) |

Fig. 2. Comparison of interception rate, precision and recall in different AEs densities

5 Conclusion

The growing cybersecurity threat requires an architectural redesign of confidential data access. We have proposed a scheme to ensure the security of sensitive data in power IoT environments against compromised devices. Our scheme is based on zero trust architecture, with the assistance of continuous perception analysis, the access behavior of the access entities in the system is continuously monitored, and the trust value is calculated in real-time. Future research will continue to refine the scheme and verify its performance.

Acknowledgments. This work is supported by the National Natural Science Foundation of China (62102312) and the State Grid Technology Project (J2021206).

References

1. Bajic, B., Rikalovic, A., Suzic, N., Piuri, V.: Industry 4.0 implementation challenges and opportunities: a technological perspective. IEEE Syst. J. (2021)
2. Mao, Y., You, C., Zhang, J., Huang, K., Letaief, B.: A survey on mobile edge computing: The communication perspective. IEEE Commun. Surv. Tutorials **19**(4), 2322–2358 (2017)
3. Wang, Y., Tao, X., Zhang, X., Zhang, P., Hou, Y.T.: Cooperative task offloading in three-tier mobile computing networks: an ADMM framework. IEEE Trans. Veh. Technol. **38**(3), 2763–2776 (2019)
4. Chim, T.W., Yiu, S.M., Hui, L.C.K., Li, V.O.K.: PASS: Privacy-preserving authentication scheme for smart grid network. In: 2011 IEEE International Conference on Smart Grid Communications, pp. 196–201. IEEE (2011)
5. Lu, R., Liang, X., Li, X., Lin, X., Shen, X.: EPPA: an efficient and privacy-preserving aggregation scheme for secure smart grid communications. IEEE Trans. Parallel Distrib. Syst. **23**(9), 1621–1631 (2012)
6. Ni, J., Alharbi, K., Lin, X., Shen, X.: Security-enhanced data aggregation against malicious gateways in smart grid. In: IEEE Global Communications Conference, pp. 1–6. IEEE (2015)
7. Yang, H., Bai, Y., Zou, Z., Zhang, Q., Wang, B., Yang, R.: Research on data security sharing mechanism of power internet of things based on blockchain. In: International Information Technology and Artificial Intelligence Conference, vol. 9, pp. 2029–2032. IEEE (2020)
8. Bogner, E.: The zero-trust mandate: never trust, continually verify. Softw. World **50**(4) (2019)
9. Gutmann, A., Renaud, K., Maguire, J., Mayer, P., Volkamer, M., Matsuura, K., Müller-Quade, J.: ZETA—zero-trust authentication: relying on innate human ability, not technology. In: IEEE European Symposium on Security and Privacy, pp.357–371. IEEE (2016)
10. Decusatis, C., Pinelli, M.: Implementing zero trust cloud networks with transport access control and first packet authentication. In: IEEE International Conference on Smart Cloud, pp. 5–10. IEEE (2016)
11. Ward, R., Beyer, B.: BeyondCorp: a new approach to enterprise security. Mag. Us-enixand Sage **39**(6), 6–11 (2014)
12. Rose, S., Borchert, O., Mitchell, S., Connelly, S.: Zero trust architecture. National Institute of Standards and Technology (2020)
13. Liu, T., Ma, Y., Jiang, H.F.: Research on power grid security protection architecture based on zero trust. Electric Power Inf. Commun. Technol. **19**(7), 25–32 (2021)
14. Samaniego, M., Deters, R.: Zero-trust hierarchical management in IoT. In: IEEE International Congress on Internet of Things (ICIOT), pp. 88–95. IEEE (2018)
15. VAST.http://www.vacommunity.org/VAST+Challenge+2013%3A+Mini-Challenge+3 (2013)
16. Yang, K., Han, Q., Li, H., Zheng, K., Shen, Z.X.: An efficient and fine-grained big data access control scheme with privacy-preserving policy. IEEE Internet Things J. **4**(2), 563–571 (2017)
17. Fang, W., Zhang, C., Shi, C., Zhao, Q., Shan, L.: BTRES: beta-based trust and reputation evaluation system for wireless sensor networks. J. Netw. Comput. Appl. **59**, 88–94 (2016)

A Parallel Implementation of 3D Graphics Pipeline

Wenjiong Fu[(✉)], Tao Li, and Yuxiang Zhang

Xi'an University of Posts and Telecommunications, 618 West Chang'an Street, Chang'an District, Xian, China
{wjfu,zhangyuxiang}@stu.xupt.edu.cn, litao@xupt.edu.cn

Abstract. Graphics pipelines include many custom accelerators which are difficult to design and not readily scalable. A graphics rendering pipeline mainly includes a front-end processor (FEP), a primitive assembler, a vertex shader stage (performing model view transform, vertex coloring, projection transform), back-face culling, 3D clipping, window transform, rasterizer, pixel shading, and fragment operations. These accelerators determine the performance of rendering. The main operation of the rasterization stage is to convert primitives into pixels. Rendering without accelerators is also viable if the SIMT (Single-Instruction Multiple-Thread) engines on a modern GPU (Graphics Processing Unit) are well-utilized. Compared with the standard TBR algorithm, this paper improves the original serial algorithm into a parallel algorithm to improve the rendering performance. A highly parallel implementation in this approach, from the very first stage of primitive assembly to fragment operations, boosts performance as experimental results indicate.

Keywords: Parallel rendering · GPU · 3D pipeline · Rasterization

1 Introduction

In recent years, with the booming in fields such as natural language processing, computer vision and autonomous driving, the GPU market performance is in short supply. Today, with the increasing market demand, the popularity of the domestic GPU market has attracted widespread attention, and many GPU startup teams have received significant amount of capital investment. However, due to the high complexity of graphics GPU design, many startup companies in China focus on the development of General-Purpose GPU [1], which does not have graphics accelerators. GPGPU processors are mainly used for accelerating AI and general-purpose computing tasks originally processed by the CPU. In addition, with the advent of the 5G era, according to the "New Moore's Law", the amount of new data generated every 18 months equals the total amount of data in the history of computing. Today, the demand for computing power for processing the super-large data volume has reached an unprecedented height and intensity, which in turns drives the GPGPU market. The design of AI (Artificial Intelligence) accelerators and GPGPU engines are not driven by graphics rendering, hence with a lack of graphics

© The Author(s), under exclusive license to Springer Nature Switzerland AG 2023
N. Xiong et al. (Eds.): ICNC-FSKD 2022, LNDECT 153, pp. 1346–1354, 2023.
https://doi.org/10.1007/978-3-031-20738-9_146

accelerator hardware. On the other hand, most computing servers do need graphics rendering capabilities. It is therefore important to implement graphics functionality on these highly parallel GPGP engines. General-purpose GPUs support for general-purpose computing libraries and languages (such as OpenCL (Open Computing Language), CUDA [2] (Compute Unified Device Architecture)) enable highly parallel software rendering.

There is previous research [3, 4] on parallel rendering using GPGPU facilities. This research demonstrates the feasibility of graphics rendering on GPGPU engines. A sorting approach to the classification of parallel rendering is given in [5]. This paper proposes further improvements on graphics rendering, including parallel primitive assembly, parallel repacking after back-face cull and 3D clipping, etc. This approach parallelizes the entire rendering pipeline.

2   GPU Rendering

The basic structure of the OpenGL ES 2.0 (OpenGL for Embedded System) graphics rendering pipeline is shown in the figure below, including front-end processing, primitive assembly, model view transformation, vertex shader, projection transformation, culling, primitive clipping, viewport transformation, rasterization, pixel shader and end Operations and other modules are finally rendered to the frame buffer [6, 7]. The specific process is shown in Fig. 1. Figure 2 shows the flow of data in the graphics pipeline. The application interface API (Application Programming Interface) of OpenGL ES 2.0 is used in this research. As a subset of OpenGL (Open Graphics Library), it is widely used in mobile handheld devices and embedded devices due to its advantages such as complete functions and multi-platform support.

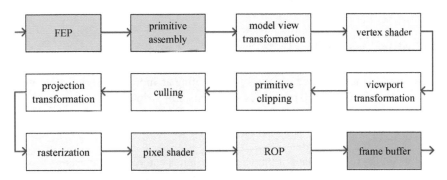

Fig. 1. A3D pipeline architecture

Figure 2 shows that the light green rectangles represent data related to the graphics pipeline, and the light-yellow rectangles represent threads. The pink arrows between the yellow and green data boxes in the figure represent the data flow of the pipeline from one stage to the next. All data in the figure has been copied from CPU to GPU. The left side of the first data frame represents the attributes of the vertex, and the right side represents the index of the vertex, and the attributes and indexes of each vertex are in a one-to-one correspondence. The second data frame primitive assembly represents the vertex indices of the first data frame as input to the second data frame for primitive assembly, where one thread assembles a primitive. The third data frame culling indicates that after the assembly of the primitives is completed, the culling operation is performed on the back or front primitives specified by the user, to reduce the amount of data in the subsequent pipelines. The fourth data frame, clipping, indicates that the clipping operation will be performed on the part beyond the front and back of the view volume. A triangle can generate up to three triangles after clipping the front and rear faces if the primitives in the view volume do not change. The fifth and sixth data frames represent the bin and tile operations of the rasterization stage. The last block diagram represents the pixel-wise processing of the ROP (raster operations pipeline) stage. The detailed implementation of each stage will be explained in the following chapters.

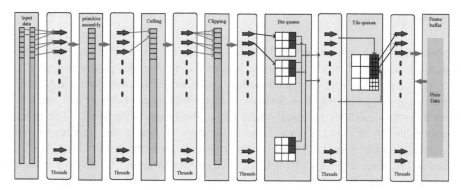

Fig. 2. A high-level diagram of our software rasterization pipeline.

3   Primitive Assembly

Element assembly refers to the process of assembling an index of a series of points into an element. In OpenGL ES2.0 primitives are points, line strips, line loop, triangle strips, triangle fans, and triangles.

The steps of the primitive drawing algorithm are as follows:

Step 1: Obtain the relevant data of the vertex attributes entered by the user.
Step 2: Copy the data from the CPU to the GPU through the cudaMemcpy function.
Step 3: Assign the data that has been copied to the GPU from global memory to share memory.
Step 4: According to the index of the assembly point of the type of the primitive, each time a vertex is assembled, the used counter of the vertex is incremented by 1.
Step 5: According to the assembled index, you can find the vertex corresponding to the primitive and its attributes. The main purpose of assembling the primitive with the vertex index is to save the storage space of share memory. Figure 3 shows the corresponding relationship between the vertex index and the triangle when the triangle strip, triangle fan and triangles are assembled. The triangle in Fig. 3a and b and the first triangle will share

an edge, but to make all triangles Both are counterclockwise, or both are clockwise, and the directions of the shared edges are different. If the vertex data is insufficient to form a triangle when assembling triangles, the vertex is not assembled.

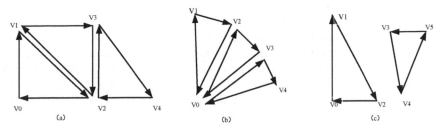

Fig. 3. (a) A triangle strips. (b) A triangle fans. (c) Independent triangles.

4   Culling

In the 3D graphics world, an object is composed of two faces. The side who's normal is the same as the direction of the camera (that is, the direction of the person's viewing angle) is regarded as the front, and the side who's normal is opposite to the direction of the camera is regarded as the back. Back face culling is to mark the back face as the face that does not need to be rendered. In OpenGL ES 2.0, the clockwise and counterclockwise determination of triangles is determined according to the order of triangle assembly in the primitive assembly stage. One way to compute this area is

$$a = \frac{1}{2} \sum_{i=0}^{n-1} x_w^i y_w^{i \oplus 1} - x_w^{i \oplus 1} y_w^i \tag{4-1}$$

where x_w^i and y_w^i are the window coordinates of x and y corresponding to each vertex in the triangle and $i \oplus 1$ is $(i + 1)$ mod n. When $a > 0$, it means that the three vertices are assembled counterclockwise (GL_CCW), and the triangle will be culled if the back face is culled. Similarly, when $a < 0$, it means that the triangle is assembled clockwise (GL_CW), if the back face is culled and the back face is culled, the triangle will not be culled.

5   3D Clipping

In the 3D clipping stage of the graphics rendering pipeline in OpenGL ES 2.0, the front and rear surfaces are mainly clipped, and the clipping of the remaining surfaces is done in the rasterization stage. In this research, the Sutherland-Hodgman algorithm is mainly used in the tailoring stage. To apply the Sutherland-Hodgman algorithm to parallel programming, each thread reads a primitive and executes the Sutherland-Hodgman algorithm once. The Sutherland-Hodgman algorithm is also called the edge-by-edge clipping method. This algorithm was proposed by Sutherland and Hodgman in 1974. The main steps of the algorithm are as follows:

Step 1: Get the vertex order of the primitive, that is, the index of the vertex.
Step 2: Find the data point corresponding to the primitive according to the index.
Step 3: Mark the triangles smaller than the near plane and larger than the far plane as deleted.
Step 4: Trim the triangles that are not marked as deleted. When trimming, first determine whether an edge of the triangle traverses the near plane or the far plane. Interpolation processing is required for the edge that traverses the near plane or the far plane. (Including vertex interpolation and attribute interpolation).
Step 5: When interpolating, first interpolate the near plane and calculate the interpolation coefficient according to the value of the near plane. See formula (5-1) and use the value of the near plane as the z value of vertex coordinate interpolation and deduce the value of the attribute from the calculation result of the interpolation coefficient. See formula (5-2) for calculation. The clipping result of the near plane is used as the input result of the far plane, and then the interpolation coefficient and vertex attribute interpolation calculation are recalculated according to the value of the far plane.

$$r = (z_1 - z_i) / (z_i - z_2) \tag{5-1}$$

$$x_i = \frac{x_1 + r * x_2}{1 + r} \tag{5-2}$$

where r represents the interpolation coefficient, z_i represents the front and back of the viewing volume, and z_1 and z_2 represent the two z values of a line segment. X_1 and x_2 represent the two x-values of a line segment. X_i represents the x-coordinate value of the interpolation point after interpolation.

In order not to change the implicit order between primitives and vertices after clipping, the primitives and vertices need to be reordered. The Hills Steele Scan algorithm is used to reorder the vertices and primitives in CUDA. The steps of the Hills Steele Scan algorithm are as follows:

Step 1: Record the number of vertices generated by each thread after clipping.
Step 2: Get the number of even digits and add the odd digits after it.
Step 3: Repeat step 2 until the last number is left.
Step4: After the calculation of step 3, the Hills Steele Scan algorithm has been completed, that is, each bit of the output array stores the sum of the first n items of the array.
Step 5: Determine the index of the clipped and blanked vertex according to the subtraction of the output array and the vertex array generated after clipping, and then reorganize the vertex according to the index.
Step 6: Assemble the vertex data generated by each thread after clipping according to the triangle fans primitives, and then reorganize the primitives according to the vertex reorganization method.

After clipping the Sutherland-Hodgman algorithm and the Hills Steele Scan algorithm of repack, the clipped vertex data and primitive data will be obtained, and the clipped vertex and primitive data will be sent to the next-level pipeline as the input parameters of the next-level pipeline. After 3D clipping, many primitives are clipped from the front and back of the view volume, which greatly reduces the processing time of rasterization.

6 Rasterization

6.1 Improved TBR Algorithm

When the data incoming to the GPU follows the pipeline to the rasterization stage, to improve the degree of parallelism, it is first necessary to divide the primitives into binning buckets, and then divide them into tiles, which is called tilling. One thread per triangle, computing the bins covered by each triangle. The method of triangle bounding box is used, that is, to find the maximum and minimum values of each triangle on the x-axis and y-axis respectively, determine the rectangular bounding box according to the found values, and then determine how many times the triangle covers according to the bounding box. Bins. Finally, look at the order in which the triangles in each bin are drawn to prevent the Z value from being wrong during the depth test. After all threads are finished, filter out the bins that are not covered by triangles and remove them. Divide each bin into 8 * 8-pixel tiles [8], and then use formula (6-1) to determine whether the tile is inside the triangle. Tiles without triangles are culled to improve performance.

The equation of an edge e is given by

$$e(x, y) = Ax + By + C = 0 \qquad (6\text{-}1)$$

where $A = \Delta y$, $B = -\Delta x$, $C = y_1 \Delta x - x_1 \Delta y$.

6.2 Triangle Rasterization

Triangle rasterization will rasterize the portion of each tile that is covered by triangles based on the previously determined tiles. The triangle mainly uses the barycentric coordinate interpolation algorithm. The triangle barycentric coordinate interpolation algorithm is the interpolation of the barycentric coordinates of any properties (position, texture coordinates, color, normal, depth, material properties...) inside the triangle. Generally, we will get the attributes on the vertices of the triangle through other steps, but we need to use the attribute value of a point inside the triangle when we continue to calculate, and the barycentric coordinates can be used to obtain a smooth transition of the value inside the triangle.

In the triangular barycentric coordinate, the coordinates of the three vertices of triangle ABC are point A (x_a, y_a), point B (x_b, y_b), point C (x_c, y_c). The formula for the area of a triangle:

$$S_{ABC} = \frac{1}{2} \begin{vmatrix} x_a & y_a & 1 \\ x_b & y_b & 1 \\ x_c & y_c & 1 \end{vmatrix} = \frac{1}{2}\left[(x_b - x_a)(y_c - y_a) - (x_c - x_a)(y_b - y_a) \right] \qquad (6\text{-}2)$$

Let the area of the triangle ABC be T, then the barycentric coordinates of the triangle are based on the ratio of the areas of the triangle and the three sub-triangles. Let the barycentric coordinates of P be (α, β, γ), then,

$$\alpha = \frac{S_{BCP}}{T}, \beta = \frac{S_{APC}}{T}, \gamma = \frac{S_{ABP}}{T} \qquad (6\text{-}3)$$

Suppose $p1(x_p \pm 1)$ is also a point in the triangle ABC, and the barycentric coordinates are $(\alpha_1, \beta_1, \gamma_1)$, then we have

$$\alpha_1 = \alpha(x_p \pm 1, y_p) = \alpha(x, y) \pm \tfrac{1}{T}\Delta y_{bc} \tag{6-4}$$

$$\alpha_1 = \alpha(x_p, y_p \pm 1) = \alpha(x, y) \mp \tfrac{1}{T}\Delta x_{bc} \tag{6-5}$$

where $\Delta y_{bc} = (y_b - y_c)$, $\Delta x_{bc} = (x_b - x_c)$, through the extension of the above formula leads to a constant increase or decrease each time in the x-axis direction or the y-direction, which can improve the calculation performance. The calculation of $\beta(x_p \pm 1, y_p)$, $\beta(x_p, y_p \pm 1)$, $\gamma(x_p \pm 1, y_p)$, $\gamma(x_p, y_p \pm 1)$ is the same as $\alpha(x_p \pm 1, y_p)$ and $\alpha(x_p, y_p \pm 1)$.

Based on the known vertex coordinates, the interpolation of vertex attributes is further studied. Triangle a1b1c1 is the projection of triangle ABC on the perspective projection plane, where the barycentric coordinates of point P are (α, β, γ), and the barycentric coordinates of point p1 are $(\alpha', \beta', \gamma')$. According to the formula of the center of gravity of the triangle:

$$1 = \alpha + \beta + \gamma \tag{6-6}$$

$$z_p = \alpha \times Z_a + \beta \times Z_b + \gamma \times Z_C \tag{6-7}$$

$$z_p' = \alpha' \times Z_a' + \beta' \times Z_b' + \gamma' \times Z_C' \tag{6-8}$$

It can be transformed from Eq. (6–6) to:

$$Z_P = \left(\tfrac{Z_p}{Z_a} \times \alpha'\right) \times Z_a + \left(\tfrac{Z_p}{Z_b} \times \beta'\right) \times Z_b + \left(\tfrac{Z_p}{Z_c} \times \gamma'\right) \times Z_c \tag{6-9}$$

From formula (6–6) and formula (6–9), we get:

$$\alpha = \tfrac{Z_p}{Z_a} \times \alpha', \beta = \tfrac{Z_p}{Z_b} \times \beta', \gamma = \tfrac{Z_p}{Z_c} \times \gamma' \tag{6-10}$$

From formula (6–6), formula (6–10) we get:

$$1 = \tfrac{Z_p}{Z_a} \times \alpha' + \tfrac{Z_p}{Z_b} \times \beta' + \tfrac{Z_p}{Z_c} \times \gamma' \tag{6-11}$$

Transform the above formula into the depth interpolation calculation formula:

$$\tfrac{1}{Z_p} = \tfrac{1}{Z_a} \times \alpha' + \tfrac{1}{Z_b} \times \beta' + \tfrac{1}{Z_c} \times \gamma' \tag{6-12}$$

The attribute interpolation formula of point p is obtained from Eqs. (6–10), (6–11)

$$l_p = \tfrac{l_a}{Z_a} \times \alpha' + \tfrac{l_a}{Z_a} \times \beta' + \tfrac{l_a}{Z_a} \times \gamma \tag{6-13}$$

Assuming that the adjacent point $q(s_p + 1, y_p)$ of the point $p(x_p, y_p)$, the point q' is the projection of the point q on the perspective projection plane, then the barycentric coordinates of the q' point have two possibilities: 1) and p' When the barycenter of the point is the same $(\alpha', \beta', \gamma')$, that is, when the barycentric coordinates of q and p are the same (p and q are the same point), the depth interpolation and attribute interpolation of point q are the same as those of point p. 2) If the barycentric coordinates of q' are $(\alpha'(x \pm 1,y), \beta'(x \pm 1,y), \gamma'(x \pm 1,y))$, in the second scenario, make an attribute The following inferences can be made when interpolating.

According to the inference of the barycentric coordinates, we get:

$$\frac{1}{Z_q} - \frac{1}{Z_p} = \frac{1}{Z_a}\left(\frac{\pm(\Delta y_{pb} - \Delta y_{pc})}{T}\right) + \frac{1}{Z_b}\left(\frac{\pm \Delta y_{ac}}{T}\right) + \frac{1}{Z_c}\left(\frac{\pm \Delta y_{ab}}{T}\right) \tag{6-14}$$

According to the inference of the barycentric coordinates, we get:

$$\frac{1}{Z_q} - \frac{1}{Z_p} = \frac{1}{Z_a}\left(\frac{\pm(\Delta x_{pb} + \Delta x_{pc})}{T}\right) + \frac{1}{Z_b}\left(\frac{\pm(-\Delta x_{ac})}{T}\right) + \frac{1}{Z_c}\left(\frac{\pm \Delta x_{ab}}{T}\right) \tag{6-15}$$

After the derivation of the above formula, it can be found that the barycentric coordinates of each point can be converted into the calculation of the increment of its adjacent coordinates. Because the pixels at any point in the triangle are adjacent to other pixels, it is only necessary to determine the barycentric coordinates of one vertex, and then the barycentric coordinates of all the pixels in the triangle can be calculated by the formula in a certain order. Similarly, vertex attribute interpolation can also be implemented using barycentric coordinate interpolation. The initial barycentric coordinates can start from any vertex. The process of moving the barycentric coordinates is the process of scanning the triangle line by line.

6.3 Early-Z

Early-Z is to advance the depth test, and early-z does the same work as the depth test. It is assigned to the depth buffer according to the depth information provided by the OpenGL API function. The specific implementation is to compare the z value of the fragment with the value of the depth buffer (the comparison parameter is improved by the API). If the current fragment is behind other fragments, it will be discarded, otherwise it will be overwritten.

7 Conclusion

After improving the TBR algorithm, the computing power of post-rasterization is greatly reduced, which saves computing resources and improves rendering efficiency. The rasterization stage advances the depth test and largely eliminates many unnecessary fragments. This operation saves storage space and reduces the computational burden for the later ROP stage. This research realizes rasterization from the software level, which can render the classic scenes of OpenGL. Figure 4. Rendering of the rendered triangle. The data in Table 1 shows that with the doubling of the number of vertices and the number of threads, the simulation running time of the system does not change much. Considering the multi-vertex data in a large scene, it will consume less time. At this level, the rendering efficiency is improved.

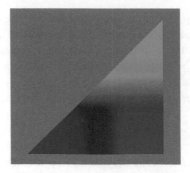

Fig. 4. Rasterized rendering result figure

Table 1. System simulation time.

Vertex num	Threads num	Simulation Time (unit: ms)
32	32	97
64	64	118
128	128	120
256	256	150

References

1. Owens, J.D., Houston, M., Luebke, D., Green, S., Stone, J.E., James, C.: GPU computing: graphics processing units powerful, programmable, and highly parallel are increasingly targeting general-purpose computing applications. Proceedings of the IEEE **96**(5), 879–899 (2008)
2. Shen, Y.: Research on the application of general-purpose GPU computing technology on high-performance computing platforms. Lanzhou University (2012)
3. Liu, F., Huang, M.-C., Liu, X.-H., Wu, E.-H.: Freepipe: a programmable parallel rendering architecture for efficient multi-fragment effects. In Proc. **I3D**, 75–82 (2010)
4. Laine, S.: HPG '11.In: Proceedings of the ACM SIGGRAPH Symposium on High Performance Graphics, pp. 79–88 (2011)
5. Molnar, S., Cox, M., Ellsworth, D., Fuchs, H.: A sorting classification of parallel rendering. IEEE Compute. Graph. Appl. **14**, 23–32 (1994)
6. Hearn, D., Pauline Baker, M., Carithers, W.R.: Computer Graphics. Translated by Cai Shijie, Yang Ruoyu. Fourth ed. Electronic Industry Press, Beijing (2014)
7. History of Graphics API Development. E-Newsletter, 2018-07-08(011). https://doi.org/10.28185/n.cnki.ndizi.2018.000382
8. Liu, Z.: Research and Implementation of tile-based triangle rasterization algorithm. Xidian University (2018)
9. Zhang, J.: Research and design of rasterization and depth pre-test unit in embedded GPU. Tianjin University (2014)

From Source Code to Model Service: A Framework's Perspective

Jing Peng[1], Shiliang Zheng[1], Yutao Li[1], and Zhe Shuai[2(✉)]

[1] HBIS Digital Technology CO., LTD., Shijiazhuang 050022, Hebei, China
{pengjing,hbzhengshiliang,hbliyutao}@hbisco.com
[2] School of Computer Science and Technology, Xidian University, Xi'an, Shaanxi, China
21031211727@stu.xidian.edu.cn

Abstract. In the current IT world, machine learning has become a powerful driving force for the development of computer science. The entire life cycle of machine learning includes data processing, code development, model training, model release, and other processes. Although there are currently mature machine learning development frameworks such as TensorFlow and PyTorch, help algorithm engineers develop quickly. However, data import, code deployment, model release and other links in the machine learning process still need to be done manually. These engineering problems distract the energy of algorithm engineering and reduce the iterative efficiency of the model. At the same time, cloud-native has become the development direction of current software systems. If machine learning can be based on cloud-native technology, it will significantly liberate productivity. Therefore, there is an urgent need to develop a one-stop machine learning platform based on cloud-native technologies.

Keywords: Machine learning · CI/CD · Workflow · Model deploy

1 Introduction

The complete life cycle of machine learning involves processes such as data preprocessing, feature engineering, model training, model evaluation, adjusting parameter, and model deployment (see Fig. 1).

It can be seen from the figure that the development process of a machine learning model is the process of continuously training the model, evaluating the model, and retraining until the prediction effect is finally achieved. In each model training iteration, it is necessary to adjust the model parameters and even the model structure so that the model under training can achieve the optimal prediction effect. Such as choosing different loss functions, classifiers, etc. At the same time, the model with a better effect will be saved and reused in other learning processes.

In the last link of the machine learning cycle, the models that meet the evaluation standards are deployed as inference services that can be called. The so-called Inference Service is to encapsulate the model and only provides some interfaces for users to input data, and obtain prediction results. There are two engineering problems in the

© The Author(s), under exclusive license to Springer Nature Switzerland AG 2023
N. Xiong et al. (Eds.): ICNC-FSKD 2022, LNDECT 153, pp. 1355–1362, 2023.
https://doi.org/10.1007/978-3-031-20738-9_147

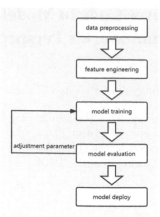

Fig. 1. Machine learning process

process of model to service. (1) The models generated by mainstream machine learning frameworks do not have the function of network services. It is necessary to use a network server or write part of the logic code of the network by yourself. (2) The model has strict requirements on the format of the input data. Still, the input data may be various heterogeneous data, such as csv, xml, txt, yaml, etc., which needs to be converted into data formats which model can resolve.

From this, we can summarize three characteristics of machine learning:

(1) Algorithm engineers need to continuously adjust parameters and even algorithm code according to the training effect of the model.
(2) The training process is complex, but there are many processes that can be reused.
(3) There are many engineering problems unrelated to the algorithm in the deployment of the model.

To solve the above problems, the idea of MLOps is put forward. The concept MLOps is proposed to introduce the technique of DevOps into the field of machine learning. The MLOps aims to support the continuous integration and continuous release of machine learning data, models and business through an automated workflow, which helps engineer develop model faster [1]. Therefore, a related tool should be developed. This tool should enable the algorithm scientist to focus on developing ML models, instead of focusing on how to run it on a production-grade system which can completely be taken over by automaton tools [2].

This article mainly solves the common engineering problems in the above machine learning inspired by the idea of MLOps. We designed and developed an end-to-end, user-friendly AI platform. Our main contributions are:

(1) Design and develop a CI/CD module that can take over the machine learning process of packaging and deploying code.
(2) Design and develop a workflow editing and execution module where users can edit and execute machine learning workflows.

(3) The model publishing module is designed and developed, and the user deploys the model as a service that can be used.

In the second 2, we introduce the industry's related research work on the AI platform and point out the shortcomings of the current research status. In the Sect. 3, we focus on the architecture of our system and the implementation of three key modules. In Sect. 4, we envision new capabilities that the system can add and future work to do.

2 Related Work

At present, relevant machine learning platforms have been developed in China. Baidu Machine Learning BML is a machine learning platform independently developed by Baidu. Relying on the machine learning algorithm library accumulated by Baidu's internal applications for many years, it can provide practical industry solutions [3]. Because the one-stop machine learning platform covers the whole process of machine learning, the following only describes the research status of some of these sub-modules.

2.1 CI/CD Tools

Continuous integration/deployment is a concept of agile development in modern software engineering. CI/CD tools are essential part of DevOps culture [4]. By closely integrating development and operation, and maintenance, agile development enables products to respond to requirements more quickly and iterate versions. CI/CD automates the work of testing and deployment in software development. Developers only need to write code and push it to the warehouse. According to the set trigger rules, developers can focus on writing logic code. At present, the mainstream CI/CD tools are Jenkins and Jenkins, but the current CI/CD tools are all used in the direction of software development, and there is no relevant case in the field of machine learning.

2.2 Workflow Engine

Although the process of machine learning consists of complex steps, many of which can be reused, we can encapsulate its code into components and organize them in sequence. It can be built quickly without writing some generic code over and over again. The Fig. 2 is an example of a machine learning workflow in which knn, roc and confusion-matrix are all common steps in machine learning. If the data transfer and work coordination between components are performed manually, it will be extremely complicated and time-consuming. For such tasks, the industry uses workflow engines, such as airflow, MLFlow, Kubeflow pipeline, and so on. Workflow enables engineers to divide complex services of building models into simpler services easily [5]. Users only need to configure, and the workflow engine can automatically coordinate the execution sequence of components and perform data transmission, log collection, and other tasks. However, the above workflow cannot be created in a visual way, and users need to write complex configuration files to describe the workflow, which is prone to format and syntax errors and is not conducive to subsequent maintenance and modification.

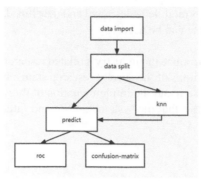

Fig. 2. Workflow example

2.3 Model Server

For model services, current mainstream machine learning framework suppliers have developed their own model servers to take over the remote invocation functions required by model services. TensorFlow launched the online native model serving system Tensorflow Serving, but in the online production environment, it is often necessary to respond to requests immediately. Tensorflow Serving, a single architecture, can no longer meet the needs of enterprises for the system, and it cannot cope with changes in traffic [6].

3 System Structure

In this paper, we design a machine learning tool. The platform includes the entire life cycle of machine learning, mainly divided into three stages:

(1) S2I (source to image) packages the algorithm as an image which can be a part of workflow.
(2) Modelflow edits and executes workflow to create model.
(3) KServe deploys the model to an inference service.

The entire system architecture can be divided into three layers (see Fig. 3).

The bottom layer is the infrastructure layer, which mainly includes the Kubernetes container orchestration cluster. Kubernetes is one of the most known open source projects for central clouds, which provides a platform for the development of the most cloud native computing technologies [7]. Kubernetes's main function is automatically managing containers and services, the greates features of which are declarative configuration and automation [8]. In addition, there is the NFS distributed file storage system. NFS can mount the folders of multiple servers to the only remote folder so that when the NFS cluster node reads and writes the folder, it is equivalent to reading and writing the same folder. When the workflow Pod is scheduled to different nodes, the uniqueness of the delivered data can be guaranteed.

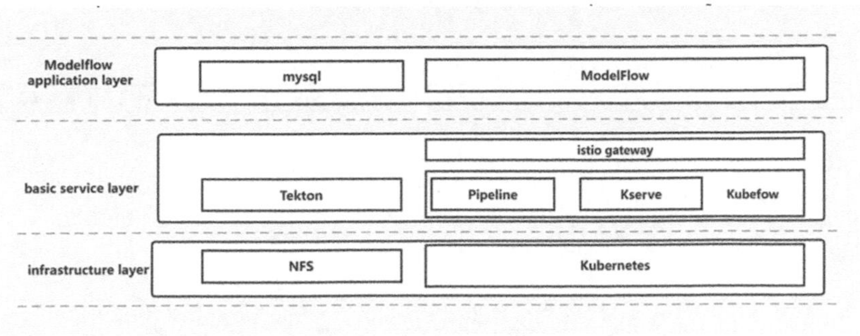

Fig. 3. Overall architecture diagram

The middle layer is the basic service layer. The open source tools Tekton, pipeline, and Kserve are deployed in the Kubernetes cluster. Pipeline is a machine learning workflow engine which provides user interface. It is a core project from kubeflow [9]. Tekton uses a separate service to access it in the form of NodePort. Pipeline and Kserve are used as part of the Kubeflow system and the both use istio gateway to expose the access entrance.

The top layer is the Modelflow application layer. The main function of the Modelflow backend is to convert the workflow diagram into a configuration file that the engine can recognize and to pass in the corresponding parameters when executing the workflow. At the same time, it provides the function of S2I and model deployment. MySQL is responsible for storing structured data that the Modelflow creates, such as information such as users, models, and workflow. The web front end of Modelflow provides the operation interface of S2I, workflow, model deploy and other functions.

3.1 S2I Architecture and Execution Process

The trigger conditions for CI/CD can be configured in the Modelflow system, which can be triggered by a timed cycle or in the form of a webhook. After the trigger condition is reached, Modelflow will send a build request to Tekton in Kubernetes to deploy the relevant Pod, and the Pod will execute the actual pulling code and packaging logic (see Fig. 4). A workflow describes the main function of Tekton. A workflow is actually a list of one or more tasks in order [10]. The pipeline is mainly divided into two steps. The first is the fetch step, which pulls the code from the git repository. Then, according to the requirement.txt file configured in the warehouse, Tekton will pull the required basic image, package the user's algorithm code into an executable image, and push it to the harbor image repository configured by the system.

3.2 Workflow Architecture and Execution Process

The existing pipeline tools are all performed by directly editing the configuration file. Directly editing the configuration file is prone to syntax errors, and the configuration file cannot visualize the workflow. A simple and user-friendly tool is needed to assist

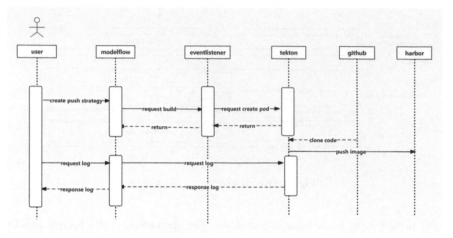

Fig. 4. S2I sequence diagram

users in editing machine learning workflow. The frontend of Modelflow will create a pipelineXml file (see Fig. 5) which describe the workflow.

```
- <userTask name="knn" id="_37_3" activiti:exclusive="true">
    - <documentation id="_3_D_3">
        - <![CDATA[
            [input] train,[output] model,[prior] _4 ,[end] true
        ]]>
    </documentation>
</userTask>
```

Fig. 5. Workflow xml example

The Modelflow backend parses the pipelineXml file to python code. The python code contains the relevant libraries of the kubeflow pipeline, which are used to compile and generate the yaml file. Finally, the python code is executed to generate the yaml file. Then Modelflow call the workflow engine interface "upipeline/apis/v1beta1/pipelines/upload" to upload yaml file. So far, the workflow is created

3.3 Model Deploy Part of the Architecture and Execution Process

In a complete machine learning life cycle, the model needs to be released as an inference service that can be called at any time. Although the existing machine learning frameworks (Tensorflow and Pytorch) have developed corresponding model server, they are mutually exclusive. That means a model server cannot be compatible with model from anther machine learning framewok. The model service of a machine learning platform should be able to adapt to all the current mainstream framework. So it is necessary to develop a model service that automatically deploys the corresponding server according to the type of model. The inference service part mainly uses the cloud native open source model server KServe. Kserve supports the current mainstream machine learning frameworks.

First, the user sends the creation requeset to Modelflow. Then, Modelflow requests Kserve in the Kubernetes cluster to create an inference service. Kserve will read the model file from the PV storage volume and create the entrance service and workload Pod. User can use the inference service from the UI of Modelflow, Modelflow will forward the request to the entrance service and return response to user.

4 Conclusion and Future Work

This paper designs a one-stop platform that can help users reduce repetitive work in machine learning work and can iteratively produce models with higher accuracy and deploy them as inference services at any time. In terms of image sharing or model sharing, there is no related function yet. In the future, we will study how to share file resources among system users.

Acknowledgement. This work is supported by the National Natural Science Foundation of China under Grant No. 62172316 and the Key R&D Program of Hebei under Grant No. 20310102D. This work is also supported by the Key R&D Program of Shaanxi under Grant No. 2019ZDLGY13-03-02, and the Natural Science Foundation of Shaanxi Province un-der Grant No. 2019JM-368.

References

1. Mei, S., Liu, C., Wang, Q., Su, H.: Model provenance management in MLOps Pipeline. In: 2022 The 8th International Conference on Computing and Data Engineering (ICCDE 2022), pp. 45–50. Association for Computing Machinery. New York, NY, USA (2022)
2. George, J., Saha, A.: End-to-end Machine Learning using Kubeflow. In: 5th Joint International Conference on Data Science & Management of Data (9th ACM IKDD CODS and 27th COMAD) (CODS-COMAD 2022), pp. 336–338. Association for Computing Machinery, New York, NY, USA (2022)
3. Yi, W.: Analysis of machine learning functions of mainstream cloud platforms in China. Sci. Technol. Commun. **11**(12), 93–94+115 (2019).https://doi.org/10.16607/j.cnki.1674-6708.2019.12.047
4. Kruglov, A., Succi, G., Vasuez, X: Incorporating energy efficiency measurement into CI\CD pipeline. In: 2021 2nd European Symposium on Software Engineering (ESSE 2021), pp. 14–20. Association for Computing Machinery, New York, NY, USA (2021)
5. Stojnić, N. Schuldt, H.: OSIRIS-SR: a scalable yet reliable distributed workflow execution engine. In: Proceedings of the 2nd ACM SIGMOD Workshop on Scalable Workflow Execution Engines and Technologies (SWEET'13), Article 3, pp. 1–12. Association for Computing Machinery, New York, NY, USA (2013)
6. Ningyuan, J., Huaxiong, Z.: Deep learning model service system based on microservices. Softw. Eng. **24**(05), 22–25 (2021). https://doi.org/10.19644/j.cnki.issn2096-1472.2021.05.006
7. Xiong, J., Chen, H.: Challenges for building a cloud native scalable and trustable multi-tenant AIoT platform. In: Proceedings of the 39th International Conference on Computer-Aided Design (ICCAD '20), Article 26, pp. 1–8. Association for Computing Machinery, New York, NY, USA

8. Patel, J., Tadi, G, Basarir, O., Hamel, L., Sharp, D., Yang, F., Zhang, X.: Pivotal Green-plum© for Kubernetes: demonstration of managing greenplum database on Kubernetes. In: Proceedings of the 2019 International Conference on Management of Data (SIGMOD'19). Association for Computing Machinery, New York, NY, USA, 1969–1972 (2019)
9. Xu, C., Lv, G., Du, J., Chen, L., Huang, Y, Zhou, W.: Kubeflow-based automatic data processing service for data center of state grid scenario. In: 2021 IEEE International Conference on Parallel & Distributed Processing with Applications, Big Data & Cloud Computing, Sustainable Computing & Communications, Social Computing & Network-ing (ISPA/BDCloud/SocialCom/SustainCom), pp. 924–930 (2021). https://doi.org/10.1109/ISPA-BDCloud-SocialCom-SustainCom52081.2021.00130
10. Mahboob, J., Coffman, J.: A Kubernetes CI/CD Pipeline with Asylo as a trusted execution environment abstraction framework. In: 2021 IEEE 11th Annual Computing and Communi-cation Workshop and Conference (CCWC), pp. 0529–0535 (2021). https://doi.org/10.1109/CCWC51732.2021.9376148

An Efficient Software Test Method for the Autonomous Mobile Robot Control Program

Chuang Cao[1]([✉]), Xiaoxiao Zhu[2], and Xiaochen Lai[1]

[1] Dalian University of Technology, Dalian 116024, China
bukita1999@mail.dlut.edu.cn, laixiaochen@dlut.edu.cn
[2] Shanghai Jiao Tong University, Shanghai 200240, China
ttl@sjtu.edu.cn

Abstract. ROS (Robot Operating System) is widely used as an open-source robotics system for various robot development. In particular, the ROS runtime is composed of several loosely coupled processes, and this loosely coupled structure is designed to allow developers to add each functional module flexibly according to the required functions of the robot. However, software testing of it is required. However, the development of a robot involves many disciplines such as mechanics, electronics, communications, and software, and as a systematic project, there is a high threshold for the average developer. For this reason. In this paper, we propose an efficient and reliable software testing method by introducing Docker to containerize the source code and runtime environment, and putting the build process into a continuous integration platform. For the components of the control software, this paper designs a unit testing scheme with functional packages as units; while for the integrated functions, a big-bang integration testing method is used to test the integrated navigation functions, and the control system is tested with the highest efficiency through an automated testing method. Finally, the test design method is carried out for the simulation experiment of ROS-based four-wheel drive autonomous robot to verify the feasibility of this scheme.

Keywords: Robot operating system · Containerization · Integration testing

1 Introduction

Robotics has become a popular research topic nowadays, and the development of a robot involves many disciplines such as mechanics, electronics, communication, software, etc. [1] As a systematic project, it is bound to take a lot of time. Especially with the rapid development and complexity of robots, the need for code reusability has become stronger and stronger, which has led to the birth of many open-source robotics systems, one of which is ROS (Robot Operating System). For most companies, the use of ROS for robot development is intended to enable robot navigation, positioning and path planning [2, 3].

© The Author(s), under exclusive license to Springer Nature Switzerland AG 2023
N. Xiong et al. (Eds.): ICNC-FSKD 2022, LNDECT 153, pp. 1363–1370, 2023.
https://doi.org/10.1007/978-3-031-20738-9_148

During the software development process, often as the application scales up, a variety of software defects tend to appear. The consequences of software defects may be only some tolerable inconvenience, or they may be destructive. In the software testing process, if the software does not work as intended, a software programmer may find that most defects are more hidden and cannot be eliminated directly through debugging during development or by their inspection. Therefore, systematic software testing is often required to find the software defects that need to be addressed and to improve the reliability of the software [4].

Recent ROS developers (especially in academia) do not pay much attention to software testing [5], resulting in software that does not meet production requirements [6, 7]. Therefore, there is an urgent need for testers to design newer testing methods for robot operating system software.

2 Methods

The specific robot targeted by the software test design is a four-wheeled autonomous mobile robot. The main hardware consists of four motor-driven wheels, robot mount, LIDAR, RGBD depth camera, and Nvidia Jetson NX development board.

In the four-wheeled mobile robot that needs to be tested in software, its main function is to perform point-to-point navigation with the constructed indoor map information to realize the autonomous movement of the mobile sampling robot on the surface. When moving along the optimal path, local obstacle avoidance can be performed; after reaching the sampling point, the sampling storage device is driven autonomously to perform the sampling task; wireless remote communication is possible during the whole working process. The system framework of the actual four-wheeled autonomous mobile robot control system is shown in Fig. 1.

2.1 Runtime Containerization

During the development of the control program for a newly designed robot, neither the developer nor the tester can directly debug it on the actual robot at the beginning. On the other hand, debugging or functional testing of the real robot during development may lead to risks and even serious accidents [8], so the robot is run in simulation, and the visualization module of the ground station PC is run on the same Ubuntu 18.04 desktop computer as the robot's upper and lower modules. The software operation, visualization interface and hardware simulation of the four-wheeled autonomous mobile robot control software are all run under the Ubuntu 18.04 environment, and the ROS version under Ubuntu 18.04 is melodic, as shown in Fig. 2.

The original runtime environment was a single physical machine with Ubuntu 18.04 installed for ROS installation and environment configuration. In order to solve the problem of long-time consuming ROS environment configuration and inconsistent development and testing environment, this paper containerizes the entire development environment architecture. The containerization of the runtime environment into several steps: building the image of the basic ROS functions, building the image of the ROS with visualization functions, and building the image of the four-wheeled autonomous mobile robot

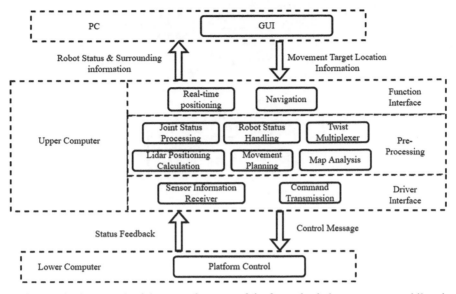

Fig. 1. System framework of the control system of the four-wheeled autonomous mobile robot

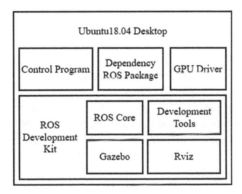

Fig. 2. Actual operation of the control system of the four-wheeled autonomous mobile robot

with source code and dependency environment. The image of ROS with visualization and the image of ROS with source code and dependencies.

2.2 Unit Testing

Since the core of ROS is a low-coupling, distributed communication mechanism, the smallest components of the system are the Nodes, each of which performs a single function. Using ROS Nodes is a good choice for developers because it allows them to more easily control the operation of the software. The four-wheeled mobile robot also uses the node feature of ROS in order to meet the intercommunication between different functions, and the specific node operation architecture is shown in Fig. 3.

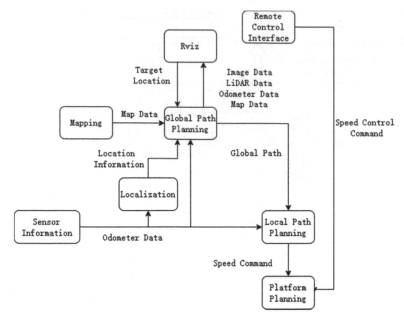

Fig. 3. ROS node architecture of the four-wheeled autonomous mobile robot

The four-wheeled autonomous mobile robot control software consists of several ROS functional packages used: joint state publisher, robot state publisher, multiplexer, adaptive Monte Carlo localization, map server and motion planner, and we perform unit tests for each of the above-mentioned functional packages. For functional testing of individual packages, we can use the "rostest" tool to run the test file or XML file or launch file in the test folder under the package to execute the ROS test functions.

After the testing of each functional unit of the four-wheeled autonomous mobile robot is completed, the integration test of the four-wheeled autonomous mobile robot is also needed. Through the integration test, we can verify the effect of each functional package operating together under the four-wheeled autonomous mobile robot system and verify the effect of using the four-wheeled autonomous mobile robot for navigation operation.

2.3 Integration Testing

For the navigation function of the control software of the four-wheeled autonomous mobile self-propelled robot in this paper, a big-bang integration test approach was used to test the reliability. The main flow of integration testing is that the test passes when it is ensured that the robot can get from the starting point to the selected endpoint. Integration testing initiates multiple nodes and tests whether they all work together as expected. Since the testing of the navigation function of the four-wheeled autonomous robot is a black-box test, a large number of test cases operating under the simulation system (including a large number of random reachable and unreachable coordinate vectors on the map) are required to ensure that the navigation test is functionally reliable. Using manual regression testing would have resulted in long test times, so an automated test solution needed to be designed for this. The test script is written in Python and the test framework is written in "unittest". 10 accessible and 10 inaccessible coordinate vectors are selected as test cases for the test system and saved in a CSV file in the test folder.

When the test script starts running, it takes about 30 s to initialize and start the simulation system, so the test script will not pass the test if it proceeds directly. Therefore, the test script waits for the initialization of the four-wheeled autonomous robot and the start-up of the simulation environment to complete. The sign of initialization and completion is that the LIDAR on the four-wheeled autonomous robot can start operating and receive scanning information. When the preparation work is completed, according to the test option of reachable or unreachable coordinate vectors, the corresponding coordinate and directional angle data need to be read into the CSV file. The coordinate vector data stored in the CSV file is partially shown in Table 1.

Table 1. Partial test data of navigation function of four-wheeled autonomous mobile robot

No.	Position-X	Position-Y	Orientation-Z	Orientation-W	Accessibility
1	2.83022	−0.0622	−0.02937	0.99956	Yes
2	6.48172	−0.8297	−0.46445	0.88560	Yes
3	5.39635	12.84637	0.24105	0.97051	No
4	−3.25528	12.91191	0.00694	0.99997	No

If the coordinates and directional angles are read as reachable points, the global path planning information should be received within 10 s to represent the system's successful planning of the corresponding path, otherwise, it means that the corresponding path is not successfully planned and the test is not passed. During the robot's navigation to the end point, the test system subscribes to the status topic information of the robot motion planner every 0.1 s. When the status topic information of the robot motion planner shows that it has arrived, the test system starts reading the current position of the robot obtained by applying the adaptive Monte Carlo localization method via LIDAR and comparing it with the coordinate vector given by the CSV file. If the difference between the values of the three-axis coordinates of the position vector and the quaternion of the angle vector is less than 0.2, the robot successfully reaches the corresponding position through the navigation system.

3 Experiment and Result

The design of the test of the source code of the four-wheeled autonomous robot mentioned in the previous chapter needs to be run in order to verify the reliability of the test design. This chapter will experiment with the design tests described in the previous chapter.

3.1 Experiment

In order to compare the efficiency of building an environment through containerization, we will compare the time to build the environment, the hard disk space occupied by the final application, the efficiency of visualization, stability and the ability to snapshot. There will be three kinds of objects for comparison, the first is to use the physical machine completely to build, the second through the Docker file generated image directly using Docker to run, the third is generated through the Docker file image into the WSL, and install the corresponding graphics card rendering tools. After testing, we can see that the second and third ways are much smaller in terms of time and hard disk space, but the second way is much worse than the first and third ways in terms of visualization because the GPU cannot be called well to assist in rendering. At the same time, we can see that the stability of WSL is not high compared with the first and the second, and there are situations such as flicker. Finally, both the second and the third are able to generate snapshots to roll back at any time with higher security, and the final results are shown in Table 2.

Table 2. Comparison table of the performance among different deployment methods

Deployment method	Set-up time	Storage occupancy (GB)	GUI performance	Stability	Snapshot
Physical	12 h	30.23	Good	Good	No
Docker	15 min	5.22	Bad	Good	Yes
WSL	30 min	10.33	Good	Bad	Yes

During the Integration testing of the four-wheeled autonomous robot, we were able to locate and identify functional errors through the existing test design. By analyzing the results displayed in the Rviz visualization interface, we found that if the robot is doing complex movements, especially during rotations, the robot's positioning itself is distorted and the map position determined by LIDAR recognition and self-localization is shifted, as shown in Fig. 4.

3.2 Discussion

Based on the results presented in Table 4.1, we can conclude that for the final deployment environment, installing the visual interface on the first physical machine with the native operating system and running the core code in Docker works best. For developers and testers, using the third installation in WSL allows them to develop and test more efficiently.

After checking and reproducing the errors, we can conclude that when the angular velocity of the robot is too large during the rotational motion, distortion of the radar reception data will occur, which is mainly related to the algorithm and the radar hardware in the simulation. Therefore, in order to fix this problem, firstly, the script strictly limits the steering angular speed of the robot so that the radar does not actively fail as much

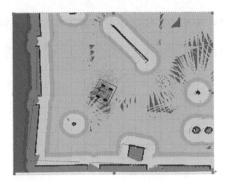

Fig. 4. The LIDAR information of the four-wheeled autonomous robot is offset

as possible. Secondly, a correction mechanism is added to the radar scan of the robot to correct the radar scan map boundaries when they differ significantly from the expected map.

4 Conclusion

This paper describes the test design of a four-wheeled autonomous mobile robot based on ROS, from containerization of the runtime environment for rapid deployment to unit testing of the functional packages used by the robot, and finally to integration testing of the robot's navigation functions. This test design not only improves the efficiency of rebuilding the runtime environment by containerizing the runtime environment but also helps testers efficiently find and correct errors in the source code of the robot. However, there is still room for improvement in this paper, such as the need to increase the number of test cases for each test function, the need to improve the generality and scalability of each test, and the need to add non-functional tests for the system. The above-mentioned elements are often the focus of other testing areas, so future research on ROS testing should also focus on these areas.

References

1. Wanjin, L., Hong, M., Qianjin, L.: A new robot sliding mode variable structure control. Mach. Tools Hydraulics **37**(6), 161–163 (2009)
2. Felter, W. et al.: An updated performance comparison of virtual machines and Linux containers. In: IEEE international symposium on performance analysis of systems and software, pp. 171–172. Singapore (2015)
3. Patton, R.: Software Testing. Sams, Indianapolis (2006)
4. Huang, M.: Robot operating system testing technology. Electronic Industry Press, Beijing (2015)
5. Kaihong, H., Xingrui, Y., Zeng Zhiwen, L., Huimin, Z.Z.: ROS-based outdoor mobile robot software system construction. Rob. Appl. **4**, 37–44 (2013)
6. Kanter, G., Vain, J.: TestIt: an open-source scalable long-term autonomy testing toolkit for ROS. In: 2019 10th International Conference on Dependable Systems, Services and Technologies, pp. 45–50. London (2019)

7. Luo, S.: Methods and practices of computer software quality assurance. Science Press, Beijing (1999)
8. Guowei, H., Wei, W.: Software reliability. Defense Industry Press, Beijing (1998)

Forward Kinematics and Singularity Analysis of an Adjusted-DOF Mechanism

Junting Fei[(✉)], Qingxuan Jia, Gang Chen, Tong Li, and Yifan Wang

School of Modern Post (School of Automation), Beijing University of Posts and
Telecommunications, Beijing 100876, China
feijunting@buptrobot.com, spacerobot@163.com, buptcg@163.com,
{tli,wangyifan}@bupt.edu.cn

Abstract. Modular robots are developing towards homogeneity, and the application of modular robots is tending to be distributed. The above work needs to be based on the kinematics and singularity analysis of one module. Firstly, the mechanical structure of an adjusted-DOF mechanism is briefly described. Next, a connection-state representation method of one module is proposed. Based on the representation, we analyze the motion transmission forms under different connection states. For kinematics analysis, we establish the kinematics model and analyze the axis of rotation involved in the motion transmission forms. Then we obtain the workspace of the mechanism under different motion transmission forms. For singularity, by establishing the explicit expression of the Jacobian matrix of the mechanism, the singularity of the mechanism can be obtained. This work provides the theoretical basis for the future application of the modular robot based on homogeneous modules.

Keywords: Modular robots · Connection-state representation · Kinematics analysis

1 Introduction

Modular robots can be reconfigured into a new topology, which gains new capabilities to adapt to the environment and meet the needs of tasks [1, 2]. Compared with the topology-fixed robot, modular robots can have strong adaptability to tasks and scenarios. Among them, the homogenization of modular robots is a major development trend [3].

According to the geometric distribution of modules, modular robots can be classified into three classes of architecture [4]: Lattice [5–8], Chain [9, 10], and Hybrid [11, 12]. Because of retaining these advantages of lattice architectures and the chain architectures, hybrid architectures have a wider application prospect.

Regarding the kinematics and singularity analysis of modular robots, Mishra et al. [13], Yun et al. [14], and Stravopodis et al. [15] obtained the kinematic parameters of the modular robot manually and established the kinematics model of the robot. However, in the analysis process, the parameters required are obtained manually, which is not convenient for the application of modular robots. In response to this problem, Feder

N. Xiong et al. (Eds.): ICNC-FSKD 2022, LNDECT 153, pp. 1371–1384, 2023.
https://doi.org/10.1007/978-3-031-20738-9_149

et al. [16] and Nainer et al. [17] automatically generated kinematic parameters of robots, such as standard and modified DH parameters, and then derived the kinematics model of robots. The above research regards modular robots as topology-fixed robots, and their kinematics analysis is derived from the conventional methods of topology-fixed robots. For robots built by modules with multiple connectors and multiple motion transmission forms, the conventional methods of topology-fixed robots will lead to complicated analysis processes and increase the period of reconfiguration.

The distributed modeling, planning, and control of modular robots are growing trends in the field of modular robot research [18]. The kinematic characteristics of robots can be analyzed based on the kinematics analysis of one module. Zhang et al. [19] deduced the mapping relationship between the parameters of modules and task-space parameters, considering the material properties of modules and the phenomenological modeling theory, then analyzed the kinematics of robots built with these modules. Yang et al. [20] and Nasibullayev et al. [21] deduced the mapping relationship between the joint parameters of modules and the pose of the end-effector through the analysis of the geometric relationship in the 3D space, considering geometric constraints contained in modules. The kinematics analysis of modular robots in the above studies is based on the kinematics analysis of one module. However, the kinematics analysis process of modules is closely related to the mechanical structure of modules, which makes the above analysis difficult to apply to other modular robots. The non-universal analysis process is not conducive to the application of modular robots.

To sum up, most of the above-mentioned literature focused on the structural design and kinematics of modular robots. There are few investigations on the universal analysis for kinematics and singularity of one module. It is well known that getting the kinematics and singularity for one module is the most fundamental understanding of a mechanism. However, when the module contains multiple connectors and multiple motion transmission forms, it will increase the complexity and difficulty of analysis for this kind of mechanism. In this paper, the kinematics and singularity of an adjusted-DOF mechanism are analyzed based on the screw theory.

2 Mechanical Structure

The proposed adjusted-DOF module (Admbot) is shown in Figs. 1 and 2. Each module mainly includes two hemispherical shells, three revolute joints modules, two docking connectors, two locking connectors, control modules, power supply modules, and other components. Each Admbot features three degrees of freedom (DOF), red axes, as shown in Fig. 3. All three DOFs are designed for continuous rotation, there are no joint limits. According to the torque output relationship, this paper regards the upper left hemispherical shell (orange) as the passive hemispherical shell and the lower right (blue) as the active hemispherical shell, as shown in Fig. 1.

The module connector usually contains two classification methods. According to the torque output attribute of connectors, they can be divided into active connectors and passive connectors. From the Geometry point of the mechanical structure, they can be divided into male and female connectors. There are an active male connector and a passive female connector distributed on each hemispherical shell of one Admbot. The connector arrangement is shown in Fig. 2. This arrangement of joints and connectors not only increases the flexibility of modules but also preserves the possibility of parallel or vertical joint axes between adjacent modules.

Fig. 1. The conceptual model of the Admbot

Fig. 2. CAD model of the Admbot

Fig. 3. Distribution of the connectors and the axis of rotation

3 Connection-State Representation

When the connector is in the connection state, it indicates that the connector has a connection with the adjacent module or environment. When the module is in the connection state, it indicates that at least one connector is in the connection state.

For the connection state of one module, this paper uses two tuples to represent, as shown in Eq. (1) and (2). Matrix I represents the number of modules, and matrix S represents the attributes and state of the connectors.

$$I = [id] \tag{1}$$

where id is the current number of modules. The number of modules is not set in stone. When modules are used to build a robot, the number will be adjusted appropriately according to the current state of the robot, which is convenient for subsequent planning and control.

$$S = \begin{bmatrix} S_1 \; S_2 \; S_3 \; S_4 \; S_5 \; S_6 \; S_7 \end{bmatrix}$$

$$\begin{cases} S_1 = \begin{bmatrix} a_1 \cdots a_{n_{surf\_con}} \end{bmatrix}^T, \; S_2 = \begin{bmatrix} b_1 \cdots b_{n_{surf\_con}} \end{bmatrix}^T, \; S_3 = \begin{bmatrix} d_1 \cdots d_{n_{surf\_con}} \end{bmatrix}^T, \\ S_4 = \begin{bmatrix} e_1 \cdots e_{n_{surf\_con}} \end{bmatrix}^T, \; S_5 = \begin{bmatrix} f_1 \cdots f_{n_{surf\_con}} \end{bmatrix}^T, \; S_6 = \begin{bmatrix} g_1 \cdots g_{n_{surf\_con}} \end{bmatrix}^T, \\ S_7 = \begin{bmatrix} h_1 \cdots h_{n_{surf\_con}} \end{bmatrix}^T \end{cases}$$

$$(2)$$

where n_{surf_con} is the total number of the connectors in a module, and n_{sub} is the total number of hemispherical shells. Each row of matrix S represents all attributes and states of a connector. Each column of matrix S represents different properties of the connector. The meaning of each column in matrix S is as follows.

(1) $a_i(i = 1, 2, ..., n_{surf\_con})$ in the first column of matrix S is the hemispherical shell number that the connector is located in, and $a_i \in \{1, 2, ..., n_{sub}\}$; (2) $b_i(i = 1, 2, ..., n_{surf\_con})$ in the second column of matrix S is the Geometry characteristics of the mechanical structure, $b_i=0$ denotes that the connector is a female connector, and $b_i= 1$ denotes that the connector is a male connector; (3) $d_i(i = 1, 2, ..., n_{surf\_con})$ in the third column of matrix S is the torque output attribute of the connector, $d_i=0$ denotes that the connector is a passive connector, and $d_i= 1$ denotes that the connector is active; (4) $e_i(i = 1, 2, ..., n_{surf\_con})$ of matrix S is the current connection state of the connector, $e_i=1$ denotes that the connector is currently connected to the environment or the adjacent module, otherwise, $e_i=0$; (5) $f_i(i = 1, 2, ..., n_{surf\_con})$ of matrix S is the module number connected to connector i, and $f_i = 0$ denotes that the connector is connected to the environment; (6) $g_i(i = 1, 2, ..., n_{surf\_con})$ of matrix S is the connector number of f_i connected to connector I; (7) $h_i(i = 1, 2, ..., n_{surf\_con})$ of matrix S is the number of the connection orientation relationship between connector i and connector g_i.

Start with the male connector on the active hemispherical shell, and number each connector in a clockwise direction. The connection state of the module shown in Fig. 2 can be represented in the above manner as

$$S = \begin{bmatrix} 1 & 1 & 1 & 0 & 0 & 0 & 0 \\ 1 & 0 & 0 & 0 & 0 & 0 & 0 \\ 0 & 1 & 1 & 0 & 0 & 0 & 0 \\ 0 & 0 & 0 & 0 & 0 & 0 & 0 \end{bmatrix} \qquad (3)$$

When the module is not in use, its number has no substantial effect. Therefore, matrix I is not reflected in Eq. (3).

The number of modules plays an important role in identifying the position in the modular robot system. During the connection process of the modular robot, the constraints contained in the mechanical structure must be satisfied. Columns 1–3 in matrix S can be used to judge the mechanical constraints when the robot is generating.

One module needs to have the ability to understand its connection state. Columns 4–6 in matrix S can fully contain the key information, such as whether there is a connection relationship between the connector and the environment or adjacent modules, and if a connection relationship exists, the module should know information about the other connector of the connection.

Due to the existence of the positioning device on the connector, there are many possible connection orientation relations. That's why column 7 exists in matrix S.

4   Motion Transmission Analysis

Some coordinate systems are established to describe the motion: coordinate system {E}, and coordinate systems {S_i}($i = 1, 2, ..., n_{surf\_con}$), as illustrated in Fig. 4. Each coordinate system is defined as follows:

(1)   Coordinate system {E} is inside the Admbot, which is used to describe the motion between the upper and lower hemispherical shell.
(2)   Coordinate systems {S_i}($i = 1, 2, ..., n_{surf\_con}$) are fixed to connectors, and they are used to describe the motion and connection orientation relationship between connectors.

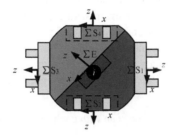

Fig. 4. Coordinate systems of one Admbot

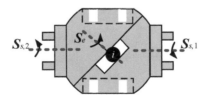

Fig. 5. Screw axes of one Admbot

The reference state of modules is defined as the state of modules when the joint angles are all 0. In the reference state, for each revolute joint in the Admbot, motion screw $S_{ei}, S_{si,1}, S_{si,2} \in \mathbb{R}^{6\times1}$ is constructed separately, where S_{ei} represents the screw axis inside the Admbot numbered i, and $S_{si,1}, S_{si,2}$ represents the screw axes at the connectors of module i.

These motion screws $\mathcal{S}_{ei}$, $\mathcal{S}_{si,1}$, $\mathcal{S}_{si,2} \in \mathbb{R}^{6 \times 1}$ can be expressed as

$$\mathcal{S}_{si,j} = \begin{bmatrix} \boldsymbol{\omega}_{si,j} \\ \boldsymbol{v}_{si,j} \end{bmatrix} \tag{4}$$

where $\boldsymbol{\omega}_{si,j} \in \mathbb{R}^{3 \times 1}$ is a three-dimensional vector, denotes the unit direction vector of a rotation; and $\boldsymbol{v}_{si,j} \in \mathbb{R}^{3 \times 1}$ is also a three-dimensional vector, and denotes the location of the rotational axis to the origin of the coordinate system fixed to the motion-input connector (Fig. 5).

4.1 Connection Enumeration

The following constraints need to be satisfied when the module is connected:

(1) Due to the limitation of the mechanical structure, the male connector can only be connected to the female connector, and the female connector can only be connected to the male connector.

(2) There must be no interference between adjacent modules connected to the same module.

A connection state that satisfies the above constraints is called a valid connection state. Strictly abiding by the above constraints, all connection states of one Admbot can be obtained by enumerating when connector 1 is used as the motion-input connector. There are 26 valid connection states, as shown in Table 1. According to the above ideas, all valid connection relationships are analyzed when connector 2, connector 3, and connector 4 are used as the motion-input connector. Finally, we get 104 valid connection states.

Table 1. Valid connection state with connector 1 as the motion input

The number of motion output	The motion output number	The number of valid connection states
1	2	2
	3	2
	4	2
2	2, 3	2^2
	2, 4	2^2
	3, 4	2^2
3	2, 3, 4	2^3

4.2 Motion Transmission Forms

The 104 valid connection states obtained in 4.1 are analyzed in turn. For example, when connector 1 is used as the motion input, and connector 2, connector 3, and connector 4 are used as the motion output, the motion transmission includes three routes as shown in Fig. 6. In Fig. 6, the pink motion transmission route passes through the joint at connector 1 and arrives at connector 2; the green motion transmission route passes through the joint at connector 1, passes through the joint inside the module, and then passes through the joint at connector 3 to reach connector 3; and the red motion transmission route passes through the joint at connector 1, and then passes through the joint inside the module to reach connector 4.

According to the above analysis, it contains a total of 12 basic motion transmission forms. According to the analysis, it can be known that the motion transfer in each connection state can be composed of one or several basic motion transmission forms. The difference between the 12 basic motion transmission forms is that the rotational axes passed through and the position offsets are different. The axes involved in these basic motion transmission forms are shown in Table 2.

Fig. 6. Motion transmission with connector 1 as the motion input and connectors 2, 3, 4 as motion output.

5   Mechanism Position Analysis

In the reference state, according to the model of modules, the pose matrixes of each coordinate system $\{S_i\}(i = 1, 2, ..., n_{surf\_con})$ in the coordinate system $\{E\}$ can be obtained, which is recorded as ${}^{E}_{S_1}A, ..., {}^{E}_{S_i}A(i = 1, 2, ..., n_{surf\_con})$. At the same time, according to the positioning device, all connection orientation relationship matrixes ${}^{S_{f_1}}_{S_{f_2}}A_1, ...$ Between connectors can be obtained.

For the Admbot, we can get

$$
{}^{E}_{S_1}A = \begin{bmatrix} -\sqrt{2}/2 & 0 & \sqrt{2}/2 & 91.02 \\ 0 & 1 & 0 & 0 \\ -\sqrt{2}/2 & 0 & -\sqrt{2}/2 & -91.02 \\ 0 & 0 & 0 & 1 \end{bmatrix}
\quad
{}^{E}_{S_2}A = \begin{bmatrix} -\sqrt{2}/2 & 0 & -\sqrt{2}/2 & -67.68 \\ 0 & 1 & 0 & 0 \\ \sqrt{2}/2 & 0 & -\sqrt{2}/2 & -67.68 \\ 0 & 0 & 0 & 1 \end{bmatrix}
$$

Table 2. Axis of rotation involved in motion transmission

Motion input connector	Motion output connector	Axes passed through	Position offsets
1	2	$S_{s,1}$	$^{S_1}_{S_2}T$
	3	$S_{s,1}, S_e, S_{s,2}$	$^{S_1}_{S_3}T$
	4	$S_{s,1}, S_e$	$^{S_1}_{S_4}T$
2	1	$S_{s,1}$	$^{S_2}_{S_1}T$
	3	$S_e, S_{s,2}$	$^{S_2}_{S_3}T$
	4	S_e	$^{S_2}_{S_4}T$
3	1	$S_{s,2}, S_e, S_{s,1}$	$^{S_3}_{S_1}T$
	2	$S_{s,2}, S_e$	$^{S_3}_{S_2}T$
	4	$S_{s,2}$	$^{S_3}_{S_4}T$
4	1	$S_e, S_{s,1}$	$^{S_4}_{S_1}T$
	2	S_e	$^{S_4}_{S_2}T$
	3	$S_{s,2}$	$^{S_4}_{S_3}T$

$$^{S_{f_1}}_{S_{f_2}}A_1 = \begin{bmatrix} -1 & 0 & 0 & 0 \\ 0 & 1 & 0 & 0 \\ 0 & 0 & -1 & 0 \\ 0 & 0 & 0 & 1 \end{bmatrix}$$

$$^{E}_{S_3}A = \begin{bmatrix} 0 & -1 & 0 & 0 \\ \sqrt{2}/2 & 0 & \sqrt{2}/2 & 91.02 \\ -\sqrt{2}/2 & 0 & \sqrt{2}/2 & 92.52 \\ 0 & 0 & 0 & 1 \end{bmatrix} \quad ^{E}_{S_4}A = \begin{bmatrix} 0 & -1 & 0 & 0 \\ \sqrt{2}/2 & 0 & -\sqrt{2}/2 & -67.68 \\ \sqrt{2}/2 & 0 & \sqrt{2}/2 & 69.18 \\ 0 & 0 & 0 & 1 \end{bmatrix}$$

$$^{S_{f_1}}_{S_{f_2}}A_2 = \begin{bmatrix} 1 & 0 & 0 & 0 \\ 0 & -1 & 0 & 0 \\ 0 & 0 & -1 & 0 \\ 0 & 0 & 0 & 1 \end{bmatrix}$$

5.1 Mechanism Analysis of the Forward Position

For an Admbot with 4 connectors, a coordinate system $\sum S_{in}$ is used to represent the coordinate system fixed to the motion-input connector of modules, and a coordinate system $\sum S_{out}$ is used to represent the coordinate system fixed to the motion-output.

In the reference state, the transformation matrix of the coordinate system $\sum S_{out}$ of modules relative to the coordinate system $\sum S_{in}$ is:

$$\underset{S_{out}}{\overset{S_{in}}{T}} = \underset{E}{\overset{S_{in}}{T}} \cdot \underset{S_{out}}{\overset{E}{T}} = \left(\underset{S_{in}}{\overset{E}{A}}\right)^{-1} \cdot \underset{S_{out}}{\overset{E}{A}} \tag{5}$$

When the screw axes involved in the motion transfer form are $S_1, ..., S_m \in \mathbb{R}^{6 \times 1}$, based on the screw theory, the homogeneous transformation matrix of the coordinate system $\sum S_{out}$ of the module relative to the coordinate system $\sum S_{in}$ is:

$$\underset{S_{out}}{\overset{S_{in}}{T}}(q) = e^{\hat{S}_1 \theta_1} e^{\hat{S}_2 \theta_2} \cdots e^{\hat{S}_m \theta_m} \underset{S_{out}}{\overset{S_{in}}{T}}(0) = \prod_{i=1}^{m} e^{\hat{S}_i \theta_i} \underset{S_{out}}{\overset{S_{in}}{T}}(0) \tag{6}$$

where $\underset{S_{out}}{\overset{S_{in}}{T}}(0)$ is the homogeneous transformation matrix of the coordinate system $\sum S_{out}$ of the module relative to the coordinate system $\sum S_{in}$ in the reference state, $\theta_1, ..., \theta_m$ are rotation angle of the screw axes involved in the motion transfer form respectively, and $\hat{S}_i$ is the antisymmetric matrix form of S_i, and $\hat{S}_i = \begin{bmatrix} \hat{\omega}_i & v_i \\ 0 & 0 \end{bmatrix} \in se(3)$.

In the connection state, due to the possibility of the motion-output connector being greater than 1, the form of motion transmission inside modules is often one or more combinations of the basic motion transmission form. When there are σ output connectors, the homogeneous transformation matrix of each coordinate system $\sum S_{out}$ of modules relative to the coordinate system $\sum S_{in}$ is

$$\begin{cases} \underset{S_{out_1}}{\overset{S_{in}}{T}}_1(q_1) = \prod_{i=1}^{m_1} e^{\hat{S}_i \theta_i} \underset{S_{out_1}}{\overset{S_{in}}{T}}_1(0) \\ \quad \vdots \\ \underset{S_{out_\sigma}}{\overset{S_{in}}{T}}_\sigma(q_\sigma) = \prod_{i=1}^{m_\sigma} e^{\hat{S}_i \theta_i} \underset{S_{out_1}}{\overset{S_{in}}{T}}_\sigma(0) \end{cases} \tag{7}$$

5.2 Workspace Analysis

Connector 1 to connector 4 is taken as the motion input in turn. Then the motion output position that can be reached is calculated when the rest except the motion-input connector are used as the motion-output connector, as shown in Figs. 7, 8, 9 and 10. In Figs. 7, 8, 9 and 10, pink, green, blue, and black are the positions that can be reached when connector 1, connector 2, connector 3, and connector 4 are used as motion output.

It can be seen from Figs. 7, 8, 9 and 10 that when connector 1 and connector 3 are taken as the motion input, a larger reachable space for the motion-output connector can be obtained. When in the following motion transmission forms, the motion-output connector position of modules can be a circular arc in three-dimensional space: (1) connector 1 is taken as the motion input, and connector 2 is taken as the motion output; (2) connector 2 is taken as the motion input, and connector 3 is taken as the motion output; (3) connector 2 is taken as the motion input, and connector 4 is taken as the motion output; (4) connector 3 is taken as the motion input, and connector 4 is taken as

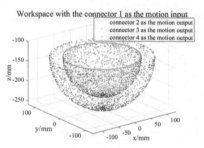

Fig. 7. Workspace with connector 1 as the motion input

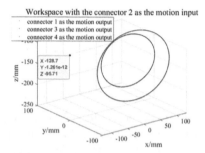

Fig. 8. Workspace with connector 2 as the motion input

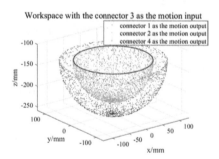

Fig. 9. Workspace with connector 3 as the motion input

the motion output; (5) connector 4 is taken as the motion input, and connector 1 is taken as the motion output; (6) connector 4 is taken as the motion input, and connector 2 is taken as the motion output.

When in the following motion transmission forms, the motion-output connector position of modules can only be a point in three-dimensional space: (1) connector 2 is taken as the motion input, and connector 1 is taken as the motion output; (2) connector 4 is taken as the motion input, and connector 3 is taken as the motion output.

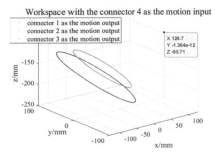

Fig. 10. Workspace with connector 4 as the motion input

6 Mechanism Flexibility Analysis

6.1 The Jacobian Matrix

According to Eq. (6), we can get the mapping relationship between the joint velocity and the end-effector velocity screw as:

$$^{s}\mathcal{V}_{st} = J_1\dot{\theta}_1 + J_2\dot{\theta}_2 + \cdots + J_m\dot{\theta}_m \tag{8}$$

where when $i = 2, ..., m$, $J_i = \text{Ad}_{e^{\hat{S}_1\theta_1}\cdots e^{\hat{S}_{i-1}\theta_{i-1}}}(S_i)$; and when $i = 1$, $J_1 = S_1$.

$\text{Ad}_{e^{\hat{S}_1\theta_1}\cdots e^{\hat{S}_{i-1}\theta_{i-1}}}(S_i) = \begin{bmatrix} R & 0 \\ [p]R & R \end{bmatrix} S_i$, $R \cdot p$ are derived by the transformation matrices

corresponding to rigid body transformations from the reference state through $\exp(\hat{S}_1\theta_1)$, ..., $\exp(\hat{S}_1\theta_1)$.

Arranging Eq. (8) into matrix form, we can get

$$^{s}\mathcal{V}_{st} = [J_1 \, J_2 \cdots J_m][\dot{\theta}_1 \, \dot{\theta}_2 \cdots \dot{\theta}_m]^T = J_s\dot{q} \tag{9}$$

Usually, in the analysis process of robot motion, the motion-output velocity is the generalized speed, that is, a six-dimensional vector composed of three-dimensional linear velocity and three-dimensional angular velocity, which is expressed as

$$^{d}\mathcal{V}_{st} = \begin{bmatrix} ^{s}\boldsymbol{\omega}_{st} \\ \dot{\boldsymbol{p}}_{st} \end{bmatrix} \tag{10}$$

According to the definition, the space motion screw at the motion-output connector of modules is

$$^{s}\mathcal{V}_{st} = \begin{bmatrix} ^{s}\boldsymbol{\omega}_{st} \\ \dot{\boldsymbol{p}}_{st} + {}^{s}\boldsymbol{\omega}_{st} \times (-\boldsymbol{p}_{st}) \end{bmatrix} \tag{11}$$

According to Eq. (10) and (11), the transformation between the motion-output connector velocity screw and the motion-output connector velocity is:

$$^{s}\mathcal{V}_{st} = \begin{bmatrix} \mathbf{I}_{3\times3} & \mathbf{0}_{3\times3} \\ \hat{\boldsymbol{p}}_{st} & \mathbf{I}_{3\times3} \end{bmatrix} {}^{d}\mathcal{V}_{st} = J_{sd}{}^{d}\mathcal{V}_{st} \tag{12}$$

Then, we can get

$$^d\mathcal{V}_{st} = J_{sd}^{-1}{}^s\mathcal{V}_{st} = J_{sd}^{-1}J_s\dot{q} = J_d\dot{q} \tag{13}$$

The Jacobian matrix is

$$J_d = J_{sd}^{-1}J_s \tag{14}$$

Through the above derivation and analysis, the space Jacobian matrix under the screw theory can be transformed into the Jacobian matrix commonly used in robot analysis.

6.2 Singularity Analysis

By dividing the space Jacobian matrix according to the rotation velocity screw and the moving velocity screw, Eq. (14) can be transformed into

$$J_d = \begin{bmatrix} \mathbf{I}_{3\times3} & \mathbf{0}_{3\times3} \\ \hat{p}_{st} & \mathbf{I}_{3\times3} \end{bmatrix}^{-1} \begin{bmatrix} J_{s\omega} \\ J_{sv} \end{bmatrix} = \begin{bmatrix} J_{s\omega} \\ -\hat{p}_{st} \cdot J_{s\omega} + J_{sv} \end{bmatrix} \tag{15}$$

where $J_{s\omega}$ is the first three rows of J_s, and J_{sv} is the last three rows of J_s.

For each basic motion transmission form, the end-effector controllable dimensions are analyzed by introducing module parameters into Eq. (15). The results are shown in Table 3. In Table 3, α, β, and γ represent rotation about the x, y, and z axes respectively, and x, y, and z represent movement along the x, y, and z axes respectively. The '$\sqrt{}$' indicates that the motion of this dimension is controllable.

Table 3. End-effector controllable dimensions under different basic motion transmission forms

Motion input	Motion output	α	β	γ	x	y	z
1	2			√	√	√	
	3	√	√	√	√	√	√
	4	√	√	√	√	√	√
2	1	√					
	3	√	√	√	√	√	√
	4	√		√	√	√	√
3	1	√	√	√	√	√	√
	2	√	√	√	√	√	√
	4			√	√	√	
4	1	√	√	√	√	√	√
	2	√		√	√	√	√
	3	√					

According to mechanism position analysis and flexibility analysis, by selecting different motion input and output terminals, the Admbot can obtain different motion transmission forms including different degrees of freedom and different offset.

7 Conclusions

In this paper, the kinematics and singularity of an adjusted-DOF mechanism are analyzed based on the screw theory. Firstly, the mechanism of the adjusted-DOF module (Admbot) is briefly described. Secondly, a connection-state representation of modules is proposed, which describes the module number and the connector properties related to the connection state through two tuples. Then, by enumerating all connection states of one Admbot, 12 basic motion transmission forms are obtained by analysis. Next, the general kinematics model of an Admbot is established, and the reachable motion-output connector position of modules under different motion transfer forms is analyzed. Finally, the space Jacobian matrix of modules under the screw theory is deduced, and the conversion between the space Jacobian matrix under the screw theory and the Jacobian matrix is established. For each basic motion transmission form, the end controllable dimension is analyzed with the help of the Jacobian matrix. The concept of the Admbot proposed in this paper, as well as the related kinematics and singularity analysis, can provide some reference value for the optimization design and kinematics research of the follow-up modular robots.

Project Support Fund. This work is supported by BUPT Action Plan to Enhance Capacity for Scientific and Technological Innovation (No. 2021XD-A10-1), National Natural Science Foundation of China (Grant No. 62173044, Grant No. 62103058), and BUPT Excellent Ph.D. Students Foundation (CX2022311).

References

1. Fukuda, T., Nakagawa, S.: Approach to the dynamically reconfigurable robotic system. J. Intell. Rob. Syst. **1**(1), 55–72 (1988)
2. Faíña, A., Bellas, F., López-Peña, F., Duro, R.: EDHMoR: Evolutionary designer of heterogeneous modular robots. Eng. Appl. Artif. Intell. **26**(10), 2408–2423 (2013)
3. Liu, S.B, Althoff, M.: Optimizing performance in automation through modular robots. In: 2020 IEEE International Conference on Robotics and Automation, pp. 4044–4050. Paris, France (2020)
4. Alattas, R.J., Patel, S., Sobh, T.M.: Evolutionary modular robotics: survey and analysis. J. Intell. Rob. Syst. **95**(3), 815–828 (2019)
5. Yim, M., et al.: Modular self-reconfigurable robot systems [grand challenges of robotics]. IEEE Robot. Autom. Mag. **14**(1), 43–52 (2007)
6. Murata, S., Yoshida, E., Kamimura, A., Kurokawa, H., Tomita, K., Kokaji, S.: M-TRAN: Self-reconfigurable modular robotic system. IEEE/ASME Trans. Mechatron. **7**(4), 431–441 (2002)
7. Kurokawa, H., Kamimura, A., Yoshida, E., Tomita, K., Kokaiji, S., Murata, S.: M-TRAN II: metamorphosis from a four-legged walker to a caterpillar. In: 2003 IEEE/RSJ International Conference on Intelligent Robots and Systems, pp. 2454–2459. Las Vegas, Nevada, United states (2003)
8. Jorgensen, M.W., Ostergaard, E.H., Lund, H.H.: Modular ATRON: Modules for a self-reconfigurable robot. In: 2004 IEEE/RSJ International Conference on Intelligent Robots and Systems, pp. 2068–2073. Sendai, Japan (2004)

9. Golovinsky, A., Yim, M., Zhang, Y., Eldershaw, C., Duff, D.: Polybot and polykinetic/spl trade/system: a modular robotic platform for education. In: 2004 IEEE International Conference on Robotics and Automation, pp. 1381–1386

10. Wei, H., Chen, Y., Tan, J., Wang, T.: Sambot: a self-assembly modular robot system. IEEE/ASME Trans. Mechatron. 16(4), 745–757 (2010)

11. Yim, M., Zhang, Y., Roufas, K., Duff, D., Eldershaw, C.: Connecting and disconnecting for chain self-reconfiguration with PolyBot. IEEE/ASME Trans. Mechatron. 7(4), 442–451 (2002)

12. Zhao, J., Cui, X., Zhu, Y., Tang, S.: A new self-reconfigurable modular robotic system UBot: Multi-mode locomotion and self-reconfiguration. In: 2011 IEEE International Conference on Robotics and Automation, pp. 1020–1025. Shanghai, China (2011)

13. Mishra, A.K., Mondini, A., Del Dottore, E., Sadeghi, A., Tramacere, F., Mazzolai, B.: Modular continuum manipulator: analysis and characterization of its basic module. Biomimetics 3(1), 3 (2018)

14. Yun, A., Moon, D., Ha, J., Kang, S., Lee, W.: Modman: an advanced reconfigurable manipulator system with genderless connector and automatic kinematic modeling algorithm. IEEE Robot. Autom. Lett. 5(3), 4225–4232 (2020)

15. Stravopodis, N.A., Moulianitis, V.C.: Rectilinear tasks optimization of a modular serial metamorphic manipulator. J. Mech. Robot. 13(1) (2021)

16. Feder, M., Giusti, A., Vidoni, R.: An approach for automatic generation of the URDF file of modular robots from modules designed using SolidWorks. In: 2022 3rd International Conference on Industry 4.0 and Smart Manufacturing, pp. 858–864 (2022)

17. Nainer, C., Feder, M., and Giusti, A.: Automatic Generation of Kinematics and Dynamics Model Descriptions for Modular Reconfigurable Robot Manipulators. In: 2021 IEEE 17th International Conference on Automation Science and Engineering, pp. 45–52

18. Stoy, K., Kurokawa, H.: Current topics in classic self-reconfigurable robot research. In: Proceedings of the IROS Workshop on Reconfigurable Modular Robotics: Challenges of Mechatronic and Bio-Chemo-Hybrid Systems

19. Zhang, Z., Wang, X., Wang, S., Meng, D., Liang, B.: Design and modeling of a parallel-pipe-crawling pneumatic soft robot. IEEE access 7, 134301–134317 (2019)

20. Yang, H.D., Asbeck, A.T.: Design and characterization of a modular hybrid continuum robotic manipulator. IEEE/ASME Trans. Mechatron. 25(6), 2812–2823 (2020)

21. Nasibullayev, I., Darintsev, O., Bogdanov, D.: In-pipe modular robot: configuration, displacement principles, standard patterns and modeling. Electromechanics Robot. 85–96 (2022)

A New Visual and Inertial and Satellite Integrated Navigation Method Based on Point Cloud Registration

Ping'an Qiao[1], Ruichen Wu[1(✉)], Jinglan Yang[2], Jiakun Shi[1], and Dongfang Yang[2]

[1] Xi'an University of Posts and Telecommunications, Xi'an 710121, China
paqiao@163.com, richardwu@stu.xupt.edu.cn, shijiakun1010@163.com
[2] The Rocket Force University of Engineering, Xi'an 710025, China
diamond_yang.sjz@foxmail.com, yangdf301@163.com

Abstract. The three navigation methods of vision, inertia and satellite have obvious complementarity in navigation accuracy, positioning method and anti-interference ability. In this paper we propose a combined visual/inertial/satellite navigation method, which can be used in both outdoors where satellites are available and indoors where satellites are unavailable where continuous geolocation can be achieved in those scenarios. First, when Global Positioning System (GPS) works normally, the absolute geographic positioning results provided by GPS are combined with the relative positioning results obtained by visual-inertial combination to obtain the coordinate conversion between the visual-inertial combined navigation reference coordinate system (world coordinate system) and the geographic coordinate system. When the GPS is interfered or the signal is lost, the relative positioning result obtained by the visual-inertial combination is converted into the geographic coordinate system by using the rotation and translation matrix and the scale factor between the world coordinate system and the geographic coordinate system, so that the result is obtained in the geographic coordinate system. Finally, this paper uses the Unmanned Aerial Vehicle (UAV) as the experimental platform to verify the method. The experimental results show that absolute geolocation results can still be obtained by using the corrected visual-inertial integrated navigation method when GPS is not available, which provides a new approach for global navigation of platforms such as UAV and Unmanned Ground Vehicle (UGV).

Keywords: VINS-Mono · GPS · SLAM · Navigation

1 Introduction

With the popularity of robots in society, how to achieve precise positioning of UAVs in various environments has gradually become a research hotspot. The GPS has low cost and high accuracy, and can provide accurate geolocation results. The disadvantage of GPS is that it is susceptible to occlusion and interference, and cannot be used indoors or in certain areas. Visual-inertial navigation uses camera imaging and Inertial Measurement Unit

© The Author(s), under exclusive license to Springer Nature Switzerland AG 2023
N. Xiong et al. (Eds.): ICNC-FSKD 2022, LNDECT 153, pp. 1385–1397, 2023.
https://doi.org/10.1007/978-3-031-20738-9_150

(IMU) data as the source of navigation information, which has stronger anti-interference ability and higher positioning accuracy. The disadvantage of Visual-inertial navigation is that its positioning results have no geographic information.

Simultaneous localization and map building method in visual navigation [1], referred to as Simultaneous Localization and Mapping (SLAM). The advantage of visual positioning is that there is no drift, and the rotation and translation can be directly measured. The advantages of monocular SLAM are low cost, simple and convenient practical application, but because the monocular camera cannot obtain depth information, which leads to the scale uncertainty of monocular SLAM. Compared with visual positioning, traditional inertial positioning has the advantages of fast response, not affected by image quality, angular velocity is generally accurate, absolute scale can be estimated, and the disadvantage is that there is zero bias, the pose divergence after the integration of low-precision sensors, and high-precision sensors are expensive. By combining the low cost and simple application of monocular SLAM with the absolute scale that can be estimated by the IMU, monocular Visual-Inertial localization makes it easier to perform SLAM. GPS and Visual-Inertial navigation and positioning method have strong complementarity, and the combined navigation of the two has gradually become a research hotspot in recent years.

In this paper, the GPS positioning results and the Visual-Inertial navigation positioning results are combined, and the GPS positioning results in a period of time are matched with the Visual-Inertial navigation positioning results to obtain the conversion relationship between the two. When GPS is unavailable, the system switches to the Visual-Inertial navigation system and geographic coordinates are obtained by coordinate transformation from the Visual-Inertial navigation system enabling the system to achieve navigation and positioning in various environments. For this reason, we call this integrated navigation method a new navigation method based on visual/inertial/satellite combination.

2   Related Work

The early research on the vision and satellite integrated navigation system is to use the visual motion parameter estimation to assist the satellite navigation system to realize the safe operation of the aircraft near the ground or during the landing [2, 3]. With the maturity of computer vision technology, it becomes more and more common to use visual information to solve navigation problems, especially when the satellite navigation signal is interfered or the navigation accuracy is insufficient [4]. In 2007, Davison et al. proposed a real-time monocular SLAM method [5]. In 2014, Engel et al. proposed a large-scale direct monocular SLAM method [6]. In 2015, Mur-Artal et al. proposed a monocular vision SLAM method based on ORB feature points [7]. In 2017, Mur-Artal et al. improved the former to a SLAM system on monocular/stereo/depth cameras [8]. In 2017, Qin et al. proposed a robust initialization method for monocular visual-inertial estimator [9]. In 2018, Ma et al. proposed a UAV video target location method that integrates monocular vision SLAM and GPS, and realized GPS location query in UAV surveillance video [10]. In 2018, Tong et al. proposed a combined positioning and navigation method of monocular visual-inertial [11], and proposed an online calibration method of monocular visual-inertial system [12]. In 2019, Qin et al. extended the former to a positioning and navigation method of monocular/stereo/inertial combination or stereo/GPS combination [13, 14]. In 2020, Feng et al. loosely combined the GNSS positioning results and the visual odometry positioning results in the Kalman filter [15], and used the visual odometry positioning results and the predicted visual odometry errors to realize the GNSS in harsh environments [16]. In 2021, Yang et al. introduced Real Time Kinematic (RTK) information in visual image tracking, obtained the conversion relationship between the visual coordinate system and the world coordinate system, and solved the scale uncertainty and trajectory drift problems, while improving the accuracy [17]. In 2021, Campos et al. improved the ORB-SLAM into a combined visual/inertial SLAM [18]. In 2022, Shen et al. proposed a tightly coupled satellite/visual/inertial fusion state estimator based on the work of Qin et al. [19].

The monocular visual-inertial combination positioning and navigation method proposed by Qin et al. is called VINS-Mono, which is a monocular visual-inertial system and belongs to the category of SLAM. The method takes monocular visual and IMU data as input and estimates the pose of the camera in the world coordinate system. At this time, the world coordinate system takes the initial position of the camera as the origin, and has physical scale, but no geographic information; because the data input is limited by the accuracy of the camera and IMU and the error of the calibration results of the device, the calculated physical scale error and camera pose error may exist.

3   Combined Navigation Method Based on Point Cloud Registration

3.1   Algorithm Overview

This paper proposes a visual/inertial/satellite combined navigation algorithm based on coordinate alignment. The overall flow of the algorithm is shown in Fig. 1. The running process of the algorithm is as follows: First the UAV vision and IMU data are processed through VINS-Mono to obtain the coordinates of the camera in the visual-inertial coordinate system, and at the same time, the GPS latitude and longitude information is converted to the geocentric coordinate system (World Geodetic System—1984 Coordinate System, WGS-84) to obtain absolute geographic coordinates. Then match GPS absolute geographic coordinates and VINS-Mono positioning data according to the timestamp alignment method. The purpose of the match is to convert coordinate data in different coordinate systems through certain rotation and translation transformations and unify them into the geocentric coordinate system. Then using the ICP algorithm [12] to calculate the optimal transformation matrix and scale factor of the geocentric coordinate system and the visual-inertial navigation coordinate system. And when GPS data is unavailable, the positioning coordinates output by VINS-Mono are similarly transformed into the geocentric coordinate system to obtain its absolute pose.

The process of timestamp alignment and matching to calculate the optimal transformation matrix and scale factor is shown in Fig. 2. The time difference between GPS data at time t_1 and time t_0' estimated by VINS-Mono is within a very small threshold c. Similarly, the difference between t_2 and t_1', t_3 and t_2', ... is also within c. Filter out the rest of the data, and then use the coordinates of the above-mentioned corresponding

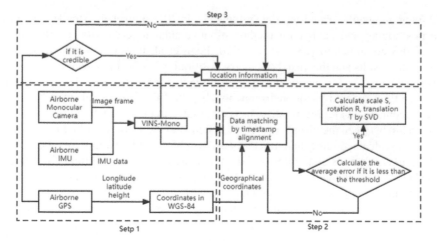

Fig. 1. Algorithm flowchart

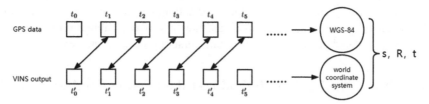

Fig. 2. Timestamp alignment and calculation of transformation parameters

moments to calculate the conversion relationship between the two coordinate systems, the rotation matrix, translation vector, and scaling scale can be obtained.

3.2 Registration Estimation Process

The world coordinate system of the VINS-Mono is the pose estimator coordinate system, and the absolute position in the coordinate system is the geocentric coordinate system. The transformation relationship between the three coordinate systems is shown in Fig. 3.

Fig. 3. Coordinate conversion diagram

The relationship between the camera coordinate system and the pixel coordinate system shown in Fig. 3 can be expressed as follows:

$$z_O \begin{bmatrix} u \\ v \\ 1 \end{bmatrix} = \begin{bmatrix} \frac{1}{dx} & 0 & u_0 \\ 0 & \frac{1}{dy} & v_0 \\ 0 & 0 & 1 \end{bmatrix} \begin{bmatrix} f & 0 & 0 & 0 \\ 0 & f & 0 & 0 \\ 0 & 0 & 1 & 0 \end{bmatrix} \begin{bmatrix} sR & t \\ 0^T & 1 \end{bmatrix} \begin{bmatrix} x_G \\ y_G \\ z_G \\ 1 \end{bmatrix} \tag{1}$$

In the formula, f is the camera focal length, u_0 and v_0 are the coordinates of the origin of the geographic coordinate system in the pixel coordinate system, dx and dy are the physical dimensions of each pixel in the direction x and direction y of the image plane. The above parameters constitute the internal parameters of the camera, which can be obtained by manual calibration. s is the scale factor, t is the translation vector, and R is the rotation matrix.

The external parameters of the camera and IMU can be obtained through calibration, including the rotation matrix R' and translation vector t' of the camera coordinate system and the IMU body coordinate system. The world coordinate system of VINS-Mono is the body coordinate system of the IMU. The camera coordinate system and the VINS-Mono world coordinate system can be converted to each other by using and through similar transformation. In this way, the process of finding the optimal s, R', t' is equivalent to finding the optimal solution of s, R, t. The output trajectory data of VINS-Mono is described as a continuous point set $Q = \{q_1, q_2, \ldots, q_n\}$ in the coordinate system; the GPS data is mapped to the WGS-84 coordinate system, and the point set is $P = \{p_1, p_2, \ldots, p_m\}$.

In the registration process, first select the trajectory points $P_m = \{p_1, p_2, \ldots, p_{200}\}$ when GPS data is available, then select N consecutive points as the registration points, and then find the corresponding points from the trajectory points $Q_m = \{q_1, q_2, \ldots, q_{200}\}$ output by VINS-Mono according to the timestamp information, and calculate the mean error between the point pairs:

$$e = \frac{1}{N} \sum_{i=1}^{N} ||p_i - q_i||^2 \tag{2}$$

When the error e is less than a certain threshold, the trajectory registration of the two groups at this time is reasonable. Otherwise, the points in the registration part of the VINS-Mono output data should be re-selected, and iteratively selects until the error between the registered point pairs is less than the threshold. Since there is a similar transformation relationship between the two, it can be obtained that:

$$q_i = sRp_i + t \tag{3}$$

If the error is defined as $e_i = sRp_i + t - q_i$, the parameter with the smallest error is the element of the desired similarity transformation matrix, that is, the scale factor, rotation matrix, and translation vector are obtained by a least-squares solution:

$$F(t) = \sum_{i=1}^{n} ||sRp_i + t - q_i||^2 \tag{4}$$

Find the partial derivative of the above formula with respect to the translation vector and set its value to 0 to solve the extreme point problem:

$$\frac{\partial F}{\partial t} = \sum_{i=1}^{n} 2(sRp_i + t - q_i) = 0 \tag{5}$$

Define two sets of point centroids as follows:

$$\bar{p} = \frac{\sum_{i=1}^{n} p_i}{n}, \bar{q} = \frac{\sum_{i=1}^{n} q_i}{n} \tag{6}$$

By formulas (4), (5) and (6), we can get:

$$t = \bar{q} - sR\bar{p} \tag{7}$$

Equation (7) can be interpreted as the optimal translation t maps the centroid of p to the centroid of q. Inserting the optimal translation t into Eq. (4), we can get:

$$F(t) = \sum_{i=1}^{n} ||sRp_i + t - q_i||^2 = \sum_{i=1}^{n} ||sR(p_i - \bar{p}) + (\bar{q} - q_i)||^2 \tag{8}$$

So far, we redefine a least squares function, which does not contain translation t, and can directly calculate the optimal rotation matrix R and scale factor s. Define two sets of parameters x_i and y_i:

$$x_i = p_i - \bar{p}, y_i = q_i - \bar{q} \tag{9}$$

According to formulas (8) and (9), the optimal rotation matrix R^* to be solved can be expressed as:

$$R^* = \arg\min \sum_{i=1}^{n} ||sRx_i - y_i||^2 \tag{10}$$

Expand the content in the summation number of the above formula, we can get:

$$(sRx_i - y_i)^T (sRx_i - y_i) = s^2 x_i^T x_i - 2y_i^T sRx_i + y_i^T y_i \tag{11}$$

From formulas (10) and (11), the minimization expression is obtained as:

$$R^* = \arg\min(s^2 \sum x_i^T x_i - 2s \sum y_i^T Rx_i + \sum y_i^T y_i) \tag{12}$$

This expression is a quadratic function, and when the minimum value is obtained at its extreme point, the optimal scale factor can be obtained:

$$s = -\left(\frac{-2\sum_{i=1}^{n} y_i^T Rx_i}{2\sum_{i=1}^{n} x_i^T x_i}\right) = \frac{\sum_{i=1}^{n} y_i^T Rx_i}{\sum_{i=1}^{n} x_i^T x_i} \tag{13}$$

Bringing s back into Eq. (12) we get:

$$\arg\min \sum_{i=1}^{n} ||sRx_i - y_i||^2 = \frac{-(\sum_{i=1}^{n} y_i^T Rx_i)^2}{\sum_{i=1}^{n} x_i^T x_i} + \sum_{i=1}^{n} y_i^T y_i \tag{14}$$

From Eqs. (10), (14), solving the least squares problem for R^* is equivalent to solving the following equations:

$$\arg\max \sum_{i=1}^{n} y_i^T Rx_i \tag{15}$$

According to the properties of the matrix trace, Eq. (15) can be transformed into:

$$\arg\max tr(Y^T RX) = \arg\max tr(RXY^T) \tag{16}$$

Next, use SVD decomposition [13] to calculate the optimal rotation matrix R^*, denote $S = XWY^T$, and perform SVD decomposition on it:

$$S = U\Sigma V^T \tag{17}$$

where U and V are both diagonal matrices. At this time, a rotation matrix B is introduced, assuming that there is an optimal solution R^*, according to the Cauchy–Schwarz inequality, there is:

$$tr(R^*S) \geq tr(BR^*S) \tag{18}$$

And from the Cauchy inequality, we get:

$$tr(AA^T) \geq tr(BAA^T)\,(BB^T = I) \tag{19}$$

Then the problem is transformed into finding a R^*, so that R^*S can be expressed as the form of AA^T, so formula (15) can be expressed as:

$$tr(RXY^T) = tr(RS) = tr(RU\Sigma V^T) \tag{20}$$

When $R^* = VU^T$, we get $R^*S = V\sum V^T$, at this time $A = V\Sigma^{\frac{1}{2}}$. So, the optimization result is:

$$R^* = VU^T \tag{21}$$

So far, the optimal similarity transformation matrix has been obtained, and the points in P are rotated and translated to obtain a new point set:

$$P' = \{P_i' = Rp_i + t | p_i < P\} \tag{22}$$

Furthermore, when the GPS positioning information is not available, the absolute position of the current UAV can be obtained by transforming the VINS-Mono track point to the WGS-84 coordinate system, thereby realizing continuous geolocation.

4   Experimental and Analysis

In the early research of visual and satellite integrated navigation system, visual navigation and satellite navigation are independent of each other, and the motion parameters of vision are only used to assist and correct satellite navigation. The methods proposed successively by Davison, Engel, Mur-Artal, Qin et al. only use vision as the only extroverted perception of the system, and their research results are local relative localization results. The experiments in this paper use the most common visual/inertial/GPS devices in the current market to experimentally verify the method in this paper under the condition of limited hardware accuracy.

4.1   Flight Dataset Experiment

The drone used in this experiment is the DJI M600. The drone is equipped with a Realsense D435i camera and a CUAV NEO V2 GNSS module. Some of the dataset thumbnails are shown in Fig. 4. The GPS position information acquisition speed of NEO V2 is 50Hz, and the GPS position is used as the ground truth of the UAV. The flight trajectory of the data set is shown in Fig. 5.

Fig. 4. Part of the aerial image

In the experiment, considering that VINS-Mono needs to meet the two requirements of real-time performance and high positioning accuracy of running results, the camera is set to 640 * 480 resolution for acquisition. After that, feature extraction is performed

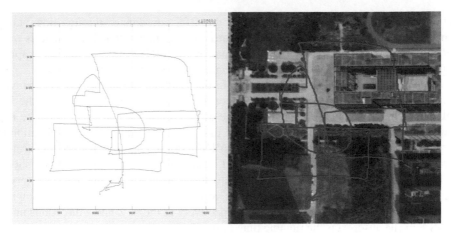

Fig. 5. Flight path under satellite image

on the camera image frame, and the IMU is pre-integrated and then processed by visual-inertial odometry. The absolute pose of the camera in the body coordinate system is obtained through back-end nonlinear optimization, and a sparse point cloud of key frame feature points is generated at the same time. The output of the VINS-Mono running dataset is shown in Fig. 6. The green line is the trajectory estimated by pure visual-inertial odometry, and the red line is the trajectory estimated by the visual-inertial information after loop detection processing.

Fig. 6. Output results from the dataset using VINS-Mono

The experimental process is divided into three stages. The first stage is the flight data processing stage. In this stage, the camera pose in the world coordinate system output by VINS-Mono is obtained, including the coordinates and direction information. The coordinates are converted into coordinates under the WGS-84 coordinate system after the coordinate conversion. The second stage is the matching stage, the coordinates output by VINS-Mono are registered with the GPS coordinates through time stamp synchronization, and the transformation matrix is calculated. The third stage is anti-GPS-interference stage, the coordinates output by the VINS-Mono system are mapped to the geocentric coordinate system one by one, and the drift error is calculated.

In the first stage, VINS-Mono needs to initialize the camera pose. For scenes with many feature points such as the teaching constructions and its surroundings, the initialization time is about ten seconds, and the estimation results are more accurate in a short period of time after initialization. The first 300 s are used for pose estimation to ensure that the positioning results output by VINS-Mono during coordinate registration are more accurate. The parameters obtained by the second stage matching are shown in Table 1. The difference between the transformed trajectory and the real trajectory in the third stage between the three axes of the geographic coordinate system is shown in Fig. 7. In the figure, the three curves are respectively in the x, y, and z directions under WGS-84. The abscissa is the running time of the dataset, in seconds, and the ordinate is the distance between the transformed trajectory and the real trajectory, in units of Meter.

Table 1. Aligned parameters from VINS-Mono and GPS trajectory using UAV data

Average time interval from data (s)	Error (m)	Hysteresis	Scale factor
0.201282	29.48	966.123	0.943245

It can be seen from Table 1 and Fig. 7 that the positioning results output by the VINS-Mono operating experimental data are converted into the positioning information under the geographic coordinate system, and the positioning error is within 100 m; there is an error of about 5.7% in the scale between geographic coordinate system and the world coordinate system of VINS-Mono; the estimated result in a short time after the initialization of VINS-Mono is more accurate, but because the positioning coordinates output by VINS-Mono will be relocated due to the loss of features, the camera will be re-positioned from the origin (occurs around the 200th second in Fig. 7), resulting in a drop in matching accuracy.

4.2 Subgrade Dataset KITTI Experiment

KITTI dataset combines visual data, IMU data, and GPS data. This dataset uses a stabilized mount, a global shutter camera, and GPS acquisition using RTK on a car. The experiments in this paper use the 2011_10_03_drive_0027_sync from KITTI for verification and comparison to illustrate the effectiveness of the experiment. The first 2000 results generated by running VINS-Mono on the KITTI are used as experimental data, the GPS coordinates are time-stamped, and the transformation matrix is calculated. The experimental results are shown in Table 2 and Fig. 8.

It can be seen from Table 2 and Fig. 8 that the positioning results output by VINS-Mono can be accurately converted into the positioning information under the geographic coordinate system by the method in this paper, and the positioning error of the KITTI dataset under the geographic coordinate system is within 40 m.

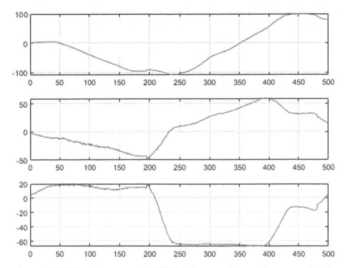

Fig. 7. Comparison of GPS trajectory and VINS-Mono trajectory between the three axes of the geographic coordinate system using UAV data

Table 2. Aligned parameters from VINS-Mono and GPS trajectory using KITTI

Average time interval from data (s)	Error (m)	Hysteresis	Scale factor
0.103629	78.566677	49	0.994545

5   Conclusion

This paper proposes a monocular visual/inertial/GPS combined navigation method. The method in this paper can use the geolocation information corrected by visual-inertial navigation for positioning when the GPS is interfered or the signal is lost in a short time. The land-based KITTI dataset and UAV experimental dataset verify the effectiveness of the method in this paper.

The method in this paper is tested on the aerial data set of a specific scene. Since it is impossible to accurately judge in which state the satellite is undisturbed and available, and in which state it is interfered or deceived, further research is needed. If the method in this paper can be practically applied in the military field, it should be able to improve the all-weather adaptability and combat level of the UAV.

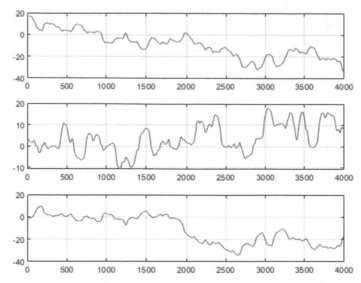

Fig. 8. Comparison of GPS trajectory and VINS-Mono trajectory between the three axes of the geographic coordinate system using KITTI

References

1. Durrant-Whyte, H., Bailey, T.: Simultaneous localization and mapping: part I. IEEE Robot. Autom. Mag. **13**(2), 99–110 (2006)
2. Saripalli, S., Montgomery, J., Sukharme, G.: Vision-based autonomous landing of an unmanned aerial vehicle. In: Processed of IEEE International Conference on Robotics and Automation, pp. 2799–2804 (2002)
3. Vergauwen, M., Pollefefeys, M., Van Gool, L.: A stere-based system for supported of planetary surface exploration. Mach. Vis. Appl. **14**, 5–14 (2003)
4. Henry, P., Krainin, M., Herbst, E. et al.: RGB-D mapping: using kinect-style depth cameras for dense 3D modeling of indoor environments. Exp. Robot. (2014)
5. Davison, A.J., Reid, I.D., Molton, N.D., et al.: MonoSLAM: real-time single camera SLAM. IEEE Trans. Pattern Anal. Mach. Intell. **29**(6), 1052–1067 (2007)
6. Engel, J., Schöps, T., Cremers, D.: LSD-SLAM: Large-scale direct monocular SLAM. European Conference on Computer Vision, pp. 834–849. Springer, Cham (2014)
7. Mur-Artal, R., Montiel, J.M.M., Tardos, J.D.: ORB-SLAM: a versatile and accurate monocular SLAM system. IEEE Trans. Rob. **31**(5), 1147–1163 (2015)
8. Mur-Artal, R., Tardós, J.D.: Orb-slam2: An open-source slam system for monocular, stereo, and rgb-d cameras. IEEE Trans. Rob. **33**(5), 1255–1262 (2017)
9. Qin, T., Shen, S.: Robust initialization of monocular visual-inertial estimation on aerial robots. In: 2017 IEEE/RSJ international conference on intelligent robots and systems (IROS), pp. 4225–4232. IEEE (2017)
10. Ma, Y., Cao, X., Ding, C., Jiang, B., Li, D., Wan, G.: A UAV video target localization method integrating monocular vision SLAM and GPS. J. Surv. Mapp. Sci. Technol. **35**(05), 497–501 (2018)
11. Qin, T., Peiliang, et al.: VINS-Mono: a robust and versatile monocular visual-inertial state estimato. IEEE Trans. Robot. (2018)

12. Qin, T., Shen, S.: Online temporal calibration for monocular visual-inertial systems. In: 2018 IEEE/RSJ International Conference on Intelligent Robots and Systems, pp. 3662–3669 (IROS). IEEE (2018)
13. Qin, T., Cao, S., Pan, J. et al.: A general optimization-based framework for global pose estimation with multiple sensors (2019)
14. Qin, T., Pan, J., Cao, S. et al.: A general optimization-based framework for local odometry estimation with multiple sensors. arXiv preprint arXiv:1901.03638 (2019)
15. Welch, G., Bishop, G.: An introduction to the Kalman filter (1995)
16. Feng, Y., Rui, T., Han, J., Furong Hou, J., Hong, J.L., Wang, X.: Navigation and positioning of binocular vision-assisted GNSS in harsh environments. Global Positioning Syst. **45**(03), 48–53 (2020)
17. Yang, K, Huang, S., Xiao, H., Luo, K.: Real-time pose estimation of UAV aerial images. Comput. Appl. Res. 1–7 (2021)
18. Campos, C., Elvira, R., Rodríguez, J.J.G., et al.: Orb-slam3: An accurate open-source library for visual, visual–inertial, and multimap slam. IEEE Trans. Rob. **37**(6), 1874–1890 (2021)
19. Cao, S., Lu, X., Shen, S.: GVINS: Tightly coupled GNSS–visual–inertial fusion for smooth and consistent state estimation. IEEE Trans. Robot. (2022)
20. Chetverikov, D., Svirko, D., Stepanov, D. et al.: The trimmed iterative closest point algorithm. In: Object Recognition Supported by User Interaction for Service Robots, vol. 3, pp. 545-548. IEEE (2002)
21. Van Loan, C.F.: Generalizing the singular value decomposition. SIAM J. Numer. Anal. **13**(1), 76–83 (1976)

Path-Following Formation Control of Multi-mobile Robots Under Single Path

Zhangyi Zhu$^{(\boxtimes)}$, Wuxi Shi, and Baoquan Li

Tiangong University, Tianjin 300387, China
2031060748@tiangong.edu.cn, shiwuxi@tiangong.edu.cn, libq@tiangong.edu.cn

Abstract. In this paper, a path-following formation control under single path is proposed for multi-mobile robots. Combining arbitrary positive uniform continuous bounded functions and barrier Lyapunov function, a possible virtual kinematic controller is designed. By utilizing the graph theory and path variable containment method, path parameter derivative is designed as additional control input to achieve the single path formation. Simulation results are given to illustrate the effectiveness of the proposed approach.

Keywords: Path following · Mobile robots · Formation control

1 Introduction

Compared with single mobile robot, multi-mobile robots can achieve higher efficiency in task execution by coordinating with each other. Over the past few decades, formation control algorithms of multi-mobile robots include leader-follow [1], virtual structure [2], behavior-based [3], artificial potential field [4], etc. Path following is one of the basic problems in motion control of mobile robot. The unique path parameter derivatives in path following often play an important role in controller design. Do [5] takes the derivative of path parameter as additional control input to synchronize formation movement. Cao [6] designs the additional control input (path parameter derivative) to achieve the desired formation and speed synchronization. Multi-mobile robots often need to achieve formation according to a single path in some narrow environments. In addition, the formation of a single path can reduce the burden of path planning. Inspired by the multi-agent containment control technology, the containing control law of path parameter derivative is designed for AUVs system in [7], realizing the control of single path following formation.

It should be noted that in most relevant researches, the underactuated control problem in kinematic was not considered. In [8], a novel path following

N. Xiong et al. (Eds.): ICNC-FSKD 2022, LNDECT 153, pp. 1398–1406, 2023.
https://doi.org/10.1007/978-3-031-20738-9_151

error model is proposed, which solves the underactuated problem, but also introduces the singularity problem into the system. Prescribed performance boundary (PPB) technology is introduced in [8] to avoid the singular points. In [9], the singularity problem is avoided by combining the barrier Lyapunov function. However, performance functions are used as constraint boundaries in above works. More parameters need to be designed in the performance function, which increases the computation load. More seriously, the exponential convergence of the performance function makes the system error limit to a narrow range. In practice, to maintain the error within this limitation is very difficult. If the error exceed the boundary, near-singularity phenomenon may be encountered [8].

Based on the above discussion, this paper proposes a single path formation control strategy based on the kinematic model of two-wheel differential mobile robots. The main contributions of this paper are as follows: 1) The singularity problems are avoided via a concise control scheme at the kinematic level. 2) Inspired by Liu et al. [7], the parameter containing update law is designed for the path parameter derivatives to perform the formation control of multi-mobile robots on a single path.

2   Preliminaries

2.1   Graph Theory

The graph theory knowledge is expounded here. A graph $\mathcal{G} = \{\mathcal{V}, \mathcal{E}\}$ contains $\mathcal{V} = \{n_1, n_2, \ldots, n_N\}$ and $\mathcal{E} = \{(n_i, n_j) \in \mathcal{E} \times \mathcal{E}\}$ represents the node set and the edge set, respectively. The element (n_i, n_j) represents the communication from node i to node j. The adjacency matrix is defined as $\mathcal{A} = [a_{ij}] \in \mathbb{R}^{N \times N}$, where $a_{ij} = 1$, if $(n_i, n_j) \in \mathcal{E}$; and $a_{ij} = 0$, otherwise. In addition, self connections are not considered, i.e., $a_{ii} = 0$. The degree matrix is defined as $\mathcal{D} = diag\{d_1, d_2, \ldots, d_N\}$, where $d_i = \sum_{j=1}^{N} a_{ij}$, $i = 1, 2, \ldots, N$. The Laplacian matrix $\mathcal{L}$ of the graph $\mathcal{G}$ is defined as $\mathcal{L} = \mathcal{D} - \mathcal{A}$.

2.2   Correlation Lemmas in Stability Proof

Lemma 1. [10] *Consider a dynamic system of the following form:*

$$\dot{z}(t) = -az(t) + bv(t), \tag{1}$$

where a, b are positive constants, and $v(t)$ is any positive function. Then, for any given bounded initial condition $z(t_0) \geqslant 0$, we have that $z(t) \geqslant 0$ for $\forall t \geqslant t_0$.

Lemma 2. [11] *For any positive constant k, the following inequality holds when $|x| < k$:*

$$ln\frac{k^2}{k^2 - x^2} \leqslant \frac{x^2}{k^2 - x^2}. \tag{2}$$

Lemma 3. [11] *For any positive constant k_b, regard $\mathcal{Z}_1 := \{z_1 \in \mathbb{R} : |z_1| < k_b\} \subset \mathbb{R}$ and $\mathcal{N} := \mathbb{R}^l \times \mathcal{Z}_1 \subset \mathbb{R}^{l+1}$ as open sets. Consider the following system*

$$\dot{\eta} = h(t, \eta), \tag{3}$$

where $\eta := [m, z_1]^T \in N$ represents the state. The function $h : \mathbb{R}_+ \times \mathcal{N} \to \mathbb{R}^{l+1}$ is piecewise continuous in t and locally Lipschitz in z_1, uniformly in t, on $\mathbb{R}_+ \times \mathcal{N}$. Suppose that there exist positive defined and continuously differentiable functions $U : \mathbb{R}^l \to \mathbb{R}_+$ and $V_1 : \mathcal{Z}_1 \to \mathcal{R}_+, l = 1, \ldots, n$, then

$$V_1(z_1) \to \infty \quad as \quad |z_1| \to k_b, \tag{4}$$

$$\gamma_1(||m||) \leqslant U(m) \leqslant \gamma_2(||m||), \tag{5}$$

with γ_1 and γ_2 as class K_∞ functions. Let $V(\eta) := V_1(z_1) + U(m)$, and $z_1(0) \in \mathcal{Z}_1$. If the inequality holds

$$\dot{V} = \frac{\partial V}{\partial \eta} \leqslant -\alpha V + \beta, \tag{6}$$

in the set $\eta \in \mathcal{N}$ and α, β are positive constants, then w remains bounds and $z_1(t) \in \mathcal{Z}_1, \forall t \in [0, \infty)$.

3   Problem Formulation

We consider a two-wheeled differential mobile robot, its kinematics model is described as follows:

$$\begin{bmatrix} \dot{x} \\ \dot{y} \\ \dot{\varphi} \end{bmatrix} = \begin{bmatrix} \cos\varphi & 0 \\ \sin\varphi & 0 \\ 0 & 1 \end{bmatrix} \begin{bmatrix} v \\ w \end{bmatrix}, \tag{7}$$

where $[x, y, \varphi]^T$ denotes the position and orientation of the robot in world coordinates, $[v, w]^T$ denotes the linear velocity and angular velocity of robot.

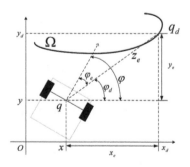

Fig. 1. Interpretation of path following errors.

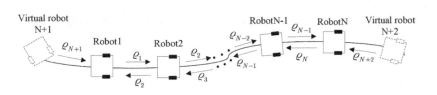

Fig. 2. Communication topology.

Figure 1 shows the interpretation of path following errors of mobile robot. Where, Ω is the desired reference path, $q = [x, y]^T$ represents the actual position of the robot, $q_d = [x_d(\varrho), y_d(\varrho)]^T$ represents the reference position, ϱ is the path parameter. φ, φ_d, φ_e represents the actual orientation, expected orientation and orientation error of the robot, respectively.

Based on the above description, the path following error model of multi-mobile robots is established [8]:

$$x_{ei} = x_{di}(\varrho_i) - x_i, \; y_{ei} = y_{di}(\varrho_i) - y_i, \; z_{ei} = \sqrt{x_{ei}^2 + y_{ei}^2}, \tag{8}$$

$$\varphi_{ei} = \varphi_i - \varphi_{di}. \tag{9}$$

where $\varphi_{di} = \arcsin\frac{y_{ei}}{z_{ei}}$.

Combining (8)–(9) and the geometric relationship in Fig. 1, we obtain:

$$\sin(\varphi_{ei}) = \frac{x_{ei}\sin(\varphi_i) - y_{ei}\cos(\varphi_i)}{z_{ei}}, \tag{10}$$

$$\cos(\varphi_{ei}) = \frac{x_{ei}\cos(\varphi_i) + y_{ei}\sin(\varphi_i)}{z_{ei}}. \tag{11}$$

From (7)–(10), the derivative of z_{ei}, φ_{ei} is

$$\dot{z}_{ei} = -\cos(\varphi_{ei})v_i + \left(\frac{x_{ei}}{z_{ei}}x'_{di} + \frac{y_{ei}}{z_{ei}}y'_{di}\right)\dot{\varrho}_i, \tag{12}$$

$$\dot{\varphi}_{ei} = w_i + \left(\frac{y_{ei}}{z_{ei}^2}x'_{di} - \frac{x_{ei}}{z_{ei}^2}y'_{di}\right)\dot{\varrho}_i + \frac{\sin(\varphi_{ei})}{z_{ei}}v_i, \tag{13}$$

where $x'_{di} = \frac{\partial x_{di}}{\partial \varrho_i}$, $y'_{di} = \frac{\partial y_{di}}{\partial \varrho_i}$.

From (12)–(13) we know, the error system will be ill-defined if $z_{ei} = 0$ and $\varphi_{ei} = \pm\frac{\pi}{2}$. Therefore, the following conditions should be satisfied when the control inputs are selected:

$$z_{ei}(t) > 0, \; |\varphi_{ei}(t)| < \frac{\pi}{2}, \quad \forall t \geqslant 0. \tag{14}$$

Next we will elaborate the path-following formation control of multi-mobile robots with single path. Assume that there are N actual mobile robots and two virtual leaders indexed by $N + 1$ and $N + 2$ in the formation queue. The communication topology between robots is shown in Fig. 2.

The communication topology could be described by a graph $\mathcal{G}$ whose Laplacian matrix $\mathcal{L}$ can be shown as

$$\mathcal{L} = \begin{bmatrix} \mathcal{L}_1 & \mathcal{L}_2 \\ 0_{2 \times N} & 0_{2 \times 2} \end{bmatrix}, \tag{15}$$

where

$$\mathcal{L}_1 = \begin{bmatrix} 2 & -1 & 0 & \cdots & \cdots & \cdots & 0 \\ -1 & 2 & -1 & 0 & \cdots & \cdots & 0 \\ \vdots & & & \ddots & & & \vdots \\ 0 & \cdots & \cdots & 0 & -1 & 2 & -1 \\ 0 & \cdots & \cdots & \cdots & 0 & -1 & 2 \end{bmatrix}_{N \times N}, \tag{16}$$

$$\mathcal{L}_2 = \begin{bmatrix} -1 & 0 & 0 & \ldots & 0 & 0 & 0 \\ 0 & 0 & 0 & \ldots & 0 & 0 & -1 \end{bmatrix}^T_{2 \times N}. \tag{17}$$

Define the path parameter error between the i-th and j-th mobile robot as

$$e_i = \sum_{j=1}^{N+2} a_{ij}(\varrho_i - \varrho_j), \tag{18}$$

where a_{ij} is the corresponding element in the adjacency matrix $\mathcal{A}$.

Define the path parameter update rate of the two virtual robots as

$$\dot{\varrho}_{N+1} = \dot{\varrho}_{N+2} = v_s, \tag{19}$$

in which $\dot{\varrho}_{N+1}(t_0) < \dot{\varrho}_{N+2}(t_0)$, and v_s is the reference speed.

Now, we set the path parameter derivative of the i-th mobile robot as

$$\dot{\varrho}_i = v_s - \chi_i, \tag{20}$$

where χ_i is a variable need to be designed.

Define $e = [e_1, \ldots, e_N]^T$ and $\chi = [\chi_1, \ldots, \chi_N]^T$, combining (16), (18)-(20), we can obtain the error dynamics of path parameters as

$$\dot{e} = -\mathcal{L}_1 \chi. \tag{21}$$

The control objective of this paper is to realize formation control for N mobile robots following a single path, and the path parameters of each actual robot are evenly distributed between the path parameters of two virtual robots.

4   Control Design

The kinematics controllers are designed as

$$v_i = \frac{k_{1i} z_{ei} + \left(\frac{x_{ei}}{z_{ei}} x'_{di} + \frac{y_{ei}}{z_{ei}} y'_{di}\right) \dot{\varrho}_i - \sqrt{\sigma_i(t)}}{\cos(\varphi_{ei})}, \tag{22}$$

$$w_i = -k_{2i}\varphi_{ei} - \left(\frac{y_{ei}}{z_{ei}^2}x'_{di} - \frac{x_{ei}}{z_{ei}^2}y'_{di}\right)v_s - \frac{\sin(\varphi_{ei})}{z_{ei}}v_i, \tag{23}$$

where k_{1i}, k_{2i} are positive constants, $\sigma_i(t) > 0$ are any positive uniformly continuous and bounded functions, i.e., $0 < \sigma_i(t) < \bar{\sigma}$, where $\bar{\sigma}$ is any positive constant.

Choose update law for χ_i as [7]

$$\begin{cases} \dot{\bar{\chi}}_i = -\bar{\chi}_i - \mu_{1i}(\xi_i - e_i) \\ \chi_i = -\mu_{2i}(\xi_i - e_i) + \bar{\chi}_i \end{cases}, \tag{24}$$

where μ_{1i} and μ_{2i} are positive constants, ξ_i is designed as

$$\xi_i = -\frac{\left(\frac{y_{ei}}{z_{ei}^2}x'_{di} - \frac{x_{ei}}{z_{ei}^2}y'_{di}\right)}{(\pi/2)^2 - \varphi_{ei}^2}\varphi_{ei} + \chi_i^{-1}k_{3i}e^T\mathcal{L}_1^{-1}e. \tag{25}$$

Theorem 1. *For system (7), with controller (22)-(23) and path parameters containing update law (24). If the initial conditions meet that $z_{ei}(0) > 0$ and $|\varphi_{ei}(0)| < \frac{\pi}{2}$, then the position error z_{ei} and the path parameters error e_i converges to zero and the orientation error φ_{ei} converges to a small neighborhood of the origin while constraint (14) can be guaranteed.*

Proof. Construct the Lyapunov function candidate as

$$V = \sum_{i=1}^{N}\left(\frac{1}{2}z_{ei}^2 + \frac{1}{2}\ln\frac{(\pi/2)^2}{(\pi/2)^2 - \varphi_{ei}^2} + \frac{1}{2}\mu_{1i}^{-1}\bar{\chi}_i^2\right) + \frac{1}{2}e^T\mathcal{L}_1^{-1}e. \tag{26}$$

The derivative of V along (12)-(13) and (22)-(24) is:

$$\begin{aligned}
\dot{V} &= \sum_{i=1}^{N}(-k_{1i}z_{ei}^2 + z_{ei}\sqrt{\sigma_i(t)} - \frac{k_{2i}\varphi_{ei}^2}{(\pi/2)^2 - \varphi_{ei}^2} - \mu_{1i}^{-1}\bar{\chi}_i^2 \\
&\quad - \mu_{2i}(\xi_i - e_i)^2) - k_{3i}Ne^T\mathcal{L}_1^{-1}e \\
&\leqslant \sum_{i=1}^{N}(-(k_{1i} - \frac{1}{2})z_{ei}^2 - k_{2i}\ln\frac{(\pi/2)^2}{(\pi/2)^2 - \varphi_{ei}^2} - \mu_{1i}^{-1}\bar{\chi}_i^2) \\
&\quad - k_{3i}Ne^T\mathcal{L}_1^{-1}e + \sum_{i=1}^{N}\frac{1}{2}\sigma_i(t) \leqslant -\alpha V + \beta,
\end{aligned} \tag{27}$$

where $\alpha = \min\{2(k_{1i}-0.5), 2k_{2i}, 2\mu_{1i}^{-1}, 2k_{3i}N, (i = 1, \ldots, N)\}$, $\beta = \sum_{i=1}^{N}\frac{1}{2}\bar{\sigma}$. To ensure the stability of system, the choice of k_{1i}, k_{2i}, k_{3i}, μ_{1i}, μ_{2i} should satisfy that $k_{1i} > 0.5$, $k_{2i} > 0$, $k_{3i} > 0$, $\mu_{1i} > 0$, $\mu_{2i} > 0$.

Let $\kappa = \frac{\beta}{\alpha}$, multiply both sides by $e^{\alpha t}$, (27) can be expressed as:

$$\frac{d}{dt}(Ve^{\alpha t}) \leqslant \beta e^{\alpha t} \tag{28}$$

Integrating (28) over $[0, t]$, obtain

$$0 \leqslant V(t) \leqslant \kappa + (V(0) - \kappa)e^{-\alpha t} \leqslant \kappa + V(0). \tag{29}$$

It can be seen from (26), (29) that z_{ei}, φ_{ei}, $\bar{\chi}_i$, e_i are bounded. Combine (8), x_{ei}, y_{ei} are bounded. From (12), (22)-(23), v_i, w_i are bounded and $\dot{z}_{ei} \in \mathcal{L}_\infty$, $\dot{e}_i \in \mathcal{L}_\infty$. Based on the Corollary of Barbalat's lemma, we obtain that $\lim\limits_{t \to \infty} z_{ei} = 0$, $\lim\limits_{t \to \infty} e_i = 0$.

It follows from Lemma 3 that $|\varphi_{ei}(t)| < \frac{\pi}{2}$, $\forall t > 0$ is ensured when the initial condition $|\varphi_{ei}(0)| < \frac{\pi}{2}$ is satisfied.

Using (12) and (22), we have

$$\dot{z}_{ei} = -k_{1i}z_{ei} + \sqrt{\sigma_i(t)}. \tag{30}$$

From Lemma 1, any selected finite initial value $z_{ei}(0) > 0$ can guarantee the constraint $z_{ei}(t) > 0$, $\forall t > 0$.

Then, from (29), we get

$$\frac{1}{2}\ln\frac{(\pi/2)^2}{(\pi/2)^2 - \varphi_{ei}^2} \leqslant \kappa + (V(0) - \kappa)e^{-\alpha t}. \tag{31}$$

Taking the exponential transformation of both sides of (31), we obtain:

$$\frac{(\pi/2)^2}{(\pi/2)^2 - \varphi_{ei}^2} \leqslant e^{2[\kappa + (V(0) - \kappa)e^{-\alpha t}]}. \tag{32}$$

It has been proved that $|\varphi_{ei}(t)| < \frac{\pi}{2}$, $\forall t > 0$. Distinctly, $(\pi/2)^2 - \varphi_{ei}^2 > 0$. So, it can be further obtained from (32):

$$|\varphi_{ei}| \leqslant 0.5\pi\sqrt{1 - e^{-2[\kappa + (V(0) - \kappa)e^{-\alpha t}]}}. \tag{33}$$

From the above equation, it can be concluded that given any $\bar{\varphi}_{ei} > 0.5\pi\sqrt{1 - e^{-2\kappa}}$, there always exists T such that for any $t > T$, it has $|\varphi_{ei}(t)| \leqslant \bar{\varphi}_{ei}$. Which means for $t \to \infty$, $|\varphi_{ei}(t)| \leqslant 0.5\pi\sqrt{1 - e^{2\kappa}}$. By choosing α and β appropriately, φ_{ei} can be made arbitrarily small. This completes the proof.

5   Simulation

In this section, three mobile robots are selected for simulation experiment to verify the effectiveness of the designed controller. The reference path is given as

$$\begin{cases} q_d = [\sin(-0.1\varrho + 1.5\pi), \cos(-0.1\varrho + 1.5\pi)], & \varrho \in [0, 10\pi) \\ q_d = [\sin(-0.1\varrho + 0.5\pi) + 2, \cos(-0.1\varrho + 0.5\pi)], & \varrho \in [10\pi, 60] \end{cases} \tag{34}$$

The initial position and orientation of robot1, robot2, robot3 are $[-1.12, 0.12, -1]^T$, $[-1.11, 0.11, -1]^T$, $[-1.10, 0.10, -1]^T$, respectively. The design parameters are chosen as $k_{1i} = 1.6$, $k_{2i} = 0.5$, $k_{3i} = 10^{-4}$, $\mu_{1i} = 0.01$, $\mu_{2i} = 0.8$, $v_s = 1$, $\varrho_i(0) = 0$, $\varrho_4(0) = 0$, $\varrho_5(0) = 3.2$, $\chi_i(0) = 0$, $\sigma_i(t) = 0.08e^{-0.05t}$.

The simulation results are shown in Fig. 3. In Fig. 3-(a), the reference path and robots motion are shown. The position errors and orientation errors are displayed in Fig. 3-(b,c), respectively. The control inputs are shown in Fig. 3-(d,e). Figure 3-(f) depicts the path parameter change of each robot.

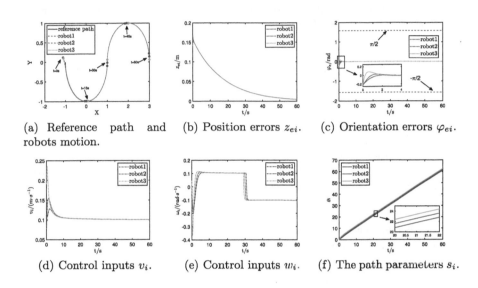

(a) Reference path and robots motion.

(b) Position errors z_{ei}.

(c) Orientation errors φ_{ei}.

(d) Control inputs v_i.

(e) Control inputs w_i.

(f) The path parameters s_i.

Fig. 3. Simulation results.

6   Conclusion

A path-following formation control strategy for multiple mobile robots with single path is proposed in this paper. Aiming at solving the singularity problem in the kinematics model, a concise and practical kinematic controller is designed. The path parameter contain control method achieves the desired formation pattern. Simulation results verify the feasibility of the proposed algorithm.

Acknowledgement. This work was partly supported by the Natural Science Foundation of Tianjin, China under Grant 20JCYBJC00180, and by the National Natural Science Foundation of China under Grant 61973234.

References

1. Dai, S.L., Lu, K., Jin, X.: Fixed-time formation control of unicycle-type mobile robots with visibility and performance constraints. IEEE Trans. Ind. Electron. **68**(12), 12615–12625 (2021)

2. Chen, X., Huang, F., Zhang, Y., Chen, Z., Zhu, S.: A novel virtual-structure formation control design for mobile robots with obstacle avoidance. Appl. Sci. **10**(17), 5807 (2020)
3. Xu, D.D., Zhang, X.N., Zhu, Z.Q., Chen, C.L., Yang, P.: Behavior-based formation control of Swarm robots. Math. Probl. Eng. (2014)
4. Chang, K., Ma, D.L., Han, X.B., Liu, N., Zhao, P.P.: Lyapunov vector-based formation tracking control for unmanned aerial vehicles with obstacle/collision avoidance. Trans. Inst. Meas. Control **42**(5), 942–950 (2020)
5. Do, K.D., Pan, J.: Nonlinear formation control of unicycle-type mobile robots. Robot. Auton. Syst. **55**(3), 191–204 (2007)
6. Cao, K.C., Jiang, B., Yue, D.: Cooperative path following control of multiple nonholonomic mobile robots. ISA Trans. **71**, 161–169 (2017)
7. Liu, L., Wang, D., Peng, Z.H.: Coordinated path following of multiple underacutated marine surface vehicles along one curve. ISA Trans. **64**, 258–268 (2016)
8. Wang, W., Huang, J.S., Wen, C.Y.: Prescribed performance bound-based adaptive path-following control of uncertain nonholonomic mobile robots. Int. J. Adapt. Control Signal Process. **31**(5), 805–822 (2017)
9. Jin, X.: Fault-tolerant iterative learning control for mobile robots non-repetitive trajectory tracking with output constraints. Automatica **94**, 63–71 (2018)
10. Wang, M., Zhang, S.Y., Chen, B., Lou, F.: Direct adaptive neural control for stabilization of nonlinear time-delay systems. Sci. China (Inf. Sci.) **53**(4), 800–812 (2010)
11. Ren, B.B., Ge, S.Z.S., Tee, K.P., Lee, T.H.: Adaptive neural control for output feedback nonlinear systems using a barrier Lyapunov function. IEEE Trans. Neural Netw. **21**(8), 1339–1345 (2010)

FPGA Prototype Verification of FlexRay Communication Controller Chip

Zejun Liu$^{(\boxtimes)}$ and Xiaofeng Yang

School of Electronics Engineering, Xi'an University of Posts and Telecommunications, Xi'an 710121, China
`1556012659@stu.xupt.edu.cn, xfyang@mail.xidian.edu.cn`

Abstract. FlexRay is high-speed, and its speed is several times higher than that of other vehicle backbone networks. It can effectively apply the simplified distributed control algorithm, implement the control mechanism with higher safety requirements, or replace the hydraulic system with an electronic control system. It can be used as the backbone network of the new generation automobile internal network. FlexRay communication controller chip makes it easy to integrate FlexRay into MCU based applications. Provide full system redundancy to meet the reliability requirements of the new safety system. MFR4310 is a mature FlexRay communication controller chip, and the function of NT4310 is compatible with MFR4310. In order to carry out the prototype verification of NT4310 chip, an FPGA prototype verification platform based on FlexRay bus architecture is developed. The process, verification environment and verification results are described. Experiments show that the platform and the corresponding verification process greatly improve the verification efficiency, shorten the whole design verification cycle, and provide a reliable guarantee for the successful chip delivery.

Keywords: FlexRay · FPGA · Prototype verification

1 Introduction

FlexRay protocol was originally developed by BMW and DaimlerChrysler based on ByteFlight protocol of BMW. With the help of FlexRay alliance, FlexRay protocol has developed rapidly and quickly become the standard of the next generation automotive network bus [1]. FlexRay communicates physically through two separate buses, each of which has a communication quantity of 10 mbit/s. The dual channel redundancy of FlexRay makes the safety of autopilot [2] higher. FlexRay is an on-board network system with high communication rate, certainty, real-time and fault tolerance. Its characteristics and mechanism meet the development needs of on-board communication network in the future. FlexRay can also be used in the distributed real-time control system with strict requirements, which can solve the application defects of the main on-board network standards of modern vehicles [3]. NT4310 is fully functionally compatible with MFR4310. NT4310 is a stand-alone FlexRay controller, which makes it easy to integrate FlexRay into MCU based applications without using embedded FlexRay controller [4].

© The Author(s), under exclusive license to Springer Nature Switzerland AG 2023
N. Xiong et al. (Eds.): ICNC-FSKD 2022, LNDECT 153, pp. 1407–1413, 2023.
https://doi.org/10.1007/978-3-031-20738-9_152

In the development of NT4310 chip, how to fully verify the design has become the key to the success of chip delivery [5]. The verification of NT4310 chip can be roughly divided into three levels: FPGA prototype, system level and P module level [6, 7]. The last two levels of verification method is to use HDL language to develop testbench and testcase to increase the incentive to the design and observe its response. However, due to the complexity of the design itself and the speed limitation of simulation verification, some functions cannot be fully verified at the simulation level, while FPGA prototype verification can make up for the deficiency of simulation verification [8, 9]. This paper mainly discusses a prototype verification platform of FlexRay communication controller chip. Most of the chip verification work is completed on this platform, which greatly improves the verification efficiency.

2   Verify Platform Structure and Function

2.1   Verification Environment Overview

The verification system completed according to the functions and verification requirements of FlexRay communication controller chip is shown in Fig. 1. The automatic verification platform includes two parts: software system and hardware system. The hardware part includes a host for debugging and two FlexRay network node verification boards. The two FlexRay node verification boards are the test node NT4310 verification board and one MFR4310 verification board as the accompanying test node. Because the function of NT4310 is compatible with MFR4310, MFR4310 verification board is used to provide function benchmarking and function benchmark judgment. A processor is implemented on both verification boards as the host of 4310 devices, it is used to run the running software of 4310. The two verification boards realize dual channel connection through FlexRay interface to verify the single and dual channels of FlexRay bus.

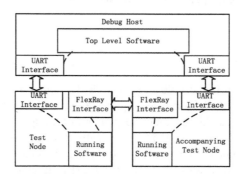

Fig. 1. Overall description of verification environment

2.2   Hardware Environment

The verification hardware environment consists of debugging host, test node and accompanying test node. The following describes the functions of the three parts in turn.

The debugging host is a PC on which the top-level debugging software is running. The debugging host provides two UART interfaces for data interaction with the test node and the accompanying test node. The debugging host and top-level software mainly realize the control of the test process and the generation of test data. The top-level planning and control of the test coverage algorithm are carried out through the top-level control software. The top-level software interacts with the tested node through driving UART, such as the distribution of test cases, the reception of test results, etc.

The test node is an FPGA verification board. The FPGA model is XC7K325T. The verification board contains the hardware environment that can meet the verification of NT4310 this time. As shown in Fig. 2, the hardware provides one UART interface and one FlexRay interface. The UART interface is used to interact with the debugging host for data, such as receiving test cases and uploading test results. The FlexRay interface connects the accompanying test node and is used to interact with the accompanying test node for test results. The NT4310 design is downloaded into FPGA, and an arm processor is integrated into FPGA as the host of NT4310 to run embedded test software.

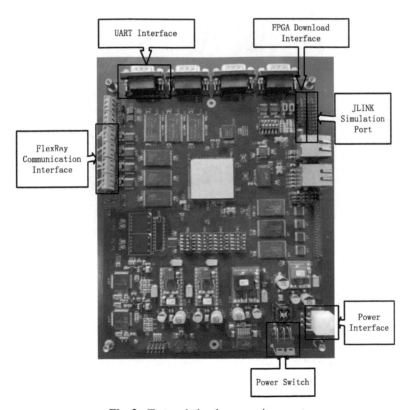

Fig. 2. Test node hardware environment

The accompanying test node is the function test board of the target device MFR4310, which includes the processor host STM32F429 for embedded operation and the accompanying test device MFR4310. The hardware is shown in Fig. 3. The hardware test board provides one UART interface and one FlexRay interface. The UART interface is used for data interaction with the commissioning host, such as receiving test cases and uploading test results. The FlexRay interface connects the test node and is used to interact with the test results of the test node.

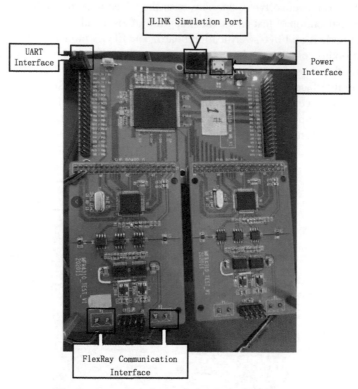

Fig. 3. Accompanying node hardware environment

2.3 Software Environment

The software system includes top-level software and running software.

The top-level software automatically deduces a full set of protocol parameters and constructs a test data pool according to the limited parameter information of the configuration file. The top-level tool first constructs and generates the test data of the test board NT4310, and generates the test data of the accompanying board MFR4310 according to the test data of the test board NT4310. After the configuration is completed, the configuration data in the data pool is sent through the serial port for testing, then wait for the board to be tested to communicate with the companion board, receive the communication data returned from NT4310 and MFR4310, and judge the result of the returned communication data. The top-level software flow is shown in Fig. 4.

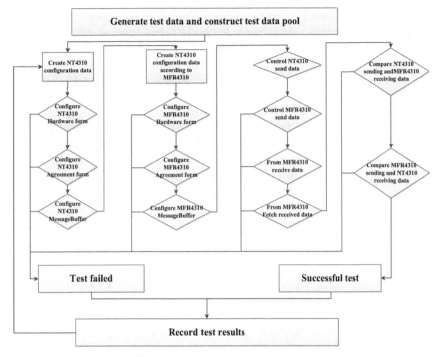

Fig. 4. Top level software workflow

Running software refers to the software directly running on the hardware platform. Its main function is to organize and manage hardware resources, provide the top-level software operation interface, respond to the operation commands of the top-level software, distribute the data issued by the top-level software to the hardware for execution, and return the execution results to the top-level software. The running machine software runs on the CPU on the test board and accompanying board. The accompanying board running software is developed and compiled using Keil MDK integrated development environment.

3   Verification Process Overview

First, input key parameter information to the top-level software. The top-level software automatically deduces a full set of protocol parameters and constructs a test data pool according to the limited parameter information of the configuration file. The top-level tool first constructs and generates the test data of NT4310, and generates the test data of MFR4310 according to the test data of NT4310. After the configuration is constructed, the configuration data in the data pool is distributed through the serial port for testing, and then the top-level tool monitors the communication results between NT4310 and MFR4310.

Running software to set NT4310 and MFR4310 to normal active state, and the two boards can send and receive normal data. When NT4310 and MFR4310 receive the data sent by the top-level software, the two boards begin to communicate. NT4310 sends data and MFR4310 receives data. Then, MFR4310 sends data and NT4310 receives data. After the two boards complete the communication, they send the communication results to the top-level software.

After monitoring the communication results between NT4310 and MFR4310, the top-level software compares the communication results. The test process will first think that the function of the MFR4310 of NXP Company is correct. If the communication result of NT4310 is different from that of MFR4310, it is considered that there is a problem with the function of NT4310. After the comparison information of the upper computer is completed, the test report is output. You can see whether the test results are correct in the test report. The test framework is shown in Fig. 5.

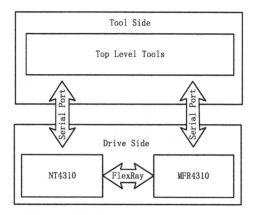

Fig. 5. Test framework

4   Conclusion

The NT4310 chip function module verification and function cross verification are carried out on the verification platform described in this paper. Using this verification platform, we can quickly and comprehensively verify the functions of the chip, find the design problems in the chip, and quickly locate and reproduce the problems, so that the problems can be solved in time, so as to speed up the verification of the chip. It has certain significance for chip localization. The test of this chip also has its limitations. Considering the time and the coverage of the test, due to the large number of functional test items and insufficient time, it is unable to cover all parameters. Chip testing is a huge task, which requires long-term and continuous testing to test the chip completely.

References

1. Kanajan, S., Abell, J.: Sensitivity analysis on FlexRay dynamic segment design parameters. Syst. Netw. Commun. 12–18 (2009)
2. Tang, Y., Meng, X., Xie, J., Huang, Z., Shenyang: Application trend of automatic driving bus technology. Electron. World **10**, 168–169 (2019)
3. Chen, T., Qin, H.: FlexRay clock synchronization analysis. Comput. Eng. **36**(14), 235–237 (2010)
4. FlexRay Consortium. MFR4310RM Rev. 2. 2008-03
5. Rashinkar, P., Paterson, P., Singh, L.: System-on-a-Chip Verification Methodology and Techniques. Springer, Norwell, MA, USA (2000)
6. Guo, M., Tian, Z., Cai, Y., et al.: Implementation of 1553B bus interface SoC verification platform. Aviation Comput. Technol. **38**(6), 99–100 (2008)
7. Zuo, H., Jin, Y.: An integrated circuit modeling and verification method based on Vera. Comput. Technol. Dev. **17**(1), 94–95 (2007)
8. Zhuang, X., Fan, Y.W.: Design of system verification platform for embedded microprocessor. Comput. Appl. Res. **24**, 240–241 (2007)
9. Wang, G., Xu, Z., Liu, Z., et al.: Research on SOC chip verification technology. Microcomput. Inf. **23**(8), 132–213 (2007)
10. Chi, H., Zhang, C., Zhao, D., Wang, Y., Tang, X.: FPGA verification of broadband power line carrier communication chip. Electron. Des. Eng. **29**(01), 128–131 (2021)
11. Lihua, D., Ben, W.: Research on IC prototype verification technology based on FPGA platform. Light Ind. Sci. Technol. **30**(11), 62–64 (2014)

Design of Extended Interface of the UAV Flight Control Computer

Qintao Wang[✉] and Xiaofeng Yang

School of Electronic Engineering, Xi'an University of Posts and Telecommunications,
Xi'an 710121, China
wangqt@stu.xupt.edu.cn, xfyang@mail.xidian.edu.cn

Abstract. With the continuous development of UAV (Unmanned Aerial Vehicle) technology, UAV flight control computer is developing towards the direction of more complex functions and smaller size. In the past, the flight control computer using only a single CPU cannot meet the development needs, so now the architecture of CPU + FPGA is usually used to design the flight control computer of UAV. In this architecture, the rich on-chip resources, large number of ports and flexible programming characteristics of FPGA are utilized to assist CPU to complete the acquisition of external data. In this study, the main use of FPGA to the CPU external interface expansion, so as to complete the external data acquisition and data acquisition chip driver work.

Keywords: FPGA · Interface expansion · Data acquisition

1 Introduction

The research of UAV began in the early twentieth century. The United States and Britain and other countries have made breakthroughs in the field of UAV research and put UAV into the later war [1, 2]. In China, the 1970s began the independent development of unmanned aerial vehicles. After more than 60 years of development, China's UAV technology is gradually mature. UAV is developing towards diversification, multi-purpose and high performance, so the research on UAV flight control computer is of vital importance. At present, MPSOC or CPU + FPGA is often used as the main control system in the research of flight control computer. Considering the cost and development cycle, most master control systems will choose the architecture form of CPU + FPGA. In this study, the main control chip of the whole system is TMS570LC4357 MCU chip produced by TI Company, and Xilinx XC6SLX100T chip is selected as FPGA chip. The main work of this study is to use the rich port resources and flexible programming characteristics of FPGA to complete the system to collect various external data [3, 4].

1.1 The Overall Structure of the System is Introduced

In this study, the overall system structure is designed based on the extended interface between FPGA and MCU, as shown in Fig. 1. MCU and FPGA communicate through

© The Author(s), under exclusive license to Springer Nature Switzerland AG 2023
N. Xiong et al. (Eds.): ICNC-FSKD 2022, LNDECT 153, pp. 1414–1422, 2023.
https://doi.org/10.1007/978-3-031-20738-9_153

EMIF bus interface. MCU accesses the register in FPGA through EMIF interface, and FPGA decodes the instructions in the register to realize the drive, data sending or reading of the external circuit connected to the expansion interface. The main extended interfaces in this study are analog quantity acquisition and output interface, digital quantity acquisition and output interface, frequency quantity acquisition interface, UART serial port and analog quantity acquisition channel selection and configuration interface, etc.

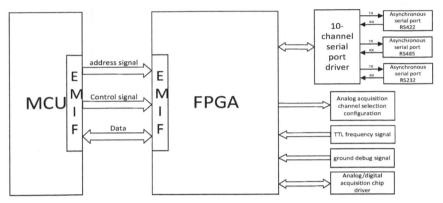

Fig. 1. Design structure of extended interface of UAV flight control system

1.2 Introduction to FPGA Design

In FPGA, the external input clock is 64 MHz, and the input reset signal is power-on reset. The reset signal is valid at low level, and the reset signal less than 45 ns needs to be filtered. First, the EMIF bus signal of CPU accessing FPGA needs to be converted into register access signal, and then the data is written into register or read out of register. The register processing module decodes the data written into the control register, and realizes the access or control of each external expansion interface according to the decoding result. The sub-modules accessed after decoding are: 10-channel UART serial port module, ground debugging signal acquisition module, TTL frequency signal acquisition module, analog quantity acquisition channel selection and configuration module, and analog and digital signal acquisition driver module. The overall detailed design structure of FPGA is shown in Fig. 2.

2 Logic Circuit Design

2.1 Design of Reset Anti-shake Module

The input reset signal in FPGA is power-on reset. When power-on reset, all functional modules are reset. Reset the active level to low. In this design, the reset signal less than 45 ns needs to be filtered, and the delay counting method is adopted. Since the reset level is low, the reset is considered complete if the high level remains above 45 ns. The

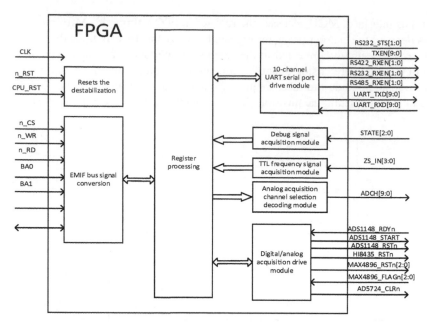

Fig. 2. FPGA detailed design structure

system reference clock of FPGA is 64M clock and one clock cycle is 15 ns. Therefore, it only needs to count the high level. When the count value is greater than or equal to 3, the reset signal can be raised.

2.2 Design of EMIF Bus Conversion Module

CPU and FPGA communicate through EMIF bus interface, and the read and write timing of EMIF bus is shown in Fig. 3a, b. EMIF bus includes the following communication interfaces: chip selection signal CS, DATA bus DATA, address bus ADDR, system clock CLK, reset signal RST, write signal WR, read signal RD. The timing sequence of the EMIF bus external interface is shown in the Fig. 3 [5]. After receiving the bus signal, the signal input from the bus is processed and converted into a simpler register access signal. So as to realize the precise access to the register. After processing, the signal that can directly access the register is: read enable, write enable, read data, write data, register address. The timing sequence of register access signal after conversion is shown in Fig. 4.

2.3 Design of Register Processing

In this module, it is mainly to decode the data written into the control register, and then access different sub-modules according to the decoding results. For UART serial port module, register processing is the interface baud rate configuration or data writing and reading in FIFO data buffer. For the ground debugging signal acquisition module, the register processing is to put the external collected signal into the corresponding

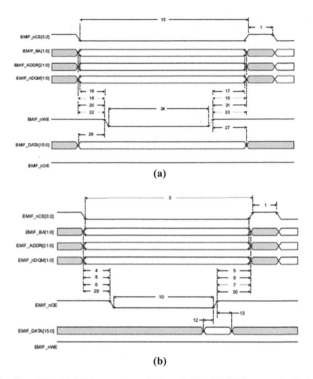

Fig. 3. a The EMIF bus writes timing, b The EMIF bus reads timing

register; For the analog quantity acquisition channel configuration module, the register processing is to send the acquisition channel selection signal to the configuration module, and then input to the external to complete the configuration of the selection channel. For TTL frequency measurement and acquisition module, register processing is to put the frequency signal value collected by the frequency quantity acquisition module into the corresponding storage register and wait for MCU to read it. For the external analog and digital quantity acquisition driver module, the register processing is to send the handshake signal between the driver module and the external driver chip to the driver module or read into the specified register.

2.4 Design of UART Extension Interface

In this design, 10 UART serial communication interfaces with the same structure are extended. Each serial port can be divided into sending module and receiving module, as shown in Fig. 6. Sending module is divided into three parts, respectively for baud rate configuration module, sending FIFO module, data packaging sending module; The serial port receiving module can also be divided into three parts. They are baud rate configuration module, receiving data processing module and receiving FIFO module. Each serial port adopts the standard UART communication protocol, and there will be 10 bits in a complete data frame, including a start bit, eight data bit, a parity, one or two

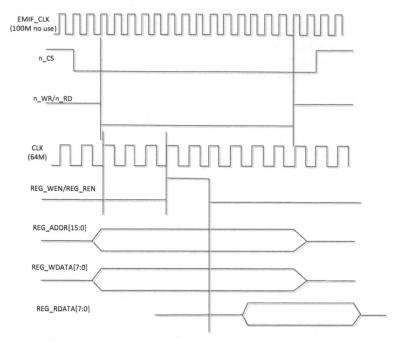

Fig. 4. Register read/write access timing

stop bits [6]. Figure 5 shows one of the data configuration formats. The interface design criteria are:

(1) The size of the FIFO data buffer for receiving and sending is 256 B.
(2) Support baud rate configuration: 9600, 19,200, 38,400, 57,600 and 115,200.
(3) Support parity configuration: odd parity, even parity and no parity.
(4) Support FIFO state query: empty, half full, full state.
(5) Support stop bit configuration: 1 bit, 2 bit.

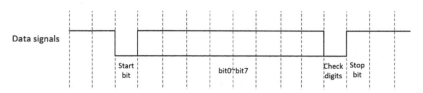

Fig. 5. UART data format

2.5 Design of Extended Interface for Frequency Acquisition

In this design, four external frequency input acquisition channels are extended. The frequency range of collection is 1 Hz ~ 20 kHz. After the external frequency is input,

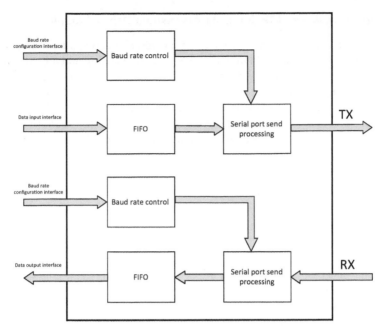

Fig. 6. UART interface driver module structure

the input signal is filtered first. After the clutter is filtered out, the signal cycle of the input frequency is calculated according to the clock frequency of the system. Because the input frequency signal is a periodic signal, the clock is counted in the high level period of frequency quantity, and the count value N can be sent to the register processing module in the low level period. Finally, the CPU calculates the collected frequency signal value according to the count value read from the register. The calculation formula is $f = \frac{f_c}{2N}$. f_c is the input clock frequency of FPGA, and f is the frequency value of the collected frequency signal.

2.6   Design of Analog Acquisition Channel Configuration Module

In this study, corresponding to the flight control system of external analog acquisition channel selection first need to determine the external acquisition chip per channel all the way to choose the configuration of the input signal values, and then to register in the configuration module processing module to send over the channel selection, configuration of coding, final output channel configuration required for the signal value choice.

2.7   Design of Signal Acquisition Module for Ground Debugging

The ground debugging signal is a level value, so the acquisition of the 3-channel ground debugging signal only needs to send the 3-channel input signal value to the register processing module.

2.8   Design of Analog and Digital Data Acquisition Driver Module

The main purpose of this module is to deal with external analog and digital data acquisition chip and CPU handshake signal. For every external analog and digital acquisition
chip, its drive signal has a pulse width requirement. If using CPU direct control, it is not
easy to control the establishment and hold time of the output signal and can not accurately collect the handshake input signal. In this module, after the register processing
module sends the CPU control signal, the module uses the delay counting method for
the control signal, so that the output of the control signal meets the control requirements
of the external acquisition chip.

3   Simulation Verification

After completing the design, the whole circuit design is modeled and simulated. By
writing Testbench programs, verify the correctness of the design and the integrity of the
function. The following is part of the simulation results. By observing the simulation
waveform and analyzing the waveform, the design simulation results all meet the design
requirements [7, 8].

Reset and shake elimination simulation: as shown in Fig. 7, after the input of external
reset signal, the system reset is completed after 3 clocks, which is consistent with the
design requirements.

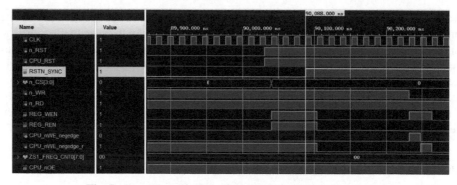

Fig. 7. Reset the vibration elimination simulation waveform

Serial port sending and receiving: according to the data sending and receiving process, the driver module is configured with baud rate and parity check, and then the DATA
buffer FIFO is read and written through emIF bus to complete the simulation of serial port
data receiving and receiving behavior. Simulation results are shown as follows (Figs. 8,
9 and 10).

Frequency acquisition: send a periodic frequency signal to the acquisition interface
in the test script, and then read the frequency meter value from the frequency storage
register. According to the value of the frequency meter read out, judge whether it is
consistent with the input periodic frequency signal to confirm whether the design is
correct. Simulation waveform is shown in Fig. 11.

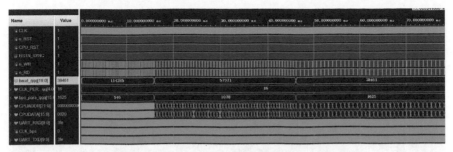

Fig. 8. Baud rate configuration simulation waveform

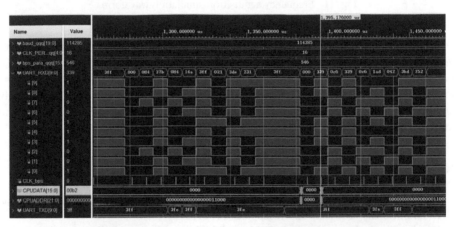

Fig. 9. Serial port data receiving simulation waveform

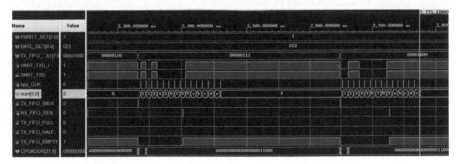

Fig. 10. Serial port data sending simulation waveform

4 Conclusion

In this paper, the design of external extended interface for data acquisition of uav flight control computer is realized based on FPGA. Through simulation and board level debugging shows that, through the FPGA implementation for the expansion of the flight control

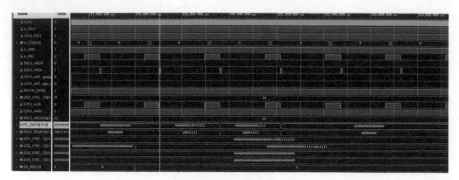

Fig. 11. Frequency acquisition simulation waveform

computer external acquisition interface, not only can easily realize the design requirements, but also a great extent, reduce the development cycle, but also can improve the integration of the whole flight control computer, the flight control computer can have more complex functions and the characteristics of smaller. The final results of this study are helpful to improve the reliability and functionality of the flight control computer and reduce the system cost.

References

1. Wang, Y.: Hardware Design and Implementation of Flight Control System of Small UAV. Henan University of Science and Technology (2019)
2. Han, B., Yi, Z., Jiang, H., et al.: Design of high-precision multi-channel real-time data acquisition system. Instrum. Tech. Sens. **9**, 42–45 (2019)
3. Shi, J., Chang, Y., Li, H.: Design of multi-channel data acquisition system based on FPGA. Electric. Autom. **38**(1), 15–18 (2016)
4. He, Z., Peng, D., Long, X.: Design of FPGA interface circuit based on multi-axis servo drive system. Mach. Tool Hydraulics **49**(05), 77–82 (2010)
5. Xuwei, Z., Wang Hong, H., Ting, L.Y., Qingrui, Y.: Design and implementation of DM642 and FPGA interface based on EMIF module. Opt. Commun. Technol. **37**(12), 35–37 (2013)
6. Jiang, Y.: Design and Implementation of an Integrated Navigation Computing Hardware Platform. Chongqing University (2019)
7. Shao, M., Li, G., Chen, M.: Interface expansion based on FPGA module in embedded NC system. Mech. Electr. Technol. **34**(03), 9–11 (2011)
8. Luo, X., Xue, Y., Zhang, L.: Electron. Meas. Technol. **44**(01), 50–54 (2010)

A Dynamic Selective Replication Mechanism for the Distributed Storage Structure

Siyi Han[✉], Youyao Liu, and Huinan Cai

School of Electronic Engineering, Xi'an University of Posts and Telecommunications, Shaanxi Xi'an 710121, China

{2002210215,alan.cai}@stu.xupt.edu.cn, lyyao2002@xupt.edu.cn

Abstract. For the characteristics of reconfigurable array processor (RAP) with large amount of data access and obvious space-time, a dynamic selective replication (DSR) mechanism of distributed storage structure is proposed. By monitoring the access of data in the cache, the array structure can replicate the data with high frequency access. Migrate the data copy to the private cache through the data path. It will be directly hit locally in the next access. Thus, the search and memory access of remote data are effectively reduced, and the line delay of remote access of the processor is reduced. In order to test the performance of the dynamic selective replication mechanism, a reconfigurable array processor with 4 × 4 PE array is designed. Experiments show that the array can achieve efficient parallel operation. Compared with the traditional static mapping distributed storage structure, the maximum latency of high-frequency data collision access is 19% lower than that of the traditional distributed cache.

Keywords: Reconfigurable computing · Selective replication · Distributed storage structure · Array processor

1 Introduction

With the development of integrated circuits entering the post Moore era and the AI era, the entire information age is in a critical period of technological change. At present, an important trend in the development of intelligent computing is how to improve system performance and reduce its power consumption [1–3]. In this context, coarse-grained architecture and reconfigurable computing emerged with the advantages of efficient computing and flexible programming [4]. At present, we have gradually stepped into the data intensive computing mode from the original computing intensive data processing mode. From the perspective of integrated circuit hardware, the performance requirements for data storage capacity and memory access data bandwidth are higher. This means that we need not only integrated circuits with strong computing power, but also more storage. That is, the ability to access memory bandwidth is more important [5]. Therefore, studying the storage structure is the key to improve the performance of the whole array processor [6].

© The Author(s), under exclusive license to Springer Nature Switzerland AG 2023
N. Xiong et al. (Eds.): ICNC-FSKD 2022, LNDECT 153, pp. 1423–1432, 2023.
https://doi.org/10.1007/978-3-031-20738-9_154

In the current processor design, in order to alleviate the problem of "storage wall" caused by the processing speed adaptation between memory and processor, designers usually use multi-level cache to improve the on-chip hit rate [7]. Reference [8] proposed a dynamic cache migration mechanism (RDMM) for reconfigurable array processors. RDMM migrates data in real time based on the parallelism and frequency of data access. Increased local access hit rate. Increase the access bandwidth of reconfigurable array processors. Literature [9] proposed a selective replication (SelRep) LLC. It can selectively copy the shared read-only data on the last level cache bank to meet the bandwidth requirements and maintain sufficient capacity to maintain the cache of shared data. Reference [10] proposed an intelligent multi hop storage structure based on distributed cache. As shown in Fig. 1. The structure can sense the storage status of candidate banks and select the appropriate target bank for the promoted data block. It improves the efficiency and effectively avoids cache pollution and other problems. It has a significant impact on the performance improvement of NUCA.

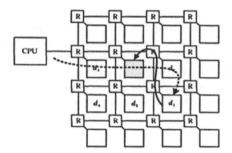

Fig. 1. Intelligent multi hop NUCA structure

In order to further improve the memory access efficiency of reconfigurable array processors and break through the four-level interconnection bus structure of traditional distributed storage structure, a reconfigurable array structure based on distributed storage mechanism is proposed. The architecture has good scalability and low hardware overhead, which satisfies the obvious spatio-temporal locality of data access of reconfigurable array processors. When the PE needs to access the storage structure, it can identify the location of the PE access target storage through instructions, and can be divided into local access, local access and global access according to the access area. Local access has absolute priority, followed by local areas with higher priority. The cross parallel access of 4 × 4 PE array to distributed storage structure is realized, which improves the memory access energy efficiency of reconfigurable array processors as a whole.

2   Research on Dynamic Selective Replication Mechanism

On chip memory structure has always been one of the important contents of reconfigurable array processor design [11]. In terms of the on-chip shared structure, each PE can access the shared memory independently through the on-chip bus structure. However, when the PES access in parallel, it is easy to cause data conflicts [12, 13]. At the same time, the on-chip shared memory has poor scalability and high failure rate. If you want to meet the memory access requirements of higher-level processors, you can only blindly increase the cache level and capacity, so as to increase the access bandwidth and the on-chip hit rate [14].

The distributed storage structure provides a private cache for each processing element. The private storage can be accessed in parallel through the on-chip interconnection structure without conflict. Figure 2 shows the distributed cache structure of the reconfigurable array processor. Different from the shared cache structure, the cache access latency at different locations is also different. When the working set is small, the local storage of the processing element is sufficient to accommodate the working set. When the working set is large, the data that cannot be placed in the local storage can be placed in the remote storage.

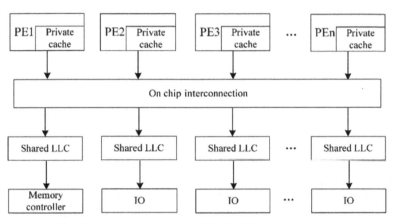

Fig. 2. Schematic diagram of distributed storage structure

2.1 Overall Structure

The overall structure of distributed storage dynamic selective replication mechanism based on reconfigurable array processor is shown in Fig. 3. The left side of the figure shows the reconfigurable array processing structure. On the right is distributed storage with dynamic selective replication mechanism. Processing element PEs can access distributed storage data through four levels of interconnection. This forms a DSR distributed storage array. DSR distributed storage and general distributed storage only need one clock cycle to return data and data feedback for local hit requests. The difference is that the DSR distributed storage designed in this paper has selective data replication behavior. That is, the replication behavior is made by balancing the cost of local loss caused by the replication behavior and the reduction of hit clock cycle. Once the data access reaches a certain frequency, local data copies will be built for remote packets. When the PE accesses the secondary data block, it can be obtained directly on the shortest path without considering the remote access path.

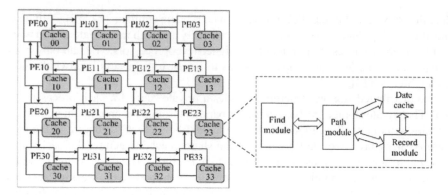

Fig. 3. Overall structure of distributed storage selective replication mechanism

The DSR-based distributed storage structure builds remote block replica files locally based on the original static address mapping. It can directly solve the long-Link delay problem of PE remote access and improve the memory efficiency. Figure 4 shows its overall access flow chart. In the case of PE00 accessing Cache10, PE00 caches the destination of the access based on the fixed four bits of the address, and then maps to its fixed Cache rows statically based on other address information. When PE00 accesses the data block again, count [1:0] plus 1, count [3:2], count [5:4], count [7:6] ~ count [31:30] are all zeroed out. Access for the first time and before replication occurs is through the four-level interconnect bus of a reconfigurable array processor. Once the count [1:0] is determined to be greater than the threshold N, a copy of the PE00 local Cache00 is immediately generated from the high frequency access data, and the valid bits are used to determine whether the data is valid or updated. When PE00 accesses the data block for the second time, it no longer needs to go to its original static mapping location to retrieve it. Instead, the local lookup module obtains a copy of the data in the local Cache based on information such as flags, which tremendously improves the performance energy efficiency of the entire system.

2.2   Read and Write Access Mechanism Flow

The distributed storage selective replication mechanism for reconfigurable array processors is designed to satisfy more efficient data access for PE arrays. While satisfying the basic parallel data block access of PE, the frequency of long path data block access is avoided. This enables the processor to access the cached data more efficiently and conveniently.

The DSR read and write access mechanism flow is as follows:

(1) DSR accepts data access capabilities, read operation addresses, and read capabilities sent from PE arrays.
(2) The destination area of PE visits is determined by the fixed four digits of the address.
(3) If it is determined that the accessed area is a local Cache, the data block is read after a local beat without bus structure. If the destination area of the data

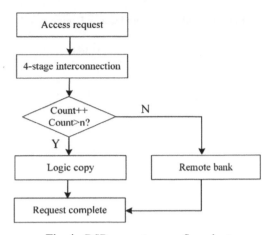

Fig. 4. DSR request access flow chart

accessed is a remote Cache, remote data access is required through a four-level fully interconnected bus structure based on its static mapping.

(4) The accessed data block detects and records the specific dynamics of the data being accessed. DSR distributed Cache storage structures are only sensitive to high frequency access.

(5) Selection of high frequency access threshold: Consider the magnitude of the processor and its access bandwidth. If the threshold N criterion is small, selective replication will occur frequently, bringing more capacity pressure to the local Cache. If the threshold N is too large, DSR replication will be difficult to occur, which adds some invalid circuit burden to on-chip chip resources.

(6) If the access frequency of a PE reaches a threshold: count > N. Start the DSR data block replication behavior to make a local copy of the data for the block.

(7) PE requests access to data blocks in the same location again. The PE's request address information is received by the lookup module. Access information is quickly searched and located by lookup tables. If the flag information for this visit is retrieved and the data status bits are valid, it means that the data block has been replicated by the DSR and a local copy of the data has been made. Get the data block and the feedback signal of the data directly in a local shot. This completes a valid read access process for the DSR distributed cache, which is similar to the read access process and is not described here.

3 Comprehensive Analysis and Performance Comparison

This paper designs a dynamic and selective replication of distributed Cache based on the distinct spatiotemporal locality and high parallelism of data access by a reconfigurable array processor. When accessing distributed Cache in DSR at low frequencies, PE arrays perform data access to distributed Cache by interconnecting buses using traditional mapping methods. During high-frequency access, the replication policy is executed by monitoring PE data access behavior to reduce long-Link data access latency.

Table 1. DSR Distributed Cache Parameters

Parameter item	Value
PE number	16
Interconnection topology	2-D mesh
Private cache	512 bit
L1 Cache number	16×16
Cache block size	16×32bit
Cache path	1
Cache groups	16

The parameters for the DSR distributed cache are shown in Table 1. In the case of local hits and low frequency data access, a four-level interconnection with a 2-dimensional mesh structure is used. That is, the PE array can access any remote Cache except the local private Cache. The 4×4 PE array in the cluster and the corresponding physically coupled 4×4 distributed Cache array form a tiled Cache storage structure that combines private and shared.

3.1   Write Access Delay Statistics Without Conflicts

Table 2. Average write access latency statistics without conflicts

Number	Local hit	Non local hit	Replica hit
5	1	2.3	2.3
15	1	2.25	1
25	1	2.13	1
55	1	2.025	1
100	1	2.01	1

Table 2 calculates the average write access latency for both local and non-local hits and replica data copy hits when PE has conflict-free access. The delay of conflict-free local access for each PE is hit directly locally without passing through the interconnect bus, regardless of the access frequency. One row is the feedback information to complete the data writing operation.

Figure 5 shows the average latency statistics for five, 15, 25, 55, and 100 write operation visits under conflict-free access conditions. In the case of non-local hits, the average write access delay is about 2.2 cycles. Because different paths to non-local Cache access result in different delays to remote access. For DSR data access, the latency is approximately the same as that of low-frequency access when no replica of the data is

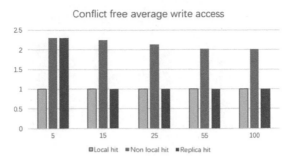

Fig. 5. Delay statistics for conflict-free write operations for each frequency

generated. The subsequent continuous data write access is obtained directly from the local, consistent with the local hit write delay.

3.2 Read/Write Access Latency Statistics with Conflict

Table 3. Add conflict average write access delay statistics

Probability (%)	Local hit	Non local hit	Replica hit
25	–	2.7	2.7
50	–	3.05	2
75	–	3.13	2
100	–	3.325	2

Table 3 calculates the average write access latency for local hits, non-local hits, and local data copy hits under four conflict probabilities.

Table 4. Add conflict average read access delay statistics

Probability (%)	Local hit	Non local hit	Replica hit
25	–	3.7	3.7
50	–	4.05	3
75	–	4.13	3
100	–	4.325	3

Figure 6 shows the average read access latency statistics with a certain conflict probability added. Similar to write access, local read access operations give absolute priority to direct hits. The number of clock cycles returned is the same for different

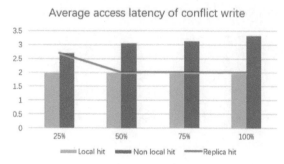

Fig. 6. Write-down latency for each conflict rate

conflict probabilities. The number of cycles to complete a non-local read operation is different due to the inconsistent access path length of the Cache for different purposes. Replica hits require only 3 clock cycles to return data. As the conflict rate increases, the number of clock cycles completed by a write operation increases. The local write operation clock delay of the copy remains unchanged (Table 4).

3.3 Performance Comparison

In order to verify the correctness of the array structure, Xilinx ZYNQ chips XC7Z045 FFG900–2 are selected for synthesis. The resource utilization of the reconfigurable array structure based on DSR distributed storage in the field-bus is compared with that in the literature [15] FS structure, the literature [16] LPAS structure, and the literature [17] as shown in Table 5.

Table 5. Comprehensive report

Logic device	Literature [15] FS	Literature [16] LPAS	Literature [17] LR2SS	This paper
LUTS	11909/218600 (<6%)	6514/218600 (<3%)	6464/218600 (<3%)	6743/218600 (<3%)
Registers	1543/437200 (<1%)	1032/437200 (<1%)	1401/437200 (<1%)	1160/437200 (<1%)
LUT-FF pairs	1460/6565 (22%)	1037/6565 (15.7%)	1300/6565 (19%)	977/6565 (14%)
Max frequency (MHz)	223	221	200	210
Bandwidth (GB/S)	7.42	7.6	6.25	7.53

The test results show that the maximum operating frequency of dynamic selective replication of distributed cache by a reconfigurable array processor is 210 MHz. The access peak bandwidth is 7.53GB/S. This architecture supports parallel read and write operations for 16 PEs with conflict-free memory access. Write and read operations are completed in one and two clock cycles, respectively. Peak bandwidth in conflict-free memory access is 1.28GB/S higher than in the literature [17] LR2SS architecture. It is similar to the peak frequency and access peak bandwidth of the [15] FS architecture with conflict-free memory access. However, it takes up less hardware resources than literature [17].

4 Conclusion

To further improve the memory access efficiency of the reconfigurable array processor and break away from the traditional static interconnection mode, a dynamic selective replication mechanism for distributed storage is proposed based on the distributed storage structure of the reconfigurable array processor. The experimental results show that the structure supports cross-parallel access to distributed Cache by a 4 × 4PE array with low conflict. Finally, the structure is prototyped using Xilinx XC7Z045 FFG900-2 development board. With conflict-free memory access, the highest frequency of parallel read-write memory access for 4 × 4 PE arrays in the cluster is 210 MHz. Peak memory access bandwidth is 7.53GB/s.

With the rapid development of in-memory computing technology [18], some design concepts of in-memory computing can be adopted in the next step. When there is a large amount of read-write access between the storage structure and the processing meta-array, memory directly undertakes some simple calculations. Further reduce the delay in data path transmission.

Acknowledgements. This work was supported by the National Natural Science Foundation of China (Nos. 61874087, 61834005, 61634004).

References

1. He, G., Ding, X., Zhou, M., Liu, B., Li, L.: Background noise adaptive energy-efficient keywords recognition processor with reusable DNN and reconfigurable architecture. IEEE Access **10**, 17819–17827 (2022)
2. Wu, J.Q., Lai, F.: Advances in emerging computing chips in the post-Moore era. Microelectronics **50**(03), 384–388 (2020)
3. Kim, S., et al.: Versa: a 36-core systolic multiprocessor with dynamically reconfigurable interconnect and memory. IEEE J. Solid-State Circuits **57**(4), 986–998 (2022)
4. Hu, S., Brandon, A., Guo, Q., et al.: Improving the performance of adaptive cache in reconfigurable VLIW processor. In: International Symposium on Applied Reconfigurable Computing, pp. 3–15. Springer, Cham (2017)
5. Lokegaonkar, I., Nair, D., Kulkarni, V.: Enhancement of cache memory performance. In: 2021 3rd International Conference on Advances in Computing, Communication Control and Networking (ICAC3N), pp. 1490–1492 (2021)
6. Liu, L., Zhu, J., Li, Z., et al.: A survey of coarse-grained reconfigurable architecture and design: taxonomy, challenges, and applications. ACM Comput. Surv. (CSUR) **52**(6), 1–39 (2019)

7. Wijtvliet, M., Waeijen, L., Corporaal, H.: Coarse grained reconfigurable architectures in the past 25 years: overview and classification. In: International Conference on Embedded Computer Systems: Architectures, Modeling and Simulation, pp. 235–244. IEEE (2017)

8. Lin, J.A., Yang, L.B., Rui, S.C., et al.: RDMM: runtime dynamic migration mechanism of distributed cache for reconfigurable array processor. Integration **72**, 82–91 (2020)

9. Zhao, X., Jahre, M., Eeckhout, L.: Selective replication in memory-side GPU caches. In: 2020 53rd Annual IEEE/ACM International Symposium on Microarchitecture (MICRO). ACM (2020)

10. Wu, J.J., Pan, X.H., Yang, X.J.: Intelligent multi-hop promotion technology for inconsistent cache. CNKI; Wan Fang (2009):9

11. Lu, T.: Exploring storage device characteristics of a RISC-V little-core SoC. In: 2021 IEEE International Conference on Networking, Architecture and Storage (NAS), pp. 1–4 (2021)

12. Guo, H., Huang, L., Lu, Y., et al.: DyCache: dynamic multi-grain cache management for irregular memory accesses on GPU. IEEE Access **6**, 38881–38891 (2018)

13. Wang, Z., Chen, X., Lu, Z., et al.: Cache access fairness in 3D mesh-based NUCA. IEEE Access **6**, 42984–42996 (2018)

14. Lian, X., Liu, Y., Yang, T., et al.: Replication strategy for spatiotemporal data based on distributed caching system. Sensors **18**(2), 222–231 (2018)

15. Jiang, L., Liu, P., Shan, R., et al.: Intra-cluster full access architecture design for array processor distributed storage. J. Xi'an Univ. Sci. Technol. **38**(004), 656–662 (2018)

16. Liu, Y.Y., Zhang, Y., Shan, R.: Design of local first access architecture for array processor distributed cache. Comput. Eng. Sci. **42**(4), 8 (2020)

17. Guo, J.L., Jiang, L., Shan, R., et al.: Design and implementation of memory structure in reconfigurable video array processor cluster. Microelectron. Comput. **34**(9), 116–120 (2017)

18. Gong, L.Q., Xu, W.D., Lou, M.: Overview of SRAM in memory computing technology. Microelectron. Comput. **38**(09), 1–7 (2021)

An Effective Algorithm for Pricing Option with Mixed-Exponential Jump and Double Stochastic Volatility

Sumei Zhang[✉], Panni Liu, and Zihao Liao

Xi'an University of Post and Telecommunications, Xi'an 710121, China
zhangsumei@xupt.edu.cn, {liupanni,liaozihao}@stu.xupt.edu.cn

Abstract. In order to combine the jump risk and stochastic volatility risk of actual asset price in the event of emergencies, the paper proposes a double stochastic volatility mixed-exponential jump-diffusion model by introducing mixed-exponential jump in the double Heston model. This paper obtains the characteristic function of logarithmic asset prices under the proposed model through Ito's formula and stochastic integration. Combing Fourier cosine series and fast Fourier transform algorithm, this paper proposes a new pricing algorithm for European options. Our algorithm truncates the integral interval of option pricing, expands the density function by Fourier COS expansion in the truncated interval, and transforms the density function into characteristic function by Fourier transform. The numerical results show that our algorithm is effective and has stable convergence for pricing European options.

Keywords: Pricing algorithm · European option · Jump risk · Stochastic volatility

1 Introduction

Affected by COVID-19 and the oil price war, many stock markets in the United States and other countries triggered a circuit breaker mechanism due to a sharp fall in 2020. This once again greatly stimulated the awareness of risk management in the financial field after the global financial crisis in 2008. European option is an important tool for risk management and risk transfer. To manage risks effectively, the first thing we should do is to price European options reasonably. The mixed-exponential jump diffusion (MEM) model [1] can perfectly fit "leptokurtic" and short-term implied volatility "smile" of real asset price distribution. The jump in the MEM model obeys a hybrid exponential distribution that can approximate any distribution, is more flexible in fitting the characteristics of the asset return distribution, and is more applicable in practice. However, it cannot fit the "volatility clustering" characteristics of real asset price distribution, nor can it perfectly explain implied volatility "smile" for longer-dated options, since the MEM model assumes that the volatility is constant. Christoffersen [2] proposed a double stochastic volatility model with two stochastic volatility factors, which can perfectly fit the long-term implied "volatility smile" and "volatility clustering", but cannot perfectly explain

N. Xiong et al. (Eds.): ICNC-FSKD 2022, LNDECT 153, pp. 1433–1440, 2023.
https://doi.org/10.1007/978-3-031-20738-9_155

implied volatility "smile" for short-dated options. Therefore, this paper considers the pricing of European options under mixed-exponential jump risk and double stochastic volatility risk.

Under the jump risk or stochastic volatility risk, it is difficult to obtain the analytical solution of option pricing, and numerical methods are usually used. Most recently, under the stochastic models, several numerical approaches for pricing European options have been studied: partial differential (integral) equation method [3–5], Monte Carlo simulation [6, 7] and transform techniques [8–10]. The method based on transform is an effective and fast numerical method with high accuracy. The paper proposes a new algorithm for pricing European options by combining Fourier cosine series and fast Fourier transform (FFT) technique. The paper is organized as follows. The pricing model is proposed in Sect. 2; The pricing algorithm is detailed in Sect. 3; Some numerical experiments are illustrated in Sect. 4; Sect. 5 concludes the paper.

2 The Pricing Model

Let $S(t)$ be the asset price which is given by the following double stochastic volatility mixed-exponential jump-diffusion model (DSVMJ):

$$
\begin{cases}
\frac{dS(t)}{S(t)} = (r - \lambda\delta)dt + \sum_{j=1}^{2} \sqrt{V_j(t)}dW_j(t) + d(\sum_{j=1}^{N(t)} (J_j - 1)), \\
dV_1(t) = \kappa_1(\theta_1 - V_1(t))dt + \sigma_1\sqrt{V_1(t)}dZ_1(t), \\
dV_2(t) = \kappa_2(\theta_2 - V_2(t))dt + \sigma_2\sqrt{V_2(t)}dZ_2(t),
\end{cases}
\tag{1}
$$

where, $\delta = E(J - 1)$, r denotes the risk-free interest rate, nonnegative constants $\kappa_j, \theta_j, \sigma_j (j = 1, 2, 2\kappa_j\theta_j \geq \sigma_j^2)$ denote the mean reversion rates, long run variance and volatility of variance $V_j(t)$ $(j = 1, 2)$ respectively. $\{W_j(t), t \in [0, T]\}, \{Z_j(t), t \in [0, T]\}$ $(j = 1, 2)$ denote all Wiener stochastic processes, $\{N(t), t \in [0, T]\}$ denotes a Poisson process with intensity parameter λ.

Assume that $Cov(dW_j(t), dZ_j(t)) = \rho_j dt (j = 1, 2)$ and other stochastic sources are independent each other. Let $S(0) = S_0, V_j(0) = V_j (j = 1, 2)$. $J_j (j = 1, 2, ...)$ is a sequence of independent identically distributed nonnegative random variables such that the density function of $Y_j = \ln(J_j)$ is given by

$$
f_Y(y) = p_u \sum_{i=1}^{m} p_i\eta_i e^{-\eta_i y}I_{y\geq 0} + q_d \sum_{j=1}^{n} q_j\hat{\theta}_j e^{\hat{\theta}_j y}I_{y<0},
\tag{2}
$$

where, $p_u \geq 0$, $q_d = 1 - p_u \geq 0, p_i, q_j \in (-\infty, \infty)$, $\eta_i > 1$, $\hat{\theta}_j > 0, \sum_{i=1}^{m} p_i = 1, \sum_{j=1}^{n} q_j = 1, i = 1, \ldots, m, j = 1, \ldots, n$. To make $f_Y(y)$ be a density function, a necessary condition is $p_1 > 0$, $q_1 > 0$, $\sum_{i=1}^{m} p_i\eta_i \geq 0$, $\sum_{j=1}^{n} q_j\hat{\theta}_j \geq 0$ and a sufficient condition is $\sum_{i=1}^{k} p_i\eta_i \geq 0$, $\sum_{j=1}^{l} q_j\hat{\theta}_j \geq 0, k = 1, \ldots m, l = 1, \ldots n$.

3 FCS Algorithm for Pricing European Option

3.1 The Characteristic Function

In order to use Fourier transform technology, we first determine the characteristic function and the cumulants of logarithmic asset price in DSVMJ.

Theorem 1 Let K denote strike price, under DSVMJ, the characteristic function of $X(T) = \ln S(T)/K$ satisfies

$$\phi(u) = \exp\{A(u) + B(u) + C(u) + D(u)\}, \tag{3}$$

where,

$$A(u) = iu \ln \frac{S_0}{K} + iurT,$$

$$B(u) = \sum_{j=1}^{2} \frac{(\kappa_j - \gamma_j)\kappa_j\theta_j T}{\sigma_j^2} - \frac{iu\rho_j\kappa_j\theta_j T}{\sigma_j} + \frac{2\kappa_j\theta_j}{\sigma_j^2} \ln \frac{2\gamma_j}{2\gamma_j + [\kappa_j - \gamma_j - iu\rho_j\sigma_j](1 - e^{-\gamma_j T})},$$

$$C(u) = \frac{iu(iu - 1)(1 - e^{-\gamma_j T})}{2\gamma_j + [\kappa_j - \gamma_j - iu\rho_j\sigma_j](1 - e^{-\gamma_j T})} V_j,$$

$$D(u) = \lambda T \left(p_u \sum_{i=1}^{m} p_i\eta_i \frac{1}{\eta_i - iu} + q_d \sum_{j=1}^{n} q_j\hat{\theta}_j \frac{1}{\hat{\theta}_j + iu} - 1 - iu\delta \right),$$

$$\gamma_j = \sqrt{[\kappa_j - iu\rho_j\sigma_j]^2 + iu(1 - iu)\sigma_j^2},$$

$$\delta = p_u \sum_{i=1}^{m} p_i\eta_i \frac{1}{\eta_i - 1} + q_d \sum_{j=1}^{n} q_j\hat{\theta}_j \frac{1}{\hat{\theta}_j + 1} - 1.$$

Proof Let $z = iu$, $dW_j(t) = \rho_j dZ_j(t) + \sqrt{1 - \rho_j^2}d\tilde{Z}_j(t)$, $j = 1, 2.\tilde{Z}_j(t)$ are Wiener stochastic processes independent of $Z_j(t), N(t)$ and J. According to Itô formula,

$$\ln \frac{S(T)}{K} = \ln \frac{S_0}{K} + (r - \lambda\delta)T + \sum_{j=1}^{2} \left\{ \rho_j \int_0^T \sqrt{V_j(t)}dZ_j(t) - \frac{1}{2}\rho_j^2 \int_0^T V_j(t)dt \right\}$$

$$+ \sum_{j=1}^{2} \left\{ \sqrt{1 - \rho_j^2} \int_0^T \sqrt{V_j(t)}d\tilde{Z}_j(t) - \frac{1}{2}(1 - \rho_j^2) \int_0^T V_j(t)dt \right\} + \sum_{j=1}^{N_T} Y_j$$

$$= \ln \frac{S_0}{K} + (r - \lambda\delta)T + \sum_{j=1}^{2} (\varsigma_{Tj} + \xi_{Tj}) + \sum_{j=1}^{N_T} Y_j,$$

so that, we have

$$\phi(u) = e^{z \ln \frac{S_0}{K} + zrT - z\lambda\delta T} E\left[\exp\left(z \sum_{j=1}^{2} (\varsigma_{Tj} + \xi_{Tj}) \right) \right] E\left[\exp\left(z \sum_{j=1}^{N_T} Y_j \right) \right]. \tag{4}$$

According to DSVMJ and reference [11],

$$E\left[\exp\left(z\sum_{j=1}^{2}\left(\varsigma_{Tj}+\xi_{Tj}\right)\right)\right]=\exp\left\{\sum_{j=1}^{2}\frac{(\kappa_j-\gamma_j)\kappa_j\theta_j T}{\sigma_j^2}-\frac{iu\rho_j\kappa_j\theta_j T}{\sigma_j}\right.$$

$$+\frac{2\kappa_j\theta_j}{\sigma_j^2}\ln\frac{2\gamma_j}{2\gamma_j+[\kappa_j-\gamma_j-iu\rho_j\sigma_j](1-e^{-\gamma_j T})}$$

$$\left.+\frac{iu(iu-1)(1-e^{-\gamma_j T})}{2\gamma_j+[\kappa_j-\gamma_j-iu\rho_j\sigma_j](1-e^{-\gamma_j T})}V_j\right\}, \qquad (5)$$

where,

$$\gamma_j=\sqrt{[\kappa_j-iu\rho_j\sigma_j]^2+iu(1-iu)\sigma_j^2}.$$

By (1) and (2), we have

$$\delta=p_u\sum_{i=1}^{m}p_i\eta_i\frac{1}{\eta_i-1}+q_d\sum_{j=1}^{n}q_j\hat{\theta}_j\frac{1}{\hat{\theta}_j+1}-1. \qquad (6)$$

By (2), we have

$$E[\exp\left(z\sum_{j=1}^{N_T}Y_j\right)]=\exp\left\{\lambda T\left(p_u\sum_{i=1}^{m}p_i\eta_i\frac{1}{\eta_i-iu}+q_d\sum_{j=1}^{n}q_j\hat{\theta}_j\frac{1}{\hat{\theta}_j+iu}-1\right)\right\}. \qquad (7)$$

Putting (5), (6) and (7) into (4), theorem 1 follows.
According to theorem 1, we have

$$c_1=\frac{1}{i}\frac{\partial\ln\phi(u)}{\partial u}\Big|_{u=0}=\ln\frac{S_0}{K}+\sum_{j=1}^{2}\frac{\theta_j-V_j}{2\kappa_j}(1-e^{-\kappa_j T})$$

$$+T\left\{r-\frac{\theta_1+\theta_2}{2}+\lambda\left[p_u\sum_{i=1}^{m}\frac{p_i}{\eta_i}-q_d\sum_{j=1}^{n}\frac{q_j}{\hat{\theta}_j}-\left(p_u\sum_{i=1}^{m}\frac{p_i\eta_i}{\eta_i-1}+q_d\sum_{j=1}^{n}\frac{q_j\hat{\theta}_j}{\hat{\theta}_j+1}-1\right)\right]\right\},$$

$$c_2=\lambda T\left[p_u\sum_{i=1}^{m}\frac{2p_i}{\eta_i^2}+q_d\sum_{j=1}^{n}\frac{2q_j}{\hat{\theta}_j^2}\right]+\sum_{j=1}^{2}\left(\prod_{0j}+\prod_{1j}e^{-\kappa_j T}+\prod_{2j}e^{-2\kappa_j T}\right),$$

where,

$$\prod_{0j}=\theta_j T+\frac{-\theta_j+V_j-\theta_j\sigma_j\rho_j T}{\kappa_j}+\frac{\sigma_j\rho_j(2\theta_j-V_j)+\frac{1}{4}\theta_j\sigma_j^2 T}{\kappa_j^2}-\frac{\sigma_j^2}{8\kappa_j^3(5\theta_j-2V_j)},$$

$$\prod_{1j}=\frac{\rho_j\sigma_j}{\kappa_j^2}(V_j-2\theta_j)+\frac{1}{\kappa_j}(1-\sigma_j\rho_j T)(\theta_j-V_j)+\frac{\theta_j\sigma_j^2}{2\kappa_j^3}+\frac{T\sigma_j^2(\theta_j-V_j)}{2\kappa_j^2},$$

$$\prod_{2j}=\frac{\sigma_j^2}{8\kappa_j^3(\theta_j-2V_j)}.$$

3.2 FCS Algorithm for Pricing European Options

Let T denote maturity date, the price of a European call option is given by

$$v(x, t) = e^{-r(T-t)} \int_R [K(e^y - 1)]^+ f(y|x) dy, \tag{8}$$

where, $x = \ln(S(t)/K)$, $y = \ln(S(T)/K)$, $f(y|x)$ is the density function.

Assume that $[a, b] \in R$, according to [12], let

$$[a, b] = [c_1 - L\sqrt{|c_2|}, \ c_1 + L\sqrt{|c_2|}], \tag{9}$$

where, L is a truncation coefficient.

Since $f(y|x)$ rapidly decays to zero with $y \to \pm\infty$, we truncate the integration range R to $[a, b]$ without losing accuracy and have

$$v_1(x, t) = e^{-r(T-t)} \int_a^b v(y, T) f(y|x) dy. \tag{10}$$

By the Fourier COS expansion, $f(y|x)$ can be written as

$$f(y|x) = \sum_{k=0}^{\infty}{}' A_k(x) \cos\left(k\pi \frac{y-a}{b-a}\right), \tag{11}$$

where,

$$A_k(x) = \frac{2}{b-a} \int_a^b f(y|x) \cos\left(k\pi \frac{y-a}{b-a}\right) dy, \tag{12}$$

so that

$$v_1(x, t) = e^{-r(T-t)} \int_a^b v(y, T) \sum_{k=0}^{\infty}{}' A_k(x) \cos\left(k\pi \frac{y-a}{b-a}\right) dy. \tag{13}$$

By interchanging the summation and integration, we have

$$v_1(x, t) = \frac{1}{2}(b-a)e^{-r(T-t)} \sum_{k=0}^{\infty}{}' A_k(x) \frac{2}{b-a} K(\chi_k(0, b) - \psi_k(0, b)),$$

where,

$$\chi_k(c, d) = \frac{1}{1 + \left(\frac{k\pi}{b-a}\right)^2} \left\{ \cos\left(k\pi \frac{d-a}{b-a}\right) e^d - \cos\left(k\pi \frac{c-a}{b-a}\right) e^c \right.$$
$$\left. + \frac{k\pi}{b-a} \left[\sin\left(k\pi \frac{d-a}{b-a}\right) e^d - \sin\left(k\pi \frac{c-a}{b-a}\right) e^c \right] \right\},$$

$$\psi_k(c,d) = \begin{cases} \frac{b-a}{k\pi}\left[\sin\left(k\pi\frac{d-a}{b-a}\right) - \sin\left(k\pi\frac{c-a}{b-a}\right)\right], & k \neq 0, \\ (d-c), k = 0. \end{cases}$$

By truncating the series summation and using Fourier transform, we have

$$v(x,t) = e^{-r(T-t)} \sum_{k=0}^{N-1} \Re\left\{\phi\left(\frac{k\pi}{b-a}; x\right)e^{-ik\pi\frac{a}{b-a}}\right\}\frac{2K(\chi_k(0,b) - \psi_k(0,b))}{b-a}. \tag{14}$$

4 Numerical Experiment

WE use the FCS algorithm to evaluate European call options under DSVMJ in the section. To test the effectiveness of the FCS algorithm, we also price the same options by the numerical integration method, and take numerical integration solutions as benchmark. We use $N = 64$ points in (14) and $L = 10$ in (9). For the numerical integration method, we use the adaptive Lobatto integration rule. Parameters are set as follows: $\sigma_1 = 0.9$, $\kappa_1 = 12$, $\theta_1 = 0.05$, $\hat{\theta}_1 = 50$, $\eta_1 = 20$, $\rho_1 = -0.5$, $V_1 = 0.05$, $\sigma_2 = 0.9$, $\kappa_2 = 16$, $\theta_2 = 0.03$, $\hat{\theta}_2 = 20$, $\eta_2 = 50$, $\rho_2 = -0.5$, $\sigma_2 = 0.9$, $V_2 = 0.02$, $p_u = 0.4$, $p_1 = 1.3$, $q_1 = 1.2$, $\lambda = 1$, $r = 0.03$, $m = 2$, $n = 2$. All numerical experiments were performed on the computer with Intel(R) Core(TM) i5-5200U CPU @ 2.20 GHz and the code is written in Matlab 9.0. We determine CPU time by averaging over 10 repeated tests. Table 1 lists the comparison of the FCS algorithm and numerical integration method for evaluating European call options under DSVMJ.

Table 1 shows that the pricing accuracy of FCS algorithm and numerical integration method is very close, and the maximum relative error of the two methods is 0.0788%. However, the FCS algorithm is obviously faster. For a given set of parameters, the FCS algorithm calculate an option price in 0.047s, while numerical integration method needs 0.065 s for the same option price. From Table 1, we can see that FCS algorithm is fast and effective.

Furthermore, by changing N, we test the convergence of the FCS algorithm. We set $N = 128$ and calculate European option prices by the FCS algorithm under $T = 0.1, 2$ and $K = 50, 100$ and take the obtained prices as the true prices. The model parameters are taken from Table 1. With the same parameter setting we calculate European option price under $N = 4, 8, 16, 32, 64$, respectively. Figure 1 reports the convergence of the FCS algorithm for pricing European call options.

Figure 1 shows that when N is small, the pricing error is large, however, when $N = 16$, the FCS algorithm presents stable convergence under each parameter setting.

5 Conclusion

Under the mixed-exponential jump risk and double stochastic volatility risk, the paper proposes a new algorithm for evaluating European options by combining with Fourier cosine series and FFT algorithm. The new algorithm truncates the integral interval of option pricing, expands the density function by Fourier COS expansion in the truncated

Table 1. Comparison of the FCS algorithm and numerical integration method for evaluating European call options under DSVMJ

T	λ	K	FCS	Numerical integration	Relative error (%)
0.5	1	70	31.4188	31.4179	0.0029
		100	8.6166	8.6170	0.0046
		130	0.8875	0.8882	0.0788
	3	70	31.4442	31.4436	0.0019
		100	8.8550	8.8556	0.0068
		130	1.0944	1.0949	0.0457
1	1	70	33.2501	33.2503	0.0006
		100	12.6433	12.6441	0.0063
		130	3.3088	3.3095	0.0212
	3	70	33.3371	33.3371	0.0000
		100	12.9831	12.9839	0.0062
		130	3.6793	3.6804	0.0299
CPU/s			0.0203	0.0590	

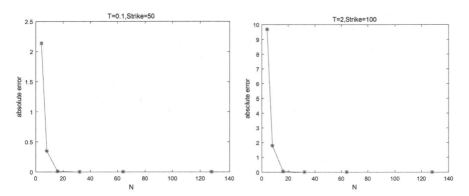

Fig. 1. The convergence of the FCS algorithm for pricing European call options

interval, and transforms the density function into characteristic function by Fourier transform, which simplifies the option pricing process. The numerical results show that our algorithm is fast, accurate and has stable convergence for pricing European options.

Acknowledgments. This work is supported the Natural Science Foundation of Shaanxi Province, China [grant number 2021JM-455].

References

1. Cai, N., Kou, S.G.: Option pricing under a mixed-exponential jump diffusion model. Manage. Sci. **57**(11), 2067–2081 (2011)
2. Christoffersen, P., Heston, S., Jacobs, K.: The shape and term structure of the index option smirk: why multifactor stochastic volatility models work so well. Manage. Sci. **55**(12), 1914–1932 (2009)
3. Yi, S.Y., Lee, K.: Numerical study for European option pricing equations with non-Lévy jumps. Appl. Anal. **100**(7), 1454–1470 (2021)
4. Kumar, A., Kumar, B.V.R.: A RBF based finite difference method for option pricing under regime-switching jump-diffusion model. Int. J. Comput. Methods Eng. Sci. Mech. **20**(5), 451–459 (2019)
5. Yousuf, M.: Numerical solution of systems of partial integral differential equations with application to pricing options. Numer. Methods Partial Differ. Equ. Int. J. **34**(3), 1033–1052 (2018)
6. Liang, Y.J., Xu, C.L.: An efficient conditional Monte Carlo method for European option pricing with stochastic volatility and stochastic interest rate. Int. J. Comput. Math. **97**(3), 638–655 (2020)
7. Jeong, D., Yoo, M., Yoo, C., Kim, J.: A hybrid Monte Carlo and finite difference method for option pricing. Comput. Econ. **53**(1), 111–124 (2019)
8. Zhang, S.M., Wang, L.H.: Fast Fourier transform option pricing with stochastic interest rate, stochastic volatility and double jumps. Appl. Math. Comput. **219**(23), 10928–10933 (2013)
9. Jackson, K.R., Jaimungal, S., Surkov, V.: Fourier space time-stepping for option pricing with Lévy models. J. Comput. Financ. **12**(2), 1–28 (2008)
10. Lord, R., Fang, F., Bervoets, F., Oosterlee, C.W.: A fast and accurate FFT-based method for pricing early-exercise options under Lévy processes. SIAM J. Sci. Comput. **30**(4), 1678–1705 (2008)
11. Scott, L.O.: Pricing stock options in a jump-diffusion model with stochastic volatility and interest rates: applications of Fourier inversion methods. Math. Financ. **7**(4), 413–426 (1997)
12. Fang, F., Oosterlee, C.W.: A novel pricing method for European options based on Fourier-Cosine series expansions. SIAM J. Sci. Comput. **31**(2), 826–848 (2008)

Data Mining and Knowledge Management for O2O Services, Brand Image, Perceived Value and Satisfaction Towards Repurchase Intention in Food Delivery

Alfred Tjandra[✉] and Zhiwen Cai

Xiamen University of Technology, Xiamen 600, China
alfredkpa@gmail.com, zwcai@xmut.edu.cn

Abstract. With the growth of big data, internet access and quality of smartphone can make a concept to combine online and offline, namely is online to offline (O2O). This research aim examined the factors to address how they influence Indonesian consumers' O2O repurchase intention. A model is quantitative method with Structural Equation Modeling (SEM) using survey data collected from 300 Grabfood users. The result indicate that perceived value is influenced by ease of use, interaction, security, entertainment and brand image. Price is not influence perceived value. Satisfaction is influenced by ease of use, price and entertainment. Interaction, security, and brand image is not influence satisfaction. Satisfaction is influenced by perceived value. Repurchase intention is influenced by perceived value and satisfaction. Experience doesn't have a moderate impact on perceived value and satisfaction, perceived value and repurchase intention, satisfaction and repurchase intention.

Keywords: Knowledge Management · O2O services · Brand image · Perceived value · Satisfaction · Repurchase intention

1 Introduction

In the era of digitalization, the internet and technology become more sophisticated and innovative. Internet and technology, especially in this pandemic, change human behavior when carrying out online activities. According to the Ministry of Communication and Informatics, in 2021, internet users in Indonesia increased by 11% from the previous year, from 175.4 million to 202.6 million users. [1]. With that development, people can quickly and conveniently access the internet from anywhere and anytime. One of the online activities which become popular is order food delivery. According to Databooks, 97% of digital spend during the corona pandemic to order food online. Followed by online delivery services by 76%. Spending on online transportation and shopping for daily necessities (online groceries) is 75% and 74%, respectively. The corona pandemic has also made most consumers frequently make online donations (54%). Another monthly digital expenditure that consumers often make is subscribing to online content

© The Author(s), under exclusive license to Springer Nature Switzerland AG 2023
N. Xiong et al. (Eds.): ICNC-FSKD 2022, LNDECT 153, pp. 1441–1452, 2023.
https://doi.org/10.1007/978-3-031-20738-9_156

platforms such as Netflix, Disney+, which reach 50%. Lastly, drug purchase (46%). See the phenomena that occur, lots of people selling food for business.

With the growth of big data, internet access and quality of smartphone, the total of consumers who spending on internet is rise. However, according to The Neilsen Global Survey of E-commerce, high shipping cost and weak logistics infrastructure in Indonesia are factors in reluctance to repurchase in e-commerce [2]. Therefore, there is a concept that is becoming a trend to solve Indonesia e-commerce logistics problems, namely is online to offline concept (O2O).

According to the phenomenon, the researchers are interested in developing research by adding two mediating variables, perceived value and customer satisfaction and experience as moderators so that they are expected to be able to answer whether perceived value and customer satisfaction mediate the perceived ease of use, interaction, price, security, entertainment, and brand image to repurchase intention under the SOR theoretical model.

2 Literature Review

2.1 Theory Stimulus-Organism-Responsible

Environmental psychology was the first discipline to employ "Stimulus-Organism-Response Model, S-O-R model", which Mehrabian and Russell proposed in 1974. The theory explicates the inter-relationship and importance regarding consumers' emotions and behaviors [3]. S represents the stimulus that causes the individual to respond, O represents the organism or the body of the reaction, and R represents the response caused by the stimulus [4].

2.2 O2O Services

O2O stands for online to offline, which refers to an offline-based business opportunity combined with the internet (online). The concept of O2O is very broad in its implementation. As long as the company can develop its online trade while still opening an offline store, it can be categorized as O2O [5]. In this study, according to [6] and [7] there are some characteristic features of O2O, namely: Ease of Use, Interaction, Price, Security, and Entertainment.

2.3 Brand Image

Brand image is built based on the impressions, thoughts, or experiences of consumers towards a brand which in turn will form attitudes towards the brand in question [8].

2.4 Perceived Value

Perceived value now comes from the customer's evaluation of the service's net worth, which is supported by the customer's appraisal of what is given (costs or sacrifice mad to get and use the service) and what is gained (benefits supplied by the service) [10].

2.5 Satisfaction

Satisfaction is defined as the feeling of pleasure or disappointment of someone that arises from comparing the product's perceived performance (or result) against their expectations. When the performance failed to fulfill their expectation, the customer will be disappointed. If their performance is appropriate to expectations, customers will be pleased [11].

2.6 Experience

Name of experience had used in the other way. Definition of experience can be disparate become two types: First, experience refers to the past (refer to knowledge and experience accumulation from time after time), and second refers to perception ongoing, sense, and direct observation [12].

2.7 Repurchase Intention

Repurchase intention is a behavior that arises because consumers feel satisfied or happy after consuming a product or service [13]. This behavior is a consumer commitment that is formed after a consumer purchases a product or service. This commitment arises because of a positive impression on consumers of a brand, and consumers are satisfied with the purchase.

3 Research Method

3.1 Hypothesis

Effect of Ease of Use on Perceived Value and Satisfaction Ease of use be indicated by the volume of use and user-application interaction [14]. Users become more habituated to a program the more they utilize it. Users also are more inclined to accept an application that they view as being simpler to use than another. Besides, According to [15] perceived ease of use has a positive and considerable impact on customer satisfaction. It means that the better-perceived ease of use, the higher their user satisfaction.

H1a: Ease of Use has a significant effect on perceived value.
H1b: Ease of Use has a significant effect on satisfaction.

Effect of Interaction on Perceived Value and Satisfaction The website's interaction tool is what determines how well it support two-way communication between users. Users can engage and participate on social media [16]. In addition, according to [16] interaction and communication among community users are crucial for building loyalty, elevate customer satisfaction, and the degree of customer loyalty is key to the long-term management of a community platform. When customers make a transaction online, customers have high satisfaction with the website.

H2a: Interaction has a significant effect on Perceived Value.

H2b: Interaction has a significant effect on Satisfaction.

Effect of Price on Perceived Value and Satisfaction Price is a factor self-directed by a perceived value which describes how to affect the price of the product against the value that will be perceived by a consumer [17]. Additionally, according to [18] perceived quality and perceived price simultaneously affected the dependent variables (satisfaction). If the price are low, customer satisfaction will also increase.

H3a: Price has a significant effect on perceived value.
H3b: Price has a significant effect on satisfaction.

Effect of Security on Perceived Value and Satisfaction Security is crucial to gaining customer's trust and sustaining user behavior, it has a significant impact on consumer decision making when they buy a product or use a variety of services [19]. Additionally, while visiting a website, user's opinions about the platform's or websites' level of security may have an impact on user's satisfaction and loyalty to the firm [20].

H4a: Security has a significant effect on Perceived Value.
H4b: Security has a significant effect on Satisfaction.

Effect of Entertainment on Perceived Value and Satisfaction The ability to satisfy audience requirements for emotional release, diversion, and enjoyment is what defines entertainment [21]. Additionally, entertainment is essential to shopping malls since it fosters an interesting and enjoyable shopping experience, which in turn may draw in customers. Consumers perceived the mall as a location for amusements as well as other activities, such as shopping, which had an emotional impact on their emotional states [22].

H5a: Entertainment has a significant effect on Perceived Value.
H5b: Entertainment has a significant effect on Satisfaction.

Effect of Brand Image on Perceived Value and Satisfaction The success of a brand image is the development of value for consumer's, a point that was further supported by target consumers group [23]. In addition, a strong brand image can influence consumers perceptions and loyalty to certain goods and services in terms of market structures that take into account both offline and online model.

H6a: Brand Image has a significant effect on Perceived Value.
H6b: Brand Image has a significant effect on Satisfaction.

Effect of Perceived Value on Satisfaction and Repurchase Intention Customers expect to receive value when they buy a goods or service, because they believe will gain more from the benefit that cost. If anything occurs after the purchase that unexpectedly lowers or raises the cost incurred or improves the benefit obtained, the perceived value is change and the customers satisfaction levels change [10]. Additionally, perceived value may predict repurchase intention more accurately than either satisfaction or quality [24].

H7: Perceived Value has a significant effect on Satisfaction.
H8: Perceived Value has a significant effect on Repurchase Intention.

Effect of Satisfaction on Repurchase Intention In regards to the O2O service, satisfaction becomes a vital determinant of customer repurchase behavior and loyalty since it can prevent customer's betrayal behavior by improving the profit of the enterprise [7].

H9: Satisfaction has a significant effect on Repurchase Intention.

Moderating Effect of Online Purchase Experience Another study [25] discovered that experience has a significant between perceived value and satisfaction. When defining their heritage sustainability strategy, heritage managers need to consider this critical issue [26].

H10: Online Purchase Experience will significant moderate the relationship between Perceived Value and Satisfaction.

Consumers assess perceived product value and design innovation, and they decide either whether to make a purchase based on past experiences and product knowledge [27]. Additionally, buy experience serves as a moderator for assessing the strength of the relations between perceived value and repurchase intention.

H11: Online Purchase Experience will significantly moderate the relationship between Perceived Value and Repurchase Intention.

When satisfied, experienced online buyers are more likely to plan to make another purchase. Experience has been demonstrated to make the affective aspect of attitude, such as satisfaction, more accessible [28].

H12: Online Purchase Experience will significantly moderate the relationship between Satisfaction and Repurchase Intention.

3.2 Research Model

Based on research objectives and literature review, the framework for thinking in this research is as follows (Fig. 1).

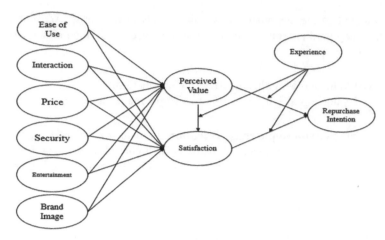

Fig. 1. Research model

4 Research Design

This research is using quantitative method which uses questionnaire survey as a research method. In this study, the questionnaire used closed questions. The questionnaire also asked about the shopper's demographic characteristics and online shopping tendencies. Respondents of the survey composed of 300 online customers who had previously shopped from grabfood and make purchases more than 3 times in 3 months. Participations was entirely voluntary.

4.1 Moderated Structural Equation Modeling

Several techniques in SEM can be used to gauge the moderating impact. The Ping method is one of the easiest methods to use and can estimate the moderating impact in SEM, which is a difficult approach. Ping claim that the latent moderating indication should be single indicator [29]. Ping advise be calculated as follows: $(x_1 + x_2) * (z_1 + z_2)$. The loading and error for the indicator of the latent product are given by the following equations [29, 30]:

$$\lambda_{x:z} = (\lambda_{x1} + \lambda_{x2})(\lambda_{z1} + \lambda_{z2}) \tag{1}$$

$$\theta_{\varepsilon x:z} = (\lambda_{x1} + \lambda_{x2})^2 VAR(X)(\theta_{\varepsilon z1} + \theta_{\varepsilon z2}) + (\lambda_{z1} + \lambda_{z2})^2 VAR(Z)(\theta_{\varepsilon x1} + \theta_{\varepsilon x2})$$
$$+ (\theta_{\varepsilon z1} + \theta_{\varepsilon z2})(\theta_{\varepsilon x1} + \theta_{\varepsilon x2}). \tag{2}$$

5 Results and Discussion

5.1 Validity and Reliability

Table 1 demonstrate that the factor loading in standardized solution is more than 0.5, the overall indicators are valid. Construct Reliability and Average Variance Extracted both have reliability values that are greater than 0.6 and 0.5, respectively. Therefore, it can be said that every variable in this study is reliable.

Table 1. Validity and reliability test result

Variable	Indicator	Standard solution	CR	AVE
Ease of use	EOU1	0.75	0.86	0.61
	EOU2	0.79		
	EOU3	0.79		
	EOU4	0.79		
Interaction	IN1	0.83	0.85	0.65
	IN2	0.74		
	IN3	0.84		
Price	PR1	0.79	0.89	0.66
	PR2	0.73		
	PR3	0.9		
	PR4	0.82		
Security	SE1	0.76	0.89	0.74
	SE2	0.91		
	SE3	0.9		
Entertainment	EN1	0.91	0.93	0.76
	EN2	0.9		
	EN3	0.81		
	EN4	0.87		
Brand image	BI1	0.8	0.86	0.61
	BI2	0.8		
	BI3	0.72		
	BI4	0.79		
Perceived value	PV1	0.78	0.85	0.58
	PV2	0.74		
	PV3	0.73		
	PV4	0.79		
Satisfaction	SA1	0.86	0.90	0.68
	SA2	0.87		
	SA3	0.8		
	SA4	0.77		
Experience	EX1	0.77	0.79	0.56

(continued)

Table 1. (*continued*)

Variable	Indicator	Standard solution	CR	AVE
	EX2	0.71		
	EX3	0.77		
Repurchase intention	RI1	0.66	0.84	0.63
	RI2	0.86		
	RI3	0.85		

5.2 Overall Model Match Test

According to the test result in Table 2 that NFI, IFI, CFI, RFI and RMSEA all fulfill the cut-off value requirements to be classified as good fit. To qualify as marginal fit, GFI and AGFI must almost completely reduce the benefit.

Table 2. Fit indices

Goodness of fit	Cut of value	Result	Notes
GFI	≥ 0.9	0.84	Marginal fit
AGFI	≥ 0.9	0.81	Marginal fit
NFI	≥ 0.9	0.94	Good fit
IFI	≥ 0.9	0.98	Good fit
CFI	≥ 0.9	0.98	Good fit
RFI	≥ 0.9	0.93	Good fit
RMSEA	< 0.08	0.044	Good fit

5.3 Hypothesis Testing

According to Table 3, the result showed that 11 hypotheses (H1a, H1b, H2a, H3b, H4a, H5a, H5b, H6a, H7, H8, H9) are accepted as describe above (T-values > 1.96). Lastly, for moderation effect, H10, H11, H12 showed that experience does not has a moderation effect (Fig. 2).

6 Conclusion

This study explores the factors that can affect customers decision to engage online to offline. The result suggest that ease of use, interaction, security, entertainment, and brand

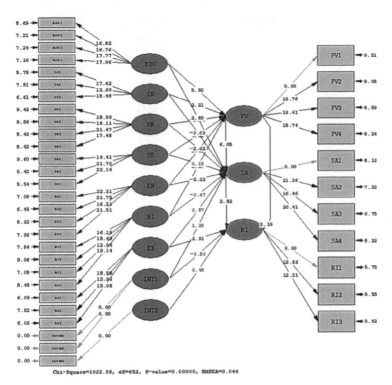

Fig. 2. Structural model of research

image are the main predictors of customers perceived value when comparing online and offline. The results also show that ease of use, price, and entertainment are the main predictors of consumers satisfaction with online to offline. The two most important variable of repurchase intention are satisfaction and perceived value. Consumer experience, which is used as a measure of moderation, doesn't have that effect.

This study also has some implications for practice. For Grabfood is better to pay more attention to how to make customers get more value from Grabfood. Because perceived value is an important variable such as ease of use, interaction, security, entertainment, and brand image. Value can increase along with quality improvement and service, which will result in higher customer satisfaction, although other factors can influence the role of perceived value.

6.1 Limitations and Future Research

There are definitely some limitations with this paper. First off, the majority of the responders are young adults from a select few Indonesian cities. The sample hardly accurately represents all users of the O2O platform. In future study, researchers will need to identify a different nation. Second, numerous publications and theoretical studies have demonstrated that satisfaction and perceived value are mediating variables. Future research is required to look at a variety of factors, including customer expectation and trust.

Table 3. Hypothesis test results

	Hypothesis	T-value	Cut-off	Notes
H1a	EOU → PV	5.92	1.96	Supported
H2a	IN → PV	2.31	1.96	Supported
H3a	PR → PV	−1.76	1.96	Not supported
H4a	SE → PV	4.52	1.96	Supported
H5a	EN → PV	2.19	1.96	Supported
H6a	BI → PV	3.57	1.96	Supported
H1b	EOU → SA	2.60	1.96	Supported
H2b	IN → SA	0.69	1.96	Not supported
H3b	PR → SA	−2.02	1.96	Supported
H4b	SE → SA	0.19	1.96	Not supported
H5b	EN → SA	−2.32	1.96	Supported
H6b	BI → SA	−0.47	1.96	Not supported
H7	PV → SA	6.05	1.96	Supported
H8	PV → RI	2.19	1.96	Supported
H9	SA → RI	2.53	1.96	Supported
H10	PV * EX → SA	1.38	1.96	Not supported
H11	PV * EX → RI	−0.80	1.96	Not supported
H12	SA * EX → RI	0.48	1.96	Not supported

References

1. Ministry of Communication and Information of the Republic of Indonesia. Netizens Increase, Indonesia Needs to Increase Cultural Values on the Internet. https://aptika.kominfo.go.id/2021/09/warganet-meningkat-indonesia-perlu-tingkatkan-nilai-budaya-di-internet/. Last accessed 21 Oct 2021
2. Priambada, A.: Is Indonesia ready to adopt the online to offline e-commerce concept (O2O)? https://dailysocial.id/post/siapkah-indonesia-mengadopsikonsep-e-commerce-online-to-offline-o2o. Last accessed 1 Jul 2022
3. Belk, R.: Situation variables and consumer behavior. J. Consum. Res. **2**(3), 157–164 (1975)
4. Sun, Q.: Research of O2O user behavior based on stimulus-organism-response model. In: Proceedings of the 2018 International Conference on Sports, Art, Education, and Management Engineering, pp. 474–479. Atlantis Press, China (2018)
5. Tiangsheng. X., Jiong. Z.: Development strategy of O2O business in China. In: Proceedings of the 2015 International Conference on Computer Science and Intelligent Communication, pp. 337–340. Atlantis Press, China (2015)
6. Chung, S., Kim, W., Bae, Y.H.: O2O trend, and future: focused on difference from each case. J. Mark. **3**(4), 49–66 (2017)
7. Che, H., Chan Lee, Y., Lei Mu, H.: An empirical study on influencing factors of O2O service repurchase intention: evidence from Meituan food service in China. Int. J. Pure Appl. Math. **120**(6), 5173–5184 (2018)

8. Yaseen, N., Tahira, M., Gulzar, A., Anwar, A.: Impact of brand awareness, perceived quality, and customer loyalty on brand profitability and purchase intention: a reseller view. J. Contemp. Res. Bus. **3**(8), 833–839 (2011)
9. Kotler, P., Keller, K.L.: Marketing Management, 14th ed. Prentice Hall, New Jersey (2012)
10. Hellier, P.K., Geursen, G.M., Car, R., Rickard, J.A.: Customer repurchase intention: a general structural equation model. Eur. J. Mark. **37**(11), 1762–1800 (2003)
11. Kotler, P., Keller, K.L.: Marketing Management, 13th ed. Erlangga, Jakarta (2008)
12. Nurrahmanto, P.A., Rahardja, R.: The effect of ease of use, shopping enjoyment, shopping experience, and consumer trust on consumer buying interest on the Bukalapak.com buying online site. Diponegoro J. Manage. **4**(2), 1–12 (2015)
13. Kusuma, P.D., Suryani, A.: The role of customer satisfaction mediates the effects of marketing mix on repurchase intention. E-J. Manage. Unud **6**(3), 1398–1424 (2017)
14. Adams, D.A., Nelson, R.R., Todd, P.A.: Perceived usefulness, ease of use, and usage of information technology: a replication. Manage. Inf. Syst. Res. Center **16**(2), 227–247 (1992)
15. Oktarini, M., Wardana, I.: The role of customer satisfaction mediates the effect of perceived ease of use and perceived enjoyment on repurchase intention. Unud Manage. E-J. **7**(4), 2041–2072 (2018)
16. Chen, S.C., Lin, C.P.: Understanding the effect of social media marketing activities: the mediation of social identification, perceived value, and satisfaction. Technol. Forecast. Soc. Chang. **140**, 22–32 (2019)
17. Buditama, W.S., Aksari, N.M.: The role of perceived quality mediates perceived price to perceived value service user Rumah Kos Jimbaran. E-J. Manage. Udayana Univ. **6**(2), 1055–1082 (2017)
18. Dapkevicius, A., Melnikas, B.: Influence of price and quality to customer satisfaction: neuromarketing approach. Sci.-Future Lithuania **1**(3), 17–20 (2009)
19. Kahar, A., Wardi, Y., Patrisia, D.: The influence of perceived usefulness, perceived ease of use, and perceived security on repurchase intention at Tokopedia.com. In: Proceedings of the 2nd Padang International Conference on Education, Economics, Business and Accounting, pp. 429–438. Atlantis Press, Indonesia (2019)
20. Wilson, N., Alvita, M., Wibisono, J.: The effect of perceived ease of use and perceived security toward satisfaction and repurchase intention. Muara J. Econ. Bus. **5**(1), 145–159 (2021)
21. Dao, W.V., Le, A.N., Cheng, J.M.: Social media advertising value. Int. J. Advert. **33**(2), 271–294 (2014)
22. Ahmad, K.: Attractiveness factors influencing shoppers' satisfaction, loyalty, and word of mouth: an empirical investigation of Saudi Arabia shopping malls. Int. J. Bus. Adm. **3**(6), 101–112 (2012)
23. Jeng, S.P.: The influences of airline brand credibility on consumer purchase intentions. J. Air Transp. Manage. **55**, 1–8 (2016)
24. Cronin, J.J., Brady, M.K., Hult, G.T.: Assessing the effects of quality, value, and customer satisfaction on consumer behavioral intentions in service environments. J. Retail. **76**(2), 193–218 (2000)
25. Amri, S., Maruf, J.J., Tabrani, M., Darsono, N.: The influence of shopping experience and perceived value toward customer satisfaction and their impacts on customer loyalty at minimarkets in Aceh. Int. Rev. Manag. Mark. **9**(4), 87–94 (2019)
26. Chen, C.F., Chen, F.S.: Experience quality, perceived value, satisfaction, and behavioral intentions for heritage tourists. Tour. Manage. **31**(1), 29–35 (2010)
27. Kim, S.J., Kim, K.H., Choi, J.: The role of design innovation in understanding purchase behavior of augmented products. J. Bus. Res. **99**, 354–362 (2017)
28. Khalifa, M., Liu, V.: Online consumer retention: contigent effects of online shopping habit and online shopping experience. Eur. J. Inf. Syst. **16**(6), 780–792 (2007)

29. Ghozali, I.: Structural equation modeling theory, concepts and applications with LISREL 9.10 program, 4th ed. Diponegoro University Publishing Agency, Semarang (2008)
30. Cortina, J.M., Chen, G., Dunlap, W.P.: Testing interaction effects in LISREL: examination and illustration of available procedures. Organ. Res. Methods **4**(4), 324–360 (2001)

Analysis of Data Loss and Disclosure in the Process of Big Data Governance in Colleges and Universities in the Smart Era

Wen Man, Yu Zhu$^{(\boxtimes)}$, and Jingyi Zhang

Air Force Engineering University, Xi'an 710051, China
308754733@qq.com, Zhangjy_1991@163.com

Abstract. In view of the problem that a large amount of data will be disclosed and shared in the process of data governance in universities in the era of wisdom, this paper focuses on the contradiction between data sharing and the risk of leakage, comprehensively analyses the subject and object of data governance, and formulates an effective and scientific data governance strategy. This paper puts forward the idea that the risk management and control of data leakage runs through every link of data governance, establishes the whole life cycle supervision process of data source, which is "acquisition, processing, cleaning, security analysis, storage, sharing", and carries out security governance through the data service mode of "collation, analysis, qualitative, authority division, integration". Through practice, it is found that this can achieve maximum disclosure of public data, desensitization of sensitive data, and life-cycle management of privacy data protection.

Keywords: Data governance · Data sharing · Leakage · Security governance

1 Introduction

Under the background of smart campus, the informatization construction of colleges and universities has entered a new stage of development. In 2018, the Ministry of Education promulgated the "Education Informatization 2.0 Action Plan" to realize the deep integration of "Internet and information technology" and educational innovation and development [1]. In 2019, the Central Committee of the Communist Party of China and the State Council issued "China's educational modernization 2035", it is proposed to build an intelligent campus and promote the whole process of education and teaching by means of information technology such as the Internet [2]. Deepening the application of big data in education and supporting the reform of education and teaching are important contents of the modernization of national education. Under the background of the new era, the rapid development of big data technology supports educational reform and innovation, and improves the scientific level of higher education decision-making. By speeding up the deployment of big data and using various big data technologies, colleges and universities can explore and release the potential value of data resources, and

N. Xiong et al. (Eds.): ICNC-FSKD 2022, LNDECT 153, pp. 1453–1462, 2023.
https://doi.org/10.1007/978-3-031-20738-9_157

establish a mechanism of "speaking with data, making decisions with data, managing with data and innovating with data" [3]. The larger the volume of data and the more dimensions, the better it can provide more potential information for users to accurately analyze and personalize services.

The governance of data means the centralization, opening, circulation, analysis and mining of data. Contradictions arise in the process of promoting data governance in colleges and universities, and the security of confidential data is greatly threatened while data sharing. Once the data governance process is not carefully managed, it will cause the leakage of confidential data and important information resources, and the harm will be irreparable. How to solve the contradiction between data sharing and confidential information security protection in data governance?

2   New Understanding of Big Data Governance in Colleges and Universities

Big data governance in universities refers to the process of collecting, cleaning, storing, deleting and protecting big data by using information technology tools according to certain standards according to the big data governance system [4]. Data governance, as a scientific tool, plays an important role in the development of higher education under the promotion of information construction in universities. Carrying out data governance can solve the problems of "low data quality, confused data flow, insufficient sharing and lack of historical data" in current data sharing, and can sort out, collect, clean, standardize and standardize the storage and application of university business system data in an orderly and standardized manner, so as to realize the effective management of school data assets and the deep sharing of data [5].

Data governance plays an important role in the process of data standardization, standardized storage and application [6]. Through organizational optimization, system innovation, data fusion and process reengineering, it can create a global data chain integrating educational administration, scientific research, students, personnel, finance and other fields, and realize the vertical connection of data between colleges and universities and the horizontal collaboration between departments [7]. But if we can organize and analyze the data of educational resources at the same time of data governance, which are the data that need to be protected, which are the data that can be made public, identify data risks through data governance, control more secure access, maximize the sharing of public data, and minimize the sharing of confidential data, we can achieve the goal of data security governance. Only in this way can we solve the contradiction between data sharing and leakage.

3   Subject and Object of Data Governance

3.1   Governance Subject

The main body of data governance mainly refers to the organization and related personnel that perform the work [8]. It is completed by the school leaders and the relevant departments responsible for the informatization work. School leaders, as the top leaders of colleges and universities, are mainly responsible for the overall management of the development of smart campus and the overall direction and overall decision-making. The leading group of informatization work mainly carries out the supervision and supervision of data management work arrangement and tasks. The Informatization Office is mainly responsible for managing relevant systems, carrying out various data management work arrangements, assigning decision-making tasks, and reporting the progress of relevant tasks. The information management center is mainly responsible for formulating technical standards, building data centers, data sharing and exchange platforms, carrying out data acquisition, cleaning, large data analysis, completing data sharing, and implementing specific instructions from superiors. Business departments are mainly responsible for providing business data to ensure the uniqueness of data sources and the authenticity and reliability of data. System developers cooperate with business departments to sort out business processes, strictly implement the requirements of national, industrial and school-level data standards, compile clear and standardized data dictionaries, constantly correct wrong data according to the problems in business process management, and provide reliable data exchange interface documents. The main body of data governance is shown in Fig. 1.

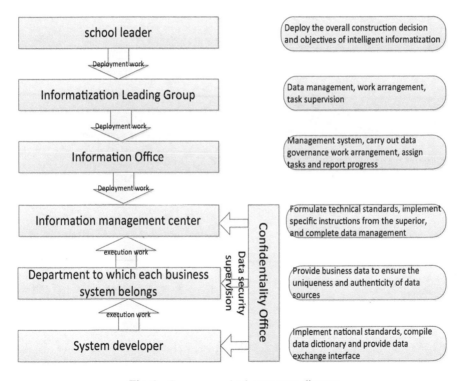

Fig. 1. Governance body structure diagram

3.2 Governance Object

The object of governance refers to the collection of data collected and stored around people, things and things in colleges and universities in the activities of teaching, scientific research and management of higher education. The collection of massive data resources in colleges and universities includes all kinds of data resources stored in application systems such as teaching, scientific research, logistics, library, educational security and educational administration [9]. As shown in Fig. 2, the overall architecture of the teaching data system in colleges and universities adopts a "four-layer" structure, which includes the basic layer, the data resource layer, the resource service layer and the data application layer.

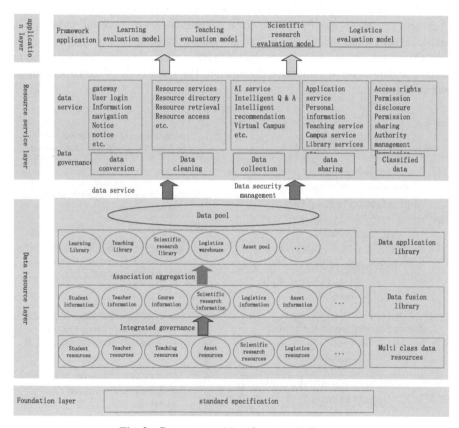

Fig. 2. Governance object framework diagram

The data base layer is the underlying unit of the data governance object, which should complete the construction of infrastructure such as data exchange platform, data storage and analysis platform, formulate relevant standards and specifications for the overall construction of university data, and unify and standardize the construction and application of data, business applications and external interfaces of application systems. It provides a basic platform for subsequent data collection, cleaning, storage and exchange.

Data resource layer, relying on the support of underlying storage devices and computing hardware, provides unified management services of university resources based on cloud computing architecture, and constructs basic databases, dynamic databases and professional databases according to student resources, teacher resources, asset resources, teaching resources and other types of data resources. Deeply sort out and clarify the logic, data source and direction of various business processes, and integrate them to form a data pool.

Resource service layer, relying on the data pool of university data resources, provides data support services based on big data technology, and provides data governance services internally, including data conversion, data cleaning, data security analysis, data exchange, data sharing, secret-related data processing, etc., to achieve the deep interaction of permissions, information and processes between multi-level and multi-level businesses; Provide portals, resource services, application services, Artificial intelligence services, etc. Do a good job of data sharing while ensuring the protection of confidential data.

Application service layer, through the service support of cloud computing and big data, constructs the typical application of university data, reconstructs the data service architecture, forms the intelligent evolution of "data-governance-value", thoroughly excavates the hidden wisdom behind the data, in order to learn evaluation model, teaching evaluation model, scientific research evaluation model and logistics. Support the construction and application of smart campus.

4 Research on the Contradiction Between Data Sharing and Risk of Leakage

Data sharing mainly refers to the sharing of teaching resources, student resources, scientific research resources, educational resources and training resources through data sharing mechanism. The governance goal of data sharing is to improve the public data sharing and exchange mechanism through the scientific establishment of rules and regulations, promote the open sharing of public data in an orderly manner, strengthen the management and control of data quality, data assets and other related links, and implement the centralized and unified management of multi-source heterogeneous data.

The management and supervision of data leakage security is the most neglected link in the data governance system. While sharing, sensitive data, user privacy, core database and other confidential information may be exposed, allowing unauthorized data to be accessed and manipulated, resulting in the contradiction between data sharing and the risk of confidential data leakage.

How to deal with data sharing without causing the risk of leakage of important confidential data? The author believes that To formulate an effective and scientific data governance strategy, to run the risk management and control of data leakage through every link of data governance, to establish the whole life cycle supervision process of data source "collection, processing, cleaning, security analysis, storage and sharing", and to organize the data through "collation, analysis, qualitative, authority division and integration". The core way to solve the contradiction is to establish the life-cycle management of maximum disclosure of public data, desensitization of sensitive data and protection of privacy data.

The big data of higher education can be divided into four categories according to the type, as shown in Table 1. It mainly consists of business data, text and unstructured data, machine data and other data.

Table 1. Data types

Serial number	Data type	Data description
1	Business data (structured data)	Data of educational administration, assets, all-in-one card, OA, network disk and other business systems
2	Text and non-institutional data	Word, Excel, txt, Audio and video data
3	Machine data	Database log, security device log, network device log, server log, etc
4	Other data	Stream data, etc

4.1 Business Data Governance and Security Management

Business data is basically structured data, so it is necessary to strengthen the confidentiality awareness of business system administrators, improve data application supervision, improve business log storage, establish data privacy and confidentiality mechanism, and develop data life cycle management mechanism. The data of the business system is classified according to the classification level, and the data assets are finally submitted to the big data platform according to the encryption of confidential data, the desensitization of sensitive data and the disclosure of public data.

How to encrypt the core confidential data of business data? Structured data basically uses a large network database to store data. Taking oracle as an example, the access of oracle database is basically based on the network protocol TCP/IP. A network session is established between the database server and the client to transmit data, and the data in the database is accessed through the network channel. The security problems faced by database mainly include three aspects: system security, network environment security and data security. The oracle database has a very complete database authentication mechanism, which controls the user's access and operation to the database by setting specific access rights for the user, including security parameters, transactions, integrity, permissions and roles [10]. To connect to the Oracle database, the user must pass the authentication before logging in. The common valid authentication methods are: external authentication and database authentication. When the security of oracle database data is in danger of intrusion, the database will automatically generate data recovery and protection defense mechanism.

The oracle database exists in the form of a file in the operating system. Before the data is accessed in the medium, the database encryption is carried out in the core part of the database system. The data generated in the process of encryption and decryption is outside the management system of the database, and the cipher text is added. The database encryption unit has the following four modes, as shown in Table 2. Through table unit encryption, attribute unit encryption, record unit encryption and database unit encryption, the database security management of structured business data is realized, and permission operation, permission access and permission application are realized [11].

Table 2. Oracle database encryption unit

Serial number	Database encryption unit	Characteristic
1	Table unit	Encrypt the whole table, encrypt different table keys, and finally form a cipher text
2	Attribute unit	Encrypt fields or domains
3	Recording unit	Equivalent to record encryption, it mainly takes the record behavior in the table as the object of encryption
4	Database element	The minimum encryption granularity is also the encryption object for recording field values

4.2 Text and Unstructured Data Governance Security Management

Unstructured data usually refers to data without a fixed display structure, such as word, excel, txt, video, audio, image and other data. According to the statistics of Gartner Group, 80% of the data today is unstructured data, which is still growing at a high speed, and unstructured data has gradually become the main body of big data. Faced with such a huge amount of data, how to ensure data security in the process of data governance? How to obtain valuable governance through effective security management means?

In view of the security management methods of text and unstructured data in the process of data governance, this subject adopts the whole life cycle data security monitoring, in-depth analysis of Word, Excel, PPT, PDF, Txt compression format files, and identification of sensitive data. The unstructured data content is identified by file basic attributes, name, type, creator, creation time, keywords, regular expressions, data fingerprints and other methods, and the sensitive file content is effectively identified based on the semantic analysis of Chinese word segmentation, so as to determine the file access rights.

When collecting and cleaning unstructured data, according to the qualitative results of data security monitoring access rights in the whole life cycle, the unstructured data management mode is established, and in the intelligent management stage, the unstructured data is structured by using AI technology combined with data warehouse technology (Extract-Transform-Load, ETL). It automatically and intelligently refines unstructured data and formulates a set of models, standards, quality and other rules applicable to unstructured data. When the unstructured data is stored, according to the data content identification, the distributed technology and HBase database are used for storage, and finally the security management, analysis, application and destruction are carried out to achieve the cleaning, classification and permission control of confidential data [12].

4.3 Machine Data

Machine data includes database logs, security equipment logs, network equipment logs, server logs and other data. With the progress of intelligent campus construction, basic network construction, network security protection system construction and business system construction are being upgraded and transformed.

Taking network equipment as an example, the operation log, hardware indicators, equipment status and other information of the equipment belong to ordinary data. In the process of governance, the online operation status of the equipment is analyzed with the big data analysis platform, so as to provide data support for operation and maintenance. The user's access log, denial log, traffic log and other information belong to confidential data, which requires access permission to protect personal privacy while doing a good job of operation and maintenance.

5 Data Governance in the Data Security Model

According to the type of education big data, the data source is sorted out, and then the data is collected and cleaned. At the same time, the data security analysis must be considered, and then the data collation, storage, exchange and analysis are completed to complete the whole data governance process. Classify by data type when combing the data source. In the process of data cleaning, check the integrity, consistency, generality, uniqueness, validity, standardization and openness of the collected data [13], establish the whole life cycle data security monitoring, conduct data security analysis, organize and store the public data according to the new data standards, and enter the big data platform for data exchange and data analysis. Establish a separate data storage space for sensitive and confidential data, allocate permissions according to data content, minimize sharing, and conduct subsequent data exchange and data analysis. As shown in Fig. 3.

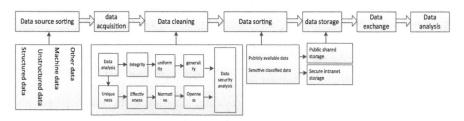

Fig. 3. Data governance process

The main functions of the data security analysis module are as follows: according to the security management strategy in the educational data system, through the scientific application of the internal data of the data platform, the security management and control of data exchange with other units are continuously strengthened to ensure that the educational data assets fully meet the four-in-one data management security standards of "storage, management, confidentiality and use". Achieve the whole process management goal of "manageable in advance, controllable in the event and check able afterwards" [5]. The data security analysis module model in the data governance process is shown in Fig. 4, which mainly includes four sub-modules: data management security module, data exchange security module, data governance security module and data sharing security module.

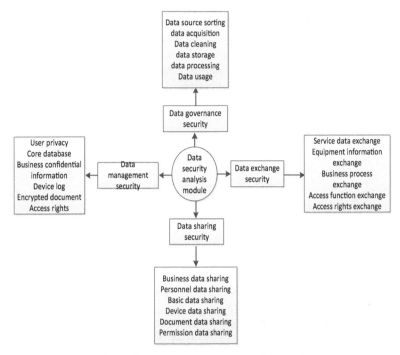

Fig. 4. Data security analysis module model

On the premise of data security guarantee, the following security measures must be taken to provide security protection capability for the data platform from the data governance level: (1) centralized security management of classified data collection; (2) unification of classified data standards; (3) hierarchical management of data authority roles with clear authority allocation; (4) safe and reliable data cleaning and storage environment; (5) Post audit of operation behavior. We will resolutely put an end to the direct connection of core data such as databases, servers and business systems, the unauthorized access to sensitive data, and the open sharing of confidential data beyond the scope.

6 Conclusion

The data governance of colleges and universities in the age of wisdom is a long-term, complex and diverse systematic project, involving the organic coordination and guarantee of every level of organizational architecture, every aspect of policy and system, and every step of technical support. In the era of advocating intelligent teaching, management and service in all colleges and universities, the big data of education covers the whole life cycle of teaching, scientific research, students, teachers, logistics and assets in colleges and universities. In view of a series of problems such as the continuous growth of current educational data, the low quality of data, the disorder of data sources, the obstacles to data sharing, the limitation of data flow, and the lack of guarantee of data security, this paper designs a set of data governance schemes that integrate data security management and solve the problems of data sharing and leakage, aiming at providing a reference for the governance and application of big data in Colleges and universities. It also guides colleges and universities to establish data standards and data security management norms while managing big data, and to enter the ecological environment of security governance, security sharing, security access, security analysis and security storage in the era of big data.

References

1. Du, Z.: Accelerate the integration of innovative development and make education informatization 2.0 a reality. China Education News, 2018-04-25 (03)
2. Xinhuanet. The Central Committee of the Communist Party of China and the State Council issued "China's Educational Modernization 2035". http://www.xinhuanet.com/2019-02/23/c_1124154392.htm
3. The State Council. Action Plan for Promoting Big Data Development. http://www.gov.cn/zhengce/content/2015-09/05/content_10137.htm
4. Wu, G., Chen, G.: University big data governance operation mechanism: functions, problems and improvement countermeasures. Univ. Educ. Sci. **6**, 34–38+66 (2018)
5. Yu, P., Li, Y.: Research on university data governance scheme from the perspective of big data. Mod. Educ. Technol. **28**(6), 60–66 (2018)
6. Dai, Y., Wang, H.: Logical framework and implementation path of data governance in higher education. Heilongjiang Higher Educ. Res. **39**(10), 41–45 (2021)
7. Dong, X., Zheng, X., Peng, Y.: Framework design and implementation of big data governance in university education. China Electron. Educ. **08**, 63–71 (2019)
8. Peng, Y., Yan, L.: Research on higher education data ecological governance system from the perspective of smart campus. China Electron. Educ. **05**, 88–100 (2020)
9. Xiong, Y., Chu, W., Cai, T., Yu, L, Tian, H.: Design and practice of big data application support system in higher education. Mod. Educ. Technol. **30**(11), 91–97 (2020)
10. Peng, X., Hu, J.: Introduction to big data under the background of the database security guarantee system. Comput. Knowl. Technol. **18**(9), 11–12 (2022)
11. Zhang, F., Zhang, M., Zhang, Z., Zhang, Y.: Distributed spatial database security mechanism study. Bull. Surv. Mapp. **01**, 41–44 (2016)
12. Duan, P.: Based on dynamic security of unstructured data block chain store. J. Jilin Univ. (Inf. Sci. Ed.) **42**(5), 595–600 (2020)
13. Liu, Z., Qu, S., Chen, S., Wang, P., Liu, D., Zhang, S.: J. Shenzhen Univ. (Sci. Technol. Ed.) **37**(S1), 139–145 (2020)

Digital Twin System Design for Textile Industry

Qingjin Wu[1], Danlin Cai[2,3,4(✉)], and Daxin Zhu[2,3,4]

[1] Laboratory and Equipment Management Office, Quanzhou Normal University, Quanzhou, China
[2] School of Mathematics and Computer Science, Quanzhou Normal University, Quanzhou, China
1795895@qq.com
[3] Key Laboratory of Intelligent Computing and Information, Quanzhou, China
[4] Fujian Provincial Key Laboratory of Data Intensive Computing, Quanzhou, China

Abstract. To address the problems of low transparency, single approach, poor real-time monitoring, lack of modeling and low production efficiency in garment production workshops, the operational logic modeling method is used to describe the operation logic of production system. By transforming the twin data of the workshop production process into corresponding workshop events, Digital Twin of the production scene is realized to digitally show the overall perspective of the production line and further reduce the production cost while improving the efficiency of textile and garment production.

Keywords: Digital Twin · Virtual simulation · Textile industry · Industrial Internet

1 Introduction

Textile and garment industry is a dominant industry in China, with a wide range, a complicated supply chains and a large number of participants. The Internet has already infiltrated into the textile and garment industry, occupying an unshakable position in Chinese economy.

Under the guidance of "Made in China 2025", the clothing industry is transforming towards automation, digitalization, networking and intelligence to realize intelligent manufacturing of enterprises. Therefore, building an intelligent production line for the clothing industry is an inevitable requirement for the manufacturing industry. With the development and popularization of computer technology, Internet technology, information technology and intelligent technology, the advanced production mode around big data and digitalization is gradually becoming the mainstream of the clothing industry development, big data is becoming more and more prominent in driving the upgrading of the garment industry, and the breadth and depth of big data extends to each production link of the textile and garment enterprises, which has become the key to the upgrading of the textile and garment industry.

The production data of textile and garment enterprises has the characteristics of big data, that is, huge amount of business data, diversified structure, diversified data sources,

N. Xiong et al. (Eds.): ICNC-FSKD 2022, LNDECT 153, pp. 1463–1470, 2023.
https://doi.org/10.1007/978-3-031-20738-9_158

and strong data timeliness. In order to obtain the management value and services brought by these data, it is necessary to analyze and mine the value of these business data. Although the level of China's textile and garment production and manufacturing has improved significantly, there are also some sore points in China's textile and garment industry.

(1) In terms of production visualization management of textile and garment workshop, some garment enterprises are still stuck in the traditional production visualization management level, simply in the form of two-dimensional charts and data reports for the management of the production site, can not guarantee the real-time production data simulation visualization monitoring, which leads to the garment production workshop into the monitoring of low transparency, single method, poor real-time, lack of models and other problems.

(2) Although textile and garment production generates massive amounts of on-site data in all links, these data are still underutilized. The garment production process is flexible and susceptible to various factors. Therefore, most textile and garment enterprises still rely on experienced managers to deploy production staff in various textile and garment production stages, resulting in low overall production efficiency.

The core issues facing the textile and garment industry are the shift to digital and intelligent manufacturing, the shift from focusing on growth rate to focusing on quality and efficiency, and the shift from production-based manufacturing to service-based manufacturing. How to promote the application of big data in all aspects of industrial production of textile and garment enterprises, and enhance the production technology innovation and efficiency improvement of industrial enterprises? How to use machine vision and other intelligent technology means to enhance the intelligence level of certain important aspects in the production process, while improving quality and efficiency? How to promote enterprises through big data, Internet of Things, cloud computing and other new information technology, to carry out the integration of industrial big data and applications to promote the accelerated development of emerging industry?

To address the above common technical problems in the textile and garment industry, this study will analyze and optimize the production line control efficiency by establishing an efficient real-time simulation model of the production environment and how to achieve it.

2 Related Technologies and Research

2.1 Digital Twin

Digital Twin is currently more widely accepted as a definition that makes full use of physical models, sensor updates, operational history, and other data to integrate multi-disciplinary, multi-physical quantity, multi-scale, and multi-probability simulation processes and complete mapping in virtual space, thus reflecting the full life-cycle processes of the corresponding physical equipment. Digital Twin is a concept that transcends reality and can be considered as one or multiple digital mapping systems of important and interdependent equipment systems.

Professor Michael Greaves of the University of Michigan introduced the concept of "digital representation of physical products", and pointed out that the digital representation of physical products should be able to represent physical products in an abstract way, and be able to test physical products under real or simulated conditions based on the digital representation. This concept is not called Digital Twin, but it has the components and functions that Digital Twin has, it creates equivalent virtual bodies of physical entities, and the virtual bodies are capable of simulating and analyzing and testing the physical entities. The theory proposed by Professor Greaves can be considered as an application of Digital Twin in the product design process.

Digital Twin, as the best way to realize the interactive integration of physical factory and virtual factory, has been highly regarded by related academic circles and enterprises at home and abroad. As shown in Fig. 1, from the connotation of Cyber Physics System (CPS) and Digital Twin, they both aim to describe the state of integration of information space and physical world, while CPS is more inclined to the verification of scientific principles and Digital Twin is more suitable for the optimization of engineering applications and more capable of reducing the cost of building complex engineering systems [1]. Fei and Meng proposed the concept of Digital Twin workshop based on Digital Twin, and analyzed from the workshop management elements, the development of Digital Twin workshop needs to go through the three stages of production elements, production activities, production control limited to the physical workshop in turn, the relative independence of physical workshop and Digital Twin workshop, and the interactive integration of physical workshop and Digital Twin workshop before it can gradually mature [2]. Tang and Lin considered Digital Twin as a way to integrate the manufacturing process of an enterprise, digitize the whole process of product from design to mainte nance, visualize the production process through information integration, form a closed loop from analysis to control to analysis, and optimize the whole production system [3]. According to Robert Plana, the most important value of Digital Twin is prediction, in case of problems in the product manufacturing process, the production strategy can be analyzed based on Digital Twin, and then production can be organized based on the optimized production strategy [4].

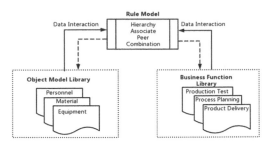

Fig. 1. Virtual interaction model framework

2.2 Intelligent Technology for the Textile and Garment Industry

Due to the complexity of the data generated by the textile and garment production process and the randomness of production planning and execution control, it is obvious that it possesses certain difficulties to realize the intelligence of the whole production process. After some years of efforts, the textile and garment industry has made obvious progress and remarkable results in several production processes. For example, Hangzhou MES printing and dyeing system performs real-time detection, automatic control and automatic distribution of process dyeing in the production process to achieve all-round management; Shandong Companion has developed a fully automatic cylinder yarn dyeing system, which realizes automatic dosing of dyestuff auxiliaries and automatic transportation of cylinder yarn at the same time, establishing an automatic and continuous production line. The intelligent system of production process, on the one hand, can realize the connection of intelligent equipment on the bottom layer, and at the same time, can provide service to the intelligent software system on the top layer, which is an essential part of the whole intelligent production model.

3   Digital Twin System Design of Textile and Garment Industry

The demand for refined production management of garment enterprises is getting higher and higher, and the amount of data has increased significantly, so the traditional management mode can no longer adapt to the requirements of the new era. In order to further improve the efficiency of enterprise management, introducing and applying new technologies such as Internet of Things, cloud computing, big data and artificial intelligence, building a kind of textile and garment industry's big data platform, and achieving the intelligent manufacturing system of textile has become inevitable. A highly intelligent Digital Twin system for the textile industry has been built. Digital Twin-based real-time simulation and modeling of the production environment has been used to visualize and monitor real-time production data, view production plan execution status, production history, and production performance. In terms of rational utilization of on-site production data, it is proposed to combine data mining technology to realize big data-based analysis and optimization methods for production line control effectiveness, and to reasonably dispatch personnel in each production link, thereby improving garment production efficiency.

3.1   Standard System of Digital Twin

Digital Twin standard system can contain the following parts [5]:

(1) Basic common standards: including terminology standards, reference framework standards and applicable guidelines, which focus on the conceptual definition, reference framework, applicable conditions and requirements of Digital Twin and provide support for the whole standard system.
(2) Key technology standards of Digital Twin: including physical entity standards, virtual entity standards, twin data standards, connection and integration standards, and service standards, which are used to standardize the research and implementation of Digital Twin key technologies.

(3) Digital Twin tool/platform standards: including two parts: tool standard and platform standard, which are used to standardize the technical requirements of hardware and software tools/platform in terms of function, performance, development and integration.

(4) Digital Twin evaluation standards: including four parts: evaluation guidelines, evaluation process standard, evaluation index standard and evaluation use case standard, which are used to standardize the testing requirements and evaluation methods of Digital Twin system.

(5) Digital Twin security standards: including physical system security requirements, functional security requirements and information security requirements, which are used to standardize the technical requirements for the safe operation of personnel and the safe storage, management.

(6) Digital Twin industry application standards: considering the technical differences of Digital Twin in different industries/fields and different scenarios, on the basis of basic common standards, key technology standards, tool/platform standards, measurement standards and safety standards, the implementation of Digital Twin in specific industry applications will be standardized.

3.2 Management and Control System

Textile and garment industry takes the control system of the production workshop as the core, as shown in Fig. 2, the architecture of the workshop control system based on Digital Twin mainly includes Digital Twin and production big data in the workshop management, with Digital Twin as the core and the control system of the manufacturing execution management system. Digital Twin is the core of the workshop management system, and the control system of manufacturing execution management system, simulation analysis system and production big data are effectively connected to form the architecture of Digital Twin workshop control system. The system adopts a star topology, which is better open and easier to expand, which facilitates the integration of new generation information technology into the workshop control system. The system is able to collect the most comprehensive manufacturing data and product data, forming industrial big data, which provides the data basis for production line intelligence through Digital Twin information model and algorithm. Digital Twin contains factory information models, basic data, business data, real-time data, and algorithms, which form a unified data source for the enterprise, reducing the difficulty of system maintenance and upgrades, and meeting the control needs of the changing intelligent factory of the future.

As shown in Fig. 2, the real-time mapping from the physical workshop to the virtual workshop is the core of realizing the 3D visualization and monitoring of the garment workshop. In order to establish a real mapping process, a workshop production system operation model is needed to accurately describe the dynamic behavior of the workshop. To this end, the system uses an operational logic modeling approach to describe the operational logic of the production system, enabling real-time mapping of logistics by transforming the twin data of the workshop production process into corresponding workshop events that drive changes in equipment status and the flow of workpieces between different workstations. At the same time, it is proposed to establish a virtual workshop using virtual model building method, and use event-driven to realize real-time mapping of logistics and model-driven to realize real-time mapping of equipment; from real-time mapping of logistics and real-time mapping of equipment combined with process flow to realize real-time mapping of products, and finally realize the front-end display of visual monitoring.

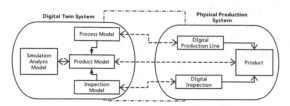

Fig. 2. Digital Twin system structure

3.3 Analysis and Optimization Method of Production Line Control Efficiency Based on Big Data

In the production process of garment enterprises, a huge amount of data is generated, including equipment status, production environment, personnel status, production data, material consumption, equipment efficiency, etc. These data exist in close connection, which makes the textile and garment production process have a certain complexity. Considering that these data have the characteristics of big data, i.e. huge volume of business data, diversified structure, diversified data sources, and strong data timeliness, this system obtains the potential management value and services of these data through the analysis and optimization method of production line control effectiveness based on big data. Thus, it gives the best production scheduling and resource adjustment plan to effectively optimize resource allocation and improve production efficiency.

In order to eliminate noisy data and unspecified data as much as possible, the system uses data pre-processing methods based on mathematical and statistical theories to improve the quality of the selected data and thus provide valuable information for garment companies. On this basis, the system combines big data association rules and temporal pattern mining methods to analyze the pre-processed data. Under the premise of fully analyzing the production scheduling state, the parameter clustering of the garment production process is clarified, and the prediction activity of the system production scheduling performance is carried out with the help of the system production scheduling prediction method. And by comparing the system regulation objectives, static and dynamic errors are determined, and the system errors are used as inputs to the fitness function, so that the steady-state errors of the system production scheduling process can be eliminated, thus completing the production scheduling performance optimization process.

4   Implementation of Production Line Control in Digital Twin

During workshop operation, production resources change dynamically. In order to form a visual real-time monitoring that can cover the whole life cycle of workshop manufacturing, the system establishes a three-tier mapping system based on twin data-driven virtual workshops in terms of logistics, equipment and products to accurately describe the dynamic behavior of the workshop. For the logistics level, the real-time state data of the workshop manufacturing process is transformed into the corresponding workshop events to drive the flow of products between different work stations to achieve logistics mapping; for the equipment level, the mapping of equipment is the smallest mapping unit of the three-layer mapping system of the virtual workshop, which realizes the real-time sensing of physical equipment actions through IoT technology, sensor technology and interface technology, and drives the virtual model based on twin data to realize the virtual synchronization of physical equipment in the virtual environment; for product level, product mapping is based on the realization of equipment level and logistics level mapping, converting product process flow, real-time location and equipment real-time status into workshop events, realizing dynamic change of product model based on event-driven approach, and finally realizing real-time mapping of product status.

As shown in Fig. 3, the realized production line control system is based on the data acquisition module and the data of the line card machine, and the signal control function of the acquisition module is used to monitor the operation of the sewing equipment on the U-shaped production line, the site operator, and the shelf on the U-shaped production line in real time. Users can grasp real-time information on the status and efficiency of personnel, sewing equipment and processes in the production process. Based on the core information of the production line such as production tempo and working hours data, sewing equipment and personnel process completion data, managers can carry out automatic payroll accounting, personnel performance evaluation, personnel skills database, production line efficiency analysis, sewing equipment failure management and prediction and warning business applications, thus helping managers to understand the weak points of the production line. Develop a reasonable improvement mechanism for weak links to improve the efficiency of the production line and save production costs.

Fig. 3. Digital Twin-based production line control system

5 Conclusion

In this study, a Digital Twin-based real-time simulation modeling method of production environment is proposed for the visualization and monitoring of textile and garment workshop production, and a big data-based analysis and optimization method of production line control effectiveness is proposed for the rational utilization of on-site production data, and a Digital Twin-based production line control system is developed. The system comprehensively reflects the whole picture of the entire sewing production line, so that managers can master the overall production situation of the workshop without entering the site, effectively improving the efficiency of garment production, the system realizes the enterprise's refined production management, and further reduces production costs while improving the quality and production efficiency of textile and garments.

Acknowledgments. The study is supported by Natural Science Foundation of Fujian Provincial Science and Technology Department (No. 2021H6037, No. 2020HZ02014), Key Projects of Fujian Provincial Education Department (No. JAT190509), 2020 Fujian Provincial new engineering research and reform practice project (No. 33), Quanzhou Science and Technology Project (No. 2021C0008R, No. 2021GZ1), 2018 Fujian provincial undergraduate teaching team project, Fujian Provincial Big Data Research Institute of Intelligent Manufacturing.

References

1. Zhang, X.: Design and Implementation of Workshop Management and Control System Based on Digital Twins. Zhengzhou University (2018)
2. Fei, T., Ying, C., Lida, X., et al.: CCIot-CMfg: cloud computing and internet of things of things-based cloud manufacturing service system. IEEE Trans. Ind. Inf. **10**(2), 1435–1442 (2014)
3. Grieves, M.: Virtually Perfect: Driving Innovative and Lean Products through Product Life cycle Management. Space Coast Press, Florida (2011)
4. Grieves, M.: Digital Twin: Manufacturing Excellence through Virtual Factory Replication, vol. 4, pp. 1–7. Florida Institute of Technology (2015)
5. Tao, F., Cheng, J., Qi, Q., Zhang, M., Zhang, H., Sui, F.: Digital twin-driven product design, manufacturing and service with big data. Int. J. Adv. Manuf. Technol. **94**(9–12), 3563–3576 (2017). https://doi.org/10.1007/s00170-017-0233-1
6. Kaszubowski Lopes, Y., Ubertino Rosso, R., Leal, A.B., et al.: Finite automata as an information model for MES and supervisory control integration. Inf. Control Probl. Manuf. **14**(1), 212–217 (2012)
7. Chao, L., Li, Q.: A Unified Decision Model for Evaluation and Selection of MES Software. **207**(6), 691–696 (2006)
8. Feng, S.C., Zhang, Y.: A Manufacturing Planning and Execution Software Integration Architecture. Globalization of Manufacturing in the Digital Communications Era of the 21st Century, pp. 363–373. Springer, US (1998)
9. Bicer, T., Wei, J., Agrawal, G.: Supporting Fault Tolerance in a Data-Intensive Computing Middleware, pp. 1–12. IEEE (2010)
10. Sandholm, T., Lai, K.: Map-reduce optimization using regulated dynamic prioritization. ACM Sigmetrics Perform. Eval. Rev. **37**(1), 299–310 (2009)

Design of FPGA Circuit for SHA-3 Encryption Algorithm

Yuxiang Zhang$^{(\boxtimes)}$, Wenjiong Fu, and Lidong Xing

Xi'an University of Posts and Telecommunications, 618 West Chang'an Street, Chang'an District, Xi'an, China

{zhangyuxiang,wjfu}@stu.xupt.edu.cn, zmy_xld@163.com

Abstract. In the age of Internet information, how to ensure the information security of users has gradually become a key subject of research. Cryptography is an important foundation for the research and protection of information security, and the proposal and design of cryptographic algorithms cannot be separated from its theoretical support. This paper studies the principle and encryption process of the SHA-3 encryption algorithm and uses the Verilog hardware programming language to realize the circuit design of the algorithm. The Virtex-5 chip of Xilinx is selected to conduct the synthesis and layout of the algorithm. The algorithm function was simulated and verified using ISE software, and the simulation result was compared with the encryption result of the official algorithm integrated in the Python standard library, so that the algorithm function was correct. In terms of performance, through the comparison with references, the hardware performance of this paper is significantly improved.

Keywords: Encryption algorithm · SHA-3 · Hash function · FPGA · Verilog

1 Introduction

In information security, cryptography plays a pivotal role, and cryptographic algorithms are the core of cryptographic technology. According to functions, cryptographic algorithms can be divided into two categories: symmetric cryptographic algorithms and asymmetric cryptographic algorithms. As an important branch of cryptography, the hash function is an irreversible one-way cryptosystem that converts any length of message input into a fixed-length digest output. The secure Hash algorithm is a family of passwords based on hash functions. In 2012, the SHA-3 encryption algorithm won a solicitation contest organized by NIST (National Institute of Standards and Technology) and became a new generation of secure hash standards.

2 The Principle of SHA-3 Encryption Algorithm

2.1 The Sponge Structure of SHA-3

According to reference, the SHA-3 encryption algorithm is a secure hash algorithm using a sealed sponge structure, which consists of the replacement function Keccak-f

N. Xiong et al. (Eds.): ICNC-FSKD 2022, LNDECT 153, pp. 1471–1478, 2023.
https://doi.org/10.1007/978-3-031-20738-9_159

and a specific filling function. As shown in Fig. 1, P is the user Input message, Z is the output message digest after hash operation, f is the permutation function used for Hash operation, r is the bit rate, which is the length of the message packet, and c is the capacity.

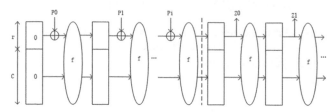

Fig. 1. Sponge structure

Before the message is input into the sponge structure, the input message must first be filled. After the packet length r is determined (r generally takes 576 bits), the input message is filled backwards so that the length of the filled message can be divided by r. According to references [1], references [2], the method of message filling is as follows: Fill the message with a "1", then fill in several "0"s, and finally fill in a "1".

The sponge structure has two processes of "absorption" and "squeezing". After the filled message is calculated in the structure, a message summary corresponding to the input message will eventually be produced.

According to reference, the calculation steps of the "absorption process", that is, the input process of the message are mainly as follows:

- After filling the input message P, divide P into i message groups according to the length of r bits in each group.
- Perform an XOR operation on the message packet P1 and the r bits of the internal initial state, and use the result of the operation as the input of the permutation function f.
- Perform an XOR operation on the r bits of the output information of the permutation function f in step 2 and the message packet P2, and use the result of the operation as the input of the permutation function f.
- Repeat the above steps iteratively, until all the message groups have been input and the permutation function operation has been output, and s is obtained.

According to reference [3], when the length l of the output message digest is less than or equal to the message packet length r, the first l bits of s are directly taken as the encrypted output result; otherwise, s is subjected to the "squeeze process" operation.

The calculation steps of the "extrusion process" are like the "absorption process", as described in the literature, the main steps are as follows:

- Take the first r bit length data of the final output data s of the "absorption process" as the output packet Z1, and input Z1 into the permutation function f.
- Take the r bits of the output of the permutation function f in step 1 as the output group Z2 and input them into the permutation function f again.

- Repeat the above steps several times.
- When the output length is the required output length, the operation is stopped, and the message summary is output.

2.2 SHA-3 State Structure

After the input message is filled, processed, and grouped, it enters the State structure, and then the message grouping in the State is output after calculation in the permutation function f. The State structure can be regarded as a three-dimensional array matrix of the follow formula represented by state [5] [5] [ω] (Fig. 2).

$$5 \times 5 \times \omega$$

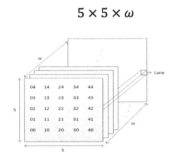

Fig. 2. State structure of SHA-3

The State structure is a three-dimensional matrix, which has three directions: X axis, Y axis, and Z axis. If the size of the matrix is b bits, the structure of State is $5 \times 5 \times b/25$.

2.3 Principle of Permutation Function F

It can be concluded from the literature [1] that the replacement function f is to iteratively operate the elements in the State structure.

The permutation function f is a core part of the entire SHA- 3 encryption algorithm. There are a total of five calculation steps for the permutation function f. The five calculation steps are: θ operation, ρ operation, π operation, χ operation, and τ operation. After five-step replacement operation, an intermediate chain value is generated as the input value of the next round of operation, and finally after 24 rounds of compression. The θ operation is to add the elements on each "column" in different positions through an exclusive OR operation, and then perform an exclusive OR with the target to be replaced, and the result of the operation will cover the replacement target.

The ρ operation is to operate the "vertical plane" in the State structure. Move the elements in each row of the "column" along the Z axis to move the bit number.

The π operation is to operate on the "horizontal" in the State structure. On a "horizontal side" of the State structure, the elements are moved by bits.

The χ operation performs AND, NOT, and XOR operations between elements on the "row" and overwrites the original elements with the results of the operations.

The basic process of the τ operation is to perform the exclusive OR operation on the first "path" element of the χ operation result, that is, state [0][0][ω] and the wheel constant.

3   Hardware Circuit Design of Sha-3 Algorithm

3.1   The Overall Circuit Design

The overall circuit module first transfers the input message to the filling module to fill the message. After the input plaintext message is filled, it is transposed with 8 bits per byte as a unit, that is, the pre-filled message is preprocessed. This step of processing is carried out in the overall circuit module.

After that, the filled and preprocessed plaintext message is then passed to the round function module to perform 24 rounds of five-step iterative calculations to obtain the calculated results. Before outputting, it is also necessary to perform output preprocessing operations in the overall circuit module, which is like filling preprocessing operations.

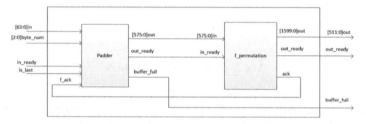

Fig. 3. Block diagram of the overall circuit module

The overall circuit module includes two sub-modules, the filling module and the round function module, and the connections inside the overall circuit module are shown in Fig. 3. The function of the filling module is to fill the input plaintext message according to the SHA-3 filling rule, and the function of the round function module is to perform multiple rounds of hash calculation on the filled plaintext message and finally output the result of the calculation.

3.2   Padder Module Design

This module is responsible for filling the input 64-bit plaintext message, and finally output the filled message.

The padding method used by this module is the multiple bit rate padding specified by the SHA-3 encryption algorithm. The 64-bit input message is filled with "1" and "1" at the end, and several "0"s are filled in the middle to make the length as the specified fill length. After the output filled message is processed in the overall circuit module, it will be passed to the round function module for hash encryption.

In this article, the length of the data after padding is 576 bits.

3.3 Round Function Module Design

The round function module is the core module in the entire circuit design. This module is responsible for performing 24 rounds of five-step iterative hashing on the input message after filling and preprocessing and outputting the result of the operation. This module contains two sub-modules: wheel constant module and wheel calculation module. The wheel constant module outputs the corresponding wheel constant value required for the round operation, and the round operation module completes the five-step iterative operation (Figs. 4 and 5).

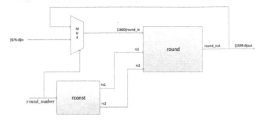

Fig. 4. Round function module circuit diagram

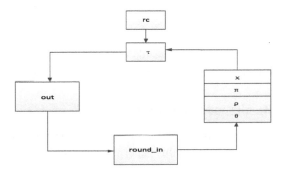

Fig. 5. Round operation module design

The round in is the input operation data. After the input operation data is calculated in the four steps of θ operation, ρ operation, π operation, and χ operation, the result of the χ operation and the round constant are subjected to τ operation, and the operation result is the output data of this round of operation. Out, and continue to use the output data as the input operation data to perform the next round of five-step operations, thereby realizing multiple rounds of operations.

4 Simulation Verification

4.1 Algorithm Function Verification

This article will give 30 sets of random decimal input data generated in Python and compare the simulation results with the encryption results of the official SHA-3 algorithm integrated with the Python library to verify the correctness of the algorithm function.

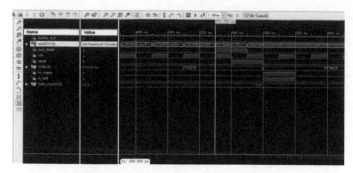

Fig. 6. Partial data simulation waveform

```
Console
This is a Full version of ISim.
Time resolution is 1 ps
Simulator is doing circuit initialization process.
Finished circuit initialization process.
ISim>
# run all
Good!
Stopped at time : 13300 ns : File "D:/Verilog/SHA-3/sha3/test_keccak.v" Line 322
ISim>
```
Console   Compilation Log   ● Breakpoints   Find in Files Results   Search Results

Fig. 7. Console output comparison success information

According to the comparison results of the test files, as shown in Fig. 6, it can be concluded that the two calculation results are completely consistent, and thus the conclusion that the function of the algorithm design in this paper is correct (Fig. 7).

4.2 Performance Results and Comparison

According to the timing analysis report, the highest operating frequency of the SHA-3 encryption algorithm implemented in this paper is 241.470 MHz (Fig. 8).

```
Timing Summary:
---------------
Speed Grade: -2

    Minimum period: 4.141ns (Maximum Frequency: 241.470MHz)
    Minimum input arrival time before clock: 1.338ns
    Maximum output required time after clock: 0.807ns
    Maximum combinational path delay: No path found
```

Fig. 8. Timing analysis report

Table 1 reports the implementation results of the proposed architecture, as well as a comparison with existing architectures. The comparison index is frequency performance.

As can be seen from the table, compared with the reference, the performance of the algorithm in this paper is significantly improved based on the same FPGA chip selection.

Table 1. Implementation results and comparisons

Reference	FPGA	MHz
[2]	V5	159
[4]	V5	189
[5]	V5	84.21
[6]	V5	118
Prop.	V5	241.47

5   Conclusion and Outlook

This article introduces the birth of the SHA-3 encryption algorithm, consults relevant literature materials to study and analyze its design ideas and encryption principles, proposes a hardware circuit design plan for the algorithm, and implements the encryption algorithm using a hardware description language. After ISE conducts comprehensive simulation, the encryption function of the algorithm is normal. Through the timing analysis report, we can see that the highest operating frequency of the algorithm is 241.470MHz, the hardware performance of this paper is significantly improved.

Keccak encryption algorithm, as the winner of Sha-3 encryption algorithm competition, has received extensive attention from the cryptography and industry, and its analysis and research are also increasing. With the continuous advancement of people's analysis and research and the continuous development of cryptography, Sha-3 encryption algorithm will be gradually optimized by technical experts, and its performance will be gradually improved. This algorithm will be applied to all aspects of our life. At the same time, as people pay more and more attention to the protection of their own information security, encryption algorithms will gradually penetrate our daily life to always protect our information security.

References

1. Tian, X.: FPGA-based SHA-3 algorithm hardware implementation optimization and system design. Xidian University, (2019). Elissa, K.: Title of paper if known, unpublished
2. Jungk, B.: Evaluation of compact FPGA implementations for all SHA-3 finalists. In: Third SHA-3 Candidate Conference (March 2012)
3. Ding, D.: Design and Implementation of Five Candidate Algorithms of SHA-3 Based on FPGA. Xidian University (2012)
4. Baldwin, B., Byrne, A., Lu, L., Hamilto, M., Hanle, N., O'Neill, M., Marnan, W.P.: FPGA implementations of the round two SHA-3 candidates. FPL **2010**, 400–407 (2010)
5. Gholipour, A., Mirzakuchaki, S.: High-speed implementation of the KECCAK Hash function on FPGA. Int. J. Adv. Comput. Sci. **2**(8), 303–307 (2012)
6. Strombergson, J.: Implementation of the Keccak Hash Function in FPGA Devices. http://www.strombergson.com/files/Keccak_in_FPGAs.pdf
7. Zhang, S., Sun, W., Zhang, J., Sun, X.: Applied Cryptography. Xidian University Press, Xi'an (2009)

8. Wang, W.: Design and Implementation of SHA-3 Algorithm Based on FPGA. PLA Information Engineering University (2017)
9. Liang, H.: Research on the Hardware Implementation of SHA-3 Hash Algorithm. Tsinghua University (2011)
10. Wang, H.: Security Analysis and Implementation of SHA-3 Standard Keccak Algorithm. Xidian University (2017)
11. Li, J.: Keccak Algorithm Hardware Implementation and Optimization. Ningbo University (2015)
12. Yang, Y., Li, Z.: FPGA Implementation of Typical Cryptographic Algorithm. Publishing House of Electronics Industry, Beijing (2017)

L-Band Broadband Dual-Polarized Dipole Phased Array

Jin Wu$^{(\boxtimes)}$, Ruiqing Guo, Yu Wang, and Heng Wen

School of Electronic Engineering, Xi'an University of Posts and Telecommunications, Xi'an 710121, China

lifewujin@xupt.edu.cn, {grq15935119823,wyf9722, w18710336696}@stu.xupt.edu.cn

Abstract. L-band radars are primarily land-based and shipborne systems for long-range military and air traffic control search applications up to 500 km. Compared with the traditional mechanical scanning radar, the phased array radar can realize the rapid scanning in a large space and find the enemy military targets in time. A wideband ±45° dual-polarized square-loop crossed dipole antenna element is designed and the operating band covers the whole L-band in this paper. Firstly, the characteristics of the element is studied, and the element has an isolation of higher than 25 dB between two orthogonal polarization. Finally, the scanning ability of the 8-element linear array in the H-plane is simulated. Seeing from the simulated results, the common bandwidth of the 8-element linear array is around 66% (active VSWR < 3) with the scanning ability of ±30° in the H-plane.

Keywords: Broadband antenna · Dual-polarization · Crossed-dipole · Wide-angle scanning · Phased array antennas

1 Introduction

Increasing the bandwidth and scanning range of phased array antennas is critical as radar and satellite systems increasingly demand dual polarization, large detection areas, and wide operation bands.

In recent years, many wide-band dual-polarized antennas have been designed for base station [1]. Dual-polarized antennas can be probably divided into two categories: patch antennas [2, 3] and crossed-dipole antennas [4, 5]. To achieve broadband dual polarization, patch antennas are generally multi-layered structures [6], so the structure of patch antennas is very complex. The crossed-dipole antennas studied in [7, 8], although these antennas achieve broadband and dual polarization, the operating frequency band does not cover the entire L-band.

In [8], a novel cast metal crossed dipole was adopted as the array element, and in order to obtain a larger gain, special metal walls around the dipole and the rear plane mirror cdge of the dipole were introduced into the design. The vertical wall of, and finally formed a 12-element linear array, but the maximum gain is only 16.3 dBi.

In this paper, a broadband ±45° dual-polarized square-loop dipole antenna element is designed, and the working frequency band covers the entire L-band, which solves the problem that the previous broadband dual-polarized antenna frequency cannot cover the entire L-band. The characteristics of the array elements are first studied to obtain the best performance, and then the scanning capability of the 8-element linear array in the H plane is studied, and the maximum gain reaches 17.01 dBi, which is an improvement over the antenna proposed in [8]. Moreover, a Wide-Angle Impedance Matching (WAIM) layer based on the concept of tight coupling is adopted, which can achieve a low active VSWR when the array is scanning at a large angle in a wide bandwidth.

2   Design and Simulation of Broadband Dual-Polarized Antenna Element

The simulation model of the element is shown in Fig. 1. This design uses electromagnetic simulation software HFSS for modeling, simulating and optimization. The proposed antenna element is mainly composed of a metal cavity, two crossed dipoles, a pair of coaxial feeders and two pairs of baluns. As shown in Fig. 1a, two crossed dipoles are placed orthogonally, resulting in dual-polarized radiation characteristics, and the parasitic patch acts as an electromagnetic wave guide. Parasitic patches can improve impedance matching performance. In [3], two directors are added to the printed quasi-Yagi antenna to increase the impedance bandwidth. The radiating and parasitic patches are located in different locations, introducing multiple resonance points to increase the bandwidth. Two pairs of baluns act as quarter impedance transformers and support structures. The metal cavity effectively improves the directivity of the radiation and makes the pattern performance of the element better. Compared with the previous square ring dipole, one of the advantages of this antenna is that parasitic small stubs are added to the square ring radiator, and the impedance bandwidth of the antenna can be further improved because four parasitic small stubs are introduced. Furthermore, when one dipole is excited, the other dipole can act as a parasitic element, which further extends its impedance bandwidth. The wide-angle impedance matching layer of the antenna is made of FR4 ($\varepsilon_r = 4.4$), which can improve the impedance matching performance when scanning at large angles.

Therefore, it can be concluded that since the two crossed dipoles are almost identical and only the position of the coaxial feeder is different, highly symmetric performance can be obtained at the two ports.

Since the performance of a single antenna and an array antenna are different, the antenna is simulated and optimized under the condition of a two-dimensional periodic boundary (simulating an infinite antenna array), the x- and y-axis directions are the periodic boundary, and the z-axis direction is the radiation boundaries, as shown in Fig. 2. After many simulation optimizations, the better performance of the element in the two-dimensional infinite array is finally obtained.

Figure 3 shows the simulation results of the active VSWR when the element is not scanning in a two-dimensional infinite array. It can be seen from the curve that the active VSWR of the dual ports are all below 3 in the whole L band.

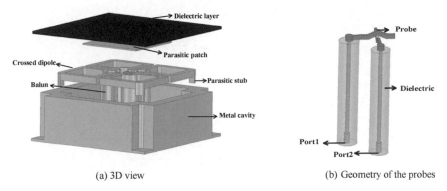

(a) 3D view   (b) Geometry of the probes

Fig. 1. Geometry of the proposed element

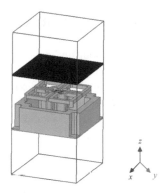

Fig. 2. Simulation model of element in 2D periodic boundaries

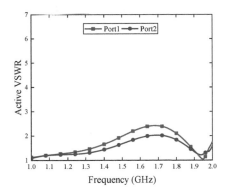

Fig. 3. Simulated active VSWR of the element when no scanning

The simulation curve of the dual-port isolation of the element in the two-dimensional periodic boundary is shown in Fig. 4. It can be seen from the figure that the isolation of the dual-port is greater than 25 dB in the whole L band.

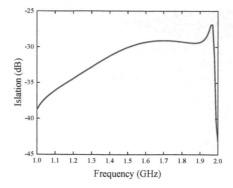

Fig. 4. Dual polarization isolation of element in periodic boundaries without scanning

3 Eight-Element Dual-Polarized Linear Array

After designing and optimizing the element under infinite array conditions using the period boundary, the scanning performance of the finite array needs to be investigated. In order to simplify and generalize the analysis, this paper analyzes the 1 × 8 array. The simulation model of the array is shown in Fig. 5. The array is infinite in the x-axis direction, set as the periodic boundary, and the y- and z-axis directions are set as the radiation boundary.

Fig. 5. Simulation model of 8-element array

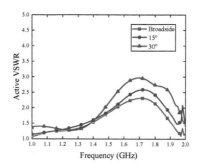

Fig. 6. Simulated active VSWR of the central element when scanning in the H plane

One of the ports of the central element scanning the active VSWR curve in the H plane, as shown in Fig. 6. It can be seen from the figure that as the scanning angle increases, the active VSWR becomes larger and is below 3 in the whole L band.

Figure 7 shows the radiation pattern performance of the array scanning at 1, 1.5 and 2 GHz. It can be seen from Fig. 7a that when the array is scanning at 1 GHz, the maximum gain is 11.06 dBi, and the maximum sidelobe level is −11.85 dB. It can be seen from Fig. 7b that when the array is scanning at 1.5 GHz, the maximum gain is 14.50 dBi, and the maximum sidelobe level is −11.97 dB. It can be seen from Fig. 7c that when the array is scanning at 2 GHz, the maximum gain is 17.01 dBi, and the maximum sidelobe level is −11.65 dB.

4 Conclusion

A dipole antenna for ±45° dual polarization wide-band wide-angle scanning array antenna is designed, and its operating band is the whole L-band. The dipole is surrounded by a metal cavity, and a parasitic patch is added on top of the dipole, which effectively improves the impedance bandwidth, and the relative bandwidth reaches 66%. The simulation results show that the dual-port isolation of the element is greater than 25 dB in the entire L band, and the active VSWR under the two-dimensional periodic boundary condition is below 3 in the whole L band. Then the element is formed into an 8-element linear array to scan in the H plane. The scanning range is ±30°, the active VSWR of all ports is below 3, the gain range is 11.06–17.01 dBi, and the maximum side lobe level is −11.65 dB. Due to the limited scientific research level of the author, the scanning range of one-dimensional linear array designed in this paper is only ±30°. In the follow-up work, how to further expand the scanning range is an important research topic.

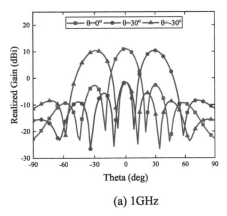

(a) 1GHz

Fig. 7. Simulated radiation patterns of the array scanning in the H plane at different frequencies

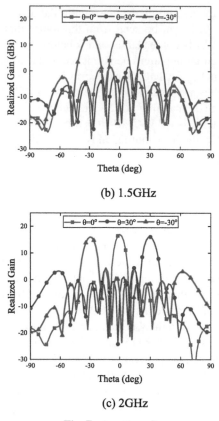

(b) 1.5GHz

(c) 2GHz

Fig. 7. (*continued*)

Acknowledgments. Supported by Shaanxi Province Key Research and Development Project (2021GY-280); Shaanxi Province Natural Science Basic Research Program Project (2021JM-459).

References

1. Deng, W.-Q., Yang, X.-S., Shen, C.-S., et al.: A dual-polarized pattern reconfigurable Yagi patch antenna for micro base stations. IEEE Trans. Antennas Propag. **65**(10), 5095–5102 (2017)
2. Mak, K.M., Lai, H.W., Luk, K.M.: A 5G wideband patch antenna with antisymmetric L-shaped probe feeds. IEEE Trans Antennas Propag. **66**(2), 957–961 (2018)
3. Wu, J., Zhao, Z., Nie, Z., et al.: Bandwidth enhancement of a planar printed quasi-Yagi antenna with size reduction. IEEE Trans. Antennas Propag. **62**(1), 463–467 (2014)
4. Tang, Z., Liu, J., Cai, Y.-M., et al.: A wideband differentially fed dual-polarized stacked patch antenna with tuned slot excitations. IEEE Trans. Antennas Propag. **66**(4), 2055–2060 (2018)
5. Yang, K.-W., Zhang, F.-S., Li, C.: Design of a novel wideband printed dipole array antenna. In: 2018 Cross Strait Quad-Regional Radio Science and Wireless Technology Conference (CSQRWC), pp. 1–3 (2018)
6. Li, M., Li, Q.L., Wang, B., et al.: A low profile dual-polarized dipole antenna using wideband AMC reflector. IEEE Trans. Antennas Propag. **66**(5), 2610–2615 (2018)

7. Wen, D.-L., Zheng, D.Z., Chu, Q.X.: A wideband differentially fed dual-polarized antenna with stable radiation pattern for base stations. IEEE Trans. Antennas Propag. 65(5), 2248–2255 (2017)
8. Rudakov, V.A., Li, Z., Sledkov, V.A., et al.: Dual-polarized dipole array with controlled beam tilt and wide radiation pattern for multi-beam antenna of base stations. In: 2021 Radiation and Scattering of Electromagnetic Waves (RSEMW), pp. 171–174 (2021)

Design and FPGA Implementation of Numerically Controlled Oscillators with ROM Look-Up Table Structure

Han Zhang⬤, Haisheng Huang(✉) ⬤, and Xin Li⬤

Xi'an University of Posts and Telecommunications, Xi'an 710121, China
{hhs,lixin}@xupt.edu.cn

Abstract. A design of numerically controlled oscillators based on Read-Only Memory locate form schematic is proposed. The working principle of NCO numerical control oscillator is introduced, and its implementation method in FPGA is analyzed in detail. Using Quartus II software to write Verilog HDL program, the NCO design is completed, and Modelsim software is called to simulate it. Finally, the performance index of NC oscillator is simply analyzed, the key parameters such as signal-to-noise ratio and spurious degree are tested, and its performance is simulated by MATLAB software. The results show that the design scheme is feasible.

Keywords: Numerically controlled oscillator · ROM look-up table structure · Field programmable gate array

1 Introduction

NCO (numerically controlled oscillator) is a digitally controlled oscillator used to generate an ideal and digitally controllable sine or cosine wave. Its implementation methods include real-time calculation method, look-up table method and so on. The sine wave sample of real-time calculation method is generated by real-time calculation. Because its calculation takes a lot of time, this method can only produce sine waves with lower frequency, and there is a contradiction between calculation accuracy and calculation time. When high-speed quadrature signal is needed, the real-time calculation method will not be realized. Therefore, the most effective and simple look-up table method is generally used in practical application. That is, first calculate the sinusoidal value of the phase according to the phase of each NCO sine wave, and store the sinusoidal value data of the phase according to the phase as the address information [1].

2 NCO Implementation Principle

NCO is mainly composed of phase accumulator and waveform generator. Its schematic is shown in Fig. 1:

© The Author(s), under exclusive license to Springer Nature Switzerland AG 2023
N. Xiong et al. (Eds.): ICNC-FSKD 2022, LNDECT 153, pp. 1486–1493, 2023.
https://doi.org/10.1007/978-3-031-20738-9_161

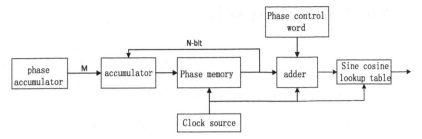

Fig. 1. NCO schematic

Its function is to convert the sum of the offset of digital LO frequency and LO frequency into phase. When a working clock arrives, the adder adds the value stored in the last cycle register to the first important parameter, then to the second key variable. At this time, the output of the adder is used as the input of the waveform generator. When the second key variable is constant, the first important case is the phase increment in one cycle; when the register overflows, the phase accumulator is cleared to indicate the completion of a cycle action.

The waveform generator is mainly used to calculate the sine and cosine values of different phases. By using the look-up table, it stores the sine and cosine signals of one cycle in the register according to the phase angle. When the steps of the input address are different, the frequency of the output sine and cosine digital signals will be different, so that the sine and cosine waveforms with different frequencies can manufactured.

Phase accumulator accumulates with its feedback value, and the high L bit of the result is used as the space of the Read-Only Memory locate form, Then read out the corresponding amplitude value from Read-Only Memory locate form to output sine and cosine waveform. If the working clock of the system is f_{clk} (sampling frequency), the frequency resolution of the sine and cosine signal generated by the NCO is:

$$\Delta f = \frac{f_{clk}}{2^N} \tag{1}$$

When the minimum n is 1 and the sampling frequency is $\frac{f_{clk}}{2}$. The phase accumulator with n bit width can divide the system clock by 2 times, and the first important parameter is. For the system clock, the phase offset increment of each clock cycle is:

$$\Delta\theta = \frac{F_{cw}2\pi}{2^N} \tag{2}$$

It is deduced from the above formula that the output frequency when the frequency control word F_{cw} is:

$$f = \frac{1}{T} = \frac{\omega}{2\pi} = \frac{\Delta\theta/t}{2\pi} = \frac{\Delta\theta \cdot f_{clk}}{2\pi} = \frac{f_{clk} \cdot F_{cw}}{2^N} \tag{3}$$

Thus, we can change the frequency of the output signal by changing.

3   FPGA Implementation of NCO

This paper mainly adopts the method of Read-Only Memory locate form to realize the generation of waveform. The core of Read-Only Memory locate form is to store the waveform amplitude value of partial cycle in a ROM. The address of waveform amplitude value in Read-Only Memory locate form can be found by the phase sequence output by aspect builder, so as to realize the generation of waveform and the conversion from phase to amplitude [2].The design implementation diagram is shown in Fig. 2.

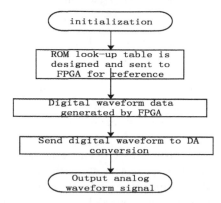

Fig. 2. NCO design and implementation block diagram structure

The implementation of NCO structure means that the preparation of Read-Only Memory locate form program and the generation of aspect builder are directly generated in FPGA. It also means that the waveform memory adopts ROM structure. In NCO, the digital phase information output by phase accumulator can be transformed into sine wave only after being searched and converted by ROM look-up table structure.

In ROM look-up table structure, the bit width of memory determines the accuracy of look-up table method, and also affects the performance of NCO and the accuracy of output waveform ROM look-up table structure is usually generated by MATLAB software simulation, generally in MIF format (Altera FPGA loading ROM files when loading MIF), so the realization of ROM look-up table method to generate MIF files is the primary operation [3].

The next step is to set the bit width and depth to generate the waveform storage file, assuming that the depth of ROM is set to 64 If the bit width of ROM is 8, it means that 64 points of sinusoidal signal cycle are quantized, and the quantization amplitude is 8 bits. The generated quantization diagram is shown in Fig. 3, and some MIF files are shown in Fig. 4.

Load the generated two MIF files into the two ROM IP cores in Quartus II. After loading the MIF file on FPGA, it means the completion of ROM look-up table structure. Using Verilog HDL language to write other NCO structure program, finally call the two ROM IP core. It is verified by Quartus II software calling Modelsim simulation software.

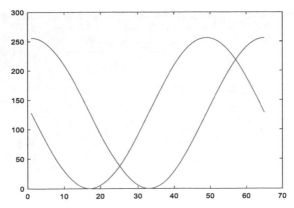

Fig. 3. Quantization diagram of 64 points in one cycle of sine cosine signal

```
WIDTH=8;
DEPTH=64;

ADDRESS_RADIX=UNS;
DATA_RADIX=HEX;

CONTENT BEGIN
      0        :        ff;
      1        :        ff;
      2        :        fe;
      3        :        fa;
      4        :        f6;
      5        :        f1;
      6        :        ea;
      7        :        e3;
      8        :        db;
      9        :        d1;
     10        :        c7;
     11        :        bc;
     12        :        b1;
     13        :        a5;
     14        :        99;
     15        :        8d;
     16        :        80;
     17        :        73;
     18        :        67;
     19        :        5b;
     20        :        4f;
```

Fig. 4. Partial MIF files

4   Simulation Results

In the waveform diagram shown in Fig. 5, RST is the system reset signal, CLK is the clock input frequency (50MHz), Phase accumulator consisting of 32 bits, the number of bits of the output signal amplitude is 8 bits (including symbol bits), the phase width is 6 bits, NCO_I and NCO_Q is the system output signal.

Waveform for the first case is generated as shown in the Fig. 6. The frequency is 0 39 MHz, the initial phase is offset by 0 degrees when the relative phase control word is 0.

Waveform for the second case is generated as shown in the Fig. 7. The frequency is 0.39MHz, the initial phase is offset by 180 degrees when the phase control word is 0.

Waveform for the third case is generated as shown in the Fig. 8. The frequency is 0.13MHz, the initial frequency is reduced by 0.5% when the relative the first parameter is 1 26MHz, the initial phase is offset by 0 degrees when the relative phase control word is 0.

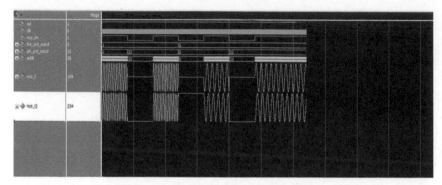

Fig. 5. Simulation waveform under Quartus II

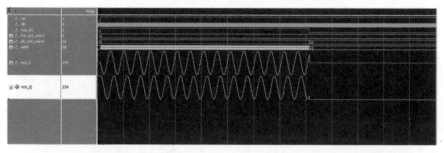

Fig. 6. Simulation waveform when the circuit main parameters change to 1 and 0

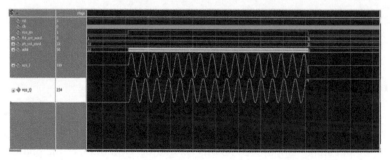

Fig. 7. Simulation waveform when the when the circuit main parameters change to 1and 32

Waveform for the fourth case is generated display the Fig. 9. The initial parameter is reduced by 0.26MHz when the initial frequency is relative to the first parameter is 1, and the initial phase is offset by 180 degrees when the initial phase relative to the phase control word is 0.

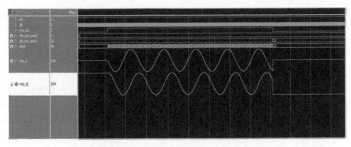

Fig. 8. Simulation waveform when the circuit main parameters change to 5 and 0

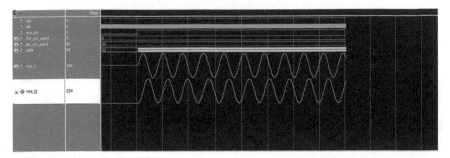

Fig. 9. Simulation waveform when the circuit main parameters change to 5 and 32

5 Error Analysis of NCO

Ideally, all the bits of the accumulator are output as the addressing bits of the address, so there will be no error. However, in practical application, it will be limited by ROM volume, cost and power consumption. As an addressed output, it is impossible to traverse all bits of the accumulator. In practical application, NCO ROM only intercepts part of the high-order output, which will introduce phase truncation error [4].

Since the output data of ROM is sinusoidal amplitude value, another error - amplitude quantization error is generated. According to Nyquist sampling theorem, the signal frequency generated by NCO cannot be greater than half of the clock frequency. In practical engineering, limited by the low-pass filter, the maximum output frequency must be smaller than 40% of the drive clock frequency. We should try to prevent spectrum repetition or all kinds of waveforms falling into the useful output frequency band to avoid interference, so we use truncation before output. The purpose of using phase truncation in hardware is to reduce the length of lookup table and reduce the amount of computation. The frequency resolution has been mentioned before, and its value is $\Delta f = f_{clk}/2^N$. It can be seen from the above formula that when the drive clock frequency is determined, the accumulator of the NCO determines the frequency resolution.

In practical engineering, the spurious of NCO is mainly caused by phase rounding error. In addition, the ROM stores the quantized value, which will also introduce some stray signals. The ratio of the main spectrum to the strongest spurious spectrum is:

$$\frac{S}{S_{spur}} \geq 6 * LdB \tag{4}$$

where, l is the addressing bits, and the relationship between the strongest spurious spectrum and the main spectrum depends on the accumulator bits and the actual interception bits. Signal to noise ratio (SNR) is used to reflect the influence of amplitude quantization error on spurious [5].

$$SNR \approx (6.02 * K + 1.76)dB \tag{5}$$

where k is the bit width of the sampling value and is a finite bit value. It shows that for each additional quantization bit, the signal-to-noise ratio will increase by about 6dB; In this design, l = k = 6. In practice, SNR formula is used to predict the signal-to-noise ratio of NCO in the worst case. The SNR and SFDR of the system are 40 dB and 54 dB respectively. The following figure is the simulation of spurious degree. The vertical axis is the amplitude, and the horizontal axis is the frequency.

ROM based look-up table structure has certain advantages, easy to use and easy to realize. It only needs to calculate the positive and residual values of the phase in advance and store the data in the table in advance. When using it, it can generate the corresponding sine and cosine waves one by one. It is the most effective and simplest method in practical application [6]. However, it also has some disadvantages. Its limitations are mainly reflected in the following aspects: first, it consumes a lot of hardware resources, For example, if 20 bit addressing data is used, the capacity of ROM table must reach at least the 20th power of 2. Therefore, it takes up a larger space, and if it is higher, it will take up a larger space, which limits the accuracy of addressing. Second, the amount of phase spurious is also large. In the ROM look-up table structure, people always hope that the higher the accuracy of the look-up table, the better. Therefore, they hope that the larger the capacity of the ROM, the better. In this way, the output of the phase accumulator can achieve better resolution and the higher the spectral purity of the signal. However, due to the limitation of the capacity of the ROM, it cannot be increased indefinitely. Third, the implementation accuracy is also low, due to the limitation of the number of bits of ROM look-up table structure and hardware memory [7]. The spurious degree simulation plot is shown in Fig. 10.

6 Conclusion

By analyzing the implementation principle and performance of NC oscillator, this paper presents a specific method to realize NCO based on FPGA ROM lookup table architecture. At the same time, the correctness of this design is verified by the simulation in Quartus II. The results show that the NCO designed by this method can output signals of various frequencies. It can control the output of different frequencies, the system sampling frequency, the depth of the table and the bit width of the storage table. It is very practical in the field of modulation signals in communication. At the same time, it can also reduce resource consumption. At the same time, this paper analyzes the

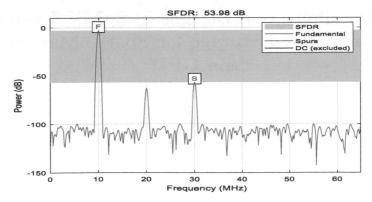

Fig. 10. Simulation diagram of clutter

shortcomings of this method. If the required precision and frequency are high, the output design requirements can be achieved by changing the algorithm or designing the practical cooperation of FPGA hardware.

References

1. Shi, J.: Design and Implementation of All-Digital Phase-Locked Loop based on Time-Average Frequency Direct Cycle Synthesis CNC Oscillator. Zhejiang University (2018)
2. Zhang, B., Song, S.: Design of CNC oscillator based on field programmable gate array. J. Xi'an Univ. Posts Telecommun. **4**, 72–75 (2017)
3. Kai, Y., Huang, H., Xing, Y.: Design of CNC oscillator circuit in GPS receiver. Inf. Technol. **9**, 58–61+66 (2017)
4. Xue, O., Zou, W.: ASIC design of NC oscillator using look-up table method. Electron. Packag. **8**, 13–15 (2012)
5. Nie, Q.: Design and implementation of direct digital frequency synthesizer based on CORDIC algorithm. Xi'an Univ. Electron. Sci. Technol. (2011)
6. Zhang, A., Zhao, P.: Design and implementation of orthogonal NC oscillator (NCO) based on FPGA. Electron. Des. Eng. 17, 149–152 (2011)
7. Li, F.: FPGA implementation of digitally controlled oscillator (NCO). Appl. Electron. Compon. **11**, 42–44 (2010)

Parallelization Designs of SpMV Using Compressed Storage for Sparse Matrices on GPU

Jianxin Wei[✉]

Information and Network Center, Hunan University of Science and Engineering,
Yongzhou, Hunan 425199, China
weijianxin@huse.edu.cn

Abstract. Sparse matrix-vector multiplication (SpMV) is one of the most basic operation in numerical and scientific computing and engineering. The efficiency of SpMV operation determines the performance of related practical applications. To improve the computational efficiency of SpMV operation, researchers need to develop accelerated programs on parallel computing platforms. Therefore, in this paper, we design the GPU-based parallelization of SpMV operation using different compressed storage formats for the input sparse matrix, including coordinate (COO) format, compressed sparse rows (CSR) format, and compressed sparse columns (CSC) format. The experimental results show that, on average, the parallel COO-, CSR-, and CSC-based SpMV algorithms achieve the speedup of 17.31, 21.61, and 19.40 over the serial SpMV algorithms, respectively.

Keywords: Compressed sparse matrix storage · GPU · Parallel · Sparse matrix-vector multiplication (SpMV)

1 Introduction

Sparse matrix-vector multiplication (SpMV) is one of the most basic operation in numerical and scientific computing and engineering. The formula of SpMV can be expressed as

$$y = A \times x, \tag{1}$$

where A represents an $M \times N$ input sparse matrix with nnz non-zeros, x represents an input dense vector with N elements, and y represents an output dense vector with M elements.

In the process of solving large-scale practical problems, such as graph computing [1], earthquake simulation [2], fluid dynamics [3], nuclear reactor simulation [4], etc., researchers usually use partial differential equations and discretize

© The Author(s), under exclusive license to Springer Nature Switzerland AG 2023
N. Xiong et al. (Eds.): ICNC-FSKD 2022, LNDECT 153, pp. 1494–1502, 2023.
https://doi.org/10.1007/978-3-031-20738-9_162

them with finite elements or finite difference methods to obtain linear equations [5]. The matrix of the linear system is very sparse, but the vectors are usually dense. For these systems of linear equations, which may have tens of billions of unknowns, iterative methods that consume less memory to obtain approximate solutions are more suitable for high-performance computing platforms than direct solutions. In the iterative method, the SpMV operation occupies most of the computation time. Therefore, the efficiency of SpMV operation determines the performance of related practical applications.

Improving the efficiency of SpMV operation has become an increasingly important problem. **Sparsity** and **large scale** are the two main challenges faced by the SpMV operation, and a series of problems in acceleration of SpMV are caused by them:

(1) **Sparsity**: One of the challenges is that different sparse matrices have differences in structure, extent of sparsity, etc., which places higher demands on load balancing in parallel algorithms. Due to the sparsity of the data, within the same sparse matrix or tensor, the number of non-zeros in each row or dimension may vary from row to dimension, which may cause irregular access to memory. Further, due to the irregular distribution of the non-zero elements of the input matrix or tensor, the size of the output matrix or tensor is difficult to determine, which causes certain problems for memory allocation.

(2) **Large scale**: Another challenges is that in real life, the scale of data related to matrices is very large, which puts forward higher requirements on the storage format of the data. Due to the limited cache size on the device, proper data partitioning is necessary. As the size of the data increases, the traditional single machine may have difficulty solving the real dilemma, so using a distributed platform is a good choice, but with it comes the problem of communication overhead.

To alleviate the above-mentioned challenges and improve the computational efficiency of SpMV operation, researchers need to develop accelerated programs on parallel computing platforms. Since it is difficult for a single-core processor to provide powerful computing capability, the current popular parallel computing platforms are equipped with multi-/many-core processors, such as multi-core central processing units (CPUs), general-purpose graphics processing units (GPGPUs), etc., or multiple computing nodes to improve computing power through parallel computing.

In addition, the storage format of the sparse matrix affects the memory access behaviors and paralleization of SpMV. Considering the parallelization of SpMV where the non-zeros in the input sparse matrix is stored by rows. We access the matrix in row order, and each thread processes the non-zero elements of the corresponding row in parallel. When the number of non-zero elements in the row differs greatly, it will cause serious thread load imbalance, resulting in performance degradation. From this we can understand that the optimizations of matrix compressed storage format, memory access behavior, and load balancing strategy have always been main points of SpMV algorithm research.

Therefore, in this paper, we design the GPU-based parallelization of SpMV operation using different compressed storage formats for the input sparse matrix, including coordinate (COO) format, compressed sparse rows (CSR) format, and compressed sparse columns (CSC) format. In addition, we analyze the performance of parallel SpMV based on different data structures of sparse matrices and various compressed storage formats.

2   Related Work

There has been a large number of researches that focus on optimizing compressed storage formats for sparse matrices. Ashari et al. [6] proposed the blocked row-column (BRC) storage format based on the CSR and CSC formats to overcome the thread divergence of parallel SpMV by blocking the rows of the matrix to processing units (warps in GPU). Zhang et al. [7] proposed the aligned CSR (ACSR) and aligned ELL (AELL) storage formats. Although compared to CSR and ELL formats, these two proposed formats need more padding elements and extra preprocessing overheads, the memory access patterns and performance of SpMV algorithm based on these two new formats can be optimized. Almasri and Abu-Sufah [8] designed the compress chunk (CCF) storage format form sparse matrices. Firstly, the computing tasks are assigned at the thread level according to the number of non-zero elements of the matrix, and each thread processes the same number of matrix sets. Then, matrix rows are added to the set until the number of non-zero elements exceeds a constant value. The third step is to sort the rows in each set according to the number of non-zero elements in each row, and rows with the same non-zero elements are collected into a bin. Finally, in each bin, the data is then grouped into chunks according to the width of vector units in cores. Besides, they used heuristics to optimize the parameters of matrix partitioning.

Quite a considerable number of researches investigate the parallelization designs of SpMV to utilize the parallel computing power of the underlying architectures. Xiao et al. [9–12] proposed an efficient method to speed up SpMV and optimize the irregular memory access behavior by performing multiplication and addition operations respectively. Fei and Zhang [13] proposed a method to dynamically adjust the parallel workloads of SpMV on GPU threads. The method uses the number of rows that each thread is assigned as a parameter for an empirical algorithm to adjust the workload for each thread. Hosseinabady and Nunez-Yanez [14] proposed an efficient streaming data engine for SpMV on FPGAs. This engine uses loop pipelining and data streaming techniques to enhance the utilization of memory bandwidth. In addition, the authors proposed an analysis model of SpMV algorithm overhead and computing platform bottleneck. Gómez et al. [15] proposed an efficient SpMV algorithm based on long vector architecture. Their implementation has fairly high utilization of the memory bandwidth of the architecture. This algorithm improves the utilization of the cache that stores the input vector x. When collecting intermediate result vectors with scatter instructions, they made instruction-level optimizations for memory access behavior, reducing unnecessary memory accesses.

3   Design

In this section, we introduce three GPU-based parallelization designs of SpMV using three compressed storage formats for sparse matrices.

3.1   GPU-Based Parallelization Design of SpMV using COO Format

COO Format. COO format is the simple compressed storage for sparse matrices. The COO format respectively stores the row id, column id, and value of each non-zero elements using three arrays.

Take the following 4×4 sparse matrix A as an example:

$$A = \begin{bmatrix} 0 & 1 & 2 & 0 \\ 3 & 0 & 0 & 0 \\ 0 & 4 & 0 & 0 \\ 0 & 0 & 0 & 5 \end{bmatrix}, \tag{2}$$

the COO format uses the array $Rows[5] = \{0, 0, 1, 2, 3\}$ stores the row ids of the 5 non-zeros in A, the array $Cols[5] = \{1, 2, 0, 1, 3\}$ stores the column ids of the 5 non-zeros in A, and the array $Vals[5] = \{1, 2, 3, 4, 5\}$ stores the values of the 5 non-zeros in A.

Parallelization Design of SpMV using COO Format. We design the GPU-based parallelization of SpMV using COO format, where nnz GPU threads are used (nnz is the number of non-zeros in the input sparse matrix A). In specific, the parallelization design includes three phases:

1. Load the three arrays $Rows[nnz]$, $Cols[nnz]$, and $Vals[nnz]$ of the input sparse matrix A and the input dense vector x from CPU host memory to GPU device memory;
2. Execute parallel SpMV on nnz GPU threads, where each GPU thread computes SpMV operation on a non-zero; Each multiplication result between a non-zero in A and the corresponding element in x is added to the corresponding element in y using atomic operation, since there exist parallel write collisions to the same element of y;
3. Return the output dense vector y from GPU device memory to CPU host memory.

3.2   GPU-Based Parallelization Design of SpMV using CSR Format

CSR Format. CSR is one of the most popular compressed storage format for sparse matrices. There are three arrays store the pointers to start and end positions of each matrix row, the column id of each non-zero, and the value of each non-zero, respectively.

Take the following 4×4 sparse matrix A shown in Eq. (2) as an example, the CSR format uses the array $PtoR[5] = \{0, 2, 3, 4, 5\}$ stores the positions of the first non-zero and end non-zero in each row in the total number of non-zeros of A, the array $Cols[5] = \{1, 2, 0, 1, 3\}$ stores the column ids of the 5 non-zeros in A by rows, and the array $Vals[5] = \{1, 2, 3, 4, 5\}$ stores the values of the 5 non-zeros in A by rows.

Parallelization Design of SpMV using CSR Format. We design the GPU-based parallelization of SpMV using CSR format, where M GPU threads are used (M is the number of rows in the input sparse matrix A). In specific, the parallelization design includes three phases:

1. Load the three arrays $PtoR[M + 1]$, $Cols[nnz]$, and $Vals[nnz]$ of the input sparse matrix A and the input dense vector x from CPU host memory to GPU device memory;
2. Execute parallel SpMV on M GPU threads, where each GPU thread computes SpMV operation on the non-zeros of one row in A; Each multiplication results between the non-zeros in a same row of A and the corresponding elements in x are added to the same element in y; In addition, in this parallelization design, there is no parallel write collisions, and atomic operations are not required;
3. Return the output dense vector y from GPU device memory to CPU host memory.

3.3   GPU-Based Parallelization Design of SpMV Using CSC Format

CSC Format. The CSC format stores the given sparse matrix by columns, and the methodology of CSC is similar with CSR. There are three CSC arrays store the pointers to start and end positions of each matrix column, the row id of each non-zero, and the value of each non-zero, respectively.

Take the following 4×4 sparse matrix A shown in Eq. (2) as an example, the CSC format uses the array $PtoC[5] = \{0, 1, 3, 4, 5\}$ stores the positions of the first non-zero and end non-zero in each column in the total number of non-zeros of A, the array $Rows[5] = \{1, 0, 2, 0, 3\}$ stores the row ids of the 5 non-zeros in A by columns, and the array $Vals[5] = \{3, 1, 4, 2, 5\}$ stores the values of the 5 non-zeros in A by columns.

Parallelization Design of SpMV using CSC Format. We design the GPU-based parallelization of SpMV using CSC format, where N GPU threads are used (N is the number of columns in the input sparse matrix A). In specific, the parallelization design includes three phases:

1. Load the three arrays $PtoC[N + 1]$, $Cols[nnz]$, and $Vals[nnz]$ of the input sparse matrix A and the input dense vector x from CPU host memory to GPU device memory;
2. Execute parallel SpMV on N GPU threads, where each GPU thread computes SpMV operation on the non-zeros of one column in A; The non-zeros in a same column of A will be multiplied with the same element in x, and the

multiplication results are further added to the corresponding element in y using atomic operation, since there exist parallel write collisions to the same element of y;

3. Return the output dense vector y from GPU device memory to CPU host memory.

4   Experimental Evaluation

4.1   Setup

We choose ten sparse matrices from SuitSparse Matrix Collection[1]. The details of the tested sparse matrices are shown in Table 1. The size of the selected sparse matrices is ranged from $21,200 \times 21,200$ to $206,500 \times 206,500$, and the number of non-zeros in the selected sparse matrices is ranged from $958,936$ to $4,344,765$.

Table 1. The selected sparse matrices.

Sparse matrices	The number of rows	The number of columns	The number of non-zeros
raefsky3	21,200	21,200	1,488,768
pdb1HYS	36,417	36,417	4,344,765
rma10	46,835	46,835	2,329,092
cant	62,451	62,451	4,007,383
2cubes_sphere	101,492	101,492	1,647,264
cop20k_A	121,192	121,192	2,624,331
cage12	130,228	130,228	2,032,536
144	144,649	144,649	2,148,786
scircuit	170,998	170,998	958,936
mac_econ_fwd500	206,500	206,500	1,273,389

We test our designed algorithms on an NVIDIA P100 GPU. In addition, the execution times are measured in million seconds.

4.2   Results

We tested the speedups of parallel SpMV algorithms respectively using the COO, CSR, and CSC formats over the serial SpMV algorithms. Figure 1 shows the speedups of parallel SpMV algorithms respectively using COO over the serial SpMV algorithm. On average, the parallel COO-based SpMV algorithm achieves the speedup of 17.31. Figure 2 shows the speedups of parallel SpMV algorithms respectively using CSR over the serial SpMV algorithm. On average, the parallel CSR-based SpMV algorithm achieves the speedup of 21.61. Figure 3 shows the

[1] https://sparse.tamu.edu/

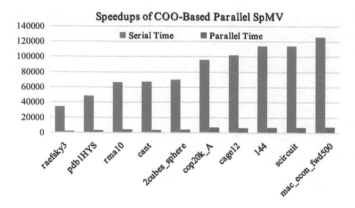

Fig. 1. Speedups of parallel SpMV algorithms respectively using COO over the serial SpMV algorithm.

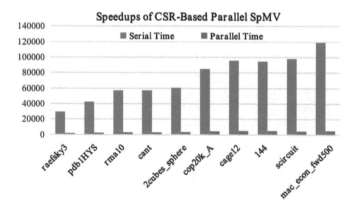

Fig. 2. Speedups of parallel SpMV algorithms respectively using CSR over the serial SpMV algorithm.

speedups of parallel SpMV algorithms respectively using CSC over the serial SpMV algorithm. On average, the parallel CSC-based SpMV algorithm achieves the speedup of 19.40.

As shown in Figs. 1, 2 and 3, we can see that the parallel algorithm using the CSR format runs the minimal execution times and obtains the highest speedups. This is because the parallel CSR-based SpMV algorithm avoids the parallel write collisions between GPU threads.

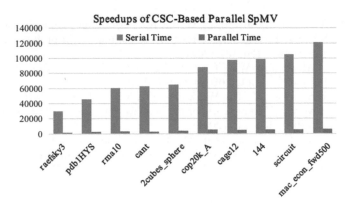

Fig. 3. Speedups of parallel SpMV algorithms respectively using CSC over the serial SpMV algorithm.

5   Conclusions

In this paper, we propose the parallelization designs of SpMV algorithms using three compressed storage formats, i.e., COO, CSR, and CSC, on GPU. The experimental results show that the parallel CSR-based SpMV algorithm can obtain the best performance.

Acknowledgement. The research was partially supported by the Program of Hunan Province (Grant No. 19C0817).

References

1. Buluç, A., Gilbert, J.R.: The combinatorial BLAS: design, implementation, and applications. Int. J. High Perform. Comput. Appl. **25**(4), 496–509 (2011)
2. Uphoff, C., Rettenberger, S., Bader, M., Madden, E.H., Ulrich, T., Wollherr, S., Gabriel, A.-A.: Extreme scale multi-physics simulations of the tsunamigenic 2004 Sumatra megathrust earthquake. In: Mohr, B., Raghavan, P. (eds.) Proceedings of the International Conference for High Performance Computing, Networking, Storage and Analysis, SC 2017, Denver, CO, USA, 12–17 Nov 2017, pp. 21:1–21:16. ACM (2017)
3. Resseguier, V., Picard, A.M., Mémin, É., Chapron, B.: Quantifying truncation-related uncertainties in unsteady fluid dynamics reduced order models. SIAM/ASA J. Uncertain. Quant. **9**(3), 1152–1183 (2021)
4. Ma, Y., Wang, Y., Yang, J.: ntkFoam: an OpenFOAM based neutron transport kinetics solver for nuclear reactor simulation. Comput. Math. Appl. **81**, 512–531 (2021)
5. Saad, Y.: Iterative Methods for Sparse Linear Systems. SIAM (2003)
6. Ashari, A., Sedaghati, N., Eisenlohr, J., Sadayappan, P.: A model-driven blocking strategy for load balanced sparse matrix-vector multiplication on GPUs. J. Parallel Distrib. Comput. **76**, 3–15 (2015)

7. Zhang, Y., Yang, W., Li, K., Tang, D., Li, K.: Performance analysis and optimization for SpMV based on aligned storage formats on an ARM processor. J. Parallel Distrib. Comput. **158**, 126–137 (2021)
8. Almasri, M., Abu-Sufah, W.: CCF: an efficient SpMV storage format for AVX512 platforms. Parallel Comput. **100**, 102710 (2020)
9. Xiao, G., Li, K., Chen, Y., He, W., Zomaya, A.Y., Li, T.: CASpMV: a customized and accelerative SpMV framework for the sunway TaihuLight. IEEE Trans. Parallel Distrib. Syst. **32**(1), 131–146 (2021)
10. Chen, Y., Xiao, G., Tamer Özsu, M., Liu, C., Zomaya, A.Y., Li, T.: aeSpTV: an adaptive and efficient framework for sparse tensor-vector product kernel on a high-performance computing platform. IEEE Trans. Parallel Distrib. Syst. **31**(10), 2329–2345 (2020)
11. Xiao, G., Chen, Y., Liu, C., Zhou, X.: ahSpMV: An autotuning hybrid computing scheme for SpMV on the sunway architecture. IEEE Internet Things J. **7**(3), 1736–1744 (2020)
12. Chen, Y., Li, K., Yang, W., Xiao, G., Xie, X., Li, T.: Performance-aware model for sparse matrix-matrix multiplication on the sunway TaihuLight supercomputer. IEEE Trans. Parallel Distrib. Syst. **30**(4), 923–938 (2019)
13. Fei, X., Zhang, Y.: Regu2D: accelerating vectorization of SpMV on intel processors through 2d-partitioning and regular arrangement. In: Sun, X.H., Shende, S., Kalé, L.V., Chen, Y. (eds.) ICPP 2021: 50th International Conference on Parallel Processing, Lemont, IL, USA, 9–12 Aug 2021, pp. 77:1–77:11. ACM (2021)
14. Hosseinabady, M., Luis Núñez-Yáñez, J.: A streaming dataflow engine for sparse matrix-vector multiplication using high-level synthesis. IEEE Trans. Comput. Aided Des. Integr. Circuits Syst. **39**(6), 1272–1285 (2020)
15. Gómez, C., Mantovani, F., Focht, E., Casas, M.: Efficiently running SpMV on long vector architectures. In: Lee, J., Petrank, E. (eds.) PPoPP'21: 26th ACM SIGPLAN Symposium on Principles and Practice of Parallel Programming, Virtual Event, Republic of Korea, 27 Feb 27–3 Mar 2021. pp. 292–303. ACM (2021)

Author Index

A

Aijun, Wen, 1045
Ananthakumar, Usha, 983
Antonic, Nemanja, 455
Auge, Daniel, 483
Aziz-Alaoui, M. A., 729

B

Bai, Wei, 1192
Bai, YuXuan, 1176
Banerjee, Parikshit, 983
Bertelle, Cyrille, 729
Bie, Wenxuan, 1056

C

Cai, Danlin, 1463
Cai, Huinan, 1423
Cai, Zhiwen, 1441
Cao, Chuang, 1363
Cao, Silei, 3
Cao, Yuwei, 739
Cappelli, Carmela, 913
Chai, Lei, 109
Chang, Chunguang, 344, 754
Chao, Zhou, 122
Chen, Chen, 210, 285, 1167
Chen, Dewang, 263, 527
Chen, Enhong, 1015
Chen, Gang, 1371
Chen, Hui, 519
Chen, Jing, 437
Chen, Lihui, 139
Chen, LiJun, 1176
Chen, Shuai, 773

Chen, Shu-Zhen, 1202
Chen, Sulin, 1002
Chen, TianRui, 939
Chen, Xiaodong, 1228, 1237
Chen, Xinqing, 519
Chen, Yan, 974
Chen, Yuxuan, 869
Chen, Zhanqi, 307
Chen, Zhimei, 386
Cheng, Xiaorui, 191, 210, 285, 1167
Cui, Yanli, 1304, 1329
Cui, Yingbao, 293, 663, 673

D

Dai, Libin, 905
Dang, Jing, 568
Dang, Siyao, 1282
Deng, Wanyu, 244, 693, 849
Ding, Ge, 377
Ding, Lijiao, 43
Ding, Pengcheng, 1024
Dong, Xiangjun, 819
Du, Huimin, 165, 682

F

Fan, Haichao, 210, 285, 1167
Fan, Lin, 361
Fan, Simeng, 353
Fan, Weinan, 648
Fan, Wenbin, 1015
Fang, Yekun, 1015
Fei, Junting, 1371
Feng, Feng, 1136
Feng, Jingyu, 1338

Feng, Xiaoshuo, 968
Feng, Zengyu, 939
Fu, Wenjiong, 1346, 1471

G
Gao, Jingyu, 1117
Gao, Shuai, 173
Gao, Xiwang, 147
Gao, Xue, 419, 428, 789
Gao, Yaqiong, 395
Gao, Ying, 621
Gao, Yue, 18
Gong, Mingliang, 798
Gong, Yunpeng, 993
Gu, Zhiyu, 947
Guan, Xinyi, 3
Guo, Kun, 1210
Guo, Ruiqing, 1479
Guo, Wenzhong, 1210
Guo, Xiaojun, 218

H
Halimu, Yeerjiang, 122
Hamila, Mohamed Elyes, 455
Han, Fei, 353
Han, Mingzhi, 584
Han, Siyi, 1423
Han, Wei, 139
Hao, Denghui, 921
Hao, Feng, 819
Hao, Liang, 781
He, Hao, 629, 1158
He, Hong, 947
He, Jie, 307
He, Ruikang, 69
Hei, Xinhong, 228, 543
Hongmin, Meng, 1045
Hou, Hongbo, 344
Hu, Xiaobing, 327, 1105
Huang, Benzun, 263, 527
Huang, Haisheng, 1282, 1486
Huang, Pu, 402
Huang, Shu, 377
Huang, Wenhua, 1338
Huang, Xiaoguang, 293, 663, 673
Huang, Yunhu, 263, 527
Huang, Zhiyong, 1067, 1228, 1237
Huang, Zihao, 1290

J
Jia, Qingxuan, 1371
Jia, Xining, 336
Jiang, Chunmeng, 447
Jiang, Huikai, 1146
Jiang, Rong, 1251, 1259

Jiang, Tao, 781
Jiang, Weijie, 1067, 1228, 1237
Jie, Jiahao, 244, 693
Jie, Wei, 1087
Jinbao, Teng, 11
Jing, Changqiang, 43
Jullapak, Rujira, 493

K
Kawahara, Daichi, 1125
Ke, Xianqun, 76
Khalid, Abdul Hanan, 455
Kita, Kenji, 1125
Knoll, Alois, 483
Kong, Weiwei, 656

L
Lai, Wenzhu, 263, 527
Lai, Xiaochen, 1363
Lei, Meng, 35, 182, 191
Lei, Yang, 656, 1056
Lekamalage, Chamara Kasun Liyanaarachchi, 139
Li, Baoquan, 1398
Li, Chao, 369
Li, Chen, 27
Li, Chi, 656
Li, Dan, 307
Li, Dong, 274
Li, Gang, 509
Li, Hongpeng, 1117
Li, Jiaqi, 519
Li, Juan, 888
Li, Laquan, 575
Li, Qi, 719
Li, Qiang, 254
Li, Qirui, 849
Li, Rongzhang, 236
Li, Ruisi, 773
Li, Shan, 377
Li, Shupeng, 447
Li, Tao, 1346
Li, Tong, 1371
Li, Xiang, 109
Li, Xianliang, 1146
Li, Xiaoge, 1146
Li, Xiaoguang, 274
Li, Xiaoning, 773
Li, Xin, 1282, 1486
Li, Xinguang, 773
Li, Xinwei, 905
Li, Yan, 361
Li, Yang, 157
Li, Yingchao, 1146
Li, Yinqiao, 896

Li, Yuan, 610
Li, Yutao, 1355
Li, Zhe, 561, 568
Li, Zhenzhen, 974
Liang, Yi, 1067
Liao, Fangting, 3
Liao, Zihao, 1433
Lin, Haoqiang, 519
Lin, Li-Wei, 1202
Lin, Luojun, 639
Lin, Meijin, 764
Lin, Qifeng, 1067
Lin, Simeng, 1095
lin, WangQiu, 1228
Lin, Xiaohui, 1210
Lin, Yatuan, 781
Lin, Yingjun, 130, 236, 1024, 1267
Ling, Xiaoxue, 754
Liu, Baoyang, 18
Liu, Changhai, 402
Liu, Chunhui, 810
Liu, Chunlei, 410
Liu, Dehua, 35, 182, 191
Liu, Deying, 974
Liu, Di, 377
Liu, Dixuan, 1304
Liu, Fan, 369
Liu, Guodong, 254
Liu, Jianhua, 1192
Liu, Jing, 157, 293, 663, 673, 896
Liu, Jixin, 109
Liu, Junxiang, 648
Liu, Kanglin, 1095
Liu, Panni, 1433
Liu, Qun, 764
Liu, Rong, 592
Liu, Tingting, 274
Liu, Tingyu, 798
Liu, Wenxi, 719
Liu, Yanhua, 519
Liu, Yi, 157, 896
Liu, Youyao, 888, 1423
Liu, Yu, 1034
Liu, Zejun, 1407
Liu, Ze-san, 377
Liu, Zhigang, 353
Long, Shun, 3
Long, Xiaolan, 773
Lu, Kuo-Liang, 1202
Lu, Yingbin, 386
Lu, Yitong, 781
Luo, Di, 707
Luo, Jing, 1136
Luo, Weina, 419, 428, 789

Luo, Xinggang, 1290
Lv, Jiajun, 93
lv, Jinrui, 509
Lv, Zeyu, 968

M
Ma, Junchi, 947
Ma, Qianguang, 719
Ma, Sugang, 1087
Ma, Yiming, 327
Ma, Yu, 165
Man, Wen, 956, 1453
Mao, Yongyi, 584, 592
Mao, Zhili, 165, 859
Matsumoto, Kazuyuki, 1125
Meng, Gong, 1185
Meng, Hong-min, 377
Meng, Kan, 1167
Meng, Xiangqing, 947
Meng, Xiangzhi, 1105
Miao, Mengmeng, 833
Ming, Yongjie, 1117
Mo, Wenxiong, 648
Mueller, Etienne, 483

N
Na, Li, 551
Niu, Jiaying, 447
Niu, Yaqiong, 930

O
Olzak, Lynn A., 798

P
Pan, Xiaoying, 475
Pang, Wenting, 536
Peng, Gaojie, 1290
Peng, Jing, 1355
Peng, Qingming, 905
Pingan, Qiao, 600
Pu, Xuetao, 1251, 1259

Q
Qian, Bai, 551
Qiao, Ping'an, 1385
Qiao, Pingan, 147, 610
Qiao, Zhe, 1176
Qiu, Jian, 629, 1158
Qiu, Sihao, 27

R
Ren, Jing, 561
Ren, Jingjing, 1274
Ren, Yulin, 859
Ruan, Heng, 1146

Ruixue, Shen, 600
Rungwachira, Petcharat, 317

S

Shan, Hongmei, 833, 1034
Shan, Li, 1045
Shang, Zhiming, 402
Shao, Xuejuan, 386
Shen, Jiandong, 201
Shen, Pei, 1117
Shen, Ruixue, 610
Shen, Tingda, 43
Shi, Jiakun, 1385
Shi, Jing, 1034
Shi, Li, 1304, 1320, 1329
Shi, Wuxi, 1398
Shi, Xiangyang, 536, 1313
Shu, Huang, 1045
Shuai, Zhe, 1355
Sihao, Chen, 551
Singh, Shubham, 983
Siyu, Jin, 1045
Song, Baoyan, 274
Song, Kang, 139
Song, Rende, 307
Song, Wei, 201
Song, Weifang, 307
Su, Jiangwen, 1237
Sun, Haian, 109
Sun, Hanlin, 1087
Sun, Hao, 410
Sun, Jiaze, 707
Sun, Jinze, 43
Sun, Jun, 122, 475
Sun, Ning, 109
Sun, Ruonan, 274
Sun, Zhengping, 968

T

Tan, Jiaxin, 575
Tan, Shyer Bin, 101
Tang, Lei, 1095
Tang, Ruizhi, 905
Thammano, Arit, 317, 493
Tian, Binhui, 53, 93
Tian, Cong feng, 465
Tian, Zhenzhou, 27, 53, 61, 69, 76, 85, 93
Tjandra, Alfred, 1441
Tuo, Shouheng, 369, 939

W

Wan, Li, 729
Wan, Ming, 274
Wang, Chenguang, 1251, 1259
Wang, DeXuan, 1015

Wang, Dongqi, 968
Wang, Fanfan, 85
Wang, Feng, 993
Wang, Hongbin, 648
Wang, Hongmin, 402
Wang, Jiapan, 833
Wang, Jingbin, 869, 1002
Wang, Jinsong, 361
Wang, Jinwei, 500
Wang, Keya, 428
Wang, Lanxin, 859
Wang, Lipo, 101
Wang, Lumeng, 61
Wang, Mei, 353
Wang, Meng, 1251, 1259
Wang, Muyu, 35, 182, 191, 210, 285
Wang, Qian, 1136
Wang, Qintao, 1414
Wang, Rui, 130, 236, 1024, 1267
Wang, Ruihao, 575
Wang, Ruixin, 1105
Wang, Shan, 410
Wang, Shenhang, 1185
Wang, Shuyan, 1219
Wang, Wei, 244, 693
Wang, Wenji, 921
Wang, Xiaodong, 293, 663, 673
Wang, Xiaoyin, 218
Wang, Xuanhong, 173
Wang, Yanyan, 1015
Wang, Yifan, 1371
Wang, Yinbo, 682
Wang, Yu, 536, 1479
Wang, Zhanmin, 543
Wang, Zhenheng, 1095
Wang, Zhenyan, 386
Wang, Zhongmin, 1087
Wang, Zhurong, 228, 543
Wang, Ziying, 1338
Wei, Bin, 930
Wei, Chenghao, 274
Wei, Cui, 1045
Wei, Jianxin, 1494
Wei, Zhen, 849
Wen, Ai-jun, 377
Wen, Gaojin, 402
Wen, Heng, 1313, 1479
Wu, Dianbin, 1228
Wu, Dunhai, 244, 693
Wu, Jin, 395, 536, 1313, 1479
Wu, Jingkai, 1067
Wu, Meiyi, 157
Wu, Peng, 509
Wu, Qingjin, 1463

Wu, Qingyue, 1077
Wu, Ruichen, 1385
Wu, Wei, 1219
Wu, Xiaokang, 719

X

Xiao, Yun, 173
Xiao, Zilong, 639
Xie, Fengjie, 739
Xie, Xuemin, 1338
Xie, Zhihua, 254
Xin, ZhengH ui, 1320
Xin, Zhenghui, 1304
Xing, Lidong, 1471
Xiong, Ning, 455, 509
Xu, Huijiao, 849
Xu, Jin dong, 465
Xu, Lanqing, 447
Xu, Tiantian, 819
Xu, Xuebin, 35, 182, 191, 210, 285, 1167
Xu, Yun, 402
Xu, Zhe-nan, 377
Xu, Zhong, 648
Xue, Wandong, 968
Xue, Yufeng, 475

Y

Yan, Jianhong, 739
Yan, Qi, 543
Yan, Xingya, 621
Yang, Dongfang, 1385
Yang, Hongxin, 274
Yang, Jia jun, 465
Yang, Jinglan, 1385
Yang, Liang, 1251, 1259
Yang, Ling, 395, 1313
Yang, Peng, 974
Yang, Xiaofeng, 1274, 1407, 1414
Yang, Xu, 519
Yang, Yuanxi, 639
Yang, Yue, 1251, 1259
Yao, Ruiling, 361
Ye, Chengfan, 1267
Ye, Yue, 896
Yi, Jiale, 165
Yin, Chunping, 764
Yoshida, Minoru, 1125
Yu, Junhe, 130, 236, 1024, 1267
Yu, Lin, 648
Yu, Tao, 130
Yu, Yuanlong, 639, 719, 1067, 1228, 1237
Yu, Zhong, 1304, 1320, 1329
Yuan, Li, 600
Yuan, Shuiquan, 1087
Yuan, Yongbin, 336, 419

Yuan, Zeduo, 3
Yuan, Zhenhua, 974
Yuwei, Qiang, 551

Z

Zeng, Jiahao, 419
Zesan, Liu, 1045
Zhan, Hongfei, 130, 236, 1024, 1267
Zhang, Bingchen, 336
Zhang, Changyou, 336, 428
Zhang, Chuanqiang, 789
Zhang, Cuiping, 53
Zhang, Han, 1486
Zhang, Hong, 18
Zhang, Jianke, 1136
Zhang, Jinggang, 386
Zhang, Jingyi, 1453
Zhang, Kejia, 353
Zhang, Lin, 1251, 1259
Zhang, Linyu, 293, 663, 673
Zhang, Qian, 833
Zhang, Qunjiao, 729
Zhang, Rong, 361
Zhang, Ru wei, 465
Zhang, Shaohong, 841, 974
Zhang, Shengli, 437
Zhang, Sumei, 1433
Zhang, Wanpeng, 1117
Zhang, Wenyu, 336, 419, 428, 789
Zhang, Xia, 165, 859
Zhang, Xiubin, 122
Zhang, Xuezhong, 905
Zhang, Yang, 673
Zhang, Yanyan, 1219
Zhang, Yin, 921
Zhang, Yingfei, 1105
Zhang, Yu-an, 307
Zhang, Yun, 1034
Zhang, Yuxiang, 1346, 1471
Zhang, Zhaoqi, 395
Zhang, Zhe, 1185
Zhang, Zongrui, 629, 1158
Zhao, Haoran, 819
Zhao, Jihong, 1290
Zhao, Jing, 930
Zhao, Long, 819, 1015
Zhao, Lv, 11
Zhao, Siyuan, 789
Zhao, Xuan-Ze, 1202
Zhao, Yong, 465
Zhenan, Xu, 1045
Zheng, Qinghai, 1067, 1228
Zheng, Shenhai, 575
Zheng, Shiliang, 1355
Zheng, Zeyao, 621

Zhong, Can, 402
Zhong, Qiyu, 841
Zhou, Hang, 327
Zhou, Jing, 228
Zhou, Jingchen, 173
Zhou, Lin, 841
Zhou, Yuan, 1015
Zhu, Daxin, 1463
Zhu, Fu, 447
Zhu, Hongmei, 764
Zhu, Jie, 956

Zhu, Tingting, 956
Zhu, Weiheng, 3
Zhu, Xiaoxiao, 1363
Zhu, YanLing, 939
Zhu, Yu, 956, 1453
Zhu, Zhangyi, 1398
Zhu, Zhaoyang, 1290
Ziwei, Guan, 11
Zong, Sha, 254
Zuo, Zhuo, 344

Printed by Printforce, the Netherlands